Ig	immunoglobulin
IK	immunoconglutinin
IL	interleukin
ISCOM	immune stimulating complex
ISG	immune serum globulin
J	joining
kDa	kilodalton
LAD	leukocyte adherence deficiency
LAK	lymphokine-activated killer (cells)
LFA	lymphocyte function-associated antigen
lpr	lymphoproliferation
LPS	lipopolysaccharide
LT	lymphotoxin (or leukotriene)
LTR	long terminal repeat
β_2M	β_2-microglobulin
MAC	membrane attack complex
M-CSF	macrophage colony stimulating factor
MDP	muramyl dipeptide
MHC	major histocompatibility complex
MLC	mixed lymphocyte culture
MLR	mixed lymphocyte reaction
mRNA	messenger RNA
N-CAM	neural cell adhesion molecule
NK	natural killer
NS	natural suppressor
NZB	New-Zealand black (mice)
NZW	New-Zealand white (mice)
PAF	platelet activating factor
PCA	passive cutaneous anaphylaxis
PFC	plaque-forming cell
PG	prostaglandin
PHA	phytohemagglutinin
PKC	protein kinase C
PPD	purified protein derivative of tuberculin
PWM	pokeweed mitogen
R	receptor (e.g., IL-2R)
RAST	radioallergosorbent test
RIA	radioimmunoassay
RNA	ribonucleic acid
S	Svedberg unit of sedimentation coefficient
SAA	serum amyloid A (protein)
SCID	severe combined immunodeficiency
SLE	systemic lupus erythematosus
SRS-A	slow-reacting substance of anaphylaxis
TAP	transporter for antigen processing
TCR	T cell antigen receptor
TdT	terminal deoxynucleotidyl transferase
TGF	transforming growth factor
Th cell	T helper cell
TIL	tumor infiltrating lymphoctyes
TK	thymidine kinase
TNF	tumor necrosis factor
Ts cell	T suppressor cell
TSH	thyroid stimulating hormone
UV	ultraviolet
VLA	very late antigen
ZAP	zeta-associated protein

Immunology

AN INTRODUCTION

Immunology

AN INTRODUCTION

Fourth Edition

Ian R. Tizard

Professor, Department of Veterinary Pathobiology

Texas A&M University

SAUNDERS COLLEGE PUBLISHING
HARCOURT BRACE COLLEGE PUBLISHERS

Philadelphia Fort Worth San Diego New York Orlando
San Antonio Toronto Montreal London Sydney Tokyo

Text Typeface: New Baskerville
Compositor: University Graphics, Inc.
Executive Editor: Julie Levin Alexander
Developmental Editor: Christine Rickoff
Managing Editor: Carol Field
Project Editor: Nancy Lubars
Copy Editor: Nanette Bendyna-Schuman
Manager of Art and Design: Carol Bleistine
Art Director: Anne Muldrow
Art Coordinator: Sue Kinney
Cover Designer: Lawrence Didona
Text Artwork: Rolin Graphics
Layout Artist: Dorothy Chattin
Director of EDP: Tim Frelick
Production Manager: Joanne Cassetti
Marketing Manager: Sue Westmoreland

Cover Credit: Nancy Kedersha, Immunogen, Inc.

Printed in the United States of America

IMMUNOLOGY: An Introduction, 4/e
0-03-004198-8

Library of Congress Catalog Card Number: 94-061384

4567890123 032 10 987654321

This book was printed on acid-free recycled content paper, containing
MORE THAN
10% POSTCONSUMER WASTE

To Claire, Robert, and Fiona

Preface

The biomedical research industry continues to produce a stream of new information about the immune system. Such a vast quantity of information makes any attempt to synthesize or simplify the material a daunting one. Indeed, some of this new information may be incorrect as it is often difficult to discern emerging patterns, identify incorrect information, and give proper weight to unanticipated results and new theories. This flood of new information increases the need for accurate textbooks while at the same time making the task of the textbook author ever more demanding.

It is no longer sufficient for an author to add new information as it appears and then to seek to integrate it into the existing knowledge base. If a book and its contents are not to grow unwieldy, strenuous efforts must be made to control the amount and quality of the new information. This inevitably eliminates material that the author perceives as unnecessary or peripheral but that others may deem important. It ensures that a text must penetrate less deeply into the subject even as the amount of available information grows. How then can immunology be addressed in a meaningful way and provide accurate and timely information, yet not succumb to oversimplification and trivialization? Sometimes new information may make an author's task simpler. As the science becomes revealed in increasing detail, new linkages, patterns, and relationships become apparent and difficult questions are answered. In many cases this increased detail provides evidence for common or shared pathways. One of the most striking features of the recent advances in immunology is its linkage with key aspects of cell biology and biochemistry. The responses of the cells of the immune system are increasingly seen, therefore, not as something separate from normal cellular functions but as modifications of basic processes.

Given the finite length of a lecture (and of a semester), it is increasingly difficult to bring immunology to undergraduates without overwhelming or hopelessly confusing them. Immunology can no longer be encompassed in a university course during a single semester, except at a very basic level. The areas of overlap between immunology, cell biology, biochemistry, and microbiology confound the issue. To make sense of it all, the student must have a good basic understanding of these other disciplines. Experience has shown, however, that this may be unrealistic. Undergraduate students commonly have a weak background in the life sciences. I hope that the changes made in this new edition address some of these issues and that it will provide a simple, yet scientifically rigorous introduction to a wonderful subject—the defense of the body.

Distinctive Features

This text is designed to provide a broad overview of immunology and, as a result, paints with broad brushstrokes. It seeks to tell the story of immunology in a logical fashion beginning with the administration of antigen and progressing eventually to the consequences of a successful immune response. Because of this goal, and to avoid disrupting the conceptual flow, the text deliberately avoids technical or experimental detail. I fully realize that immunology is an experimental discipline. All the information in this text is derived from carefully conducted experiments. It is not, however, my object to provide this detail but to give the student a concept of the scope of the discipline, its vocabulary, its implications, and its role in medicine and everyday life; in short, to provide a broad, overall view. This is especially important in immunology where some topics are complex and a student easily is overwhelmed by the volume of material available. Those who are stimulated by this presentation will be able to turn to many excellent texts to review experimental details at their leisure. This book is designed to instruct at a somewhat more basic level than other immunology textbooks.

Instructional Features

Basic Chapters. Because of the uneven academic background of students studying immunology for the first time, three completely new chapters describing basic principles have been provided. The first, Chapter 2, is on microbiology and infectious diseases. It presents a brief overview of the agents that invade the body and against which the immune responses are normally directed. The second new chapter, Chapter 3, reviews some selected aspects of basic biochemistry including gene function and protein structure. The third, Chapter 7, is a review of relevant concepts in cell biology including exocytosis, cell surface structures, receptors, signal transduction, cell adhesion molecules, and oncogenes.

Key Concepts. The key concepts encompassed in each chapter are summarized at the beginning. These are designed to ensure that students recognize the main thrust of each chapter. Students should not go far wrong if they learn these first.

Key Words. To further assist students in studying, a list of key words has been provided for every chapter. This should also help students to focus their attention on the terms of greatest importance and identify the most important features of the text.

Illustrations. The number of illustrations has been significantly increased, and many of the earlier ones extensively revised to enhance student comprehension of key concepts.

Methods. As discussed previously, immunology is an experimental science. In order to ensure that students realize this, yet without interfering with the flow of the text, a limited number of key experimental procedures have been outlined in boxes throughout the book. This is not designed to provide full experimental details but simply to provide a flavor of the methods employed.

"A Closer Look." There are many fascinating ramifications of immunology that cannot be encompassed in a text of this nature. Nevertheless, I have provided many examples of these through the use of boxes throughout the text. These boxes complement the text and hopefully will enrich student comprehension and appreciation of this subject.

Questions. Previous editions were characterized by a small number of thought-provoking questions at the end of each chapter. Many of these required not only factual knowledge but also provoked philosophical or speculative analysis. These are now complemented by the addition of carefully chosen multiple choice questions with answers for self-testing. These have been student-tested and should provide excellent support for the learning process.

References. References have been expanded and updated. They have been carefully selected to provide a mixture of key research papers and reviews. They should provide interested students with an opportunity to expand their knowledge of a specific area, should they so wish.

Glossary. I have already alluded to the complex and specialized terminology that has evolved to describe the molecules and processes involved in immunology (some call it immunobabble). This is compounded by the close relationship between immunology and medicine and the need, on occasion, to use medical terminology as well. In order to assist the beginning student, an extensive glossary has been provided and words defined in the glossary have been so identified in the text.

Organization. One contentious issue among teachers of immunology has been the order in which topics should be presented to students. I have discussed this at length with my colleagues, including many of those who have reviewed this book. No consensus has been reached. There is no single correct order of topics that all immunologists will accept. I therefore use an organization that I have found most satisfactory in teaching immunology for over twenty years. The chapter sequence in the first half of the book follows the course of an immune response from the initial administration of antigen to the final consequences of the immune response, defense of the body. The simple linear progression is complicated by the branching of the immune responses into the cell-mediated and antibody-mediated arms. There is also a need to digress to review features such as the structure of the lymphoid organs. The second half of the book describes the body's defenses against infections and the consequences of a failure in these defenses. This is followed by a description of the abnormal features of the immune system, immunologic diseases, and hypersensitivity.

New Features

The flood of new information has required the publication of this new edition only three years after the previous one. Most areas have been updated or rewritten. Some of the major changes in this edition are listed below. The information provided is current up to mid-1994.

- Three new chapters on basic microbiology, biochemistry, and cell biology have been added.
- A large number of new illustrations have been provided and many of the older ones extensively updated and clarified.
- The chapter on antigen processing has been completely rewritten to reflect the central role of MHC molecules in the process. This includes the function of MHC class Ib molecules, the binding of peptides to MHC molecules, the structure of the MHC binding site in both class I and class II molecules, the role of transporter proteins and proteasomes.
- The chapter on cytokines has been completely rewritten. These are now described on the basis of their cellular source. The chapter includes new information on the newest cytokines, especially IL-8, IL-12, IL-13, IL-14, and IL-15, as well as the cytokine receptor families.
- The chapter on the response of B cells to antigen has been greatly changed. This includes the latest information on the structure of the B-cell antigen receptor complex, B-cell signal transduction, the development and apoptosis of B cells within germinal centers, IgD knockout animals, CD40 and its significance, bcl-2 and protection against apoptosis, and B-cell subpopulations.
- The chapter on T-effector cell functions now includes a greatly enlarged discussion on superantigens, mechanisms of cytotoxicity and their relationship to apoptosis, fas, and other TNF-like inducers of cytotoxicity.
- The chapter on AIDS has been significantly revised to include updates on the global status of AIDS, new ideas on the pathogenesis of the disease, and AIDS vaccines. The possibility of HIV components acting as superantigens is discussed as is the possible role of CD26 as a coreceptor for HIV.
- The process of apoptosis is considered in much greater detail than before. This includes its role in cell development, in immunity, and in cytotoxity. The role of oncogenes such as fas and bcl-2 in apoptosis is also considered.
- Signal transduction and proteins such as ras and FcϵRIγ are described and their significance explained. The role and significance of costimulating pathways involving molecules such as CD28, CD40, CD19, CR2, B7, and CD40-L are also described.
- New information has been provided on the important cell surface molecules up to CD130.
- Nitric oxide and its functions are described in much greater detail than before.
- Updated information is provided on the development and function of $\gamma\delta$ T cells.
- The events occurring in lymph nodes when responding to antigen and the roles of apoptosis, somatic mutation, and cell selection are described.
- The significance of different T-cell subpopulations and helper-cell imbalances in infectious diseases is examined.
- The chapter on complement includes significant updates on mannose-binding protein and on complement deficiencies.
- The chapter on grafting includes the new information on the mechanisms of action of the immunosuppressive cytophilins such as cyclosporin and FK 506.
- The latest ideas on the origin and functions of NK cells are described and their relation to T cells is clarified.
- There is an update on positive and negative selection in the thymus and the components that regulate this process.
- The existence of suppressor cells continues to be questioned, and their presumed function is examined in detail.
- The evidence for a protective role of IgA inside cells is reviewed. Intraepithelial lymphocytes are described and their functions discussed.
- New information in the chapter on vaccines includes updates on recombinant vaccines and the use of naked DNA as a vaccine.
- The section on cell adhesion molecules, their expression, and regulation is expanded as is lymphocyte circulation and its control. This includes recent advances in our understanding of cell adherence to endothelium.
- The latest information on the control of the allergic reactions by Fcϵ receptors and the IgE network as well as helper-cell and cytokine imbalances in allergies is explained.
- The newest information on the structure and functions of blood group antigens is described.
- The chapter on autoimmunity now contains a section on multiple sclerosis. In addition, the newest developments in regard to molecular mimicry are described.

Acknowledgments

As always it is a pleasure to publicly acknowledge my enormous debt to those colleagues who assisted materially in the preparation of this book. Individuals who reviewed the manuscript and provided critical and insightful recommendations included:

Sharon D. Bramson, The College of Staten Island
Dale J. Erskine, Lebanon Valley College
Julian B. Fleischman, Washington University in St. Louis
J. C. Hennings, South Dakota State University
Ronald Humphrey, Prairie View A & M
David J. Hurley, South Dakota State University
Laura Jenski, Indiana University–Purdue University
Robert I. Krasner, Providence College
Paul A. LeBlanc, The University of Alabama
Pat Lord, James Madison University
Michael Lynes, University of Connecticut at Storrs
Louise B. Montgomery, Marymount University
Carol Park, Ohio Wesleyan University
Thomas C. Peeler, Susquehanna University
Dr. Charles Pfau, Renssalaer Polytechnic
Gene Scalrone, Idaho State University
Cynthia V. Sommer, University of Wisconsin

It is a special pleasure to thank Dr T. K. Hunt of the University of California at San Francisco for his hospitality during a sabbatical leave where the great bulk of writing this edition was undertaken. To this I must add my thanks to my graduate students and staff at Texas A&M University for their tolerance and patience while I ignored their needs as I concentrated on the book.

The staff of Saunders College Publishing have been unusually patient, professional, and supportive. I am especially grateful to Julie Levin Alexander, Biology Editor; Christine Rickoff, Developmental Editor; and Nancy Lubars, Project Editor.

Most importantly, I must express my ongoing appreciation for the support and tolerance of my wife Claire.

Ian Tizard
College Station
September 1994

Contents

CHAPTER 11

The Helper T-Cell Responses 139

CHAPTER 12

Lymphokines and Cytokines 155

CHAPTER 13

The Nature of Antibodies 170

CHAPTER 14

The Response of B Cells to Antigen 192

CHAPTER 28

Inflammation 432

CHAPTER 29

Allergies: Type I Hypersensitivity 454

CHAPTER 30

Other Hypersensitivities 468

CHAPTER 31

Autoimmunity: Breakdown in Self-Tolerance 488

CHAPTER 1

General Principles of Immunology

Lady Mary Montagu *A remarkable woman who, as the wife of the English Ambassador at Constantinople, learned how the Turks protected their children against smallpox by the process of variolation. Not only did she have her children successfully variolated in the face of opposition from her compatriots but, through her friendship with the wife of King George II, encouraged the English aristocracy to try it as well. This support from the highest levels of English society ensured that the process was adopted with enthusiasm in Britain. (Courtesy of the National Library of Medicine.)*

CHAPTER OUTLINE

Historical background
Antibody-mediated immune responses
Cell-mediated immune responses
Tolerance
Immune response mechanisms
Current problems in immunology

CHAPTER CONCEPTS

1. The idea that individuals could become resistant to an infectious disease is recent, and the mechanisms of this immunity have been clarified only over the past 100 years.
2. The immune system has two major tasks. It must protect individuals against infectious agents invading the body from outside, and at the same time, it must prevent the development of abnormal cells within the body.
3. Molecules that induce an immune response are called antigens.
4. There are two types of immune responses. One, mediated by proteins called antibodies, is responsible for resistance to infectious agents found in body fluids. The other, mediated by cells called lymphocytes, is responsible for the destruction of abnormal cells. Such abnormal cells include cancer cells and virus-infected cells.
5. The immune system does not usually react against normal body components. It is therefore said to be ''tolerant'' of self-components.

The animal body with its warmth, moisture, cells, and rich supplies of nutrients represents an ideal habitat for the growth of bacteria, fungi, viruses, and other parasites. If an animal is to survive and function, it must defend itself effectively against invasion by these microorganisms. Failure to do so will inevitably result in death from overwhelming infection. The body system that defends us against the constant assault by microorganisms is called the immune system. Immunology is the science that studies the immune system, how it works, and the way in which we can stimulate the immune system to protect against disease.

Because of its importance to an animal's survival, the immune system must be very effective. Even a temporary breach in the body's defenses must be closed immediately. For this reason, the immune system has many components. Some are optimized to defend against a single invader while others are directed against a great variety of infectious agents. Some are designed to exclude infectious agents and so prevent them from getting a "toehold" in the body. Others are directed against invaders within the body and even against organisms hiding within cells. Considerable redundancy is built into the system so that several defense mechanisms are active against a single invader. Evolutionary pressures have ensured that the defenses of the body are as effective as possible.

One remarkable feature of the immune system is that it possesses a memory. Thus it has the ability to remember previous encounters with specific microorganisms and to respond even more effectively to these organisms on second or subsequent encounters. As a result, recovery from an attack of many infectious diseases confers resistance to subsequent attacks of the same disease. Despite this remarkable efficiency, the immune system is not always completely effective in excluding invaders, and not all diseases result in subsequent immunity. For example, most readers of this book will have suffered from the common cold within the past year or so. You will probably recollect having suffered from many such infections. It is therefore not surprising that until recently most people did not accept that recovery from some infections confers resistance to subsequent attacks of the same disease. Nevertheless, even the ancient Greeks suspected that those persons who survived one attack of the plague would not suffer the disease a second time. It would have required an act of great courage to test such a radical idea, and it is unlikely to have been widely accepted.

HISTORICAL BACKGROUND

Of the great diseases of humankind, smallpox was one of the most feared. Not only did it kill huge numbers of people, but also survivors were scarred for life with disfiguring pockmarks. The Chinese determined, about the beginning of the eleventh century, that persons who survived an attack of smallpox would not get the disease a second time. It therefore became accepted practice in ancient China to infect young children with smallpox. Scabs from the pocks on the skin of an infected person were either put up the nose of an infant or rubbed into a scratch. Children who survived the resulting disease became resistant to smallpox; those who died saved their parents the trouble and expense of raising them, only to have them die from the disease later. As the Chinese gained experience with this procedure, they found that the mildest disease occurred when the smallpox scabs used were selected from mildly affected donors. As a result, they eventually succeeded in reducing the hazards of the procedure to fairly low levels.

News of this method of protecting children against smallpox gradually spread westward along the caravan routes of Central Asia and eventually reached Turkey. In 1718 Lady Mary Montagu, the wife of the English ambassador in Constantinople, decided that the technique should be used on her children. Her own face is said to have been scarred by smallpox, so she had a special interest in the disease. However, her portrait does not show this. Her chaplain tried to dissuade her on the grounds that the technique would be ineffective in Christians. Nevertheless, Lady Montagu persisted, and her children were successfully protected.

News of this technique, by now called **variolation** ("variola" is the Latin word for smallpox), spread rapidly to England, where Queen Anne's son and heir had recently died from smallpox. The British ruling classes were thus acutely aware of the disease and adopted variolation with enthusiasm. The procedure was adapted with extraordinary rapidity in the American colonies, where George Washington ordered all his troops to be inoculated in 1776. Nevertheless, variolation remained a hazardous procedure since lethal smallpox occasionally resulted from the process.

In 1774 an English farmer named Benjamin Jesty used dried scab material from a case of cowpox to variolate his children in the belief that this would protect them against smallpox. Cowpox is a skin disease of cattle that is caused by a virus closely related to smallpox virus. It is also a very mild disease in humans. In 1798 Edward Jenner, a physician, learned about this use of cowpox from one of his patients and decided to investigate the matter further. Jenner conducted many experiments and eventually confirmed that exposure to dried cowpox scabs could indeed safely protect humans against smallpox. He, unlike Jesty, published his results and received the credit for the discovery. Jenner's technique of **vaccination** ("vacca" is the Latin word for cow) rapidly replaced variolation as the preferred method of protecting humans against smallpox. Variolation and vaccination were so successful in reducing smallpox mortality that a population explosion resulted. This, in turn, had major social consequences, since it provided labor for the new factories of the Industrial Revolution. In 1958, the World Health Organization decided to eradicate smallpox by means of mass vaccination. This was so effective that by 1980 smallpox had become the first infectious disease to be eradicated from the earth.

In spite of the discovery of vaccination against smallpox in the late eighteenth century, it was not suspected that this type of approach could be used to protect against other infectious diseases, although a few unsuccessful attempts were made to use cowpox to cure unrelated skin diseases. It was not until almost 100 years later that another significant advance occurred.

In 1879 in France, Louis Pasteur was studying the bacterium that causes a disease of chickens called fowl cholera. (The bacterium is now called *Pasteurella multocida.*) Pasteur possessed a culture of this bacterium that, when injected into chickens, consistently caused an infection that killed them. One afternoon he told his assistant, Charles Chamberland, to infect some birds with the culture. Since it was late in the day and he was about to go on vacation, Chamberland decided to postpone the experiment until he returned. As a result, the chickens eventually received an injection of the bacterial culture that had remained on the bench for several weeks. The inoculated chickens remained healthy (Fig. 1–1). Pasteur then decided to inject these chickens with a second dose of bacteria from a fresh bacterial culture. To Pasteur's surprise, the birds survived this second dose without becoming ill. Pasteur, with remarkable insight, recognized that this phenomenon was identical in principle to vaccination. By injecting his chickens with the aged culture of bacteria (a **vaccine**), he had protected them against disease caused by a fresh culture of the same organism.

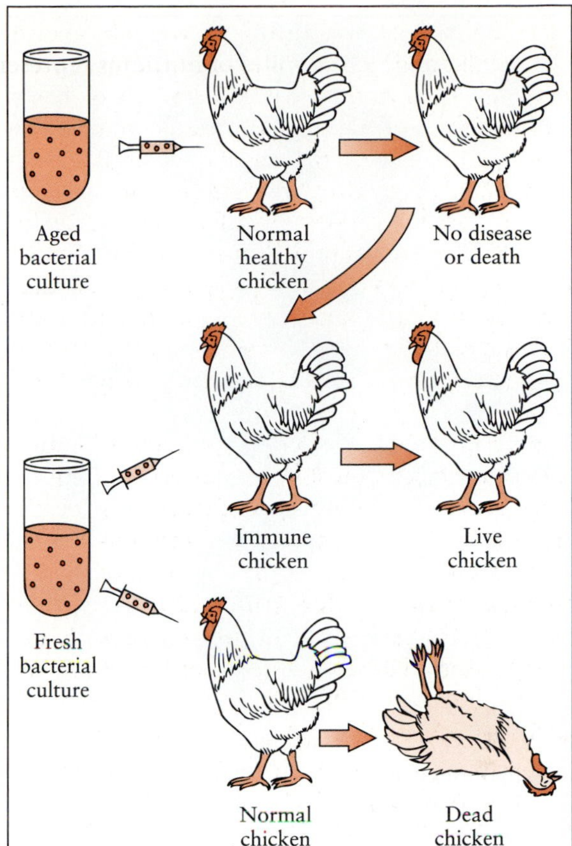

Figure 1–1 Pasteur's fowl cholera experiment. An aged culture of the bacterium *Pasteurella multocida* not only failed to kill chickens but also rendered them resistant to subsequent infection by a fresh culture of *P. multocida.* This fresh culture would kill unimmunized chickens.

Once he had established the general principle of vaccination, Pasteur tried to apply it to other infectious diseases. He first produced a vaccine against anthrax, a disease caused by a bacterium called *Bacillus anthracis.* Pasteur found that he could not make this organism safe by aging it on the laboratory bench, but he could do so by growing it at an unusually high temperature. Pasteur then conducted a public experiment that convincingly showed that administration of his heated anthrax culture would protect sheep, cattle, and goats against a subsequent lethal dose of anthrax bacteria. A few years later he developed an effective vaccine against rabies. He thus showed that the general principles of vaccination applied to diseases other than smallpox, and this approach could be used to protect animals and humans against other infections. Louis Pasteur can therefore be considered the founder of the science of immunology.

Pasteur, although remarkably successful in developing effective vaccines, had little concept of the mechanisms involved. He suggested that the organisms in the vaccine removed essential nutrients from the body and thus prevented the subsequent growth of the disease-causing agent. It was in Berlin about ten years later, in 1890, that Emil von Behring and Shibasaburo Kitasato demonstrated that the protection induced by vaccination was not due to removal of nutrients but was associated with the appearance of protective factors in the blood (Fig. 1–2). They called these factors **antibodies.** Within a few years Paul Ehrlich had proved that antibodies could protect animals against foreign toxins other than those found in bacteria (for example, the toxin ricin from the castor bean), and another German, Richard Pfeiffer, had shown that antibodies could clump and then destroy *Vibrio cholerae* bacteria, the cause of cholera in humans.

Figure 1–2 Emil von Behring discovered that immunity was due to the presence of factors in blood serum that he called antibodies. *(Culver Pictures.)*

CLARIFICATION

Serum and Plasma

If fresh blood is allowed to clot in a tube, the clot will contract and express a clear yellow fluid called **serum.** If blood is drawn into an anticoagulant so that it will not clot and is then centrifuged so that its cells are sedimented, the clear fluid is called **plasma.** Serum differs from plasma in that it does not contain clotting proteins such as fibrinogen.

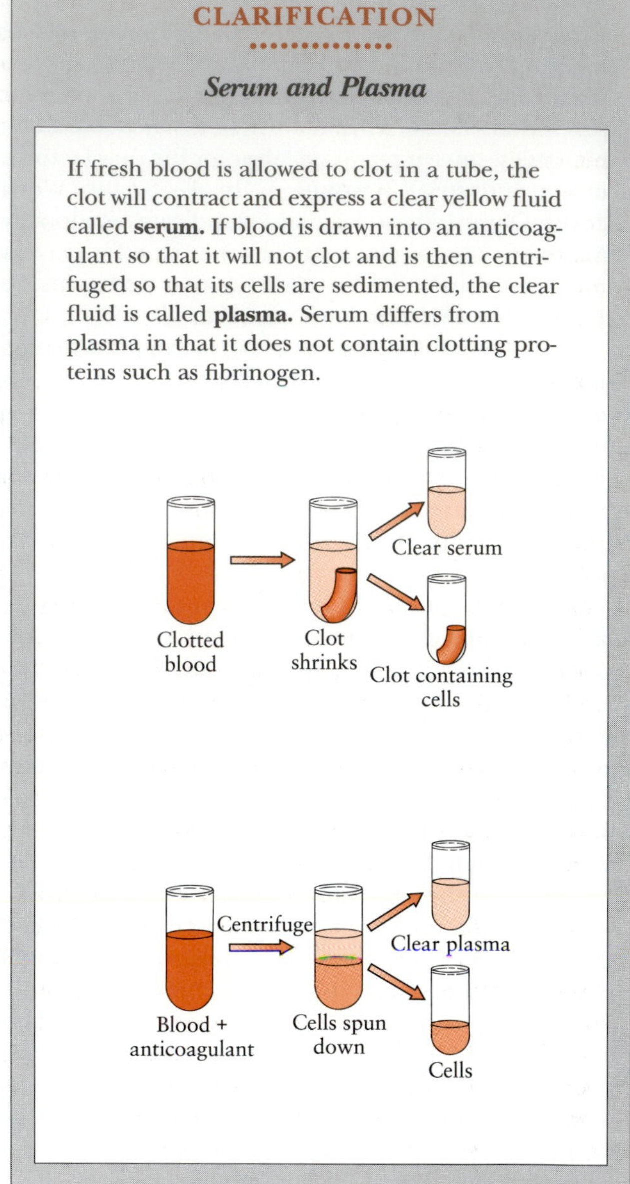

The demonstration that antibodies could make a suspension of bacteria clump was soon applied by Isidore Widal to the diagnosis of enteric disease such as typhoid fever. Thus serum from an infected individual would make the bacteria clump, whereas serum from an unaffected individual would not.

In 1894 Emile Roux showed that patients suffering from diphtheria could be cured by an injection of serum from the blood of a horse immunized against diphtheria. (The horse serum thus contained antibodies to diphtheria toxin.) This fostered hopes that such therapy could be used to treat other diseases. Unfortunately, none succeeded as well as that for diphtheria. It was shown, however, that serum from a horse immunized by repeated injections of tetanus toxin could be used to protect a human against tetanus for a few weeks. This technique, known as **passive immunization,** is still used to protect persons against tetanus after they have received a deep wound. The horse serum contains antibodies against tetanus toxin.

Further studies on the properties of antibodies led in 1893 to the observation that heating an antibody-containing serum to 56° C caused the loss of its ability to kill bacteria. Since antibodies are not destroyed by heating, another component must be responsible for this activity. The heat-labile component that acts with antibodies to kill bacteria is now called **complement.**

The first major controversy to engulf the emerging science of immunology was triggered by the Russian investigator Eli Metchnikoff in 1882. Metchnikoff demonstrated, first in invertebrates and later in mammals, that certain cells could "eat" foreign material. He proposed that these cells, which he called **phagocytes** (eating cells), were the body's main defense against invading microorganisms. Furthermore, he suggested that antibodies were of little significance! This radical proposal upset a lot of people and provoked violent controversy. The conflict was only resolved in 1904, when Almroth Wright in England and Joseph Denys in Belgium showed that antibodies could coat bacteria and promote their destruction by phagocytes (Chapter 5).

The history of immunology since the beginning of the twentieth century is complex but can be outlined by summarizing the Nobel prizes awarded to immunologists. The first Nobel Prize in medicine was awarded to Emil von Behring in 1901 for his work on the production of antibodies against toxins (antitoxins). In 1905 Robert Koch was awarded the prize for his studies on tuberculosis; one of his major contributions to our knowledge of this disease was his discovery of the **tuberculin** reaction, an immunologically mediated inflammatory skin reaction that is a very useful diagnostic test. It is described in detail in Chapter 18. The Nobel Prize of 1908 was shared by Ehrlich and Metchnikoff. Paul Ehrlich's contributions to immunology were diverse; he determined the time course of the immune response described later in this chapter and developed several important concepts, including the idea that the immune system would not normally mount a response against normal body components. In 1913 Charles Richet won the prize for his discovery of **anaphylaxis.** While trying to immunize dogs against a toxin from sea anemones, Richet and his colleague Paul Portier observed that very small nontoxic doses of this toxin made some dogs collapse and die. We now know that this reaction is caused by certain antibodies. This dangerous immunological response to small amounts of foreign material is called anaphylaxis, meaning "without protection." Anaphylaxis is described in Chapter 19. Jules Bordet, the discoverer of complement, won the prize in 1919.

Karl Landsteiner was awarded a Nobel Prize in 1930 for his demonstration of **blood groups,** the complex carbohydrates found on the surface of red cells. These molecules are important in immunology because they can stimulate a life-threatening immune response if blood is transfused into the wrong person. Landsteiner's research eventually resulted in the development of successful blood transfusion procedures (Chapter 30). For the next 30 years, the progress of immunology appeared to slow. It was not until 1960 that Peter Medawar and Macfarlane Burnet jointly won the prize for their studies of the ways in which the immune system recognizes foreign material while avoiding reacting against itself. Burnet developed the first rational explanation of this ability of the immune system to identify foreign material—the **clonal selection theory,** which Medawar later confirmed experimentally (Chapter 14). The concept of clonal selection lies at the core of modern immunology.

In 1972 two protein chemists, Rodney Porter and Gerald Edelman, received the prize for demonstrating the chemical structure of antibody molecules and, incidentally, showing how they bound to foreign material (Chapter 15). One third of the 1977 prize was awarded to Rosalyn Yalow for her studies on the technique of **radioimmunoassay.** By combining the use of radioisotopes with antibodies that bind specifically to hormone molecules, Yalow showed that it was possible to develop remarkably sensitive tests to detect low levels of these hormones in body fluids (Chapter 17).

The 1980 Nobel Prize in Medicine was shared by three immunologists who had made significant contributions in immunogenetics. George Snell developed genetically defined mouse strains and determined the existence of genes that control the process of graft rejection. Jean Dausset developed techniques to identify the molecules found on the surface of human cells. This enables transplantation surgeons to match organ donors with recipients so that the immune response leading to graft rejection is minimized. Baruj Benacerraf studied the inheritance of an animal's ability to make antibodies to defined molecules. In this way, he and his co-workers discovered that genes regulate an individual's ability to respond to a specific foreign molecule and so determine resistance and susceptibility to disease (Chapter 8).

The 1984 Nobel Prize in Medicine was awarded to Niels Jerne, Georges Köhler, and César Milstein. The prize was awarded to Köhler and Milstein for discovering the principles of the production of **monoclonal antibodies** and to Jerne for his provocative theories on the specificity, development, and control of the immune system. Monoclonal antibodies are described in Chapter 14, and some of Jerne's ideas are discussed in Chapter 22.

The 1987 Nobel Prize in Medicine was awarded to Sutsumu Tonegawa for his molecular studies that showed how antibodies were generated so that they could combine with a vast array of foreign molecules. These studies are described in Chapter 15.

The 1990 prize was awarded to Donnall Thomas and Joseph Murray for their pioneering work on organ transplantation. Murray was the first surgeon to perform a successful organ transplant when he grafted a kidney between identical twins. Thomas was the first to perform a successful bone marrow transplant in a patient who had first been irradiated to remove leukemia cells. Organ grafting is discussed in Chapter 19.

The remarkable recent growth in immunology is well demonstrated by the award of the Nobel prizes. Only three immunologists won the prizes between 1919 and 1972. In contrast, between 1980 and 1990 nine prizes were awarded for immunology research. The spectacular growth of immunology continues to the present.

ANTIBODY-MEDIATED IMMUNE RESPONSES

Antibodies are proteins that appear in serum following exposure of an animal to foreign substances such as infectious agents. Those foreign substances that can provoke the production of antibodies are called **antigens.** Antibodies can combine specifically with the antigen that stimulated their production and hasten its destruction. Antibodies directed against a specific antigen are found only in animals that have previously encountered that same antigen. The blood of a normal person contains a great variety of antibodies. These antibodies are present because each of us is constantly exposed to many different foreign antigens from our environment.

A good example of a "typical" antigen is the toxin produced by the bacterium *Clostridium tetani.* This bacterium, which grows in deep wounds, produces a highly toxic protein—tetanus toxin. This toxin poisons nerve cells and so causes the disease called tetanus. Because it is a foreign protein, tetanus toxin acts as an antigen and provokes an immune response in animals. It is clearly much too poisonous to be used as a vaccine. When treated with formaldehyde, however, tetanus toxin loses its toxic activities while still remaining a powerful antigen. This treated toxin, called tetanus **toxoid,** can be safely used to vaccinate individuals. The antibodies provoked by injection of tetanus toxoid will protect an individual against tetanus.

If tetanus toxoid is injected into an animal, antibodies against the toxoid are produced in response. If serum containing these antibodies is mixed with tetanus toxoid solution, a cloudy precipitate will develop as a result of the combination of tetanus toxoid and the antibodies. These antibodies attach to tetanus toxin so that it can no longer bind to nerve cells and is thus rendered nontoxic. It is by means of this **neutralization** that antibodies protect animals from the lethal effects of tetanus infection.

The antibodies produced by an animal in response to a foreign organism are able to protect the animal only against that organism. Antibodies can bind only to the antigen that induces their production. Thus, in the preceding example, antibodies produced in response to tetanus toxoid will protect only against tetanus. They have no effect on any other disease.

It is not difficult to follow the course of antibody production following the injection of a single dose of tetanus toxoid. Blood samples can be taken at intervals after the injection and the serum separated. The amount of antibody in the serum can be estimated by measuring either the amount of precipitate formed on the addition of toxoid or, alternatively, the ability of the serum to neutralize tetanus toxin. Both methods yield similar results.

Following injection of a single dose of tetanus toxoid into an individual who has never before been exposed to this antigen, antibodies cannot be detected in blood until several days have elapsed. This is called the **lag period.** The length of the lag period depends on the method used to detect the antibodies, but it is never shorter than three to four days. Once antibodies appear in the blood, their level climbs, reaching a maximum 10 to 14 days after exposure to the antigen (Fig. 1–3); they then decline rapidly to undetectable levels within a few weeks. Because the amount of antibody produced during this **primary immune response** is small, the level of protection against tetanus is also low. A single dose of tetanus toxoid cannot therefore be relied on to give effective protection against tetanus. It is therefore necessary to give a second dose of toxoid.

Following injection of a second dose of tetanus toxoid, the lag period lasts for only two or three days. When antibodies appear in the blood, they climb rapidly to high levels, before declining slowly over a period of months. This **secondary immune response** is specific in that it can be induced only by exposure to

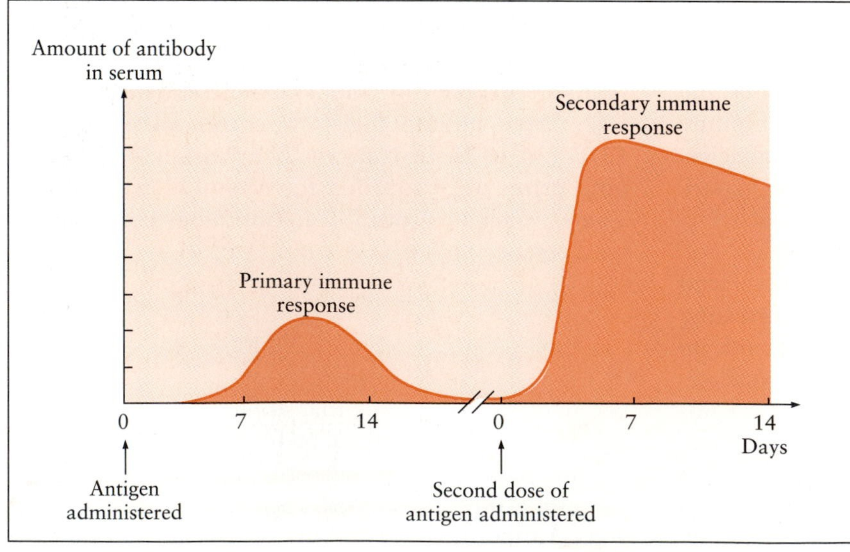

Figure 1–3 A typical time course of the immune responses to two injections of an antigen as measured by serum antibody levels. Note the very different characteristics of the primary and secondary immune responses.

an antigen identical to the first. A secondary response may be provoked many months or years after first exposure to an antigen, although its size tends to decline as time passes. A secondary response may also be provoked in animals in which the primary response was so weak as to be undetectable. It is clear therefore that the antibody system has the ability to "remember" a previous exposure to an antigen. For this reason, the secondary immune response is sometimes known as an **anamnestic, or memory, response** ("anamnesis" is the Greek word for memory).

Administration of a third dose of the same antigen to an animal provokes an immune response with an even shorter lag period and higher, more prolonged antibody levels. However, repeated injections of an antigen do not lead to higher and higher antibody levels indefinitely. The level of antibodies in blood is well regulated and tends to stabilize even after multiple doses of antigen are administered.

The stimulation of resistance to infectious disease through repeated injections of microbial antigens forms the basis of most current vaccination techniques. Thus tetanus toxoid is normally given to children in four doses to ensure immunity to tetanus. The interval between each of the first three doses is eight weeks, and the fourth dose is given one year after the third dose. Booster doses given at ten-year intervals are recommended to maintain high levels of protective antibody. The effectiveness of this procedure is demonstrated by the rapid decline in the number of cases of tetanus since vaccination was introduced (Fig. 1–4).

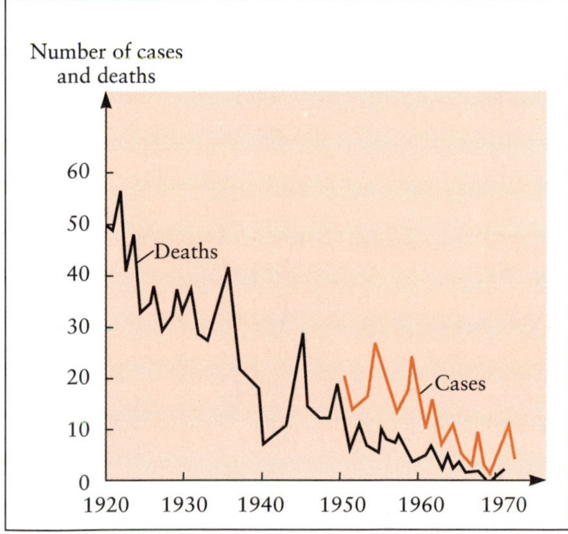

Figure 1–4 The progressive decline in the number of deaths from and cases of tetanus in Canada from 1920 to 1970. The major drop that occurred around 1940 is attributed to the introduction of tetanus toxoid. *(Used with permission of the Laboratory Centre for Disease Control, Canada.)*

CELL-MEDIATED IMMUNE RESPONSES

The production of antibodies in response to a foreign antigen is responsible for the development of resistance to many infectious agents. It is not, however, the whole story. Despite their potency, antibodies alone cannot protect the body against all invaders. Some bacteria such as those that cause tuberculosis can enter cells and hide from antibody molecules. Viruses also grow within cells and cannot be eliminated by antibodies. In these cases a second type of immune response has to be employed. The infected cells have to be destroyed by special killer cells called **lymphocytes.** This process is called **cell-mediated immunity** and is best demonstrated by the rejection of a foreign organ graft.

It was not until the development of modern surgical techniques that it became possible to successfully transplant organs or tissues between individuals. The first attempts at transplantation almost inevitably failed (although see Fig. 1–5). For example, if a piece of skin is surgically transplanted from one individual to a second, unrelated individual, the initial events will be encouraging. At first, the grafted skin will appear healthy, and blood vessels will connect the graft to the underlying tissues. After about one week, however, the graft will become pale and die a few days later. If the dead graft is not removed, it will fall off. The destruc-

Figure 1–5 Saints Cosmos and Damian, two early Christian martyrs, are supposed to have been the first to overcome the cell-mediated immune response when they successfully transplanted a leg from a black donor to a white recipient. *(The Bettmann Archive.)*

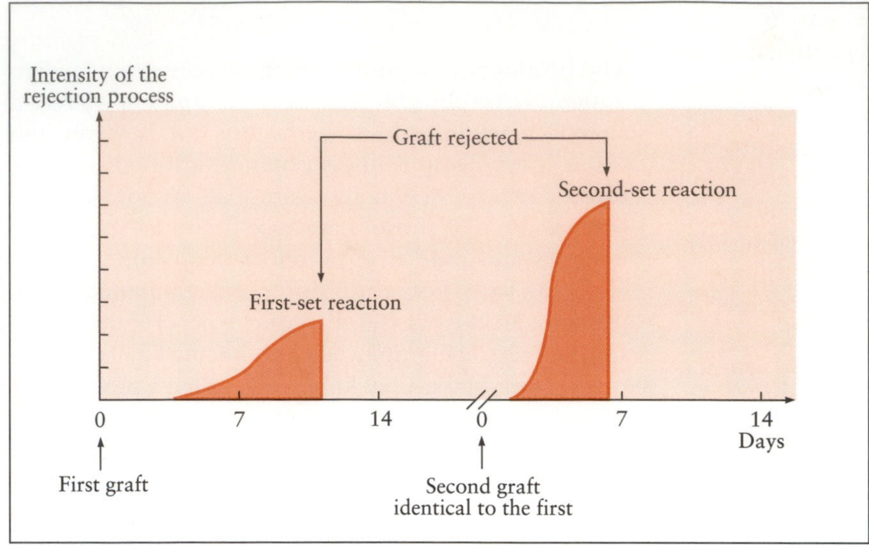

Figure 1–6 The time course of the immune response to a foreign skin graft as shown by the intensity of the rejection process. Note the similarities with Figure 1–4, especially the shortened lag period and the increased intensity of a second-set reaction.

tion of a skin graft in this way is due to damage to the connecting blood vessels as a result of an immune response against the graft. The rejection of a graft by an unsensitized person in this way is known as a **first-set reaction** (Fig. 1–6).

If a second graft from the same donor is placed on the same recipient, the grafted skin usually survives for no longer than one or two days before an immune response causes its destruction. During the rejection process, the underlying tissue becomes severely inflamed. This intense rejection process is known as a **second-set reaction.** Experience has shown that unless the immune response is suppressed, tissue grafts between unrelated individuals are rapidly destroyed. Untreated grafts survive only when the donor and the recipient are genetically identical, as in the case of identical twins.

Transplantation of organs between individuals is an artificial procedure that has no counterpart in nature. Animals did not evolve immune responses of this type just to frustrate transplantation surgeons. What, then, is the reason for the rapid rejection of foreign grafts? One clue is the finding that persons who have received drugs to suppress graft rejection are more likely than normal to develop cancer. This has led to the concept that the rejection of foreign grafts reflects the existence of a system that can recognize and destroy cells that differ from normal. Such abnormal cells may include not only the cells in a foreign organ graft, but also cells infected by viruses, cells altered by chemicals, and cancer cells.

The process of graft rejection is a specific form of immune response. A second-set reaction occurs only when the second graft is identical to the first. If the second graft is from a new donor, only a first-set reaction is provoked. The rejection process also possesses a memory, as shown by the fact that a second-set reaction can be induced months or even years after the first graft is rejected. Graft rejection differs significantly from the immune response to tetanus, how-

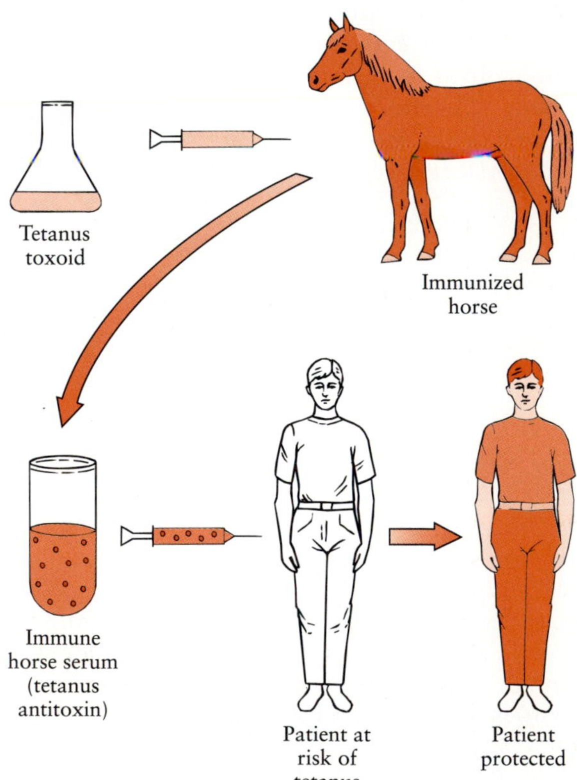

Figure 1–7 The passive transfer of immunity to tetanus by means of antibody. Antibodies against tetanus toxin (tetanus antitoxin) made in a horse make a human recipient temporarily immune to tetanus. Immunity to tetanus must therefore be mediated by antibodies in the horse serum.

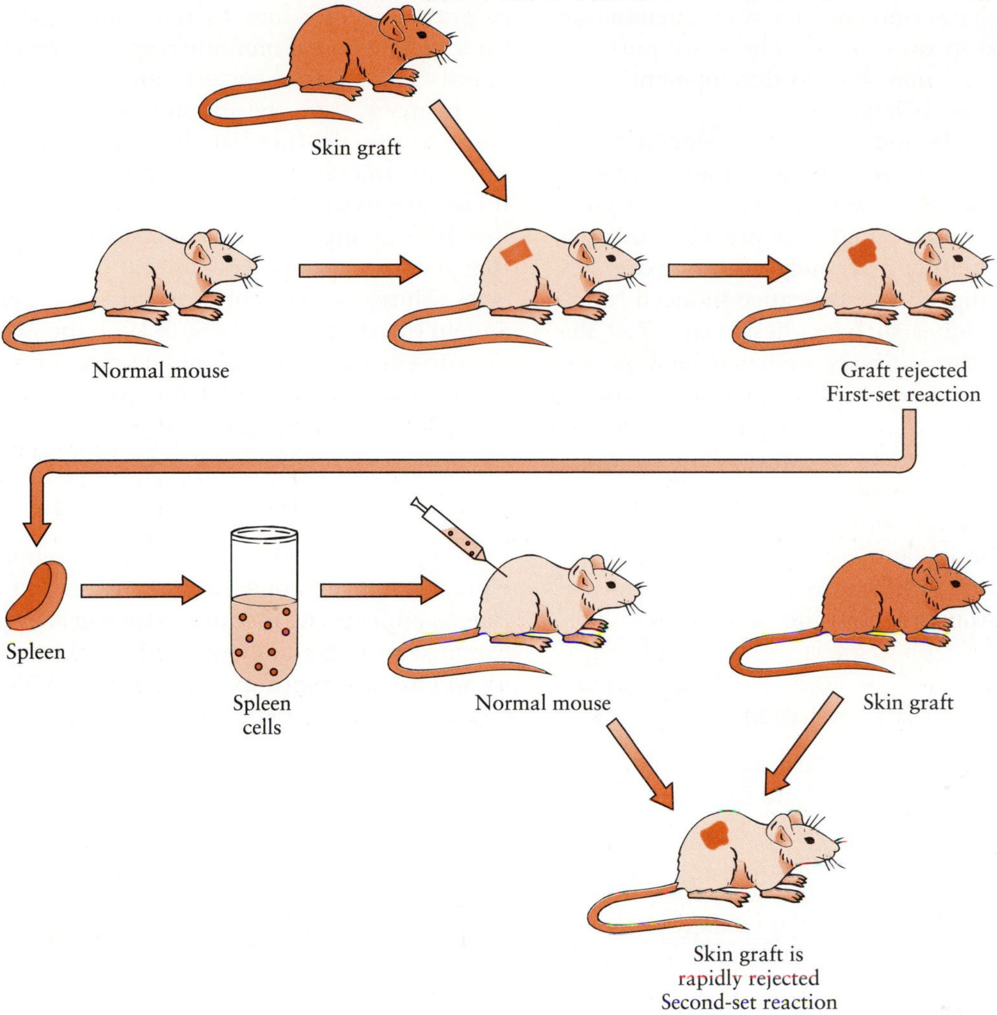

Figure 1–8 The adoptive transfer of immunity to a foreign skin graft by cells. In this case, spleen cells taken from a mouse that had rejected a skin graft from another mouse strain will transfer immunity to the graft. A skin graft placed on the adoptively immunized mouse will be rapidly rejected in a second-set reaction. Rejection of foreign grafts must therefore be a cell-mediated reaction.

ever, in that the rejection process cannot be transferred by giving serum from a sensitized person to a normal person; graft rejection therefore is not caused by antibodies (Fig. 1–7). If, however, spleen cells are taken from an animal that has rejected a graft and these are administered to an unsensitized, identical animal, then the recipient of the cells will develop the ability to mount a second-set response (Fig. 1–8). Thus the ability to reject grafted tissue can be transferred by the spleen cells. The cells found in the spleen that are responsible for graft rejection are called **lymphocytes.** Lymphocytes able to cause graft rejection are readily obtained form the spleen, lymph nodes, or blood of a sensitized individual. The rejection of foreign tissue grafts is therefore a very different type of immune response from that mediated by antibodies. Graft rejection is one example of **cell-mediated immunity.**

TOLERANCE

The production of antibodies against tetanus toxoid and the rejection of a graft from an unrelated person occur because both the toxoid and the graft are recognized by the body as foreign. This in turn implies that a mechanism exists that permits the body to identify foreign material and distinguish this from its own cells. Indeed, healthy, normal individuals do not usually mount an immune response against themselves and are therefore said to be self-tolerant. If the body loses this ability to discriminate, it may mount immune responses against its own tissues. These will cause tissue destruction and may lead to development of an **autoimmune disease** (Chapter 31).

Tolerance may be induced experimentally by administering foreign antigens to very young animals, whose immature immune systems respond by becoming tolerant. Tolerance can also be provoked in adult animals by giving either extremely small or extremely large doses of antigen. The tolerance induced by experimental techniques such as these may affect the cell-mediated or the antibody-mediated immune system or both. Tolerance is specific for the inducing antigen and, like the other forms of immune response, will fade unless boosted by reexposure to the same antigen. In the absence of reexposure, tolerance is gradually lost, and the animal regains the ability to respond to the foreign antigen. Self-tolerance is therefore maintained by continuous exposure to normal body components throughout life.

IMMUNE RESPONSE MECHANISMS

As described previously, there are two basic forms of immune response, the antibody- and the cell-mediated responses. Immunological tolerance is a special form of these two basic responses (Fig. 1–9).

The two basic immune responses serve complementary functions. Thus the cell-mediated responses detect and eliminate abnormal cells such as virus-infected or chemically modified cells. In contrast, the antibody-mediated responses protect against non-cell-associated invaders, such as bacteria or parasites. This distinction is not absolute. Antibodies can contribute to graft rejection and to immunity against viruses, while cell-mediated immune responses can participate in resistance to many bacterial and parasitic infections.

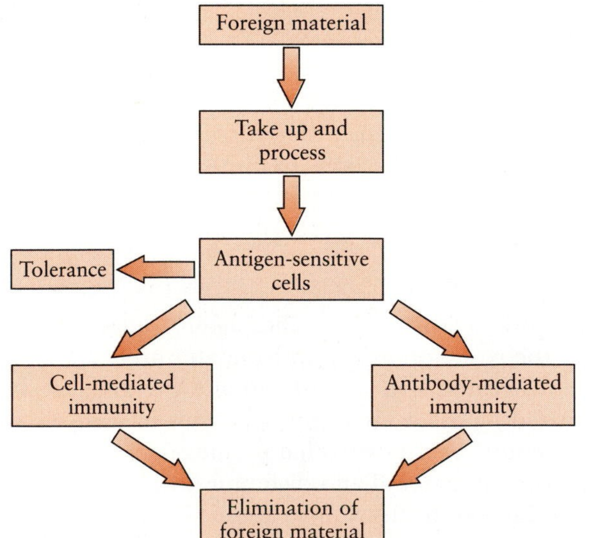

Figure 1–9 The essential features of the immune response.

In many ways, the immune system may be compared to a totalitarian state in which citizens who conform are tolerated but foreigners are expelled and those citizens who "deviate" are eliminated. Although this analogy must not be pursued too far, it is clear that such regimes possess several essential characteristics. These include border police to repel outsiders and an internal police force to keep the populace under surveillance and to eliminate dissidents at the first trace of dissent. A state of this type also requires an identification system so that foreigners not possessing the necessary identification can be readily detected and dealt with.

Similarly, when antigen enters the body, it must be trapped and processed. If it is recognized as foreign, this information must be processed in such a way that an appropriate immune response is triggered and the foreign material eliminated. In the next chapter we will discuss the properties required for an antigen to stimulate an immune response.

CURRENT PROBLEMS IN IMMUNOLOGY

The science of immunology has moved at an astonishingly rapid pace during the past decade. New discoveries are being made at an ever more rapid rate. Advances in cell biology, genetics, and biochemistry have all combined to increase our understanding of immunological processes. Advances in protein and nucleic acid characterization enable proteins and their genes to be identified very rapidly. Our ability to selectively "knock-out" genes has provided us with a unique ability to determine the true importance of individual molecules. Notwithstanding this phenomenal progress, many key questions remain to be answered, and serious disease problems mediated by immunological processes remain to be solved. Some areas of basic biochemistry such as the role of carbohydrates in glycoprotein function have scarcely been touched. Significant gaps exist in our understanding of **signal transduction** processes. Overconfident attempts to use **cytokines** for therapeutic purposes have revealed their intrinsic toxicity and a failure to appreciate the true complexity of the regulatory processes of the immune system.

Immunology, of course, has had significant triumphs. The remarkable decline in the significance of infectious diseases in developed countries is almost entirely attributable to the efforts of the early immunologists such as Pasteur, Koch, and von Behring. The elimination of smallpox and the control of diseases such as poliomyelitis and measles have had a significant impact, especially in developed countries. New drugs have made organ grafting a routine procedure and provided new hope to many who would have died prematurely. Simple immunological procedures have controlled major causes of infant mortality such as **hemolytic disease** of the newborn. New immunological tests have enabled physicians to diagnose diseases with speed and accuracy.

Nevertheless, many major disease problems remain to be solved, and students contemplating a career in immunology can be assured of many exciting opportunities. Unquestionably, the most important of these diseases is **AIDS,** a virus disease that results in the destruction of the immune system. Although a major effort to control AIDS has been under way for many years, solutions to the problems of AIDS are not readily apparent at this time. Indeed, it is difficult to be optimistic about the prospects for its control in the near future.

Cancer continues to be responsible for about a quarter of the deaths in developed countries. Chemotherapy, radiation, and surgery save many patients, although success rates are climbing only very slowly or not at all. Tumor immunology has been an area slow to develop and yield practical results. The mechanism of action of the cells that protect us against cancer (**natural killer [NK] cells**) is still poorly understood. Nevertheless, immune responses against tumor cells do occur, and some innovative immunological treatments of cancer are showing encouraging results. About 25% of the population of the United States claim to suffer from some form of **allergy.** This can range from mild discomfort to severe life-threatening asthma or shock. Although the overall immunological basis for allergies has been known for many years, progress in controlling or preventing these diseases has been very slow.

Five percent of adults in developed countries suffer from one or more autoimmune diseases. These include juvenile-onset diabetes mellitus, which requires daily injections of insulin in order for a patient to lead a normal life; rheumatoid arthritis, a painful crippling disease that causes enormous amounts of suffering; and multiple sclerosis, a disease of the central nervous system that affects a significant number of young adults. All these diseases as well as many other autoimmune conditions require considerably more study before they can be successfully treated or prevented.

Even in the areas where immunology has had its greatest successes, such as vaccination and organ grafting, there is still much room for improvement. In the case of vaccines, too few children, even in the United States, are receiving appropriate vaccines. In Third World countries, the problem is even more acute. Highly innovative techniques such as incorporation of antigen genes into fruits and vegetables may enable vaccines to reach almost everyone. Parasitic diseases, although of minor importance in developed countries, are a major threat to health and economic development in much of the world. Many diseases, including the major parasite-mediated ones, malaria, schistosomiasis, trypanosomiasis, and leishmaniasis, remain uncontrolled and cause enormous problems worldwide. New diseases are emerging that will have to be controlled by immunological techniques. Unavoidable adverse reactions to vaccines have resulted in problems of liability and slowed the development of vaccines in the United States. In the case of organ grafting, major issues that are emerging include the purchasing of organs and the problems of obtaining an adequate supply of fresh tissue.

Advances in immunology will continue to yield dividends in the form of improved health care and the alleviation of suffering. Immunological methods of contraception may contribute to reducing the growth of the world's population; animal vaccination can increase the available food supply.

KEY WORDS

Anamnestic response p. 7
Antibody(ies) p. 3
Antigens p. 5
Autoimmune disease p. 10
Cell-mediated immunity p. 7
Clonal selection theory p. 5
Complement p. 4
Cowpox p. 2
Paul Erhlich p. 5
First-set reaction p. 8
Fowl cholera p. 3
Immunogenetics p. 5
Edward Jenner p. 2
Lag period p. 6
Lymphocytes p. 7
Memory response p. 7
Eli Metchnikoff p. 4
Lady Mary Montagu p. 2
Neutralization p. 6
Passive immunization p. 4
Louis Pasteur p. 3
Phagocytes p. 4
Plasma p. 4
Primary immune response p. 6
Radioimmunoassay p. 5
Secondary immune response p. 6
Second-set reaction p. 8
Serum p. 4
Smallpox p. 2
Tolerance p. 9
Toxoid p. 6
Vaccination p. 2
Vaccine p. 3
Variolation p. 2
Emil von Behring p. 3

QUESTIONS

1. The basic biological function of the cell-mediated immune system is to
   a. cause delayed hypersensitivity
   b. destroy abnormal cells
   c. combat virus infections
   d. combat intracellular bacteria
   e. reject skin grafts
2. The primary immune response is characterized by
   a. induction by one dose of antigen
   b. a long lag period
   c. low levels of antibody produced
   d. rapid decline
   e. all of the above
3. Vaccination using cowpox (vaccinia) was first described by
   a. Louis Pasteur
   b. Lady Mary Montagu
   c. Emil von Behring
   d. Edward Jenner
   e. Paul Ehrlich
4. Antibodies are
   a. foreign substances that stimulate an immune response
   b. a form of vaccine
   c. serum proteins that protect the body
   d. a type of cell that protects the body
   e. a virus that causes disease
5. A secondary immune response
   a. occurs after several months
   b. has a very short lag period
   c. is mediated by cells
   d. only occurs against toxoids
   e. can be induced by one high dose of antigen
6. Rejection of a foreign skin graft is an example of
   a. destruction of virus-infected cells
   b. tolerance
   c. antibody-mediated immunity
   d. a secondary immune response
   e. a cell-mediated immune response
7. Graft rejection occurs because
   a. transplantation is bad for the body
   b. grafts resemble foreign organisms
   c. the body destroys foreign cells
   d. grafts act as infectious agents
   e. grafts invade the body
8. Tolerance is a
   a. failure to make any antibodies
   b. defect in the cell-mediated immune system
   c. method of inducing autoimmunity
   d. method of preventing autoimmunity
   e. loss of resistance to infections
9. Antibodies are necessary to
   a. help destroy invading bacteria
   b. destroy foreign grafts
   c. help develop tolerance
   d. cause autoimmunity
   e. produce toxoids
10. The first successful vaccine was against
    a. cowpox
    b. smallpox
    c. fowl cholera
    d. anthrax
    e. rabies
11. Why do you think it took so long for the concept of immunity to infectious diseases to become established in the West?
12. What features of smallpox made its eradication possible whereas we have made little progress in eradicating other infectious diseases such as tuberculosis and influenza?
13. If the immune system is so efficient, why do we continue to suffer from infectious and parasitic diseases?
14. Describe the time course of a typical antibody response to tetanus toxoid. Outline a vaccination schedule that makes effective use of this information.
15. Antibodies to tetanus made in horses can be used to protect an individual against tetanus. What might be the disadvantages of this form of immunization?
16. Why should an individual be able to reject and destroy grafted tissues from another individual of the same species? Should a pregnant animal be able to reject its fetus? If not, speculate why.
17. Is it possible to confer cell-mediated immunity on an unsensitized recipient by transferring cells? Is this a practical method of immunization, and what problems may be encountered when conducting this type of experiment?
18. Speculate on the consequences if the immune system were to lose its ability to distinguish self-antigens from foreign ones.

Answers: 1b, 2e, 3d, 4c, 5b, 6e, 7c, 8d, 9a, 10b

SOURCES OF ADDITIONAL INFORMATION

Series in Immunology

Many publishers produce a series of review texts in immunology. These include *Advances in Immunology* (Academic Press, New York), *Immunological Reviews* (Munksgaard, Copenhagen), *Annual Review of Immunology* (Annual Reviews, Palo Alto, California), and *Seminars in Immunology* (Academic Press, London).

Immunology Journals

There are a vast number of immunology journals of varying degrees of general usefulness. In writing this text, I have found that the most useful journal was *Nature,* followed by *Science, Journal of Immunology, Cell, Immunology Today, Journal of Experimental Medicine, Proceedings of the National Academy of Sciences, New England Journal of Medicine, FASEB Journal, Hospital Practice, Infection and Immunity,* and *Immunology.*

CHAPTER 2

Infections and Diseases

Robert Koch *Born in Germany in 1843, Robert Koch shared with Louis Pasteur the early key discoveries in bacteriology and antibacterial immunity. Indeed, he, in effect, devised the basic principles of modern microbiology and the rules for showing that an organism actually caused disease. He was the first to identify and isolate the causal agents of major diseases such as tuberculosis, cholera, and anthrax. His studies of tuberculosis were especially important since he was the first to show that it was an infectious disease. He developed the tuberculin skin test, an immunological technique that is still one of the key diagnostic methods for this disease. Koch was awarded the Nobel prize in 1905 for his studies on tuberculosis. He died in 1910.*

CHAPTER OUTLINE

CHAPTER CONCEPTS

1. Many microorganisms would invade the animal body to make use of its rich resources, if they were not prevented by the immune system.
2. The major groups of microorganisms that can cause disease are bacteria, fungi, viruses, protozoa, and helminths.
3. Not all bacteria cause disease. Many colonize the body and assist in normal body functions as members of the normal flora.
4. Viruses can only grow inside the cells of their host. They may kill those cells, alter their function, or make them divide uncontrollably to cause cancer. Some viruses can hide inside cells without causing disease.
5. Parasites such as protozoa or helminths rarely benefit from the death of their animal. Thus they may not cause disease unless they occur in the wrong tissues or in excessive numbers.
6. If, for some reason, the immune system is suppressed, then resistance to infectious agents may be severely impaired, and serious disease may result.

This chapter reviews some basic aspects of microbiology. The first section describes the microorganisms that cause disease. We then look at the major groups of disease-causing agents—bacteria, fungi, viruses, protozoa, and helminths. The text focuses on the structure of these organisms as well as how they cause disease. These are both immensely relevant to the ways in which the immune response defends the body against invasion. Finally, we briefly review the effects of suppression of the immune system on resistance to infectious disease.

ORGANISMS THAT CAUSE DISEASE

The animal body is a warm, nutrient-rich environment that is very attractive to microorganisms. If the body was not actively defended, it would be rapidly invaded by many different organisms. This can be readily seen if a piece of raw meat is placed in the open. It begins to decompose, change color, and smell as it is invaded by organisms attracted to the protein-rich environment. The organisms that seek to invade the body and use its many resources for themselves are a complex mixture of bacteria, fungi, viruses, and other parasites (Fig. 2–1).

Microorganisms can be divided into two major groups based on their cellular structure. One group, the **eukaryotic organisms**, consists of cells that contain a distinct nucleus with a well-defined nuclear membrane and characteristic organelles such as mitochondria. In this group are included organisms such as the protozoa and the fungi. The second group, the **prokaryotic organisms**, have cells whose genetic material is free within the cytoplasm. As a result, they do not contain a distinct nucleus with a nuclear membrane. They also lack obvious cellular organelles. The best examples of prokaryotes are the bacteria (Fig. 2–2). Prokaryotes and eukaryotes both contain two types of nucleic acid, DNA and RNA.

Viruses are organisms that do not exist as cells (Fig. 2–3). Unlike the prokaryotes or eukaryotes, they contain only a single type of nucleic acid (either DNA or RNA). Because of their lack of metabolic enzymes, viruses can grow only within cells. Some viruses such as the bacteriophages grow inside prokaryotic cells, whereas others grow inside eukaryotic cells, including the cells of the animal body. As with other organisms, some viruses are pathogenic and others may be able to integrate themselves inside a cell so that the cell's function is unimpaired and disease does not result.

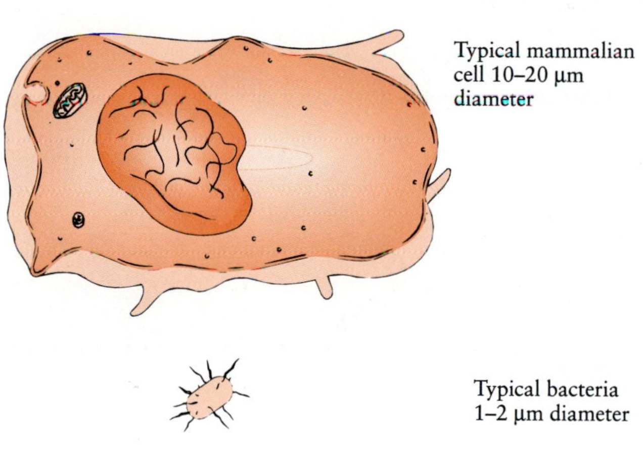

Figure 2–1 Some pathogenic microorganisms and their size relative to a typical mammalian cell. Rickettsiae are difficult to see under a light microscope, whereas viruses cannot be seen by light microscopy.

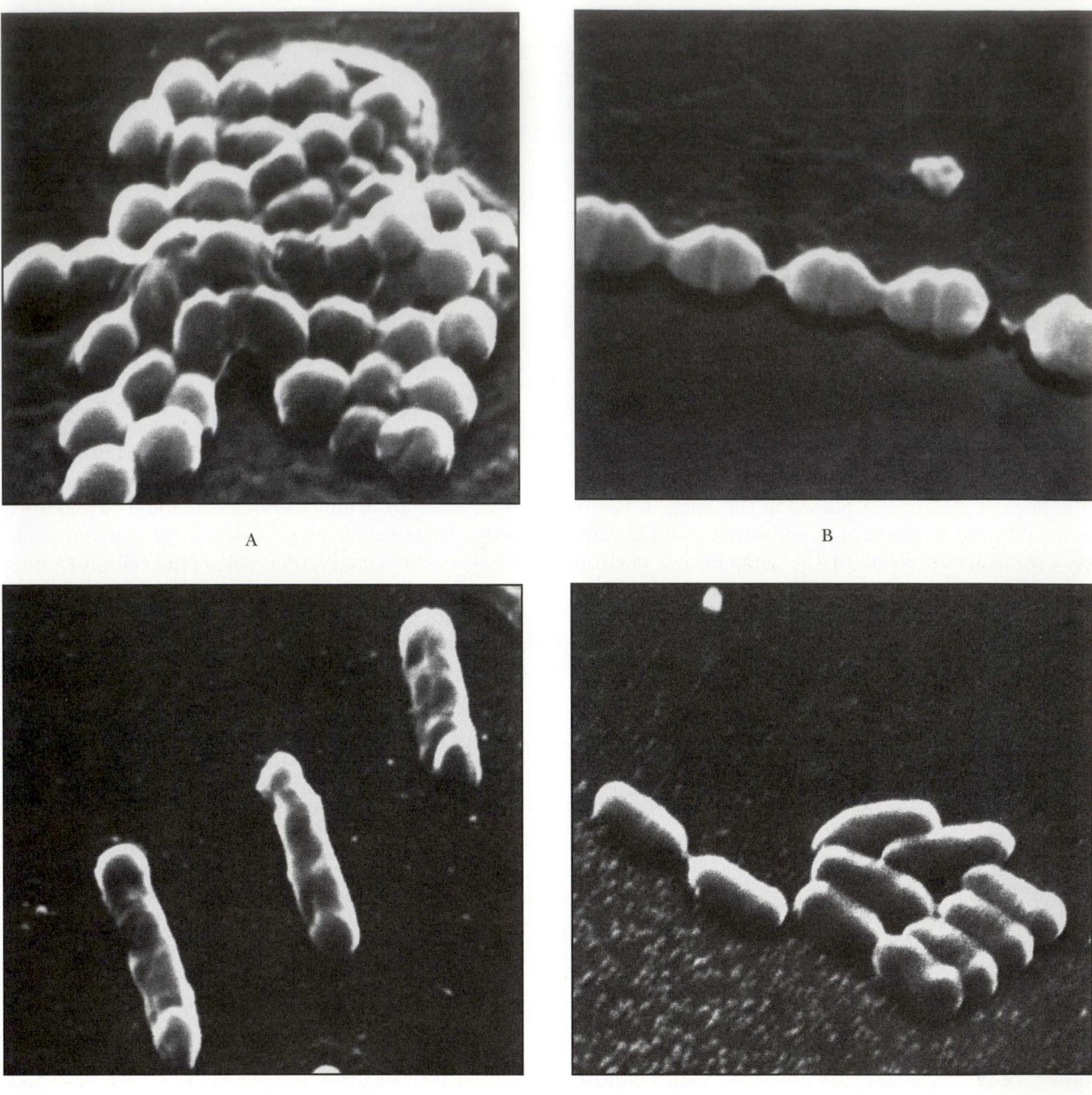

Figure 2–2 Scanning electron micrographs of some typical bacteria. A, *Staphylococcus aureus.* B, *Streptococcus agalactiae.* C, *Salmonella typhimurium.* D, *Escherichia coli.* (All magnifications ×8000)

Normal healthy animals may carry many of these integrated viruses within their cells.

Another type of infectious agent is even smaller than the viruses. This is called a prion. Although the nature of prions is still debated, evidence is accumulating that they lack nucleic acid and may consist of a protein that can copy itself! Prions cause rare neurological diseases in animals. These include scrapie in sheep and kuru and Creutzfeldt-Jakob disease in humans.

The microorganisms that are of interest to us are those that are found in association with animals since these may on occasion be harmful and cause disease. An organism that can cause disease is said to be **pathogenic**. It is important to point out, however, that only a small proportion of the world's microorganisms are associated with animals and that only a very small proportion of these are pathogens.

Although many organisms are associated with the animal body, usually growing on its surfaces, very few

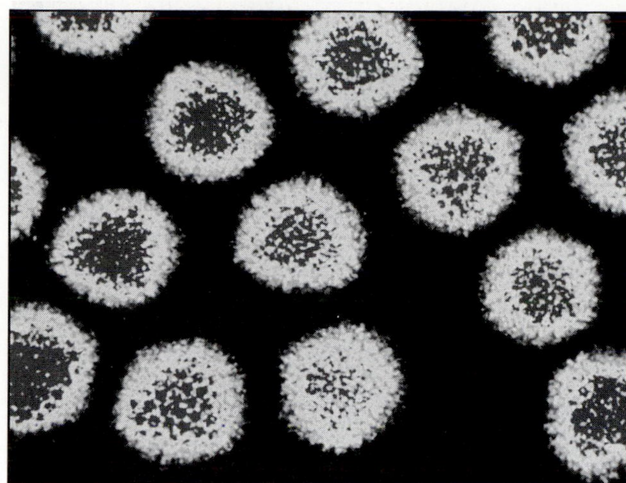

Figure 2–3 Electron micrograph of Adenovirus. (© Omikron/Science Source /Photo Researchers, Inc.)

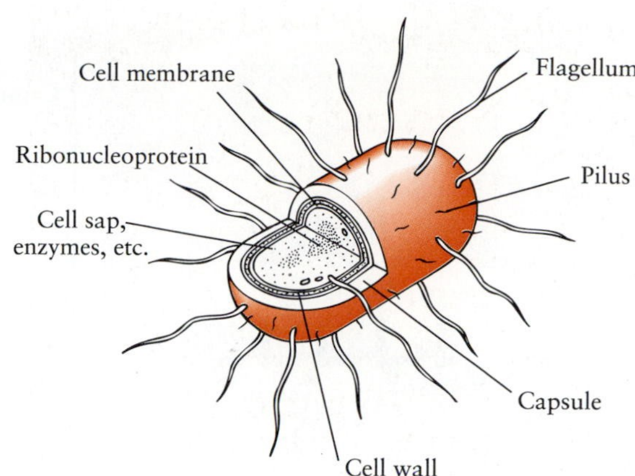

Figure 2–4 The structure of a typical bacterium. This bacillus is motile as shown by its flagella. It is surrounded by a capsule. Its pili help it to attach to cells and may also play a role in mating.

have the ability to invade tissues deeply and damage them. These pathogens vary greatly in their ability to cause disease. This ability is termed **virulence**. Thus a highly virulent organism has a greater ability to cause disease than an organism with low virulence. If an organism can cause disease almost every time it invades a healthy individual, even in low numbers, then it is a **primary pathogen**. Examples of primary pathogens include the influenza virus, and AIDS virus (human immunodeficiency virus, HIV), and *Salmonella typhi*, the cause of typhoid fever. Other pathogens may be of such low virulence that they will cause disease only if administered in very high doses or if the immune defenses of the body are impaired. These are **opportunistic pathogens**. Examples of opportunistic pathogens are those that invade AIDS patients whose immune systems have been destroyed. They include bacteria such as *Mycobacterium intracellulare* and parasites such as *Pneumocystis carinii*. These organisms rarely, if ever, cause disease in normal healthy individuals.

BACTERIA

Bacteria are spherical or rod-shaped prokaryotic organisms with a simple basic structure (Fig. 2–4). They consist of a cytoplasm containing the essential elements of the cell surrounded by a cell membrane. The cell membrane is in turn covered by a cell wall that, in some bacteria, is enclosed by a capsule.

Classification

Bacteria are classified according to certain key properties. The first of these is shape. Thus bacteria may form small spherical bodies called cocci. Other bacteria are rod-shaped and are called bacilli. Some bacteria are covered by a thick wall of polysaccharide or protein called a capsule. Capsules protect bacteria against destruction in the body (Chapter 5), and antibodies directed against the capsule may therefore protect an infected animal. Bacteria also have several surface structures. Thus many have flagella extending from their surface. Flagella (singular: flagellum) are fine, threadlike structures that can be moved by the bacteria. This repeated movement can make the bacteria move through their surrounding fluid. Bacteria may also possess pili (singular: pilus). These are short projections that cover the surfaces of some bacteria. They are of several different types with different functions. Thus some pili help bacteria to attach to surfaces and can be important in helping bacteria invade the body. Other pili are used to transfer genetic material between bacteria in a form of mating.

The second key property that is used to classify bacteria is the nature of their cell wall (Figs. 2–5 and 2–6). The structure of the bacterial cell wall determines just how the organisms will stain when treated with certain dyes. The stain used to determine this is called Gram's stain, after Hans Christian Gram, the Danish microbiologist who invented this technique in 1884. Gram found that bacteria stained in one of two different ways when treated with the dye called crystal violet followed by iodine. (The iodine acts as a mordant and ensures that the dye is firmly attached to the surface.) Some bacteria, called gram-positive bacteria, take up the crystal violet blue dye so strongly after iodine treatment that the dye cannot be washed out by an organic solvent such as alcohol or acetone. Thus when examined under the microscope, the bacteria appear purple. In contrast, in other bacteria, called

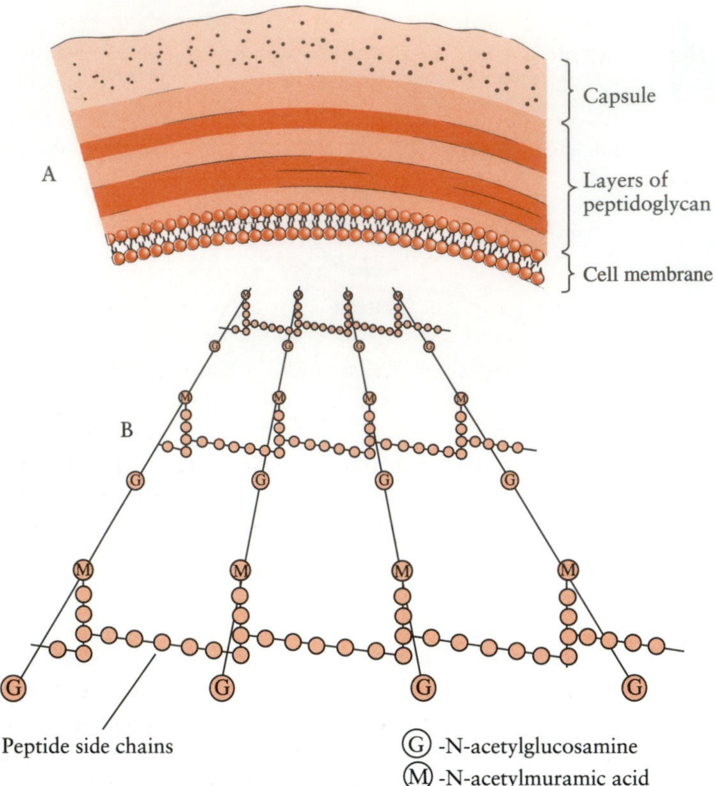

Figure 2–5 A, The structure of the cell wall of a typical gram-positive bacterium. B, The structure of the cell wall peptidoglycan.

gram-negative bacteria, the crystal violet can be washed out by alcohol or acetone treatment. These bacteria can then be counterstained using a dye of a different color such as safranine. Under the microscope, these bacteria appear pink. Almost all pathogenic bacteria can be classified as either gram-positive or gram-negative.

In addition to the two major features, shape and staining properties, bacteria can be classified according to the nutrients that they use for growth, their

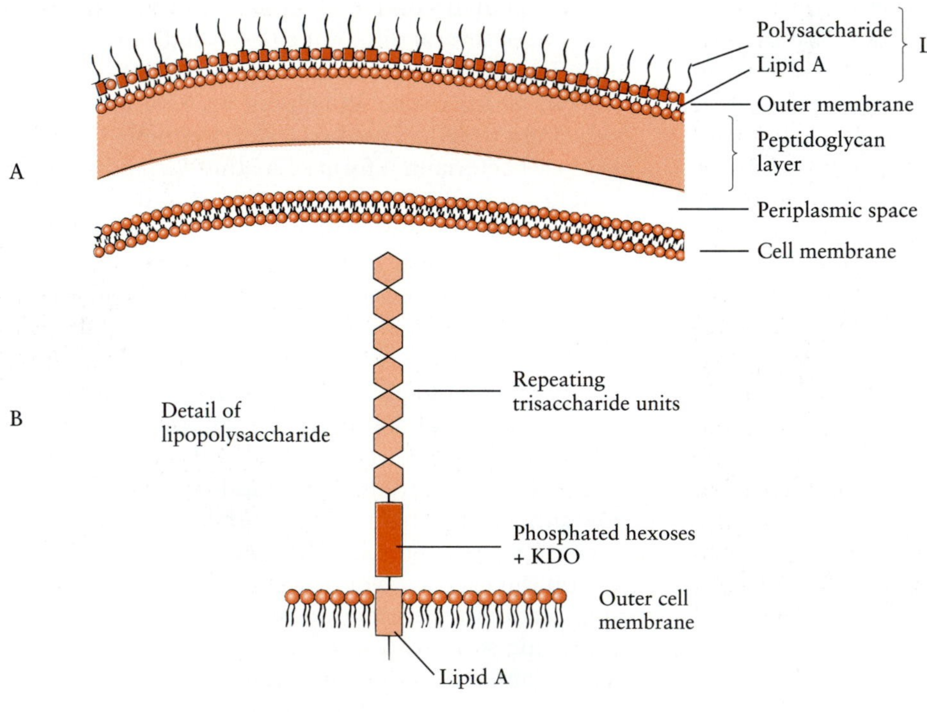

Figure 2–6 A, The structure of the cell wall of a typical gram-negative bacterium. B, The outer component of the bacterial wall is an oligosaccharide composed of repeating trisaccharides or tetrasaccharides. The core of the structure consists of various hexoses and 2-keto-3-deoxyoctonic acid (KDO). This structure is also known as endotoxin.

oxygen requirements, their ability to form spores, and, most basically, the relationship of their DNA to that of other bacteria.

As noted, gram-positive and gram-negative bacteria have very different structures in their cell walls. The cell wall of gram-positive bacteria is largely composed of peptidoglycans (chains of alternating *N*-acetylglucosamine and *N*-acetylmuramic acid cross-linked by short peptide side chains) (Fig. 2–5); the cell wall in gram-negative bacteria has, in addition to a much thinner peptidoglycan layer, an outer membrane composed of a polysaccharide-lipid-protein structure. The polysaccharide component consists of an **oligosaccharide** attached to a lipid (lipid A) and to a series of repeating trisaccharides (Fig. 2–6). Animals may recognize these trisaccharides as foreign and make antibodies against them. Some bacteria are classified according to this antigenic structure. For example, the salmonellae have been classified into about 2000 species on this basis. Their polysaccharide antigens are called O antigens. The cell wall lipopolysaccharides of gram-negative bacteria induce toxic responses when injected into an animal and thus are also called **endotoxins**.

Some forms of bacteria differ significantly from the main group described previously. Thus spirochetes are long coiled organisms that have a thin flexible cell wall (Fig. 2–7). Although they are gram-negative, they stain so poorly that they must usually be visualized with other special stains or with a special darkfield microscope. Another important group of atypical bacteria is the Rickettsia and their relatives (Fig. 2–8). These are very small, nonmotile coccobacilli that also stain poorly with Gram's stain. Unlike other bacteria, they grow only inside cells. They are often spread by biting arthropods such as lice or ticks.

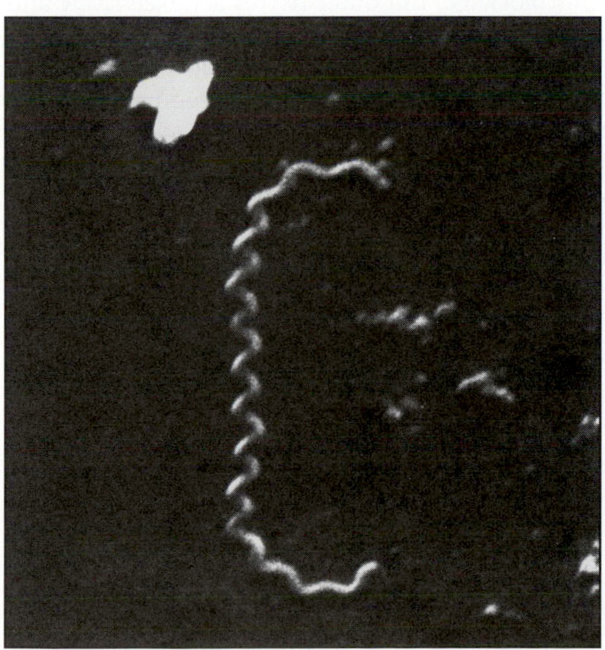

Figure 2–7 A scanning electron micrograph of *Leptospira pomona*. A typical spirochete.

How Bacteria Cause Disease

Bacteria are well adapted to growing in the warm, wet, nutrient-rich environment of the animal body. Indeed, many grow in large numbers on the body surface, where they do no harm. They are usually prevented from invading the body itself by the immune system. Nevertheless, on occasion, bacteria can enter the body and cause disease. They do this by several mechanisms. Some important diseases caused by bacteria are listed in Table 2–1.

Tissue Invasion

Some bacteria may invade the body and release enzymes that cause tissue damage. For example, they secrete enzymes such as hyaluronidase, collagenase, and elastase. These enzymes can split the connective tissue structural molecules, hyaluronic acid, collagen, and elastin, respectively. As a result, the extracellular tissues open up and allow the bacteria to move readily between cells. Other bacteria can secrete enzymes that disrupt fibrin, the major component of blood clots, and still others release coagulases that have the opposite effect and cause blood to clot. Organisms that can grow within blood clots may find coagulases of

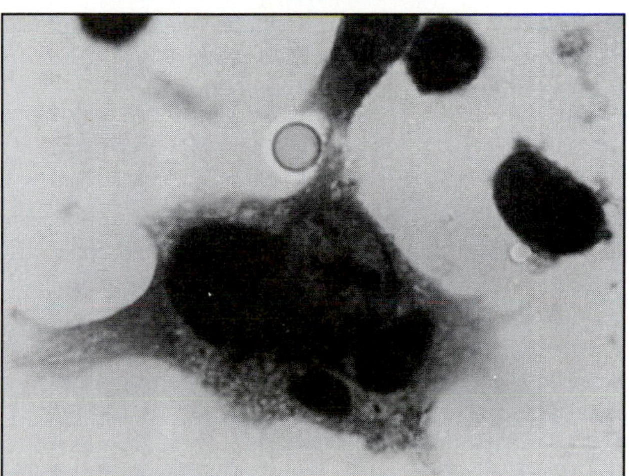

Figure 2–8 A cell infected by the rickettsial organism, *Chlamydia psittaci*. The large dark bodies within the cell are composed of masses of organisms. *(Courtesy of Dr. D.K. Winsor, Jr.)*

Table 2–1 Some Important Bacterial Diseases and Their Causes

Bacteria	Disease
Streptococcus pyogenes	Pharyngitis, septicemia, wound infections
Streptococcus pneumoniae	Pneumonia, meningitis, otitis
Haemophilus influenzae	Meningitis, pneumonia
Corynebacterium diphtheriae	Diphtheria
Mycoplasma pneumoniae	Pneumonia
Bordetella pertussis	Whooping cough
Legionella pneumophila	Pneumonia
Mycobacterium tuberculosis	Tuberculosis
Neisseria meningitidis	Meningitis
Neisseria gonorrhoeae	Gonorrhea
Clostridium tetani	Tetanus
Staphylococcus aureus	Toxic shock syndrome, wound infections, diarrhea
Escherichia coli	Diarrhea
Shigella sonnei	Diarrhea
Salmonella typhi	Typhoid fever
Vibrio cholerae	Diarrhea (cholera)
Chlamydia trachomatis	Venereal disease, trachoma
Borrelia burgdorferi	Lyme disease
Treponema pallidum	Syphilis

great assistance in generating sites for their growth. Once they invade the tissues, bacteria can cause damage by consuming nutrients that would otherwise be used by body cells. Thus some bacteria growing in tissues can use oxygen, proteins, or other nutrients and will eventually starve nearby cells.

Although most bacteria that invade the body usually do so by penetrating the spaces between cells, some bacteria can get into cells (see Fig. 25–5). These intracellular bacteria may multiply inside cells, where they are sheltered from many of the defense mechanisms of the body—a significant advantage. A special form of cell-mediated immune response is required to destroy these bacteria.

Exotoxins

Invading bacteria commonly release toxic molecules as they grow in tissues. These toxic molecules are called **exotoxins**. They consist of proteins formed in the bacterial cytoplasm and either secreted by the living organism or released when the bacteria die and break up. The best example of a disease caused by an exotoxin is tetanus. Tetanus is caused by the release of a powerful toxin from the anaerobic, gram-positive bacillus *Clostridium tetani* (Fig. 2–9). Because *C. tetani* is a strict anaerobe, it grows only in tissues where the oxygen tension is very low. Thus it is found in deep penetrating wounds where tissue death has resulted in a loss of the local blood supply. Tetanus toxin, also called tetanospasmin, is released by the bacteria and is activated by proteolytic enzymes in the tissues. It travels from the site of bacterial growth along nerves to the spine. There the toxin interferes with the activities of inhibitory neurons (these are the nerve cells that regulate the activities of other cells). As a result of this blockage, the other nerves act excessively and make the muscles that they connect to contract excessively. This "tetanic spasm" can lead to paralysis. If the tetanus toxin paralyzes the respiratory muscles, death from asphyxiation may result. Other organisms that produce exotoxins include *Clostridium botulinum* and *Clostridium perfringens,* the causative agents of two forms of food poisoning. The toxin of *C. botulinum* blocks nerve transmission and can cause death as a result of respiratory paralysis. Botulinum toxin is very potent, since as few as eight molecules of the toxin can block nerve transmission. The exotoxin of *C. perfringens* is an enzyme that disrupts cell membranes and so causes tissue destruction.

Endotoxins

The walls of gram-negative bacteria are formed by a complex of polysaccharides, lipids, and proteins. These wall components are toxic to mammals and are called **endotoxins** (Fig. 2–6). Most of the toxic properties of endotoxins are due to a specific component called lipid A. Lipid A exerts its toxic effect by stimulating the release of cellular proteins called cytokines.

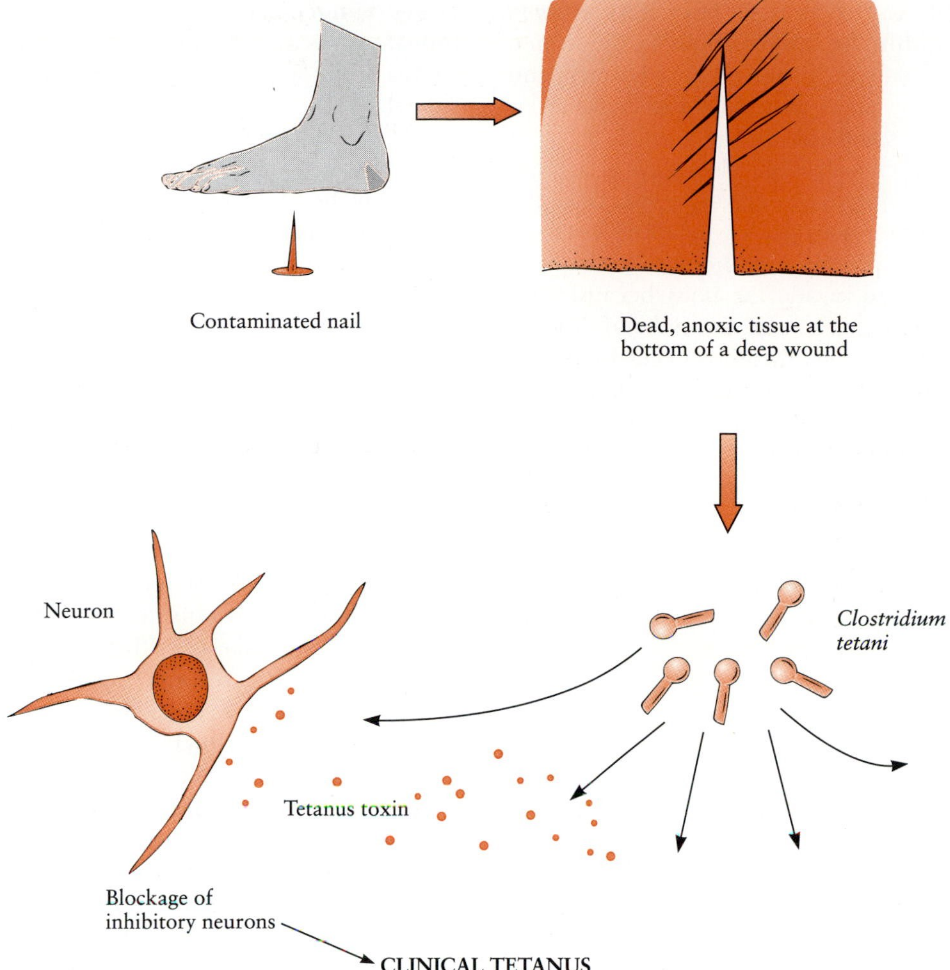

Figure 2–9 The pathogenesis of tetanus. For details, see text.

The cytokines cause a fever, malaise, hypotension, and vomiting. This condition is called **septic shock** (see Fig. 12–1, Chapter 12). In sufficiently high doses, endotoxin causes multiple organ failure and death.

The Normal Flora

Although most bacteria live free in the environment, many grow in close association with the animal body. Every animal carries a dense, complex microbial population, or flora, on its surfaces. Once a mammal is born and leaves the sterile environment of the uterus, it becomes exposed to environmental organisms. Some of these permanently colonize the body within a few days. Some of these bacteria are so adapted to life on an animal that they are restricted to certain anatomic areas. For example, viridans streptococci colonize the mouth and throat, lactobacilli colonize the intestines of breast-fed infants, and coliform organisms colonize the intestines of bottle-fed infants. The **normal flora** that eventually establishes itself on the skin and in the nose, digestive tract, and urogenital tract is relatively stable. Its composition depends on local environmental factors such as pH, oxygen tension, and moisture. This normal flora can be of significant benefit to an animal. Thus in the intestine these bacteria may synthesize vitamins such as vitamin K, riboflavin, or biotin. They may also detoxify potential carcinogens and contribute to the metabolic breakdown of certain foods. This is important in herbivores such as horses, cattle, and sheep. Mammals do not make the enzymes that can break down complex carbohydrates such as cellulose in plants. Thus much plant structural material cannot be used as a source of energy. Herbivores therefore rely on their intestinal flora to digest plant material and make its breakdown products available to the animal. They have evolved large chambers within the intestinal tract—the rumen in cattle and the cecum in horses—where bacteria and protozoa can digest plant materials and make their breakdown products available to the animal.

The normal microbial flora also plays a defense role, since by occupying the microenvironment of the body surface, it excludes other, less adapted microorganisms. Thus many potential pathogens fail to colonize body surfaces and invade the body because they are excluded by the presence of the normal flora. If we destroy the normal body flora, by excessive **antibiotic** treatment, for example, we run the risk of allowing other, pathogenic organisms to replace them.

In individuals whose body defenses are severely impaired, as happens in AIDS patients, for example, some of the normally nonpathogenic organisms of the normal flora may act as opportunistic pathogens and invade the body to cause disease.

FUNGI

Although we think of fungi as being large plantlike organisms such as mushrooms or molds growing on stale bread, there are also microscopic fungi that can invade the body and cause disease. The fungi are distinctly different from the bacteria insofar as they are eukaryotic organisms (Fig. 2–10). They also differ from the bacteria in that they may exhibit sexual reproduction and have a complex morphology. They have very complex cell walls consisting of carbohydrates. Fungi can invade the body by the inhalation of fungal spores into the lung or by invasion of tissues by spores or fungal fragments as a result of trauma. Some fungal infections such as those due to *Candida albicans* are especially common in patients with a defective immune system. Some, such as the ringworm fungi, invade only the superficial layers of the skin. Others such as histoplasma and cryptococcus may, however, invade deeper tissues such as the lungs.

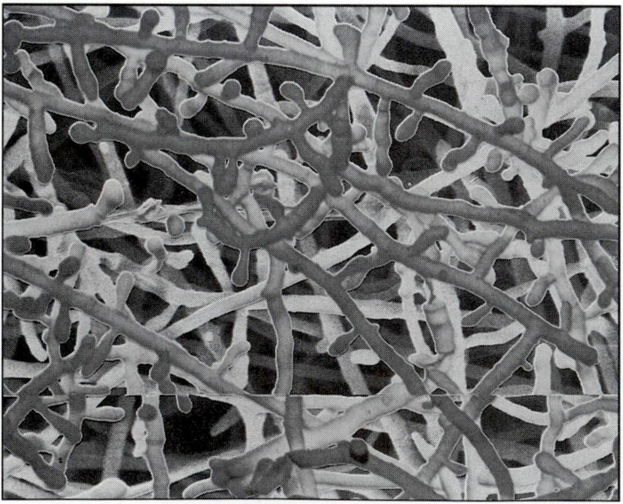

Figure 2–10 Scanning election micrograph (SEM) of the fungus *Trichophyton tonsurans,* a causative agent of the skin condition *tinea* or ringworm. Magnification × 275. (© E. Guecho-CNRI/Science Photo Library/Photo Researchers, Inc.)

VIRUSES

Viruses differ from bacteria in many ways. They are much smaller than bacteria, so small that they cannot be seen under the light microscope (Fig. 2–11). Their structure is also very different in that they consist simply of a nucleic acid core surrounded by protective layers of protein. Most important, viruses, being so simple, cannot grow outside living cells. They need to invade cells and take over the cellular enzymes in order to survive. Viruses are thus obligate intracellular parasites, or parasites of the genome. Viruses are classified by the type of nucleic acid they contain—either DNA or RNA—and then by their shape, size, and the arrangement of their protein subunits. The protective layer of protein is termed the **capsid**, and the complete structure of nucleic acid and capsid is called the **nucleocapsid**. Some viruses may also be surrounded by an envelope containing lipoprotein and glycoproteins. The complete viral structure is called a **virion**.

Virions are the nonreproducing, transmissible form of a virus. To reproduce themselves, viruses must invade a cell. As a result of virus infection, a cell may be destroyed (through cell lysis), or alternatively, the virus genome may integrate itself into the cell genome. This integration may alter cellular functions

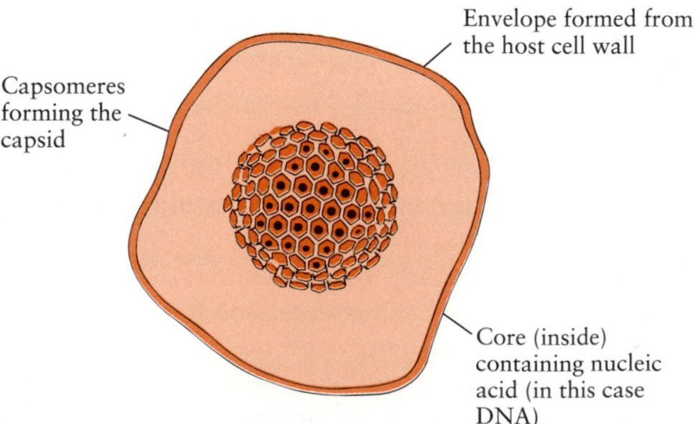

Figure 2–11 The structure of a typical enveloped virus. In this case a herpesvirus.

Table 2–2 Some Important Animal Viruses and the Diseases They Cause

DNA Viruses	Virus	Disease
Poxviruses	Variola major	Smallpox
Herpesviruses	Herpes simplex	Mouth and genital ulcers
	Herpes zoster	Chicken pox
	Epstein-Barr virus	Infectious mononucleosisa
Papovaviruses	Papillomavirus	Warts
Hepadnaviruses	Hepatitis B	Liver disease
Parvoviruses	Canine parvovirus	Diarrhea
RNA Viruses		
Picornaviruses	Poliovirus	Poliomyelitis
	Rhinoviruses	Common cold
Togaviruses	Rubella virus	German measles
Flaviviruses	Yellow fever virus	Yellow fever
Rhabdoviruses	Rabies virus	Rabies
Orthomyxoviruses	Influenza virus	Influenza
Paramyxoviruses	Measles virus	Measles
	Mumps virus	Mumps
Retroviruses	Human immunodeficiency virus	AIDS

such as the surface properties or growth habits of the cell. Some virus-infected cells may become cancerous (Table 2–2).

Virus Multiplication

The first step in the invasion of a cell by a virus occurs when the virus binds to the cell surface. This is called adsorption. The virus first binds to specific receptors on the cell surface. These are not receptors designed for the convenience of viruses, but they have a physiologic function in normal cells. Thus the rabies virus binds to the receptor for acetylcholine, a neurotransmitter. The Epstein-Barr virus (the cause of infectious mononucleosis) binds to a receptor for C3, one of the proteins of the complement system. Rhinoviruses that cause the common cold bind to cell surface adhesion proteins called integrins. The bound virus is then taken into the cell. Once inside a cell the virus capsid breaks open so that its nucleic acid is released into the cell cytoplasm—a process called uncoating. Once the virus genome is uncoated it begins the process of replication (Fig. 2–12). First, the host cell DNA, RNA, and protein synthesis are inhibited so that only the viral genetic information is processed. The site where this happens differs between viruses and depends on their nucleic acid. If the virus, for example, a herpesvirus, contains DNA as its genetic material, then this viral DNA is replicated so that its amount in the cell increases. The new viral DNA is then transcribed into viral messenger RNA, and this RNA is translated into new viral proteins. The host cell also replicates the viral nucleic acid so that large quantities of viral DNA are produced. This viral DNA is packaged inside the capsid so that a complete new virion is formed. If the

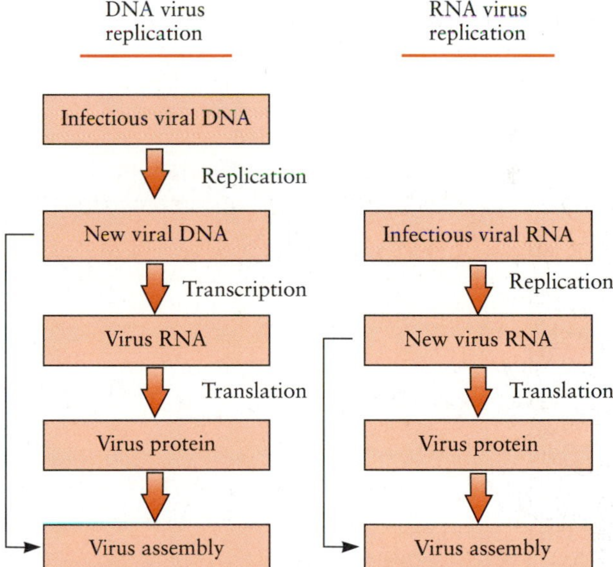

Figure 2–12 The major steps in virus replication. DNA viruses replicate in a manner very similar to that of mammalian cells. Thus they transcribe their DNA to RNA, which is then translated into protein. Some RNA viruses, in contrast, can replicate without using DNA at all. They can replicate their RNA and translate this into protein without using DNA. Once sufficient new nucleic acid and capsomere protein are made, new virions can be assembled.

virus is unenveloped, then the infected cells disintegrate and the virions are released into the environment. If the virions are enveloped, then they leave the cell by budding through the cell surface. The cell membrane that encloses them serves as the new envelope. The released virions may thus spread to nearby cells and invade them.

If a virus contains RNA rather than DNA as its nuclear material, its replication takes a slightly different course. For most RNA viruses such as influenza or poliomyelitis virus, cellular DNA is not used. In poliovirus infection for example, the virus single-stranded RNA (the "positive" strand) is used as a template to synthesize a complementary "negative" strand of RNA. These "negative" strands are then used to generate new "positive" strands that can then be translated into viral proteins. Some viruses contain double-stranded RNA where they use only one of the strands generated during replication. In other RNA viruses the RNA from the infecting virus may be used to make new RNA that can be directly translated into viral proteins.

A different replication mechanism is employed in the case of some RNA tumor viruses and immunodeficiency viruses (Fig. 2–13). These viruses are called **retroviruses** since their RNA is first reversely transcribed into DNA. This is done by means of an enzyme called a **reverse transcriptase**. As a result, new viral DNA is formed that moves into the cell's nucleus and is then integrated into the host cell genome as a provirus. This proviral DNA can then be transcribed into RNA as well as being able to replicate itself. The proteins and RNA can then be packaged into complete new virions.

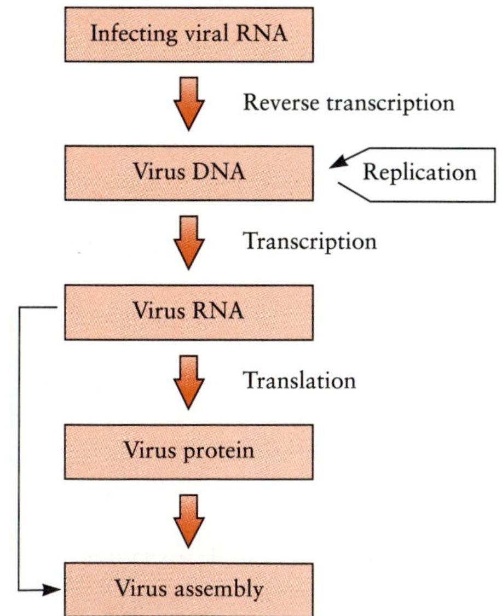

Figure 2–13 The major steps in the replication of a retrovirus. These viruses use an enzyme called a reverse transcriptase in order to generate new DNA from their RNA. This is a major exception to the rule that information flows from DNA to RNA. Once viral DNA is made it can be integrated into the host cell genome where it is replicated to produce more DNA or transcribed into RNA to make viral protein.

Virus Diseases

The most obvious effect of virus infection on a cell is its destruction. The cells round up, die, and lyse, releasing new virions that can infect nearby cells. Small virions are readily released by cells in this way, although large virions such as poxviruses may remain within the ghosts of infected cells. Sooner or later, after virus infection, a cell's outer membrane will be altered. This alteration can result in several changes. For example, some viruses make cells fuse with each other to form giant masses of cytoplasm containing hundreds of nuclei called **syncytia**. In other cases, especially with enveloped viruses, alterations may occur in the surface proteins of an infected cell. These changes may be detectable by the immune system and trigger an immune response.

Some viruses may infect cells yet not make infectious virus. This happens, for example, if the virus multiplication cycle is arrested. This arrest may not be permanent. The virus multiplication may resume some time (several weeks to years) later. This stage of arrested development is called latency and is a feature of herpesvirus infections.

Virus genes may become permanently integrated into the host cell genome, and the converse—cellular genes becoming part of the viral genome—may also occur. In the first case, these endogenous viral genes can code for new proteins that will be expressed on otherwise normal cells. They can play a role in controlling the immune system (Chapter 18). In the second case, cellular genes involved in signal transduction and cell division may become incorporated in the viral genome. These are called **oncogenes.** A cell infected with a virus containing oncogenes may divide uncontrollably to form a cancer (Chapter 7).

Not all cells in the body are equally susceptible to virus infection. Usually, rapidly dividing cells are more susceptible to infection than nondividing or slowly dividing cells. Viruses may also attack specific cell targets. Thus the AIDS virus (HIV) can only bind and invade cells carrying a protein on their surface called CD4. In this case, CD4 acts as a specific receptor for the virus, allowing it to bind and invade the cell. Similar receptors are found on other cells so that specific viruses generally tend to invade only certain cell populations. For example, influenza viruses tend to invade cells in the respiratory system, and hepatitis viruses invade liver cells.

PROTOZOA

Protozoa are eukaryotic parasites that occur in a wide variety of shapes and sizes (Fig. 2–14). Many may be capable of living in the environment such as in water or soil, and others live in invertebrate hosts, especially ticks and insects. When they invade the body, protozoa may be found in many tissues. Some, such as the trypanosomes (the cause of African sleeping sickness), circulate freely in the bloodstream. Others, such as *Giardia* and *Cryptosporidium*, attach to the surface of the intestine, and some, such as *Toxoplasma*, are intracellular parasites that can grow readily within any nucleated cell (Table 2–3).

Protozoan diseases have many manifestations reflecting the varied ways that protozoa invade tissues. Some have a relatively simple pathogenesis. For example, *Toxoplasma* grows inside cells, and as the number of parasites increases, the parasitized cell eventually bursts open. The released parasites can then move to a new cell and invade it. Plasmodia, the cause of malaria, cause red cell destruction and anemia. Other protozoa may release toxins that cause cell destruction or even provoke immune responses that destroy parasitized cells.

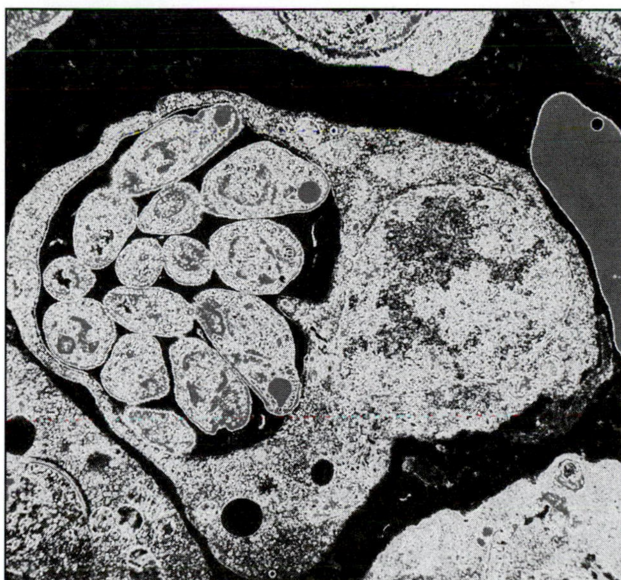

Figure 2–14 Transmitting election micrograph of malaria infecting blood. (© Omikron/Science Source/Photo Researchers, Inc.)

Table 2–3 Some Important Protozoan Diseases

Organism	Disease
Plasmodium falciparum	Malaria
Trypanosoma rhodesiense	Encephalitis (sleeping sickness)
Leishmania major	Skin lesions (leishmaniasis)
Toxoplasma gondii	Fetal death (toxoplasmosis)
Trichomonas vaginalis	Vaginitis (trichomoniasis)
Giardia lamblia	Diarrhea (giardiasis)
Entamoeba histolytica	Diarrhea (amebiasis)
Cryptosporidium muris	Diarrhea

HELMINTHS

The largest of the foreign organisms that may invade the body are the helminth parasites. These are worms, either roundworms (nematodes), tapeworms (cestodes), or flukes (trematodes). As parasites, these organisms usually derive no benefit from killing their host, who acts as a source of nutrients as well as shelter. Most helminth parasites rarely cause disease, especially if present in small numbers. Only when unusual numbers of helminths enter an individual, provoke a severe tissue reaction, or become lost and end up at a critical organ such as the brain or eye is disease or death likely to occur. Diseases such as schistosomiasis (also called bilharziasis) result from the host's aggressive attempts to destroy the parasite rather than any direct actions of the parasite itself.

THE IMPORTANCE OF THE IMMUNE SYSTEM

The primary function of the immune system is to protect the body against invasion by microorganisms. Normally, this system works well and individuals remain free of clinical disease for much of their lives. If, however, the immune system fails, the body immediately becomes more susceptible to infectious or parasitic diseases. If the defect in the immune system is sufficiently severe, the consequent infectious disease may be lethal. An excellent example of this is AIDS, in which the HIV wipes out the immune system, and the diseases occurring as a result of secondary infections by opportunistic pathogens are invariably fatal. In other situations, the defect in the immune system may be relatively minor and the resulting disease may be fairly mild. Thus stress may result in decreased resistance to the common cold viruses. Clearly, the best defense against infectious diseases is to have a fully functioning immune system.

KEY WORDS

Adsorption p. 23
Bacillus (bacilli) p. 17
Bacterium (bacteria) p. 17
Capsid p. 22
Capsule p. 17
Cestode p. 25
Coccus (cocci) p. 17
DNA virus p. 23
Endotoxin p. 20
Eukaryotic organism p. 15
Exotoxin p. 20
Flagellum (flagella) p. 17
Fungus (fungi) p. 22
Gram's stain p. 17
Helminth p. 25
Immune system p. 25
Intracellular bacteria p. 20
Latency p. 24
Lipid A p. 20
Lipopolysaccharide p. 19
Nematode p. 25
Normal flora p. 21
Nucleocapsid p. 22
Oncogene p. 24
Opportunistic pathogen p. 17
Pathogenesis p. 21
Pathogenic organism p. 16
Peptidoglycan p. 19
Pilus (pili) p. 17
Primary pathogen p. 17
Prion p. 16
Prokaryotic organism p. 15
Protozoa p. 25
Replication p. 23
Retrovirus p. 24
Reverse transcriptase p. 24
Rickettsia p. 19
RNA virus p. 24
Septic shock p. 21
Spirochete p. 19
Syncytium (syncytia) p. 24
Tetanus toxin p. 20
Trematode p. 25
Uncoating p. 23
Virion p. 22
Virulence p. 17
Virus p. 22

QUESTIONS

1. Which one of the following is eukaryotic?
   a. bacteria
   b. viruses
   c. fungi
   d. rickettsiae
   e. prions
2. The outer coat of budding viruses is the
   a. nucleoprotein
   b. capsid
   c. envelope
   d. cell surface antigens
   e. all of the above
3. Which of the following are obligate intracellular organisms?
   a. bacteria
   b. viruses
   c. protozoa
   d. helminths
   e. fungi
4. A pathogenic organism is one that
   a. does not cause disease
   b. grows within cells
   c. stimulates an immune response
   d. causes disease
   e. is gram-positive
5. An exotoxin is a
   a. toxin that works outside the body
   b. protein toxin with specific activity
   c. bacterial lipopolysaccharide
   d. viral subunit
   e. product of a rickettsial organism
6. A virus particle is a
   a. capsid
   b. nucleocapsid
   c. capsomere
   d. viroid
   e. virion
7. A retrovirus is a virus that
   a. uses DNA as its nuclear material
   b. uses a reverse transcriptase
   c. causes immunosuppression
   d. has a cell surface envelope
   e. causes cancer in animals
8. Protozoa are parasites that
   a. use RNA as their nuclear material
   b. are exclusively found in blood
   c. are eukaryotic organisms
   d. are gram-negative organisms
   e. cause severe skin disease
9. Immunosuppression causes increased susceptibility to
   a. protozoan disease
   b. viral disease
   c. bacterial disease
   d. rickettsial disease
   e. all of the above
10. Proteoglycans are major cell wall components in
    a. gram-positive bacteria
    b. protozoa
    c. helminths
    d. gram-negtive bacteria
    e. rickettsia
11. List the types of microorganisms that can cause disease in animals. Indicate whether they are eukaryotic or prokaryotic.
12. What are the major differences in structure between gram-positive and gram-negative bacteria? How do these differences influence the diseases caused by these organisms?
13. List the ways in which bacteria may cause disease. Why

do bacteria cause disease? Does this help the organism in any way?

14. In what ways can the normal bacterial flora be of benefit to an animal?
15. Outline the ways by which viruses can cause tissue damage or otherwise cause disease.
16. How do retroviruses differ from other viruses in their replication cycle?
17. Outline the types of infections that might arise in an animal whose immune system fails to function.

Answers: 1c, 2c, 3b, 4d, 5b, 6e, 7b, 8c, 9e, 10a

SOURCES OF ADDITIONAL INFORMATION

Baron, E.J., Peterson, L.R., and Finegold, S.M. Barley and Scotts Diagnostic Microbiology, 9th ed. Mosby, St. Louis, 1994.

Davis, B.D., Dulbecco, R., Eisen, M.N., and Ginsberg, M.S. Microbiology, 4th ed. J.B. Lippincott Co. Philadelphia, 1990.

Jawetz, E., et al. Medical Microbiology, 18th ed. Appleton and Lange, Norwalk, Connecticut, 1989.

Joklik, W.K., Willett, H.P., Amos, D.B., and Wilfert, C.M. eds. Zinsser Microbiology, 19th ed. Appleton and Lange, Norwalk, Connecticut, 1988.

Koneman, E.W., et al. Color Atlas and Textbook of Diagnostic Microbiology, 3rd ed. J.B. Lippincott Co. Philadelphia, 1988.

Lenz, T.L. The recognition event between virus and host cell receptor: A target for antiviral agents. J. Gen. Virol., 71:751–766, 1990.

Prusiner, S.B., Collinge J., Powell J., and Anderton B., eds. Prion Diseases of Humans and Animals. Ellis Morwood, New York, 1992.

Scott, A. Pirates of The Cell. Basil Blackwell, Oxford 1985.

CHAPTER 3

Key Biochemical Processes

James D. Watson *Born in Illinois in 1928, James Watson shared the Nobel Prize for Medicine with two British scientists, Francis Crick and Maurice Watkins, in 1962. The prize was awarded for their roles in elucidating the structure of DNA, the key molecule of heredity. Watson and Crick developed a model of the molecule by trial and error based on very early results from X-ray crystallography performed by Watkins. Once they arrived at the structure, they immediately realized how the DNA molecule could carry information and code for the amino acid sequence of proteins. James Watson heads this chapter in recognition of the key role of DNA in all life processes, not just in immunity.*

CHAPTER OUTLINE

CHAPTER CONCEPTS

1. All the information needed to manufacture any of the body's proteins, including those important in immunity, is contained in the nucleotide sequence of a cell's DNA. To make a protein, the DNA is first transcribed into RNA. The RNA is then translated into a peptide chain.
2. The shape of a peptide chain is determined by only its amino acid sequence.
3. Certain basic structures are observed in many proteins. These include the α-helix and the β-pleated sheet.
4. Proteins may have carbohydrate side chains added to them in the endoplasmic reticulum or the Golgi apparatus. Glycoproteins formed in this way are characteristically located on cell membranes.

In this chapter, we review in a very simple manner protein synthesis and structure as they apply to immunology. We first review the process by which protein structure is inherited and briefly describe DNA transcription and RNA translation. We then look at basic features of protein structure, focusing especially on how the precise shape of protein molecules is determined. Finally, since many of the proteins involved in the immune responses have carbohydrates attached to them, we review the key features of protein glycosylation. All of the key molecules that play a role in immunology are proteins. These include all the cell surface receptors as well as soluble proteins such as **antibodies** and **cytokines.** An understanding of the mechanisms by which proteins are produced is critical for the student of immunology.

BASIC GENETIC MECHANISMS

Proteins consist of chains of assorted amino acids linked by peptide bonds. Twenty major amino acids are available for use, so the number of possible combinations of amino acids in a peptide chain is truly enormous. The precise order in which amino acids are assembled along a peptide chain is determined by the genes that code for each specific protein. These genes, composed of DNA, are first transcribed into RNA. The RNA in turn is translated into protein.

DNA Transcription

DNA molecules are very long polymers containing thousands of deoxyribonucleotides (Fig. 3–1). There are four kinds of nucleotide, each containing a different base—cytosine, thymine, adenine, and guanine. The nucleotides are joined in a sequence that is characteristic of each organism. Each nucleotide is linked to the phosphate groups on its neighbors through the 3′ (pronounced "3-prime") and 5′ sites on its deoxyribose ring. Each 3′ site binds to its neighbor's 5′ site. Because of this, a nucleic acid chain has polarity. One end has a free 3′ end, the other has a free 5′ end. (When printing nucleotide or gene sequences, it is usual to place the 5′ end on the left and the 3′ end

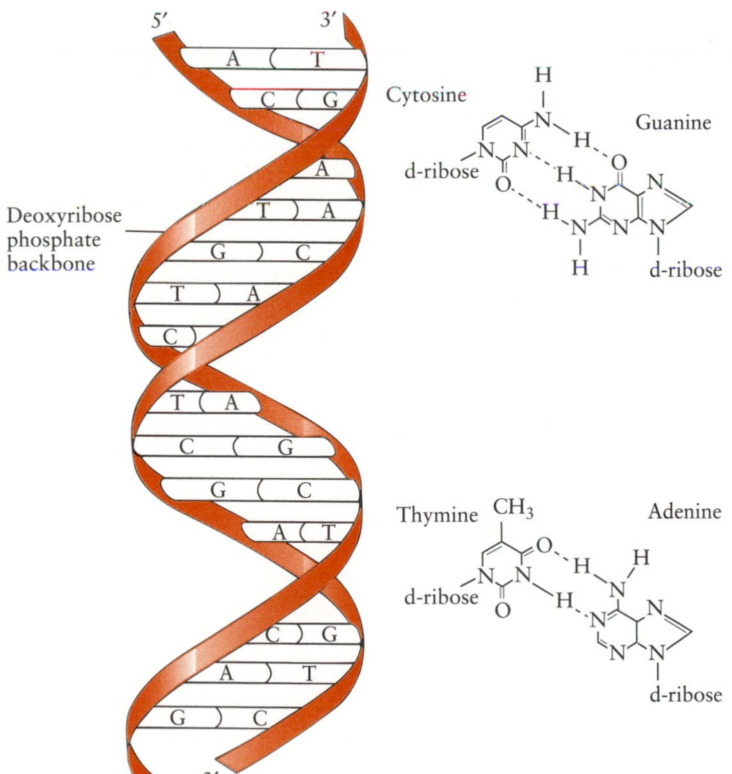

Figure 3–1 The structure of the DNA double helix. Because of their structure, guanine bases can bind only to cytosine and vice versa. Likewise, thymine bases can bind only to adenine. This highly specific pairing of nucleotides is the key to the genetic code.

on the right.) DNA is found within the chromosomes in the cell nucleus. When a cell divides, an exact copy of the DNA is made so that each daughter cell carries a complete set of DNA. A DNA molecule consists of two DNA chains entwined around a single axis—the so-called double helix. Within the double helix, the two chains are connected by hydrogen bonds between purine and pyrimidine bases. These bases are positioned at right angles to the sugar-phosphate backbone. The hydrogen bonds can form pairs only between adenine and guanine or between thymine and cytosine. As a result, when a new DNA molecule is being synthesized, one chain is used as a template on which the bases of the new chain are assembled. Because of the strict pairing rules, the new chain must form a complementary sequence to the template chain. The arrangement of the four bases along the nucleic acid molecule is thus faithfully replicated. This sequence of bases is also important because it encodes precise information regarding the proteins of the body.

To synthesize a protein, the information about its specific amino acid sequence must be made available to the cell. The DNA in the nucleus contains this information, which is in the form of a code. Each set of three bases (a codon or triplet) codes for a specific amino acid. For example, the triplet GCA (guanine-cytosine-adenine) codes for the amino acid arginine, and the triplet TGA (thymine-guanine-adenine) codes for the amino acid threonine. Thus as one reads along a DNA chain, one can read off the corresponding sequence of amino acids. It should be noted that to make sense of the genetic code the nucleotide triplets must be read correctly. If one or two nucleotides are

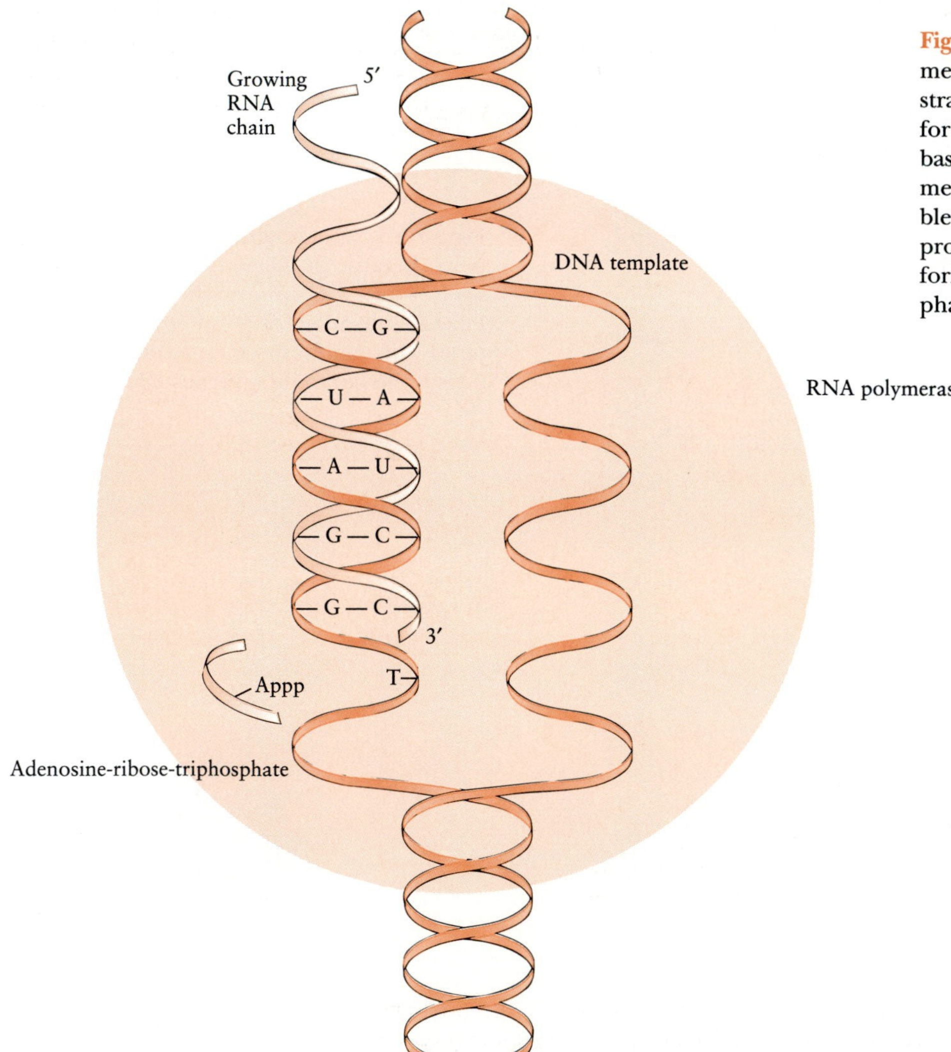

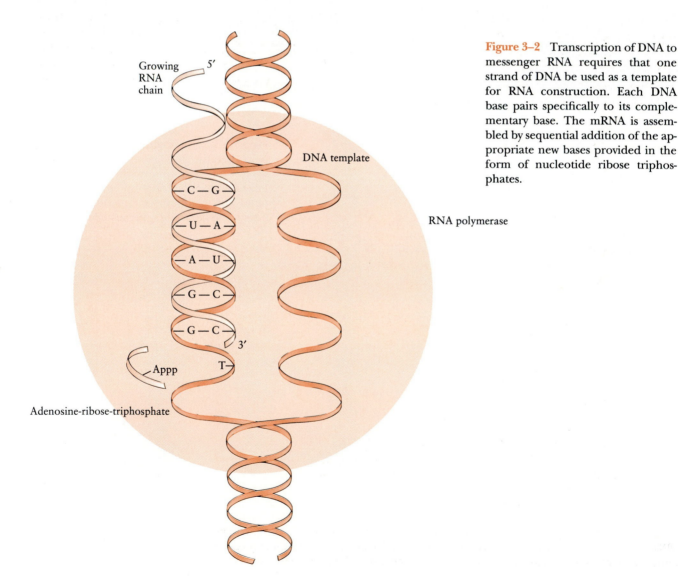

Figure 3–2 Transcription of DNA to messenger RNA requires that one strand of DNA be used as a template for RNA construction. Each DNA base pairs specifically to its complementary base. The mRNA is assembled by sequential addition of the appropriate new bases provided in the form of nucleotide ribose triphosphates.

deleted from the DNA for some reason, then each triplet will be read incorrectly and the resulting message will be useless. This is known as a frame shift.

The first step in making a new protein is the **transcription** of the DNA chain within the nucleus to an RNA chain in the cytoplasm. The DNA is transcribed to messenger RNA (mRNA) by employing enzymes called RNA polymerases (Fig. 3–2). Just as a faithful complementary copy of the DNA can be made when a cell is dividing, so too can a corresponding mRNA chain be made when protein synthesis is initiated. The RNA is similar to the DNA structure except that it consists of long strings of ribonucleotides. In these the sugar is ribose, not deoxyribose as in DNA, and the four bases are adenine, guanine, cytosine, and uracil. Thus the DNA codon TGA is transcribed to ACU in mRNA.

RNA Translation

Once transcribed, the RNA undergoes processing in the nucleus. The mRNA is then carried to the cytoplasm, where it cooperates with the other forms of RNA in ribosomes to make peptide chains. The mRNA is used as a template by ribosomes for the translation of the mRNA sequence into the amino acid sequence of a protein. Thus a cell may contain many mRNA molecules, each coding for a different peptide chain. Ribosomes are small, compact ribonucleoprotein particles that coordinate the **translation** of mRNA to protein (Fig. 3–3). Proteins are made at groups of ribosomes called polyribosomes. Within the polyribosomes, the bases in the mRNA chain are read sequentially by transfer RNA (tRNA). Transfer RNAs are highly folded strands of RNA that contain 70 to 95 nucleotides. Each tRNA contains a nucleotide triplet called an anticodon that is complementary to the triplet of the mRNA—the codon. Each of the 20 amino acids normally used for protein construction has one or more corresponding tRNAs that will bind it. The tRNA binds the amino acid, carries it to the ribosomes, and acts as a device that translates the genetic code contained in the mRNA nucleotide sequence into the amino acid sequence of the proteins. As the mRNA chain is read, each nucleotide triplet binds an amino acid. For example, the triplet ACU is linked to threonine. The amino acids are linked in turn to form a peptide chain. Thus a peptide chain grows as each new amino acid is added. The sequence of the peptide chain depends on the nucleotide sequence in the mRNA, which is usually a faithful copy of the cell's own DNA.

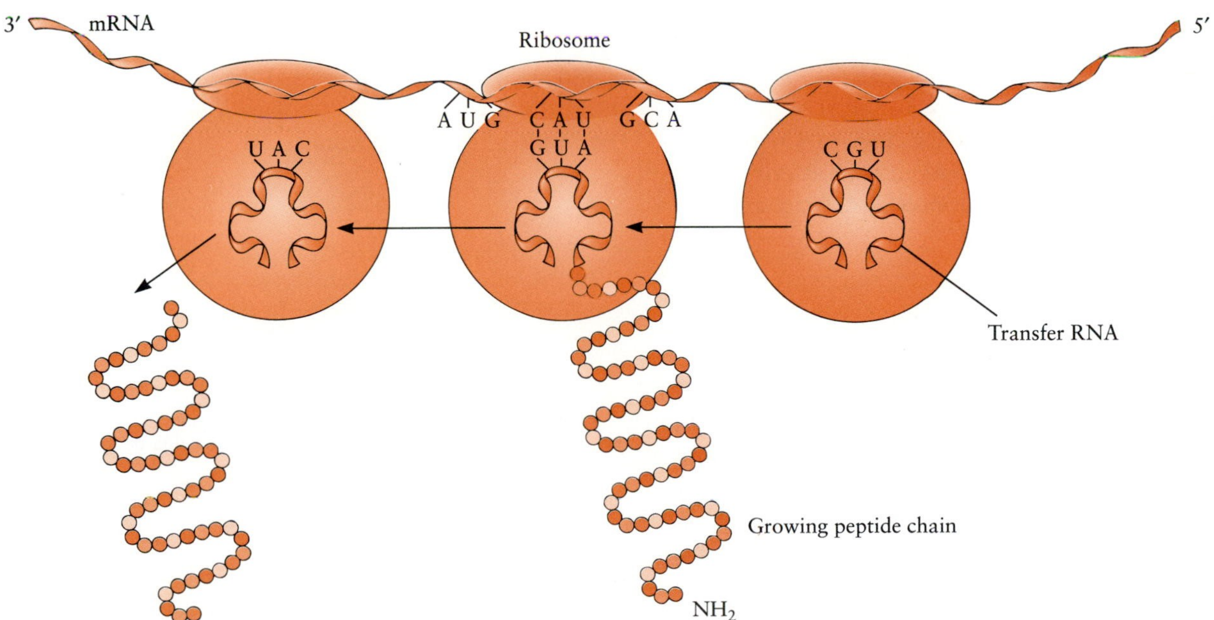

Figure 3–3 Translation of mRNA into a peptide chain requires the use of specific transfer RNA molecules each attached to an amino acid molecule. The mRNA codon sequence determines which tRNA anticodon will attach the next amino acid in the peptide chain.

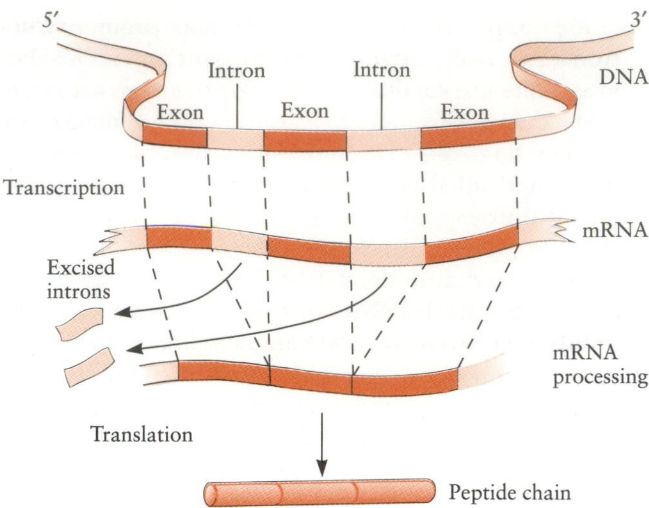

Figure 3–4 DNA coding for a protein molecule contains both expressed segments (exons) and inserted segments (introns). The introns must be excised before translation occurs. This is usually done once an mRNA transcript is made.

Introns and Exons

For many years, it was believed that there was a direct correspondence between the DNA, the mRNA, and the final polypeptide chain product and that a single DNA chain was eventually translated into a colinear polypeptide chain. Once it became possible to sequence mammalian DNA, it was found that genes were, in fact, interrupted sequences of nucleotides coding for single polypeptide chains. Within genes for individual proteins, there are segments of DNA that do not code for peptides (Fig. 3–4). These noncoding DNA segments are deleted from the mRNA after transcription. They are called **introns,** or intervening sequences. The expressed gene segments are called **exons.** The mRNA is processed so that the introns are removed and the exons are joined in the correct order through a process known as **splicing.** In Chapter 15 we will see that the presence of introns is critical for the proper assembly of antibodies and T-cell antigen-receptor molecules.

PROTEIN STRUCTURE

As we have seen, the sequence of amino acids in a protein is determined only by the nucleotide sequence coding for it in a cell's DNA. Twenty amino acids are normally used for mammalian proteins and these can be arranged in any possible order. When a peptide chain is synthesized on a ribosome, the growing chain begins to fold (Fig. 3–5). Each amino acid has a characteristic side chain that interacts with its neighbors in a unique fashion. Thus the sequence of amino acids determines the precise shape of the peptide chain.

Protein Shape

Proteins can be broadly classified on the basis of their shape. Thus there are two major classes—the globular proteins, whose polypeptide chains tend to be folded into compact round shapes, and fibrous proteins, which are long, stringy molecules. Although we will encounter both types, most of the proteins important in immunology are globular. These molecules are soluble in aqueous solutions and diffuse readily. Fibrous proteins are much less soluble and usually have a structural function. They include collagen and elastin in skin. Proteins may also contain carbohydrate side chains **(glycoproteins)** or be associated with lipids (lipoproteins). Glycoproteins are commonly closely associated with cell membranes. Lipoproteins are commonly found free in the circulation or associated with the lipids of cell membranes.

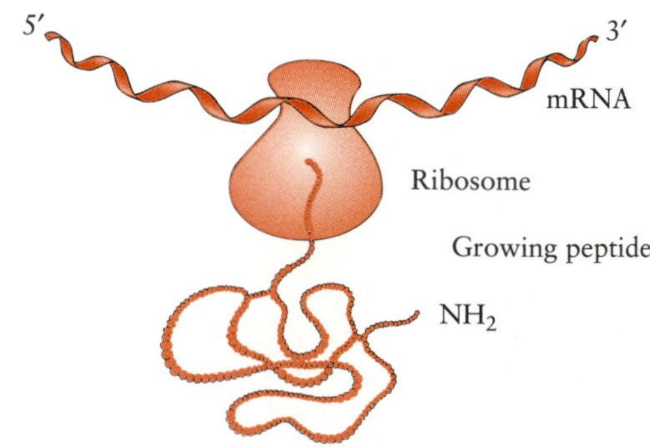

Figure 3–5 The growing peptide chain emerging from the ribosome is free to assume its most energetically favorable configuration. This may turn out to be a very complex shape.

Since polypeptide chains contain amino acids linked by single bonds, it might be expected that polypeptides might be very flexible and not assume a fixed configuration. This is not the case, since proteins tend to assume a very stable configuration. In the 1940s Linus Pauling showed that the —C—N— bond joining two peptides is short and unable to rotate freely. It resembles a double bond and acts as if it was in a rigid plane. Because of these constraints, polypeptide chains tend to assume stable configurations and, especially in globular proteins, may be tightly folded. The precise way in which a protein is folded is important for its biological activity. Certain rules are fairly readily apparent. Four forces stabilize protein structure (Fig. 3–6). First, hydrogen bonding occurs between side chains in adjacent folds of the chain. Thus as the chain folds, two amino acids may come together in space and bond. Likewise, ionic attractions between acidic and basic side chains result in inter- or intrachain bonding. Third, hydrophobic interactions occur as hydrophobic side chains tend to associate in the interior of the molecule. Finally, covalent cross-links may form between adjacent loops. These covalent bonds are much stronger than the other, noncovalent bonds and ensure that the proteins assume a stable conformation. In some situations in immunology, such as when antigen and antibody bind, two proteins reversibly attach to each other. In these cases no covalent bonds form.

Because of its configuration, proline produces a 90-degree "bend" when inserted in a polypeptide chain (Fig. 3–7). Because proline can rotate freely around its peptide bonds, the effect of closely spaced proline residues is to produce a "universal joint" around which the protein chains can swing freely. Cysteine residues, since they contain a reactive sulfydryl group, tend to react with other cysteine residues to form disulfide bonds. If the two cysteines are found on the same peptide chain, these form intrachain di-

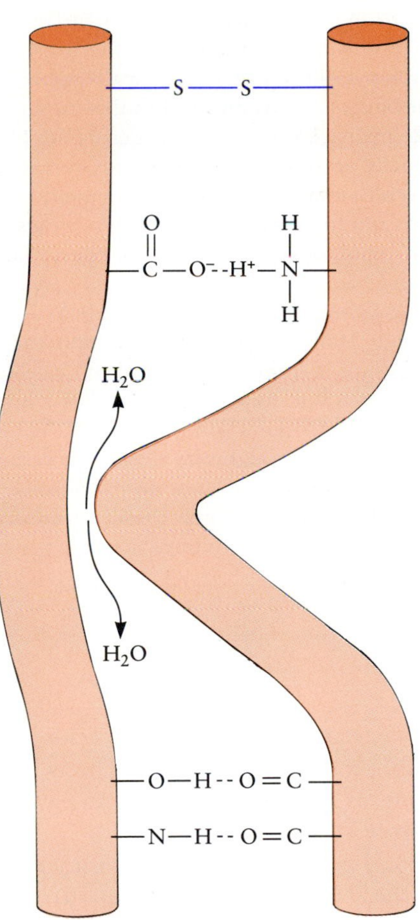

Figure 3–6 The stability of a protein structure is determined in large part by both covalent and noncovalent bonding between adjacent peptide chains. These include, from top to bottom, covalent bonding by means of disulfide bonds, ionic bonding between amino acids with oppositely charged side chains, hydrophobic bonding between chains positioned so as to exclude water molecules, and hydrogen bonding between adjacent —OH, —NH, and —C=O groups.

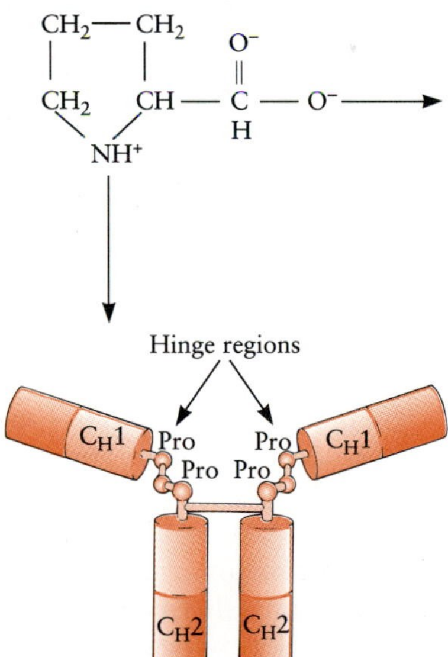

Figure 3–7 Because of its structure, proline causes a peptide chain to bend sharply. Since peptide bonds, although directionally stable, are free to rotate, several proline residues in a row serve as a hinge region around which parts of a protein can rotate freely. This is seen in an antibody molecule, for example.

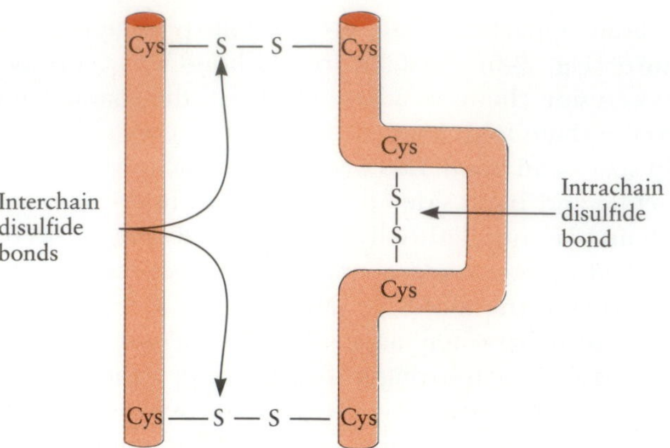

Figure 3–8 A major stabilizing force within proteins is the presence of covalent bonds between peptide chains. Disulfide bonds are most important in this respect. These bonds formed by cysteine molecules may be between two separate chains (interchain bonds) or between regions in a single chain (intrachain bonds).

sulfide bonds. If found on different peptide chains, they are called interchain disulfide bonds (Fig. 3–8).

By combination of 20 amino acids in random order, an infinite number of arrangements is possible. Nevertheless, proteins are not randomly scrambled sequences. Certain sequences are of much more use than others, and certain patterns emerge on analysis. The amino acid sequence of amino acids is called the **primary structure** of a protein. The **secondary structure** is the formation of secondary structures such as α-helices and β-pleated sheets described subsequently. The term **tertiary structure** designates how the peptide chains of proteins are folded into compact structures. The **quaternary structure** is defined by the way in which different peptide chains are packed together to form the complete protein.

A CLOSER LOOK

Antigen–Antibody Binding

One of the key features of the immune system is the highly specific binding between certain receptor molecules and foreign proteins. Thus the proteins called antibodies and certain cell receptors called T cell antigen receptors or TCRs can bind with exquisite specificity to antigens or antigen fragments. This binding is mediated by noncovalent interactions as described in the text. Thus the major bonds formed between an antigen and its antibody are hydrophobic. When antigen and antibody molecules come together, they exclude water molecules from the area of contact. This exclusion frees some water molecules from constraints imposed by the proteins and results in a favorable gain in entropy; the complex is, therefore, energetically stable. Other major bonds involved in this binding include hydrogen bonds. Hydrogen bonds are normally present between proteins and water molecules in aqueous solution, so the binding of antigen to antibody by hydrogen bonds normally requires relatively little net energy change. Electrostatic bonds formed between oppositely charged amino acids may contribute to antigen–antibody binding, but the charge on many protein groups is commonly neutralized by electrolytes in solution. As a result, the relative importance of electrostatic bonds is unclear. Van der Waals forces, although very weak, may become collectively important when two large molecules come into contact and may therefore contribute to antigen–antibody binding.

Folding Patterns

Two major folding patterns are found in proteins that account for much of their secondary structure. One is the α-helix and the other is the β-pleated sheet. These are found in many molecules of immunological interest. The simplest conformation of a peptide chain with rigid peptide bonds is an α-helix (Fig. 3–9). In this, the peptide backbone is wound around an axis and the amino acid side chains protrude outward. The helix forms because it permits hydrogen bonds to form between the H atom on the —NH group and the —NH group and the O atom on the carbonyl group of every fourth amino acid. Thus each coil of the helix allows several hydrogen bonds to form and so makes a very stable structure. Some amino acids are not compatible with α-helix formation. (These include proline, which, as described earlier, causes the chain to bend sharply.)

A second important conformation seen in peptide chains is the β-pleated sheet (Fig. 3–10). In this structure, two or more peptide strands are arranged side by side in a series of pleats. The strands may be parallel, antiparallel, or mixed. In this conformation, hydrogen bonds form between adjacent amino acids. The side chains of each amino acid protrude from the plane of the sheet. Parallel β-sheets may sometimes curve inward to form a cylindrical "barrel." Loops often connect α-helices and β-pleated sheets.

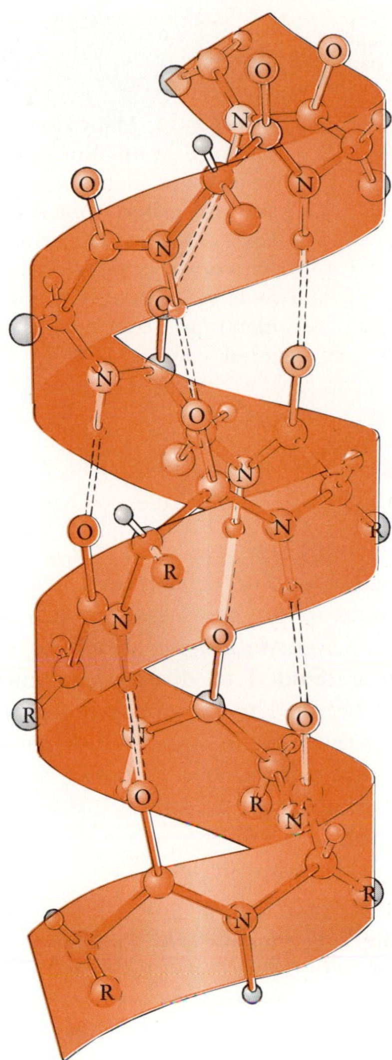

Figure 3–9 One of the most stable structures assumed by a peptide chain is the α-helix. Hydrogen bonds between side chains stabilize the structure.

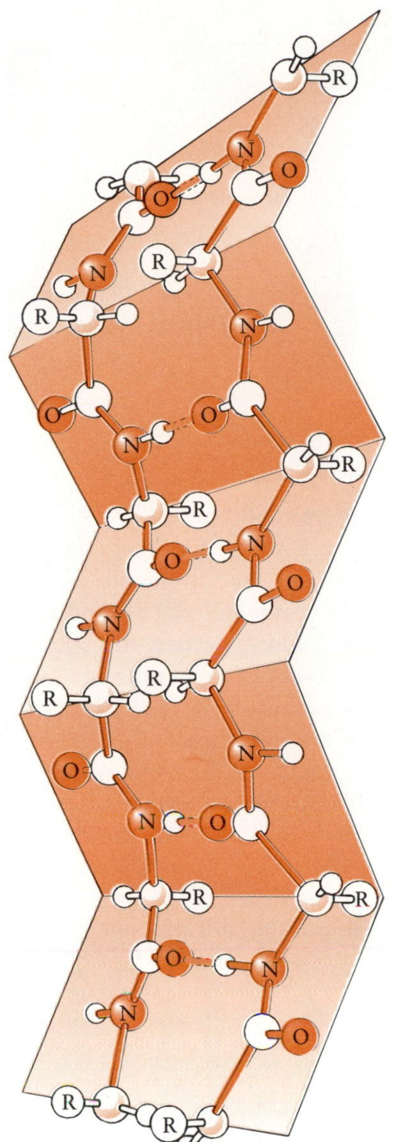

Figure 3–10 The other stable structure formed by peptide chains is the β-pleated sheet.

Another feature of normal protein structure is a **domain.** When it became possible to determine the amino acid sequence of many proteins, it became apparent that certain sequences containing about 100 amino acids were used repeatedly, both within a single protein molecule and in different proteins. The widespread use of these sequences suggests that proteins are constructed of modules. Each module forms a clearly identifiable region within a protein. These modules are called domains. Each is usually coded for by a single exon and may have a well-defined function. For example, in proteins that bind to cell surfaces, one domain consists of a sequence of hydrophobic amino acids. As a result, it can penetrate the lipid bilayer on the cell surface. Other domains may be responsible for the structural stability of a protein or for its biological activities such as signaling to other molecules. Thus in antibody molecules there is a domain that is used to bind specifically to an antigen while other domains are responsible for their biological functions. The presence of similar domains in proteins of dissimilar function suggests that they had a common origin. Proteins may be assigned to certain protein families or superfamilies based on their domain structure. For example, members of the **immunoglobulin superfamily** all contain one or more characteristic immunoglobulin domains.

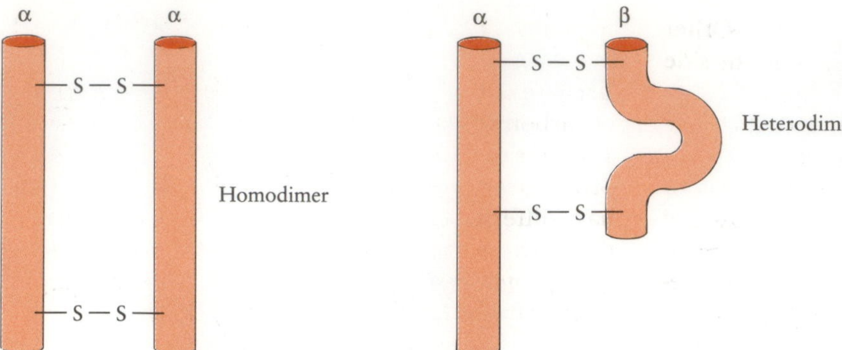

Figure 3–11 Many key proteins of importance in immunology are dimers. Some have identical peptide chains and are called homodimers. Others have dissimilar chains and are thus heterodimers. Larger polymeric proteins (e.g., trimers, tetramers, pentamers) are also commonly encountered.

Protein Assemblies

A protein may be made of a single peptide chain or be a complex consisting of several chains (oligomeric protein). When a protein molecule is made from two identical peptides, it is called a **homodimer.** If two different peptides come together, the structure is a **heterodimer** (Fig. 3–11). An example of a homodimer of importance in immunology is the CD28 molecule, and an example of a heterodimer is an MHC class I molecule, which consists of an α-chain associated with a molecule called β_2-microglobulin. Many of the cell surface receptors important in immunology are heterodimers in which one chain is responsible for binding another molecule while the other chain is responsible for sending a signal to the cell.

GLYCOPROTEINS

Some protein molecules contain significant amounts of carbohydrate. These are called **glycoproteins.** They are commonly found on cell surfaces, whereas soluble proteins found in the cell cytoplasm are usually not glycosylated. Glycosylation is a result of the biosynthetic sorting needed to direct the protein to the plasma membrane. The carbohydrate side chains may help protect the protein against destruction or help in specific recognition. The carbohydrates on membrane glycoproteins are found exclusively on the outside of the cell.

Carbohydrates are linked to proteins as they are formed. They may be linked to the amino group on the side chain of an asparagine residue. These are called N-linked oligosaccharides and are added to pro-

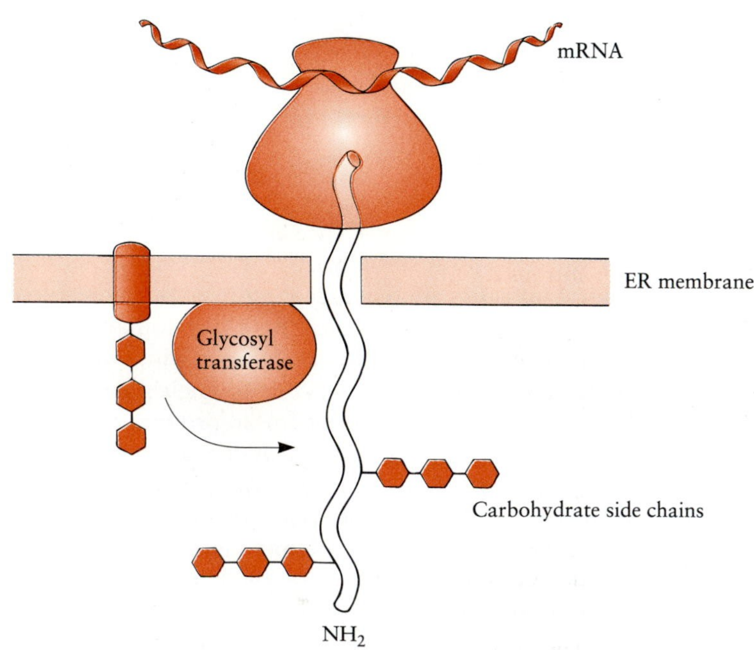

Figure 3–12 If a newly formed protein is destined for export or is to act as a cell surface receptor, it must be glycosylated. Oligosaccharide side chains are attached to the growing peptide chain as it emerges from the ribosome and is extruded into the lumen of the rough endoplasmic reticulum. An enzyme called a glycosyl transferase attaches the assembled carbohydrate. This side chain may subsequently be shortened.

teins in the endoplasmic reticulum (Fig. 3–12). Other carbohydrates may be linked to OH groups on the side chain of serine, threonine, or tyrosine residues. These are called O-linked oligosaccharides, and they are formed exclusively in the Golgi apparatus.

Glycosylation is important in sorting out protein traffic within a cell. Thus glycoproteins made in the rough endoplasmic reticulum are enclosed in transport vesicles derived from the smooth endoplasmic reticulum and are then delivered to the Golgi apparatus, where sorting occurs. Glycoproteins to be secreted are packaged in secretory vesicles that fuse with the plasma membrane and release their contents by exocytosis. Nonsecretory glycoproteins are packaged in other regions of the Golgi apparatus and delivered in vesicles that fuse with internal organelles such as lysosomes. The molecular features of each glycoprotein determine which path they take.

Isoforms and Glycoforms

Because of the possibilities for variation in the amount of carbohydrate bound to a protein as well as differing amounts of trimming of the carbohydrate side chains, proteins with an identical amino acid sequence may have differing carbohydrate content and biological activity. These are called **glycoforms.** Several important glycoproteins in the immune system occur as mixtures of glycoforms.

Another source of variation in protein structure results from minor variations in the amino acid sequence of a specific protein. Provided that these variations do not induce drastic changes in a protein's shape, these changes may result in only minor changes in the biological properties of a molecule. These variations in sequence that are found in all members of a species are called **isoforms.**

KEY WORDS

α-helix p. 34
β-pleated sheet p. 34
Codon p. 31
Domain p. 35
DNA p. 29
Exon p. 32
Fibrous protein p. 32
Frame shift p. 31
Globular protein p. 32
Glycoform p. 37
Glycoprotein p. 36
Glycosylation p. 36
Golgi apparatus p. 37
Heterodimer p. 36
Homodimer p. 36
Hydrogen bond p. 33
Hydrophobic interaction p. 33
Immunoglobulin superfamily p. 35
Intron p. 32
Ionic attraction p. 33
Isoform p. 37
N-linked oligosaccharide p. 36
Oligomeric protein p. 36
O-linked oligosaccharide p. 37
Polyribosome p. 31
Primary structure p. 34
Proline p. 33
Quaternary structure p. 34
Ribosome p. 31
RNA p. 31
Endoplasmic reticulum p. 37
Secondary structure p. 34
Splicing p. 32
Superfamily p. 35
Tertiary structure p. 34
Transcription p. 29
Translation p. 31

QUESTIONS

1. Transcription occurs when
   a. the amino acid sequence in a protein is converted to a corresponding nucleotide sequence in RNA
   b. the nucleotide sequence in mRNA is converted to a corresponding sequence in DNA
   c. the nucleotide sequence in mRNA is converted to a corresponding amino acid sequence in protein
   d. the nucleotide sequence in DNA is converted to a corresponding amino acid sequence in protein
   e. the nucleotide sequence in DNA is converted to a corresponding sequence in mRNA
2. Translation occurs when
   a. the amino acid sequence in a protein is converted to a corresponding nucleotide sequence in RNA
   b. the nucleotide sequence in mRNA is converted to a corresponding sequence in DNA
   c. the nucleotide sequence in mRNA is converted to a corresponding amino acid sequence in protein
   d. the nucleotide sequence in DNA is converted to a corresponding amino acid sequence in protein
   e. the nucleotide sequence in DNA is converted to a corresponding sequence in mRNA
3. Introns are segments of DNA that
   a. are expressed as genes
   b. regulate mRNA production
   c. code for long sequences of amino acids
   d. are inserted between expressed genes
   e. are derived from endogenous viruses
4. Disulfide bonds can be formed
   a. between chains as well as within a single chain
   b. between glycine and serine residues
   c. between two different peptide chains only
   d. between peptides and nucleic acids
   e. within residues on a single peptide chain only
5. The primary structure of a protein refers to its

a. molecular weight
b. carbohydrate side chains
c. polymerization between subunits
d. amino acid sequence
e. three-dimensional configuration

6. One of the common basic features of protein structure is a(n)
a. binding groove
b. transmembrane domain
c. α-helix
d. variable region
e. convalently linked lipid

7. A well-defined structural region within a protein is called
a. a region
b. a domain
c. the N-terminus
d. a variable region
e. a β-pleated sheet

8. When many protein molecules make use of the same basic structural unit, this is called a(n)
a. superfamily
b. gene complex
c. oligopeptide
d. heterodimer
e. polypeptide

9. Glycoproteins are proteins that
a. contain lipids
b. contain oligosaccharides
c. contain interchain bonds
d. stimulate the immune system
e. are found only in bacteria

10. Protein molecules that differ in a few amino acid residues are called
a. polymers
b. isoforms
c. glycoforms
d. heterotypes
e. isotypes

11. Describe the process of DNA transcription.

12. Describe the process of RNA translation.

13. What are introns and exons? How may the use of introns be of any benefit to an organism?

14. What bonds are involved in stabilizing protein structure? How could these bonds act in binding two separate protein molecules?

15. Using the index of this book, list some of the important roles of α-helices and β-pleated sheets in molecules of immunological interest.

16. What are glycoproteins? Where are they found? What is the significance of protein glycosylation?

Answers: 1e, 2c, 3d, 4a, 5d, 6c, 7b, 8a, 9b, 10b

SOURCES OF ADDITIONAL INFORMATION

Campbell, M.K. Biochemistry. Saunders College Publishing, Philadelphia, 1991.

Devlin, T.M., ed. Textbook of Biochemistry, with Clinical Correlations, 3rd ed. Wiley-Liss, New York, 1992.

Lehniger, A.I., Nelson, D.L., and Cox, M.M. Principles of Biochemistry, 2nd ed. Worth, New York, 1993.

Rothwell, N.V. Understanding Genetics, a Molecular Approach. Wiley-Liss, New York, 1993.

Werner, R. Essential Biochemistry and Molecular Biology, 2nd ed. Elsevier, Amsterdam, 1992.

CHAPTER 4

Antigens and Antigenicity

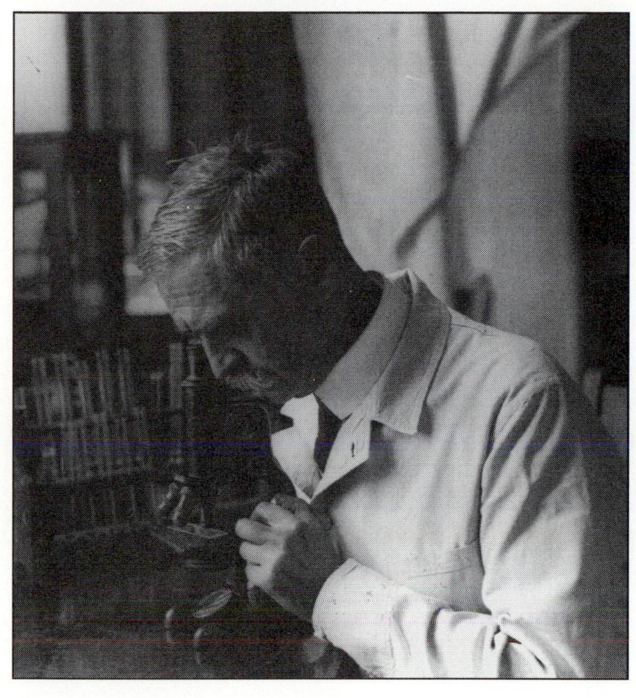

Karl Landsteiner *Born in Vienna, Austria in 1868, Karl Landsteiner made many of the key discoveries in relation to antigens and antigenicity. Thus, in the 1920s he discovered haptens and in a series of elegant experiments showed the remarkable specificity of antibodies for antigens. Landsteiner also discovered blood groups, the antigenic molecules located on the surface of red blood cells. He showed how these acted as antigens following incompatible blood transfusions. This discovery finally permitted blood transfusions to be made safely and as a result transformed modern surgery. He was awarded the Nobel Prize in Medicine in 1930 for his discovery of blood groups.*

CHAPTER OUTLINE

CHAPTER CONCEPTS

1. The most effective foreign antigens are large, complex molecules such as proteins.
2. Only selected portions of foreign molecules are recognized. Those portions recognized by the immune system are called epitopes.
3. Small molecules that do not normally stimulate an immune response will do so if they are chemically linked to a large protein molecule. These small molecules are called haptens.
4. Studies with haptens have shown that antibodies can distinguish between very similar molecular structures.
5. Many foreign molecules can provoke an immune response. These include proteins from infectious agents and foreign cells, as well as drugs and vaccines.

In this chapter we examine the properties of antigens, the foreign material that actually triggers the immune system to respond. We first look at the properties that determine whether a foreign substance can act as an antigen. Then we examine antigens in more detail to determine what special features are recognized by the immune system. These special features tell us much about how the body recognizes invaders as well as about the structure of biological molecules. Finally, we briefly review some of the common antigens that we will encounter as we study immunology.

Since the function of the immune system is to protect the body against foreign invaders, it is fairly obvious that mechanisms must exist for cells to recognize these invaders. Somehow, the body must recognize a foreign substance in order for an immune response to be provoked. In this chapter, we discuss just what it is about foreign material that triggers immunity. Although we usually consider infectious microorganisms as the major threats to the body's integrity, it should be pointed out that we are exposed to a great variety of foreign material in everyday life and not all of it is threatening. Our most obvious exposure is to food. We consume quantities of foreign protein, carbohydrate, and fat. These foods are generally not a threat to the body and are not usually regarded as foreign. Likewise, large inert foreign bodies such as metal bone pins or plastic heart valves fail to provoke an immune response. We must therefore consider the features required of a foreign molecule in order for it to be recognized by the cells of the immune system, act as an antigen, and provoke an immune response.

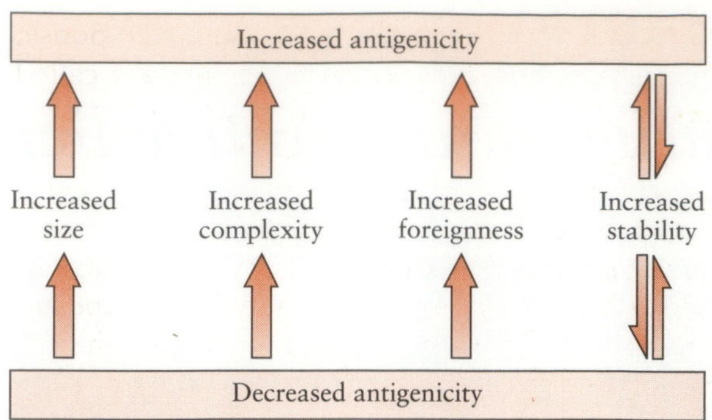

Figure 4–1 The major factors that determine antigenicity. Increased size, complexity, and foreignness all promote antigenicity. Physical stability has a more complex effect since very inert or very unstable molecules are poor antigens.

ESSENTIAL FEATURES OF ANTIGENS

There are two major restrictions on antigenic molecules. First, and more important, the molecules must be recognized as foreign. Second, because of the processing antigens must undergo, there are physical and chemical limitations on the types of foreign molecules that can stimulate the immune system. The most effective antigens are large, rigid, chemically complex molecules that can be degraded fairly readily to soluble fragments that the immune system can recognize.

Factors That Influence Antigenicity (Fig. 4–1)

Molecular Size. In general, large molecules are better antigens than small molecules (Fig. 4–2). For example, hemocyanin, a very large protein from the blood of invertebrates (6.7×10^3 kDa), is a potent antigen. Serum albumin from other mammals (69 kDa) is a fairly good antigen but may also provoke tolerance. The hormone angiotensin (1031 Da), however, is a poor antigen. The record for minimal antigenic size is held by p-azobenzene-arsenate trityrosine (750 Da), which has been used to provoke antibodies in guinea pigs and rabbits. Very small molecules may, however, bind to large proteins, and the resulting

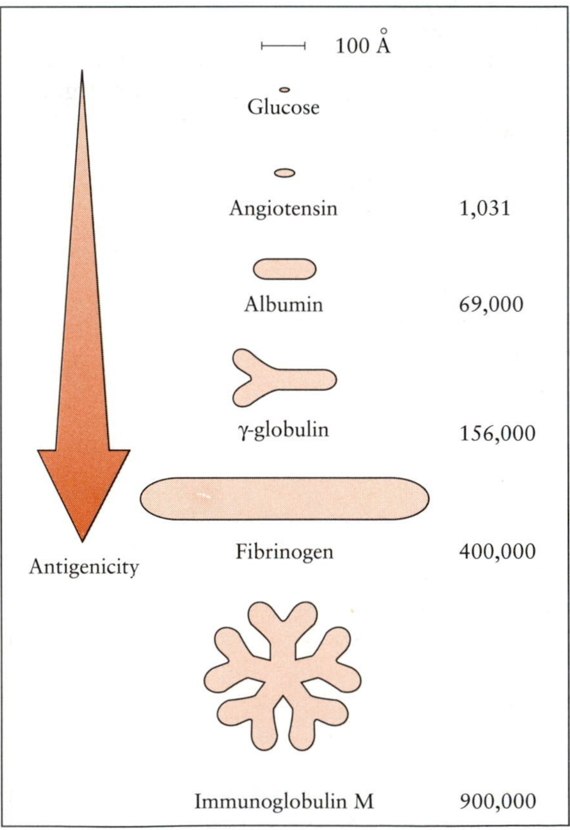

Figure 4–2 The relative sizes of some important antigens. Molecules the size of angiotensin or smaller are usually very poor antigens.

complex may then provoke an immune response. When this happens, these small molecules are called **haptens.**

Structural Stability. To recognize a molecule or part of a molecule as foreign, the cells of the immune system must recognize its specific shape. Consequently, highly flexible molecules that have no fixed shape are poor antigens. For example, gelatin, a protein well known for its structural instability (this is why it can wobble), is a poor antigen unless it is stabilized by the incorporation of tyrosine or tryptophan molecules, which cross-link the peptide chains. Similarly, flagellin, the major protein of bacterial flagella, is a structurally unstable, weak antigen. Its stability and hence its antigenicity are greatly enhanced by polymerization. Starch and other simple repetitive polysaccharides are poor antigens because they do not assume a stable configuration. For the same reason, proteins are better antigens than large, repeating polymers, such as the lipids, carbohydrates, and nucleic acids.

Degradability. The cells of the immune system recognize small molecular fragments and soluble antigens. If a molecule cannot be broken up or solubilized, then it cannot act as an antigen. For example, stainless steel pins and plastic joints are commonly implanted in the body without triggering an immune response. The lack of antigenicity of metals or large, inert organic polymers, such as the plastics, is due to their inertness. They cannot be fragmented and processed to a form suitable for triggering an immune response. Conversely, since immune responses are antigen driven, foreign molecules that are very rapidly destroyed on entering the body may not provide sufficient stable antigen fragments to stimulate an immune response.

Another example of the importance of fragmentation in antigen processing is seen when using copolymers of D–amino acids. D–amino acids do not occur naturally in mammals. Because mammalian enzymes cannot degrade them, they are metabolically inert. Peptides made from D–amino acids are therefore very poor antigens. If, however, a few short peptides consisting of L–amino acids are inserted into a D–amino acid polymer, the resulting molecule is a good antigen. The presence of the L–amino acids allows the peptide to be broken into fragments that can be recognized by the cells of the immune system.

Foreignness. The cells whose function is to respond to antigen (antigen-sensitive cells) are selected in such a way that they do not usually respond to normal body components. They will respond, however, to foreign molecules that differ even in minor respects from those usually found within the body. The elimination of cells that react to normal body components and the subsequent development of tolerance to these components occur because these cells are exposed to self-antigens at an immature stage in their development (usually early in fetal life). Self-reactive immature cells exposed to self-antigens are killed. If antigen-sensitive cells are not exposed to an antigen when immature, tolerance to that antigen will not develop. For example, the sperm-forming cells in the testes are separated from the rest of the body by a tissue barrier. As a result, the cells of the immune system do not normally encounter sperm. These cells are therefore not tolerant to sperm antigens. If the tissue barrier is broken down by injury or infection, sperm antigen may reach the bloodstream, where the antigen-sensitive cells will regard it as foreign and mount an immune response against it. The development of antisperm antibodies is a common sequel to vasectomy as a result of leakage of sperm antigens into the tissues. On a smaller scale, mitochondria are not normally exposed to antigen-sensitive cells in the circulation. When extensive cell destruction occurs—as, for example, following a heart attack—mitochondria are exposed to the cells of the immune system and antimitochondrial antibodies may develop. These can be detected in serum several weeks later.

Foreign molecules differ in their ability to stimulate an immune response. This property is called **immunogenicity.** The immunogenicity of a molecule depends to a great extent on its degree of foreignness. The greater the difference between a foreign antigen and an animal's own antigens, the greater will be the intensity of the immune response. For example, a kidney graft from an identical twin will be readily accepted because its proteins are identical to those on the recipient's own kidney. A kidney graft from an unrelated human will be rejected in about two weeks unless drugs are used to control the rejection. A kidney grafted from a chimpanzee to a human will be rejected within a few hours despite the use of drugs.

EPITOPES

Complex foreign particles, such as bacteria, fungi, viruses, and foreign cells, readily provoke an immune response following injection. These particles are a complex mixture of proteins, glycoproteins, polysaccharides, lipopolysaccharides, lipids, and nucleoproteins. The response to such a foreign particle is therefore a mixture of many simultaneous immune responses against each of the foreign molecules.

A single large molecule such as a protein can also be shown to have regions against which the immune

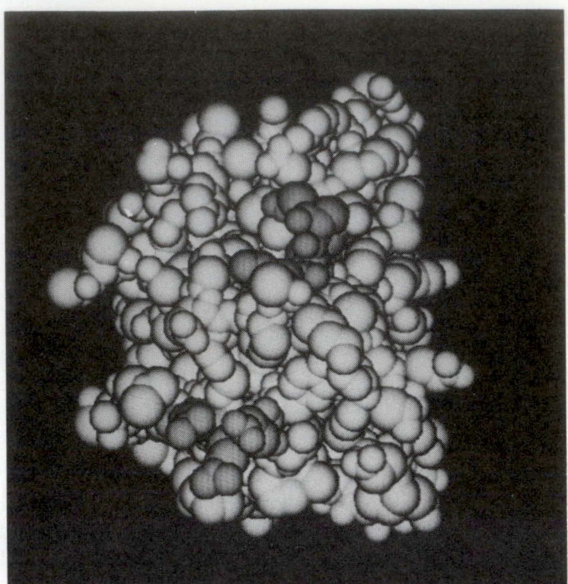

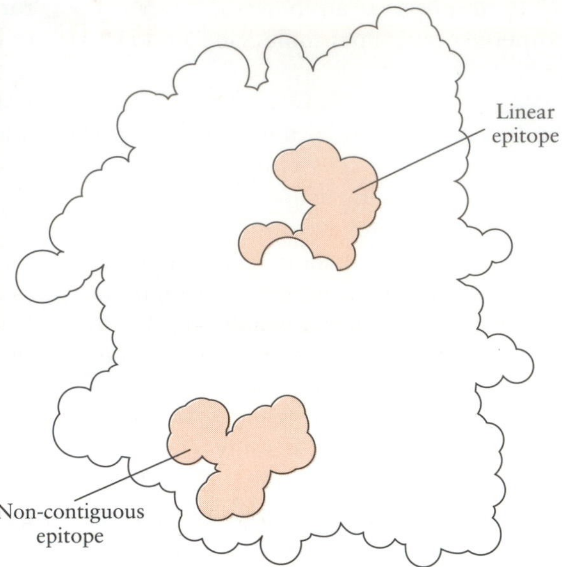

Figure 4–3 A molecular model of the plant protein "thaumatin" showing two types of epitope on its surface. The upper epitope is formed by three linear amino acids. The bottom epitope is formed by amino acids located at different points in the peptide chain. They come together on the surface of the molecule as a result of the folding of the chain. This type of noncontiguous epitope will be destroyed if the peptide chain is broken by proteolytic enzymes. The reader will note several other prominent sites on the surface of the molecule that could also serve as epitopes. *(Courtesy of Dr. Scott Linthicum.)*

responses are directed and with which antibodies will bind. These molecular sites that stimulate immune responses are termed **epitopes,** or **antigenic determinants** (Fig. 4–3). As with the intact molecule, the most intense immune responses are directed against epitopes that are most "foreign." As a result, some are much more immunogenic than others. Thus mice immunized with the enzyme lysozyme, obtained from chickens, preferentially respond to a single favored epitope, and the remainder of the molecule is virtually nonimmunogenic. Such epitopes are said to be **immunodominant.** In general, the number of epitopes on a molecule is directly related to its size, and there is usually about one epitope for each 5 kDa of a protein. For this reason, large molecules are usually more potent antigens than small molecules. When we describe a molecule as foreign, we are implying that it contains epitopes that are not found on self-antigens. The immune system recognizes and responds to such foreign epitopes.

Haptens

Two features of a foreign molecule determine how an immune response is generated: its immunogenicity and its **antigenicity.**

Immunogenicity, as described earlier, is the ability of a molecule to elicit an immune response. This is determined both by the way in which the epitopes on a molecule are processed and presented to the cells of the immune system and by the number of antigen-sensitive cells available to react to that epitope. Antigenicity, in contrast, is the ability of a molecule to be recognized by antibodies or lymphocytes. Any molecule that is immunogenic must also be antigenic, but the reverse need not be true. This apparent paradox is best explained by using the following as an example.

If a small organic compound is chemically linked to a large antigenic molecule, such as a protein, the complex can be used to immunize an animal (Fig. 4–4). When a protein molecule modified in this way is used to immunize an animal, the modified protein

O = C
C − CH_2 − CH_2 − CH_2 − CH_2 − NH − [ring] − NO_2
N − H
NO_2

Figure 4–4 Dinitrophenol can be chemically linked to a lysine side chain on a peptide. This new structure forms a unique epitope and thus acts as a hapten.

METHODOLOGY

How to Determine the Size of an Epitope

Several techniques have been employed to determine the average size of an epitope. One of the most elegant was that of Elvin Kabat. Kabat made antibodies to dextran, a long glucose polymer. He then used various short glucose polymers to block the combination between dextran and its antibodies. He found that molecules containing one or two glucose units were ineffective as blockers, molecules containing three glucose units had slight blocking activity, those containing four units were consistently good blockers, but for maximum inhibition, molecules containing six units (hexamers) were needed (see figure). Polymers larger than the hexamer were no more effective than the hexamer. Thus, in this case, six linear glucose residues probably defined the upper size of the antigen-binding site on antidextran antibodies. Similar experiments using defined polypeptides indicate that epitopes on proteins contain about six to eight amino acids.

Kabat, E.A. The upper limit for the size of the human antidextran combining site. J. Immunol., 84:82–85, 1960.

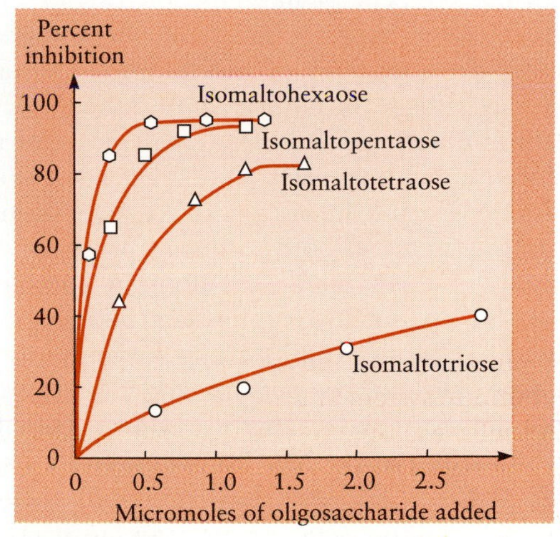

Isomaltotriose, a trimer, has a maximal inhibiting activity of 40%. Isomaltoteraose, containing four sugars, can inhibit dextran binding to its antibody by 80%. Maximum inhibition was obtained with the hexaose suggesting that at least six isomaltose units are required to fill the antigen-binding site completely. *(From Kabat, E. A., J. Am. Chem. Soc., 76:3709, 1954. Copyright 1954, American Chemical Society.)*

is processed and antibodies are generated against its epitopes. Although many of these antibodies are directed against unaltered epitopes on the original protein, some are directed against the new epitopes formed by the small organic molecule. Small molecules or chemical groups that can function as epitopes when bound to other molecules in this way are called **haptens** (in Greek "haptein" means to grasp or fasten), and the antigenic molecule to which they are attached is called the **carrier** (Fig. 4–5). By using haptens of known chemical structure, it is possible to study the interaction between antibodies and epitopes in great detail. For example, antibodies raised against one hapten can be tested for their ability to bind to other, structurally related haptens. Haptens can be very small indeed. For example, it is possible by careful immunization procedures to induce antibody formation against metal ions such as mercury. Some individuals may be unlucky enough to develop allergies against metals such as nickel.

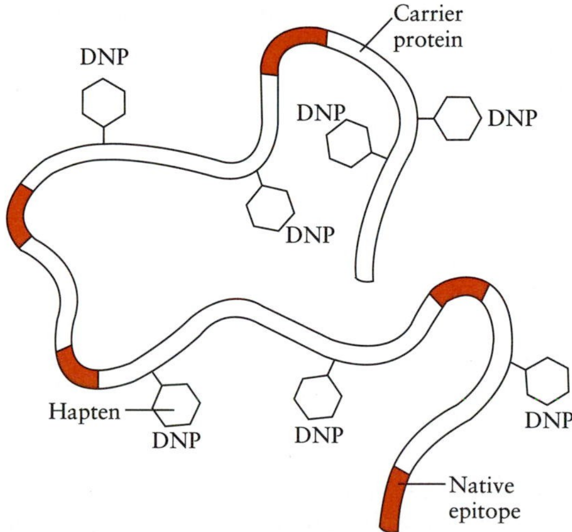

Figure 4–5 Two types of epitope are found on a hapten-protein conjugate formed using a hapten such as DNP. They are unchanged native epitopes found on the protein molecule and haptenic epitopes formed by the DNP.

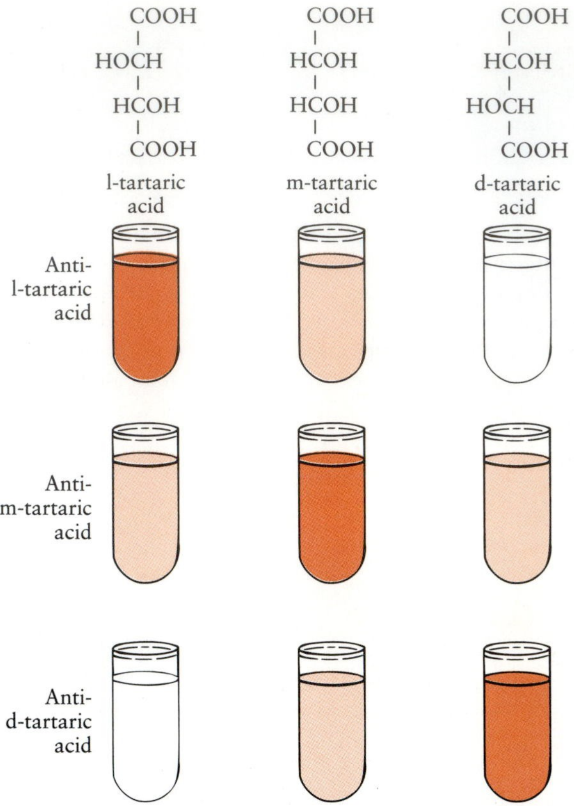

Figure 4–6 The ability of specific antisera to distinguish between the three stereoisomers of tartaric acid. Each antiserum reacts most strongly with the tartaric acid hapten that induced its formation as shown by the intensity of shading.

Simple tests have shown that any alteration in the shape, size, or charge of the hapten reduces its ability to bind antibodies directed against the unmodified hapten. Even very minor structural modifications such as the difference between haptens that are stereoisomers usually result in a significant alteration in the shape of the hapten and therefore influence its ability to bind antibody (Fig. 4–6). Since there is an enormous variety of potential haptens, and since each hapten can provoke its own specific antibodies, it follows that animals must be able to generate an extremely large number of different antibody molecules.

Protein Epitopes

Because the antibody responses against protein molecules are directed against epitopes of a specific shape, the way in which the peptide chain is folded plays a critical role in determining the shape of these epitopes and the specificity of antiprotein antibodies. For example, the protein lysozyme has an intrachain **disulfide bond** that folds the peptide chain into a loop (Fig. 4–7). This loop peptide forms an important epitope. If the loop is removed so that its disulfide bond remains intact, then the detached loop retains its ability to combine with antibodies to lysozyme. If, however, the disulfide bond is reduced so that the loop is opened, antibodies will no longer bind to the loop peptide. This and other similar experiments confirm that it is the shape of an epitope, not its amino acid sequence, that determines antibody specificity.

The reader must not assume, however, that this rule applies to other components of the immune system. As will be described in subsequent chapters, some cells recognize fragments of foreign protein. In these cases the shape of the epitope is of little significance. The amino acid sequence is what is recognized.

Examples of Hapten-Carrier Conjugation

Although the concept of haptens and carrier molecules provides the basis for much of our knowledge concerning the specificity of the immune response, haptens may also be of clinical importance. For example, the antibiotic penicillin is a small molecule that is not immunogenic by itself. Once degraded within the body, however, it forms a very reactive ''penicilloyl'' group, which can bind to serum proteins, such as albumin, to form penicilloyl-albumin conjugates (Fig. 4–8). The penicilloyl hapten may be recognized as a foreign epitope and provoke an immune response, especially in those persons who have a genetic predisposition to make antibodies against penicillin. These antibodies can mediate an allergic response to penicillin that can be hazardous or life-threatening (Chapter 29). Highly sensitive persons must ensure that they avoid exposure to penicillin.

A second example of a naturally occurring reactive chemical that binds spontaneously to normal proteins and so acts as a hapten is the toxic component of poison ivy *(Rhus radicans)*. The resin of this plant, a mixture of complex catechols known as urushiol, binds to any protein it comes into contact with, including the skin proteins of anyone who rubs against the plant. The skin cells modified in this way are then regarded as foreign and attacked by lymphocytes in a manner similar to the rejection of a skin graft. Destruction of these altered skin cells by a cell-mediated immune response causes blister formation and the development of a very uncomfortable skin rash known as **allergic contact dermatitis** (Chapter 30).

CROSS-REACTIVITY

Identical or similar epitopes are sometimes encountered in unrelated molecules. This is known as a **cross-reaction.** Cross-reactions are of two types. One type is

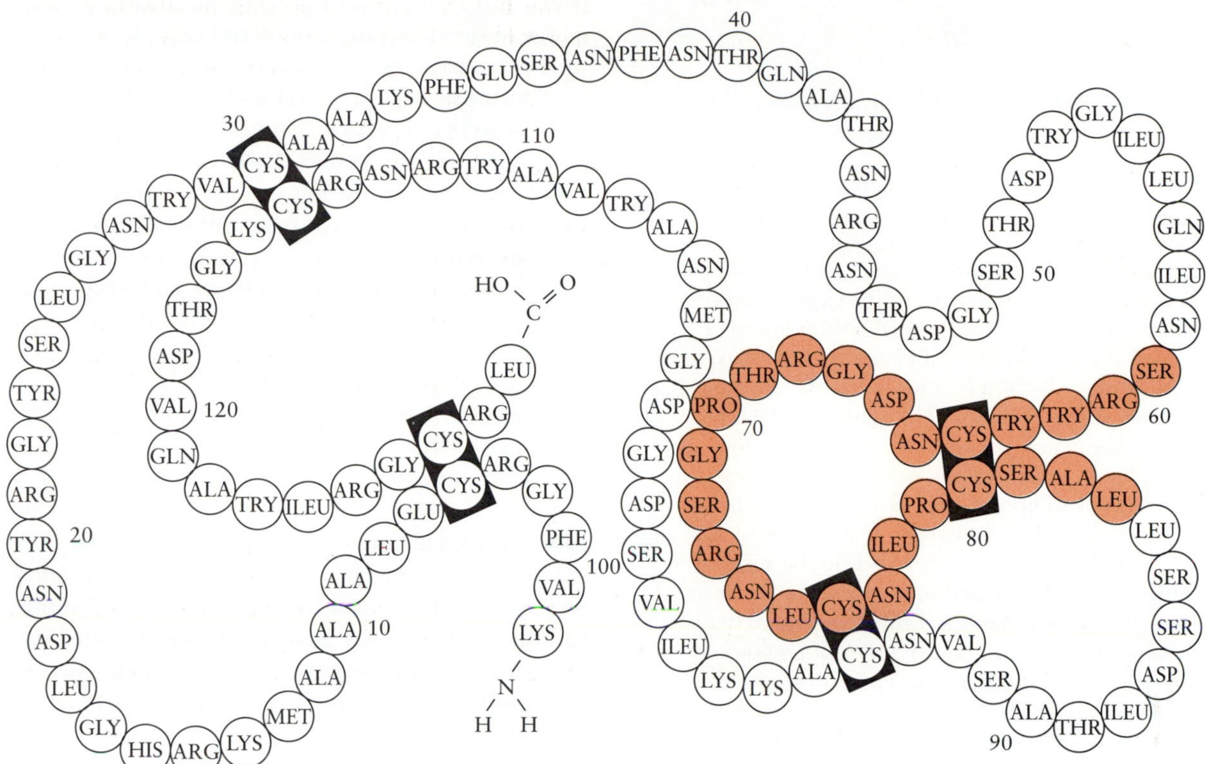

Figure 4–7 The structure of lysozyme from hen egg white. The region of the loop peptide is shaded. This loop functions as an epitope. Antibodies bind to the loop peptide only if the loop structure remains intact. If the loop is broken, it no longer binds to these antibodies, confirming that antibodies recognize the conformation of an epitope rather than its sequence. *(From Arnon, R., and Sela, M., Proc. Natl. Acad. Sci. U S A, 62:164, 1969. With permission.)*

unpredictable and results from a coincidental similarity between totally unrelated molecules. The other type results from the structural similarities between related molecules from different species.

There are several good examples of cross-reactions of the first type. Many bacteria possess cell wall polysaccharides that are also found on mammalian red blood cells. For example, some intestinal bacteria have blood group glycoproteins A and B on their cell walls (Chapter 30). When these glycoproteins are absorbed through the intestinal wall into the bloodstream, the animal mounts an immune response against the foreign epitopes. Thus blood group glycoprotein A is foreign to a person of blood group B. This person therefore produces anti-A antibodies in response to the bacterial antigens in spite of never having been exposed to red blood cells of blood group A. The reverse is also the case, so that a person of blood group A has anti-B antibodies in his or her blood. Cross-reacting antibodies of this type are called **heterophile antibodies.**

Another example of cross-reactivity between antigens from unrelated sources is seen in two bacteria, the relatively avirulent *Proteus vulgaris* (strain OX19) and the highly dangerous *Rickettsia typhi,* the cause of typhus. Antibodies to *R. typhi* react with *P. vulgaris.* Because of this cross-reactivity it is possible to diagnose

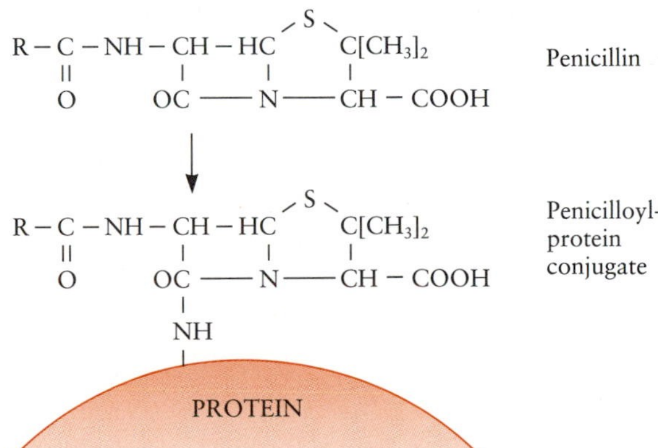

Figure 4–8 Penicillin can function as a hapten. When degraded by enzymes, penicillinic acid is formed. This reacts with protein amino groups to form a penicilloyl-protein complex. It is this complex that stimulates antibody formation in most individuals who are allergic to penicillin.

A CLOSER LOOK

Vaccines Against Heart Disease?

Since antibodies can be raised against small organic molecules by using them as haptens, is it possible to make antibodies against cholesterol? If so, could these antibodies then control the development of heart disease? In experiments where cholesterol esters linked as haptens to protein carriers were used to vaccinate rabbits, a significant reduction in atherosclerotic plaque formation was observed. The rabbits had made antibodies against the cholesterol hapten. It is probable that the antibodies bound to cholesterol linked to lipoproteins and enhanced their removal from the circulation. Unfortunately this cholesterol-lowering effect could not be sustained, despite repeated immunization. Studies have also shown that many humans possess natural antibodies to cholesterol that can be boosted by appropriate vaccination. While these are exciting results, considerably more basic research must be undertaken before the practicality of this approach is established.

Bailey, J.M. Cholesterol vaccines. Science 264: 1067, 1994

typhus simply by detecting the appearance of antibodies to *P. vulgaris* in a patient's serum, thus avoiding the necessity of exposing laboratory wokers to the very dangerous *Rickettsia.*

The second type of cross-reactivity, that between related proteins, can be demonstrated in many different biological systems. One example is the method used to detect relationships between animal species. Thus antisera to human serum cross-react well with chimpanzee and gorilla sera but react much more weakly with sera from other monkeys (Fig. 4–9). Presumably, this reflects the presence of shared epitopes on these proteins. The more closely related the proteins are, the more epitopes they share. This is thus a useful tool in determining evolutionary relationships.

SOME SPECIFIC ANTIGENS

Although many antigens are described and discussed throughout this book, some general comments can be made here.

As mentioned earlier, proteins are the best antigens because of their size and structural complexity. Almost all proteins larger than 1 kDa are antigenic. Many of the major antigens of microorganisms, such as the clostridial toxins, bacterial flagella, viral capsids, and protozoan cell coats, are all proteins. Other important antigenic proteins include snake venoms, serum proteins, milk and food proteins, hormones, and even antibody molecules themselves. The cell surface antigens responsible for antigen presentation and graft rejection, known as histocompatibility molecules, are also proteins.

Simple polysaccharides, such as starch or glycogen, are usually not good antigens because they are rapidly degraded within cells and, in addition, do not form structurally stable epitopes. More complex carbohydrates, however, may be of immunological importance, especially if bound to proteins or lipids. These include the major cell wall antigens of gram-negative bacteria, the blood group antigens of red blood cells, and some glycoprotein side chains on nucleated cells (see Fig. 3–12). Many of the so-called natural antibodies found in the serum of unimmunized animals are directed against polysaccharide epitopes and probably develop as a result of exposure to glycoproteins or carbohydrates from the normal intestinal flora or from food.

Lipids tend to be poor antigens because of their wide distribution, relative simplicity, structural instability, and rapid metabolism. Nevertheless, when linked to proteins or polysaccharides, they may function as haptens.

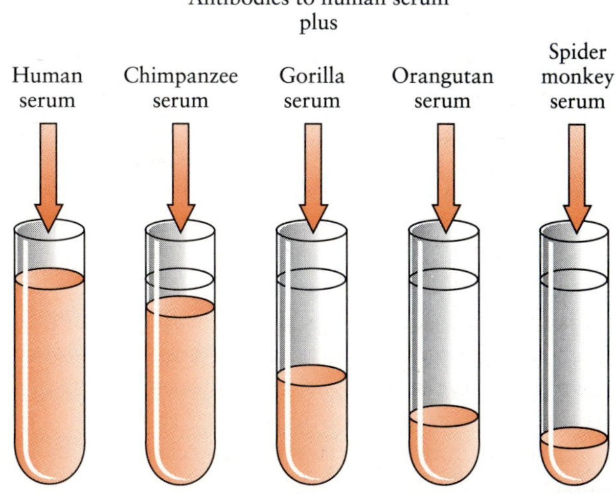

Figure 4–9 When antigen and antibody combine, they can form a precipitate (Chapter 17). This graph shows the quantity of precipitate obtained when rabbit antihuman serum is permitted to react with serum from various other primates. The amount of precipitate is presumably a reflection of the degree of antigenic similarity between humans and these other primates.

CLARIFICATION

Some Commonly Used Antigens

The major considerations used in selecting antigens for immunological research are that they be cheap, pure, and available in large quantity.

Bacterial Lipopolysaccharide. Usually derived from *Escherichia coli*, it is also known as endotoxin.

Bovine Serum Albumin. A soluble white protein from bovine blood with reasonable antigenicity.

Flagellin. A protein derived from the flagella of motile salmonellae. Commonly polymerized to improve its antigenicity and stability.

Hemocyanin. A very potent protein antigen in mammals. It is the blood pigment of the giant keyhole limpet *Megathura crenulata.*

Immunoglobulin G. Derived from human blood, this protein is an excellent antigen in experimental animals.

Lysozyme. This protein is derived from egg white. Because its complete structure is well known, it is useful for careful investigations into the structure of protein epitopes.

Sheep Red Blood Cells. Plentiful and inexpensive, washed sheep red blood cells are potent antigens and used in a wide variety of immunological techniques. Purists make certain that the sheep are of a specific blood group to insure reproducibility.

Tetanus Toxoid. Formalin-treated tetanus toxin. A good protein antigen, also used to protect against tetanus.

TGAL. A multichain synthetic antigen in which tyrosine and glutamic acid residues are attached to alanine side chains on a lysine backbone. Useful for analyzing the immune response to defined epitopes.

Nucleic acids are poor antigens because of their relative simplicity and flexibility and also because they are very rapidly degraded. Nevertheless, antinucleic acid antibodies can be produced by artificially stabilizing and linking them to an immunogenic carrier. It is also noteworthy that autoantibodies to nucleic acids are a characteristic feature of several important autoimmune diseases of humans and other animals. The most important of these is called systemic lupus erythematosus (Chapter 31).

Cell Surface Antigens

The surface of every mammalian cell consists of a mosaic of glycoprotein molecules immersed in a fluid lipid bilayer. Most of these glycoproteins are immunogenic when injected into an animal of another species or even into a different animal of the same species.

Early attempts to transfer blood between unrelated individuals usually met with disaster. The transfused cells were often rapidly destroyed even though the recipient had never before received a transfusion. Investigation revealed that the problem was due to the presence of naturally occurring antibodies to complex carbohydrates and glycoproteins on the surface of red blood cells, known as **blood groups.** The most important of these are the antigens of the ABO system (Chapter 30).

Nucleated cells, such as leukocytes, possess hundreds of proteins on their surface. These proteins are good antigens and readily provoke an immune response (such as graft rejection) if transferred into a genetically different individual of the same species. Some of these proteins are much more potent than others in provoking an immune response and are called the major **histocompatibility molecules.** These antigens are of such importance in immunology that they deserve a complete chapter of their own (Chapter 8).

When advances in immunology made it possible to make monoclonal antibodies (Chapter 14) against specific epitopes, it was soon found that lymphocytes possessed many different surface proteins, each of which possessed many distinct epitopes. To classify these lymphocyte antigens, a numbering system was established that clusters molecules with similar epitopes. Thus all **monoclonal antibodies** that detected the epitopes on a single antigen were assigned to a numbered **cluster of differentiation**(CD). In most cases, a defined CD denotes a protein of specific function. For example, the protein called CD4 is associated with cells (lymphocytes) that "help" the immune response, whereas CD8 is found on cells that suppress the immune response.

Autoantigens

In some situations (not always abnormal), antibodies or lymphocytes are directed against normal body components. Antigens that induce this **autoimmunity** are called **autoantigens.** They can include hormones, such as thyroglobulin; structural components, such as basement membranes; complex lipids, such as myelin; intracellular components, such as the mitochondrial proteins, nucleic acids, or nucleoproteins; and cell membrane antigens, especially hormone receptors. The production of these **autoantibodies** and the consequences of this production are discussed in detail in Chapter 31.

KEY WORDS

Allergic contact dermatitis p. 44
Antigen p. 40
Antigenic determinant p. 42
Antigenicity p. 42
Autoantibodies p. 48
Autoantigen p. 48
Blood group p. 47
Carrier p. 43
Cluster of differentiation (CD) p. 47
Cross-reaction p. 44
Disulfide bonds p. 44
Epitope p. 41
Foreignness p. 41
Hapten p. 42
Heterophile antibody p. 45
Immunodominant p. 42
Immunogenicity p. 41
Lysozyme p. 44
Penicillin allergy p. 44
Penicilloyl group p. 44
Poison ivy p. 44
Tolerance p. 41
Urushiol p. 44

QUESTIONS

1. Which one of the following is not immunogenic?
   a. keyhole limpet hemocyanin
   b. bovine serum albumin
   c. bacterial lipopolysaccharide
   d. glucose
   e. DNA
2. Cross-reactions may be due to
   a. nonspecific antibodies
   b. dissimilar epitopes on antigens
   c. similar epitopes on antigens
   d. chemical reactions with antigens
   e. genetic similarities between species
3. An antigenic determinant is about the size of
   a. 1 amino acid residue
   b. 5 amino acid residues
   c. 20 amino acid residues
   d. 50 amino acid residues
   e. 100 amino acid residues
4. Which one of the following properties of a molecule does *not* influence its antigenicity?
   a. conformation
   b. complexity
   c. color
   d. stability
   e. molecular weight
5. In penicillin allergy, the penicillin acts as a(n)
   a. antibody
   b. carrier
   c. hapten
   d. antigen
   e. adjuvant
6. A hapten is a(n)
   a. carbohydrate side chain
   b. amino acid side chain
   c. small molecule attached to a protein
   d. large protein attached to a small molecule
   e. antibiotic
7. The ability of an antibody that binds blood group antigens to also react with some *Escherichia coli* bacteria is an example of
   a. cross-reactivity
   b. autoimmunity
   c. passive immunity
   d. transfusion reaction
   e. none of the above
8. Which of the following is most likely to induce a strong immune response?
   a. glycoprotein
   b. phospholipid
   c. glycolipid
   d. simple carbohydrate
   e. nucleic acid
9. An autoantigen is a(n)
   a. antigen from bacteria
   b. self-antigen
   c. artificial antigen
   d. carbohydrate antigen
   e. nucleic acid
10. Plastics are not good antigens because they are
    a. degraded too fast
    b. synthetic
    c. toxic

d. degraded too slowly
e. complexed with proteins

11. List the properties of an ideal antigen. How might you alter the proteins in a vaccine to ensure maximum immunogenicity?
12. Which of the following are not normally antigenic in humans: sodium chloride, bovine myoglobin, a glutaraldehyde-treated pig heart valve, aspirin, a contact lens, *Vibrio cholerae*? Explain your answers.
13. How might you reconcile the observations that antigens are usually structurally stable molecules yet antibodies bind to flexible epitopes?
14. When nickel clasps were used to fasten underwear, some people developed allergic reactions where the metal touched the skin. Explain how someone might develop an immune respone to a metal.
15. Complex antigens are taken up by cells and partially degraded before they are presented to the cells of the immune system. Can you reconcile this degradation with the fact that antibodies recognize antigen in its native conformation?
16. Find out more about "heat-shock" proteins. What makes them special antigens?

Answers: 1d, 2c, 3b, 4c, 5c, 6c, 7a, 8a, 9b, 10d

SOURCES OF ADDITIONAL INFORMATION

Allison, C.A. Mode of action of immunological adjuvants. J. RES., 26:619–630, 1979.

Arnon, R., and Van Regenmortel, M.H.V. Structural basis of antigenic specificity and design of new vaccines. FASEB J., 6: 3265–3274, 1992.

Atassi, M.Z. Precise determination of the entire antigenic structure of lysozyme. Immunochemistry, 15:909–936, 1978.

Borek, F. Immunogenicity. Frontiers of Biology Series. Elsevier North-Holland, Amsterdam, 1972.

Green, N., et al., Immunogenic structure of the influenza virus hemagglutinin. Cell, 28:477–487, 1982.

Kabat, E.A. Some configurational requirements and dimensions of the combining site on an antibody to a naturally occurring antigen. J. Am. Chem. Soc., 76:3709–3716, 1954.

Katz, M.E., et al. Immunological focusing by the mouse major histocompatibility complex: Mouse strains confronted with distantly related lysozymes confine their attention to very few epitopes. Eur. J. Immunol., 12:535–540, 1982.

Landsteiner, K. The Specificity of Serological Reactions. Harvard University Press, Cambridge, Massachusetts, 1945.

Marx, J.L. Do antibodies prefer moving targets? Science, 226: 819–821, 1984.

Mitchison, N.A. The carrier effect in the secondary response to hapten-protein conjugates. II. Cellular cooperation. Eur. J. Immunol., 1:18–22, 1971.

Sela, M. Antigenicity: Some molecular aspects. Science, 166: 1365–1374, 1969.

Sela, M., ed. The Antigens. Academic Press, New York, 1973.

Wilson, I.A., et al. The structure of an antigenic determination in a protein. Cell, 37:767–778, 1984.

Wylie, D.E., et al. Monoclonal antibodies specific for mercuric ions. Proc. Natl. Acad. Sci. U S A, 89:4104–4108, 1992.

CHAPTER 5

Destruction of Foreign Material—The Myeloid System

Elie Metchnikoff *Born in Russia in 1845, Elie Metchnikoff was a zoologist by training. In 1882 he observed that the mobile cells found in starfish larvae moved toward and surrounded a foreign body, such as a rose thorn. Metchnikoff, with remarkable insight, reasoned that these mobile cells had a defensive function and acted to protect the body against invasion. He coined the term phagocyte and showed that mammalian white blood cells were phagocytic, a process described in this chapter. He devoted the rest of his life to developing this theory in the face of considerable opposition from the medical establishment. The prolonged debate concerned the significance of phagocytosis versus destruction by antibodies. However, when Almroth Wright identified opsonins, the controversy died down. Metchnikoff shared the Nobel Prize for Medicine with Paul Ehrlich (Chapter 31) in 1908 and died in 1916.*

CHAPTER OUTLINE

CHAPTER CONCEPTS

1. Many bacteria or other foreign particles that enter the body are taken up and destroyed by neutrophils in a process called phagocytosis. Neutrophils respond rapidly to the presence of foreign particles.
2. The phagocytosis of bacteria by neutrophils is promoted by opsonins. Important opsonins include antibodies and complement. Neutrophils possess receptors that bind foreign particles coated with these proteins and trigger their ingestion and destruction.
3. Neutrophils kill ingested bacteria by means of reactive oxygen metabolites generated through a respiratory burst, as well as through the activities of lysosomal enzymes.
4. Neutrophils destroy ingested material and do not process it for the immune system. As a result, it will not provoke an immune response.
5. Eosinophils are specialized phagocytic cells that can destroy parasites.
6. Basophils are granulocytes that are found in low numbers in blood and participate in allergic reactions.

In this chapter we examine the cells that provide the first line of defense in the body. These cells, called neutrophils, can trap and destroy many invading organisms and other foreign particles. We first study their origin and their structure. We then look in some detail at how they recognize that invading microorganisms are present and how they migrate toward invaders and trap them. We will examine how they can kill and destroy the invaders and what happens to them afterwards. We will also briefly review some related blood cells that also assist in defending animals against microbial invasion.

Animals must permit the free access of nutrients and oxygen to the body while excluding dangerous organisms. To do this, they employ several systems that can trap and then kill organisms that succeed in entering tissues. These trapping systems employ cells that engulf and destroy foreign material through a process known as **phagocytosis** (Greek for eating by cells).

The Russian scientist Elie Metchnikoff discovered phagocytosis in 1882. He showed that mobile cells found in the body cavity of starfish larvae engulfed foreign particles. He promptly suggested that this was a general phenomenon and that phagocytic cells were essential components of the defense system of all animals. Indeed, phagocytosis of invading organisms represents a primitive, nonspecific defense system. It can be considered the first line of defense in protecting the animal body against invasion. If the neutrophils are successful in preventing microbial invasion, then it may not be necessary for the body to activate its main defenses, the immune system, at all.

The phagocytic cells **(phagocytes)** of mammals belong to two complementary systems. The **myeloid system** consists of cells that are rapidly phagocytic but incapable of sustained effort. The cells of the **mononuclear-phagocytic system,** in contrast, act more slowly but are capable of repeated phagocytosis. Mononuclear phagocytic cells also process antigen so that it can stimulate an immune response. In this chapter we describe the properties of the cells of the myeloid system.

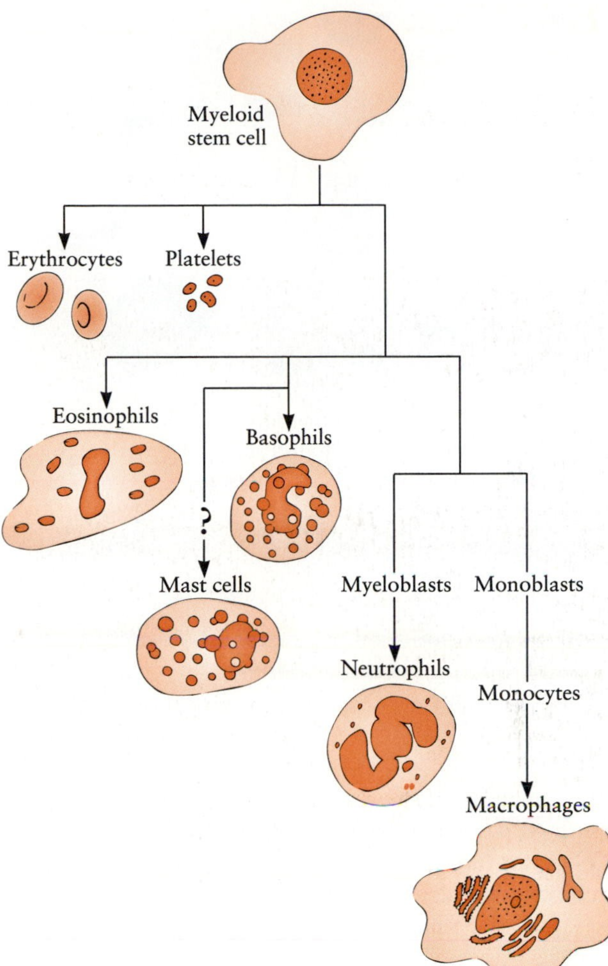

Figure 5–1 All phagocytic cells originate from myeloid stem cell precursors in the bone marrow. These include the myeloid cells discussed in this chapter and macrophages and monocytes, described in the next chapter.

THE MYELOID SYSTEM

The cells of the myeloid system are derived from the bone marrow (''myelos'' is Greek for bone marrow) (Fig. 5–1). All the cells of the system possess a cytoplasm that is filled with granules, so they are collectively called **granulocytes.** These cells also contain an extensively lobulated, irregular nucleus so they are described as ''polymorphonuclear'' as opposed to the single, rounded nucleus characteristic of ''mononuclear'' cells. The three types of granulocytes are distinguished by the staining property of their granules. Cells whose granules take up basic dyes such as hematoxylin (a blue dye) are called basophils; those whose granules take up acidic dyes such as eosin (a red dye) are called eosinophils; and those that take up neither basic nor acidic dyes are called neutrophils.

NEUTROPHILS

The most important cell type of the myeloid system is the **polymorphonuclear neutrophil granulocyte,** otherwise known as the **neutrophil.** Neutrophils are formed in the bone marrow, migrate to the bloodstream, and about 12 hours later move into the tissues, where they die. Their life span is only about a day. Neutrophils are the most abundant **leukocytes** (white

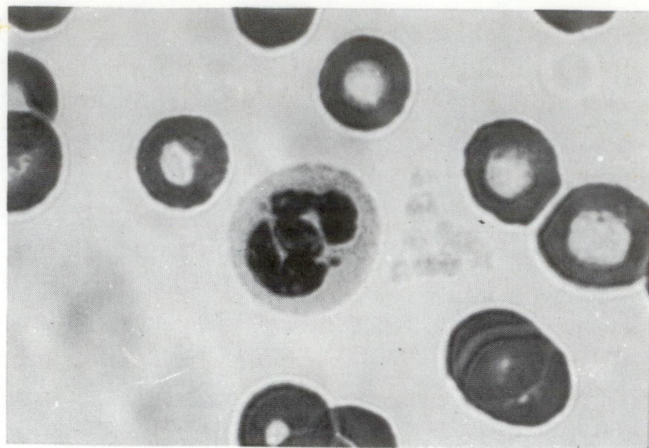

Figure 5–2 A neutrophil polymorphonuclear granulocyte in a blood smear (Wright-Giemsa stain; original magnification ×1400). *(From Bellanti, J. A. Immunology II. W. B. Saunders, Philadelphia, 1979. With permission.)*

blood cells) in blood, constituting about 60% to 75% of the blood leukocytes in humans.

Structure of Neutrophils

When suspended in blood, neutrophils are round cells about 12 μm in diameter (Fig. 5–2). They possess a finely granular cytoplasm, at the center of which is an irregular, sausage-like or segmented nucleus (Fig. 5–3). The chromatin in the nucleus is compacted and assumes this segmented shape since these cells are no longer able to divide. Electron microscopy shows that neutrophils contain two types of cytoplasmic granule. The primary granules are electron-dense structures that contain bactericidal enzymes such as myeloperoxidase and **lysozyme;** neutral proteases, such as elastase; and acid hydrolases, such as β-glucuronidase and cathepsin B. The secondary granules, which are not electron-dense, contain enzymes such as lysozyme and collagenase and the iron-binding protein lactoferrin. Mature neutrophils also have a small Golgi apparatus and some mitochondria but very few ribosomes or rough endoplasmic reticulum (Fig. 5–4). Most of the proteins necessary for their function are transcribed and translated during their development in the bone marrow. Thus they cannot synthesize large quantities of protein. Their cytoplasm also contains large amounts of glycogen. This glycogen is a source of glucose that can be used for anaerobic glycolysis. As a result, neutrophils can remain functional in damaged tissues where oxygen tension is low, such as at sites of bacterial invasion.

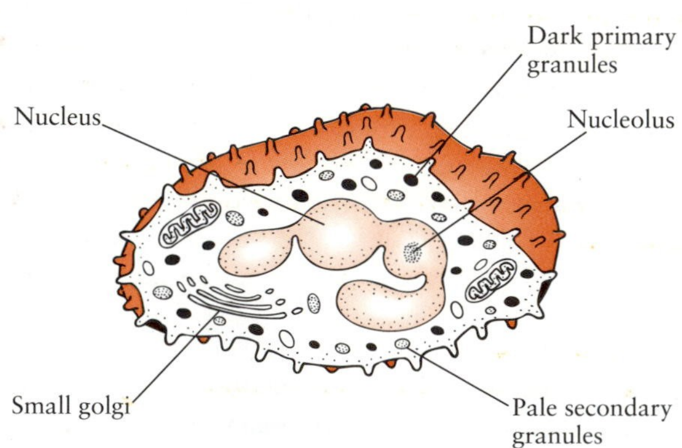

Figure 5–3 The major structural features of a neutrophil polymorphonuclear granulocyte. Note the highly irregular nucleus, indicating that this is not a dividing cell, as well as a lack of rough endoplasmic reticulum, suggesting that very little new protein synthesis will occur.

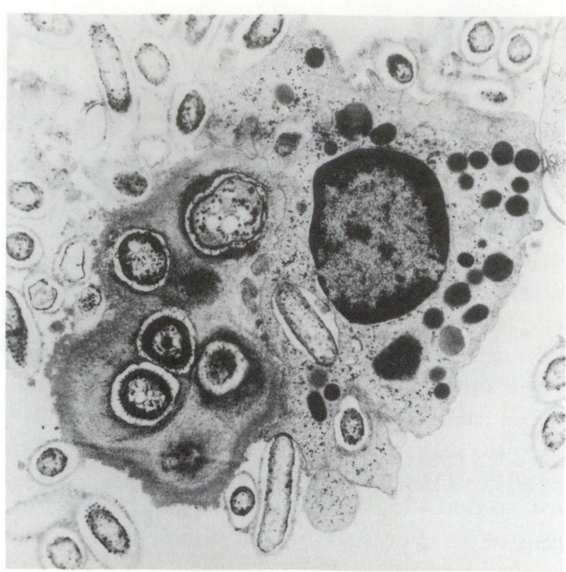

Figure 5–4 A transmission electron micrograph of a neutrophil phagocytosing bacteria. Several of the bacteria are already enclosed within phagosomes. *(Courtesy of Dr. Scott Linthicum.)*

Functions of Neutrophils

Phagocytosis

The major function of neutrophils is the capture and destruction of foreign organisms through **phagocytosis.** Although a continuous process, phagocytosis can be divided into four discrete stages: chemotaxis, adherence, ingestion, and digestion (Fig. 5–5).

Chemotaxis

Chemotaxis is the directed movement of neutrophils under the influence of external chemical gradients. Thus neutrophils are attracted toward the source of certain chemicals. Since neutrophils can crawl but cannot swim, they must attach to a surface before they

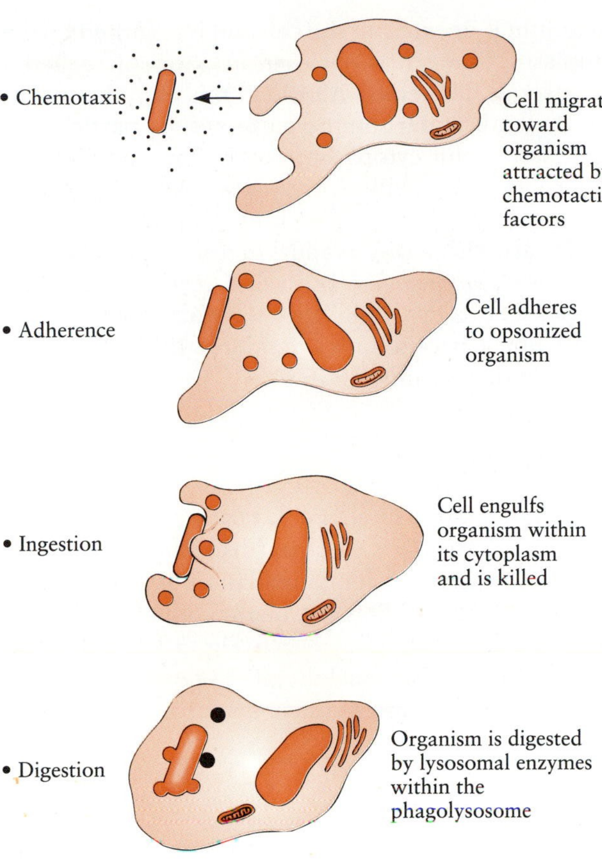

• Chemotaxis
Cell migrates toward organism attracted by chemotactic factors
• Adherence
Cell adheres to opsonized organism
• Ingestion
Cell engulfs organism within its cytoplasm and is killed
• Digestion
Organism is digested by lysosomal enzymes within the phagolysosome

Figure 5–5 The four basic steps in the phagocytic process.

can respond to chemotactic stimuli. Circulating neutrophils are triggered to leave the bloodstream as a result of local changes in blood vessel walls. In tissues where cells are damaged as a result of bacterial invasion, there is an increase in adhesiveness of both neutrophils and vascular endothelial cells. In the vessel walls, the endothelial cells are stimulated to express surface adhesive proteins called **integrins** and **selectins.** (These are discussed in Chapter 7.) The neutrophils therefore stick to these endothelial cells before migrating into tissues.

Bacterial invasion and tissue destruction result in the production of many chemotactic factors. One of the most important of these is a small peptide called C5a, generated by activation of the **complement** system. (The complement system is a set of enzymes found in the blood that help destroy invading organisms. Complement system proteins are denoted by the letter C followed by a number. See Chapter 16.) Another important chemotactic molecule is a small protein called **interleukin-8** (IL-8). IL-8 is produced by many cells, including macrophages, fibroblasts, epithelial cells, and endothelial cells. Small lipid molecules, such as the leukotrienes released by blood platelets, can also attract neutrophils. Most important, invading bacteria release peptides containing a unique amino acid, formylated methionine, that is not produced by mammalian cells. Neutrophils are strongly attracted to peptides containing formylated methionine and, as a result, move rapidly toward sites of bacterial invasion.

All the principal neutrophil chemoattractants trigger a change in cell surface electrical potential, an enhancement of plasma membrane fluidity, and a rapid increase in intracellular calcium levels. As a result, the cells align themselves along the concentration gradient and crawl toward the source of the chemotactic material. They correct their direction continuously, indicating that they can sense gradients across their length through their cell surface receptors. Large doses of chemotactic agents may desensitize neutrophils and thus enable them to leave sites of microbial invasion once they have completed phagocytosis.

Adherence and Opsonization

Normally, bacteria are free to float away when they encounter a neutrophil suspended in blood plasma. If, however, a bacterium is lodged in tissues, or trapped between a neutrophil and another cell surface and thus prevented from floating away, it can be readily ingested. This process is known as surface phagocytosis. Once a neutrophil encounters a bacterium, it must bind it firmly. This adherence does not happen spontaneously, since both cells and bacteria have a negative charge called the ζ (zeta) potential and so repel each other. The negative charge on the bacterium must be neutralized by coating it with a positively charged protein. Examples of such charged proteins include antibody molecules and a protein called C3b (the third component of complement, considered in Chapter 16). Bacteria coated with antibody or C3b therefore have a reduced surface charge, enabling them to stick to negatively charged neutrophils. Neutrophils also have specific cell surface receptors for antibody and C3b. As a result, bacteria coated with these molecules can adhere to neutrophils through their receptors (Fig. 5–6). Molecules that coat bacteria in this way and so promote phagocytosis are called **opsonins.** This word is derived from the Greek word for sauce, implying perhaps that they make the bacterium tastier for the neutrophil.

Antibodies, the major products of the immune system, are by far the most effective opsonins. They bind to foreign bacteria and link them to receptors on neutrophils or macrophages, and they provoke ingestion by these cells. Various complement components, especially C3b, are almost as effective. C3b is deposited on the bacterial surface and binds to cell receptors, although it may not trigger ingestion.

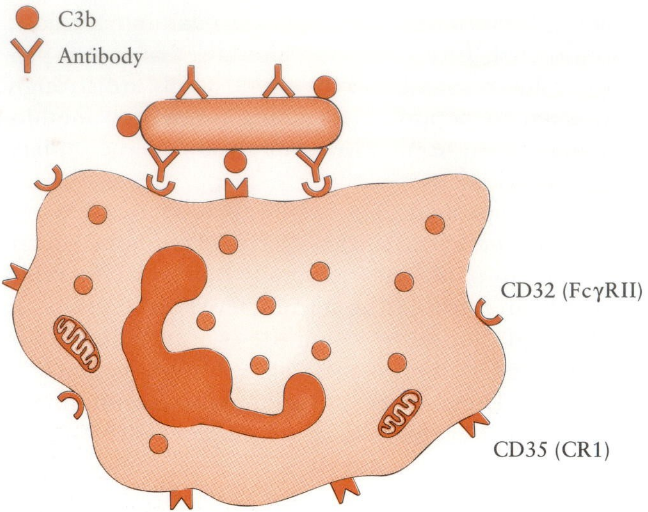

Figure 5–6 The role of antibody receptors (CD32) and complement receptors (CD35) in the opsonization of bacteria for phagocytosis by neutrophils. Antibody- or complement-coated bacteria binding to these receptors stimulate phagocytosis.

Ingestion

As neutrophils crawl toward a chemotactic source, a pseudopod advances first, followed by the main portion of the cell. When the pseudopod meets a bacterium, the pseudopod flows over and around it (Fig. 5–7). Once bound firmly to the neutrophil surface, a bacterium is drawn into the cell and is engulfed by the cytoplasm, where it is enclosed in a vacuole called a phagosome. The ease with which this engulfment occurs depends in part on the nature of the particle surface. Neutrophil cytoplasm readily flows over hydrophobic (water-repellent) surfaces. Bacteria with a waxy, hydrophobic coat, such as *Mycobacterium tuberculosis,* are therefore readily ingested. In contrast, *Streptococcus pneumoniae,* the cause of lobar pneumonia, possesses a hydrophilic carbohydrate capsule. It is poorly phagocytosed unless rendered hydrophobic by a coating of antibodies or C3b.

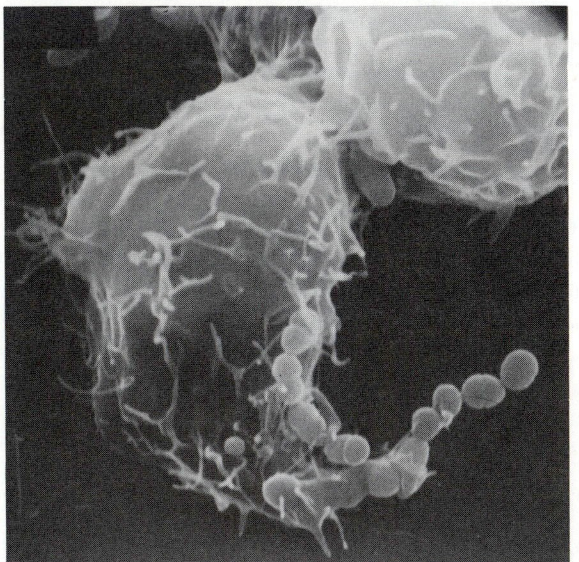

Figure 5–7 A scanning electron micrograph of a neutrophil polymorphonuclear granulocyte in the process of ingesting a chain of streptococci. Note how the neutrophil cytoplasm appears to flow over the bacteria (original magnification ×5000).

Destruction

Once ingested by a neutrophil, a bacterium is exposed to several destructive mechanisms. The most important of these involves lethal oxidation of the bacterial lipids and proteins. This is brought about by a process known as the **respiratory burst** (Fig. 5–8). Other destructive mechanisms include the use of lysosomal enzymes and the release of nitric oxide.

The Respiratory Burst

Within seconds of binding to a foreign particle, neutrophils increase their oxygen consumption nearly 100-fold. This increase is due to the rapid assembly and activation of a cell surface enzyme called NADPH-oxidase. NADPH-oxidase is a multicomponent enzyme that forms a transmembrane electron transport chain with cytosolic NADPH as the electron donor and oxygen as the electron acceptor. When a bacterium binds to a neutrophil Fc receptor, the oxidase is activated (Fig. 5–9). Activated NADPH-oxidase thus converts NADPH (the reduced form of NADP, nicotinamide-adenine dinucleotide phosphate) to $NADP^+$ in the presence of oxygen. The oxygen accepts a single donated electron, resulting in the generation of the superoxide anion (the dot in $.O_2^-$ denotes the presence of an unpaired electron). This gives the molecule a net negative charge.

$$NADPH + 2O_2 \xrightarrow{\text{NADPH-oxidase}} NADP^+ + H^+ + 2.O_2^-$$

The $NADP^+$ generated by the NADPH-oxidase accelerates the pentose phosphate pathway, a metabolic pathway that converts sucrose to a pentose and CO_2, and releases energy for use by the cell. The two molecules of $.O_2^-$ interact spontaneously (dismutation) to generate one molecule of H_2O_2 under the influence of the enzyme superoxide dismutase.

$$2{\cdot}O_2^- + 2H^+ \xrightarrow{\text{Superoxide dismutase}} H_2O_2 + O_2$$

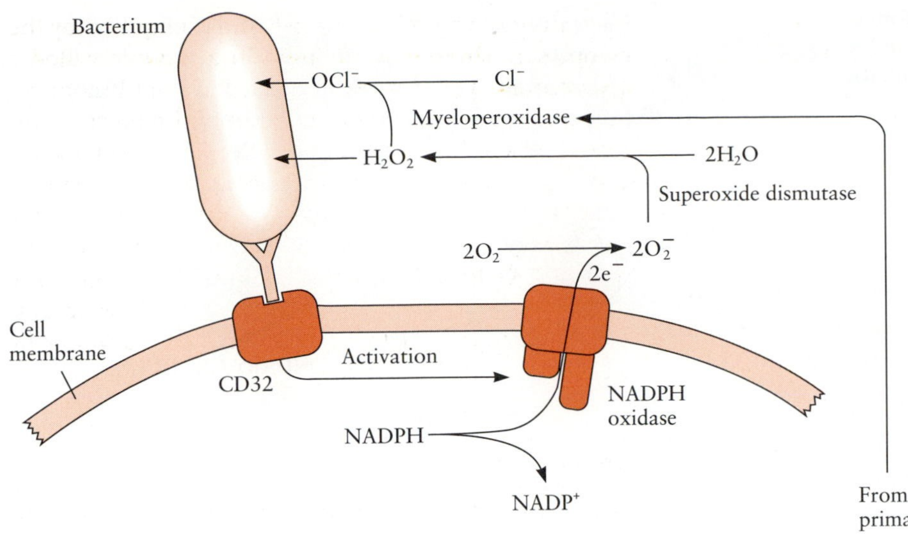

Figure 5–8 The neutrophil respiratory burst. Bacteria binding to CD32 trigger activation of NADPH-oxidase. As a result, a chain of reactions is triggered, resulting in bacterial killing by hydrogen peroxide and hypochloride ions.

This is a rapid reaction, so there is usually very little free superoxide anion around. H_2O_2, in contrast, is a stable oxidant. The hydrogen peroxide may then be converted to other bactericidal compounds through the action of myeloperoxidase, the most significant respiratory burst enzyme in neutrophils. Myeloperoxidase is found in large amounts in the primary granules. It catalyzes several oxidative reactions, but the most important one is that between hydrogen peroxide and intracellular halide ions (Cl^-, Br^-, I^-, or SCN^-) to produce hypohalides:

$$H_2O_2 + Cl^- \xrightarrow{\text{Myeloperoxidase}} H_2O + OCl^-$$

Human neutrophils can generate H_2O_2 for up to 3 hours after triggering, and this reacts with plasma Cl^- to generate OCl^-. Because of its reactivity, OCl^- does not accumulate but rapidly disappears in multiple reactions. It is a very powerful oxidant that rapidly attacks many biological molecules. OCl^- kills bacteria by oxidizing their proteins and enhances the bactericidal activities of the lysosomal enzymes. (Remember HOCl is the active ingredient of household bleach and is commonly used to prevent bacterial growth in swimming pools.)

Hydroxyl radicals and singlet oxygen, both very potent bactericidal agents, may also be generated by O_2^-, H_2O_2, and a metal catalyst, thus

$$H_2O_2 + O_2^- \longrightarrow 2OH^{\cdot} + {}^{\cdot}O_2$$

There is no evidence, however, that neutrophils actually do this in vivo. Ferric iron, the most likely catalyst, is bound by neutrophil lactoferrin, and H_2O_2 is rapidly used up by myeloperoxidase so that none is available for this reaction.

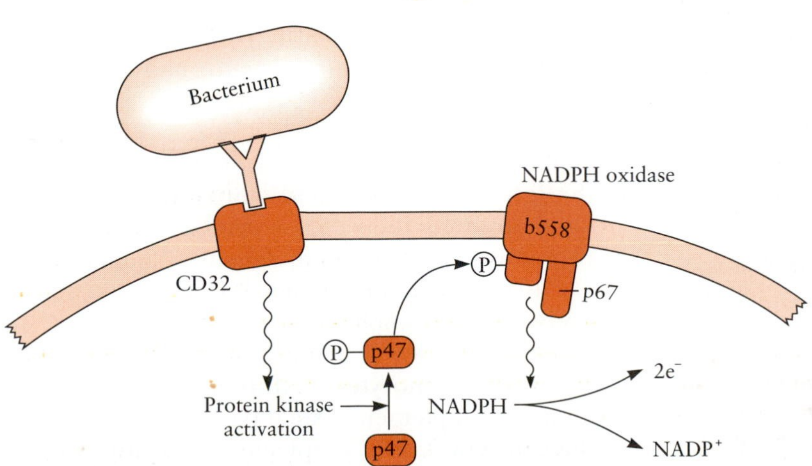

Figure 5–9 The mechanism of activation of NADPH oxidase. Binding of an antibody-coated particle to CD32 leads to activation of a protein kinase. This enzyme phosphorylates the p47 subunit of the oxidase, so activating it.

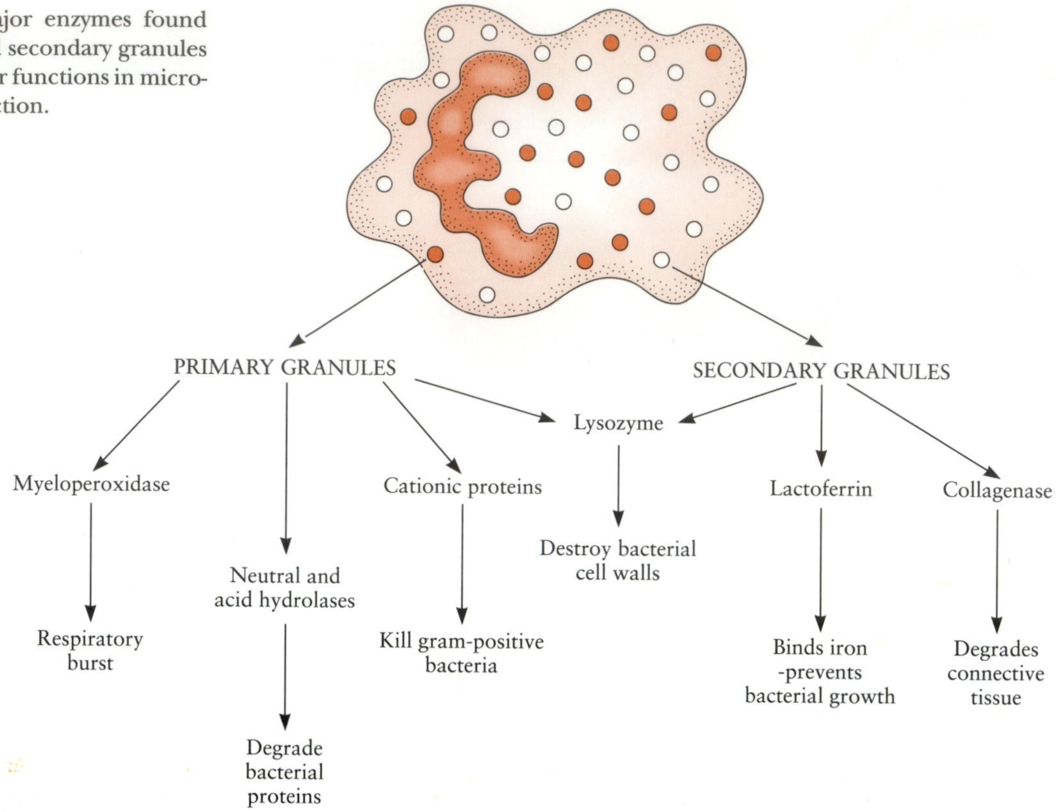

Figure 5–10 The major enzymes found within the primary and secondary granules of neutrophils and their functions in microbial killing and destruction.

Lysosomal Enzymes

Another mechanism of bacterial destruction involves their digestion by **lysosomal enzymes.** This is especially important in tissues where the oxygen tension is low and the respiratory burst cannot function effectively. As a bacterium is drawn into the cytoplasm and enclosed in a **phagosome,** the primary granules (or **lysosomes**) of the neutrophil migrate through the cytoplasm, fuse with the phagosome, and release their enzymes into it. The vacuole is then known as a phagolysosome. The enzymes released into the phagolysosome may destroy bacterial walls and kill some bacteria, although their main function is probably to digest dead organisms (Fig. 5–10). **Lysozyme** can destroy the cell walls of some nonpathogenic gram-positive bacteria (see Fig. 2–5). Lactoferrin, by binding iron, may also prevent bacterial growth. However, some bacteria, such as *Brucella abortus* and *Listeria monocytogenes,* can multiply within phagocytic cells even in the presence of lysosomal enzymes.

Neutrophil Surface Proteins

Neutrophils carry many glycoprotein receptors on their surface (Fig. 5–11). Among the most important are the integrins of the CD18 (β_2) family. These are three heterodimers consisting of one protein, CD18, linked to a second protein, either CD11a, CD11b, or CD11c. CD11a/CD18 and CD11b/CD18 play important roles in the adhesion of granulocytes to blood vessel walls in inflammation. CD11b/CD18 is also a receptor for some components of the complement system (hence it is also called complement receptor 3 and is discussed further in the chapter on complement). When stimulated by chemoattractants, neutrophils show an increased adhesiveness to other cells that may be due to increased expression of CD11b/CD18. CD11c/CD18 is also found on granulocytes and macrophages, where it binds a breakdown fragment of C3. The integrins are described in detail in Chapter 7.

The glycoprotein called FcγRII or CD32 is a receptor for antibody found on neutrophils (and macrophages). It binds to the Fc region of antibody molecules and is therefore called an **Fc receptor.** It has a low binding strength **(affinity)** for antibody molecules of the IgG class. FcγRII binds antibody-coated particles and promotes their phagocytosis and destruction. Fc receptors are described in detail in Chapter 10.

CR1 (complement receptor-1, CD35) is the major receptor for the complement component called C3b. It is found not only on neutrophils but also on other granulocytes, monocytes, erythrocytes, and B cells. Binding of complement-coated particles to CR1 leads to their phagocytosis. Complement receptors are described in detail in Chapter 16.

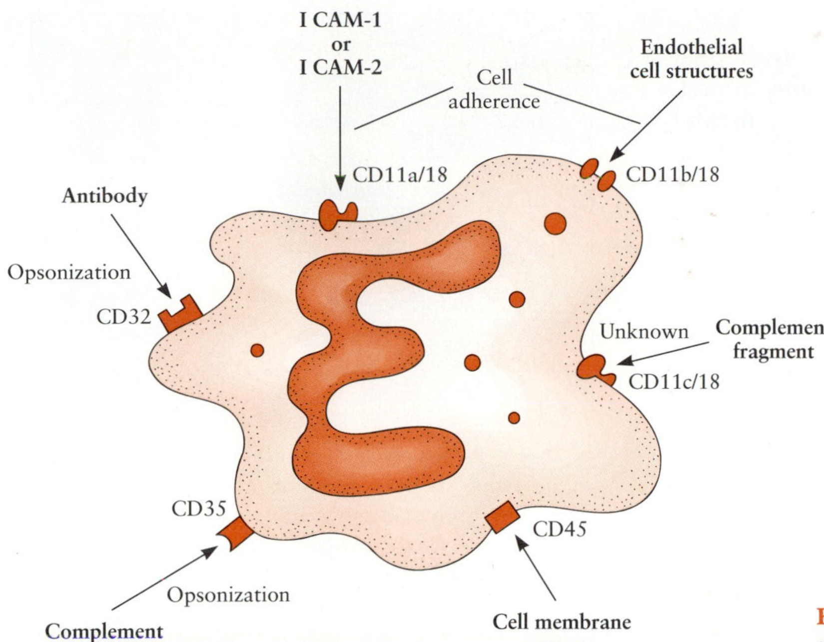

Figure 5–11 The major receptors found on the surface of neutrophils, the structures that they bind to, and their functions.

Fate of Neutrophils

Neutrophils possess a limited reserve of energy in the form of glycogen that cannot be replenished. They are therefore active immediately after being released from the bone marrow but are rapidly exhausted and can ingest only a limited number of bacteria. Because of their lack of stored energy sources, even inactive neutrophils survive for less than a day before dying in a process called **apoptosis** (Chapter 18). The dying neutrophil is usually ingested and destroyed by macrophages before it can release its lysosomal enzymes and cause inflammation or tissue damage. Macrophages recognize these dying neutrophils as a result of changes in their cell membrane lipids.

Thus, neutrophils may be considered a first line of defense, moving rapidly toward foreign material and destroying it promptly but being incapable of sustained effort. The second line of defense is the mononuclear-phagocytic system. Since neutrophils usually destroy all ingested foreign material, they cannot prepare antigen for presentation to antigen-sensitive cells.

EOSINOPHILS

The second type of polymorphonuclear granulocyte is the eosinophil, so called because its cytoplasmic granules stain intensely with eosin, a red dye (Fig. 5–12). Like neutrophils, eosinophils are formed in the bone marrow. Eosinophils leave the bone marrow in a relatively immature state and probably move to the spleen, where they reach maturity. They then spend a short period circulating in the bloodstream. Their half-life in the circulation is only about 30 minutes. Eosinophils subsequently migrate into the tissues, where they have a half-life of about 12 days. As a result, for every eosinophil seen in the blood, there are about 500 stored in tissues. The proportion of eosinophils among the blood leukocytes is highly variable since their level rises if the animal is parasitized. Normal values range from 2% to 5% in healthy humans.

Like neutrophils, eosinophils are phagocytic cells that can ingest and destroy foreign material. The granules of the eosinophil, however, contain large quan-

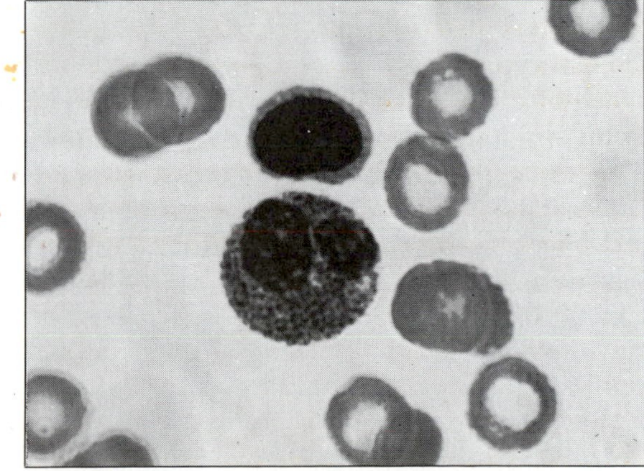

Figure 5–12 An eosinophil granulocyte and a lymphocyte from peripheral blood (Wright-Giemsa stain; original magnification ×1400). *(From Bellanti, J. A. Immunology II. W. B. Saunders, Philadelphia, 1979. With permission.)*

tities of acid phosphatase and peroxidase. The eosinophil peroxidase is chemically distinct from that found in neutrophils and uses bromide ions in preference to chloride, so producing OBr^-. Eosinophil peroxidase is more efficient than neutrophil peroxidase in killing certain organisms. In addition, each large eosinophil granule contains in its crystalline core a protein called major basic protein (MBP). MBP is highly toxic for invading parasitic worms (helminths). The major function of eosinophils is probably to destroy helminths.

Phagocytosis by eosinophils is similar to the process in other cells. Eosinophils are attracted to areas of activity by tissue damage and by specialized eosinophil chemotactic factors. The most potent of these are two tetrapeptides (Val-Gly-Ser-Glu and Ala-Gly-Ser-Glu) secreted by mast cells. (Mast cells are large connective tissue cells that play a major role in allergies. See Chapter 29.) Nonspecific chemotactic factors for eosinophils include C5a, C567, histamine, and some metabolites of arachidonic acid. Eosinophils possess both antibody and complement receptors. Their antibody receptors are generally fewer and of lower affinity than the receptors on neutrophils. However, the number of antibody receptors is enhanced by the presence of eosinophil chemotactic factors.

Once a particle is bound to the receptors on an eosinophil, it triggers a respiratory burst that may exceed that of neutrophils. Like neutrophils, eosinophils can phagocytose small particles, but they are much more suited to extracellular destruction of large parasites. For this reason, eosinophils can extrude their granules into the surrounding fluid. Thus when they stick to parasites, they extrude MBP and other toxic proteins that can destroy the worm's cuticle. Because MBP has also been identified in basophils, it has been suggested that eosinophils and basophils are closely related and have common functions. Eosinophils and their function are described in detail in Chapter 28.

BASOPHILS

The least numerous of the granulocytes, the basophils, are so called because their cytoplasmic granules stain intensely with basic dyes, such as hematoxylin (Fig. 5–13). Basophils constitute about 0.5% of blood leukocytes; they are not normally found in extravascular tissues. They may, however, infiltrate tissues under the influence of cells called lymphocytes. In the tissues, basophils provoke inflammation since their granules contain vasoactive amines such as histamine and serotonin. Basophils, together with eosinophils and mast cells, are functionally related since all have the ability to promote acute inflammation.

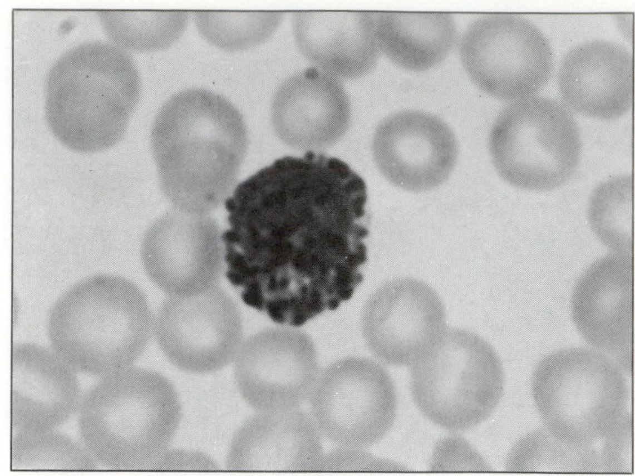

Figure 5–13 A basophil granulocyte from peripheral blood (Wright-Giemsa stain; original magnification ×1400). Note the large, intensely staining granules. *(From Bellanti, J.A. Immunology II. W. B. Saunders, Philadelphia, 1979. With permission.)*

PLATELETS

Blood platelets can adhere to and engulf some bacteria. This is not a true phagocytic process, but the bacteria may enter the cytoplasmic canaliculi of the platelets. Bacteria attached to platelets can be removed from the circulation and destroyed by the cells of the mononuclear-phagocytic system. It is also possible that bacteria hidden within clumps of aggregated platelets may be protected from circulating antibodies and complement and so survive.

KEY WORDS

Lysosome p. 52
Metchnikoff, Elie p. 51
Mononuclear-phagocytic system p. 51
Myeloid system p. 51
Myeloperoxidase p. 55
NADPH-oxidase p. 54
Neutrophil p. 51
Nitric oxide p. 54
Opsonin p. 53
Peroxidase p. 58
Phagocyte p. 51
Phagocytosis p. 52
Phagolysosome p. 56
Phagosome p. 54
Polymorphonuclear cell p. 51
Primary granule p. 52
Respiratory burst p. 54
Secondary granule p. 52
Superoxide anion p. 54
Superoxide dismutase p. 54
Surface phagocytosis p. 53

QUESTIONS

1. The predominant type of leukocyte in the blood is the
   a. monocyte
   b. eosinophil
   c. basophil
   d. neutrophil
   e. lymphocyte
2. The movement of neutrophils under the influence of external chemical gradients is called
   a. endocytosis
   b. chemotaxis
   c. phagocytosis
   d. chemolysis
   e. exotaxis
3. Which of the following is a potent chemotactic factor for neutrophils?
   a. C-reactive protein
   b. interleukin-8
   c. arachidonic acid
   d. complement
   e. immunoglobulin
4. For which one complement protein do neutrophils contain a receptor important in phagocytosis?
   a. C1a
   b. C2a
   c. C3b
   d. C4a
   e. C5b
5. What is an opsonin?
   a. a chemotactic factor
   b. a type of granulocyte
   c. a chemokine
   d. a substance that enhances phagocytosis
   e. a lysosomal enzyme
6. The key cell surface molecule that initiates the respiratory burst is
   a. hypohalide
   b. myeloperoxidase
   c. superoxide dismutase
   d. NADPH-oxidase
   e. singlet oxygen
7. A receptor that binds immunoglobulin (antibody) to a cell surface is called a(n)
   a. Fc receptor
   b. complement receptor
   c. integrin
   d. CD molecule
   e. selectin
8. When neutrophils die, they do so by a process called
   a. necrosis
   b. phagocytosis
   c. diapedesis
   d. endocytosis
   e. apoptosis
9. One function of eosinophil major basic protein is to
   a. destroy invading bacteria
   b. destroy helminth parasites
   c. cause acute inflammation
   d. cause asthma
   e. destroy normal tissue
10. Integrins are
    a. opsonins
    b. chemotactic factors
    c. cell surface adhesive proteins
    d. complement components
    e. antibody receptors
11. Why might the body require two phagocytic cell systems—the myeloid system and the mononuclear-phagocytic system? What advantages are there in having two systems?
12. List the most important enzymes found inside neutrophils. What might happen if a neutrophil released its enzymes into healthy tissues? Does this happen normally?
13. The respiratory burst generates some highly toxic free radicals with the potential to cause severe tissue damage. How would you suggest that the body prevent this?
14. Speculate on the suggestion that the three forms of granulocytes (basophils, neutrophils, and eosinophils) are really very similar in function and that their staining properties exaggerate their differences while minimizing their similarities.
15. In the first decade of the twentieth century, physicians often attempted to stimulate phagocytic cell activity to enhance resistance to disease. In fact, George Bernard Shaw wrote a play about it. Is this a valid medical concept? How might you stimulate the phagocytes?
16. Describe the fate of a bacterium that gains access to the bloodstream via a skin wound from a thorn. What will eventually happen to the thorn if it remains embedded deep in the tissues?

Answers: 1d, 2b, 3b, 4c, 5d, 6d, 7a, 8e, 9b, 10c

SOURCES OF ADDITIONAL INFORMATION

Babior, B.M. Oxygen-dependent microbial killing by phagocytes N. Engl. J. Med., 298:659–668, 721–726, 1978.

Baggiolini, M., Boulay, F., Badwey, J.A., and Curnutte, J.T. Activation of neutrophil leukocytes: Chemoattractant receptors and respiratory burst. FASEB J., 7:1004–1010, 1993.

Baggiolini, M., and Dewald, B. The neutrophil. Int. Arch. Allergy Appl. Immunol., 76(suppl. 1):13–20, 1985.

Boxer, G.J., Curnutte, J.T., and Boxer, L.A. Polymorphonuclear leukocyte function. Hosp. Pract., 20:69–90, 1985.

Geisow M. Pathways of endocytosis. Nature, 288:434–436, 1980.

Ginsburg, I., and Lahav, M. Are bacterial cells degraded by leukocytes in vivo? An enigma. Clin. Immunol. News, 4:147–153, 1983.

Jaffe, E.A., and Mosher, D.F. Synthesis of fibronectin by cultured human endothelial cells. J. Exp. Med., 147:1779–1791, 1978.

Jordison-Boxer, G., Curnutte, J.T., and Boxer, L.A. Polymorphonuclear leukocyte function. Hosp. Pract., 20:69–90, 1985.

Kishimoto, T.K., et al. The leukocyte integrins. Adv. Immunol. 46:149–210, 1989.

Rossi, F., and Patriarcha, P., eds. Biochemistry and function of phagocytes. *In* Advances in Experimental Medicine and Biology, vol. 141. Plenum Press, New York, 1982.

Savill, J., Fadok, V., Henson, P., and Haslett, C. Phagocyte recognition of cells undergoing apoptosis. Immunol. Today, 14:131–136, 1993.

Snyderman, R., and Goetzl, E.J. Molecular and cellular mechanisms of leukocyte chemotaxis. Science, 213:830–837, 1981.

Yoshida, T., and Torisu, M., eds. Immunobiology of the Eosinophil. Elsevier Biomedical, New York, 1983.

CHAPTER 6

The Mononuclear-Phagocytic System

Emil R. Unanue *Born in Cuba in 1934, Emil Unanue has been responsible for many of the key discoveries relating to antigen processing by macrophages. For example, he first demonstrated that antigen processing was first necessary for T cells to recognize an antigen. He did this by fixing macrophages in paraformaldehyde at intervals after ingestion of antigen. Antigen could activate T cells when macrophages were fixed one hour after phagocytosis but not if fixed at the time of antigen exposure. He and his colleagues also demonstrated that T cell activation also depended on the presence of IL-1 derived from macrophages as well as on the presence of MHC class II molecules.*

CHAPTER OUTLINE

CHAPTER CONCEPTS

1. Macrophages are large phagocytic cells found throughout the body. They possess a single, rounded nucleus and extensive cytoplasm.
2. Macrophages phagocytose particles in a similar manner to neutrophils, but they can do so repeatedly and live much longer.
3. Macrophages are secretory cells. They produce proteins that influence inflammation, healing, and the body's response to infection.
4. The most important of these proteins are the cytokines. They include interleukin-1, interleukin-6, interleukin-12, and tumor necrosis factor–α.
5. Macrophage-derived cytokines can cause a fever and are responsible for many of the clinical signs of infection.
6. Macrophages in the spleen, liver, and bone marrow are largely responsible for removing foreign particles such as bacteria from the bloodstream.

In this chapter we look in detail at the properties of the phagocytic cells called macrophages. Macrophages are of major importance in immunity since they are actively phagocytic. Their function is to trap foreign material and then process it so that it can be recognized by the cells of the immune system. We first look at the structure of these cells in different areas of the body. We then examine their function, discussing their ability to phagocytose and destroy foreign material as well as their role in tissue repair. We then study some of the proteins that macrophages secrete, since these proteins are important in transmitting signals to other cells and influencing the response of the body to infection. Finally, we look at the fate of foreign material that enters the body by different routes and how it is destroyed by strategically located macrophages.

If an animal is intravenously injected with a suspension of carbon particles, such as those found in India ink, the particles are rapidly removed from the blood by phagocytic cells distributed throughout the body. Ludwig Aschoff, the German scientist who discovered this phenomenon, believed that these cells collectively formed a body system, which he called the **reticuloendothelial system.** Subsequent investigations showed, however, that this was not the case. The so-called reticuloendothelial system was a mixture of cells and systems. Some of the cells that take up carbon particles do so inadvertently and not as their prime function; these include epidermal cells, intestinal epithelial cells, and vascular endothelial cells. Some are neutrophils of the myeloid system. Nevertheless, among the cells that take up particulate matter there are professional phagocytes whose major function is the removal of foreign material and cell debris from tissues and blood. These professional phagocytes form a well-delineated body system called the mononuclear-phagocytic system.

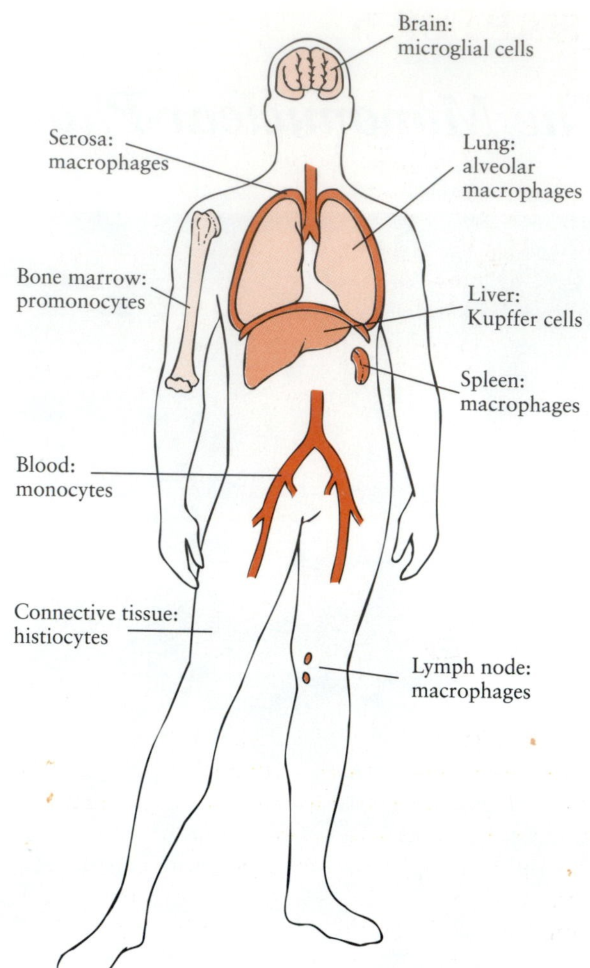

Figure 6–1 The location of the mononuclear-phagocytic system and its cells.

THE MONONUCLEAR-PHAGOCYTIC SYSTEM

The **mononuclear-phagocytic system** consists of a single population of cells called **macrophages** with similar morphology, common function, and common origin. Macrophages usually have a single, rounded nucleus and can therefore be readily distinguished from neutrophils. They are avidly phagocytic—hence they are mononuclear phagocytes. In contrast to neutrophils, macrophages are capable of sustained, repeated phagocytic activity. This is, however, just one of their functions. They also process antigen so that it can stimulate an immune response; they secrete proteins that amplify the immune response; they control inflammation and healing; and they remove dead, dying, and damaged tissue.

Macrophages are widely distributed throughout the body (Fig. 6–1). Immature macrophages found in the bloodstream are called **monocytes.** They constitute about 5% of the total blood leukocyte population. Mature macrophages are found in connective tissue, where they are known as **histiocytes;** those found lining the sinusoids of the liver are called **Kupffer cells;** those in the brain are **microglia;** and those in the lungs are **alveolar macrophages.** Many are found in the sinusoids of the spleen, bone marrow, and lymph nodes. Regardless of their name or location, they are all macrophages and all are part of the mononuclear-phagocytic system.

The Structure of Macrophages

Macrophages assume a wide variety of shapes in response to their environment. In suspension, however, they are round cells, about 14 to 20 μm in diameter. They have abundant cytoplasm, at the center of which is a single nucleus that may be round, bean-shaped, or

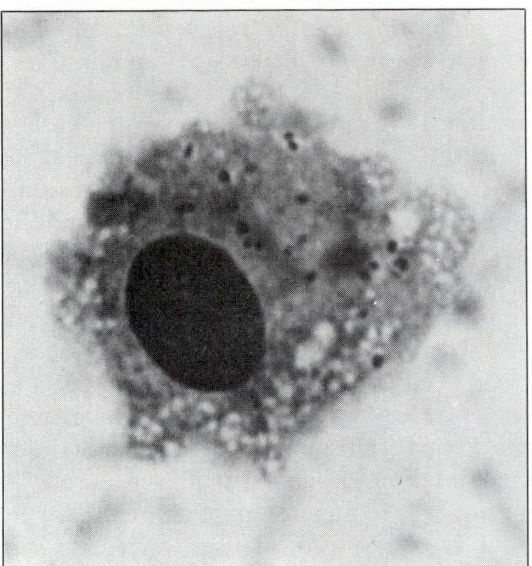

Figure 6–2 A typical alveolar macrophage. The small dark dots within the cytoplasm are ingested bacteria (original magnification ×900). *(Specimen provided by Dr. B. Wilkie.)*

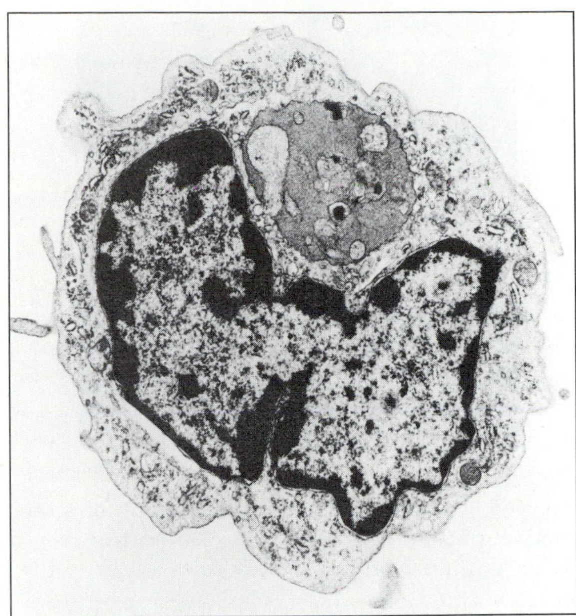

Figure 6–4 A transmission electron micrograph of a rabbit monocyte. This cell contains a large inclusion body of unknown significance. The rough endoplasmic reticulum is readily apparent. *(Courtesy of Dr. Scott Linthicum.)*

indented (Figs. 6–2 and 6–3). The perinuclear cytoplasm contains mitochondria, many lysosomes, some rough endoplasmic reticulum, and a Golgi apparatus, indicating that they can synthesize and secrete proteins (Fig. 6–4). The peripheral cytoplasm is usually devoid of organelles and is in continuous movement, forming and reforming veil-like ruffles (Fig. 6–5). Macrophages stick to glass surfaces, on which they spread by sending out long cytoplasmic filaments (Fig. 6–6). Some macrophages show variations from this basic structure. Peripheral blood monocytes tend to have round nuclei, which elongate as the cells mature; alveolar macrophages rarely possess rough endoplasmic reticulum, but their cytoplasm is full of granules; the microglia of the central nervous system have rod-shaped nuclei and very long cytoplasmic processes that are lost when the cell is stimulated into activity by tissue damage.

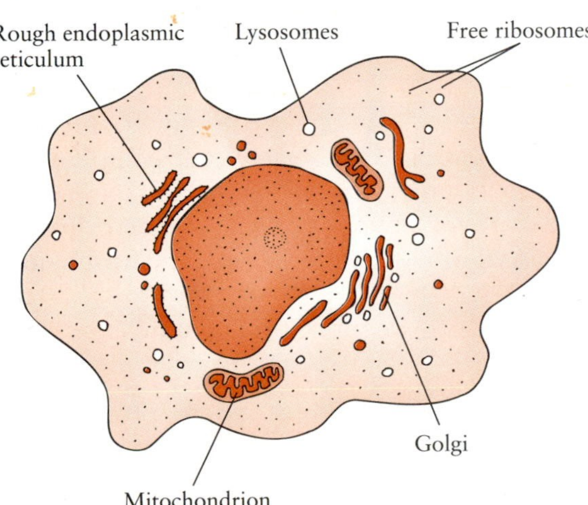

Figure 6–3 The major structural features of a macrophage. Note the extensive Golgi apparatus and rough endoplasmic reticulum, indicating that this cell can synthesize and secrete proteins.

The Life History of Macrophages

All macrophages originate from bone marrow stem cells known as monoblasts. Monoblasts develop into promonocytes and promonocytes develop into mono-

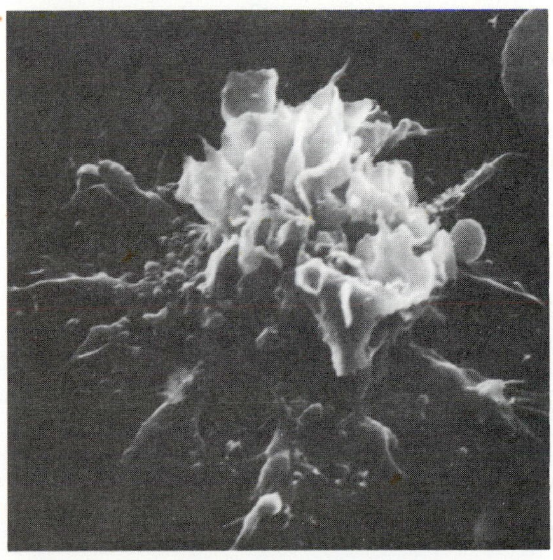

Figure 6–5 A scanning electron micrograph of a guinea pig peritoneal macrophage, showing extensive veil-like ruffles (original magnification ×4000).

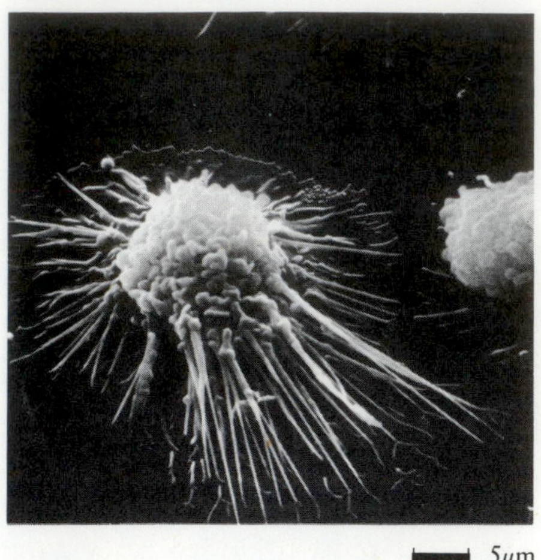

Figure 6–6 A scanning electron micrograph of a human monocyte spread on a glass surface. *(Biophoto Associates.)*

cytes, all under the influence of proteins called colony-stimulating factors. Monocytes then enter the bloodstream. It takes about six days for a promonocyte to develop into a mature monocyte (Fig. 6–7). After about three days, monocytes enter tissues and develop into macrophages. Some of these macrophages may be able to divide. Thus macrophages may be derived both from monocytes and from local proliferation. In the lung, where macrophages are found in the alveoli, most of these cells arise by local proliferation and only about 30% are derived from monocytes. On the other hand, in the liver, the macrophages (called Kupffer cells) are almost entirely dependent (90%) on blood monocytes for replenishment. Blood monocytes are actively recruited into inflamed tissues where they differentiate into macrophages. Macrophages may also divide in inflamed tissues. Macrophages are relatively long-lived cells, replacing themselves at a rate of about 1% per day unless activated by inflammation or tissue damage. Macrophages may live for a long time after ingesting chemically inert particles, such as the carbon injected in tattoo marks, although several may fuse to form **giant cells** in an attempt to eliminate the foreign material. In some circumstances, such as in smokers, macrophages may carry the particles to the lungs or intestines and from there into the bronchi or the intestinal lumen, from which they are eliminated. Some foreign particles are toxic to macrophages. For example, asbestos particles kill macrophages after phagocytosis. When released from dead cells, the particles must be repeatedly rephagocytosed, leading to the release of lysosomal enzymes and reactive oxygen metabolites. This results in chronic tissue destruction,

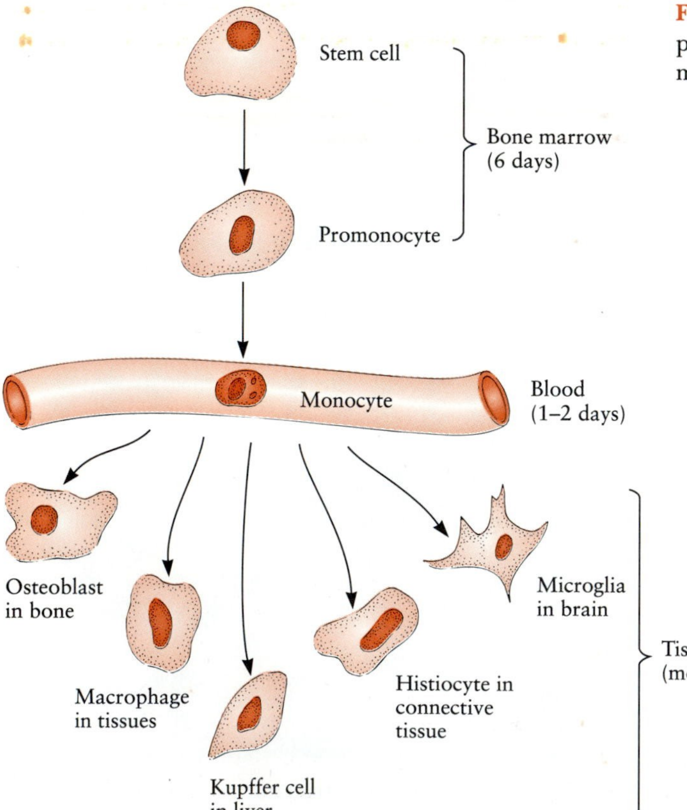

Figure 6–7 The progressive maturation of mononuclear phagocytic cells from a bone marrow stem cell through a monocyte stage in blood.

inflammation, and scar formation and eventually causes severe lung damage, a condition known as **asbestosis.**

The Functions of Macrophages

Phagocytosis

Phagocytosis by macrophages is similar to the process in neutrophils. Macrophages are attracted not only to bacterial products such as the *N*-formylmethionyl peptides and the products of immune reactions such as C5a and some cytokines, but also to factors released by damaged cells and extracellular matrix. These include fragments of collagen, elastin, and fibrinogen. Neutrophils, on dying, release elastase and collagenase and so generate monocyte chemotactic factors. Neutrophils are thus the martyrs of the immune system: they reach and attack foreign material first, and in dying they attract macrophages to the site of microbial invasion. Macrophages destroy antigen by both oxidative and nonoxidative mechanisms. The **respiratory burst** is especially intense in activated macrophages. An important role for macrophages is the removal of dead and dying cells. For example, the macrophages that line the splenic sinusoids remove aged red blood cells.

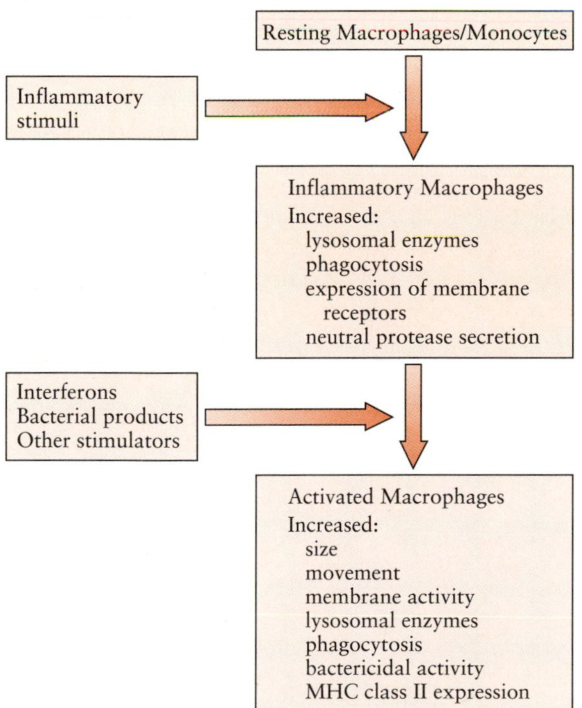

Figure 6–8 The activation of mononuclear phagocytic cells is a multistep process. Thus monocytes first become inflammatory macrophages and are subsequently fully activated. (See also Fig. 18–14.)

Macrophage Activation

Unstimulated monocytes and macrophages are relatively quiescent cells. When stimulated appropriately however, they can be rapidly activated. Several levels of activation are recognized. Thus when monocytes first move into inflamed tissues, they develop increased levels of lysosomal enzymes, enhanced phagocytic activity, increased expression of antibody, complement, and transferrin receptors, and increased neutral protease secretion. These cells are called inflammatory macrophages (Fig. 6–8). Inflammatory macrophages can be stimulated further to become **activated macrophages** as a result of exposure to bacterial products and proteins called interferons. Activated macrophages have a greatly enhanced ability to kill bacteria and some tumor cells by releasing reactive nitrogen metabolites. In contrast, their ability to secrete many proteins is reduced (Chapter 15).

Generation of Reactive Nitrogen Metabolites

When macrophages are activated by exposure to foreign particles, or chemotactic agents, synthesis of an enzyme called nitric oxide synthase is induced. This enzyme oxidizes the terminal guanido nitrogen atom of L-arginine, leading to the production of NO (**nitric oxide**) and citrulline (Fig. 6–9). Other reactive oxides of nitrogen such as NO_2, NO_2^-, N_2O_3, and NO_3^- as well as nitrosamines and nitrosothiols may also be produced. These oxides of nitrogen are toxic. They form

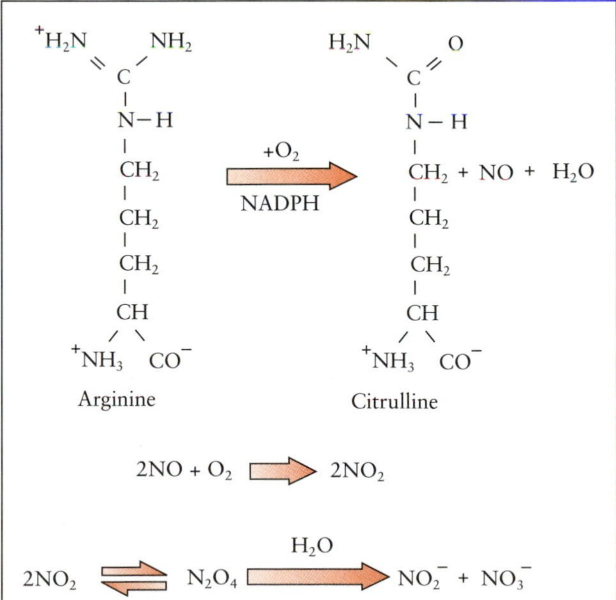

Figure 6–9 The production of bactericidal oxides of nitrogen, especially nitric oxide derived from the conversion of arginine to citrulline by nitric oxide synthase.

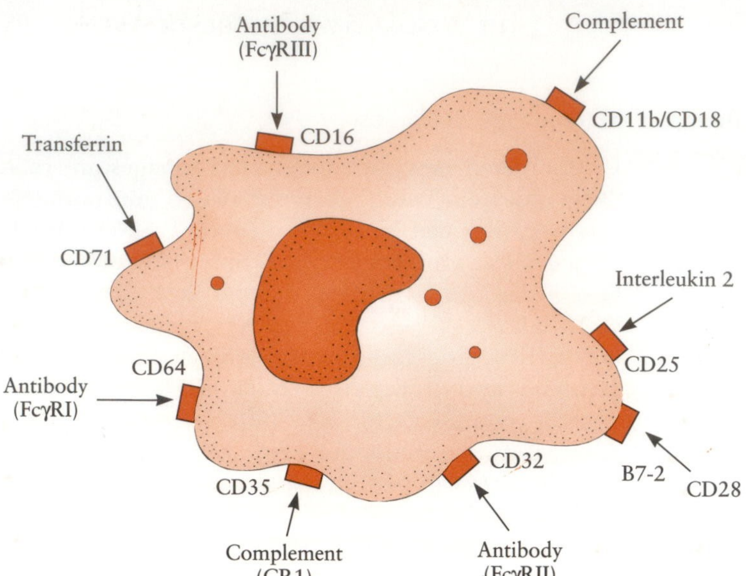

Figure 6–10 The major macrophage surface receptors and the molecules that they bind.

complexes with iron containing enzymes and thus inhibit oxidative metabolism of ingested bacteria and intracellular parasites. Macrophage-derived NO can inhibit lymphocyte and tumor cell DNA synthesis. Alternatively, NO can react with superoxide (O_2^-) to form $ONOO^-$, the peroxynitrite anion. This decays to form the highly toxic hydroxyl radical

$$ONOO^- + H^+ \rightleftharpoons OH + NO_2$$

(Neutrophils may also produce NO but only if activated. Thus NO is not produced by blood neutrophils but can be found in neutrophils in inflammatory exudates.)

When foreign material persists for long periods within the body, macrophages may accumulate in large numbers around the persistent material and look like epithelium on histological examination. These cells are therefore called **epithelioid cells.** Epithelioid cells are usually packed closely together and hence are polygonal. They possess abundant cytoplasm containing many lysosomes and much endoplasmic reticulum. Epithelioid cells may fuse to form multinucleated giant cells when attempting to enclose particles too large to be ingested by a single cell. Epithelioid cells and giant cells are a prominent feature of **tubercles,** the persistent inflammatory lesions that develop in individuals suffering from tuberculosis. In this case they result from the body's attempt to remove persistent *Mycobacterium tuberculosis* in the tissues.

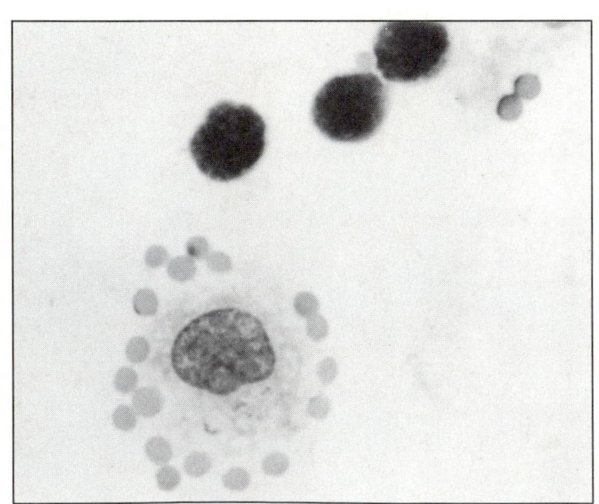

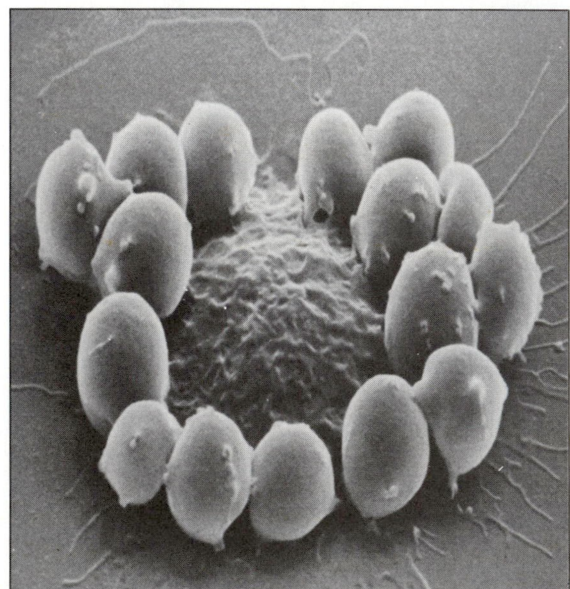

Figure 6–11 Light and scanning electron micrographs of guinea pig macrophages that have been allowed to settle on a glass coverslip for 1 hour before fixation. Each cell is surrounded by a rosette of erythrocytes coated with antibody and attached to the cell by Fc receptors (original magnification: top, ×900; bottom, ×1700). *(From Tizard, I. R., Holmes, W. L., and Parapally, N. P. J. Reticuloendo. Soc., 16:225–231, 1974. With permission.)*

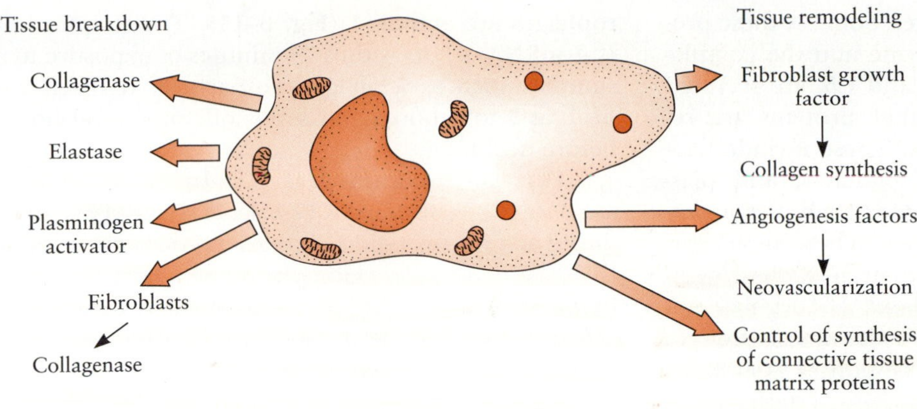

Figure 6–12 The role of macrophages in wound healing. They are involved both in the breakdown and removal of damaged tissues as well as in remodeling of newly healed tissues.

Macrophage Surface Receptors

Macrophages carry many proteins on their surface some of which are receptors for antibodies or complement (Fig. 6–10). For example, FcγRI (CD64) is a 75-kDa glycoprotein expressed on monocytes and macrophages and to a lesser extent on neutrophils. (Like other antibody receptors, it binds to the Fc region of antibody molecules and so is an **Fc receptor**—FcγRI.) FcγRI binds both free and antigen-bound antibody molecules with high affinity (Fig. 6–11). Its expression is enhanced up to 20-fold by activation of the macrophage with **interferon-γ.** Human macrophages also carry two other antibody receptors called FcγRII (CD32) and FcγRIII (CD16). These have a much lower affinity for antibody, although FcγRII binds aggregated antibodies and antigen–antibody complexes. (You can read more about these CD molecules in Chapter 10.)

Mouse and human macrophages also have receptors for the complement component called C3b and some of its breakdown products. Thus CR1 (CD35) is the major receptor for C3b. The integrin CD11b/CD18 is also a macrophage receptor for fragments of C3b. These receptors permit antigen particles coated with complement to bind to macrophages. (You can read more about complement in Chapter 12.)

Macrophages and Wound Healing

Macrophages are necessary for the proper healing of wounds (Fig. 6–12). When they arrive at a wound, macrophages first act in a similar manner to neutrophils. That is, they phagocytose and kill bacteria through the secretion of oxygen metabolites and lysosomal enzymes. However, macrophages also secrete elastase and collagenase that break down connective tissue. Macrophages also regulate the secretion of collagenase by fibroblasts by releasing interleukin-1. Once the damaged tissue is removed, the macrophages participate in tissue remodeling by acting as a source of growth factors for fibroblasts and stimulate these fibroblasts to secrete collagen. They also secrete molecules that promote the growth of new blood vessels. Macrophage participation is thus essential for the proper healing of wounds.

MACROPHAGES AS SECRETORY CELLS

Macrophages contain a Golgi apparatus and rough endoplasmic reticulum and can synthesize and secrete proteins (Fig. 6–13). Indeed, almost 100 different pro-

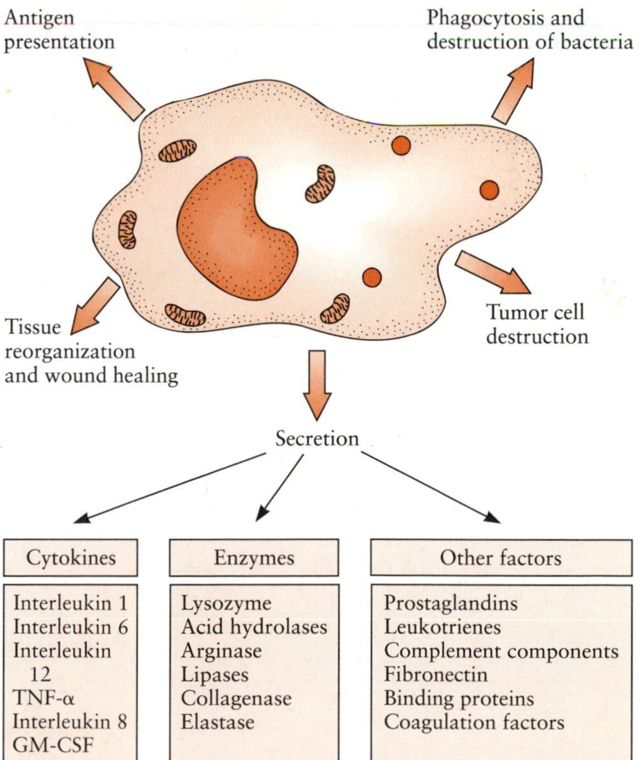

Figure 6–13 Some of the secretory products of macrophages as well as their major functions.

teins are secreted by macrophages. Some of these proteins, such as the enzyme lysozyme and the complement components C2, C3, C4, and C5, are secreted continuously (Chapter 16). Other proteins are released only during phagocytosis. These include lysosomal proteases, collagenases, elastases, and plasminogen activators as well as lipids such as platelet-activating factor and **leukotrienes.** These molecules cause tissue damage and inflammation. Macrophages also secrete four important proteins that regulate immune responses. These are interleukin-1 and interleukin-12, which promote immune responses by enhancing lymphocyte responses to antigen; interleukin-6, which influences the body's overall response to infection; and tumor necrosis factor–α, which destroys cancer cells. As might be expected, their net effect is complex. It must be pointed out, however, that although macrophages are the major source of these proteins, these proteins can also be secreted by a variety of other cell types.

Interleukin-1

Although many cell types can produce interleukin-1 (IL-1), such as dendritic cells, lymphocytes, endothelial cells, fibroblasts, and keratinocytes, the major producer of IL-1 is the activated macrophage. IL-1 is produced spontaneously by macrophages in small quantities and in much larger amounts when the macrophages are activated (Fig. 6–14). Transcription of IL-1 mRNA occurs within 15 minutes of exposure to a stimulus such as a phagocytic particle. It reaches a peak at 3 to 4 hours and levels off for several hours before declining.

There are two forms of IL-1, IL-1α and IL-1β. They are both glycoproteins of 17 kDa. IL-1β is usually produced at a 10- to 50-fold higher level than the α form. IL-1β is secreted by cells, while IL-1α remains membrane bound. Membrane-bound IL-1α is biologically active and can therefore act on cells that come into direct contact with the macrophage.

Interleukin-1 is a typical **cytokine** in that it acts on a great variety of cell types to produce many different effects (Fig. 6–15). When first identified, it was called lymphocyte-activating factor because of its effects on **lymphocytes.** (Lymphocytes are the small round cells that mediate the immune response. They are of two major types, T cells and B cells.) Thus IL-1 is necessary for the successful initiation of some forms of immune response. It binds to its receptor on certain lymphocytes (specifically the T-cell subpopulation called T helper 2 [Th2] cells) and induces them to secrete a mixture of proteins such as interleukin-4, interleukin-5, and interleukin-10. In addition to playing a central role in the immune responses, interleukin-1 acts on many other cells. For example, when you get an infection, macrophages attempting to destroy the invading microorganisms release interleukin-1. This acts on the

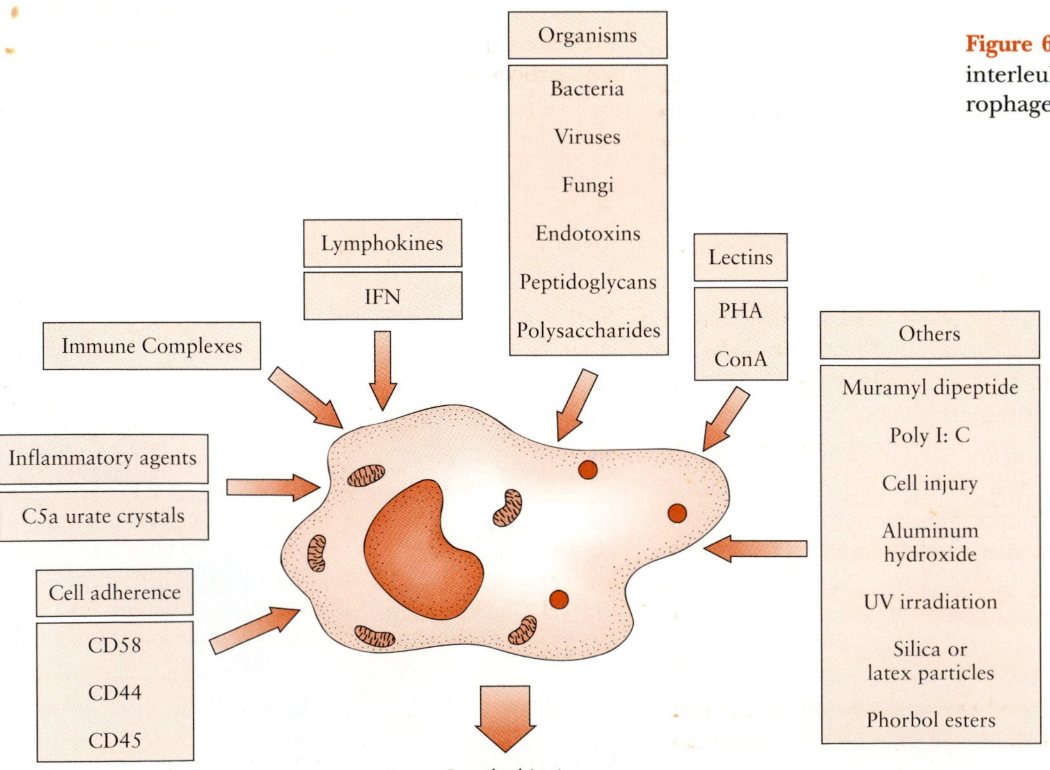

Figure 6–14 The inducers of interleukin-1 synthesis by macrophages.

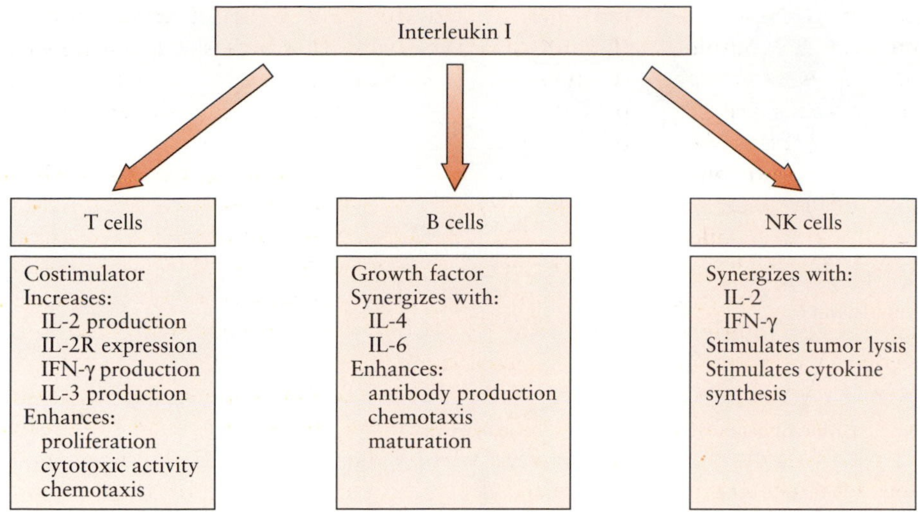

Figure 6–15 The effects of interleukin-1 on the cells of the immune system.

brain to cause a fever and its associated effects, lethargy, malaise, and lack of appetite. Thus it is the IL-1 that makes you feel "ill." IL-1 acts on muscle cells to mobilize a pool of amino acids. It acts on liver cells to induce the production of many new proteins. These new proteins, called **acute-phase proteins,** all assist in the defense of the body. Thus interleukin-1 plays a central role in the body's response to infection. These systemic effects of interleukin-1 are discussed in detail in Chapter 28.

Two receptors for IL-1 (IL-1R) of 60 and 80 kDa have been cloned. (They are also called CDw121 a and b.) These are type I transmembrane proteins that are homologous in their extracellular domains but very different in their cytoplasmic domains (see Chapter 7). It is believed that their **signal transduction** is mediated through the sphingomyelin signaling pathway involving a ceramide-activated protein kinase. Very few IL-1 receptors need to be bound for signal transduction to occur.

Interleukin-1 Receptor Antagonist

Interleukin-1 has very significant biological activities that must be carefully controlled. This is done by an interleukin-1 receptor antagonist (IL-1RA). IL-1RA is a glycoprotein that belongs to the same family as IL-1 and is produced by the same cells. Its only known function is to bind to IL-1 receptors, blocking IL-1 and preventing signal transduction. IL-1RA is important in many diseases. Thus it reduces mortality in septic shock, reduces graft-versus-host disease, and reduces the severity of malaria and arthritis.

Interleukin-6

Interleukin-6 (IL-6) is a 26-kDa glycoprotein produced not only by activated macrophages but also by fibroblasts, endothelial cells, **mesangial cells,** keratinocytes, and lymphocytes (Fig. 6–16). IL-6 acts on a variety of cell targets, including lymphocytes, fibroblasts, and hepatocytes. It acts on lymphocytes to promote antibody synthesis and other activities. IL-6, like IL-1, plays an important role in the body's response to infection. It acts with IL-1 to stimulate the release of acute-phase proteins from hepatocytes. It also causes a fever, although it is much less potent than IL-1 or TNF–α in this regard. The IL-6 receptor (IL6R or CD126) is a heterodimer of 80 and 130 kDa.

Interleukin-12

Interleukin-12 (IL-12) is a 75-kDa heterodimeric glycoprotein produced by macrophages and other antigen-processing cells (B cells and dendritic cells) (Fig. 6–17). Its receptor is a single protein of 110 kDa found

> **A CLOSER LOOK**
>
> ***Viruses Can Inhibit IL-1 Production***
>
> For IL-1β to be produced, its inactive precursor must be cleaved by a specific protease. This enzyme, called IL-1β convertase, cleaves the precursor, releasing the mature cytokine. In the cowpox virus, one of the viral genes codes for a very specific inhibitor of IL-1β convertase. As a result, cowpox virus can effectively suppress the IL-1β response to infection. It is recognized that cowpox virus can effectively suppress inflammation in some experimental systems. As a result, the virus can infect more cells and cause more tissue damage.

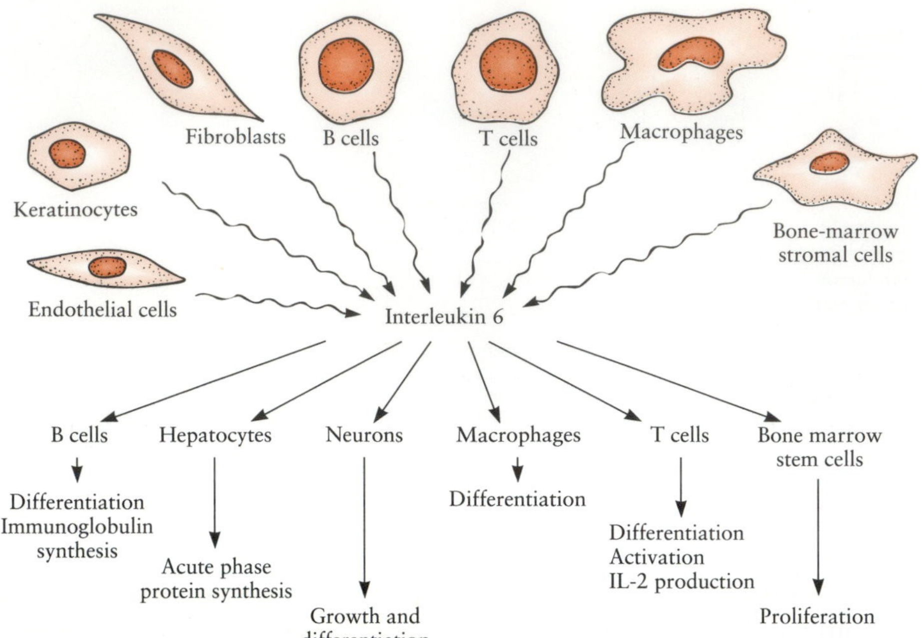

Figure 6–16 The origins and targets of interleukin-6.

on certain lymphocytes (T cells and NK cells). IL-12 plays a comparable role to IL-1 in costimulating helper T cell activity. Thus IL-1 is the major costimulator of the helper T cell subset called Th2 cells, and IL-12 is the costimulator of the helper T cell subset called Th1 cells (Chapter 11). IL-12 acts on these cells to enhance their secretion of proteins such as IL-2 and interferon-γ as well as the interleukin-2 receptor. It also stimulates their cytotoxic activities and thus promotes their antiviral and anticancer functions. (You can read more about IL-12 in Chapter 12.)

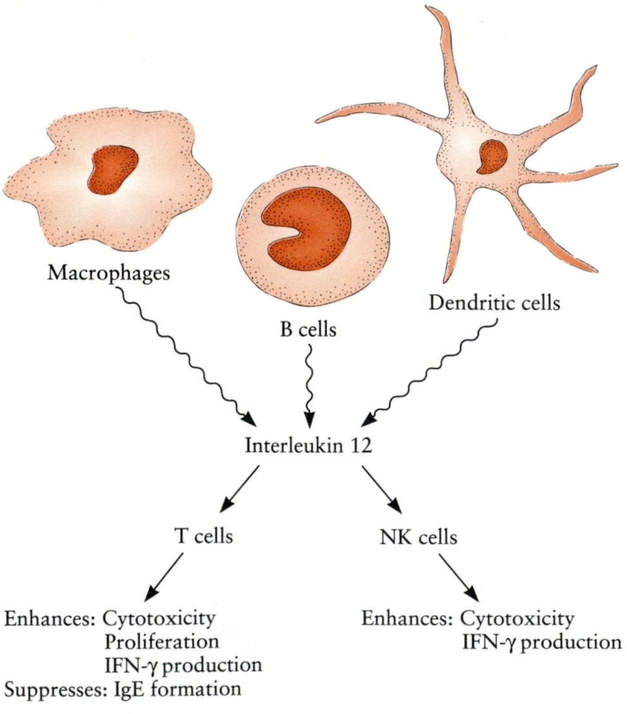

Figure 6–17 The origins and targets of interleukin-12.

Tumor Necrosis Factor–α

Tumor necrosis factor–α (TNF–α) is a 17-kDa protein produced primarily by macrophages or monocytes. Its production is triggered mainly by gram-negative bacteria, but it can also be produced in response to many other stimuli (Fig. 6–18). TNF–α can enhance proliferation of lymphocytes. It stimulates IL-1 and IL-6 production in macrophages and acts as a growth stimulator for fibroblasts. As its name implies, it can kill tumor cells.

Individuals infected with gram-negative bacteria such as *Escherichia coli* or *Salmonella typhi* may develop a lethal syndrome with fever, rigors, muscle pains (myalgia), headache, hypotension (low blood pressure), kidney, liver, and lung injury, and eventually death. This is called **septic shock.** All these effects are mediated by the release of TNF–α and other cytokines by endotoxin-stimulated macrophages. Animals exposed to chronic, sublethal doses of TNF–α lose weight and become anemic and depleted of protein. The weight loss occurs because TNF–α makes fat cells lose their stored lipids. TNF–α is also responsible for the severe wasting seen in individuals suffering from cancer, AIDS, or chronic parasitic and bacterial diseases. TNF–α acts directly on the temperature-regulating center of the hypothalamus to cause a fever. It causes

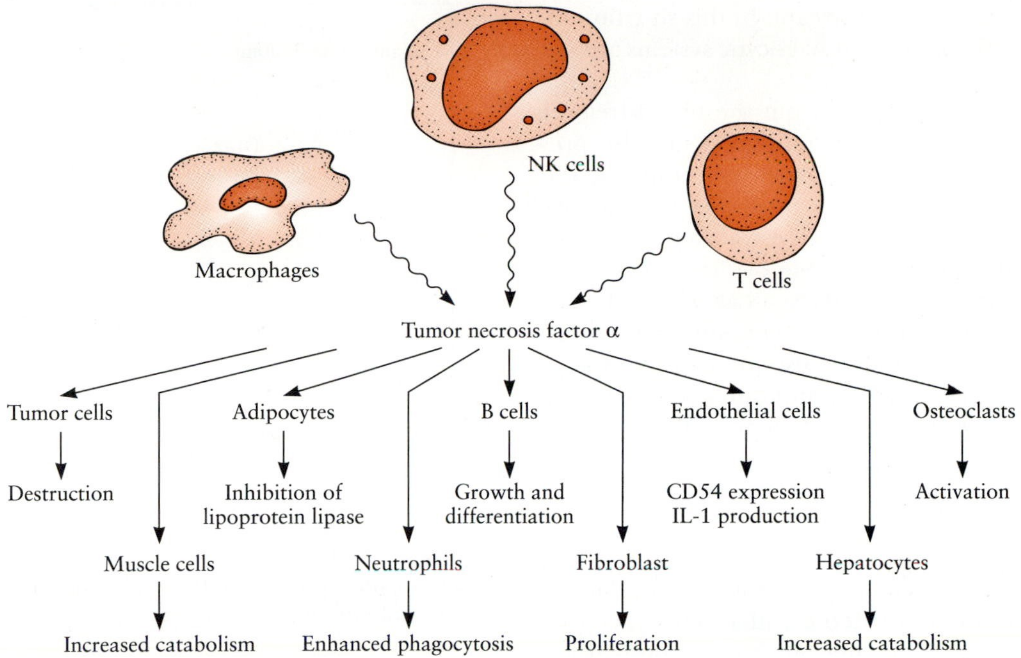

Figure 6–18 The origins and targets of tumor necrosis factor–α.

hepatocytes to release acute-phase proteins. It also enhances macrophage and neutrophil chemotaxis and increases their phagocytic and cytotoxic activities.

TNF–α kills some human tumor cell lines and causes **necrosis** in the center of some implantable tumors in vivo. The mechanism of TNF-induced cell lysis is not well understood, but the binding of TNF–α to its receptor is interpreted by some cells as a signal to die. (See apoptosis, Chapter 18; you can read more about TNF in Chapter 12.)

FATE OF FOREIGN MATERIAL WITHIN THE BODY

Particles Given Intravenously

If suspended particles, such as bacteria, are injected intravenously into an animal, they will eventually be trapped and removed by the macrophages that line the blood sinusoids of the liver, spleen, and bone marrow (Fig. 6–19) as well as by blood neutrophils. The spleen is a more effective filter than the liver but, being a much smaller organ, traps less material. There is also a major difference in the types of particle removed by the liver and spleen. Particles coated with antibody tend to be preferentially removed in the spleen. Splenic macrophages have receptors for antibody (FcγRI) and strongly bind antibody-coated particles. In contrast, particles coated with complement are preferentially removed by the Kupffer cells in the liver. Kupffer cells possess complement receptors (CR1). Some injected particles may also be trapped as they pass through the capillary beds of the lungs. The rate of clearance of particles from the circulation is regulated by the presence of opsonins such as fibronectin. If an animal is injected intravenously with a very large dose of colloidal carbon, **fibronectin** will be depleted and other particles (such as bacteria) will not

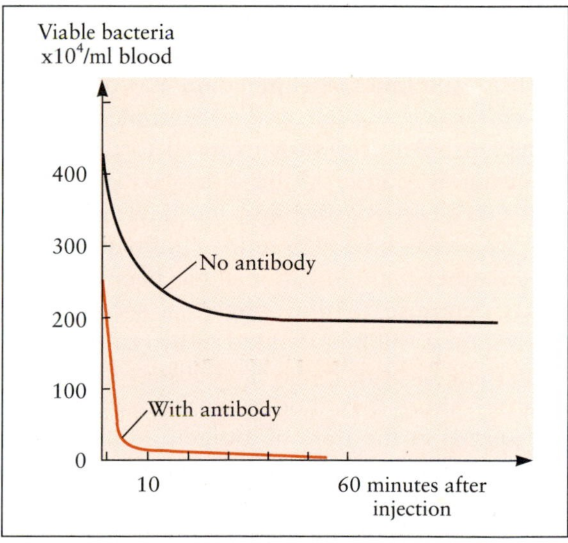

Figure 6–19 The clearance of bacteria from the bloodstream (in this case *E. coli* in swine) in the presence and absence of antibodies against *E. coli.*

be cleared from the bloodstream. In this situation the myeloid and mononuclear-phagocytic systems are said to be blockaded.

Clearance of organisms from the bloodstream is greatly enhanced if specific antibodies are also present, since opsonization increases the trapping efficiency of neutrophils and macrophages. If antibodies are absent or the bacteria possess an antiphagocytic polysaccharide capsule, the rate of their clearance is decreased. Some compounds, such as bacterial endotoxins, estrogens, and simple lipids, stimulate macrophage activity and therefore also increase the rate of bacterial clearance. Steroids and other drugs that depress macrophage activity also depress the clearance rate.

Soluble Antigens Given Intravenously

Unless carefully treated, protein molecules in solution tend to aggregate spontaneously. If a solution of a soluble protein antigen is injected intravenously, these aggregates are rapidly removed by neutrophils and macrophages. The unaggregated molecules remain in solution and are distributed evenly through the animal's blood. Molecules that are sufficiently small (less than 100 kDa) are also distributed through the extravascular tissue fluids. Once distributed, the antigen is treated like other body proteins and catabolized, resulting in a slow but progressive decline in antigen concentration. Within a few days, however, the animal begins to mount an immune response to the antigen. Antibodies are produced that combine with the antigen to form **immune complexes.** These immune complexes are rapidly cleared from the circulation by phagocytic cells. In this way all the antigen is rapidly and completely eliminated (Fig. 6–20).

This triphasic clearance pattern of distribution, catabolism, and **immune elimination** may be modified under certain circumstances. For example, if the animal has not been previously exposed to the antigen, the immune response will be a primary one. In this case, it takes between five and ten days before antibodies are produced and immune elimination occurs. If, on the other hand, the animal has been primed by previous exposure to the antigen, a secondary immune response will be mounted in two to three days, and the stage of progressive catabolism will therefore be relatively short. If antibodies are already circulating in the animal at the time of antigen administration, immune elimination will occur immediately, and no phase of catabolism is seen. If the injected material is not antigenic or if the immune response does not occur, because of either tolerance or immunosuppression, catabolism will continue until all the material is eliminated.

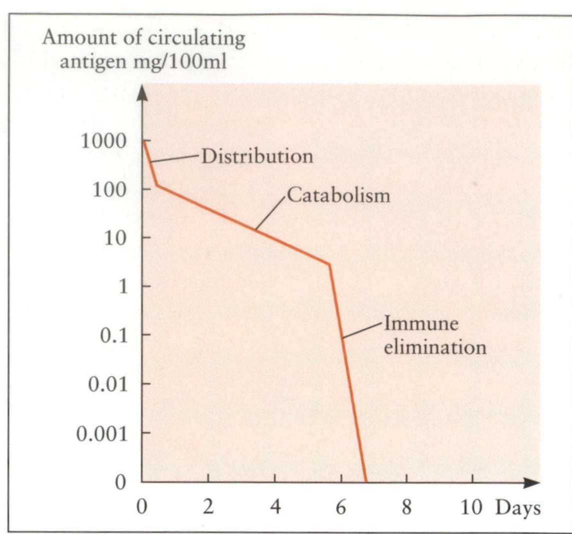

Figure 6–20 The clearance of a soluble antigen from the bloodstream, showing the three phases: distribution, catabolism, and immune elimination.

Fate of Antigen Administered by Other Routes

When insoluble or aggregated antigen is injected into a tissue, some slight damage is bound to occur. As a result, phagocytic cells (first neutrophils, then macrophages) migrate toward the injection site under the influence of chemotactic factors released from the damaged tissue. These cells phagocytose the injected material. The antigen taken up by macrophages is processed and eventually stimulates an immune response. Antibodies and complement (Chapter 16) interact with the antigen, generating more chemotactic factors that attract still more phagocytic cells and in this way hasten the final elimination of the offending material. In an animal already sensitized to the antigen, a web of specialized dendritic cells called **Langerhans cells** within the skin may trap immune complexes and present the antigen directly to lymphocytes. For this reason intradermal injection of antigen may be more effective in stimulating an immune response than subcutaneous inoculation.

Soluble antigen injected into a tissue is redistributed by the flow of tissue fluid through the lymphatic system. It eventually reaches the bloodstream, so its final fate is similar to intravenously injected material. Any aggregated material present is phagocytosed either by blood neutrophils, tissue macrophages, or by the macrophages and dendritic cells of lymph nodes through which the tissue fluid flows.

Normally, antigens passing through the intestinal tract are broken down by digestive enzymes to small, nonantigenic fragments. However, some antigenic molecules may remain intact and pass through the intestinal epithelium. Bacterial polysaccharides and those antigens that associate with lipids are especially effective in this respect, since they are absorbed in fat droplets. Antigens that enter the bloodstream from the intestine are promptly removed by the macrophages of the liver, whereas those antigens entering the intestinal lymphatics are trapped in the mesenteric lymph nodes. The relatively large size of these lymph nodes in animals testifies to their activity in this respect.

The fate of inhaled antigenic particles depends on their size. Large particles (greater than 3 μm in diameter) are deposited on the mucus layer overlying the respiratory epithelium from the trachea to the terminal bronchioles. These particles are then removed from the respiratory tract by the flow of mucus toward the pharynx or by coughing. Most particles that reach the alveoli are phagocytosed by alveolar macrophages, which carry them to the bronchoalveolar junction; from there they are also removed from the lung by the flow of mucus and by the cilia on the respiratory epithelial cells. Nevertheless, some antigens may be absorbed from the alveoli. Small particles absorbed in this way are cleared to the draining lymph nodes, whereas soluble antigens enter the blood vessels and are therefore distributed throughout the body. When large quantities of particles are inhaled, as occurs in workers exposed to industrial dusts and in cigarette smokers, the alveolar macrophage system may be temporarily blockaded and the lungs rendered more susceptible to invasion by microorganisms.

KEY WORDS

Alveolar macrophage p. 62
Antibody (Fc) receptor p. 67
Asbestosis p. 65
Clearance p. 71
Complement receptor p. 67
Epithelioid cell p. 66
Fever p. 69
Fibronectin p. 71
Giant cell p. 64
Histiocyte p. 62
Integrin p. 67
Interleukin-1 (IL-1) p. 68
Interleukin-1 receptor antagonist (IL-1RA) p. 69
Interleukin-6 p. 69
Interleukin-12 p. 69
Kupffer cell p. 62
Langerhans cells p. 72
Macrophage p. 62
Microglia p. 62
Monocyte p. 62
Mononuclear-phagocytic system p. 62
Nitric oxide p. 65
Reticuloendothelial system p. 62
Septic shock p. 70
Spleen p. 71
Tumor necrosis factor–α p. 70
Wound healing p. 67

QUESTIONS

1. Macrophages produce
   a. IL-2 and IL-3
   b. IL-2 and IL-1
   c. IL-4 and IL-1
   d. IL-1 and IL-3
   e. IL-12 and IL-1
2. Bacteria are cleared from the bloodstream mainly by
   a. lung capillaries
   b. spleen and liver macrophages
   c. circulating neutrophils
   d. marginating neutrophils
   e. renal capillaries
3. Particles given intravenously are mainly trapped in the
   a. lung
   b. thymus
   c. lymph nodes
   d. spleen
   e. tonsils
4. The constituent cells of the mononuclear-phagocytic system are
   a. eosinophils
   b. basophils
   c. neutrophils
   d. lymphocytes
   e. macrophages
5. The form of macrophage lining the sinuses of the liver is the
   a. histiocyte
   b. Kupffer cell
   c. monocyte
   d. astrocyte
   e. hepatocyte
6. Macrophages can be activated by
   a. neutrophils
   b. antibodies
   c. complement
   d. interferons
   e. all of the above
7. Activated macrophages can kill intracellular organisms by producing
   a. nitric oxide
   b. antibodies
   c. proteases
   d. interferons
   e. interleukins

8. The key protein produced by macrophages on exposure to microorganisms and other stimuli is
   a. complement
   b. integrins
   c. interleukin-1
   d. interleukin-2
   e. nitric oxide
9. Septic shock is mediated by which macrophage-derived product?
   a. tumor necrosis factor–α
   b. nitric oxide
   c. singlet oxygen
   d. interleukin-12
   e. prostaglandins
10. FcγRI are receptors for
   a. interleukins
   b. antibodies
   c. complement
   d. TNF
   e. nitric oxide synthase
11. Why is the concept of a reticuloendothelial system not considered valid? Is the mononuclear-phagocytic system a more legitimate concept? Why?
12. Compare the bactericidal mechanisms of neutrophils and macrophages. How do these different mechanisms reflect functional differences between these cell populations?
13. What sequence of events takes place when a small piece of sterile carbon is inserted under the skin in a tattoo mark? How does this response differ from that due to a crystal of asbestos?
14. What happens when a bacterium such as *E. coli* is ingested by a macrophage? How does this differ from the fate of a different bacterium, *M. tuberculosis*? Are these differences reflected in the diseases caused by these bacteria?
15. Macrophages can be selectively destroyed in the body by the administration of colloidal silica. Discuss the potential consequences of total or partial loss of functional macrophages. Which might be worse—no macrophages or no neutrophils?
16. Draw a table comparing the properties of IL-1, IL-12, and IL-6.
17. Do molecules such as interleukin-1 or tumor necrosis factor have any effect on wound healing? Could this be exploited therapeutically?

Answers: 1e, 2b, 3d, 4e, 5b, 6d, 7a, 8c, 9a, 10b

SOURCES OF ADDITIONAL INFORMATION

Allison, A.C. Macrophage activation and nonspecific immunity. Int. Rev. Exp. Pathol., 18:304–346, 1978.

Babior, B.M. Oxygen-dependent microbial killing by phagocytes. N. Engl. J. Med., 298:659–668, 721–726, 1978.

Bogdan, C., Vodovotz, Y., and Nathan, C. Macrophage deactivation by interleukin 10. J. Exp. Med., 174:1549–1555, 1991.

Cadena, D.L., and Gill, G.N. Receptor tyrosine kinases. FASEB J. 6:2332–2337, 1992.

Dinarello, C.A. Biology of interleukin 1. FASEB J., 2:108–115, 1988.

Fiorentino, D.F., et al. IL-10 inhibits cytokine production by activated macrophages. J. Immunol., 147:3815–3822, 1991.

Geisow M. Pathways of endocytosis. Nature, 288:434–436, 1980.

Jaffe, E.A., and Mosher, D.F. Synthesis of fibronectin by cultured human endothelial cells. J. Exp. Med., 147:1779–1791, 1978.

Johnston, R.B. Current concepts: Immunology. Monocytes and macrophages. N. Engl. J. Med., 318:747–752, 1988.

Linsley, P.S., and Ledbetter, J.A. The role of the CD28 receptor during T cell responses to antigen. Ann. Rev. Immunol., 11:191–212, 1993.

Mandel, T.E., et al. Long-term antigen retention by dendritic cells in the popliteal lymph node of immunized mice. Immunology, 43:354–362, 1981.

Metlay, J.P., Purè, E., and Steinman, R.M. Control of the immune response at the level of antigen presenting cells: A comparison of the function of dendritic cells and B lymphocytes. Adv. Immunol., 47:45–116, 1989.

Mocking, W.G., and Golde, D.W. The pulmonary alveolar macrophage. N. Engl. J. Med., 301:580–587, 639–645, 1979.

Nathan, C. Nitric oxide as a secretory product of mammalian cells. FASEB J., 6:3051–3061, 1992.

Nathan, C.F., Murray, H.W., and Cohn, Z.A. The macrophage as an effector cell. N. Engl. J. Med., 303:622–626, 1980.

Scott P. IL-12: Initiation cytokine for cell-mediated immunity. Science, 260:496–497, 1993.

Snyderman, R., and Goetzl, E.J. Molecular and cellular mechanisms of leukocyte chemotaxis. Science, 213:830–837, 1981.

Webb, D.S.A., et al. LFA-3, CD44 and CD 45: Physiologic triggers of human monocyte TNF and IL-1 release. Science, 249:1295–1297, 1990.

CHAPTER 7

Cellular Interactions

Harold Varmus *Harold Varmus was born in New York in 1939. He and Dr. J. Michael Bishop were awarded the 1989 Nobel Prize for their discovery that normal cellular genes can malfunction and so trigger the development of cancer. These investigators had been working on the ways in which some chicken viruses could cause cancer. They found that certain genes in the virus, called oncogenes, were responsible for the uncontrolled cancer cell growth. Subsequently they found that these viral oncogenes were actually modified forms of genes found in normal cells. This was a key finding in the fight against cancer as well as a major contribution to our understanding of cell biology.*

CHAPTER OUTLINE

CHAPTER CONCEPTS

1. Cells use glycoproteins on their surface to communicate with other cells in the body.
2. Cell surface proteins are in a state of flux as the cell surface is constantly renewed through endocytosis and exocytosis.
3. Cells go through a predictable cycle as they proceed from one mitosis to another.
4. Some cell surface proteins are receptors for ligands that transmit signals to cells. These signals are then transduced through several biochemical pathways so that a message is eventually transmitted to the cell nucleus. In the nucleus new genes are turned on by transcription factors so that their products are expressed and cell function is altered.
5. Many of the proteins involved in signal transduction are coded for by genes called proto-oncogenes. If these mutate, they can cause uncontrolled cell division—cancer.
6. Cells of the immune system bind to other cells through cell adhesion proteins that belong to four families—the immunoglobulin superfamily, the cadherin family, the integrin family, and the selectin family.

In this chapter we review some basic aspects of cell biology that are especially important in immunology. The immune response results from cells interacting with each other through receptors. We first briefly review the nomenclature of cell surface proteins. Then we examine the dynamics of the cell surface and how proteins are moved to and from the surface by the processes of exocytosis and endocytosis. We also discuss the key features of the cell cycle at this stage. The next section of the chapter deals with cell membrane receptors and how they signal to the cell how to turn cell functions on and off. The relationship of signal transduction to cancer formation is also discussed here. The last section of the chapter reviews the properties of the cell surface proteins involved in the adherence of cells to surfaces and other cells.

Natural membranes consist of a fluid phospholipid bilayer. Embedded within, or attached to, this phospholipid bilayer are the cell surface proteins. Through these proteins a cell communicates with its environment, receiving and transmitting signals. In the case of the immune system these signals are almost exclusively chemical. Most of these responses require a series of distinct steps. First, a **ligand** must bind to a **receptor** located on the outside of a cell. Second, information that the receptor is occupied must be transmitted to the inside of the cell. Third, genes must be selectively turned on or off. Finally, the cell must respond. This response can be divided into the primary events that occur immediately after the receptor is occupied and the secondary events that follow from this. In this chapter we review the basic properties of cell surface receptors and their mechanisms of action.

CELL SURFACE PROTEINS

All nucleated cells have a great variety of proteins associated with their surface membranes. These can be attached in many ways. Some are loosely attached to the surface by hydrophobic bonding and may be easily removed. Others are integral membrane proteins. Integral membrane proteins are of five major types (Fig. 7–1). Type I transmembrane proteins are single-pass polypeptides that have their C-terminus in the cytoplasm and their N-terminus outside the cell. The domain that passes through the cell membrane has a hydrophobic sequence that enables it to bind to the membrane lipids. These are the most common forms of cell membrane protein on leukocytes. Type II transmembrane proteins have the opposite orientation to type I proteins. Their N-terminus is in the cytoplasm, and the C-terminus is extracellular. The third form of membrane protein is covalently attached to the cell membrane by a glycosyl-phosphatidylinositol (GPI) anchor. An example of this is the Thy-1 molecule found on immature T cells. The other two types of integral transmembrane proteins pass through the cell membrane several times. One type passes through four times; the other type passes through seven or more times. Both have both hydrophilic and hydrophobic regions and orient themselves in the lipid bilayer so that the hydrophobic regions are associated with the interior of the membrane and the hydrophilic regions protrude into the aqueous tissue fluid (Fig. 7–2). Some proteins found on the surface of cells of the immune system may be found on many types of cells, and others may be specific for the cells' function and state of development. Thus some are found only on neutrophils, and some only on macrophages or lymphocytes.

Many apparently unrelated cell surface proteins are in fact united by the use of common domains. Several families of receptor proteins have been identified where the individual members respond to different ligands and elicit different responses. These form "superfamilies" of structurally related but functionally diverse proteins (see Fig. 13–15). The development of superfamilies reflects modular construction of proteins within each family. In other words, the body arranges a limited number of domains in many different ways to generate a great diversity of receptors. Another important feature of the proteins found on cell surfaces is that they are glycosylated. The carbohydrate side chains of the cell surface proteins are all located on their extracellular domains.

CELL SURFACE DYNAMICS

The outer cell membrane of animal cells is continually being exchanged with intracellular membranes through the movement of membrane-lined vesicles. The vesicles may move in either direction between the different membranes. These vesicles are formed by pinching off invaginations from the membrane to form small spherical bodies. Those formed by the outer cell membrane are called **endosomes,** and the process is called **endocytosis** (Fig. 7–3). Vesicles may also be transported to the outer cell membrane, where they fuse with the outer membrane in the process of **exocytosis.** Both endocytic and exocytic pathways involve the selective transfer of membrane components between membranes within the cell by vesicles that pinch off from one membrane and then fuse with another.

The Endocytic Pathway

Endocytosis involves the formation of invaginations in the plasma membrane that are then pinched off to form cytoplasmic vesicles. This process, called **pino-**

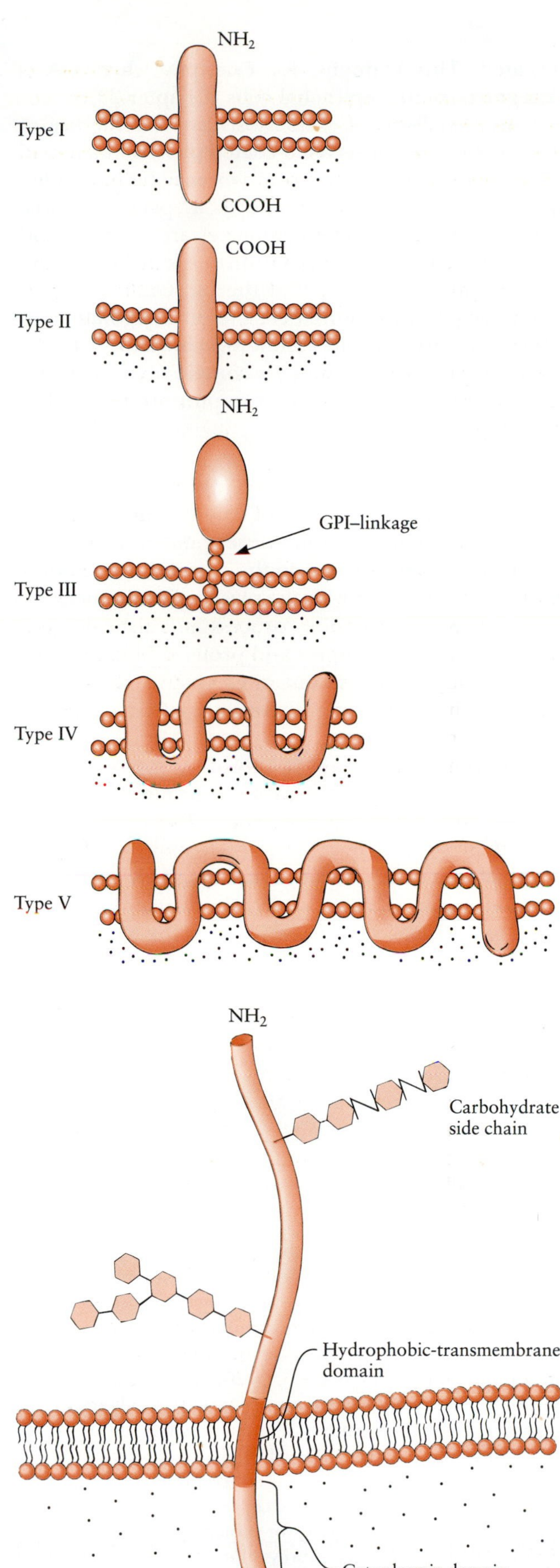

Figure 7–1 Cell surface proteins may be grouped according to their method of attachment to the cell membrane. Integral membrane proteins pass completely through the lipid bilayer once and are thus exposed to both the cytoplasm and the extracellular environment. Most are type I proteins, in which the amino terminus is at the extracellular end and the carboxyl terminus is in the cytoplasm. Type II integral membrane proteins have a similar structure but their orientation is reversed. A third type is attached directly to the membrane lipid by a short phosphatidyl inositol linker chain. The fourth and fifth types of cell membrane protein pass through the cell membrane either four or seven times, exposing several sites to the cytoplasm and the extracellular fluid.

Figure 7–2 A typical type I single-pass integral membrane glycoprotein showing the key features of these molecules. Note the hydrophobic domain that is embedded within the lipid membrane and the presence of carbohydrate side chains on the extracellular domains. The cytoplasmic domain often binds cytoplasmic or membrane-associated proteins involved in signal transduction.

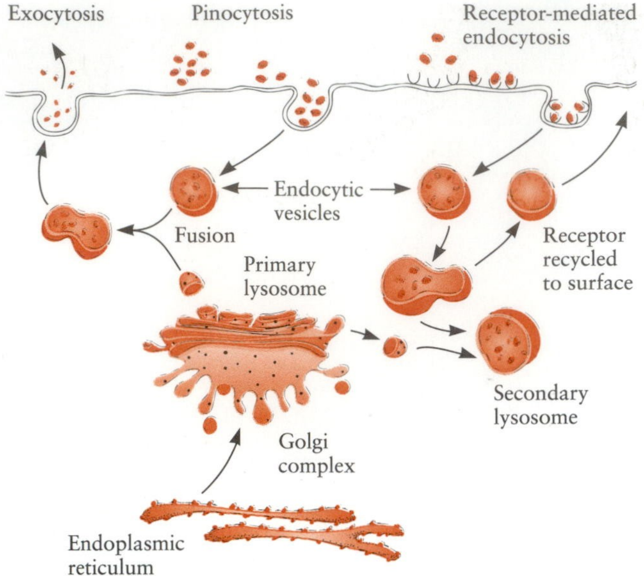

Figure 7–3 Cell membranes and their associated proteins are constantly being recycled. Membranes are interiorized by endocytosis and are returned to the cell surface by exocytosis.

cytosis, effectively internalizes portions of cell membrane, any proteins or receptors on the membrane, and any ligands attached to those receptors. The fate of these receptors and their ligands varies after endocytosis. In some cases, as when soluble antigen binds to B-cell antigen receptors, both receptor and ligand are directed to the lysosomes (Chapter 14). In other cases the vesicles may be transferred to another region of the outer cell membrane, where the ligand is then released. This happens, for example, when IgA is transferred across epithelial cells (Chapter 23).

Another form of endocytosis involves the use of specialized structures called **coated pits.** After a ligand binds to its receptor, the complex may be moved laterally through the membrane to a specialized area called a coated pit. The complexes are retained and concentrated in these pits. Clathrin is a protein whose subunits form the surface of the pit. Once sufficient complexes have accumulated, the pit invaginates and eventually forms a cytoplasmic vesicle called a receptosome. The receptosome may contain only the ligand as both the clathrin and the receptor are recycled to the surface.

The Exocytic Pathway

Exocytosis is a mechanism whereby material enclosed in a cell vacuole is passed to the extracellular fluid by fusion of the vacuole with the plasma membrane. Exocytosis is both a secretory process and a mechanism for replenishing the lipids and proteins of the plasma membrane. As pointed out in Chapter 3, the precise routing of proteins in the exocytic pathway depends on their glycosylation.

THE CELL CYCLE

Normal cells divide through mitosis. This process of division, however, takes a fairly short time during the cells' reproductive cycle and is called the M phase (M = mitosis). The much longer period between one M

A CLOSER LOOK

A Note on Cell Surface Molecule Nomenclature

Cell surface molecules are classified by the **CD system.** Each of these molecules—and there are now 130—is given a number, CD4, CD8, CD16, and so on. (A provisional designation, CDw, is given to incompletely characterized molecules.) The CD system was introduced to replace a haphazard collection of abbreviations and acronyms, such as B7, CEA, and VCAM-1. Some of the CD molecules, such as CD4 and CD8, play a well-recognized role in cell functions. Others do not have a known function at this time.

Both systems of nomenclature have disadvantages. It would be almost impossible for students to memorize the functions of many arbitrary CD molecules. On the other hand, meaningless letters and numbers are equally difficult to remember.

In this text I have compromised by using two basic principles. First, the common name will be used if it is well accepted or describes its function. Examples include FcαR (CD89), IL-6R (CD126), and L-selectin (CD62L). Second, if the CD designation has only recently replaced the common name and is not yet in wide usage, for example, B7 (CD80).

CD nomenclature will be used for molecules where the designation is well accepted, such as CD8 and CD4. It will also be used for molecules that have an irrational or arbitrary abbreviation.

phase and the next is called interphase. Several key events occur during interphase that prepare the cell for division. For example, the nuclear DNA of a cell must be duplicated. This period of DNA synthesis is called the S phase (S = synthesis). There is usually an interval between the end of the M phase and the beginning of the S phase known as the G_1 phase (G = gap) and another interval between the end of the S phase and the beginning of the next M phase called the G_2 phase.

Cells of different types or in different states of activation have division cycles that can vary enormously in their length. Some divide very rapidly, as in the early embryo, and others divide very slowly, as in liver cells. This variation is a result of changes in the length of G_1. Indeed, some cells divide so slowly that they appear to be no longer proliferating. Some slowly dividing cells may pause in G_1 for weeks or years. These nonproliferating cells are then said to be in G_0. The termination of the G_0 state and progression to S phase can be triggered by a variety of stimuli depending on the cell type involved. In this book, much of the discussion in later chapters centers on the triggering of cell proliferation by foreign substances.

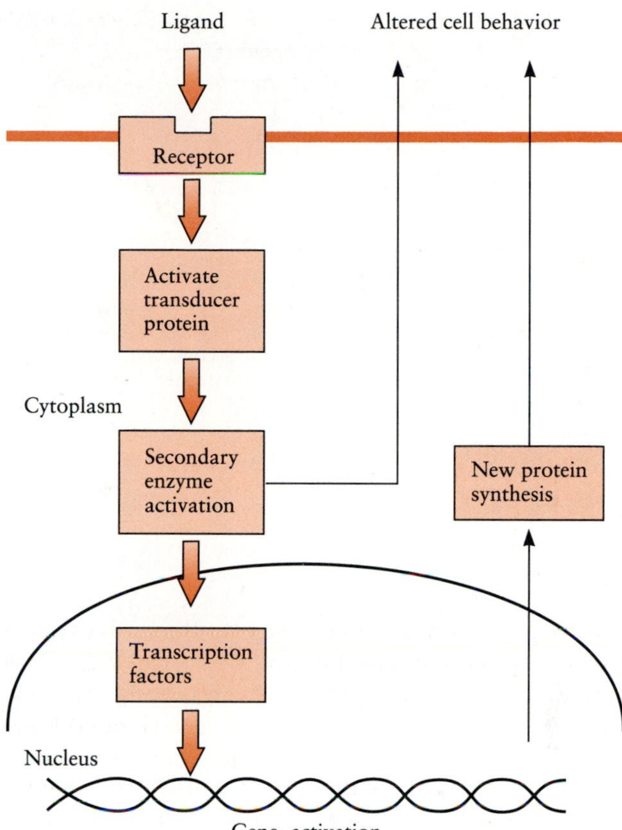

Figure 7–4 The basic features of the signal transduction pathways.

CELL SURFACE RECEPTORS AND SIGNAL TRANSDUCTION

Immunology is largely concerned with interactions between cells. Cells must be able to sense what is occurring in their surroundings and react accordingly. For example, neutrophils must be able to detect the presence of nearby bacteria and migrate toward them by chemotaxis. These processes require that cells possess many different receptors on their surface membrane and that once these receptors bind to their **ligands,** they transmit signals to the cell nucleus (Fig. 7–4). This conversion of an extracellular signal into a series of intracellular events is called **transduction.** The key components of these signal transduction pathways include activation of a transducer protein by the receptor, secondary activation of other enzymes, generation of new transcription factors, and gene activation leading to altered cell behavior. Most cell surface receptor proteins belong to one of four classes based on their mode of action (Fig. 7–5).

Channel-linked receptors that use transmitter-gated ion channels are involved in rapid signaling between nerve cells. Thus the receptor itself is a channel, and binding of the agonist opens that channel, allowing ions to pass through it. Channel-linked receptors are found in inflammatory and immune cells, but their roles are unclear.

A second class of receptors consists of proteins that have a cytoplasmic domain that acts as a tyrosine-specific **protein kinase** that becomes activated when the ligand binds. Thus when the ligand binds to the extracellular domain, a conformational change occurs that activates the intracellular kinase domain. These kinases add a phosphate group to tyrosine residues on other cytoplasmic or membrane proteins or even the receptor itself (autophosphorylation). The phosphorylation modifies the protein structure and function and initiates a series of changes in cellular activities. Many reactions of the immune system operate through this type of receptor (especially through protein kinases of the *src* family). A related family of receptors consists of those that are not themselves tyrosine kinases but that can activate cellular tyrosine kinases noncovalently associated with the receptors. This type of receptor is also widely employed in the cells of the immune system.

The third major class of receptors is associated with one of a group of membrane-bound GTP-binding proteins, called **G-proteins.** (One of the most important of these is called ras [Chapter 11].) G-proteins act as chemical switches. When inactive, they bind guanosine diphosphate (GDP). When active, they bind guanosine triphosphate (GTP). Thus once these receptors bind their ligand, a conformational change in

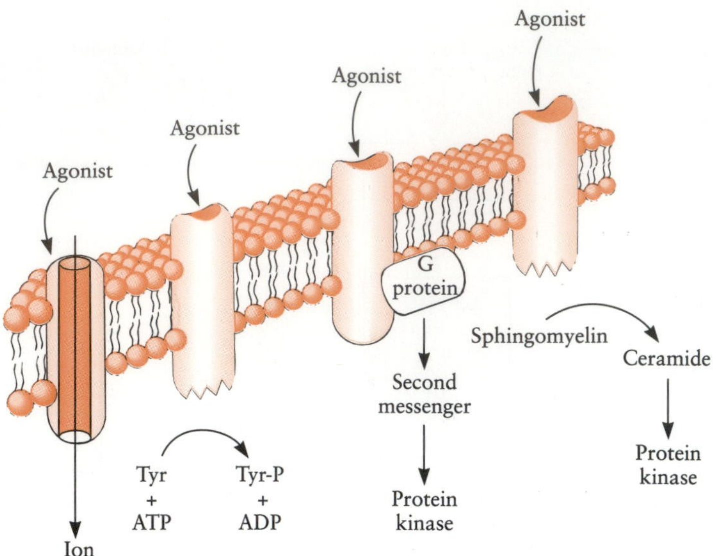

Figure 7–5 Four major mechanisms by which receptor signals are transduced. Tyrosine kinases may be integral membrane proteins in which binding to the receptor activates a tyrosine kinase site in the cytoplasmic portion of the molecule. Alternatively, the receptor protein may be coupled to a separate cytoplasmic tyrosine kinase.

the receptor–G-protein complex results in the G-protein losing its GDP and binding GTP (Fig. 7–6). The activated G-protein can then act on other substrates, causing their activation. The GTP is very rapidly hydrolyzed to GDP by the G-protein's own GTPase activity, so the G-protein is then turned "off." More than 20 G-proteins have been identified. They are all αβγ heterotrimers, although there is considerable diversity between the subunits. All G-protein–linked receptors are closely related, single polypeptides that pass through the membrane several times (see Fig. 7–1). The ligand binds to multiple sites on the receptor.

The targets of G-proteins can include ion channels, enzymes such as adenylate cyclase, phospholipases C, and some protein kinases. Modulation of adenylate cyclase activity alters the intracellular concentration of cyclic adenosine monophosphate (cAMP). cAMP is a second messenger that regulates a wide variety of intracellular activities. Most important, increasing cAMP increases the activity of a cAMP-dependent protein kinase (protein kinase A), which then phosphorylates specific protein substrates.

There are at least seven **isoforms** of phospholipase C that can be activated by G-proteins, although only the γ isoform interacts with tyrosine kinases. When activated by a G-protein, phospholipase C cleaves the membrane-bound lipid phosphatidylinositol 4,5-bisphosphate (PIP_2) into two second messengers, inositol trisphosphate and diacylglycerol (Fig. 7–7). Inositol trisphosphate binds to intracellular receptors, releasing Ca^{++} from internal stores and so increasing the concentration of intracellular free Ca^{++}. These calcium ions can activate many different proteins, including calmodulin. The diacylglycerol remains in the plasma membrane and along with calcium activates an enzyme called protein kinase C.

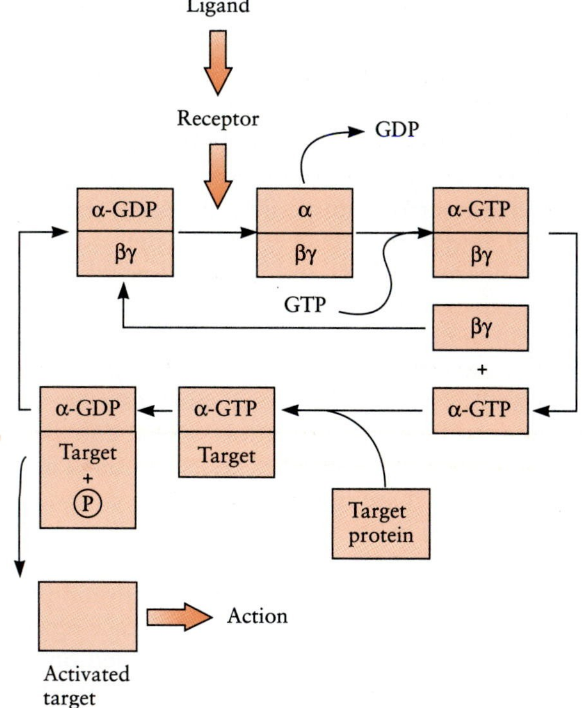

Figure 7–6 The process by which a G-protein consisting of three units, α, β, and γ, acts as a signal transducer. In summary, binding of agonist to the inactive molecule causes GDP to dissociate and GTP to bind to the α component. As a result, the complex dissociates and the α chain binds to the target molecule phosphorylating and hence activating it. Once activated, the target dissociates, the GTP hydrolyzes to GDP, and the complete G-protein complex reassociates.

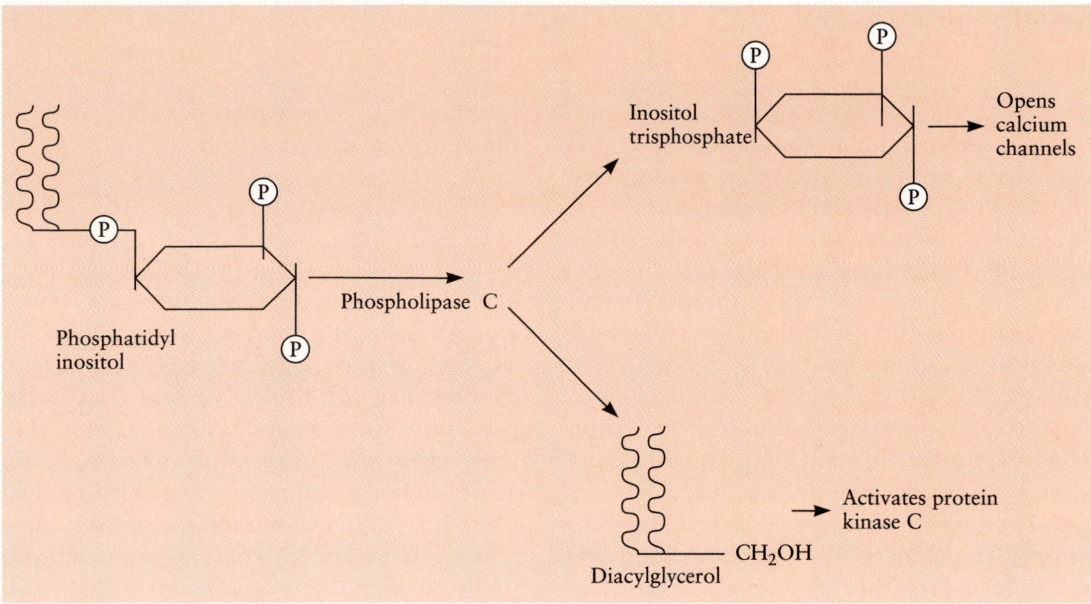

Figure 7–7 Phospholipase C acts on phosphatidylinositol biphosphate to generate both inositol trisphosphate and diacylglycerol.

A fourth class of receptor mediates signal transduction through the sphingomyelin pathway. Ligand binding activates a neutral sphingomyelinase, which then hydrolyzes sphingomyelin in the cell membrane to ceramide. The ceramide then acts as a second messenger. It stimulates a serine threonine protein kinase that phosphorylates cellular proteins. These in turn generate transcription factors such as NF-κB. This mechanism of signal transduction is employed by the receptors for interleukin-1 (IL-1) and tumor necrosis factor–α (TNF–α).

Protein Phosphorylation

Phosphorylation is the most important form of reversible modification of proteins. All of the described signal transduction systems involve the use of a high-energy compound (for example, GTP or ATP) to modify a protein or lipid and send a signal to a cell. Cell growth, cell division, as well as other critical processes are regulated by protein phosphorylation. Protein kinases enzymatically phosphorylate serine, threonine, and tyrosine (Fig. 7–8).

$$\text{Protein} + \text{ATP} \xrightarrow{\text{Protein kinase}} \text{Protein-P} + \text{ADP}$$

In some proteins only one amino acid is phosphorylated; in others multiple amino acids are phosphorylated. The phosphorylated and nonphosphorylated proteins have different functional properties. These may be due to conformational changes resulting from phosphorylation, from a change in location within a cell, from enzymic activities, or from interactions with

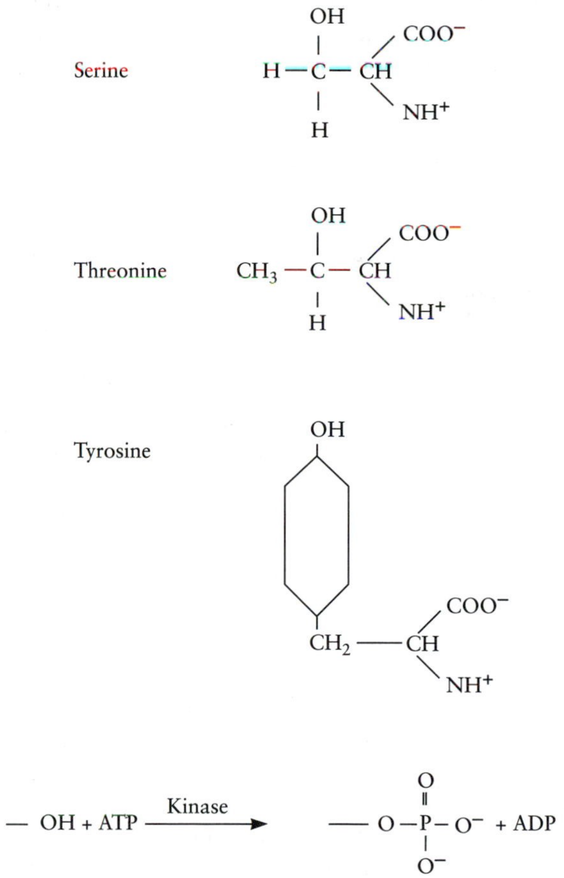

Serine: $H-C(OH)(H)-CH(COO^-)(NH^+)$

Threonine: $CH_3-C(OH)(H)-CH(COO^-)(NH^+)$

Tyrosine: $HO-C_6H_4-CH_2-CH(COO^-)(NH^+)$

$$-OH + ATP \xrightarrow{\text{Kinase}} -O-P(=O)(O^-)-O^- + ADP$$

Figure 7–8 The three amino acids that are subject to phosphorylation by protein kinases are serine, threonine, and tyrosine. Tyrosine kinases are a distinctly different group of enzymes from the serine/threonine kinases.

other cellular components. Many enzyme activities are regulated by phosphorylation in a process in which the phosphorylation of serine or threonine activates the enzyme and dephosphorylation has an opposite effect. Phosphorylation of the three key amino acids, serine, threonine, and tyrosine, plays a critical role in regulating many cellular functions. When phosphorylated proteins are examined, about 90% of the phosphate is attached to serine and about 10% to threonine. Only about 1/2000 of the phosphate is linked to tyrosine. Thus tyrosine phosphorylation is a rare event, although it is a key mechanism of almost all the signal transduction pathways described in this book.

The protein kinases are a large family of enzymes with more than 50 members. They can be divided into two subsets based on whether they phosphorylate serine and threonine or alternatively tyrosine. A third subset that phosphorylates histidine or lysine may also exist. Their activity is commonly regulated by second messengers or hormones such as cAMP, calcium ions, diacylglycerol, double-stranded RNA, and insulin. Diacylglycerol specifically activates a family of serine/threonine kinases called protein kinase C. Some protein kinases are not regulated in this way. Many of these "independent" kinases are tyrosine kinases. Independent kinases are, however, regulated by other factors such as receptor binding and conformational changes.

Transcription Factors

Protein kinases generated by extracellular signals, in turn, generate DNA-binding proteins called transcription factors. These proteins control gene expression by binding to specific sites on promoter regions. These promoter regions are found upstream of major genes and act as regulatory elements. A gene will only be effectively transcribed when its promoter is correctly stimulated. Thus the binding of transcription factors to promoter regions turns on protein synthesis. Several different transcription factors must usually bind to a promoter region in order for its downstream target gene to be activated. Some transcription factors are widely distributed, being found in many different cell types. Others, in contrast, are restricted to specific cell types.

A good example of a widespread transcription factor is NF-κB. NF-κB is a heterodimeric protein that exists in an inactive form in the cytoplasm bound to an inhibitor called I-κB. When a cell is appropriately stimulated, second messengers such as the protein kinases can release NF-κB from I-κB. The NF-κB can then enter the nucleus and bind to its specific binding site on enhancer sequences and thus turn on the synthesis of certain proteins. NF-κB is important in activating the genes for antibody formation. Other transcription factors such as NF-AT play a role in activating the genes for IL-2 in T cells.

ONCOGENES

The discovery that some viruses could cause tumors in animals made it important to learn how they did so. The genes responsible were identified by molecular dissection of the viral DNA that transforms infected cells into cancer cells. These oncogenic (cancer-causing) retroviruses were found to carry genes that are responsible for the excessive growth of the host cell. They were called viral **oncogenes.** Each oncogene was named after the virus where it originated. Thus *src* was found in a chicken sarcoma virus, *myb* was found in myeloblastosis virus, *rel* in reticuloendotheliosis virus, *fms* in feline sarcoma virus, and so on. Some time later, virologists found that very similar genes occurred in normal cells. In fact, the viral oncogenes turned out to be variants of normal cellular genes found in the genome of the host animal. These normal cellular genes are called **proto-oncogenes.** It now appears that the viral oncogenes have arisen as a result of the capture of the cellular proto-oncogenes by the virus. To distinguish the normal cellular genes from their viral counterparts, the viral genes are prefixed by *v*, as in *v-src, v-rel,* and *v-fms.* In contrast, the cellular genes or proto-oncogenes are prefixed by *c*, as in *c-src, c-rel,* and *c-fms.* Some oncogenes were not first identified in viruses but found in human cancer cells as a result of a search for cellular oncogenes related to the viral ones. Thus *lck* was first identified in a human lymphoma cell line. As might be expected, proto-oncogenes code for a great variety of different proteins. Nevertheless, despite their large number, they have a limited range of activities (Table 7–1). Some code for growth factors. For example, *sis* codes for platelet-derived growth factor. Other oncogenes code for growth factor receptors. Thus *erb* codes for a form of the epithelial growth factor receptor. Many of the proto-oncogenes code for components of the receptor signal transduction path-

Table 7–1 Example of Oncogene Functions

Growth factors	*sis, hst*
Growth factor receptors with tyrosine kinase activity	*erbB, ros, fms, kit, met, neu*
Cytoplasmic tyrosine kinases	*abl, src, fgr, syn, lck, lyn*
G-proteins	*Ha-ras, N-ras, gsp*
Serine/threonine kinases	*mos, cot, pls*
Transcription factors	*myc, myb, fos, jun, rel*
Other functions	
Angiotension receptor	*mas*
Inhibitor of apoptosis	*bcl-2*
Apoptosis signal	*fas*

way. Thus there are three families of oncogenes that code for tyrosine kinases. The *src* family includes *src, yes, fgr, fyn, lck, lyn, hck,* and *tkl.* The other two families are the *abl* family and the *fes/fps* family. One feature of the *src* tyrosine kinase family is the presence of a tyrosine whose phosphorylation regulates the protein's kinase activity. Thus addition of phosphate to the tyrosine may inhibit the enzyme's activity, and its removal has the opposite effect. Other proto-oncogenes code for serine/threonine kinases, especially the members of the *raf* family, while members of the *ras* family code for G-proteins. Finally, there is a group of proto-oncogenes whose products localize in the nucleus. These nuclear proto-oncogenes code for transcription factor. Among the nuclear proto-oncogenes are *c-myc, c-myb, c-fos, c-erb, c-rel,* and *c-jun.* Activation of the *c-myc* gene in a B-cell tumor called Burkitt's lymphoma is discussed in Chapter 20.

It is clear that all the oncogene products play a role in the pathways leading to cell division. It is not difficult to see that if control of their production is lost, as occurs in some virus infections, they may cause uncontrollable cell division—cancer.

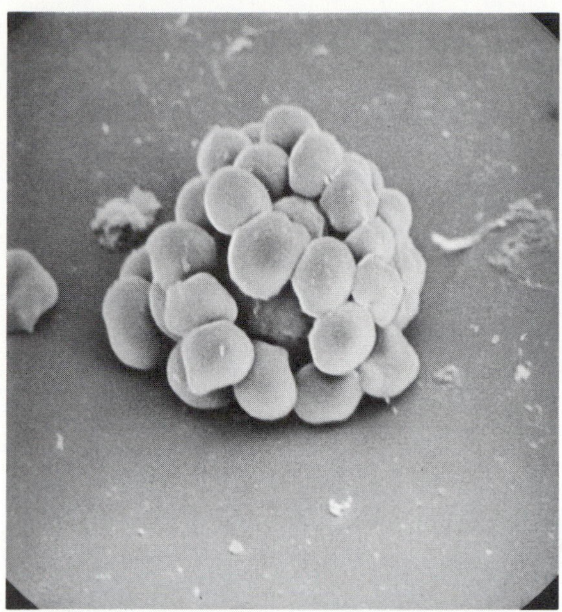

Figure 7–9 This scanning electron micrograph shows a "rosette" formed by allowing antibody-coated sheep erythrocytes to bind to receptors for antibody on a lymphocyte. The lymphocyte is almost totally hidden by the red cells. Original magnification was 22,500 times.

CELL ADHESION MOLECULES (CAMs)

Adherence between cells and between cells and connective tissue molecules is mediated by cell membrane proteins. These adhesion molecules belong to four glycoprotein families. The **immunoglobulin superfamily,** the cadherin family, the **integrin** family, and the **selectin** family.

The Immunoglobulin Superfamily

The proteins of the immunoglobulin superfamily are the major regulators of lymphocyte function, and more than 35% of cell surface proteins belong to this family. All members of the superfamily share, as a common building block, a structure called an immunoglobulin domain (see Fig. 13–13). This domain consists of a peptide chain of about 100 amino acids folded over into two β-pleated sheets and stabilized by a single intrachain disulfide bond. (Its structure is discussed in detail in Chapter 13.) Different members of the family have varying numbers of these immunoglobulin domains. The members of the superfamily fall into two functional groups. One group consists of proteins specialized for specific antigen recognition. Its members include the immunoglobulins, the MHC molecules, and the T-cell antigen receptor proteins (see Fig. 13–15). The other group includes cell surface receptor proteins that are not antigen specific. These proteins include the proteins called CD2 (LFA-2), CD58 (LFA-3), ICAM-1 (CD54), ICAM-2 (CD102), VCAM-1 (CD106), B7-1 (CD80), CD3, CD4, and CD8. Many of these proteins form adhesive pairs. Thus CD2 binds to CD58; B7-1 to CD28, CD4 to MHC class II molecules; CD8 to MHC class I molecules; and ICAM-1 to CD11a/CD18.

CD2, also called LFA-2 (leukocyte function antigen-2) is a 54-kDa single-chain glycoprotein. Approximately 40,000 molecules of CD2 are found on resting T cells. (Details of T cells are provided in Chapter 10.) Activated T cells have up to 250,000 CD2 molecules per cell. The ligands of CD2 include proteins called CD58 (LFA-3), CD48, and CD59 found on other cells. Activated T cells can adhere to red cells through CD2 and CD58 to form "rosettes" (Fig. 7–9). CD2–CD58 binding can augment the T-cell response to antigen since CD58 is found on antigen-presenting cells such as macrophages. Their linkage enhances the adherence of the two different cells, increasing recognition of antigen by the T-cell antigen receptor while at the same time stimulating the macrophage to secrete IL-1. CD2–CD58 binding also occurs between thymocytes and thymic epithelial cells (Chapter 9) and between cytotoxic T cells and their target cells (Chapter 18).

ICAM-1, also known as CD54, is the ligand for CD11a/CD18 and CD43 (Fig. 7–10) (Chapter 5). (ICAM stands for **inter**cellular **a**dhesion **mo**lecule.) In the absence of inflammation, ICAM-1 is normally found only on dendritic cells and B cells. Cytokines, however, such as interferon-γ, tumor necrosis factor–α, or interleukin-1, cause the appearance of ICAM-1 on other cells within a few hours. Thus inflammation

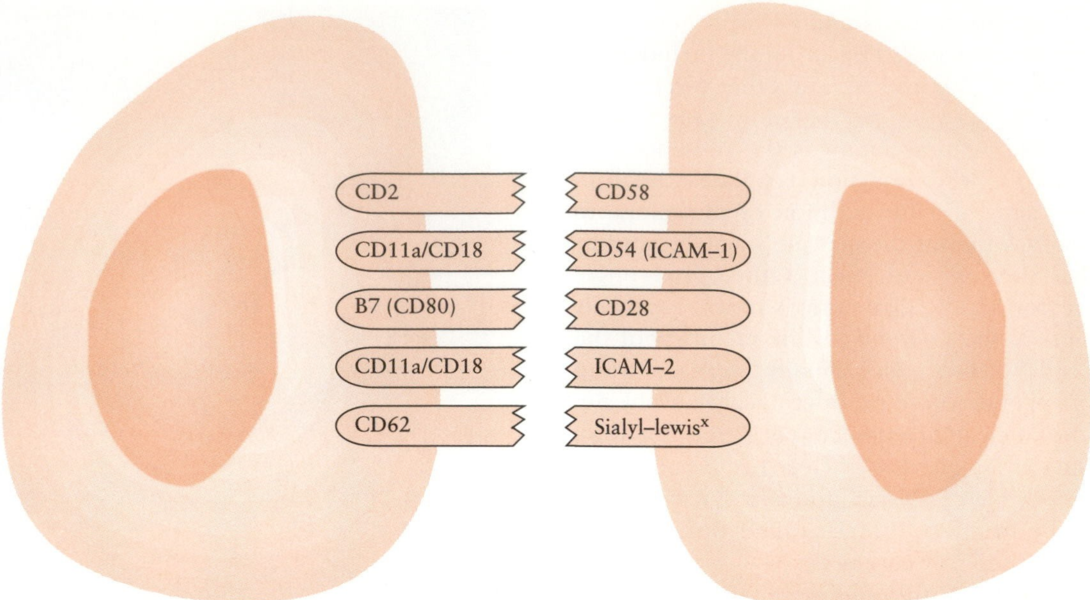

Figure 7–10 Some of the major protein-receptor interactions that bind interacting cells of the immune system. Although there may be considerable redundancy in the process, as a general rule, the tighter the binding between cells, the more efficient is the interaction.

induces ICAM-1 production on capillary endothelial cells and so permits lymphocytes and monocytes to adhere and move into inflamed tissues (Chapter 28). It is ICAM-1 that is responsible for the migration of T cells into areas of inflammation (so-called delayed hypersensitivity reactions, Chapter 18). ICAM-1 is also the receptor for rhinoviruses, the cause of the common cold. Thus rhinoviruses can penetrate T cells after first binding to ICAM-1.

ICAM-2 (CD102) is an adherence protein structurally related to ICAM-1 but has only two immunoglobulin domains as opposed to ICAM-1, which has five. It also binds CD11a/CD18. Unlike ICAM-1, ICAM-2 is expressed on unstimulated endothelial cells and is not increased during inflammation.

CD58 is another ligand for CD2. Since CD2 is restricted to T cells whereas CD58 is widely distributed on many cell types, it is suggested that CD58 facilitates T-cell binding to any cell that is to undergo surveillance (Chapter 20). In the thymus, immature T cells adhere to thymic epithelial cells by the CD2–CD58 linkage.

Another important adherence protein is CD28. CD28 is expressed on almost all T cells. The ligands for CD28 are proteins called B7-1 (CD80 or BB1) and B7-2 found on antigen-presenting cells such as macrophages and B cells. B7-2 is also a member of the immunoglobulin superfamily. Linkage of CD28 to B7-2 is a critical event in T-cell activation.

The Cadherin Family

This family consists of several calcium-dependent transmembrane proteins involved in cell adhesion. They include L-CAM, E-cadherin, N-cadherin, and P-cadherin. They are essential for normal tissue formation since they set up strong physical bonds between cells. They may also be involved in regulating cellular motility and migration. No cadherins have been found on leukocytes.

The Integrin Family

Integrins are transmembrane proteins that mediate adherence between cells or between cells and extracellular matrix proteins such as collagen. In so doing they can also modulate cellular activities. They are all noncovalently linked heterodimers consisting of one α and one β chain (Fig. 7–11). At least 14 α chains and eight β chains have been identified. They are divided into subfamilies, such as β_1, β_2, and β_3, each with a common β chain binding with one of several α chains (Fig. 7–12). About 20 $\alpha\beta$ heterodimers have been identified, but this number continues to grow. Integrin binding activity depends on divalent cations.

The β_1 (CD29) Subfamily. The β_1 subfamily is the largest of the integrin families. These integrins have been called VLA (very late antigen) integrins because some of them appear on activated T cells two to

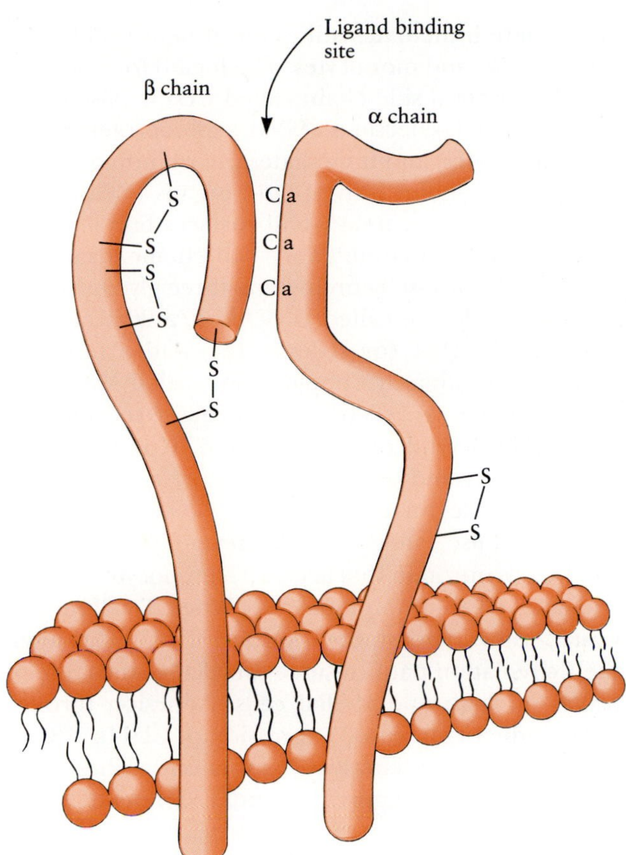

Figure 7–11 The structure of a typical integrin. It consists of a noncovalently associated transmembrane heterodimer. As a typical cell membrane protein, the extracellular domains are glycosylated and bind the appropriate ligand. The short cytoplasmic domains are attached to the cytoskeleton. The α chain contains calcium-binding areas (Ca).

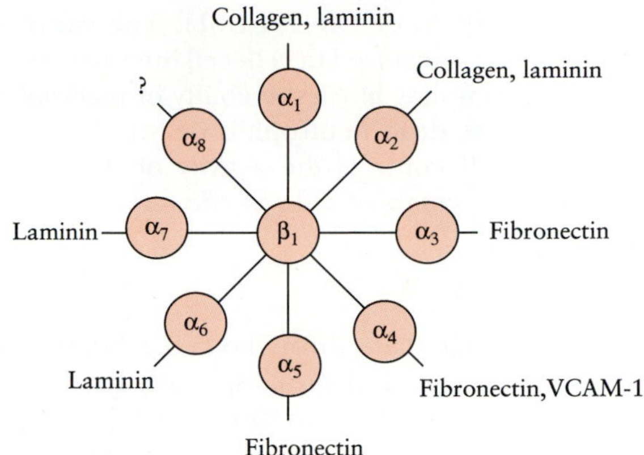

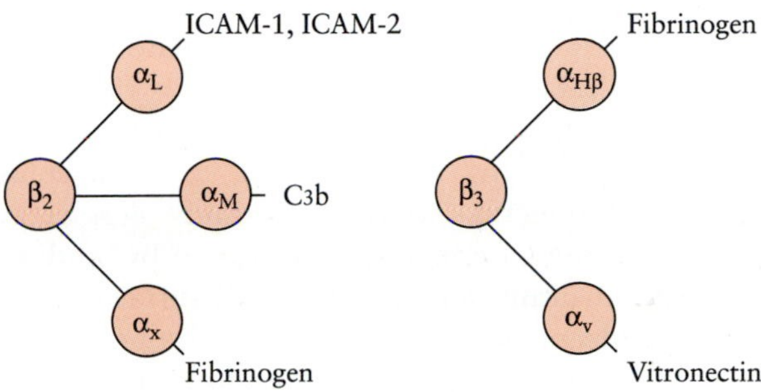

Figure 7–12 The leukocyte integrins. Each is a heterodimer composed of an α and a β chain. A limited number of β chains can associate with several different α chains to create a family of binding proteins of different specificities.

four weeks after antigen stimulation. However, most cell types express one or more β_1 integrins constitutively (that is, they produce small amounts all the time). The β_1 chain is common to all members of the subfamily and is also called CD29 (see Fig. 7–12). Each member has a different α chain (also called CD49a, b, c, d, e, f, g, or h). β_1 integrins are expressed on many different cell types and bind them to extracellular matrix proteins such as fibronectin, laminin, and collagen. For example, $\beta_1\alpha_1$ and $\beta_1\alpha_2$ bind collagen, $\beta_1\alpha_4$ binds fibronectin, and $\beta_1\alpha_6$ binds laminin. $\beta_1\alpha_4$ is found on lymphocytes and monocytes, where it binds to a protein of the immunoglobulin superfamily called CD106 (also called VCAM-1, vascular cell adhesion molecule-1). CD106 is a 110-kDa protein induced by inflammatory mediators such as IL-1 and TNF–α on endothelial cells. $\beta_1\alpha_5$ is found on a subpopulation of T cells and is a fibronectin receptor. Resting T cells express these integrins at low levels. Activation of T cells increases this expression and enables them to bind to fibronectin or laminin. This binding acts as a costimulator of T-cell function and thus promotes their ability to proliferate. Memory T cells have three to four times as many integrins on their surface as do naive T cells.

The β_2 Subfamily. The β_2 integrins consist of three heterodimers: CD11a/CD18—the α chain is also called CD11a and the β chain is called CD18 ($\beta_2\alpha_L$ or LFA-1); CD11b/CD18 ($\beta_2\alpha_M$, Mac-1, or CR3); and CD11c/CD18 ($\beta_2\alpha_x$, p150, 95, or CR4). They are found only on leukocytes (Chapter 5). When T cells interact with antigen-presenting cells, the initial adherence between the T cells and presenting cell is mediated by both CD11a/CD18 and CD2. These two molecules bind strongly to ICAM-1 and CD58, respectively, on the presenting cell. The CD11a/CD18 adherence reaction is Mg^{2+} dependent and temperature dependent.

The β_3 Subfamily. The members of the β_3 subfamily of integrins are called cytoadhesins. One is a vitronectin receptor ($\beta_3\alpha_{IIb}$ or CD51/61), the other is a platelet receptor ($\beta_3\alpha_v$ or CD41/CD61). The vitronectin receptor may be involved in cell–cell interactions since antibodies to it may block the ability of macrophages to phagocytose dead neutrophils or lymphocytes. Vitronectin itself controls the activity of the terminal portion of the complement cascade.

The Selectin Family

Lymphocytes circulate throughout the blood system and migrate through tissues, especially lymphoid organs (Chapter 9). The circulatory pattern of these cells is regulated by the interaction between molecules on the surface of lymphocytes and receptors on vascular endothelial cells. Many molecules that are found on lymphocytes or **endothelium** and regulate this traffic belong to a family of adhesion proteins called **selectins.** Three glycoproteins are currently recognized as belonging to this family. These are P-selectin (CD62P), L-selectin (CD62L), and E-selectin (CD62E). All mediate leukocyte adhesion by binding to the carbohydrate side chains of cell surface glycoproteins (**addressins**). P-selectin, a 140-kDa protein, is found in the secretory granules of endothelial cells in capillaries. When these are activated by thrombin or histamine, the P-selectin is distributed to the cell surface, where it mediates adhesion of neutrophils, activated T cells, and monocytes. The ligand for P-selectin is a glycoprotein side chain called CD15s (also called sialyl Lewis^x). L-selectin is a 90-kDa glycoprotein found on lymphocytes, and it mediates their binding to high endothelial venules in lymphoid organs (Chapter 8). It is also found on neutrophil surfaces but is rapidly shed by them on stimulation with chemoattractants. The ligands for L-selectin include three glycoproteins: a 50-kDa molecule called GlyCAM-1, a 90-kDa molecule called CD34 (or sialomucin), and a molecule found in intestinal lymphoid tissue called MAdCAM-1. E-selectin is a 115-kDa glycoprotein receptor for neutrophils found on endothelium. It is rapidly induced by IL-1 and other inflammatory agents. The ligand for E-selectin is also CD15s.

CD44 (also called H-CAM, Hermes, or Pgp-1) is a 90-kDa glycoprotein that acts as a lymphocyte receptor and as a cell adhesion molecule. When T cells are activated, they increase their surface CD44. This alters their circulation pattern since the CD44 binds to molecules on high endothelial cells. Adhesion through CD44 induces increased expression of CD11a/18 as a result of protein kinase C activation. Adhesion to monocyte CD44 triggers IL-1 and TNF release.

KEY WORDS

Addressin p. 86
B7 (CD80) p. 84
Cadherin p. 84
CD2 p. 83
CD11/CD18 p. 85
CD28 p. 84
CD44 p. 86
CD54 p. 83
Cell adhesion molecule p. 83
Cell cycle p. 78
Channel-linked receptors p. 79
Clathrin p. 78
Cluster of differentiation (CD) p. 78
Coated pit p. 78
Cyclic AMP p. 80
Diacylglycerol p. 80
Domain p. 83
Endocytosis p. 76
Endosome p. 76
Exocytosis p. 78
G-phase p. 79
G-protein p. 79
ICAM-1 p. 83
Immunoglobulin superfamily p. 83
Inositol trisphosphate p. 80
Integral membrane protein p. 76
Integrin p. 84
Ligand p. 76
M-phase p. 78
Oncogene p. 82
Pinocytosis p. 76–77
Protein kinase p. 82
Proto-oncogene p. 82
Receptor p. 76
Receptosome p. 78
S-phase p. 79
Signal transduction p. 79
Selectin p. 86
Src oncogene p. 82
Superfamily p. 83
Transcription factors p. 82
Tyrosine kinase p. 82

QUESTIONS

1. The protein that lines coated pits is called
   a. immunoglobulin
   b. complement
   c. clathrin
   d. integrin
   e. selectin
2. One of the major target molecules linked to G-proteins is
   a. adenylate cyclase
   b. *c-fos*
   c. complement receptors
   d. lipopolysaccharide
   e. immunoglobulin
3. The phase of the cell cycle immediately preceding mitosis is called
   a. S phase

b. G_1 phase
c. M phase
d. G_2 phase
e. T phase

4. What residues are usually phosphorylated by the protein kinase that is directly linked to receptor proteins?
   a. serine
   b. tyrosine
   c. histidine
   d. threonine
   e. proline
5. The normal cellular equivalent of a viral oncogene is called a
   a. cyto-oncogene
   b. proto-oncogene
   c. cellular oncogene
   d. nuclear oncogene
   e. pro-oncogene
6. The receptor glycoproteins that bind foreign antigen belong to the
   a. immunoglobulin superfamily
   b. selectin superfamily
   c. integrin superfamily
   d. receptor superfamily
   e. cytokine superfamily
7. The β_2 family of integrins contains which of the following proteins?
   a. CD29 and CD18
   b. CD16 and CD32
   c. CD11 and CD18
   d. CD11 and CD42
   e. CD16 and CD18
8. Selectins are adherence proteins that bind to
   a. cadherins
   b. adhesins
   c. integrins
   d. immunoglobulins
   e. addressins
9. The major tyrosine kinase associated with lymphocyte receptors is called
   a. *fos*
   b. *lck*
   c. *ras*
   d. *erb*
   e. *fos*
10. The uptake of fluid by cells is through a process called
    a. pleocytosis
    b. exocytosis
    c. pinocytosis
    d. endocytosis
    e. phagocytosis
11. Describe and illustrate the key features of an integral cell membrane glycoprotein.
12. How are cell membranes recycled? Describe the endocytic and exocytic pathways.
13. List the four major types of cell surface receptor. Outline the mechanisms by which they transduce signals.
14. What does a protein kinase do? How do protein kinases assist in modifying cellular activities?
15. Describe the basic features of glycoproteins of the immunoglobulin superfamily.
16. Describe the integrin family. What are its functions? How does the structure of an integrin determine its functions?

Answers: 1c, 2a, 3d, 4b, 5b, 6a, 7c, 8e, 9b, 10c

SOURCES OF ADDITIONAL INFORMATION

Avers, C.J. Molecular cell biology, Addison-Wesley, Reading, MA, 1986.

Barklay A.N., et al. The Leukocyte Antigen Facts Book. Academic Press, London, 1993.

Clement, L.T. Isoforms of the CD45 common leukocyte antigen family: Markers for human T-cell differentiation. J Clin Immunol., 12:1–10, 1992.

Etzioni, A. Adhesion molecules in host defense. Clin. Diagn. Lab. Immunol., 1:1–4, 1994.

Horejsf, V. Leukocyte surface glycoproteins—from qualitative analysis to functional understanding. *In* Gahmberg, C.G., et al., eds. Leukocyte Adhesion. Basic and Clinical Aspects, pp. 17–32. Elsevier Science Publishers, Amsterdam, 1992.

Hynes, R.O. Integrins: Versatility, modulation, and signalling in cell adhesion. Cell, 69:11–25, 1992.

Mackay, C.R., and Imhof, B.A. Cell adhesion in the immune system. Immunol. Today, 14:99–102, 1993.

Mathias, S., et al. Activation of the sphingomyelin signaling pathway in intact EL-4 cells and in a cell-free system by IL-1β. Science, 259:519–522, 1993.

Öbrink, B. Cell adhesion—an overview. *In* Gahmberg, C.G., et al., eds. Leukocyte Adhesion. Basic and Clinical Aspects, pp. 3–13. Elsevier Science Publishers, Amsterdam, 1992.

Pardi, R., Inverardi, L., and Bender, J.R. Regulatory mechanisms in leukocyte adhesion: Flexible receptors for sophisticated travellers. Immunol. Today, 13:224–230, 1992.

Pigott, R., and Power, C. The Adhesion Molecule Facts Book. Academic Press, London, 1993.

CHAPTER 8

Antigen Processing and Histocompatibility Antigens

Baruj Benacerraf *Baruj Benacerraf was born in Venezuela in 1920 but has worked primarily at Harvard University. He has played a prominent role in many different areas of immunology. Thus during his studies on phagocytosis he was the first to show that cell-mediated immune responses were directed against epitopes that required processing by cells. He identified the first Fc receptors on macrophages, clarified the mechanism of antibody affinity maturation, and helped to confirm the correctness of the clonal selection theory. He was awarded the Nobel Prize in 1980 for his role in demonstrating that MHC genes could control the ability of an animal to mount an immune response to a specific epitope. This led to our current understanding of the way in which antigens are processed and bound to MHC molecules.*

CHAPTER OUTLINE

CHAPTER CONCEPTS

1. Foreign antigens must be specially processed before they can be presented to the antigen-sensitive cells of the immune system.
2. The most important antigen processing cells are macrophages, dendritic cells, and B cells.
3. Antigen fragments generated inside these cells are bound to specialized receptors called MHC molecules in the cell cytoplasm and then transported to the surface of antigen-presenting cells.
4. Foreign, phagocytosed antigens, called exogenous antigens, are fragmented in endosomes and the fragments attached to MHC class II molecules.
5. Foreign molecules synthesized within a cell (as in a virus-infected cell) are attached to MHC class I molecules while they are being synthesized.
6. A third class of MHC molecules (MHC class III) are not involved in antigen processing but code for molecules involved in the complement and other systems.

In this chapter we look at the two ways in which foreign antigens are processed prior to presentation to the cells of the immune system. We first look at the processing of foreign material that has been phagocytosed by cells such as macrophages. Other cell types also involved in this processing are reviewed. Then we look at the specialized protein receptors that antigens must bind to. These receptors have a complex structure as well as an unusual gene arrangement that we must examine. Finally, we study how foreign antigen fragments bind to these receptors. Some foreign material can originate inside infected cells. This too must be processed and bound to special receptors. This process and the molecules involved are described in the second half of the chapter.

The foreign antigens that trigger an immune response are of two distinct types. First, there are the antigens derived from organisms such as bacteria that, when they enter the body, are phagocytosed by cells such as macrophages. These antigens may also be found free in the circulation for a time. These are called **exogenous antigens** and are processed by specialized antigen processing cells such as macrophages. A second type of antigen is actually made within the body. Thus if a virus invades a cell and takes over its biosynthetic processes, then new viral proteins are formed within the infected cell. These are called **endogenous antigens.** Endogenous antigens are presented to antigen-sensitive cells by the cells in which they are formed. This presentation of antigen to the cells of the immune system is a key step in mounting an immune response.

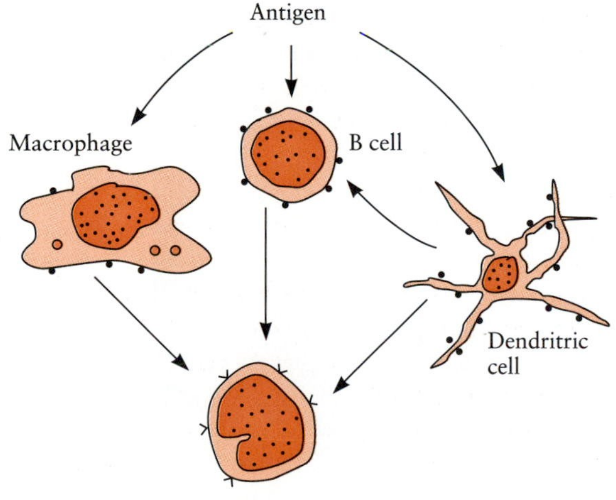

Figure 8–1 The three major groups of antigen-presenting cells. Presentation of antigen by B cells is regulated by helper T cell–derived interleukins. Presentation by macrophages is regulated by T cell–derived interferon-γ.

PROCESSING OF EXOGENOUS ANTIGEN

If all foreign material that gets into the body were totally ingested, digested, and destroyed by phagocytic cells, there would be no need and no stimulus for an immune response. This may in fact be a common occurrence. Nevertheless, some antigen must persist to stimulate antigen-sensitive cells and initiate an immune response. Three cell populations have the ability to process this exogenous antigen and present it appropriately. These are **macrophages, dendritic cells,** and **B cells** (Fig. 8–1). The relative importance of each of these cell types is unclear. It probably depends on the nature of the antigen, the route by which it enters the body, and whether the body has been exposed to it previously. All three antigen-presenting cell types also secrete IL-1 and IL-12, although their relative proportions may vary.

ANTIGEN PROCESSING CELLS

Macrophages

Macrophages are the most accessible and best understood of the antigen-presenting cells. Their key properties have been described in the previous chapter. They are probably of greatest importance in processing antigen that has not been encountered previously by the body. Thus macrophages can phagocytose organisms in the absence of antibody for opsonization. They are probably of less importance when there are preexisting antibodies, since these greatly increase the efficiency of antigen processing by dendritic cells and B cells.

Dendritic Cells

Antigen processing by macrophages is inefficient, since much of the endocytosed antigen is destroyed by lysosomal proteases. An efficient alternative pathway of antigen presentation involves antigen uptake by a specialized population of mononuclear cells, collectively called dendritic cells.

Dendritic cells are located throughout the body, but especially in lymphoid organs. They have many long, filamentous cytoplasmic processes called dendrites (Fig. 8–2). They also have lobulated nuclei and a clear cytoplasm containing characteristic granules called Birbeck granules. All dendritic cells carry complement and antibody receptors on their surface but are poorly phagocytic. Several types of dendritic cells are found in the body. These include **Langerhans cells** located in the skin and interdigitating cells and follicular dendritic cells in lymphoid tissues.

Figure 8–2 A scanning electron micrograph of a dendritic cell from a guinea pig lymph node. Note its extensive array of dendrites that trap antigen (×4000).

In lymphoid tissues (e.g., spleen and lymph nodes), dendritic cells form an extensive interdigitating web that efficiently traps antigen while at the same time allowing cell interactions and movement. There are two populations of these cells, interdigitating cells in the T-cell areas (see Chapter 9 on the structure of lymphoid tissues) and follicular dendritic cells in the B-cell areas. Interdigitating cells are present in all lymphoid tissues, including the thymus. Follicular dendritic cells are confined to the follicles in the B-cell areas where they form clusters with lymphocytes.

Langerhans cells are dendritic cells found in the epidermis of the skin. They trap and present antigen that penetrates the skin. This includes topically applied antigen such as the resins of poison ivy or intradermally injected antigens such as mosquito saliva. As a result, Langerhans cells influence the development of skin allergies such as **delayed hypersensitivity** and **allergic contact dermatitis** (Chapter 30). Langerhans cells can leave the skin and colonize lymph nodes, where they become interdigitating cells. Veiled cells are the dendritic cells found in the lymph that flows from tissues into lymph nodes (efferent lymph). They may represent a transition form between Langerhans cells and interdigitating cells.

Dendritic cells present antigen in two ways. In an unprimed animal (i.e., an animal that has not previously been exposed to the antigen), antigen presentation is a passive process. The dendritic cells simply provide a surface on which antigen can be presented. In animals that have previously been exposed to an antigen and so possess antibodies, the antigen and antibody combine to form antibody–antigen complexes (also called **immune complexes**). Follicular dendritic cells in the B-cell areas take up these immune complexes on their surface and then shed them in round, beaded structures from their processes (Fig. 8–3). These immune complex bodies (also called **iccosomes**) subsequently attach to B cells. The B cells ingest the antigen and, after processing, present it to

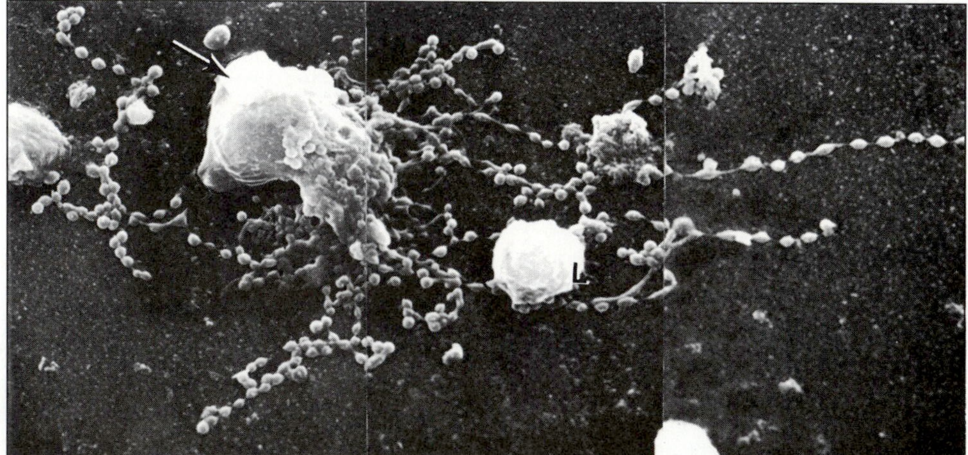

Figure 8–3 Scanning electron micrograph of a follicular dendritic cell with "beaded" dendrites. The beads, which are known to be coated with immune complexes, are termed "iccosomes" (immune-complex–coated bodies). The arrow indicates the cell body from which the dendrites emanate. L is a lymphocyte attached to the dendrites (original magnification ×3700). *(From Szakal, A.K., et al. J. Immunol., 134:1353–1354, 1985, © 1985 American Association of Immunologists. Reprinted with permission.)*

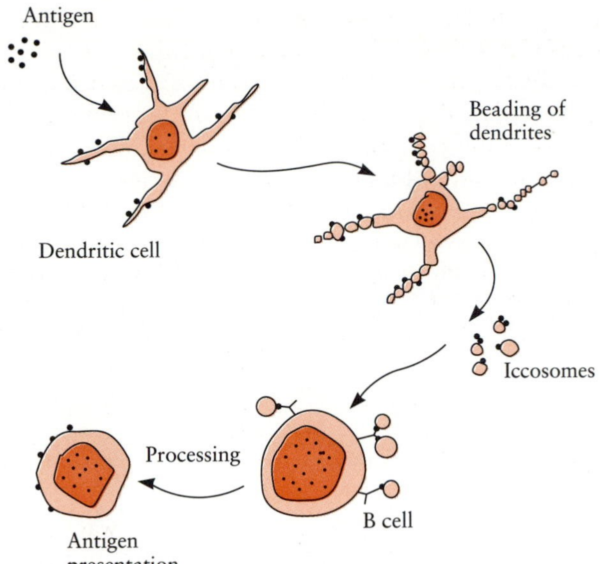

Figure 8–4 The suggested way in which iccosomes deliver antigen to B cells for processing.

antigen-sensitive T cells (Fig. 8–4). Dendritic cells can retain antigen on their surface for more than three months. Antigen processed by dendritic cells is a potent stimulant for T cells—about 10,000 times more efficient than unbound antigen.

B Cells

One type of lymphocyte called the **B lymphocyte** or B cell may also be an effective and important antigen-presenting cell. This is described in Chapter 14 but may be summarized here. B cells bind whole antigen molecules by means of their antigen receptors. They then ingest and process it and present the processed antigen, in association with specific antigen-presenting molecules to T cells. B cells are especially effective in presenting antigen to memory T cells and Th1 cells since they secrete interleukin-12. The importance of B cells as antigen-presenting cells can be demonstrated by showing that T-cell responses are seriously impaired in B-cell–deprived animals.

Other Antigen-Presenting Cells

Antigen may be presented to lymphocytes, albeit poorly, by many cell types. These include eosinophils, vascular endothelial cells, and skin keratinocytes. For example, vascular endothelial cells can also take up antigen, synthesize IL-1, and, under the influence of interferon, express antigen-molecules on their surface. Even skin keratinocytes can produce factors similar to IL-1, express antigen-presenting molecules, and present antigen to T cells.

THE MAJOR HISTOCOMPATIBILITY COMPLEX

Antigen processing requires not only the fragmentation of antigen molecules inside cells, but also the linkage of these fragments to appropriate antigen-presenting molecules. These antigen-presenting molecules are called **histocompatibility** or **MHC molecules.** They are in fact specialized receptor glycoproteins inherited in a gene complex called the **major histocompatibility complex (MHC).** The absolute need for antigens to be presented bound to MHC molecules is called **MHC restriction.**

All mammals possess a major histocompatibility complex in their genome. In humans the MHC is located on the short arm of chromosome 6; in mice it is found on chromosome 17. The human MHC contains about 3.5 megabases, which is two to three times larger than the mouse MHC and about the same size as the genome of the bacterium *Escherichia coli.*

The major histocompatibility complex contains three classes of genes (Fig. 8–5). Class I genes code for MHC molecules found on the surface of most nucleated cells (and on red cells in some species). The class I genes can be divided into those that are highly

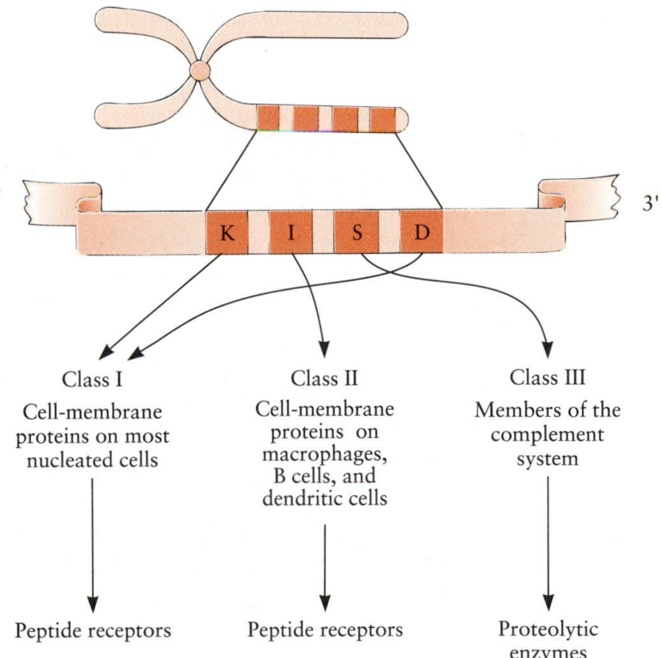

Figure 8–5 The three major classes of genes found within the mouse major histocompatibility complex and the functions of their gene products.

Table 8–1 Comparison of MHC Class I and Class II Structure

	Class I	Class II
Loci include	HLA-A, B, and C H-2-K, D, and L	HLA-D (DP, DQ, DR, DO, and DN) H-2-I (IA and IE)
Detected by	Serology	Serology and mixed lymphocyte reaction*
Distribution	Most nucleated cell	B cells, macrophages, and other dendritic cells
Function	Present antigen to cytotoxic T cells	Present antigen to helper T cells
Result	T-cell–mediated cytotoxicity	T-cell–mediated help

*A mixed lymphocyte reaction occurs when lymphocytes from two individuals with different MHC class II molecules are cultured together. Each population stimulates the other to divide.

polymorphic, called class Ia genes, and those that show very little **polymorphism,** called class Ib genes. (Polymorphism refers to inherited structural differences between proteins.) A second class, called class II genes, codes for a different set of polymorphic MHC molecules found only on the surface of macrophages, dendritic cells, and B cells. Because both class Ia and class II MHC molecules act as receptors that bind antigen fragments and present antigen to lymphocytes, they effectively regulate the immune responses (Table 8–1). Class III genes code for a mixture of proteins with a wide variety of functions that are not directly linked to antigen presentation. Some of these class III genes code for **complement** proteins. The complement system is a set of serum proteins responsible for protection against microbial invasion.

Each MHC studied so far contains all three classes of genes, although their number and arrangement within the complex vary widely (Table 8–2). The collective name given to the proteins produced by the MHC also varies between species. In humans these molecules are called HLA (human leukocyte antigen); in dogs they are called DLA; in rabbits, RLA; in guinea pigs, GPLA; and so on. In some species, histocompatibility antigens were identified as blood group antigens before their actual function was recognized. In these cases, the nomenclature is anomalous. Thus in the mouse the MHC is called H-2; in the rat it is called RT1; and in chickens it is called B.

Table 8–2 The Recognized HLA Alleles

Locus	Genetically Recognized Alleles	Serologically Recognized Alleles
A	25	14
B	32	24
C	11	7
DR	39	19
DQ	21	7
DP	23	7

MHC CLASS II PROTEINS

The human MHC class II region (called HLA-D) is 1.1 megabases in size and is located on the centromeric side of the class I and III regions on chromosome 6. Within the HLA-D region are five loci called DN, DO, DP, DQ, and DR as well as the genes for proteins involved in transporting antigen fragments across the endoplasmic reticulum. In the mouse class II gene region, two loci are recognized. They are called IA and IE.

Class II molecules are found constitutively on the professional antigen-presenting cells (B cells, dendritic cells, or macrophages), or they may be induced (as on T cells, keratinocytes, and vascular endothelial cells). The number of class II molecules on the cell surface is enhanced in rapidly dividing cells and in cells treated with interferon (Chapter 12).

Structure

MHC class II molecules are glycoproteins, each consisting of two noncovalently linked polypeptide chains called α and β (Fig. 8–6). The α chains are 31 to 34 kDa, and the β chains are 25 to 29 kDa. Each chain consists of two extracellular domains, a connecting peptide, a transmembrane domain, and a cytoplasmic domain. A third chain, a γ or invariant chain (also called CD74) of 32 kDa, is associated with intracellular class II molecules. The γ chain is not coded for by genes located within the MHC, although its expression is modulated by both interferon-γ and interleukin-4.

Gene Arrangement

Within the MHC class II region of humans, five gene loci are arranged in the order DP, DN, DO, DQ, and DR (Fig. 8–7). Within each locus, the genes for α chains are designated A and the genes for β chains are

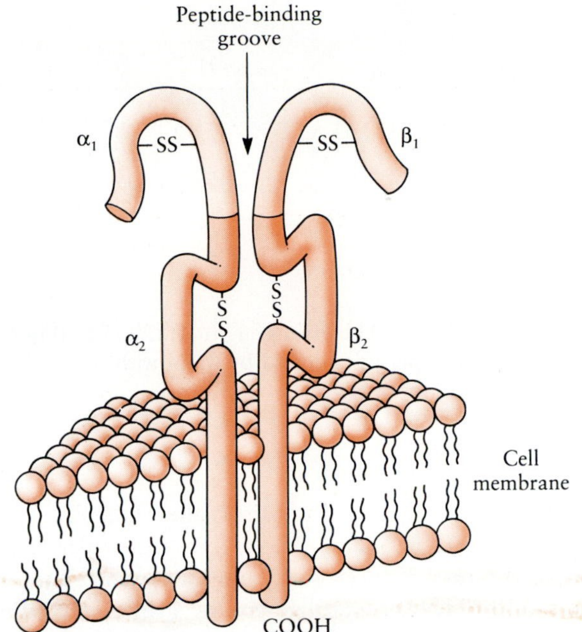

Figure 8–6 A simplified view of the structure of a MHC class II molecule showing its heterodimeric structure.

called B. The DN locus contains only a single A gene and DO contains only a single B gene, but protein products of these genes have never been found. In contrast, the DP locus contains two A and two B genes. Of these, DPA2 and DPB2 are **pseudogenes** that cannot form a protein product. The DQ locus contains two A genes and three B genes. Of these, DQB3 is a pseudogene while DQB2 and DQA2 are not pseudogenes but their protein products have not been identified. In the DR locus, only two B genes are present

A CLOSER LOOK

Haplotypes and MHC Restriction

The entire set of alleles inherited in one chromosome is called a haplotype. In mice, the haplotypes of inbred strains are given a letter designation. Thus the MHC alleles in A strain mice have the designation H-2^a. In strain C57/B1 mice the haplotype is H-2^b, and in DBA/2 mice the haplotype is H-2^d, and so on. Thus in strain A mice all the MHC regions have the "a" designation (i.e., K^a, I^a, S^a, D^a.) When recombinants are made between different inbred mice, the MHC of these animals may contain regions of different haplotypes. Thus a recombinant between strain A and DBA/2 might have the haplotype K^a I^a S^d D^d.

These recombinant mice have been of tremendous importance in demonstrating the functional role of MHC molecules. For example, the key cellular interactions of the immune system—between T cells and B cells, between T cells and antigen-presenting cells, and between cytotoxic T cells and their targets—have been shown to be MHC restricted. This means that for each of these interactions to occur successfully, it is necessary for one of the interacting cells to recognize the MHC of the other. Thus cytotoxic T cells will kill only virus-infected target cells of the same MHC class I haplotype. In contrast, macrophages can only successfully present antigen to T cells of the same MHC class II haplotype.

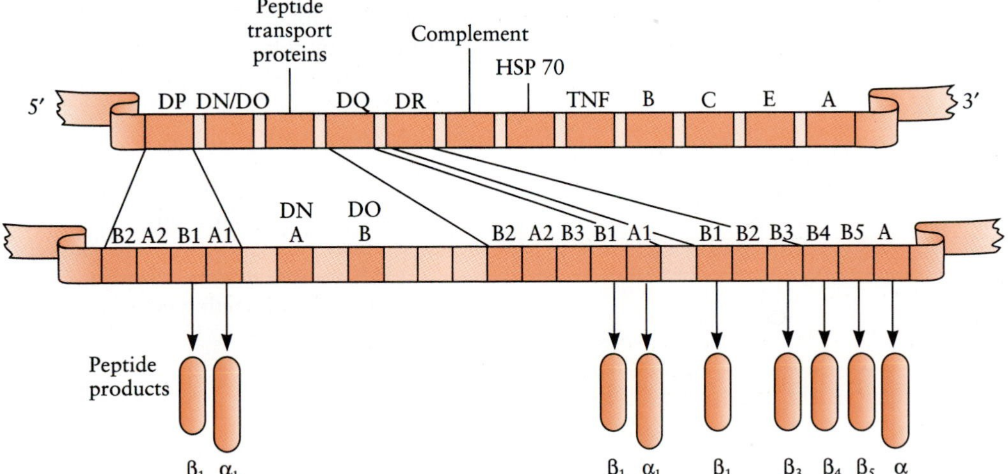

Figure 8–7 The arrangement of the genes within the human MHC as well as the peptides coded for by the polymorphic MHC class II genes, DP, DQ, and DR. Usually only two class II genes are functional and expressed in each locus. In the DR region, however, there are multiple functional B genes but only two are present in most haplotypes.

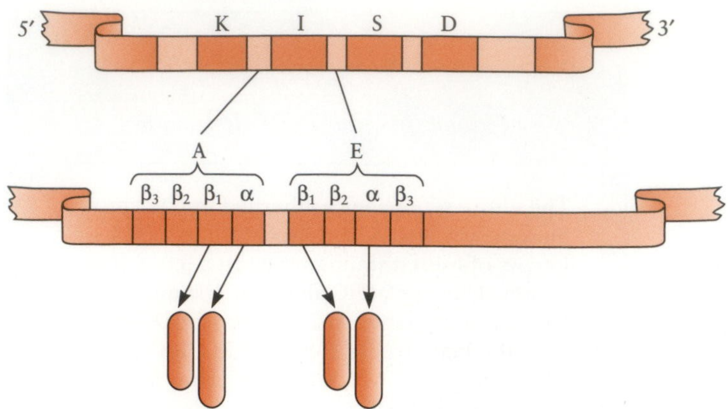

Figure 8–8 The arrangement of the class II genes and the peptides coded by them within the MHC of the mouse (H-2).

in most individuals, although the number varies. DRB2 is a pseudogene; the other genes are functional. There is a single DRA gene. The DR, DQ, and DP β chains associate primarily with the α chains of their own family. The large number of pseudogenes may serve as donors of nucleotide sequences that can be used in generating additional class II polymorphism.

The genes for the transporter proteins involved in antigen peptide transport across the endoplasmic reticulum are located between the DP and DQ gene loci.

There are eight class II genes at two loci in the I region of the mouse. In IA there are Aβ3, Aβ2, Aβ1, and Aα. In IE there are Eβ1, Eβ2, Eα, and Eβ3 arranged in that order on chromosome 17 (Fig. 8–8). Aβ2 and Eβ2 are transcribed but not translated while Aβ3 is a pseudogene. The IA locus is homologous to DQ, and IE is the **homolog** of DR. The mouse class II region is 600 kilobases in size—just over half the size of the human class II region.

All class II genes have six exons, namely signal sequence, α or β1, α or β2, connecting peptide, transmembrane region, and cytoplasmic region (Fig. 8–9). These correspond to the domains in the class II protein.

Polymorphism

Class II molecules show great polymorphism. That is, their amino acid sequence shows great variability. These sequence variations are restricted to the α1 and β1 domains. Within these domains, the polymorphic amino acids are concentrated in three to four discrete **hypervariable regions.** X-ray crystallography has demonstrated that the outermost domains of each MHC class II chain (α1 and β1) fold together to form an open-ended groove (Fig. 8–10). This groove functions as an antigen-binding site. The hypervariable regions are located almost exclusively in the walls of the groove, and they determine the shape of that groove. The shape of the groove as determined by their amino acid sequence dictates the ability of each class II molecule to bind processed antigen fragments and thus determines the ability of an animal to respond to a specific epitope.

A β-pleated sheet composed of eight antiparallel strands forms the floor of the groove, and two α-helices form its walls. The overall shape of the groove is determined by the conserved peptides of the α and β domains, while the polymorphic residues are in a po-

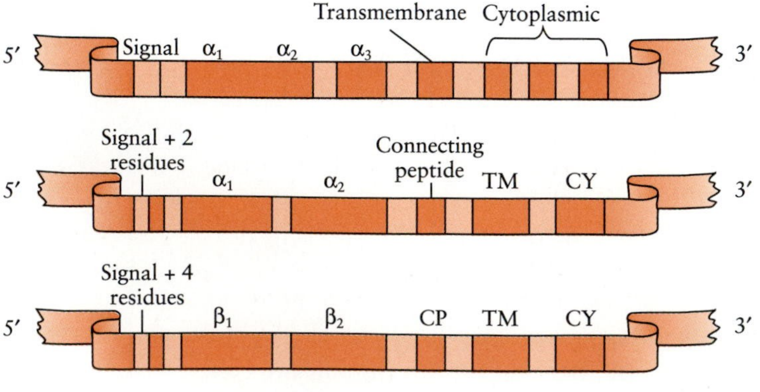

Figure 8–9 The MHC peptide chain exon structure. *Top,* MHC class I α chain genes. *Center,* MHC class II α chain genes. *Bottom,* MHC class II β chain genes. (TM, transmembrane domain; CY, cytoplasmic domain; CP, connecting peptide.)

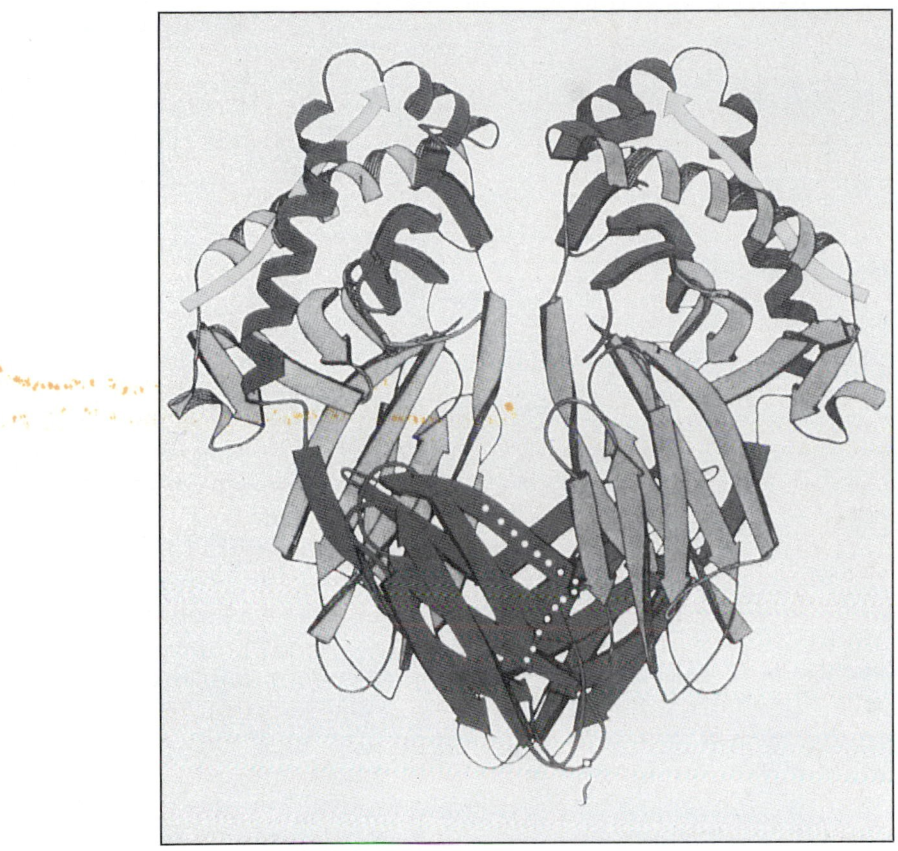

Figure 8–10 The three-dimensional structure of an MHC class II molecule, HLA-DR1. This molecule forms a dimer of the αβ heterodimer. The α chains are lighter shades; the β chains are darker. The location of the peptide-binding grooves is shown by the arrow at the top of each unit. *(From Brown, J.H., et al. Nature, 364:38, 1993. With permission.)*

sition to bind to the side chains of the bound peptides. The groove can hold a peptide of 12 to 24 amino acids as a straight, extended chain that projects out of both ends of the site. There is a "pocket" in one wall near the end of the groove that can accommodate an amino acid side chain. X-ray crystallography also suggests that class II molecules may exist in a paired form (i.e., a dimer of dimers), so that two peptides may be presented in close association. Since the peptide–MHC class II linkage may be fairly weak, this dimerization may help ensure that an adequate signal reaches the T cells.

In humans, DOB and DRA (Eα in the mouse) are the least polymorphic, while the DRB genes and DQB1 (Aβ and Eβ in mice) are most polymorphic (Figs. 8–11 and 8–12). This MHC class II polymorphism results from multiple point mutations and gene conversion. In **gene conversion,** small blocks of genetic material are exchanged between different class II genes in a nonreciprocal fashion. The size of the transferred

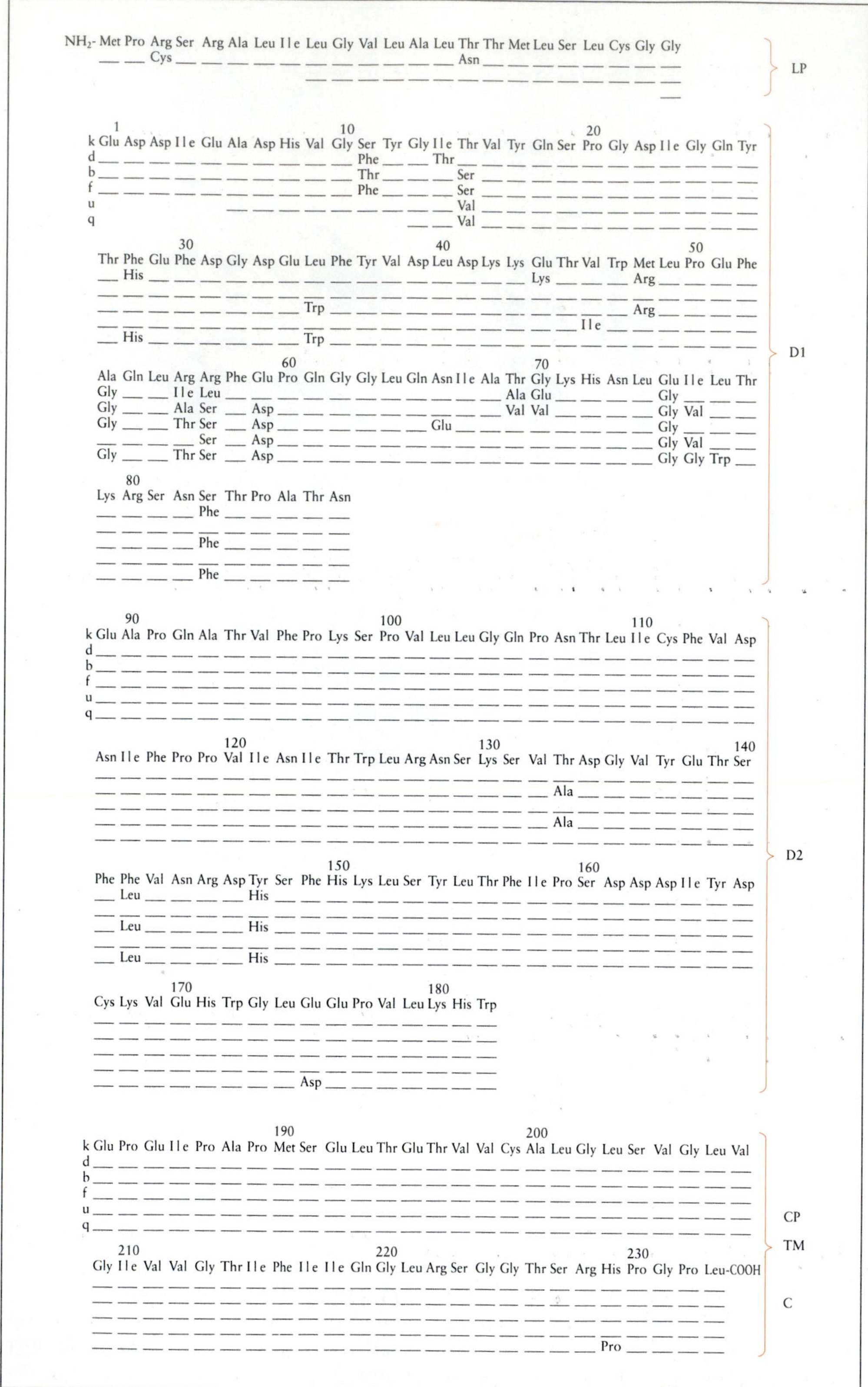

Figure 8–11 The complete amino acid sequences of six allelic Aα chains from mice. Dashes indicate locations where the amino acid residues are identical. Note that there are several highly variable positions scattered throughout the sequence but no hypervariable regions. *(From Benoist, C.O., et al. Cell, 34:169, 1983. With permission.)*

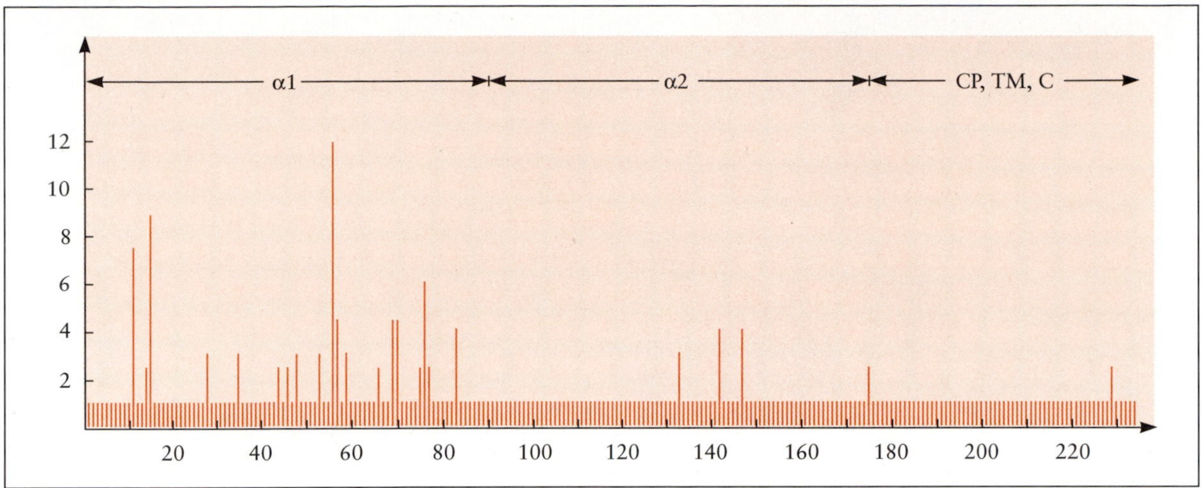

Figure 8–12 This diagram presents the sequences from Figure 8–10 in a graphical format known as a Kabat-Wu plot. For each position in the peptide chain, a variability index is calculated thus:

$$\text{Variability index} = \frac{\text{The number of amino acids}}{\text{The frequency of the most common amino acid}}$$

Thus if six amino acids are found equally in one position, the variability index for that position will be 6/0.16 = 36. The plot shows only occasional scattered variable positions. The reader may wish to compare this with Figures 11–4 and 13–5. *(From Benoit, C.O., et al. Cell, 34:169, 1983. With permission.)*

blocks ranges from 5 to 95 nucleotides. The donated blocks may be derived from nonpolymorphic class II genes and **pseudogenes** as well as from the polymorphic class II genes. Blocks of nucleotides may be obtained from the same or different chromosomes. Single nucleotide substitutions and reciprocal recombinations occur as well as gene conversion. As a result of the use of all these mechanisms, MHC genes have the highest mutation rate of any germ-line genes yet studied. This rate is 10^{-3} mutations per gene per generation in mice. This high mutation rate implies that there must be significant advantages to be gained by having very polymorphic MHC genes.

Function

The presentation of exogenous antigen is the prime function of MHC class II molecules. Although all macrophages can phagocytose foreign particles, only those that express MHC class II molecules can process this material in such a way that it can stimulate an immune response. The MHC class II molecules can bind fragments of the ingested antigen and present them to antigen-recognizing cells (Fig. 8–13). These antigen-recognizing cells are a subpopulation of lymphocytes called **helper T cells** since they stimulate the immune response. Helper T cells can only recognize a foreign antigen and respond to it if it is physically bound to a MHC class II molecule.

Macrophages carrying MHC class II molecules predominate in the spleen, thymus, and liver. Most peritoneal macrophages, however, are normally MHC class II negative but express MHC class II molecules when activated. If macrophages are treated with antibodies against MHC class II molecules so that they disappear from the cell surface, they lose their ability to initiate an immune response. Similarly, because the macrophages of newborn mice lack MHC class II molecules, these animals are effectively immunodeficient. Only when MHC class II molecules appear on their macrophages do they acquire the ability to mount an immune response. Macrophages regulate the dose of antigen presented on their surfaces and so prevent the inappropriate development of tolerance. Helper T cells are triggered into responding to antigen by exposure to a narrow range of antigen doses. If antigen is presented to T cells without being linked to a MHC class II molecule, they may be turned off and tolerance may result (Chapter 21).

There are several steps in exogenous antigen processing by macrophages. First, the antigen must be

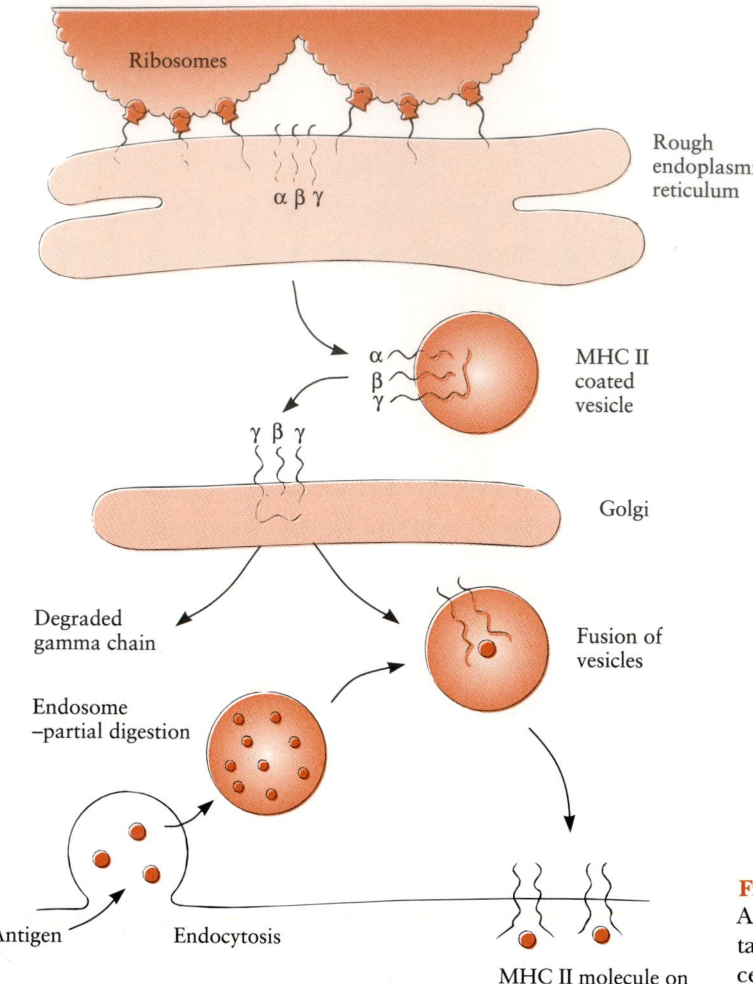

Figure 8–13 The processing of exogenous antigen. Antigen is fragmented in endosomes and then attached to MHC class II molecules for transport to the cell surface. The γ chain protects the assembled MHC molecule until it binds antigen.

phagocytosed or, if in solution, pinocytosed and taken into phagosomes. Phagosomes then fuse with granules containing acidic proteases. Ingested peptides are broken down by lysosomal enzymes into fragments about 10 to 20 amino acids long. The endosomes containing the antigenic peptides then fuse with other endosomes carrying newly synthesized MHC class II molecules. The fused vesicles continue to move toward the cell surface while the antigen peptides bind to the MHC groove. Once the vesicles reach the cell surface, they fuse with the cell membrane and the MHC–peptide complex is presented on the cell surface in such a way that it can be recognized by receptors on the T cell. Occasionally, some short antigen peptides bind directly to MHC molecules on the macrophage surface without being endocytosed.

During the assembly and export of MHC class II molecules, they are associated transiently with the invariant chain (γ). The γ chain binds to the antigen-binding groove on MHC class II molecules as they are formed. The presence of the γ chain prevents the MHC class II molecule from being immediately transported to the cell surface. (Unlike most transmembrane proteins that are expressed minutes after passing through the Golgi apparatus, MHC class II molecules are retained for several hours.) The γ chain redirects the MHC molecule to special endosomes, where it can be loaded with antigen fragments. In the endosomes the γ chain is degraded by proteases, leaving a vacant groove to bind the foreign peptides and at the same time permitting the complex to move to the cell surface.

Since helper T cells must be stimulated for most immune responses to occur, MHC class II molecules effectively determine whether an immune response will occur in response to any epitope. Class II molecules specifically bind some, but not all, peptide antigens created during antigen processing and thereby select the epitopes that are to be presented to T cells. The peptide–MHC complex is the signal that activates

T cells through their antigen receptor. As a result, a helper T cell recognizes an antigen only if it is presented by a class II molecule.

PROCESSING OF ENDOGENOUS ANTIGEN

Some antigens that trigger an immune response are not first ingested by macrophages or other antigen processing cells but actually originate within the cell itself—endogenous antigens. Good examples of endogenous antigens are the new proteins made by a cell when taken over by a virus. These proteins are handled in a different manner than exogenous antigens in that after fragmentation they are bound to MHC class I molecules and transported to the cell surface. Antigen bound to MHC class I antigens triggers a response by a population of lymphocytes called **cytotoxic T cells,** not helper T cells. Cytotoxic T cells destroy virus-infected cells. The key difference lies in the nature of the MHC molecules used to present the antigen to the T cells.

Unlike the highly polymorphic MHC class II molecules, the MHC class I molecules fall into two groups: a polymorphic group of proteins called MHC class Ia molecules, and a group that is very much less polymorphic called class Ib molecules.

MHC CLASS Ia PROTEINS

Structure

All class Ia molecules consist of a single peptide chain of 45 kDa called α that is noncovalently linked to a much smaller chain called β2-microglobulin (the β chain) and bound to a cell surface (Fig. 8–14). The α chain contains three extracellular domains—α1, α2, and α3, each about 100 amino acids long and containing a single intrachain disulfide bond, as well as a transmembrane domain, and a cytoplasmic domain. The cytoplasmic domain is coded for by one to three exons so that the total number of exons coding for a MHC class Ia α chain is six to eight (see Fig. 8–9). The β chain consists of a single extracellular domain.

Class Ia molecules are found on the surface of all nucleated cells except those that are highly differentiated, or primitive, cells in early embryonic life. They are found in highest concentration on lymphocytes and macrophages. They are not found on human red blood cells but are on mouse red blood cells. Some tissues such as myocardium and skeletal muscle show little or no expression of class Ia molecules. Class Ia molecules are usually synthesized at a greater rate than is required by the cell and, as a result, may be found free in tissue fluid and serum.

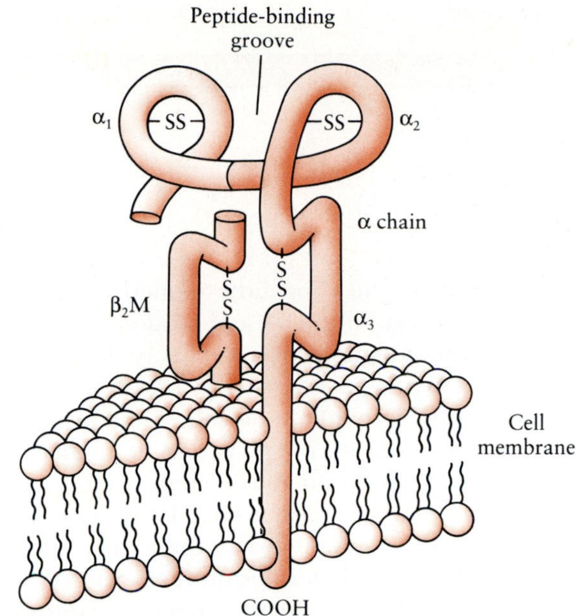

Figure 8–14 A simplified view of the structure of a MHC class I antigen. The actual three-dimensional structure of a MHC class I antigen is seen in Figure 8–17.

Gene Arrangement

Each mammal species studied has two or three MHC class Ia loci. For example, in humans there are three class Ia loci, called A, B, and C. The class Ia loci in the mouse H-2 complex are called K and D (and in some strains, L) (Fig. 8–15). The number of identified class Ia genes or gene fragments varies greatly between species. Thus rats have over 60, and pigs as few as 6 genes or gene fragments. There are about 30 genes coding for class Ia α chains in mice. Yet in this species only two or three polymorphic α chains have a known function.

Polymorphism

In human populations, as many as 23 α chain **alleles** can be found at the A locus, 49 α chain alleles at B, and 8 α chain alleles at C (Table 8–3). This polymorphism is a result of variations in the amino acid sequence of the α1 and α2 domains. These alleles of MHC class Ia molecules differ in many positions, not just one or two amino acid residues. About half of the 182 amino acids in the α1 and α2 domains are variable. Certain clusters of amino acid differences occur, such as at positions 62–83 in α1, 105–116 in α2, and 177–194 in α2–α3. Usually, however, only two or three different amino acids are found at each position and one is usually predominant. The α3 domain as well as the transmembrane and cytoplasmic domains are

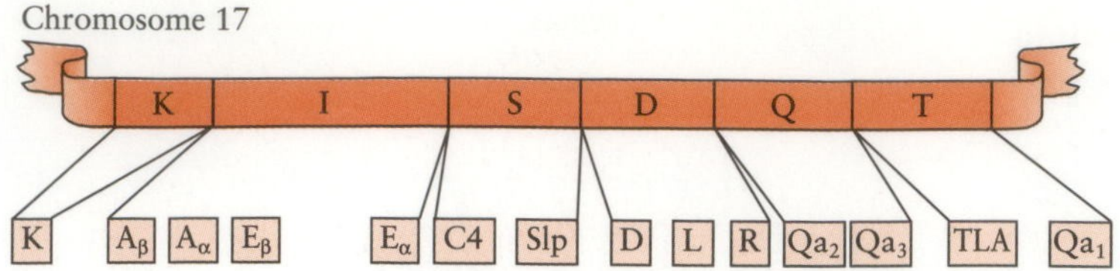

Figure 8–15 The overall arrangement of the H-2 gene complex in the mouse. The I and S regions contain class II and III genes, respectively. The remainder of the region codes for class I molecules.

highly conserved and show little sequence variation. The polymorphism of the α1 and α2 domains occurs as a result of several molecular mechanisms, including point mutation and gene conversion.

The β2-microglobulin molecule found associated with class Ia molecules is not polymorphic and its function is unclear. Mice genetically engineered to have no β2-microglobulin appear to be almost normal. In the absence of β2-microglobulin, the animals do not express MHC class I molecules on their cell surfaces. As a result, they do not present antigen to cytotoxic T cells and have very low levels of cytotoxic T cells. They do not absorb maternal antibody through their intestines. (The absorption of antibody across epithelial brush border membranes is by a β2-microglobulin–associated receptor.) Presumably, these animals could be unusually susceptible to virus infections.

Function

One function of the T-cell–mediated immune response is the identification and destruction of abnormal cells such as those infected by viruses. To destroy these cells, cytotoxic T cells must respond to fragmented viral proteins expressed on the surface of infected cells. They respond to these fragments only if they are first bound to the antigen-binding groove of the MHC class Ia molecules on the target cell. Cytotoxic T cells recognize this virus protein–class Ia molecule complex. This can be demonstrated experimentally by showing that cytotoxic T cells destroy virus-infected target cells only if the T cells and their targets possess identical MHC class I molecules (Chapter 18). These cells are thus said to be MHC restricted. The T cell can recognize foreign epitopes only in as-

Table 8–3 The Recognized HLA Specificities*

HLA-A	HLA-B	HLA-C	HLA-DR	HLA-DQ	HLA-DP
A1	B7	Cw1	DR1	DQw5 (w1)	DPw1
A2	B8	Cw2	DR'BR'	DQw6 (w1)	DPw2
A3	B13	Cw3	DR4	DQw2	DPw3
A11	B14	Cw5	DR7	DQw4	DPw4
A24 (9)	Bw65 (14)	Cw6	DRw8	DQw7 (w3)	DPw5
A25 (10)	Bw62 (15)	Cw7	DR9	DQw8 (w3)	DPw6
A26 (10)	B18	Cw11	DRw10	DQw9 (w3)	DP'Cp63'
A29 (w19)	B27		DRw11 (5)		
A30 (w19)	B35		DRw12 (5)		
A31 (w19)	B37		DRw13 (w6)		
A32 (w19)	B38 (16)		DRw14 (w6)		
Aw33 (w19)	B39 (16)		DRw15 (2)		
Aw68 (28)	B40		DRw16 (2)		
Aw69 (28)	Bw41		DRw17 (3)		
	Bw42		DRw18 (3)		
	B44 (12)		DRw52a		
	Bw46		DRw52b		
	Bw47		DRw52c		
	B49 (21)		DRw53		
	B51 (5)				
	Bw52 (5)				
	Bw57 (17)				
	Bw58 (17)				

*The suffix *w* in the name of an allele indicates a provisional designation assigned at an international workshop. Once the existence of the allele is confirmed beyond doubt, the *w* designation is removed.

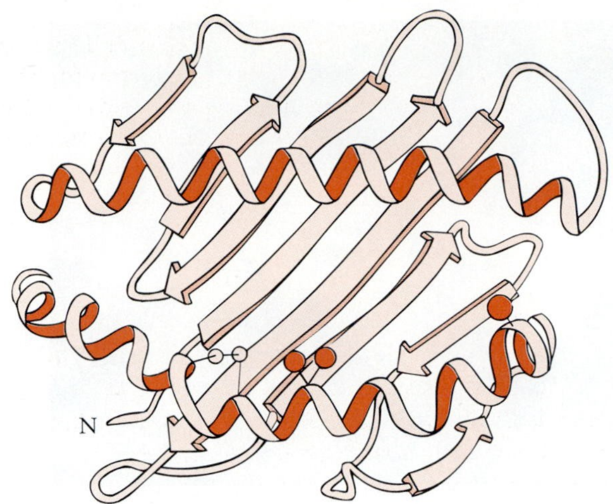

Figure 8–16 A view (from "above") of the antigen-binding groove on a MHC class I molecule. The floor of the groove is formed by an extensive β-pleated sheet. The walls of the groove are formed by two parallel α-helices. This structure is formed by the folding of the α1 and α2 domains. *(Reprinted by permission from Dr. P.J. Bjorkman and Nature, Vol. 320, p. 506. Copyright © 1987 MacMillan Magazines Limited.)*

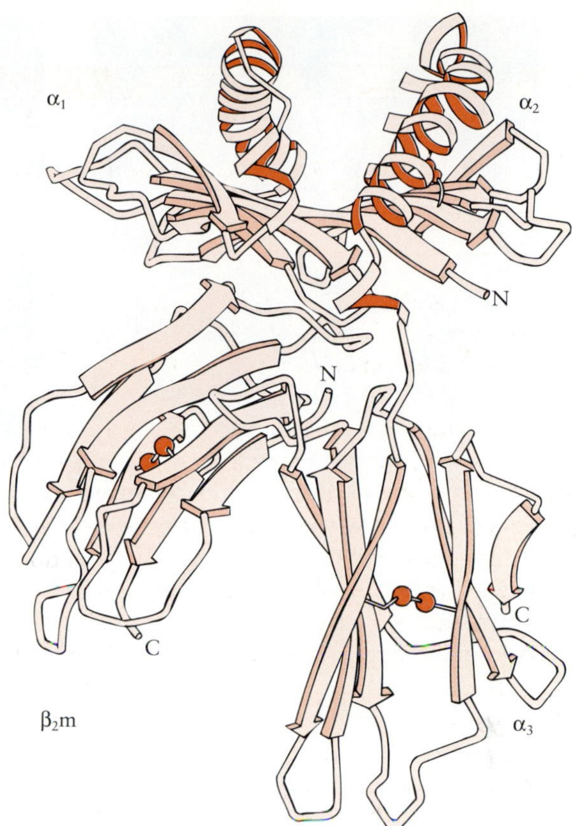

Figure 8–17 A schematic three-dimensional view of the complete structure of HLA-A2 derived by x-ray crystallography. The antigen-binding groove at the top is formed by the α1 and α2 domains, and the α3 domain binds to the cell membrane. The β chain (β2-microglobulin) has no direct role in antigen binding. *(Reprinted by permission from Dr. P.J. Bjorkman and Nature, Vol. 320, p. 506. Copyright © 1987 MacMillan Magazines Limited.)*

sociation with self-MHC molecules. The association of class Ia molecules with viral antigens also controls the amount of antigen that can be effectively presented to a T cell and thus prevents saturation of T-cell receptors by excessive exposure to viral proteins.

X-ray crystallography studies show that the conformation of the MHC class Ia α chain is similar to the conformation of MHC class II molecules. Thus it too is folded in such a way that a large groove is formed on its outermost surface. This groove is formed by the folding of the α1 and α2 domains (Fig. 8–16). The floor of the groove is formed by a β-pleated sheet, and the sides are formed by two α-helices (Fig. 8–17). This groove, however, differs from the groove on class II molecules in that it is closed at each end with deep pockets. As a result, bound peptides cannot project out of the ends of the groove. These pockets are conserved, and tyrosine and threonine residues form hydrogen bonds with the C-terminus of bound peptides so that they all bind with the same orientation. There is a deep polymorphic pocket in the middle of the groove that binds one of the peptide's side chains and plays a major role in specific binding. Other shallow polymorphic pockets in the groove may also bind the peptide's other side chains but play a relatively minor role in specific binding. Because the groove is closed at each end, MHC class Ia molecules can only bind peptides containing 8 to 10 amino acids. Indeed, to do this, these peptides must bulge out in the middle (Fig. 8–18). However, overall the grooves on class II and class Ia molecules function in a very similar manner. The α3 domain attaches this receptor-like structure to the cell membrane, and the β2-microglobulin component probably stabilizes the structure.

MHC class Ia molecules bind peptides obtained from intracellular sources. The processing of these peptides is very different from that of the peptides associated with class II molecules. Cells continually break down and recycle proteins. As a result, abnormal proteins are removed, regulatory peptides are not allowed to accumulate, and amino acids are made available for other purposes. The mechanisms that degrade most cell proteins usually require, as a first step, that the protein molecules be marked by linking them to the peptide ubiquitin (Fig. 8–19). This linkage requires several enzymes and ATP. Once linked to ubiquitin, the peptide is rapidly hydrolyzed by one of two very large proteolytic complexes, or **proteasomes,** in the cytoplasm. These proteasomes are complex structures having about 30 components with three or four peptidase activities. (At least some of these constituents are coded for by genes located within the MHC

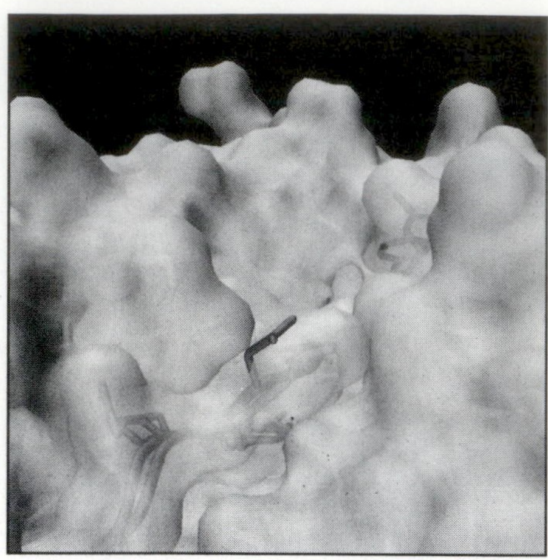

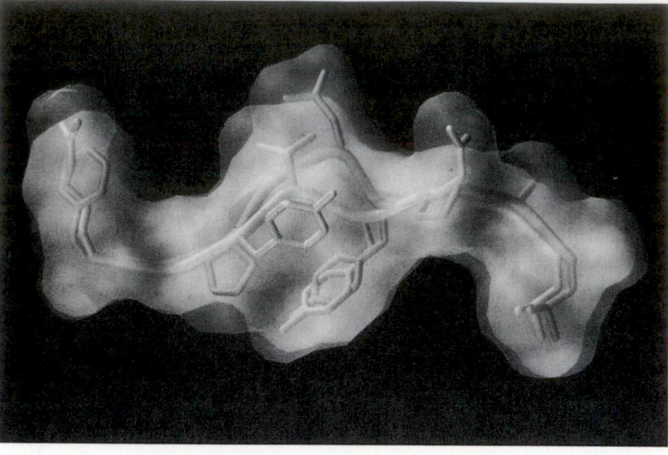

Figure 8–18 A peptide, eight amino acids long, fits into a groove in the mouse MHC class I molecule (*left*). The right panel shows how two related peptides are configured when they fit into the MHC-binding groove. *(From Baringa, A. Science, 257:881, 1992. With permission.)*

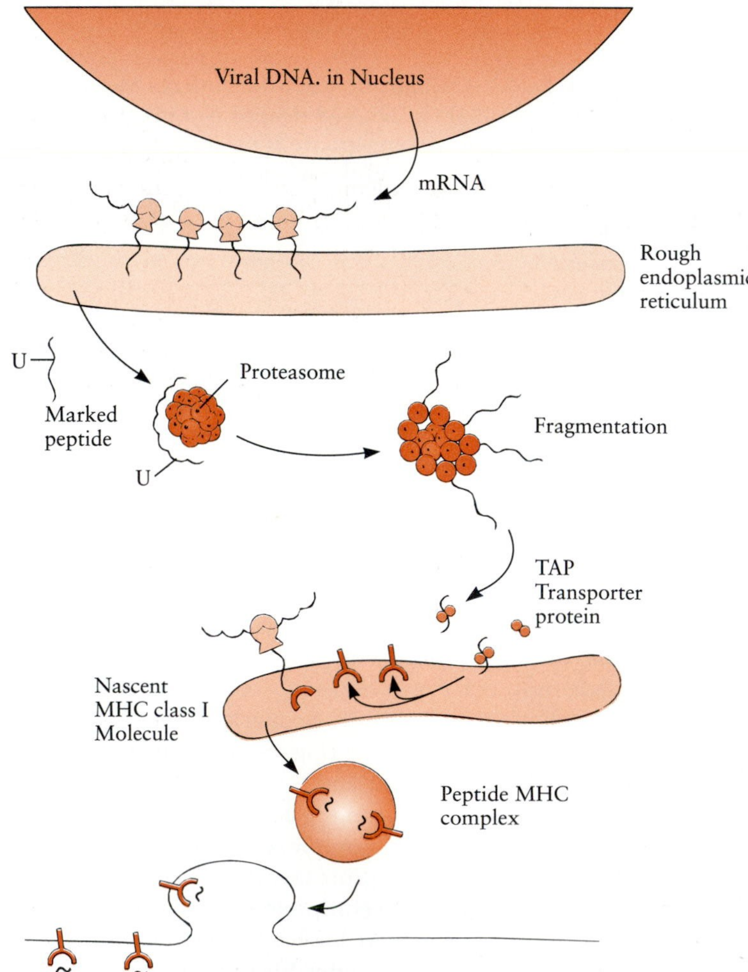

Figure 8–19 The processing of endogenous antigens. Newly synthesized peptides are marked by attachment to ubiquitin (U), degraded by proteasomes, and their fragments transported to the newly forming MHC class I molecule, where they are inserted in the antigen-binding groove. The complexes are then transported to the cell surface for presentation to antigen-sensitive cells.

class II region.) They probably produce multiple peptides simultaneously from the same protein molecule. As the proteasomes digest cytosolic proteins, they also generate antigenic peptides for presentation by MHC class Ia antigens.

As a protein is degraded, peptides are rescued from further breakdown by attachment to transporter proteins. The genes for these transporter proteins are encoded in the MHC class II region, adjacent to the genes for proteasome subunits. Two transporter proteins called TAP-1 and TAP-2 have been identified. These form a heterodimer (called TAP—**t**ransporter for **a**ntigen **p**rocessing) that acts as a peptide pump and carries the antigen fragments from the cytoplasm across the endoplasmic reticulum and into its lumen, where the MHC class Ia molecules are located. The peptides are then placed in the groove of the MHC molecule, which they bind with fairly high affinity. It is believed that the MHC class Ia molecules are assembled around the peptides. The MHC class Ia molecules subsequently carry the peptides to the cell surface for recognition. When the MHC Ia–peptide complex reaches the cell surface, the bound peptides are displayed for many hours. Some empty MHC molecules may reach the surface, but these tend to be unstable and fall apart. Polymorphism in the transporter proteins determines which peptides are loaded on an MHC class Ia molecule. Thus these transporter molecules can also determine which peptides will and will not trigger an immune response.

Neither the proteasome nor the TAPs or MHC molecules can distinguish peptide fragments from normal cellular constituents or from abnormal constituents such as viral proteins. The ability to distinguish these is a function of the T cells that encounter the MHC-peptide complex on the cell surface. It has been calculated that a cell expressing 10^5 copies of one class I molecule may present as many as 10^4 different bound peptides.

EVOLUTIONARY SIGNIFICANCE OF MHC BINDING

The binding site of an MHC class Ia or II molecule acts as a fairly nonspecific receptor. The groove can actually bind many peptides because it binds tightly to the peptide backbone rather than to its amino acid side chains. Nevertheless, structural constraints limit the efficiency of binding of each allele, and one molecule cannot possibly bind all peptides made available to it. As a result of these constraints, it is likely that only one or two peptides from an average protein will be able to bind to any given MHC molecule. Increasing the number of MHC loci in an individual will clearly increase the number of peptides that can be bound and presented. Because most individuals are heterozygous at their MHC, each individual normally expresses six class Ia molecules (two each coded for by the HLA-A, B, and C loci). The number of MHC types is not larger, since the presence of more MHC molecules would require **negative selection** of much larger numbers of self-reactive T cells. (Negative selection is the process by which lymphocytes that react with self-antigens are destroyed in the thymus (Chapter 21). Nevertheless, the efficacy of antigen binding by each MHC allele must be a limiting factor in resistance to infectious diseases. Thus six MHC class Ia molecules probably represent a reasonable compromise between maximizing recognition of foreign antigen and minimizing recognition of self-antigens (Fig. 8–20). Homozygosity at one of the three class Ia loci reduces the available number of MHC molecules from six to five, a drop of 16%, and results in a major loss in immune capability. Homozygosity is therefore something to be avoided.

Some class Ia loci may code for a very large number of alleles. Thus H-2K in the mouse can contain more than 100. If the purpose of having polymorphism is to ensure heterozygosity at each locus, this is clearly far in excess of the body's needs. Thus although heterozygosity is important, there must be another reason for having such an enormous diversity of MHC class Ia molecules. The most likely reason is to protect the population as a whole from disease. Because of extensive polymorphism, most individuals in a population will have a unique set of class Ia molecules, and these will be able to bind a unique diversity of epitopes. As a result, when a new infectious disease strikes, it is likely that at least some individuals will be able to bind and present the new microbial epitopes. These individuals will be able to respond to the agent and so be immune.

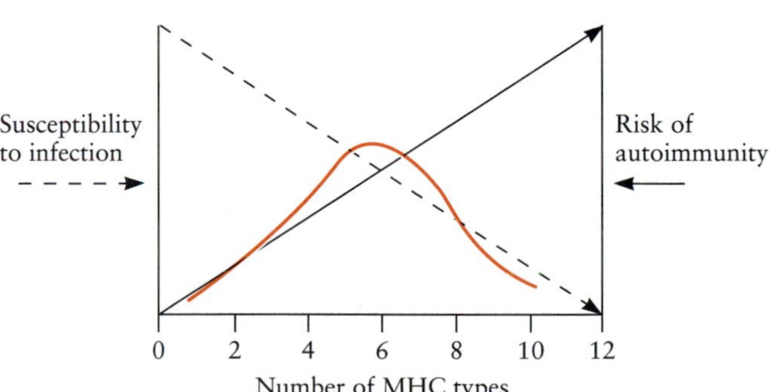

Figure 8–20 The optimal number of MHC molecules is a balance between the need to respond to many different invading microorganisms and the need to avoid autoimmunity. The optimal number of MHC molecules appears to be approximately six, the same number that is found in nature.

No single MHC allele has been found to be present at a very high frequency in a population. In other words, no single MHC molecule confers major advantages on an individual (although linkage disequilibrium, described later, suggests that some allelic combinations may be advantageous). This reflects the futility of the host's attempting to match invading organisms in antigenic variability. A microorganism will always be able to mutate to evade the immune response very much faster than a mammalian population can develop resistance. Any mutations in the MHC may increase resistance to one organism; they may, at the same time, decrease resistance to another. It is more advantageous for each member of a population to be different at the MHC class Ia locus so that an organism spreading through a population will have to adapt anew to each host. It is thus desirable for the population to contain as many different MHC molecules as possible.

Highly adaptable social animals such as humans or mice with large populations through which disease can spread rapidly show extensive class Ia polymorphism. In contrast, low-density solitary species such as whales or cats have much less polymorphism. It is also of interest to point out the case of the cheetah, which is essentially monomorphic at its class Ia and II loci as a result of a recent population bottleneck. Because of this lack of MHC diversity, a single lethal infectious disease has the potential to cause the cheetah's extinction.

If a certain MHC molecule—for example, HLA-A1—occurs in 16% of the population and another antigen—for example, HLA-B8—occurs in 10% of the population, then the combination of A1 + B8 would be expected to occur in 1.6% of the population (10% × 16%). In fact, the A1 + B8 combination occurs in European populations at a frequency of 8.8%. This combination, A1 and B8, is said to be in **linkage disequilibrium.** There are two possible causes of this. One is that these specific alleles might have evolved so recently that their distribution is not yet random. Alternatively, there may exist some unknown factor that gives an advantage to individuals carrying this combination, perhaps resistance to disease. Other examples of linkage disequilibrium include A3 and B7 that occur in 2.8% of the population rather than the expected 1.3% and A29 and B12 that occur in 3.4% rather than the expected 1%.

MHC CLASS Ib PROTEINS

The nonpolymorphic MHC class Ib molecules have a restricted tissue distribution; some may be secreted, and some are not encoded by genes located within the MHC.

Structure and Gene Arrangement

The class Ib gene loci in mice can be divided into three clusters designated Q, T, and M. These clusters are found 3′ to the D and L class I loci (Fig. 8–21). There are approximately ten Q alleles, 23 T alleles, and eight M alleles. They code for proteins found on the surface of regulatory and immature lymphocytes and on hematopoietic cells. These proteins each consist of a membrane-bound α chain of 44 kDa closely related to the α chain of class Ia molecules, associated with β2-microglobulin. Their extracellular conformation is similar to that of class Ia molecules, and they do have an antigen-binding groove. They characteristically have short cytoplasmic domains, and some may be anchored to the cell membrane by a GPI anchor.

In humans there are six class Ib loci—HLA-E, F, G, H, J, and X. The last three are pseudogenes. HLA-E and X are located between HLA-A and C and expressed on a variety of cells. HLA-F is differentially expressed in resting T cells and skin cells. A second family of class Ib–like molecules called CD1 is found on another chromosome.

Function

As a result of their lack of polymorphism, class Ib molecules can bind to only a limited range of peptides. Nevertheless, it is now believed that they can act as

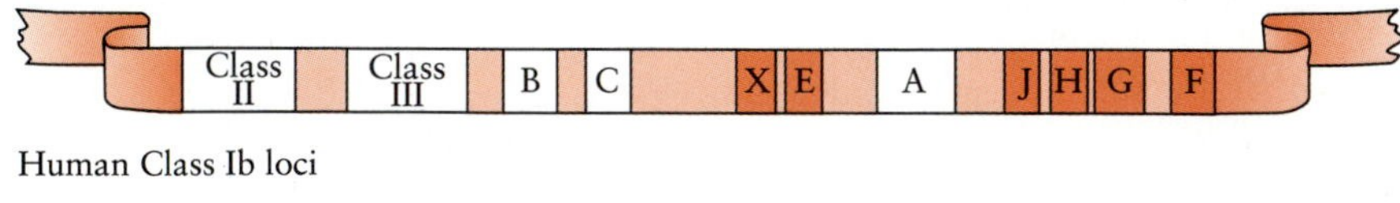

Human Class Ib loci

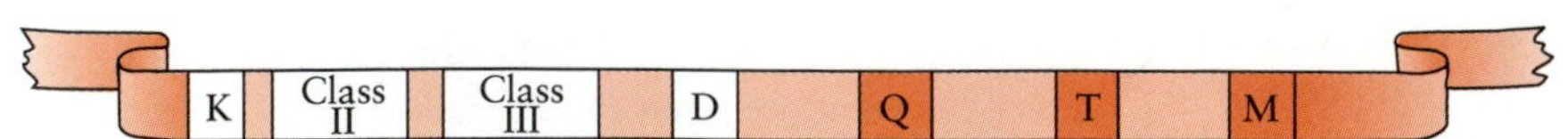

Mouse Class Ib loci

Figure 8–21 The location of the class Ib loci in humans and mice. In humans, H, J, and X are pseudogenes. Note that the closely related CD1 molecules are coded on another chromosome.

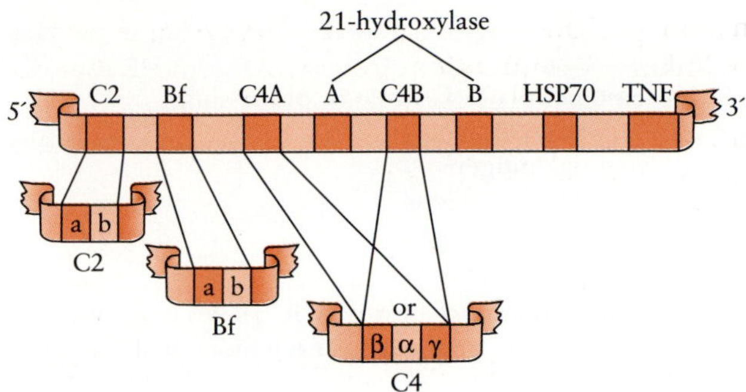

Figure 8–22 The structure of the MHC class III gene region in humans. C2, B, and C4 are components of the complement system. The significance of the presence of the 21-hydroxylase gene in this region is unknown.

receptors for specific, commonly encountered microbial antigens. For example, the mouse class Ib molecule known as M3 appears to bind specifically to exogenously processed peptides that have an N-formyl-methionine residue at their amino terminus. As pointed out in Chapter 5, these structures are commonly encountered on bacteria. M3 can also present a peptide from *Listeria* to cytotoxic T cells. CD1b, a human class Ib–like molecule, can present exogenous mycobacterial antigens to some T cells.

MHC CLASS III PROTEINS

The remaining genes located within the major histocompatibility complex can be grouped as class III genes (Fig. 8–22). They code for proteins with a great variety of functions. Several of these genes code for proteins that belong to the complement system. (The complement system is described in detail in Chapter 16.) In addition, there are other loci that code for the enzymes 21-hydroxylase A and B, tumor necrosis factor (TNF–α and β), and heat shock protein 70 (HSP70). All these are classified as MHC class III loci.

In mice, a gene locus called S codes for two proteins collectively known as SS. There are two major proteins in this class. One is called C4, the fourth component of mouse complement. The other protein is called Slp (sex-linked protein). Mice that make both proteins are SS-high. Mice that make only C4 are SS-low. The production of Slp is sex linked in some strains of mice; that is, it is found only in males since in these strains the Slp gene is regulated by testosterone. However, in other SS-high strains, testosterone is not a requirement and Slp is present in mice of both sexes. The genes for C4 and Slp are discrete but closely linked, and both proteins have a similar structure. C4 and Slp are discussed in more detail in Chapter 16. Factors B and C2 (two components of the complement system) are also coded for by genes located within the S region.

Humans also possess four genes for complement components in their class III region—two for C4 and one each for factors B and C2. The first two class III genes identified were believed to be blood group antigens known as Chido and Rogers (named after the individuals in whom they were first detected). It was subsequently shown that these antigens are plasma proteins adsorbed onto red blood cells. Chido and Rogers antigens are in fact variants of the C4 molecule (C4A and C4B, respectively).

KEY WORDS

Antigen-binding groove p. 94
B cell p. 91
β2-microglobulin p. 99
C4 p. 105
Class I genes p. 99
Class II genes p. 92
Class III genes p. 105
Complement p. 105
Cytotoxic T cells p. 99
Dendrite p. 89
Dendritic cell p. 89
Endogenous antigen p. 99
Endosome p. 98
Exogenous antigen p. 89
Follicular dendritic cell p. 90
Gene conversion p. 95
Haplotype p. 93
Helper T cell p. 97
Histocompatibility molecule p. 91
HLA p. 92
Hypervariable region p. 94
Iccosome p. 90
Immune complex p. 90
Interdigitating cell p. 90
Invariant chain p. 98
Langerhans cell p. 90
Linkage disequilibrium p. 104
Macrophage p. 89
Major histocompatibility complex p. 91
MHC class Ia molecule p. 99
MHC class Ib molecule p. 104
MHC class II molecule p. 92
MHC-restriction p. 91
Point mutation p. 95
Polymorphism p. 94
Processing p. 89
Proteasome p. 101
Pseudogene p. 93
Transporter protein p. 103
Ubiquitin p. 101
Veiled cell p. 90

QUESTIONS

1. Exogenous antigen processing mainly occurs within
   a. T lymphocytes
   b. neutrophils
   c. plasma cells
   d. macrophages
   e. none of the above
2. Class I MHC molecules are found on
   a. B cells and macrophages
   b. erythrocytes, B cells, and T cells
   c. T cells only
   d. all nucleated cells
   e. neutrophils, T cells, and B cells
3. Class II MHC molecules contain
   a. one γ and one δ chain
   b. one α chain and one β chain
   c. two light and two heavy chains
   d. one γ chain
   e. β2-microglobulin
4. Endogenous antigens are antigens that are
   a. adsorbed onto cell surfaces
   b. phagocytosed by macrophages
   c. recognized by B cells
   d. recognized by CD4 positive cells
   e. synthesized within cells
5. Many of the genes within the MHC are pseudogenes—what is their function?
   a. These form complete gene products.
   b. These do not form functional gene products.
   c. These code for MHC class II molecules.
   d. These code for complement components.
   e. These code for immunoglobulins.
6. An individual's HLA haplotype is
   a. a chromosome segment
   b. one allele at each MHC class I locus
   c. one allele at each MHC class II locus
   d. the individual's complete set of MHC alleles
   e. the individual's class I antigen type
7. Nonpolymorphic MHC gene products are called
   a. MHC class III molecules
   b. MHC class Ia molecules
   c. MHC class Ib molecules
   d. MHC class II molecules
   e. cell surface molecules
8. Proteasomes play a key role in processing
   a. exogenous antigens
   b. bacterial antigens
   c. cell surface antigens
   d. endogenous antigens
   e. MHC class II peptides
9. When the prevalence of two MHC alleles is significantly greater than the prevalence of each individual allele, this is called
   a. MHC restriction
   b. linkage disequilibrium
   c. allotypic restriction
   d. independent assortment
   e. codominance
10. β2-Microglobulin is an integral part of
    a. immunoglobulin M
    b. MHC class II molecules
    c. MHC class I molecules
    d. T-cell antigen receptor
    e. B-cell antigen receptor
11. Why do animals need histocompatibility molecules? Could we function without them? Speculate on the clinical consequences of a total lack of class I MHC molecules.
12. Describe the structure of the class II MHC molecules. How does this structure relate to the function of these molecules?
13. Why do we have both class I and class II MHC molecules? Why not have just one type?
14. What is the function of the class III MHC molecules? Is this a legitimate class of proteins? Speculate as to why these genes are closely associated with those of the class I and II molecules.
15. How might you identify the class II MHC molecules on lymphocytes of a prospective organ donor?
16. A protein called interferon enhances the amount of MHC molecules on the surface of cells under immunological attack (e.g., the cells of an organ graft). What might be the benefits of this?
17. Outline a breeding program for any domestic animal species that is designed to eventually increase its resistance to infectious disease.

Answers: 1d, 2d, 3b, 4e, 5b, 6d, 7c, 8d, 9b, 10c

SOURCES OF ADDITIONAL INFORMATION

Amigorena., et al. Transient accumulation of new class II MHC molecules in a novel endocytic compartment in B lymphocytes. Nature, 369:113–119, 1994.

Benacerraf, B. The role of MHC gene products in immune regulation. Science, 212:1229–1238, 1981.

Bjorkman, P.J., et al. Structure of the human class I histocompatibility antigen, HLA-A2. Nature, 329:506–512, 1987.

Bodmer, J.G., Marsh, S.G.E., and Albert, E. Nomenclature for factors of the HLA system, 1989. Immunol. Today, 11:3–10, 1990.

Brown, M.G., Driscoll, J., and Monaco, J.J. MHC-linked low-molecular mass polypeptide subunits define distinct subsets of proteasomes. J. Immunol., 151:1193–1204, 1993.

Buus, S., et al. The relation between major histocompatibility complex (MHC) restriction and the capacity of Ia to bind immunogenic peptides. Science, 235:1353–1358, 1987.

Collins, D.S., Unanue, E.R., and Harding, C.V. Reduction of disulfide bonds within lysosomes is a key step in antigen processing. J. Immunol., 147:4054–4059, 1991.

Daar, A.S., et al. The detailed distribution of MHC class II antigens in normal human organs. Transplantation, 38:293–298, 1984.

Fremont, D.H., et al. Crystal structures of two viral peptides in complex with murine MHC class I H-2K^b. Science, 257: 919–927, 1992.

Geraghty, D.E, et al. The HLA class I gene family includes at least six genes and twelve pseudogenes and gene fragments. J. Immunol., 149:1934–1946, 1992.

Goldberg, A.L., and Rock, K.L. Proteolysis, proteasomes and antigen presentation. Nature, 357:375–379, 1992.

Grusby, M.J., et al. Mice lacking major histocompatibility complex class I and class II molecules. Proc. Natl. Acad. Sci. U S A, 90:3913–3917, 1993.

Hedrick, S.M. Dawn of the hunt for nonclassical MHC function. Cell, 70:177–180, 1992.

Hood, L., Steinmetz, M., and Goodenow, R. Genes of the major histocompatibility complex. Cell, 28:685–687, 1982.

Kurlander, R.J., et al. Specialized role for a murine class-Ib MHC molecule in prokaryotic host defenses. Science, 357: 678–679, 1992.

Lee, J., and Trowsdale, J. Molecular biology of the major histocompatibility complex. Nature, 304:214–215, 1983.

Lew, A.M., et al. Class I genes and molecules: An update. Immunology, 57:3–18, 1986.

Loss, G.E., and Sant, A.J. Invariant chain retains MHC class II molecules in the endocytic pathway. J. Immunol., 150: 3187–3197, 1993.

Madden, D.R., et al. The three dimensional structure of HLA-B27 at 2.1 Å resolution suggests a general mechanism for tight peptide binding to MHC. Cell, 70:1035–1048, 1992.

McDevitt, H.O. The HLA system and its relation to disease. Hosp. Pract. [Off.], 20:57–72, 1985.

Monaco, J.J. A molecular model of MHC class I–restricted antigen processing. Immunol. Today, 13:173–179, 1992.

Mouritsen, S., et al. MHC molecules protect T cell epitopes against proteolytic destruction. J. Immunol., 149:1987–1993, 1992.

Neefjes, J.J., and Ploegh, H.L. Intracellular transport of MHC class II molecules. Immunol. Today, 13:179–183, 1992.

Nepom, G.T. The effects of variations in human immune-response genes. N. Engl. J. Med., 321:752–753, 1989.

O'Brian, S.J., et al. Genetic basis for species vulnerability in the cheetah. Science, 227:1428–1434, 1985.

Pamer, E.G., Bevan, M.J., and Lindahl, K.F. Do nonclassical, class Ib MHC molecules present bacterial antigens to T cells? Trends Microbiol., 1:35–38, 1994.

Pullen, J.K., et al. Structural diversity of the classical H-2 genes: K, D, and L. J. Immunol., 148:958–967, 1992.

Robertson, M. Proteasomes in the pathway. Nature 253:300–301, 1991.

Rosa, F., and Fellous, M. The effect of gamma-interferon on MHC antigens. Immunol. Today, 5:261–262, 1984.

Spies, T., et al. A gene in the human major histocompatibility complex class II region controlling the class I antigen presentation pathway. Nature, 348:744–747, 1990.

Steinmetz, M. The major histocompatibility complex: Organization and evolution. Clin. Immunol. News, 7:134–137, 1986.

Stern, L.J., et al. Crystal structure of the human class II MHC protein HLA-DR1 complexed with an influenza virus peptide. Nature 368:215–221, 1994.

Stroynowski, I. Molecules related to class I major histocompatibility antigens. Annu. Rev. Immunol., 8:501–530, 1990.

Wurst, W., Benesch, C., and Drabent, B. Localization of heat shock protein 70 genes inside the rat major histocompatibility complex close to class III genes. Immunogenetics, 30: 46–49, 1989.

CHAPTER 9

The Lymphoid Organs

Jacques Miller *Despite its large size, the function of the thymus was unknown until relatively recently. Dr. Jacques Miller working in Australia was the first to demonstrate that the thymus played a significant role in the development of the immune response. In 1961 he published his studies on neonatal thymectomy in mice. These animals showed a loss of lymphocytes from the bloodstream and were unable to reject skin grafts from allogeneic mice indicating a loss of cell-mediated immune responses. Thus with a series of simple experiments he established, for the first time, the role of the thymus and laid the foundations of modern cellular immunology.*

CHAPTER OUTLINE

CHAPTER CONCEPTS

1. The organs in which lymphocytes develop are called primary lymphoid organs. These include the thymus, the bursa of Fabricius, and some Peyer's patches in the intestine.
2. Lymphocytes that undergo development in the thymus are called T cells and are responsible for the cell-mediated immune responses. Lymphocytes that undergo development in the bursa of Fabricius or equivalent tissue are called B cells and are responsible for the antibody-mediated responses.
3. The organs in which mature lymphocytes reside and in which they encounter antigen are called secondary lymphoid organs. Secondary lymphoid organs include the lymph nodes, spleen, tonsils, and lymphoid tissues in the intestine, the lungs, and other body surfaces.
4. Removal of the thymus from a newborn animal impairs the cell-mediated immune responses. Removal of the bursa or some Peyer's patches impairs antibody formation.
5. T lymphocytes circulate between the blood, tissues, and secondary lymphoid organs.
6. The structure of the lymphoid organs provides an optimal environment for cell interaction and the development of immune responses.

In this chapter we examine the structure of the major lymphoid organs of the body. In these organs the cellular interactions that lead to the immune response occur. We first look at the bone marrow, the source of all lymphoid cells. Then we look at the structure and function of the thymus and the bursa of Fabricius, where lymphocytes develop and are selected so that they will not attack normal body components. We then study the structure and function of lymph nodes and the spleen, where much of the body's defense occurs. This includes examination of the way in which lymphocytes circulate. The chapter concludes with a brief look at other lymphoid organs.

Although antigen is trapped and processed by macrophages and dendritic cells, the mounting of an immune response is a function of **lymphocytes.** Lymphocytes are the small round cells that are the predominant cell type in organs such as the spleen, lymph nodes, and thymus. Lymphocytes are extraordinarily complex cells. Their major function is to mount immune responses following exposure to appropriately presented antigen. These responses occur within lymphoid organs, which must therefore provide an environment in which efficient interaction between lymphocytes, antigen-presenting cells, and antigen may occur (Fig. 9–1).

Lymphoid organs may be classified on the basis of their role in generating lymphocytes, in regulating the production of lymphocytes, and in providing an environment for the optimal interaction between antigen and antigen-sensitive cells (Fig. 9–2).

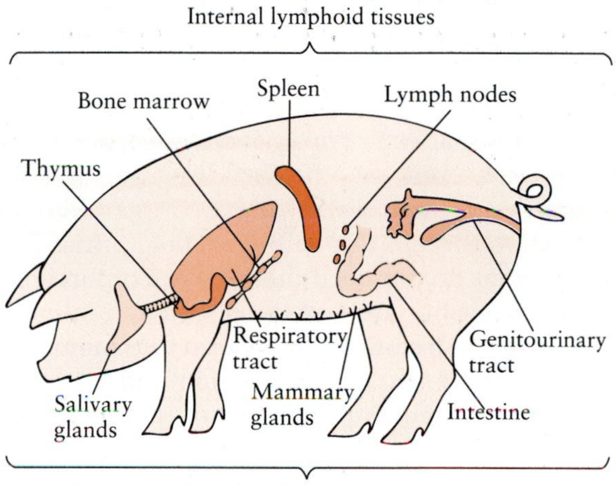

Figure 9–1 The lymphoid tissues of a mammal. In general, the differences in structure between different species of mammal are functionally insignificant.

SOURCES OF LYMPHOID CELLS

Bone Marrow

All the cells that participate in the immune responses originate from a population of **stem cells** that both reproduce themselves and give rise to the specialized cell lines, neutrophils, macrophages, and lymphocytes. Animals or humans that have received a high dose of radiation over their entire body will die as a

Figure 9–2 The different roles of the lymphoid tissues in the development and function of the immune system.

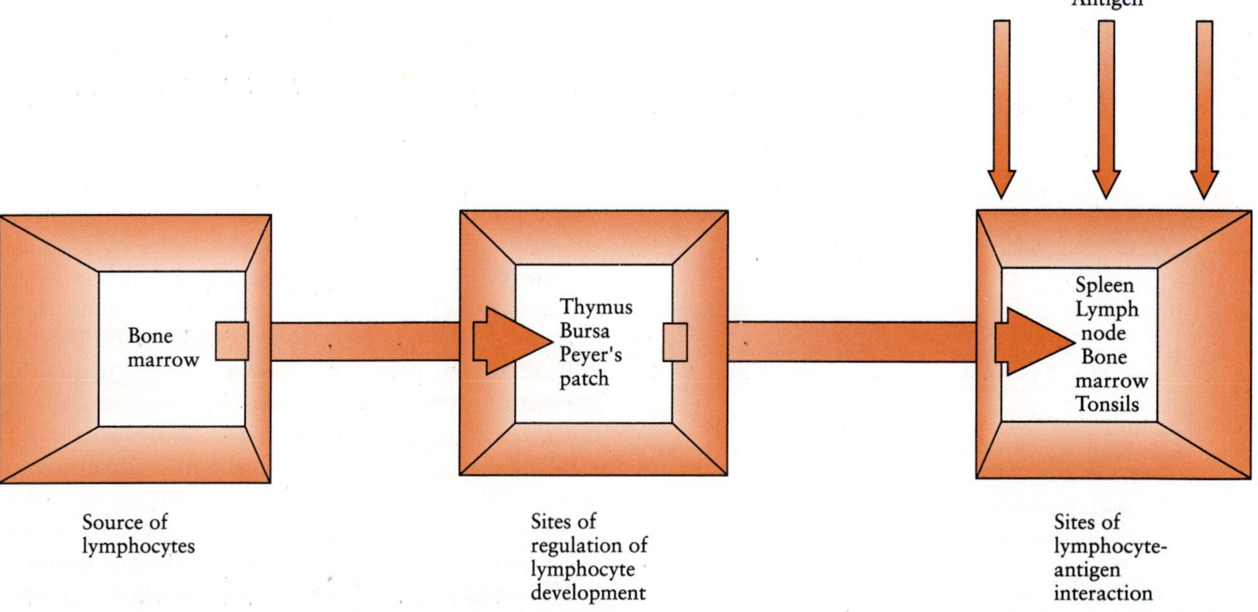

result of a failure to replenish their blood cells. The radiation kills their stem cells, and no new lymphocytes or other leukocytes can therefore be made. The condition can be treated by a bone marrow transplant. Indeed, only grafted bone marrow cells can prevent the death of lethally irradiated animals. A single stem cell of the correct type is theoretically capable of generating all the cells necessary to restore the functions of the other lymphoid organs.

Not only is the bone marrow the source of all blood cells, including lymphocytes, but like the spleen, liver, and lymph nodes, it contains many macrophages that can remove foreign particles from the blood. The bone marrow is also a major source of antibodies. It consists of a meshwork of reticular tissue divided into two distinct compartments, a **hematopoietic** compartment and a vascular compartment (Fig. 9–3). The hematopoietic areas contain precursors of all the blood cells, including macrophages and lymphocytes, and are enclosed by a layer of reticular cells. The lymphocytes are found either scattered among the other hematopoietic cells or grouped in discrete clusters. Macrophages are commonly associated with these lymphocyte clusters. In older individuals, the reticular cells may become so loaded with fat that the hematopoietic tissue is masked and the marrow has a fatty, yellow appearance. The vascular compartment of the bone marrow consists of blood sinuses lined by endothelial cells and crossed by reticular cells and macrophages. It is here that bacteria are trapped.

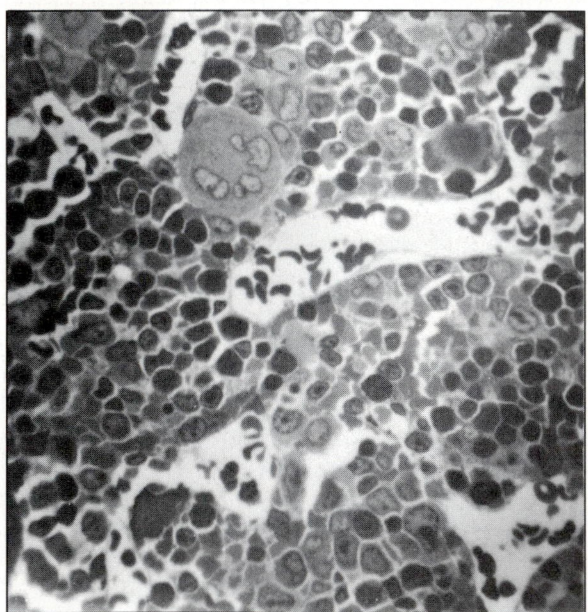

Figure 9–3 An Epon thin section of bone marrow. Note the large vein and the large multinucleated cell, which is a megakaryocyte. (Megakaryocytes give rise to platelets.) The other cells are myeloid, lymphoid, and erythroid precursors. *(Reprinted by permission of the publisher from J.J. Oppenheim, D.L. Rosenstreich, and M. Potter, eds. Cellular Functions in Immunity and Inflammation. © 1981, Elsevier Publishing Co., Inc.)*

PRIMARY LYMPHOID ORGANS

Primary lymphoid organs regulate the production and differentiation of lymphocytes. There are three of these organs—the thymus and Peyer's patches, found in both mammals and birds, and the bursa of Fabricius, found only in birds. These organs all arise early in fetal life either from outgrowths at the ectoendodermal junctions or from endoderm. For example, the thymus arises from the endoderm of the third and fourth pharyngeal pouches, the bursa develops from the cloacodermal junction, and Peyer's patches arise in the intestinal wall. Newly formed lymphoid stem cells from the yolk sac, the fetal liver, and the bone marrow first migrate to the primary lymphoid organs via the bloodstream (Table 9–1). These primary lymphoid organs do not need the presence of foreign antigens to grow but are well developed and fully functional in germ-free animals. (Germ-free animals are born aseptically by cesarean section and raised in a germ-free environment. They are thus subject to very little antigenic stimulation. The only foreign antigens that enter their body are derived from their sterilized feed.)

Thymus

The thymus is a pale, lobulated organ located in the thoracic cavity between the lungs (Fig. 9–4). The size of the thymus can vary considerably. Its relative size (relative to total body weight) is greatest in the newborn, and its absolute size is greatest at puberty. After puberty, the thymus atrophies, and its cortex is gradually replaced by fatty tissue, but remnants of the thymus do persist until old age. In addition to age-related shrinkage, the thymus shrinks rapidly in response to stress, so that in individuals dying after prolonged illness it may be abnormally small.

Structure. The thymus consists of lobules of loosely packed epithelial cells covered by a connective tissue capsule (Fig. 9–5). The outer part of each lobule, the **cortex,** is densely infiltrated with rapidly dividing lymphocytes called **thymocytes.** In the inner part, the **medulla,** there are fewer thymocytes and the epithelial cells are clearly visible. Within the medulla are round bodies called thymic (Hassall's) corpuscles, whose function is not known. The blood supply to the thymus is derived from arteries that enter through connective tissue septa and run as arterioles along the cortico-

Table 9–1 Comparison of Primary and Secondary Lymphoid Organs

	Primary Lymphoid Organ	Secondary Lymphoid Organ
Origin	Ectoendodermal junction or endoderm	Mesoderm
Time of development	Early in embryonic life	Late in fetal life
Persistence	Involutes after puberty	Persists in adults
Effect of removal	Loss of lymphocytes Loss of immune responses	No or minor effects
Response to antigen	Unresponsive	Fully reactive
Examples	Thymus, bursa, some Peyer's patches	Spleen, lymph nodes

medullary junction. The capillaries that arise from these arterioles enter the cortex and loop back to the medulla. These cortical capillaries have an abnormally thick basement membrane and a continuous outer layer of epithelial cells. Their walls thus form a barrier that may prevent circulating antigens from entering the thymic cortex.

Function. The function of the thymus may be demonstrated by removing it and studying the consequences. **Thymectomy** in adults has no immediate obvious effect, but if the thymus is removed from newborn rodents, its consequences are immediately apparent (Table 9–2).

Neonatal Thymectomy. Circulating lymphocytes drop to very low levels in neonatally thymectomized

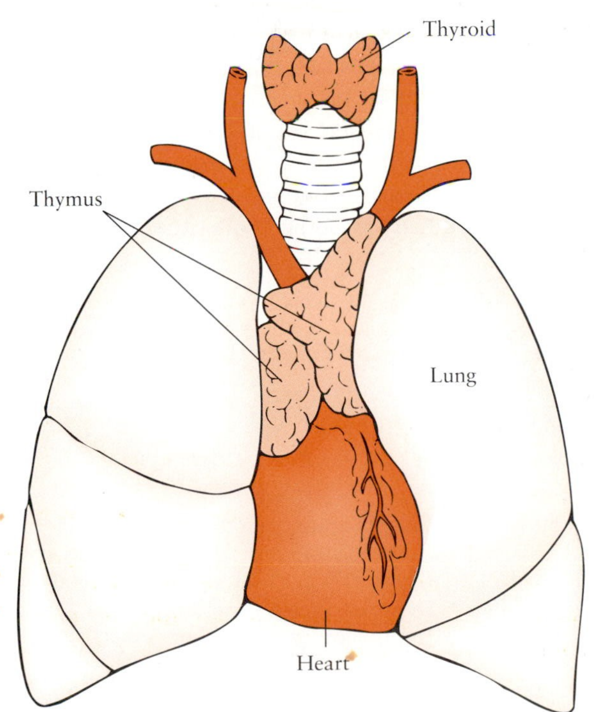

Figure 9–4 The location of the thymus in the chest of a child.

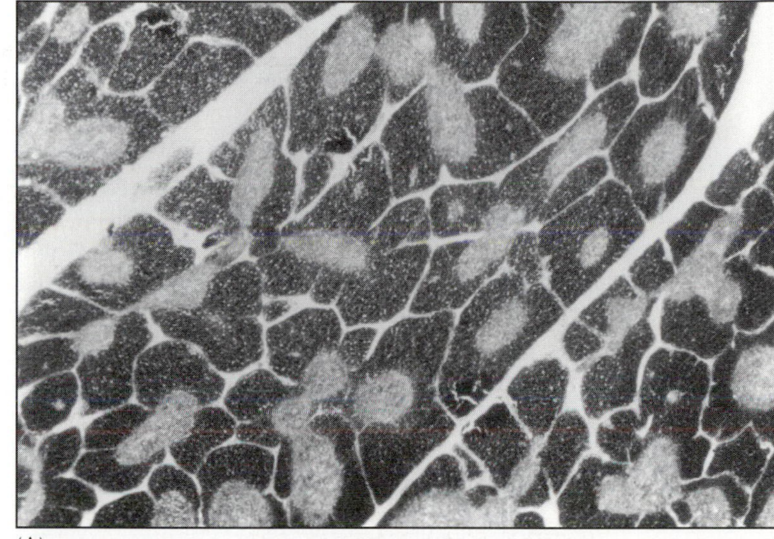

(A)

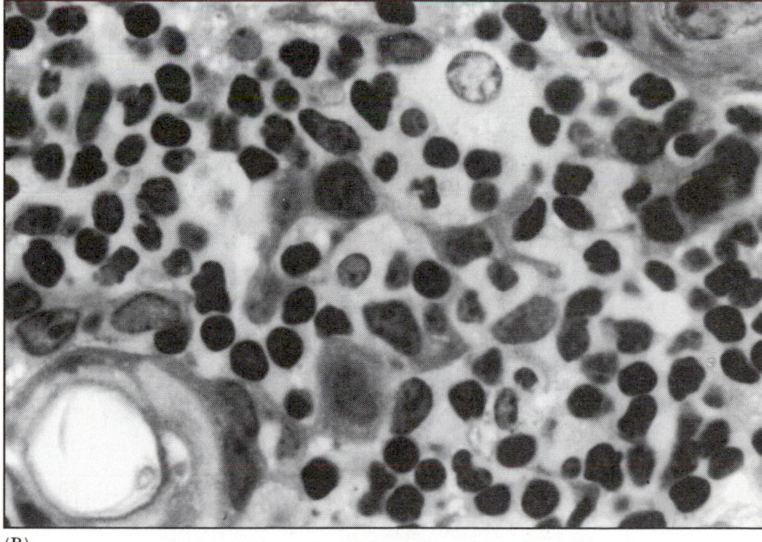

(B)

Figure 9–5 *A*, A section of monkey thymus. Each lobule is divided into a cortex rich in lymphocytes and hence darkly staining and a paler medulla, consisting mainly of epithelial cells (×10). *B*, A high-power view of the medulla of a monkey thymus, showing several pale-staining epithelial cells with cytoplasmic processes surrounded by many dark-staining thymocytes (×1000).

Table 9–2 The Effects of Neonatal Thymectomy and Bursectomy

Function	Thymectomy	Bursectomy
Numbers of circulating lymphocytes	↓↓↓	—
Presence of lymphocytes in T dependent areas	↓↓↓	—
Graft rejection	↓↓↓	—
Presence of lymphocytes in T independent areas	↓	↓↓↓
Plasma cells in lymphoid tissues	↓	↓↓↓
Serum immunoglobulins	↓	↓↓↓
Antibody formation	↓	↓↓↓

mice. As a result, the mice become unusually susceptible to infections (Fig. 9–6). These animals cannot reject foreign organ grafts because they are unable to mount cell-mediated immune responses. The effects of neonatal thymectomy on antibody production are variable. Thus the antibody response to antigens such as bovine serum albumin or sheep red cells is significantly inhibited. On the other hand, the response to the carbohydrate from the capsule of *Streptococcus pneumoniae* (pneumococcal polysaccharide) is unaffected. For this reason, antigens such as pneumococcal polysaccharide are classified as **thymus-independent** antigens to distinguish them from the **thymus-dependent** antigens such as sheep red cells and bovine serum albumin. Thymus-dependent antigens induce an immune response only in the presence of cells that mature within the thymus.

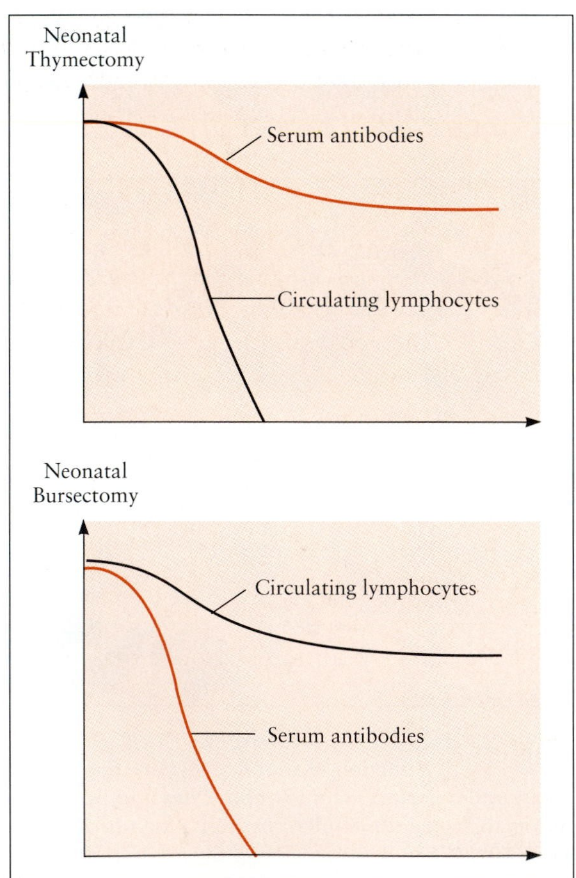

Figure 9–6 The effects of neonatal thymectomy and bursectomy on antibody- and cell-mediated immune responses.

Adult Thymectomy. In contrast to neonatal thymectomy, surgical removal of the thymus from adult mice produces no immediately obvious results. If, however, thymectomized animals are observed for several months, their circulating lymphocytes decline progressively and their ability to mount cell-mediated immune responses is gradually reduced. Thus the adult thymus must remain functional, but there is a reservoir of thymus-derived lymphocytes that must be exhausted before the immunosuppressive effects of adult thymectomy become apparent (Table 9–3).

From studies on the effects of thymectomy we now know that lymphocytes leaving the bone marrow are attracted to the neonatal thymus by hormones secreted by thymic epithelial cells. The immature lymphocytes (called thymocytes) colonize the cortex and divide rapidly. Most of the new cells produced die within the cortex, but the survivors (about 3% of the total in rodents) migrate to the medulla, mature, and eventually leave the thymus to colonize the **secondary lymphoid organs.** These thymus-derived lymphocytes are called **T lymphocytes,** or **T cells.**

The T cells released by the thymus into the circulation must participate in immune responses by responding to foreign antigens, yet they must not respond to normal body constituents. This is

Table 9–3 Loss of T-Cell Function in Mice Thymectomized as Adults

Time after Thymectomy	Effect
One hour	Disappearance of circulating thymic hormones
One week	Disappearance of T cells from blood
One month	Depressed responses to mitogens Depression of suppressor-cell activity Depletion of T cells in spleen
One year	Depression of antibody formation to T-dependent antigens Delay in skin allograft rejection

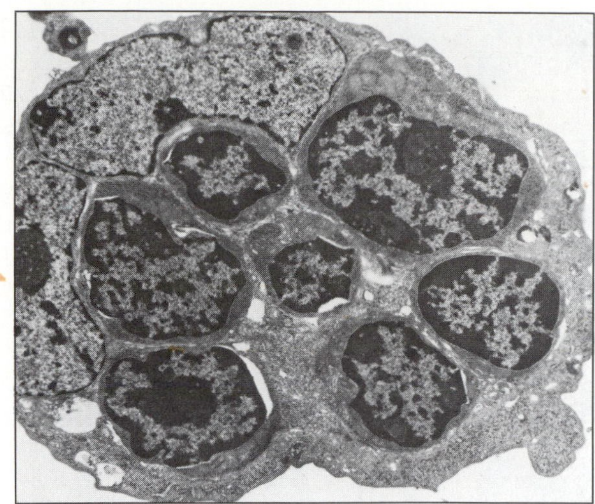

Figure 9–7 A thymic nurse cell. Within the cytoplasm can be seen seven lymphocytes. It is believed that, in contrast to their name, these cells are actually sites of thymocyte death (apoptosis). *(Courtesy of Dr. H. Wekerle.)*

accomplished by selection of the thymocytes within the thymus. The selection process, which is described fully in Chapter 21, is summarized here. The thymic epithelial cells function as antigen-presenting cells to the thymocytes. First, only those thymocytes that have functioning receptors that can bind self-MHC molecules are selected to proliferate, a form of **positive selection.** Then thymic epithelial cells present "self"-antigens (in other words, normal body constituents) in association with MHC antigens. If the thymocyte responds to a self-antigen, it is destroyed—a process of **negative selection.** The other thymocytes—those that presumably cannot respond to self-antigens but can respond to foreign antigens—survive. This selection process may involve the destruction of thymocytes by apoptosis within nurse cells (Fig. 9–7). Nurse cells are so large that they can engulf up to 50 lymphocytes at one time.

Thymic Hormones. If mouse thymic tissue is placed in a chamber whose walls are permeable to proteins but not to cells and this chamber is implanted in a neonatally thymectomized mouse, then the immune function of the mouse is partially restored. The thymus must therefore be able to secrete hormones and function as an endocrine gland. Thymic epithelial cells release several hormones, which are all polypeptides and, at present, do not seem to be related. Unlike the cytokines described in Chapter 12, the thymic hormones are not well defined and their immunological significance remains obscure. They are all peptides ranging from 1 to 15 kDa and include thymosins, thymopoietins, thymic humoral factors, thymulin, and thymostimulins. Thymic epithelial cells also produce interleukin-1, a protein discussed at length in Chapters 6 and 28.

Skin

A subpopulation of T cells resides in the skin. These cells interact with keratinocytes and may undergo some maturation within the epidermis, suggesting that the skin may have thymus-like activity. However, the skin also functions as an effective antigen-trapping barrier and skin T cells may serve a local defense function.

Bursa of Fabricius

Structure. The bursa of Fabricius is an organ found in birds but not in mammals. It is a hollow sac about 1 cm in diameter located just above the cloaca and connected to it by a duct (Fig. 9–8). The bursa reaches

Figure 9–8 The bursa of Fabricius is a round structure located just above the cloaca of a chick. Its lumen connects directly with the cloaca. This bursa has been opened to show the folds inside.

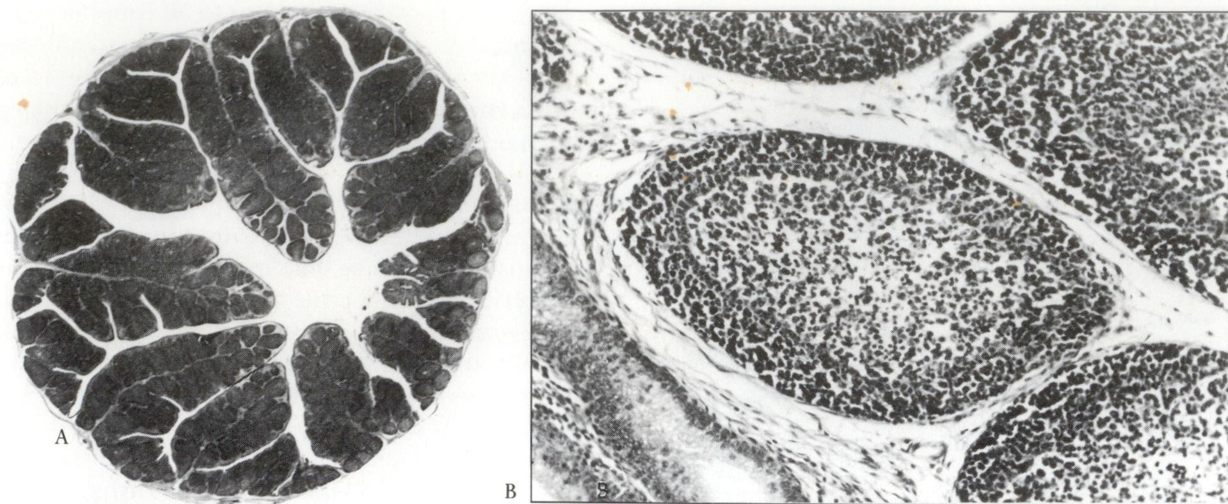

Figure 9–9 The structure of the bursa of Fabricius. *A,* Low-power micrograph showing the bursa of the 14-day-old chick (×5). *B,* A single follicle (×360). *(Courtesy of Dr. S. Yamashiro.)*

its greatest size about one to two weeks after the chick has hatched, and then it gradually atrophies. Inside the bursa, folds of epithelium extend into the lumen, and scattered through these folds are follicles of lymphoid cells (Fig. 9–9). Each follicle is divided into a cortex and medulla. The cortex contains lymphocytes, **plasma cells** (mature, antibody-producing lymphocytes), and macrophages. At the boundary between these regions (corticomedullary junction) there is a basement membrane and capillary network, on the inside of which are epithelial cells. These medullary epithelial cells are replaced by lymphocytes toward the center of the medulla, so that the center of the follicle appears to consist solely of lymphocytes. Each follicle is directly connected with the epithelium covering the surface of the fold.

Function. The bursa may be removed surgically (**bursectomy**). Bursectomized birds show a slight drop in their circulating lymphocytes, but they stop producing antibody and lose their antibody-producing plasma cells. Since bursectomized birds can still reject foreign skin grafts, however, this operation appears to have little effect on the cell-mediated immune responses.

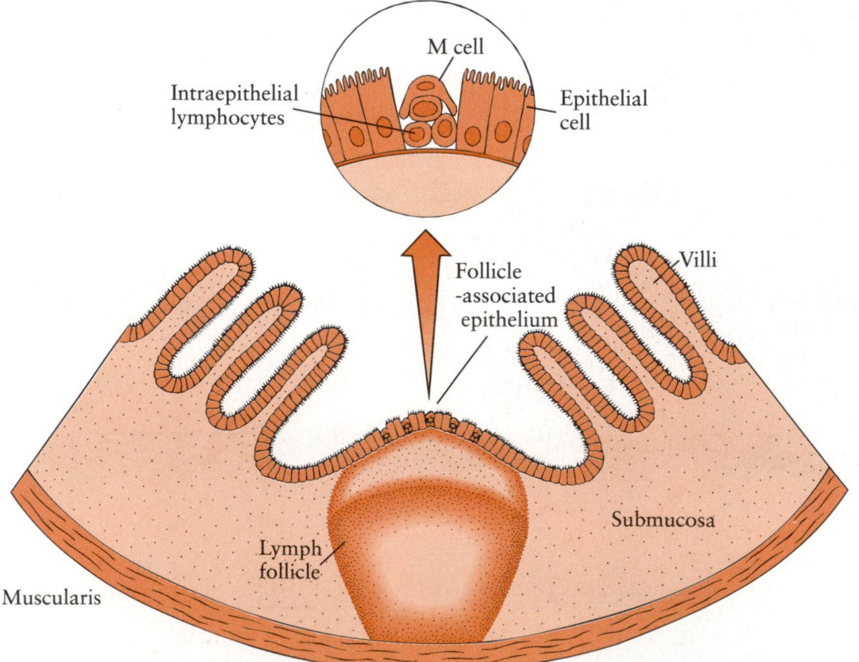

Figure 9–10 The major structural features of a Peyer's patch. The lymph follicle is covered by a specialized epithelium containing intraepithelial lymphocytes and antigen-binding M cells.

Bursectomized birds are more susceptible than normal to bacterial infections. These results have been interpreted to suggest that the bursa is a primary lymphoid organ, whose function is to serve as a maturation and differentiation site for the lymphocytes of the antibody-forming system. These cells are called **B lymphocytes,** or **B cells.** Several hormones have been extracted from the bursa. The most important of these is bursin, a tripeptide (lys-his-glycylamide). Bursin activates B cells but not T cells.

Peyer's Patches

Peyer's patches (Fig. 9–10) are areas of lymphoid tissue located in the wall of the intestine, and in some species, such as sheep, these Peyer's patches have a similar function to the bursa.

Two distinct types of Peyer's patch have been described in the sheep intestine. One type is found in the jejunum, and the other is found in the ileum and cecum (Fig. 9–11). The two types of Peyer's patch have

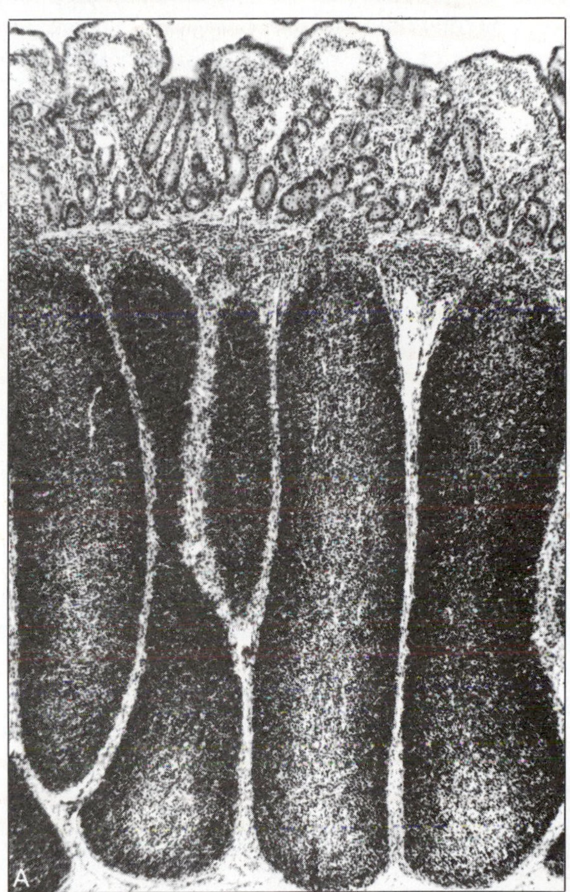

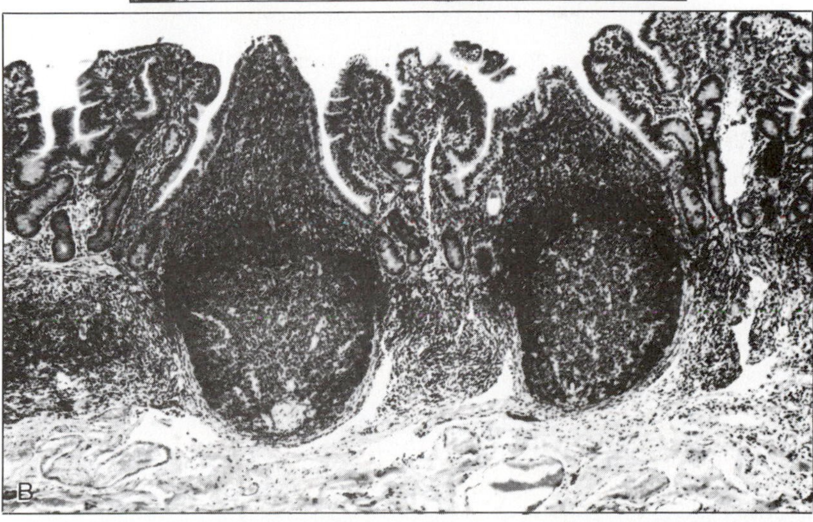

Figure 9–11 The structure of the two types of Peyer's patch in sheep. *A,* An ileocecal Peyer's patch at 8 weeks of age. *B,* A Peyer's patch from the jejunum, also at 8 weeks (×32). *(From Reynolds, J.D., and Morris, B. Eur. J. Immunol., 13:631, 1983. With permission.)*

very different structures. Ileocecal Peyer's patches consist of densely packed lymphoid follicles, each separated by a connective tissue sheath, and contain only B cells. Jejunal Peyer's patches, however, have pear-shaped follicles separated by extensive interfollicular tissue and up to 30% of their lymphocytes are T cells. Jejunal Peyer's patches persist throughout the animal's lifetime and function as secondary lymphoid organs. Ileocecal Peyer's patches, in contrast, reach their maximum size and maturity before birth, at a time when they are shielded from external antigens. They disappear by about 15 months of age and cannot be detected in adult sheep. The ileocecal Peyer's patches collectively form the largest lymphoid tissue in six-week-old lambs, constituting about 1% of total body weight (about the same size as the thymus). It has been calculated that the ileocecal Peyer's patch can produce about 3.6×10^9 lymphocytes per hour, but most of these are destroyed and only about 0.2×10^9 lymphocytes per hour are released into the circulation. Surgical removal of the ileocecal Peyer's patch leads to a loss of circulating B cells and a failure of antibody production. The bone marrow of lambs contains many fewer lymphocytes than the bone marrow of laboratory rodents, so the ileocecal Peyer's patches are probably a significant source of B cells in this species. The ileocecal Peyer's patches are therefore primary lymphoid organs (at least in lambs) and serve a similar function to the bursa of Fabricius. Indeed, the bursa may be no more than a specialized Peyer's patch.

SECONDARY LYMPHOID ORGANS

In contrast to the thymus, ileocecal Peyer's patches, and bursa, the **secondary lymphoid organs** arise from mesoderm late in fetal life and persist through adult life. They grow in response to antigenic stimulation and thus are poorly developed in germ-free animals. Removal of individual secondary lymphoid organs does not significantly affect an animal's ability to respond to antigen. The secondary lymphoid organs include the spleen, lymph nodes, bone marrow, tonsils, and jejunal Peyer's patches as well as isolated lymphoid nodules in the gastrointestinal, respiratory, and urogenital tracts. The secondary lymphoid organs are rich in macrophages and dendritic cells that trap and process antigens and in T and B lymphocytes, which mediate the immune responses. The anatomical structure of these organs is designed to facilitate antigen trapping and to maximize opportunities for processed antigen to be presented to antigen-sensitive cells.

Lymph Nodes

Structure. Lymph nodes are round or bean-shaped structures placed on lymphatic vessels in such a way that they can filter out any foreign material carried in the lymph (Figs. 9–12 and 9–13). Lymph nodes consist of a fibrous reticular network filled with lymphocytes, macrophages, and dendritic cells. Lymphatic sinuses penetrate the node. A subcapsular sinus is located immediately under the connective tissue capsule of the node; other sinuses pass through the node but are most prominent in the medulla. Afferent lymphatic vessels (those flowing into the node) enter the node around its circumference, and efferent lymphatics (those flowing out of the node) leave from a depression (or hilus) on one side. The blood vessels supplying a lymph node also enter and leave via the hilus.

The interior of a lymph node is divided into a **cortex** on the outside, a central **medulla,** and an ill-defined area between these two regions called the **par-**

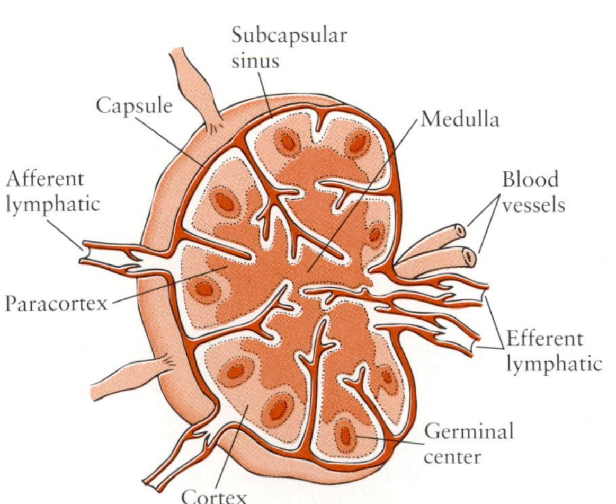

Figure 9–12 The major structural features of a typical lymph node.

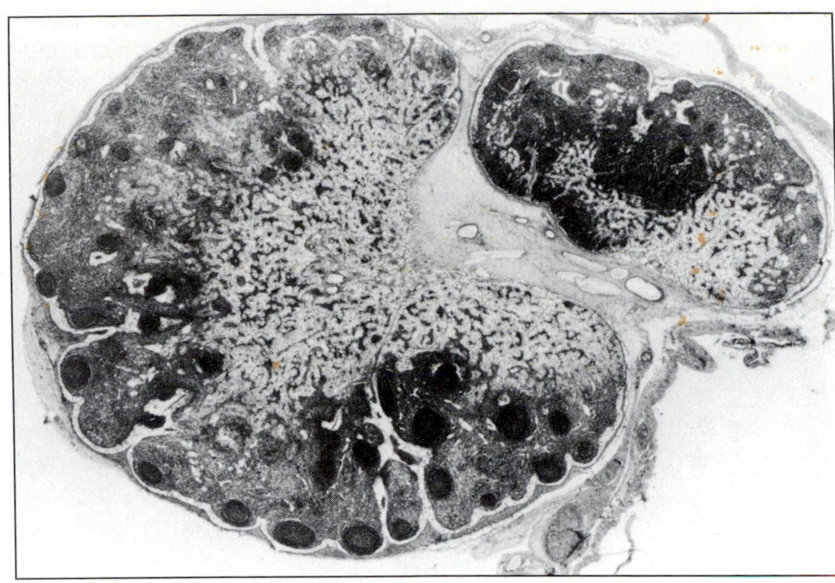

Figure 9–13 A histological section of a lymph node showing the features labeled in Figure 6–13 (×12). *(Courtesy of Dr. W.E. Haensly.)*

acortex. The cells in the cortex are predominantly B cells and are arranged in nodules (Fig. 9–14). Before exposure to antigen, these nodules are termed primary follicles. In lymph nodes that have been stimulated by antigen, the cells within primary follicles expand to form structures known as **germinal centers;** a follicle containing a germinal center is known as a secondary follicle. Germinal centers are sites where B cells undergo a process called somatic mutation. In this process the cells' ability to bind antigen changes at random. Those cells that increase their ability to respond to an antigen leave the germinal center to colonize other secondary lymphoid organs. Those cells whose ability to respond to antigen is reduced undergo **apoptosis** and are removed by macrophages. Germinal centers are also the site of immunoglobulin **class switching** and memory cell formation (Chapter 14). Three cell populations are recognized in germinal centers. The cells closest to the paracortex have small dark-staining nuclei. At the other end, the cells closest to the cortex have large pale nuclei (Fig. 9–15). A follicular mantle of small lymphocytes forms a

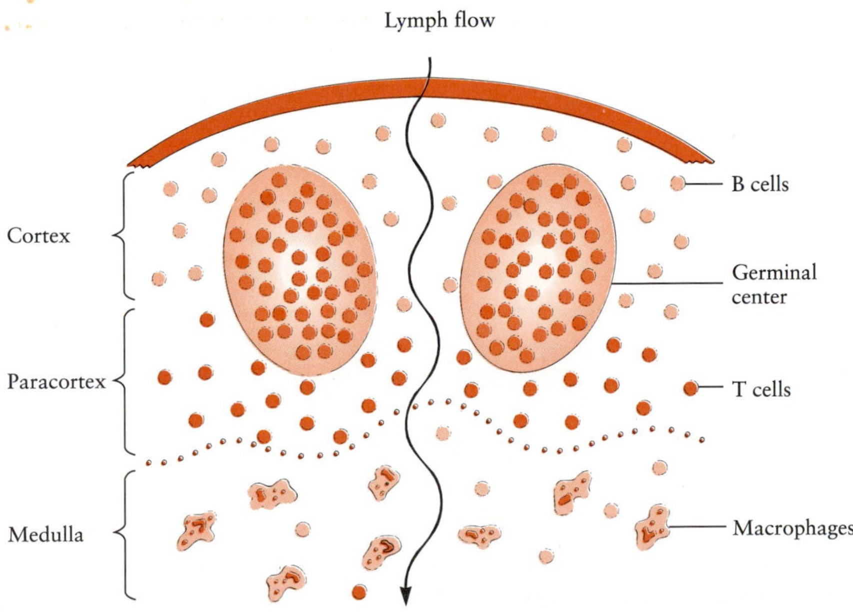

Figure 9–14 The distribution of B cells, T cells, and macrophages within a lymph node.

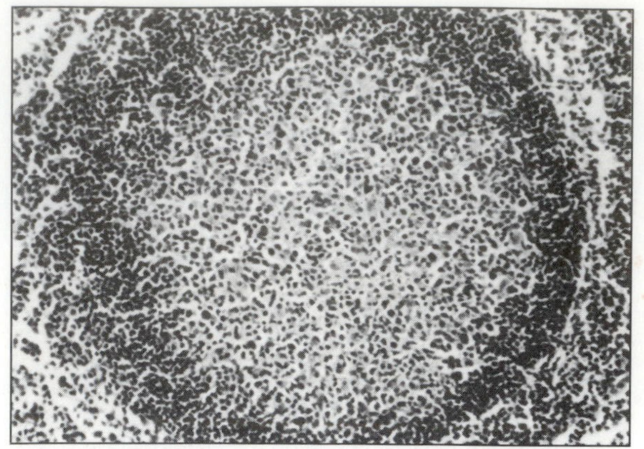

Figure 9–15 A germinal center in the cortex of a cat lymph node (×120). *(Courtesy of Dr. W.E. Haensly.)*

cap over the outer pole of the germinal center. A few T cells are found in the cortex, in a region immediately surrounding each germinal center.

T cells are mainly located in the paracortex. In neonatally thymectomized or congenitally athymic animals, lymphocytes disappear from the paracortex. The cells found in the medulla include B cells, macrophages, reticulum cells, and plasma cells. They are arranged in cellular cords between the lymphatic sinuses.

Lymphocyte Circulation. The thoracic duct is the major lymphatic vessel of the body. Lymphatics from the lower body join to form the duct, and it empties into the vena cava. Lymph flows through the thoracic duct at about 500 ml per hour, and it contains about 1×10^8 lymphocytes per milliliter. If a tube is inserted

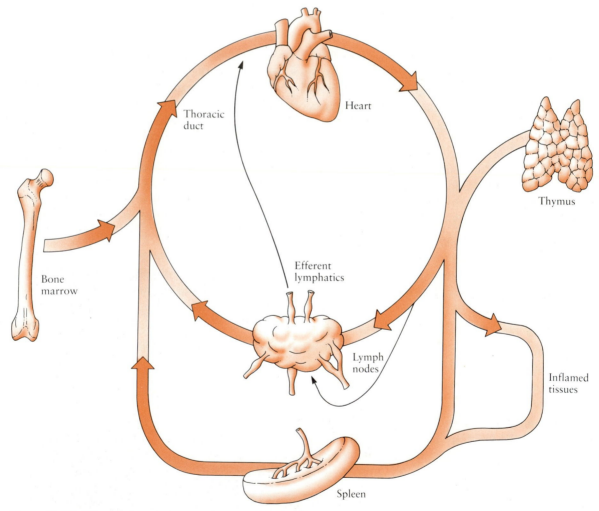

Figure 9–16 The circulation of lymphocytes. Lymphocytes found in blood may be newly formed cells passing from the bone marrow to the primary lymphoid organs or cells passing from the primary to the secondary lymphoid organs. Most of these cells, however, recirculate between the lymph nodes and the bloodstream. Lymphocytes may leave the blood to enter inflamed tissues or to pass through tissues in the afferent lymph flow.

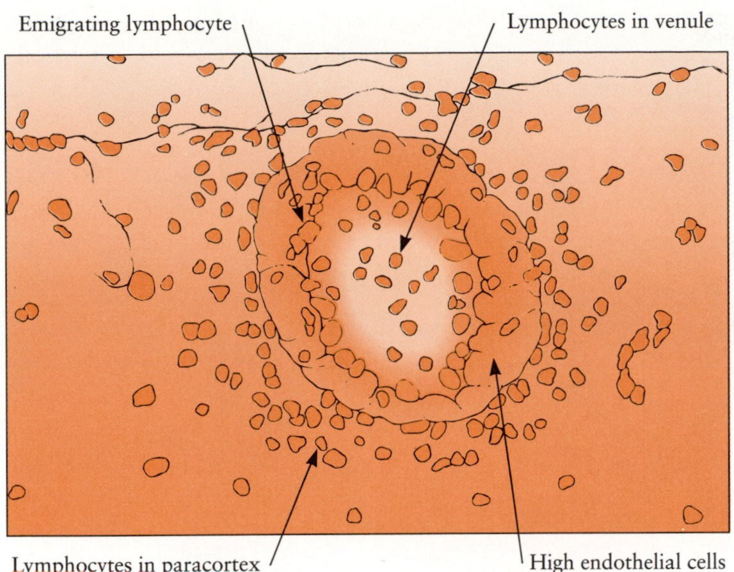

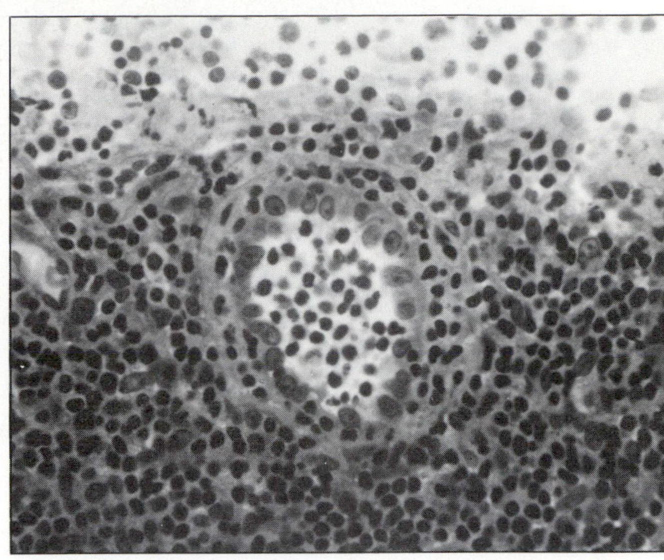

Figure 9–17 A section of a human tonsil showing a high endothelial venule (×900).

into the thoracic duct and the lymph removed from the body, the lymphocytes in the blood eventually drop to very low levels and cells disappear from the paracortices of lymph nodes. The rapid depletion of T lymphocytes by this technique indicates that the thoracic duct cells must normally recirculate back to lymph nodes by way of the blood. In fact, the lymphocytes that enter the vena cava from the thoracic duct spend an average of only 30 minutes in the blood before returning to lymphoid organs, so about 500×10^9 lymphocytes travel through the blood daily (Fig. 9–16). These cells leave the bloodstream by two routes. Virgin T cells (i.e., T cells that have not yet encountered antigen) bind to the **endothelium** of the venules in the paracortex of the lymph node. These venules are called **high endothelial venules** because they possess an extremely tall endothelium (Figs. 9–17 and 9–18). The number and length of high endothelial venules are variable and controlled by local immune activity. Thus stimulation of a lymph node by the presence of antigen results in a rapid increase of its high endothelial vessels. If a lymph node is protected from antigen, its high endothelial vessels tend to disappear. Circulating **virgin** lymphocytes adhere to high endothelial cells and then migrate into the node by passing between these cells. In contrast, memory T cells leave the bloodstream via conventional blood vessels and are then carried to lymph nodes via tissue fluid (the afferent lymph). All the migrating T cells percolate through the lymph node and then reenter the lymphatic circulation via the efferent lymph vessels. Ninety percent of lymphocytes in the efferent lymph are derived from cells entering the node through high endothelial venules, and 6% enter by way of afferent lymph. A small proportion of circulating T cells recirculate between blood and the lymphoid tissues in the lung or intestine. These cells are specifically involved in the development of immune responses at body surfaces. As a result of the continuing recirculation of cells between lymphoid organs and blood, most lymphocytes found in peripheral blood are T cells. The lymphocytes in the blood account for only 2% of the total lymphocyte population of the body, so minor alterations in the migration of lymphocytes through lymphoid organs may therefore result in major changes in the number of blood lymphocytes.

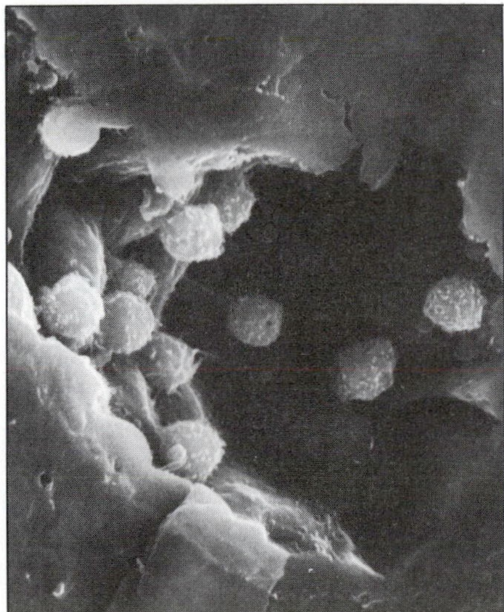

Figure 9–18 Scanning electron micrograph showing T cells within the lumen of a high endothelial venule. *(Courtesy of Becker, Ph.D./Custom Medical Stock.)*

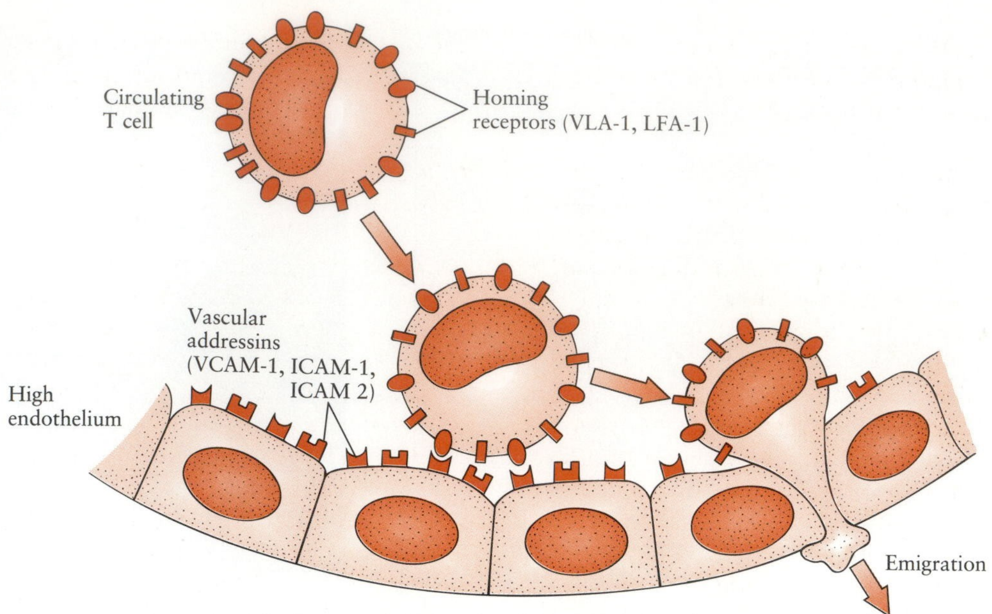

Figure 9–19 A schematic diagram showing how circulating lymphocytes bind to vascular addressins on high endothelium by means of their homing receptors.

The control of lymphocyte recirculation is an important feature of the immune system. Lymphocytes must be able to leave blood vessels for various reasons. Thus to reenter lymph nodes, lymphocytes must bind to high endothelial cells (Fig. 9–19). To enter tissues where microbial invasion is occurring, lymphocytes must bind to the endothelial cells of inflamed blood vessels. This binding between lymphocytes and vascular endothelial cells is mediated by lymphocyte cell surface proteins called **homing receptors** that bind to proteins on endothelial cells called **addressins.** The homing receptor–addressin system ensures that specific lymphocytes leave the blood vessels in appropriate locations. The addressins are heavily glycosylated proteins. Distinct forms of addressin are found on vascular endothelial cells in different lymphoid tissues such as lymph nodes, Peyer's patches, spleen, or lung-associated lymphoid tissues as well as on vascular endothelium in inflamed tissue (Fig. 9–20). For example, GlyCAM-1 (CD34) is an addressin found on high endothelial venules in lymph nodes, MAdCAM-1 is an addressin found only in mucosal lymphoid tissues such as Peyer's patches, and sialomucin is found on many different endothelial cells. The ligands of these addressins, the homing receptors, are all adherence proteins of the selectin family (Chapter 7). Thus GlyCAM-1, sialomucin, and MAdCAM-1 all bind L-selectin (CD62L).

The binding of the lymphocyte through its homing receptors to the addressins causes the cells to bind weakly. This effectively slows down the fast-moving lymphocyte. Once it has slowed down, the lymphocyte is then firmly bound to the endothelial cells through a second set of adhesion molecules, the integrins (Chapter 7).

Response of Lymph Nodes to Antigen. Antigen deposited in tissues is carried by the flow of tissue fluid to local lymph nodes. Its fate within these nodes depends on whether the animal has been previously exposed to the antigen. Lymph nodes contain two separate antigen-trapping systems. One system uses the macrophages in the lymph node medulla. The other system involves dendritic cells found in secondary fol-

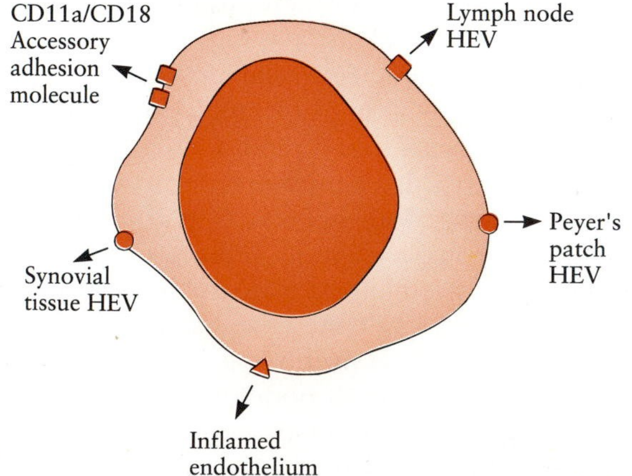

Figure 9–20 A schematic diagram showing the major homing receptors known to be present on lymphocytes. In practice, a single cell may express only one of these receptor populations at a time.

licles in the lymph node cortex. Dendritic cells have extensive cytoplasmic processes that form a web through which antigen must pass as it flows through the cortex.

Antigen entering the body for the first time is phagocytosed by macrophages situated in the lymph node medulla. These antigen-carrying macrophages migrate to the paracortex, where they encounter helper T lymphocytes. Once stimulated by the antigen, these T cells migrate to the cortex and trigger the B cells' response to the antigen. B cells respond by division, and their antibody-producing progeny cells migrate to the medulla. Some of the antibody-producing cells are also released into the efferent lymph and are carried to other lymph nodes downstream. Several days after antibody production is first observed in the medulla, germinal centers appear in the cortex as a result of proliferation of B cells within primary follicles. Antigens that do not stimulate antibody production do not usually cause germinal center formation. The function of germinal centers is to provide a focus for the production of the memory cells that are required for a secondary response. Germinal centers also provide an environment in which B cells can undergo **somatic mutation** and thus increase the ability of the antibodies produced to bind antigen.

On second exposure to antigen, antigen molecules are trapped by adherence to antibody-coated dendritic cells in the cortex. In a secondary response, the germinal centers tend to become less obvious as the activated memory cells migrate from the cortex to the medulla and out in the efferent lymph. Once this stage is completed, the germinal centers redevelop. The movement of cells within the lymph node is necessary to ensure that antibody-producing cells are kept well away from antigen-sensitive cells. This segregation prevents an immediate inhibition of the immune response through the **negative feedback** exerted by antibody (Chapter 22). When responding to antigens that stimulate a cell-mediated rather than an antibody-mediated immune response, the T-cell–rich paracortex produces large pyroninophilic cells. These large cells (known as **lymphoblasts**) (see Fig. 18–6) give rise in turn to more small lymphocytes, which participate in the cell-mediated immune responses (Chapter 18).

Spleen

Just as lymph nodes filter lymph, so the spleen filters blood. The filtering process removes both antigenic particles and aged blood cells. The spleen also stores red blood cells and platelets and undertakes red blood cell formation in the fetus. This organ is also of great immunological significance since it is a major site for the production of antibodies and effector T cells. The spleen is divided into two compartments: the red pulp is for the formation and storage of red blood cells and for antigen trapping, and the white pulp is where the immune response occurs. The white pulp is discussed here.

Structure. The white pulp of the spleen consists of masses of lymphocytes closely associated with the spleen's blood vessels (Fig. 9–21). Vessels that enter the spleen travel through muscular trabeculae before entering its functional areas. Immediately on leaving the trabeculae, each arteriole is surrounded by a thick layer of lymphocytes called a periarteriolar lymphoid sheath. The arteriole eventually leaves this sheath and, after branching, opens either directly or indirectly into venous sinuses that drain into the splenic venules in the red pulp. The periarteriolar lymphoid sheath consists of T cells and is thus depleted following neonatal thymectomy. However, scattered through the sheath are primary follicles containing B cells. On antigenic stimulation, these follicles develop germinal centers and so become secondary follicles. Each follicle is surrounded by a layer of T cells in what is known as the mantle zone. The white pulp as a whole is separated from the red pulp by a marginal sinus and by a marginal zone of cells, many of which are macrophages.

Response of the Spleen to Antigen. Intravenously administered antigen is trapped, in part, in the spleen, where it is taken up by macrophages situated in the marginal zone and lining the sinusoids of the red pulp. These macrophages carry the antigen to the primary follicles of the white pulp. After a few days, an-

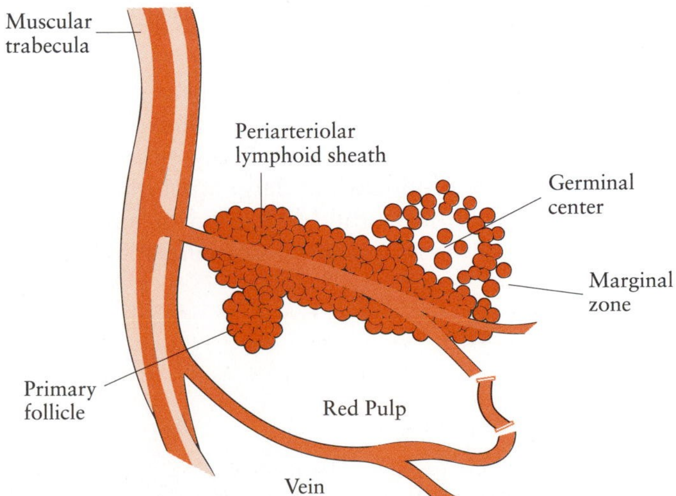

Figure 9–21 A diagram showing the structure of the spleen.

tibody-producing cells migrate from the primary follicles and colonize the marginal zone and the red pulp, so antibody production is first detected in these regions. Germinal center formation occurs within the primary follicles within a few days. In an animal already possessing circulating antibody, the antigen is trapped by dendritic cells within the secondary follicles. As in a primary immune response, the antibody-producing cells migrate from these follicles to the red pulp and to the marginal zone, where antibody production largely occurs. Some antibody may also be produced within the secondary follicles.

Lymphocyte Trapping

When antigen enters the spleen or lymph nodes, it triggers lymphocyte trapping. Lymphocytes that normally pass freely through these organs get trapped so that they cannot leave and the organ begins to swell. The trapping process probably occurs as a result of the interaction between antigen and macrophages, leading to the release of factors that influence the movement of lymphocytes. Trapping concentrates antigen-sensitive lymphocytes close to sites of antigen accumulation and so increases the efficiency of the immune responses. After about 24 hours, the lymph node releases the trapped cells and shows increased cellular output for about seven days. Toward the end of this time, many of the released cells are antibody producers or memory cells.

Bone Marrow

Although its scattered nature makes it difficult to measure, the bone marrow is the largest mass of secondary lymphoid tissue in the body. If antigen is given intravenously, much of it becomes trapped not only in the liver and spleen but also in the bone marrow. During a primary immune response, however, the antibodies produced are derived only from the spleen and lymph nodes. Toward the end of that response, memory cells apparently leave the spleen and colonize the bone marrow. When a second dose of antigen is given, the bone marrow produces very large quantities of antibody. Up to 70% of the production of antibody to some antigens may be derived from the bone marrow (Fig. 9–22).

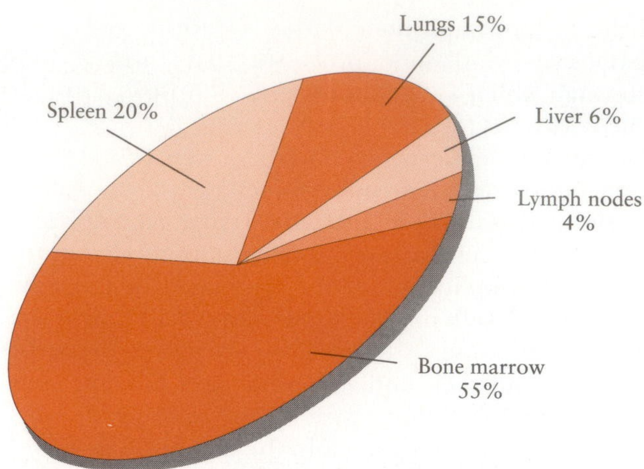

Figure 9–22 The relative contribution of different organs or tissues to antibody production following the administration of antigen intravenously. Compare this with Figure 10–4.

Other Secondary Lymphoid Organs

In addition to being derived from the spleen, lymph nodes, and bone marrow, antibodies are produced in lymphoid nodules scattered throughout the body, particularly in the digestive, respiratory, and urogenital tracts.

A specialized population of lymphocytes is located within the epithelium of the intestine, especially the follicle-associated epithelium of the Peyer's patches, and may be involved in controlling immune responses to food antigens (Fig. 9–10). Antigen administered orally may stimulate the intestinal lymphoid tissues if it is taken up by specialized cells (M cells). As a result, sensitized lymphocytes leave the intestine, circulate in the bloodstream, and then colonize surfaces throughout the body. Antigenic stimulation of one part of the intestine may thus provoke antibody formation throughout the digestive tract as well as in the lung, the mammary glands, and the urogenital tract (Chapter 23). Inhaled antigen stimulates local antibody production in the lymphoid tissues of the respiratory tract and, if absorbed into the bloodstream, provokes a systemic immune response.

KEY WORDS

QUESTIONS

1. High endothelial venules of lymph nodes are important for
   a. antigen trapping
   b. antigen presentation
   c. lymphocyte emigration from blood to lymph node
   d. phagocytosis of foreign material
   e. production of plasma cells
2. Neonatal bursectomy in birds causes
   a. sharp reduction in circulating lymphocytes
   b. cancer in chicks
   c. impaired rejection of skin grafts
   d. thymic atrophy
   e. reduction in serum immunoglobulin levels
3. Lymph nodes perform the following functions *except*
   a. trapping antigen on dendritic cell processes
   b. pumping returning lymph
   c. exposing trapped material to macrophages
   d. exposing trapped material to lymphocytes
   e. filtering lymph
4. B lymphocytes in the spleen are largely found in
   a. the marginal zone
   b. the red pulp
   c. the periarteriolar lymphoid sheath
   d. the follicular areas
   e. all of the above
5. The thymus is
   a. a primary lymphoid organ
   b. a secondary lymphoid organ
   c. a reticuloendothelial organ
   d. a lymphoreticular organ
   e. all of the above
6. In which area of the lymph node are T cells mainly found?
   a. subcapsular
   b. sinusoids
   c. primary follicles
   d. germinal center
   e. paracortex
7. Where in the lymph node are memory cells for antibody production generated?
   a. subcapsular
   b. sinusoids
   c. primary follicles
   d. germinal center
   e. paracortex
8. Following a single intravenous injection of antigen, antibody formation primarily occurs in the
   a. Peyer's patches
   b. lymph nodes
   c. spleen
   d. thymus
   e. bone marrow
9. Neonatal thymectomy causes an immediate
   a. loss of cell-mediated immune responses
   b. drop in peripheral blood lymphocytes
   c. loss of antibody-mediated immune responses
   d. increased susceptibility to disease
   e. all of the above
10. Germinal centers of lymphoid tissues largely contain
    a. macrophages
    b. neutrophils
    c. T cells
    d. B cells
    e. plasma cells
11. Predict the effect of surgical removal of the following from a newborn mouse: the spleen, a lymph node, the thymus.
12. Animals raised in an environment totally free of microorganisms are called gnotobiotic animals. Can you predict the effects of this gnotobiotic state on the development of the primary and secondary lymphoid organs? Will a gnotobiotic animal develop a normal immune system?
13. Outline the events that occur in a draining lymph node following a single subcutaneous injection of a vaccine.
14. Are B cells ever found in peripheral blood? What are they doing there?
15. How does the different circulation pattern of T and B cells reflect their functional differences? What are the advantages of having circulating T cells?
16. What evidence is there to suggest that at least some Peyer's patches have a similar function to the bursa of Fabricius? Is it fair to say that the bursa is simply a modified Peyer's patch?
17. What is the importance of the bone marrow in an adult? What are the consequences of destruction of the bone marrow? How might you treat an individual who lacks a functional bone marrow?
18. Splenectomy in the adult human results in the person's becoming very susceptible to overwhelming bacterial infection. Explain this.

Answers: 1c, 2e, 3b, 4d, 5a, 6e, 7d, 8c, 9e, 10d

SOURCES OF ADDITIONAL INFORMATION

Aguilar, L.K., et al. Thymic nurse cells are sites of thymocyte apoptosis. J. Immunol. 152:2645–2651, 1994.

Anderson, N.D., et al. Specialized structure and metabolic activities of high endothelial venules in rat lymphatic tissues. Immunology, 31:445–473, 1976.

Audhya, T., et al. Tripeptide structure of bursin, a selective B cell differentiating hormone of the bursa of Fabricius. Science, 231:997–999, 1986.

Barclay, A.N. The organization of B and T lymphocytes in lymph nodes. Immunol. Today, 3:330–331, 1982.

Cahill, R.N.P., et al. Two distinct pools of recirculating T lymphocytes: Migratory characteristics of nodal and intestinal T lymphocytes. J. Exp. Med., 45:420–428, 1977.

Duijvestijn, A., and Hamann, A. Mechanisms and regulation of lymphocyte migration. Immunol. Today, 10:23–28, 1989.

Edelson, R.L., and Fink, J.M. The immunologic function of skin. Sci. Am., 252:46–53, 1985.

Ekino, S., et al. The bursa of Fabricius: A trapping site for environmental antigens. Immunology, 55:405–410, 1985.

Friedman, H., ed. Thymus factors in immunity. Ann. N. Y. Acad. Sci., 249:1–547, 1975.

Gowans, J.L., and Knight, E.J. The route of recirculation of lymphocytes in the rat. Proc. R. Soc. Lond. [Biol.], 159:257–282, 1964.

Hay, J.B., and Hobb, B.B. The flow of blood to lymph nodes and its relation to lymphocyte traffic and the immune response. J. Exp. Med., 145:31–44, 1977.

Knowles, D.M., and Holck, S. Tissue localization of T-lymphocytes by the histochemical demonstration of acid α-naphthyl acetate esterase. Lab. Invest., 39:70–76, 1978.

Kubai, L., and Auerbach, R. A new source of embryonic lymphocytes in the mouse. Nature, 301:154–156, 1983.

Low, T.L.K., Mu, S.-K., and Goldstein, A.L. Complete amino acid sequence of bovine thymosin V4: A thymic hormone that induces terminal deoxynucleotidyl transferase activity in thymocyte populations. Proc. Natl. Acad. Sci. U. S. A., 78: 1162–1166, 1981.

Owen, R.L., and Nemanic, P. Antigen processing structure of the mammalian intestinal tract: An SEM study of lymphoepithelial organs. Scanning Microsc., 11:367–378, 1978.

Reynolds, J.D. Evidence of extensive lymphocyte death in sheep Peyer's patches. I. A comparison of lymphocyte production and export. J. Immunol., 136:2005–2010, 1986.

Reynolds, J.D., and Morris, B. The effect of antigen on the development of Peyer's patches in sheep. Eur. J. Immunol., 14:1–6, 1984.

Scollary, R.G., Butcher, E.C., and Weissman, I.L. Thymus cell migration. Quantitative aspects of cellular traffic from the thymus to the periphery in mice. Eur. J. Immunol., 10:210–218, 1980.

Weiss, L. Cells and Tissues of the Immune System: Structure, Functions, Interactions. Foundations of Immunology Series. Prentice-Hall, Englewood Cliffs, New Jersey, 1972.

Wekerle, H., Ketelsen, V.P., and Ernst, M. Thymic nurse cells, lymphoepithelial cell complexes in murine thymuses. Morphological and serological characterization. J. Exp. Med., 151:925–944, 1980.

CHAPTER 10

Lymphocytes

Gus Nossal *Dr. G.J.V. Nossal is an eminent Australian immunologist whose studies have focused on the events occurring in lymph nodes following administration of antigen. Thus he demonstrated that antigen was either trapped on macrophages or on the surface of dendritic cells. He showed that this antigen, when appropriately presented, caused some lymphocytes to progressively develop into plasma cells. He was also the first to formally demonstrate the one cell-one antibody rule (that each antibody-producing cell produces antibody of a single specificity).*

CHAPTER OUTLINE

CHAPTER CONCEPTS

1. The two major types of lymphocytes, T cells and B cells, cannot be distinguished by any structural features.
2. B cells can be identified by the presence of antibodies (immunoglobulin) on their surface. They account for about 20% of blood lymphocytes.
3. T cells can be identified by the characteristic proteins on their surface. They account for about 70% of blood lymphocytes.
4. T and B cells can be distinguished by the presence, on their surface, of many different receptors, enzymes, or adherence molecules. Many of these proteins are classified by the CD system.
5. Lymphocytes may be provoked to divide in response to carbohydrate-binding proteins of plant origin called lectins.
6. Cell surface antigens can be readily analyzed by labeling with fluorescent monoclonal antibodies and examining the cells with a flow cytometer.

In this chapter we examine the basic features of the cells called lymphocytes. Lymphocytes are the key cells of the immune system since they are the ones that recognize and respond to foreign antigens. We look at their structure, which provides very little useful information about their function. We then study their cell surface proteins, which are much more informative. These proteins enable us to distinguish the major populations of T and B cells. We also review their surface receptors and their enzymes. The proteins called lectins that can stimulate certain lymphocyte populations to divide will also be described. The chapter concludes with an examination of the flow cytometer, an instrument that enables immunologists to study lymphocyte surface properties in detail.

Lymphocytes are small round cells, ranging from 7 to 15 μm in diameter. The average human has about 10^{10} of them. Each lymphocyte contains a large round nucleus that stains intensely and evenly with dyes such as hematoxylin. It is surrounded by a thin rim of cytoplasm containing some mitochondria, free ribosomes, and a small Golgi apparatus (Figs. 10–1 and 10–2). Their structure provides no clue as to their complex role within the body (Fig. 10–3). Lymphocytes are found in the blood as well as in the lymphoid organs described in the previous chapter. They account for about 20% to 30% of blood leukocytes. In total they form an organ about the same size as the brain in humans (Fig. 10–4).

The function of lymphocytes cannot be determined by their structure. Lymphocytes can, however, be characterized by their surface receptors, their cell surface proteins, their ontogeny (where and how they develop), their enzymes, and their response to molecules that provoke cell division (**mitogens**).

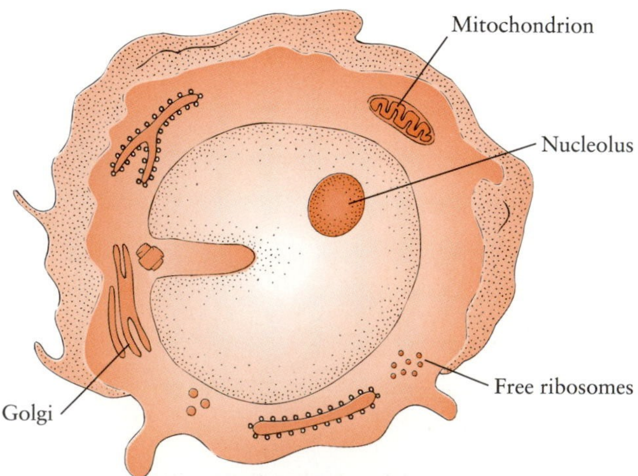

Figure 10–1 The basic structure of a lymphocyte.

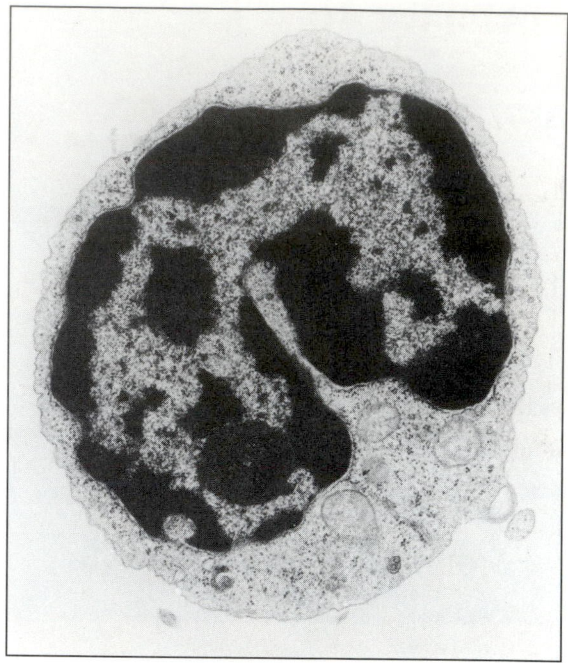

Figure 10–2 A transmission electron micrograph of a lymphocyte. *(Courtesy of Dr. Scott Linthicum.)*

T AND B LYMPHOCYTES

Lymphocytes that develop within the thymus are known as **T cells** (Fig. 10–5). T cells leave the thymus and accumulate in the paracortex of lymph nodes (see Fig. 9–14), in the periarteriolar lymphoid sheath of

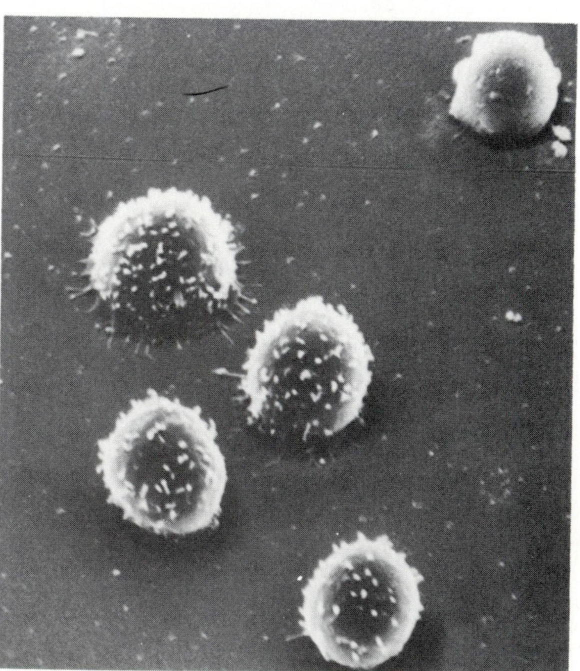

Figure 10–3 A scanning electron micrograph showing lymphocytes derived from a mouse mesenteric lymph node. (Original magnification ×1500.)

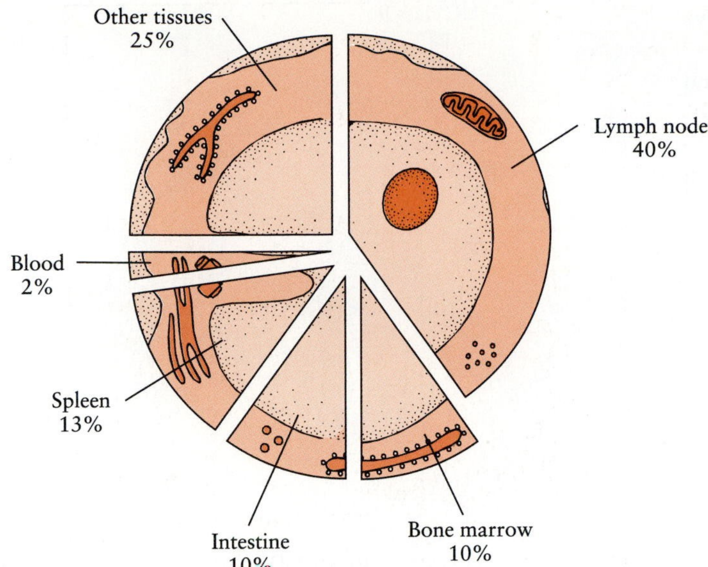

Figure 10–4 The distribution of lymphocytes within the body. Note that blood lymphocytes account for only 2% of the total and thus changes in these cells may not accurately reflect changes in the total lymphocyte population. It is interesting to compare this with Figure 9–22, which shows the organs in which antibodies are made.

the spleen (see Fig. 9–21), and in the interfollicular areas of the Peyer's patches. All these areas are depleted of cells in neonatally thymectomized animals. T cells account for about 70% of the lymphocytes in blood (Table 10–1). Some of these T cells recirculate through the high endothelial venules of the lymph node paracortex and the lymphatic circulation (Chapter 9).

B cells originate in the bone marrow but probably mature in intestinal lymphoid tissues or the bone marrow before migrating to the secondary lymphoid organs. They are found in the cortex of lymph nodes, in follicles within the Peyer's patches and spleen, and in the marginal zone of the white pulp of the spleen. They account for about 20% of blood lymphocytes.

T cells probably survive for six months to as long as ten years in humans. B cells, in contrast, have a

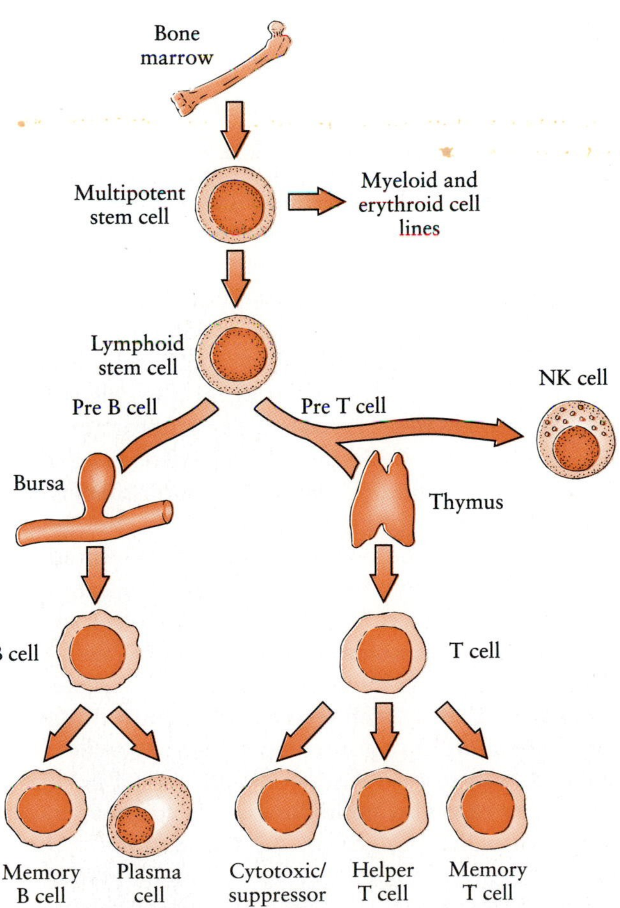

Figure 10–5 The origin and development of B, T, and NK lymphocytes. The bursa is, of course, only present in birds. In mammals its function is served by Peyer's patches or the bone marrow.

Table 10–1 Human Blood Lymphocytes

Cells	Range (%)
B cells	
Ig^+	16–28
IgG*	4–12.7
IgA	1–4.3
IgM	6.7–13
IgD	5.2–8.2
T cells	
E rosettes (CD2)	65–75
T (FcγR)	13–15
T (FcμR)	50–60
CD4/CD8 ratio	1.8–2.2

*These are the names given to immunoglobulin classes. They are described in Chapter 13.

variable short life span; in mice this ranges from five to seven weeks.

The small proportion of blood lymphocytes that are neither T nor B cells are **natural killer (NK) cells.** NK cells are large lymphocytes with an extensive cytoplasm and many cytoplasmic granules. They form a distinct third population of lymphocytes.

LYMPHOCYTE SURFACE MOLECULES

As discussed in Chapter 7, the antigenic molecules located on the surface of lymphocytes are known as CD molecules and listed numerically. In addition to CD molecules, lymphocytes carry histocompatibility molecules and specialized receptors for foreign antigens (TCRs and BCRs).

Antibodies. B cells carry antibody (or **immunoglobulin**) molecules on their surface that act as highly specific antigen receptors. They are discussed in detail in Chapters 13 and 14. T cells do not carry surface immunoglobulins, so the detection of these molecules on lymphocytes provides an excellent method of distinguishing between the two major types of lymphocyte.

Histocompatibility Molecules. Most B cells carry MHC class II molecules on their surface. In contrast, less than 1% of normal T cells are MHC class II positive, but after stimulation by mitogens, up to 70% express these proteins. Mouse thymocytes, but not peripheral T cells or B cells, express TL molecules. These TL molecules are nonpolymorphic MHC class Ib proteins coded for by the T loci in the MHC class I region (Chapter 8). Because the expression of TL is regulated not only by regulatory genes but also by the activities of the murine leukemia viruses, TL^- mice may develop TL^+ leukemias. Mouse lymphocytes also express Qa proteins. Qa-1 is a nonpolymorphic MHC class Ib molecule found on about 60% of T cells. Qa-2 is found on T cells as well as on a subpopulation of B cells. Qa-3 is found on a subpopulation of peripheral T cells. The genes for the TL proteins and the three Qa proteins as well as some minor histocompatibility antigens constitute the Tla gene complex.

TCR. The antigen receptor complex on T cells is called the TCR (**T cell antigen receptor**). This receptor complex is described in detail in the next chapter on the response of T cells to antigen. In addition to its antigen-binding and signal transduction peptides, this receptor contains other proteins, including CD4 or CD8.

CD4, a 55-kDa cell surface glycoprotein, is expressed on about 65% of blood T cells in humans,

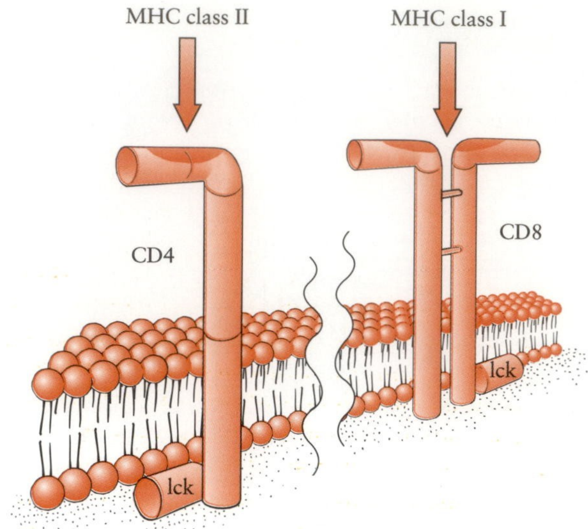

Figure 10–6 The structure of CD4 and CD8, major components of the T cell antigen receptor (TCR). Both are attached to the *lck* tyrosine kinase.

where it acts as a receptor for MHC class II molecules (Fig. 10–6). It is a single peptide chain linked to the *lck* tyrosine kinase in the cell membrane. CD4 is the receptor for human immunodeficiency virus (HIV), the cause of acquired immune deficiency syndrome (AIDS). Cells that possess CD4 on their surface are therefore susceptible to invasion by HIV.

CD8 is a disulfide-linked heterodimeric glycoprotein of 34 kDa found on about 30% of T cells in blood (see Fig. 10–6). It is associated within the cell with the *lck* tyrosine kinase. CD8 acts as a receptor for MHC class I molecules by binding to their α3 domain. Binding between CD8 and MHC class I molecules is necessary if a T cell is going to respond to the endogenous antigen fragments bound to the MHC molecule. When appropriately stimulated, T cells bearing CD8 can attack and destroy virus-infected and other altered cells. They are therefore called **cytotoxic T cells.** The CD8α chain alone can bind MHC class I molecules and costimulate a T-cell response. (By convention, the peptide chains of oligomeric proteins are called α, β, γ chains and so on.) A small proportion of normal T cells carry CD8 α–α homodimers and function apparently normally. The role of the CD8β chain is therefore unclear. The CD8α–β heterodimer binds MHC class I molecules with a higher affinity than the α–α homodimer. It has also been shown that CD8β chain undergoes structural changes when T cells are activated. It is likely, therefore, that the CD8β chain has a regulatory function.

Changes in Surface Protein Expression

Lymphocytes do not express the same surface proteins at all stages in their life cycle. Which protein is expressed depends on the stage of maturity of the cell as well as its state of activation and differentiation. For example, immature human T cells carry CD9 and CD10. As the cells mature within the thymus, CD9 is lost and the cells acquire CD4, CD6, and CD8. Mature thymocytes then split into two subpopulations; one population is $CD4^+CD8^-$. These cells promote immune responses and are called **helper T cells.** They account for about 65% of peripheral blood T lymphocytes. The other T-cell subpopulation becomes $CD4^-CD8^+$. These cells kill other cells and may suppress the immune response; they account for about 30% of peripheral blood T cells and are called **cytotoxic T cells.** The ratio of $CD4^+$ to $CD8^+$ cells in blood may be used to estimate lymphocyte function. An elevated CD4 count implies increased immune reactivity as helper cells predominate, while a high CD8 level implies depressed immune reactivity as a result of excessive cytotoxic cell activity.

LYMPHOCYTE SURFACE RECEPTORS

Proteins are located on cell surfaces to serve a physiological function (Table 10–2). Some are enzymes, some are transport proteins, and many are receptors. All cells use specific receptors to communicate with their environment or with other cells. The receptors found on lymphocytes may be classified according to their function. Thus lymphocytes require receptors that recognize antigen. They need receptors to recognize antigen-presenting cells as well as receptors for the many molecules that regulate lymphocyte responses. In the previous chapter we described the receptors that mediate cell adherence and regulate lymphocyte circulation pathways.

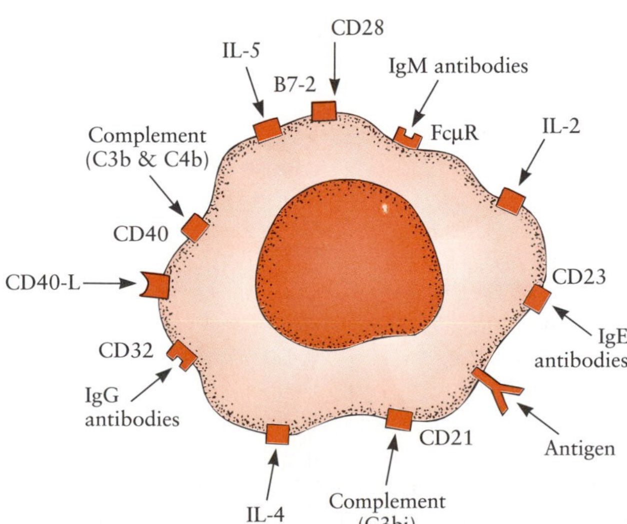

Figure 10–7 The major receptors found on the surface of B lymphocytes.

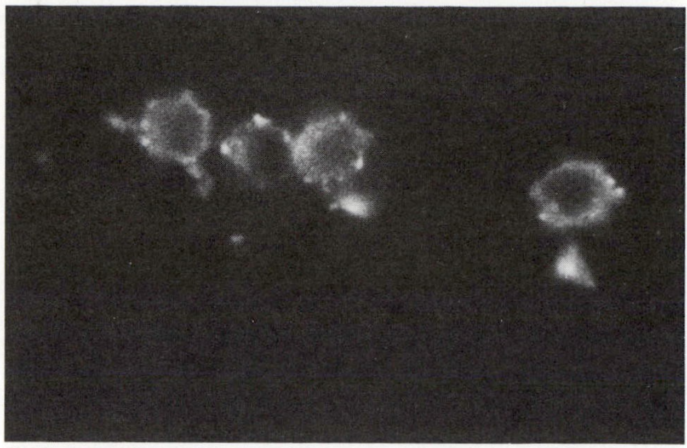

Figure 10–8 Immunoglobulin on the surface of lymphocytes as revealed by the use of the fluorescein-labeled antiserum. The thin rim of fluorescence marks the presence of immunoglobulin. *(Courtesy of Dr. D.H. Lewis.)*

Antigen Receptors

On B cells (Fig. 10–7) the antigen-binding receptors (BCR) consist of antibody molecules together with a set of other proteins that serve as signal transducers (Fig. 10–8). Because of their importance and complexity, B-cell antigen receptors are discussed in detail in Chapter 14. On T cells the most important receptors are also the receptors for antigen (TCR) (Fig. 10–9). Their structure is discussed in the next chapter.

Lymphokine Receptors

B cells possess receptors for many regulatory proteins (**lymphokines**), including **interleukin-2** (IL-2), interleukin-4 (IL-4), and interleukin-5 (IL-5). Binding of the correct ligand to these receptors is required for a successful B-cell response to antigen. Interleukin-2 receptors are also found on T cells. The IL-2 receptors are protein heterodimers that can combine in different ways to generate three receptors with different affinities for IL-2—low, intermediate, and high. The major peptide chain in the IL-2 receptor is called CD25. Its function is described in Chapter 12.

Antibody Receptors

Antibody-coated particles such as red blood cells adhere to lymphocytes. This adherence is mediated through receptors that bind the antibodies to the lym-

Table 10–2 Selected CD Molecules That Serve as Receptors for Defined Ligands

CD	Other Names	MW (kDa)	Cell Location	Ligand
1	T6	49/12	thymocytes, dendritic cells, some B cells	TCR
2	LFA-2	67	T cells, some B cells	CD58, CD59, and CD48
3		16–28	T cells	Antigen
4	T4	59	T cell subset, thymocytes, monocytes	MHC class II
5	T1, Ly-1	67	T cells, some B cells	CD72
8	T8	32	T cell subset, thymocytes, NK cells	MHC class I
11	LFA-1	180	leukocytes	ICAM-2, CD54
14		55	macrophages, granulocytes	Lipopolysaccharide
15s	Sialyl-Lewis^X	multiple	granulocytes	CD62 (selectins)
16	Fcγ RIII	50–65	NK cells, granulocytes, macrophages	IgG
18	LFA-1β	95	leukocytes	
21	CR2	145	B cells, some T cells, dendritic cells	CD23, C3d, and EBV
22	BL-CAM	130/140	B cells	CD45RO
23	Fcγ RII	45–50	B cells, macrophages, eosinophils	IgE, CD21
25	IL-2Rα, Tac	55	activated T and B cells	IL-2
28		45/45	T cells	CD80 (B7, B70)
29	Integrin β_2	130	leukocytes, platelets	Extracellular matrix
32	FcγRII	40	macrophages, granulocytes, B cells, eosinophils	IgG
35	CR1	160–250	granulocytes, monocytes, B cells, NK cells, erythrocytes	C3b
36		90	macrophages, platelets	Thrombospondin
40		50	B cell, dendritic cell, macrophage	CD40-L
41	Integrin α	120/23/110	platelets	Fibrinogen
43	Sialophorin	90–100	leukocytes (not B cells)	CD54
44	Pgp-1, HCAM	90	T cells, leukocytes	Hyaluronate, collagen
45		190–220	leukocytes, T cells	CD22
48	Blast-1	47	lymphocytes, thymocytes	CD2
49	Integrin-α_1	170	leukocytes, platelets	Extracellular matrix
51	VNR-α	125/25	platelets, endothelial cells	Vitronectin
54	ICAM-1	90	endothelial cells	CD11, CD43
58	LFA-3	65	most cells	CD2
59		18–20	most cells	CD2
61	Integrin β_3	110	platelets	
62	Selectins	75–150	platelets, endothelial cell	CD15s, Sialyl Lewis^X
64	FCγRI	75	monocytes, stimulated granulocytes	IgG
70		55–170	activated T and B cells	CD27
71	T9	95	activated T and B cells, macrophages	Transferrin
72		43/39	B cells	CD5
73		69	B and T cell subsets, endothelial cells	Fibronectin
75			B cells, T cell subset	CD22
80	B7	60	B cells	CD28
87		50–65	myeloid cells, activated T cells	Urokinase plasminogen activator
88	C5aR	42	myeloid cells	C5a
89	FcαR	55–75	granulocytes, monocytes	IgA
91		600	myeloid cells	α2-macroglobulin
105	TGFR	95	endothelial cells	TGF-β1 and TGF-β3
115	M-CSFR	150	macrophages and their precursors	M-CSF
w116	GM-CSFR	75–85	granulocytes, monocytes, eosinophils	GM-CSF
117	SCFR, c-kit	145	stem cells	stem cell factor
118	IFNα/βR			IFN-α/β
w119	IFNγ-R	90	lymphocytes, monocytes	IFN-γ
120a	TNFR-I	55	most cells	TNF
120b	TNFR-II	75	myeloid cells, activated T and B cells	TNF
w121a	IL-1RI	80	thymocytes, fibroblasts, keratinocytes, endothelial cells	IL-1
w121b	IL-1RII	68	macrophages, B cells	IL-1
122	IL-2Rβ	75	T cells, activated B cells, NK cells, monocytes	IL-2
w124	IL-4R	140	lymphocytes, fibroblasts, endothelial cells, stem cells	IL-4
126	IL-6R	80	B cells, plasma cells, epithelial hepatocytes	IL-6
w127	IL-7R	75	stem cells, T cells, monocytes	IL-7
w128	IL-8R	58–67	leukocytes, keratinocytes	IL-8
w130	IL-6R	130	many	IL-6/IL-11

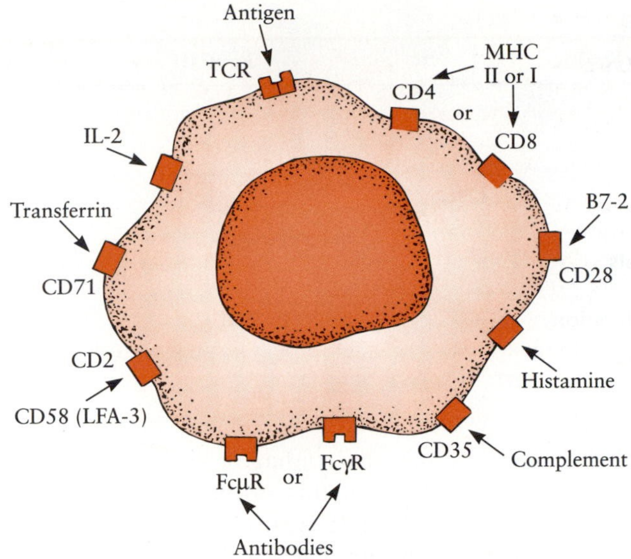

Figure 10–9 The major receptors found on the surface of T lymphocytes.

phocytes. (To be more specific, the receptors bind the "Fc" regions of antibody molecules. For this reason they are called Fc receptors, or FcR for short. The meaning of the term "Fc" can be found in Chapter 13. The reader might wish to return to this section after reading Chapter 13.) These Fc receptors are specific for different antibody heavy chains. Thus FcγR specifically binds immunoglobulin G molecules. Three classes of FcγR have been described in human leukocytes (Fig. 10–10). They are FcγRI (otherwise called CD64), FcγRII (CD32), and FcγRIII (CD16) (Table 10–3).

FcγRI (CD64) is a glycoprotein of 72 kDa that is expressed on monocytes and macrophages and, to a much lesser extent, on neutrophils and eosinophils. (It is not found on lymphocytes but is mentioned here for the sake of completeness.) FcγRI binds free and antigen-bound IgG1 and IgG3 with high affinity. Its expression is enhanced 20-fold by interferon-γ. As with many other cell surface receptors, while the receptor binds the ligand, in this case IgG, a second peptide chain acts as a signal transducer. This chain is a homodimer of 20 kDa found also in FcεRI, FcγRIII, and γδ TCR. It is called FcεRIγ.

FcγRII (CD32) is mainly found on B cells, granulocytes, and macrophages and is a single-chain glycoprotein of 40 kDa. The number of FcγRII receptors varies according to the degree of stimulation of a B cell. Free Fc fragments, presumably acting through these receptors, can either act as B-cell mitogens (in other words, can make them divide) or suppress B-cell functions. The FcγRII glycoprotein occurs in at least six **isoforms** that are homologous in their extracellular domains but differ significantly in their cytoplasmic domains. These isoforms have different cellular distributions and functions. For example, FcγRIIb1 is found on B cells, where it does not influence endocytosis but inhibits cell function through serine phosphorylation. In contrast, FcγRIIb2, found on macrophages, promotes endocytosis and stimulates their release of cytokines.

FcγRIII (CD16) consists of two glycoproteins of 50 to 80 kDa that bind IgG with low affinity. They are expressed on granulocytes, NK cells, and macro-

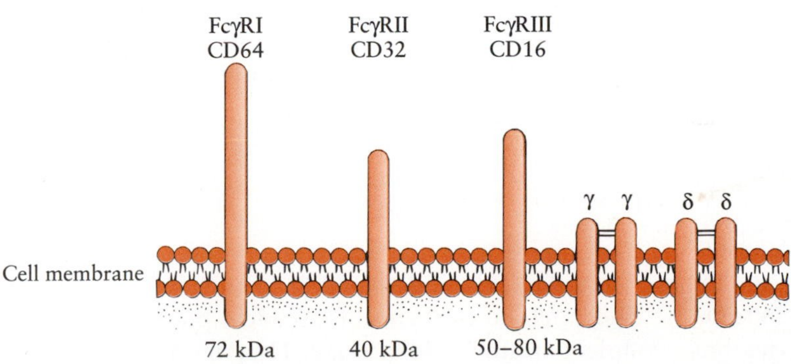

Figure 10–10 Three major Fcγ receptors. FcγRI and FcγRIII may be associated with the dimeric CD3 γ or δ molecules which presumably serve a signal transducing function.

Table 10–3 Fc Receptors

Property	FcγRI	FcγRII	FcγRIII
CD designation	65	32	16
Molecular weight	75 kDa	39–48 kda	50–65 kDa
Cells	monocytes macrophages	B cells macrophages granulocytes eosinophils	NK cells granulocytes macrophages
Affinity	high	medium	low
Function	phagocytosis	B cells-inhibition macrophages-phagocytosis	NK cells-ADCC granulocytes-phagocytosis

phages but not on B cells. The forms found on granulocytes and NK cells differ functionally. Thus the receptor on NK cells can mediate antibody-dependent cellular cytotoxicity (ADCC, Chapter 18), whereas FcγRIII on neutrophils cannot. FcγRIII is complexed with a dimer consisting of either or both of the γ or ζ chains of CD3 (i.e., either γ–γ, ζ–ζ, or γ–ζ). These are required for signal transduction. It may also be associated with the γ chain of FcϵRI (FcϵRIγ), another signal-transducing molecule.

Activated but not resting B cells carry a receptor of 60 kDa that binds the Fc region of IgM (FcμR). It is not found on T cells, monocytes, or granulocytes. IgM binding may enhance lymphocyte responses to antigens and mitogens. In addition, about 15% of human peripheral blood T cells have FcμR, and other T cells may have FcαR (CD89), FcδR, or FcϵRI (CD23). FcαR (CD89) is a 50- to 70-kDa glycoprotein, and T cells that carry it have helper activity specific for IgA.

FcϵRI is a high-affinity IgE receptor expressed on mast cells and discussed in Chapter 29. FcϵRII, in contrast, is a receptor expressed weakly on a few freshly isolated resting B cells but is markedly increased after activation by exposure to interleukin-4. It therefore increases after T cell–B cell interactions. FcϵRII is also found on platelets, eosinophils, macrophages, natural killer cells, dendritic cells, and possibly even T cells. When a B cell is activated, it can release soluble FcϵRII. This soluble FcϵRII can bind to nearby B cells and so permits them to be activated. Soluble FcϵRII can also bind to IgE-producing B cells and so regulate the production of IgE. (For more details, see Chapter 29.)

Complement Receptors

B cells possess receptors for activated complement components. (You may wish to return to this section after reading Chapter 16.) About 50% to 70% of mature B cells have a complement receptor called CR1 (CD35), which binds C3b and C4b. CR1, a 250-kDa glycoprotein, is found not only on lymphocytes but also on neutrophils, macrophages, and eosinophils. T cells possess CR1, but this probably becomes expressed only when the T cells are activated by antigen.

CR2 (CD21), another complement receptor on B cells, is a 145-kDa glycoprotein that binds C3d and C3bi. CR2 acts as a receptor for the Epstein-Barr virus, the cause of infectious mononucleosis. CR2 is closely associated with the B-cell surface immunoglobulins that serve as antigen receptors, and it plays a role in the triggering of the B-cell response to antigen. CR2 is also a ligand for FcϵRII and thus plays a role in regulating the B-cell IgE response.

Other Lymphocyte Receptors

Histamine Receptors. T cells possess two types of histamine receptor called H1 and H2. Exposure of T cells to histamine reduces their activity by stimulating H2 receptors and raising intracellular cyclic AMP levels. Stimulation of T-cell H1 receptors depresses intracellular cyclic AMP and thus promotes T-cell activity. Likewise, drugs that block H2 activity such as cimetidine may act as immunostimulants.

Transferrin Receptors. The cell surface receptor for transferrin, CD71, is a 100-kDa disulfide-linked homodimer. Transferrin receptors are not found on resting lymphocytes; their expression is induced by mitogenic stimulation, and they permit iron uptake by T cells. This iron uptake is necessary for cell division. As a result blockage of transferrin binding by monoclonal anti-CD71 inhibits division of stimulated T cells.

LYMPHOCYTE ENZYMES

Both mouse and human T cells possess lysosomal enzymes that can be detected by histochemical staining. They include β-glucuronidase, acid phosphatase, and alpha naphthyl acid esterase. They are usually concentrated in a single small round body (a **lysosome**) in the cytoplasm. B cells have no characteristic enzyme-staining patterns.

Three lymphocyte enzymes—5′-nucleotidase, adenosine deaminase, and purine nucleotide phosphorylase—are all important in immunity since a deficiency of any of them leads to defective immune function. They are involved in purine metabolism. 5′-Nucleotidase is found in many cells, including T and B cells. Adenosine deaminase is present in higher concentrations in T cells than in B cells. Purine nucleotide phosphorylase can be found in the cytoplasm of T cells but not B cells.

Several lymphocyte surface proteins are also enzymes (Table 10–4). Thus CD10 (CALLA) is a type II integral membrane protein on B cells that acts as a neutral endopeptidase, CD26 is a peptidase found on activated T cells, CD73 is a 5′-nucleotidase, and CD45 is a phosphotyrosine phosphatase. The cytoplasmic tails of CD4 and CD8 are associated with the *lck* tyrosine kinase.

CD45 identifies a family of cell membrane glycoproteins found on all blood cells with the exception of erythrocytes and platelets. (They are also called leukocyte common antigens.) Up to 10% of the surface of a lymphocyte may be composed of one or more members of this family, and CD45 therefore is a major cell surface component. The CD45 molecules are all receptor phosphotyrosine phosphatases that span the cell membrane and are required for normal signal transduction and lymphocyte activation (page 147). There are four isoforms of CD45 (CD45RA, CD45RB, CD45RC, and CD45RO) coded for by a single gene. They result from differential splicing of several exons. This results in specific expression of different isoforms in various cell types as well as at different stages of differentiation. CD45RA associates with LFA-1, CD45RO with CD2, and CD45RC with CD4 and CD8. They may also associate with Thy-1 and the TCR. On B cells, CD45 associates with the antigen receptor, with CD22, and with MHC class II molecules. The CD45RA$^+$ population contains virgin, unprimed T cells. CD45RO, in contrast, is a marker of stimulated and memory T cells. Thus virgin T cells carry CD45RA, but memory T cells carry CD45RO. Both populations respond to phytohemagglutinin in the presence of monocytes, but only CD45RO$^+$ cells produce interleukin-2 (IL-2) and express IL-2 receptors (IL-2R). The two cell populations appear to have different triggering requirements and also circulate in a different manner. For example, CD45RO$^+$ cells enter lymph nodes via high endothelial vessels and CD45RA cells enter through afferent lymphatics.

OTHER LYMPHOCYTE SURFACE PROTEINS

Some surface proteins, although present in significant amounts on lymphocyte surfaces, do not have a well-defined function at this time. Two of the most interesting are CD1 and Thy-1 (CDw90). CD1 consists of four polypeptides of 43 to 49 kDa named CD1a, b, c, and d. They are noncovalently attached to β2-microglobulin and thus bear a marked resemblance to MHC class Ib molecules, although their genes are not linked to the MHC. They are expressed on thymocytes and a subpopulation of developing B cells as well as on Langerhans cells and dendritic cells. Their function is unknown.

Thy-1 (CDw90) was the first surface marker that could be used to identify T cells. Subsequently, it was found that its expression varied between different species. Thus in mice it is found on both thymocytes and peripheral blood T cells. In rats it is found on thymocytes but not on peripheral blood T cells. A human Thy-1 equivalent has been detected on brain cells and some early thymocytes but not on mature T cells. A 19-kDa glycoprotein attached by a glycophosphatidylinositol anchor, Thy-1 is found on the surface of brain cells (astrocytes), epidermal cells, fibroblasts, and peripheral blood T lymphocytes. It is a member of the immunoglobulin superfamily. In mice Thy-1 is found

Table 10–4 Some CD Antigens with Recognized Enzymic Activities

CD	Other Names	MW (kDa)	Cell Location	Enzymic Activity
10	CALLA	100	lymphoid stem cells, CLL cells, granulocytes	neutral endopeptidase
13		150	monocytes, granulocytes, high endothelium	aminopeptidase
26		120	activated T and B cells, macrophages	exopeptidase
45	T200	180–200	leukocytes	tyrosine phosphatase
46	MCP	66/56	hematopoietic cells, non-hematopoietic cells	cofactor for factor I
55	DAF	70	hematopoietic cells	complement inactivator
59		18–20	most cells	complement inactivator
73		69	B cell subset, T cell subset	5′-nucleotidase

in large amounts on immature thymocytes (there are about 10^6 molecules per cell); the amount on T cell surfaces declines rapidly once T cells leave the thymus. Thy-1 is distributed evenly through the T-cell subsets, although some intraepithelial T-cell populations are Thy-1 negative. Anti-Thy-1 in the mouse can activate T cells independently of the T-cell antigen receptor and triggers a rise in the concentration of Ca^{2+} in cells. Thus Thy-1 may be part of an alternative pathway of T-cell activation, like human CD2.

LYMPHOCYTE MITOGENS

Certain natural compounds have the ability to make lymphocytes divide (Fig. 10–11). The most important of these compounds is a family of proteins called **lectins,** which are usually isolated from plants. Examples of lectins include **phytohemagglutinin** (PHA), obtained from the red kidney bean (*Phaseolus vulgaris*); **concanavalin A** (Con A), obtained from the jack bean (*Canavalia ensiformis*); and **pokeweed mitogen** (PWM), obtained from the pokeweed plant (*Phytolacca americana*).

These lectins can bind sugars specifically. For example, phytohemagglutinin binds N-acetyl-D-galactosamine, and concanavalin A binds α-D-mannose and α-D-glucose (Table 10–5). The binding of lectin to lymphocyte membranes stimulates nucleoside incorporation, phospholipid synthesis, DNA synthesis, and cell division. Not all lymphocytes respond equally well to all lectins. PHA primarily stimulates T-cell division, although it has a slight effect on B cells. Con A is also a T-cell mitogen, and PWM acts on both T and B cells and induces some B cells to synthesize immunoglobulin.

Other Lymphocyte Mitogens. Although the plant lectins are the most efficient lymphocyte mitogens, mitogenic activity is also found in other unexpected sources. For example, an extract from the snail *Helix pomata* stimulates T cells. Lipopolysaccharide from gram-negative bacteria stimulates B cells to divide, but the presence of T cells is required for an optimal response. Other important B-cell mitogens include neutral proteases, such as trypsin, and Fc fragments of immunoglobulins. BCG vaccine (bacille Calmette-Guérin, an avirulent strain of *Mycobacterium bovis* that is used as a vaccine against tuberculosis) is a T-cell mitogen, although it may exert a slight effect on B cells.

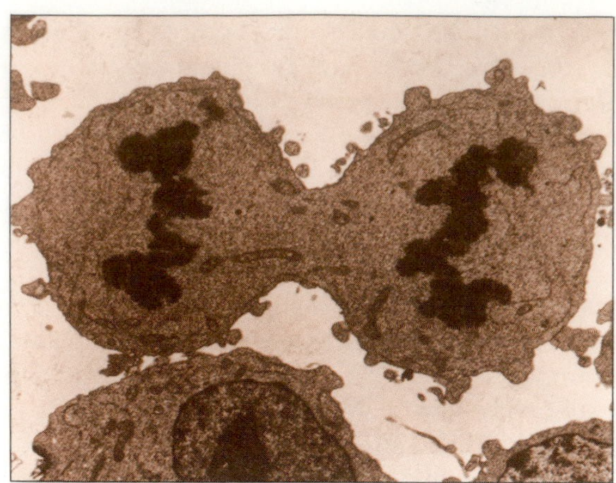

Figure 10–11 Two T cells in the act of dividing. T-cell division can be triggered by antigen or by mitogens. *(Courtesy of Visuals Unlimited/David M. Philips.)*

These mitogens can be used to assist in the differentiation of T and B cells and, by measurement of the response provoked, demonstrate the ability of the T-cell and B-cell systems to respond to nonimmunological stimuli.

THE FLOW CYTOMETER

It is possible to analyze the proteins on cell populations in great detail and with very high efficiency using an instrument called a flow cytometer (Fig. 10–12). In this instrument a suspension of cells flows through a

Table 10–5 Some Lymphocyte Mitogens

Lectin	Carbohydrate Specificity	Target
Phytohemagglutinin	N-acetyl-D-galactosamine	T(B)
Wheat-germ agglutinin	N-acetylglucosamine, sialic acid	T
Concanavalin A	α-D.-mannose, α-D-glucose	T
Lentil lectin	α-D.-mannose, α-D-glucose	T
E. coli endotoxin	α-D.-mannose	B
Helix pomata (garden snail)	N-acetylgalactosamine	T
Pokeweed mitogen	Not carbohydrate-specific	T,B
Lipopolysaccharide	Not carbohydrate-specific	B

METHODOLOGY

How to Measure Mitogenicity

To measure the effect of lymphocyte mitogens, lymphocytes are grown in tissue culture. Lymphocytes can be obtained directly from the spleen in small rodents or, more usually, from peripheral blood. The lymphocytes are grown in tissue culture for at least 24 hours before the mitogen is added. Once this is done, they begin to divide, synthesize new DNA, and take up any available nucleotides from the medium. It is usual to incorporate in the tissue-culture fluid a small quantity of thymidine labeled with the radioactive isotope of hydrogen, tritium (^3H). The thymidine is incorporated only into the DNA of cells that are dividing. About 24 hours after exposure to tritiated thymidine, the cultured cells are separated from the tissue-culture fluid, by either centrifugation or filtration, and their radioactivity is counted in a liquid scintillation counter. The amount of radioactivity in the lectin-treated cells may be compared with that in an untreated lymphocyte culture. This ratio is known as the stimulation index. As an alternative to the use of tritiated thymidine, it is possible to use a radiolabeled amino acid such as ^{14}C-leucine. Uptake of this compound is an indication of increased protein synthesis by the cells.

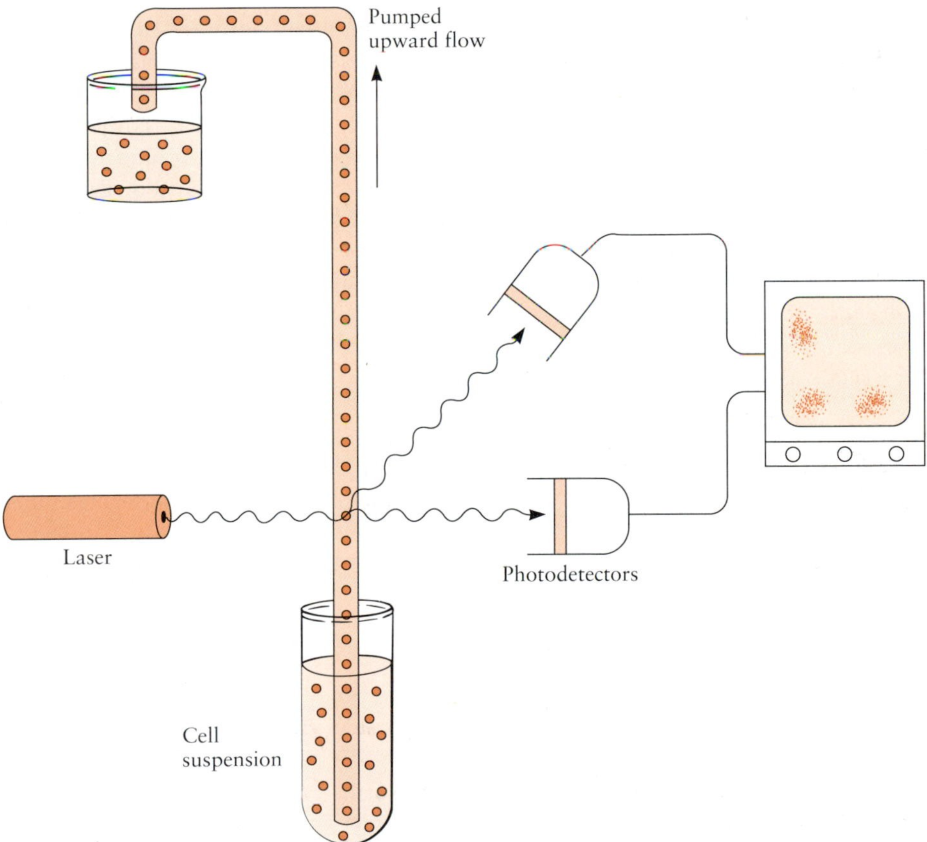

Figure 10–12 The basic features of the flow cytometer. A thin stream of cells is drawn through a fine capillary tube and passes across a laser beam. The light is scattered as each cell passes through the beam. Side and forward light scatter are detected and measured. These can be used to characterize a cell population. If the cell components, such as surface proteins, are stained using a fluorescent dye, even more details about cell populations can be determined.

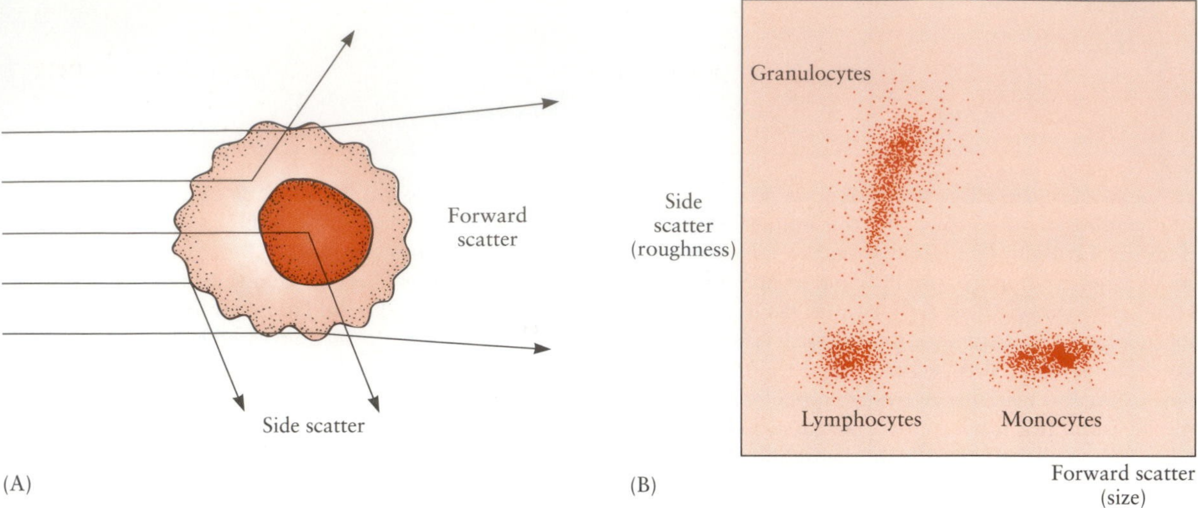

Figure 10–13 (A) Forward and side light scatter caused by a cell in a light beam. (B) The pattern observed on the flow cytometer screen when unlabelled blood cells are examined. Monocytes, being relatively large and smooth, show little side scatter. Granulocytes are small rough cells while lymphocytes are small and smooth.

very fine-bore tube so that the cells pass through in single file. A laser beam is directed through the cell stream, and the effects of the cells on the light beam are measured. Thus the scatter of the light beam in a forward direction can be used to give a measure of a cell's size. The light scattered to the side by a cell gives a measure of a cell's surface roughness and internal complexity. A combination of these two parameters can be used to identify all the cells in a blood sample (Fig. 10–13).

The flow cytometer can, however, be used to measure much more than this. Thus if a subpopulation within a cell suspension is labeled with a monoclonal antibody (Chapter 14) conjugated to a fluorescent dye, then this subpopulation can be automatically characterized and counted. Two fluorescent labels can be analyzed simultaneously, and it is possible to use the flow cytometer to follow sequential changes in the composition of mixed cell populations (Fig. 10–14).

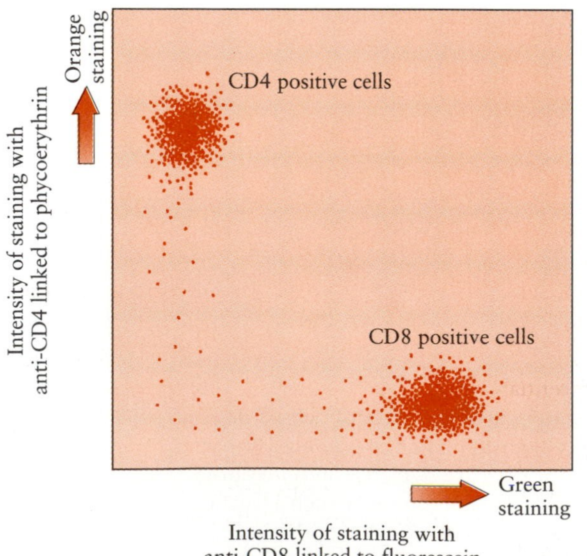

Figure 10–14 The pattern seen on a flow cytometer screen when a stained lymphocyte preparation is examined. In this case the cells are treated with two fluorescent antibody preparations. One is anti-CD4 linked to the orange fluorescent dye phycoerythrin. The other is anti-CD8 linked to the green dye fluorescein. Thus both types of cell can be discriminated by their fluorescence.

KEY WORDS

QUESTIONS

1. T cells constitute approximately which of the following percentages of peripheral blood lymphocytes?
   a. 10%
   b. 50%
   c. 70%
   d. 100%
   e. 0%
2. On the basis of morphology, T and B cells can be distinguished by
   a. size differences
   b. microvillous projections on cell surfaces
   c. secretory granules
   d. cytoplasmic structures
   e. none of the above
3. The FcεRII molecule on B-cell surfaces binds
   a. C3b
   b. IgE
   c. IgA
   d. CD3
   e. CD8
4. The CD45 molecule is a
   a. tyrosine kinase
   b. serine kinase
   c. phosphotyrosine phosphatase
   d. complement receptor
   e. Fc receptor
5. One characteristic cell surface molecule largely restricted to human thymocytes is
   a. CD4
   b. CD8
   c. TCR
   d. CD45
   e. Thy-1
6. One potent T-cell mitogen is
   a. phytohemagglutinin
   b. complement
   c. immunoglobulin
   d. endotoxin
   e. interleukin-1
7. The flow cytometer is an instrument that
   a. separates blood cells
   b. detects different leukocyte populations
   c. measures T-cell responses
   d. measures antibody responses
   e. diagnoses AIDS
8. CD8 is a receptor for
   a. antigen
   b. MHC class II molecules
   c. antibody
   d. TCR
   e. MHC class I molecules
9. In addition to T cells and B cells, there is a third distinct type of lymphocyte called
   a. MHC cells
   b. IgG cells
   c. Langerhans cells
   d. astrocytes
   e. NK cells
10. The CD4 molecule is
    a. a heterodimer
    b. a receptor for MHC class II molecules
    c. part of the BCR
    d. a complement receptor
    e. a tyrosine kinase
11. Do you think it possible that lymphocyte-erythrocyte rosettes could form in vivo? Could this rosetting phenomenon have any biological significance?
12. Given a suspension of lymphocytes, how would you differentiate the T cells from the B cells?
13. What functions do Fc receptors serve on T cells and on B cells?
14. Would you expect to find a perfect correspondence between the lymphocyte surface proteins in humans and mice? Why?

15. What is a lectin? What effect do lectins have on lymphocytes? Is this of any biological significance, or is it purely a coincidental finding?
16. Which receptors on lymphocyte surfaces act to suppress lymphocyte function when triggered? Could you suggest which of these are most important and why?
17. Do you think that we may eventually be able to catalog all lymphocyte surface molecules? What could we do with this information?

Answers: 1c, 2e, 3b, 4c, 5e, 6a, 7b, 8e, 9e, 10b

SOURCES OF ADDITIONAL INFORMATION

Albelda, S.M., and Buck, C.A. Integrins and other cell adhesion molecules. FASEB J., 4:2868–2880, 1990.

Casabó, L.G., et al. T cell activation results in physical modification of the mouse CD8β chain. J. Immunol., 152:397–404, 1994.

Clark, E.A., and Ledbetter, J.A. Leukocyte cell surface enzymology: CD45 (LCA, T200) is a protein tyrosine phosphatase. Immunol. Today, 10:225–227, 1989.

Clement, L.T. Isoforms of the CD45 common leukocyte antigen family: Markers for human T-cell differentiation. J. Clin. Immunol., 12:1–10, 1992.

Coico, R.F., et al. IgD-receptor-positive human T lymphocytes. I. Modulation of receptor expression by oligomeric IgD and lymphokines. J. Immunol., 145:3556–3561, 1990.

Gordon, J., et al. CD23: A multifunctional receptor lymphokine? Immunol. Today, 10:153–157, 1989.

Haynes, B.F., et al. CD44—A molecule involved in leukocyte adherence and T-cell activation. Immunol. Today, 10:423–428, 1989.

Hemler, M.E. VLA proteins in the integrin family: Structure, functions and their role on leukocytes. Annu. Rev. Immunol., 8:365–400, 1990.

Howard, F.D., et al. A human T lymphocyte differentiation marker defined by monoclonal antibodies that block E-rosette formation. J. Immunol., 126:2117–2122, 1981.

Knapp, W., et al., eds. Leukocyte typing IV: White cell differentiation antigens. International Workshop and Conference on Human Leukocyte Differentiation Antigens. Oxford University Press, Vienna, Austria, 1989.

Koopman, G., et al. Triggering of the CD44 antigen on T lymphocytes promotes T cell adhesion through the LFA-1 pathway. J. Immunol., 145:3589–3593, 1990.

Moller, G., ed. B cell differentiation antigens. Immunol. Rev., 69:5–159, 1983.

Nadler, L.M., et al. Characterization of a human B cell–specific antigen (B2) distinct from B1. J. Immunol., 126:1941–1947, 1981.

Ohno, T., et al. Biochemical nature of an Fcμ receptor on human B-lineage cells. J. Exp. Med., 172:1165–1175, 1990.

O'Neill, G.J., and Parrott, D.M.V. Locomotion of human lymphoid cells. I. Effect of culture and Con A and T on non-T lymphocytes. Cell. Immunol., 33:257–267, 1977.

Rowlands, D.T., and Daniele, R.P. Surface receptors in the immune responses. N. Engl. J. Med., 293:26–32, 1975.

Scholl, P.R., and Geha, R.S. Physical association between the high affinity IgG receptor (FcγRI) and the γ subunit of the high-affinity IgE receptor (FcεRIγ). Proc. Natl. Acad. Sci. U. S. A. 90:8847–8850, 1993.

Springer, T.A. Adhesion receptors of the immune system. Nature, 346:425–433, 1990.

Thomas, M.L. The leukocyte-common antigen family. Annu. Rev. Immunol., 7:339–369, 1989.

Van de Winkel, J.G.J., and Capel, P.J.A. Human IgG Fc receptor heterogeneity: Molecular aspects and clinical implications. Immunol. Today, 14:215–221, 1993.

Webb, D.S.A., et al. LFA-3, CD44, and CD45: Physiologic triggers of human monocyte TNF and IL-1 release. Science, 249: 1295–1297, 1990.

Weissman, I.L., and Cooper, M.D. How the immune system develops. Sci. Am. 269:65–71, 1993.

CHAPTER 11

The Helper T-Cell Responses

***Macfarlane Burnet** was born in Australia in 1899 and died in 1985. His contributions to immunology were mainly theoretical. Thus he was the first to recognize the then revolutionary concept that the ability to be immune to disease was not inherited but developed gradually in the embryo. He postulated that during this gradual development tolerance to self antigens would develop and predicted that exposure to self-antigens in the embryo would cause tolerance. He went on to propose the Clonal Selection theory, in the 1950s. Burnet was also an excellent virologist and was the first to isolate the agent of Q fever. He was awarded the Nobel Prize in 1960.* (UPI/Bettman)

CHAPTER OUTLINE

CHAPTER CONCEPTS

1. Helper T cells have many identical receptors on their surface that recognize fragments of foreign antigen in association with a MHC class II molecule on the surface of an antigen-presenting cell.
2. T-cell antigen receptors (TCRs) are heterodimeric protein complexes closely associated with either CD4 or CD8 molecules. In helper T cells, the antigen receptor is associated with CD4.
3. There are two major types of TCR, each with two different peptide chains. Each TCR is bound to the cell membrane so that its antigen-binding site faces outward.
4. For a T-cell response to occur, helper T cells must recognize a foreign antigen bound to a MHC class II antigen in the presence of a costimulator and in association with a second signal mediated through CD28.
5. There are two major subpopulations of helper T cells that secrete different mixtures of interleukins and stimulate different types of immune response.
6. Th1 cells secrete interleukin-2 and interferon-γ. These cells act to "help" mainly cell-mediated immune responses.
7. Th2 cells secrete a mixture of cytokines, including IL-4, IL-5, and IL-10. These cells act to "help" mainly antibody responses.

In this chapter we examine the cells that recognize and respond to processed exogenous antigen. These cells are called helper T cells. We first look in detail at the receptors that they use to recognize the antigen and at the complex set of other proteins that are used to transmit a signal to the T cell. Antigen presentation by itself is not usually sufficient to trigger a T-cell response. Other costimulators are also needed, so we next examine these. We then look at the results of all this signaling and how T cells respond to antigen. You will find out how there are two types of helper T cell. These cells secrete proteins called lymphokines, which trigger activities in other cells. Finally, the chapter concludes with a brief review of the way in which T cells develop.

A key step in a successful immune response is the presentation of antigen to cells that can recognize it and respond appropriately. These responding cells are the T and B lymphocytes. B cells, as we shall see later, can recognize intact soluble antigen molecules free in body fluids. T cells, in contrast, can recognize only selected fragments of antigen. As described in Chapter 8, exogenous antigen is fragmented within antigen-presenting cells, bound to a MHC class II molecule, and presented on the cell surface. This recognition is said to be **MHC restricted.** The antigen–MHC complexes are recognized by a population of T lymphocytes called **helper T cells.** Once appropriately stimulated, helper T cells provide an essential stimulus for the other components of the immune response to occur (Fig. 11–1). Helper T cells are therefore required to initiate immune responses by binding processed fragments of foreign antigen and, as a result of receptor signaling, to secrete proteins that will regulate the responses of other lymphocytes to that antigen.

ANTIGEN PRESENTATION

Antigen-Presenting Cells

Helper T cells respond to exogenous antigens processed by macrophages, dendritic cells, and B cells. It has been calculated that an antigen-presenting cell carries about 2×10^5 MHC class II molecules that are capable of presenting peptides to T cells. A T cell can be activated by exposure to 200 to 300 peptide–MHC complexes. Thus it is possible for one antigen-presenting cell to present many foreign antigens simultaneously.

The T cell must recognize the antigen fragment in association with an appropriate MHC molecule through its receptor complex. T cells that are going to develop into cytotoxic cells recognize endogenous antigen presented in association with a MHC class I molecule. Helper T cells, in contrast, recognize exogenous antigen in association with a MHC class II molecule. As a result of binding between a specific T-cell antigen receptor and its ligand, an antigen fragment located within the groove on a MHC molecule, a signal is transmitted to the cell. This signal stimulates T cells to proceed from the resting stage of the cell cycle into the G_1 stage. At the same time the genes for cytokines and their receptors are transcribed and translated.

THE T-CELL ANTIGEN RECEPTORS

Antigen Recognition

Each T cell possesses about 50,000 to 100,000 identical receptors for antigen. These receptors are called TCRs. The TCRs are clonally distributed so that each T-cell **clone** expresses only one type of TCR. TCRs are

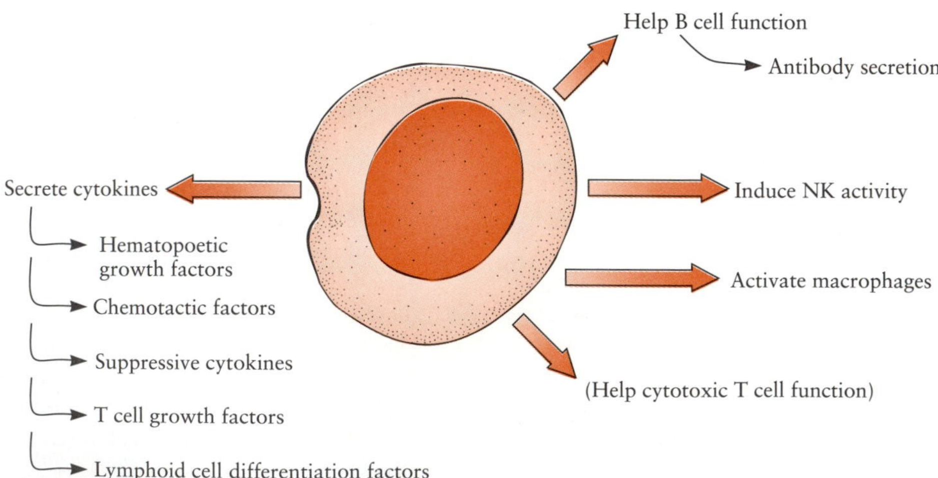

Figure 11–1 The central role of the helper T-cell (CD4+) response in promoting both antibody and cell-mediated immune responses.

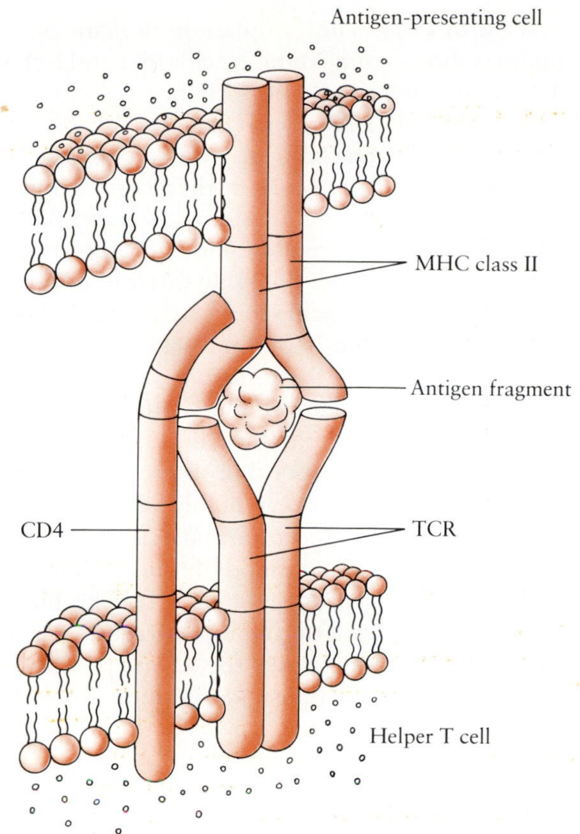

Figure 11–2 A simplified view of the antigen receptor complex on helper T cells. CD4 is required to bind to a MHC class II molecule in order to optimize antigen presentation.

cell membrane–associated protein complexes (Fig. 11–2). Unlike the antigen receptors on B cells, TCRs are not secreted into body fluids.

The antigen binding components of TCRs are glycoprotein heterodimers. Two major classes of TCR have been identified. Between 1% and 3% of T cells in blood have TCRs that consist of γ and δ chains (TCR γ/δ). The remaining T cells in blood have TCRs consisting of α and β chains (TCR α/β). These TCR heterodimers are always membrane bound and always associated with additional glycoproteins, either with CD3 and CD4 or with CD3 and CD8. As discussed later, the CD3 glycoproteins play a role in signal transduction, while CD4 and CD8 are coreceptors that help the T cells bind to antigen-presenting cells (Table 11–1).

The four peptide chains found in TCRs—α, β, γ, and δ—all have a similar overall structure, although their sizes vary (Fig. 11–3). In humans the α chain ranges from 43 to 49 kDa, and the β chain ranges from 38 to 44 kDa. The γ chain is 40 to 55 kDa, and the δ chain is 40 kDa. The genes for the α and δ chains are located together on chromosome 14. The genes for the β and γ chains are at separate locations on chromosome 7. Each of the four TCR chains consists of four **domains**—two extracellular domains, a transmembrane domain, and a cytoplasmic domain. In α/β receptors, the two peptides are linked by an interchain disulfide bond. In γ/δ receptors the interchain bonding is variable and depends on the isoform of γ chain employed. The N-terminal extracellular domain of each peptide has a highly variable amino acid sequence and therefore forms a variable region (V region). The other domains have a relatively constant amino acid sequence. The V region contains about 100 amino acids. It has two conserved cysteine residues that fold the chain to form an **intrachain** disulfide bond. The constant extracellular domain of the TCR consists of four regions of about 138 to 179 amino acids in length coded for by four separate exons (Chapter 15). This domain contains three cysteine residues, two of which are in the first exon and fold the chain over, forming an intrachain disulfide bond. The

Table 11–1 The TCR–CD3 Receptor Complex

Peptide Chain	Function	Molecular Weight (kDa)
TCR α	Recognition of Ag and MHC	43–49
β		38–44
γ	Recognition of antigen	40–55
δ		40
CD3 γ	Signal transducer	21–28
δ	Signal transducer	20–28
ϵ	Signal transducer	20–25
ζ	Signal transducer	16
η	Signal transducer	22
ω	Signal transducer	23
CD4	MHC II receptor	55
CD8	MHC I receptor	34
lck	Tyrosine kinase	56
ZAP	Tyrosine kinase	70

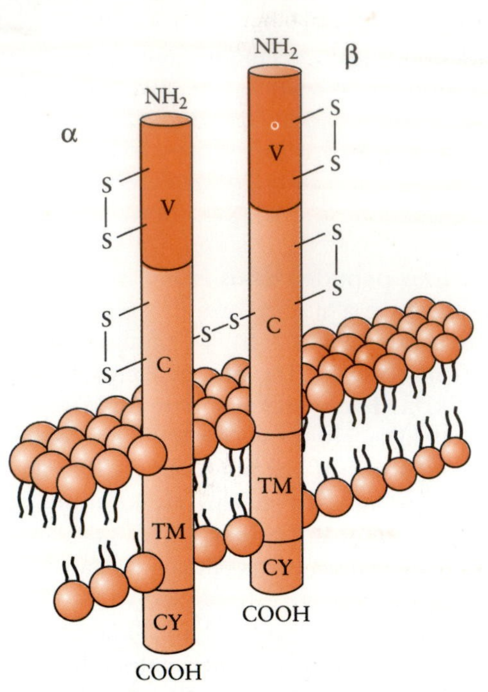

Figure 11–3 The basic domain structure of α/β T-cell antigen receptors. (V, variable segment; D, diversity segment; C, constant segment; TM, transmembrane domain; CY, cytoplasmic domain.)

third cysteine residue forms the **interchain** disulfide bond with the other chain of the heterodimer. The transmembrane domain consists of 20 hydrophobic amino acids, although there is a highly conserved lysine residue in the middle of the membrane that may interact with CD3. The cytoplasmic domain of TCR peptides is short—only 5 amino acids in α and δ chains and 15 amino acids in β and γ chains.

Because each TCR consists of two linked chains, the V regions of each chain are located close together at the end of the molecule farthest away from the T-cell surface. These two V regions (α1 and β1) collectively form the antigen-binding site. The shape of this antigen-binding site is highly variable as a result of the different amino acid sequences in the V regions. The specificity of the TCR for its antigen is determined by the shape of these regions.

When the variability in the sequence of the V regions of the TCR is examined in detail, it is found that there is one area around position 100 where the amino acid sequence is very highly variable indeed. This is the part of the receptor that actually binds antigen. For this reason, it is called the complementarity-determining region (CDR) (Fig. 11–4). The binding site is a shallow groove on the top surface of the molecule between the α-helices of the α1 and β1 domains. When the groove is occupied by a peptide, the surface of the complex is fairly flat.

Gamma and Delta T Cells

About 1% to 3% of peripheral blood T cells have a TCR consisting of γ and δ chains. Three isoforms of human TCR γ/δ are recognized (Fig. 11–5). One isoform has a 40-kDa γ chain disulfide-linked to the δ chain. In a second isoform, a 55-kDa γ chain is noncovalently linked to the δ chain, and in a third form, a 40-kDa γ chain is noncovalently linked to the δ chain. These isoforms result because two Cγ exons

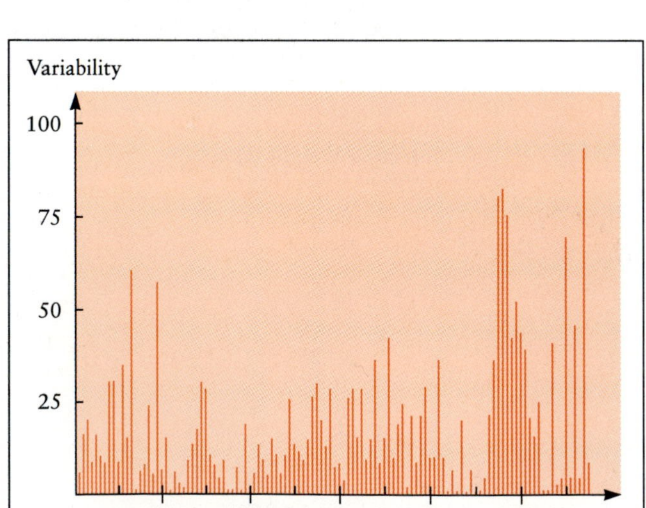

Figure 11–4 This is a Kabat-Wu plot of the V domain of a TCR α chain. Each bar represents an amino acid residue. The height of each bar indicates the variability of the amino acid residue at that position. One region, around position 100, shows major sequence variability. As will be discussed in Chapter 15, this is where two gene segments join. *(Calculated from data in Caccia, N., et al. in The T Cell Receptors, Edited by T. W. Mak, Plenum, New York, 1988.)*

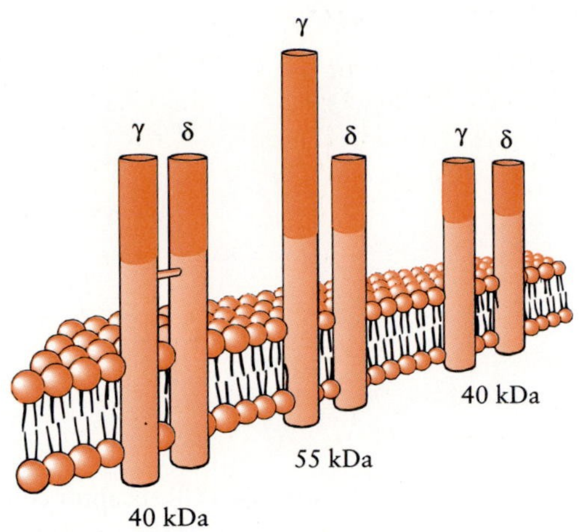

Figure 11–5 There are three different forms of γδ TCR that differ either in size of the γ chain or in the presence of an interchain disulfide bond between γ and δ chains.

(Cγ1 and Cγ2) are used by γ/δ T cells. Only the Cγ1 exon can form intrachain disulfide bonds with the δ chain.

The γ/δ TCR is mainly found on immature T cells (Fig. 11–6). Thus the first T cells to colonize the embryonic thymus carry the γ/δ TCR. These cells emigrate from the thymus to colonize the skin and reproductive organs. In the mouse, γ/δ T cells are found in the intestinal wall and epithelium, where they constitute 1% to 15% of intestinal $CD3^+$ cells. They are found in high numbers in the mammary gland and irregularly scattered in the dermis and in the basal layer of the epidermis into the skin. This is not the case in humans, where γ/δ T cells are evenly distributed through tissue sites in the same location as α/β cells. γ/δ T cells do not develop into α/β cells.

γ/δ T cells are cytotoxic and can lyse various target cells such as chicken red blood cells, cells infected with mycobacteria, and some leukemic cells. Their numbers are increased in measles, in leprosy, and in mice infected with *Listeria monocytogenes* or influenza virus. They may represent a first line of defense against certain bacteria since a high proportion recognize bacterial heat shock proteins (Chapter 25). Some γ/δ T cells interact with hematopoietic cells of both T and non-T lineages and regulate their differentiation and activation. They may also have a role in NK cell antigen recognition or in suppressor T cell function. Anti-CD3 antibodies can induce γ/δ T cells to become cytotoxic and secrete IL-2, suggesting that they are relatively normal T cells. Most reports on γ/δ T cells indicate that they are MHC unrestricted like NK cells. Although γ/δ TCR may be associated with certain MHC class Ib molecules, growing evidence suggests that activation of γ/δ T cells does not

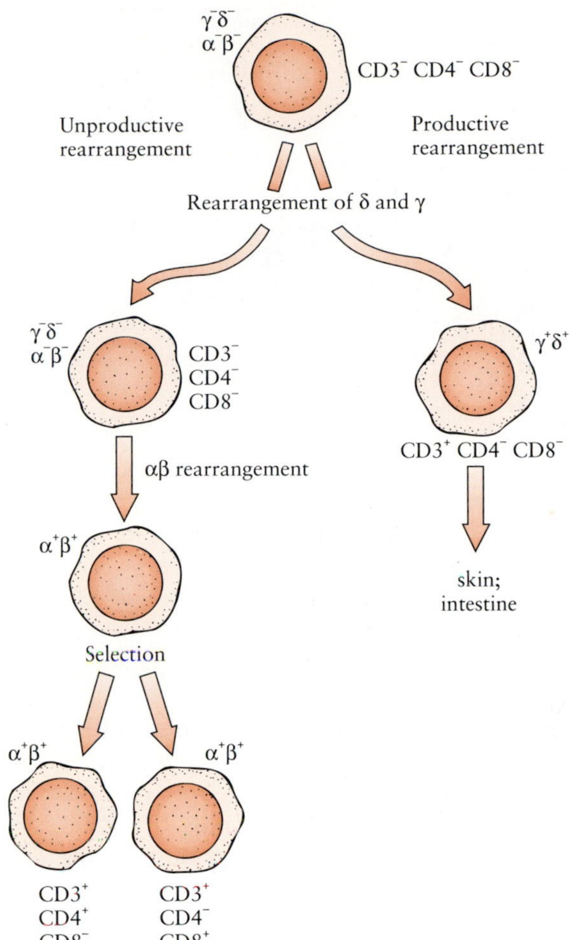

Figure 11–6 The development of T cells showing the initial attempt to generate a γδ receptor followed by formation of an αβ receptor.

require antigen processing. Indeed δγ TCRs tend to be structurally more similar to antibodies (immunoglobulins) and, like them, can recognize free, unprocessed antigens.

THE ACCESSORY MOLECULES

CD3

T-cell antigen receptors are always membrane bound and noncovalently associated with a set of four invariant glycoproteins collectively called CD3. CD3 consists of two dimers, γ-ϵ and δ-ϵ (Fig. 11–7). It has a total size of 66 kDa. The TCR β chain is directly linked to CD3γ-ϵ. The TCR α chain is linked to CD3δ-ϵ. Each CD3 protein contains an N-terminal extracellular domain, a transmembrane domain, and a cytoplasmic domain. When the TCR complex is being assembled on the endoplasmic reticulum of the T cell, it is tran-

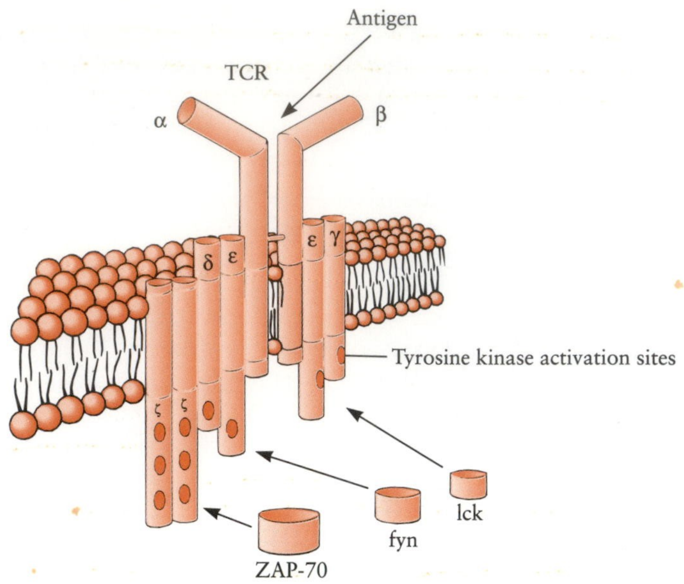

Figure 11–7 The structure of a TCR receptor complex. The CD3 dimers, δ and ε, as well as the ζ–ζ dimer have activation sites that interact with and activate cytoplasmic tyrosine kinases. Three of these tyrosine kinases are ZAP-70, fyn and lck. Ten to twenty percent of TCRs possess δ-η heterodimers instead of δ-δ homodimers.

siently associated with a 28-kDa protein called CD3ω (omega), whose function is unknown. It has been suggested that participation of CD3ω is necessary for proper assembly of the CD3 complex. The CD3 complex is required for signal transduction from the TCR molecule to an intracellular tyrosine kinase.

ζ (Zeta) and η (Eta) Chains

Two additional peptide chains are closely associated with TCR and CD3. These are called ζ and η. The ζ and η chains are structurally and genetically distinct from the CD3 chains. The η peptide is generated by alternative splicing of exons from the ζ gene locus. All T cells carry on their surface a mixture of ζ homodimers and heterodimers. Eighty percent to 90% of TCRs carry a 32-kDa ζ–ζ **homodimer.** The remaining 10% to 20% carry 38-kDa ζ–η heterodimers. The ζ–ζ homodimer plays a role in assembly and expression of the complex as well as signal transduction through tyrosine kinase activation. The ζ–η structure is important in coupling the TCR to inositolphospholipid hydrolysis.

FcεRIγ Chain

It has generally been assumed that signal transduction by γ/δ TCR was mediated by the same peptides as α/β TCR. This is not the case. Although α/β TCRs employ ζ and η chains, most γ/δ TCRs employ the γ chain first associated with the high-affinity IgE receptor (FcεRI) as one unit of its TCR–CD3 complex (this molecule is described in Chapter 29). FcεRIγ occurs as both a homodimer and a heterodimer associated with the TCR ζ chain. Stimulation of γ/δ TCR results in its rapid phosphorylation. Thus it is believed to be a signal-transducing component.

CD4 and CD8

CD4 is a single-chain glycoprotein of 55 kDa. CD8 is a disulfide-linked heterodimer of a 34-kDa subunit (Fig. 10–6). Both are members of the immunoglobulin superfamily. Either CD4 or CD8 is found on mature T cells, although immature T cells may express both. About 65% of human circulating T cells are CD4 positive, and the remaining 30% are CD8 positive. Both CD4 and CD8 are closely associated with the TCR molecule. Their function is to determine the class of MHC molecule that is recognized by a T cell. CD4 is an elongated peptide that can extend from the T-cell surface to bind to a site on the MHC class II molecule. The two N-terminal domains of the CD4 molecule form a binding site for MHC class II molecules. T cells bearing CD4 will thus bind only to antigen associated with these MHC molecules. The N-terminal domains of the CD8 α and β chains in contrast bind specifically to MHC class I molecules. As a result cells bearing CD8 bind only to antigen presented on the surface of virus-infected or other altered cells. These two molecules, CD4 and CD8, thus form receptors that bind conserved structures on the MHC molecules. Antibodies against CD4 block the helper effect of $CD4^+$ cells (a MHC class II restricted reaction), and antibodies against CD8 block the cytotoxicity of $CD8^+$ cells (a MHC class I restricted reaction).

The TCR complexes and CD4 or CD8 molecules are normally unlinked and diffuse independently through the T-cell membrane unless they are brought together when both recognize the MHC–peptide complex. At their normal density on the T-cell surface, CD4 and CD8 do not bind strongly to MHC antigens. However, engagement of the TCR with antigen activates the CD8 so that its avidity for the MHC molecule is increased. The mechanism of this is unknown. The importance of CD4 and CD8 lies in signal transduction. Stimulation of a T cell by an antigen-presenting cell is potentiated about 100-fold when CD4 or CD8 is associated with the TCR. In receptors containing CD4 or CD8 the threshold ligand concentration required to trigger a T-cell response drops 100-fold. The binding of CD4 with MHC class II or of CD8 with MHC class I clearly has a major role in regulating T-cell responses.

COSTIMULATORS

Soluble Factors

The recognition of antigen alone, even when associated with a MHC molecule, is insufficient by itself to trigger a T-cell response. Indeed, the recognition of antigen alone may lead to tolerance through induction of an incomplete T-cell activation signal. In addition to antigen and MHC, other signals are needed to induce an optimal response. One such signal requires adhesion between T cells and antigen-presenting cells. Another signal is generated by soluble mediators such as interleukin-1 and interleukin 12.

IL-1 is secreted by macrophages in response to a wide variety of stimuli, including phagocytosis of foreign particles (see Fig. 6–14). Its release can also be induced when a T cell comes into contact with a macrophage. The T cell probably induces the macrophage to synthesize IL-1 through signals transmitted via the antigen/MHC/TCR complex. T cells may also induce release of interleukin-1 indirectly through lymphokines such as tumor necrosis factor (Chapter 12).

IL-1 may be secreted into the tissue fluid or remain bound to the macrophage surface. Membrane-bound IL-1α, together with properly presented antigen, can stimulate division of adherent T cells. The effects of the two stimuli, antigen and IL-1α, are additive. IL-1 activates a ceramide-activated protein kinase through the sphingomyelin signaling pathway (Fig. 11–8). As a result, the T-cell genes for the transcription factors jun and NF-κβ are activated. The jun protein acts in conjunction with other transcription factors induced by antigen to activate the genes for IL-4.

B cells, which are also antigen-presenting cells, secrete IL-1 too, although they produce much lower levels of IL-1 much more slowly than do macrophages. IL-2 and, to a lesser extent, IL-4 stimulate B-cell IL-1 release, although neither does this to macrophages. Tumor necrosis factor (TNF) does not stimulate B-cell IL-1 release.

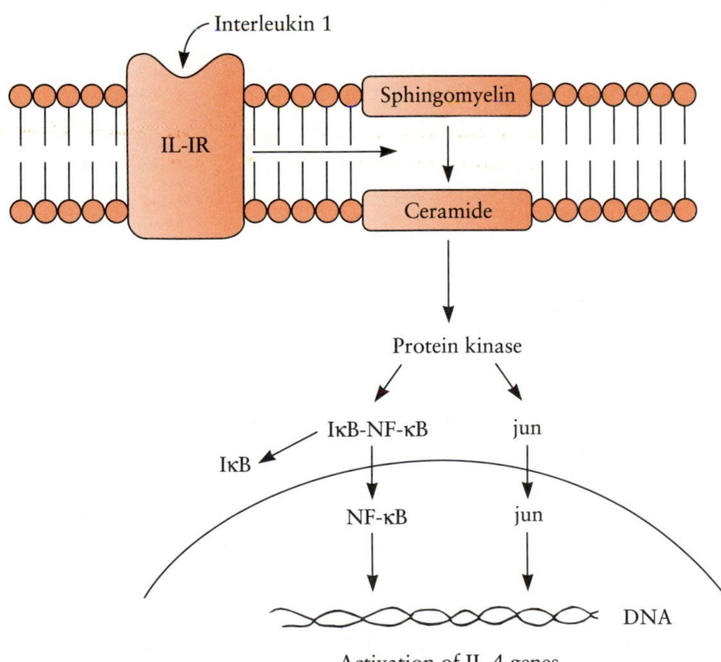

Figure 11–8 Interleukin-1 acts through a ceramide-sensitive protein kinase to generate the transcription factors NF-κB and jun among others.

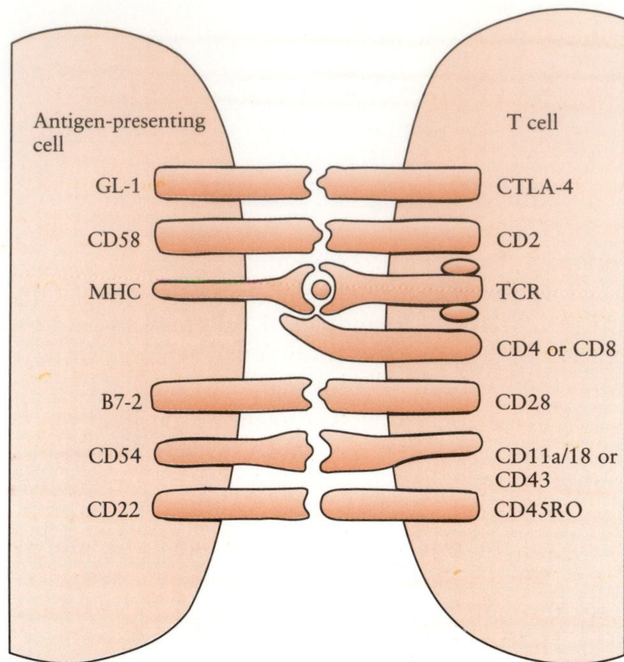

Figure 11–9 Some of the molecules that participate in the interactions between an antigen-presenting cell and a helper T cell.

Cell Adherence Molecules

In addition to the effects of antigen and soluble costimulators such as IL-1, optimal T-cell responses require that adhesion occurs between the T cell and the antigen-presenting cell. This cell adhesion requires binding through proteins such as CD2 and CD58, CD43 and ICAM-1, CD5 and CD72, and CD11a/18 and CD54 (Fig. 11–9). It is believed that this adhesion helps stabilize and support the TCR–antigen bonding and increases the strength of the stimulus transmitted to the T cell. By themselves, these adherence molecules will not turn on a T cell.

Costimulatory Signals

CD28-B7 and Related Interactions. When a T cell encounters an antigen-presenting cell for the first time, the T cell surface receptor CD28 binds to its ligand, a 70-kDa glycoprotein called B7-2 (or B70) on the antigen-presenting cell (Fig. 11–10). B7-2 is expressed on activated macrophages, activated B cells, and dendritic cells. Once the T cell is activated, another higher-affinity ligand for B7-2, called CTLA-4, is expressed. CTLA-4 binds B7-2 with a 20-fold greater affinity than CD28. CD28 and CTLA-4 are both 44-kDa glycoprotein homodimers belonging to the immunoglobulin superfamily. CTLA-4 also binds a B-cell surface glycoprotein of 100 kDa called GL-1. GL-1 is expressed on activated but not resting B cells. The binding of CD28 to B7-2 or of CTLA-4 to B7-2 or GL-1 is required for a successful T-cell response since blockage of these reactions effectively inhibits immune responses and results in the T cell becoming tolerant to the antigen. Stimulation of CD28 or CTLA-4 by some monoclonal antibodies can dramatically enhance T-cell responses to antigen. These antibodies significantly increase expression of IL-2Rα (CD25) and the production of IL-2, IL-3, IL-4, interferon γ (IFN-γ), GM-CSF, and TNF-α. IL-2 production is increased 30-fold. It is suggested that CD28 stimulation prevents degradation of intracellular messengers. In

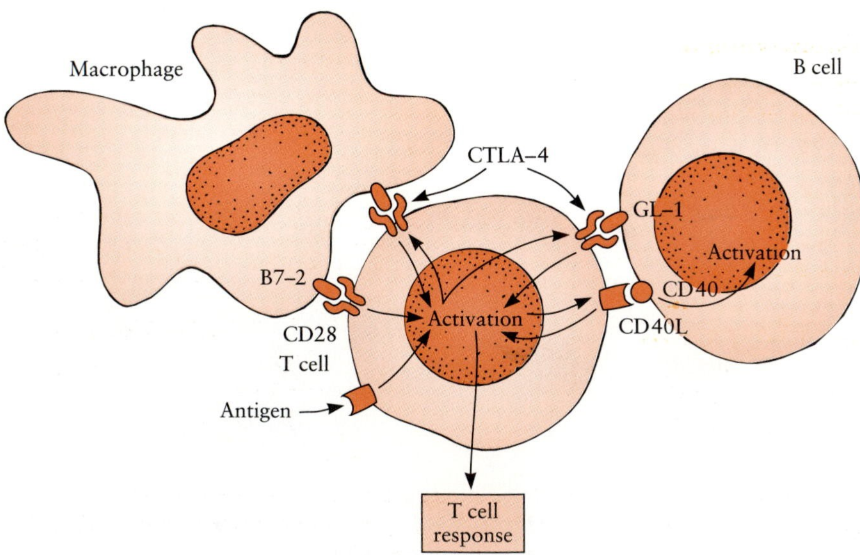

Figure 11–10 The accessory signal molecules of the B7-CD28 pathway. Macrophages use a molecule called B7-2 to bind to CD28 on the resting T cell with moderate affinity. In contrast, activated B cells use a closely related protein called GL-1 to bind to T cell CTLA-4, a protein related to CD28. CTLA-4 is expressed only on activated T cells.

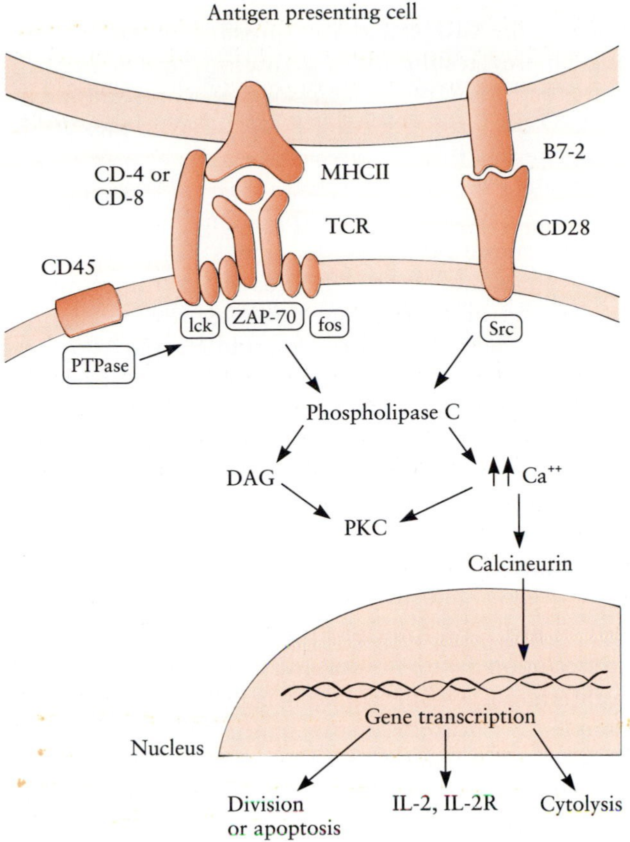

Figure 11–11 The signal transduction pathway of the TCR and CD28 complexes.

the absence of CD28 costimulation, an inhibitor is produced that prevents subsequent proliferation by the T cell. This inhibited state is called **anergy** and involves blockage of both IL-2 gene transcription and responsiveness to IL-4.

CD40 and Its Ligand. CD40 is a 50-kDa glycoprotein expressed constitutively on B cells, dendritic cells, and thymic epithelial cells and is induced on activated macrophages. It binds to a counterreceptor, the CD40-ligand (CD40-L), a 33-kDa type II transmembrane glycoprotein expressed on activated $CD4^+$ T cells. Thus antigen binding to a TCR induces expression of CD40-L. The interaction of CD40–CD40-L induces B cell growth and antibody production. However, it also stimulates the T cell as well and makes it divide in the presence of IL-2. Thus this pathway permits selective expansion of $CD4^+$ T cells after interaction with CD40 bearing antigen-presenting cells. The CD40–CD40-L system is closely linked to the CD28 system described above. CD28-induced cell help is delivered via CD40-L while cross-linking of CD40-L promotes expression of B7-2.

SIGNAL TRANSDUCTION

The first event that follows antigen binding to the TCR is transmission of a signal from the antigen-binding TCR α and β chains to the signal-transducing chains of CD3 and the ζ chain (Fig. 11–11). This results in activation of three cytoplasmic tyrosine kinases. Two of the tyrosine kinases involved have been identified as fyn and lck (Chapter 7). The lck tyrosine kinase is a unique tyrosine kinase expressed only in T cells and NK cells. It is a 56-kDa peptide bound to the cytoplasmic face of the cell membrane and noncovalently associated with the cytoplasmic domains of CD4 and CD8α, although it can act independently of these molecules. (It should also be pointed out that there is much more lck associated with CD4 [90% of cellular lck] than with CD8 [10%].) A third tyrosine kinase called ZAP-70 (ζ-associated protein of 70 kDa) also plays an important role in this process and may be the most important of the three. The CD45 **isoforms** have phosphotyrosine phosphatase activity and can also interact with lck. CD45 removes phosphate from a tyrosine in lck, resulting in activation of the kinase. Thus CD45 also regulates T-cell responses to antigen (Fig. 11–12).

These tyrosine kinases phosphorylate and so activate an isoform of cell membrane phospholipase C. The phospholipase C hydrolyzes phosphatidylinositol bisphosphate to produce the two second messengers, inositol trisphosphate and diacylglycerol. The inositol trisphosphate then releases calcium ions from intra-

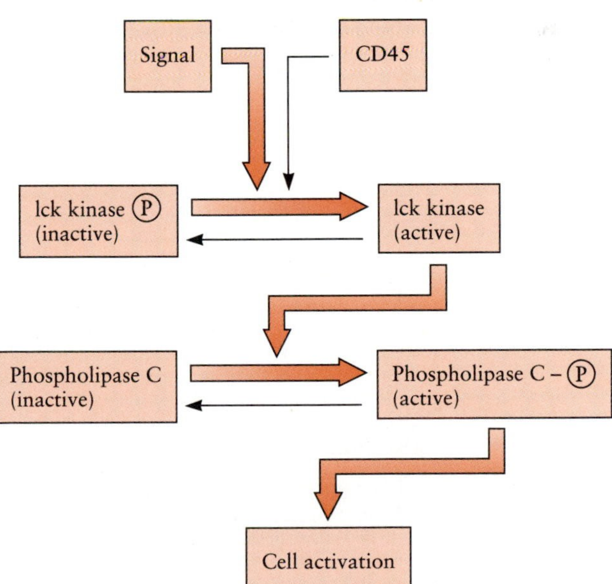

Figure 11–12 The regulation of T-cell responses requires the action of CD45 to activate tyrosine kinases before they can act as signal transducers.

cellular organelles and opens transmembrane channels, allowing Ca^{2+} to enter the cell. As a result, there is a substantial increase in cytosolic free Ca^{2+} levels. This rise in intracellular calcium provides one of two signals required for cell activation. The intracellular calcium also causes **protein kinase C** to translocate from the cytoplasm to the cell membrane, where it is activated by diacylglycerol. The lck tyrosine kinase also activates a protein kinase called erk, which then activates additional protein kinases.

The increase in intracellular calcium leads to the activation of calmodulin-regulated enzymes involved in lymphokine gene expression. One of the most important of these is calcineurin, a phosphatase that acts on a transcription factor that regulates IL-2 gene expression.

One of the key signaling pathways that appears to be critical to normal cell function involves the G-protein ras. Like other **G-proteins,** ras acts as a chemical switch located on the inner surface of the cell surface membrane. The activity of ras proteins is regulated by two other proteins (Fig. 11–13). Thus GTPase-activating proteins (GAPs) turn the ras protein off by accelerating the hydrolysis of GTP to GDP. Another class of proteins called guanine nucleotide releasing factors (GRFs) directly activate ras. They catalyze the release of bound GDP from ras and so permit GTP to bind. Thus the relative activities of GRFs and GAPs acting on ras determine its activation state.

Ras is a key protein in a major common pathway of signaling from the cell surface to the nucleus. Thus in the case of T cells, membrane-bound ras is activated by stimulation of the TCR–CD3 complex. Phosphorylation of the GRF called Vav (possibly by the lck tyrosine kinase) results in its activation, and it in turn causes activation of ras. Once activated, ras activates a related proto-oncogene called raf, which in turn phosphorylates a protein kinase and passes signals to the cell nucleus through phosphorylation of transcription factors that regulate gene expression.

The activation of protein kinases also turns on certain genes within the T cell. Thus several transcription factors known to regulate gene expression are induced and activated. These include NF-κB, AP-1, AP-3, and NF-AT. This gene activation occurs in waves. Thus two of the most important of the molecules activated within 15 to 20 minutes are fos and jun. Fos is induced by stimulation of the TCR by antigen, and jun is induced by the binding of IL-1. The fos and jun proteins bind to form a heterodimer called AP-1. AP-1 is a transcription factor required to induce a second wave of gene activation for IL-2, IL-2R, IL-3, IL-4, IL-5, IL-6, myc, and interferon γ (Fig. 11–14). (Myc is a transcription factor important in cell activation and proliferation.) Appropriate triggering of the TCR complex also results in the synthesis and release of GM-CSF, IFN-α, and TNF-α. The IL-2 and IL-2R genes are switched on first. Slightly later the IL-3 gene is switched on followed by a burst of lymphokine production involving IL-4, IL-5, and IL-6. This is followed by cell division under the control of IL-4 and IL-2. The genes for serine esterases, which probably have a role in cytotoxicity, are activated in a third wave, several days after stimulation by antigen (Fig. 11–15). The net result of all this is that the T cell is driven from the G_0 phase into the G_1 phase, enlarges, and enters the cell cycle. Finally, a group of very late activating (VLA) molecules is produced 7 to 14 days after stimulation.

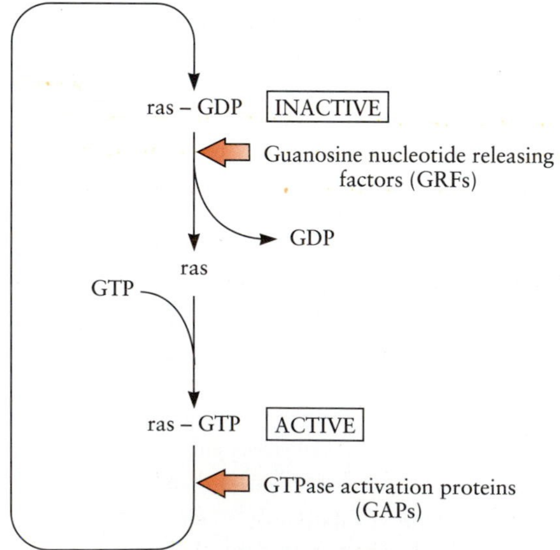

Figure 11–13 The role of ras in signal transduction. The activation of ras is controlled by GRFs and GAPs.

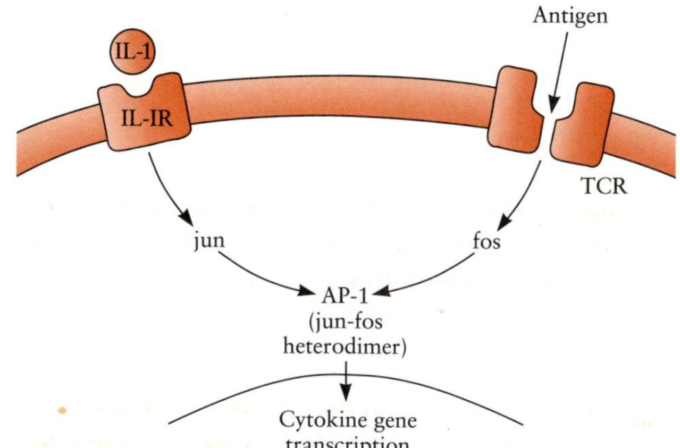

Figure 11–14 The interaction of the IL-1 and TCR pathways leads to the production of the transcription factor AP-1 and cytokine production.

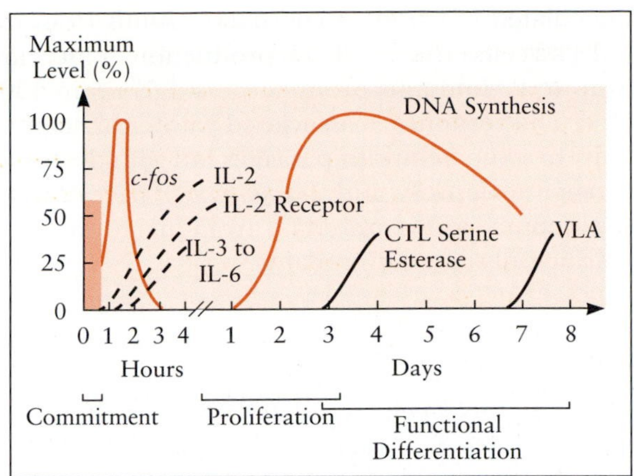

Figure 11–15 The sequence of events in T-lymphocyte activation after antigen-MHC triggering. Commitment refers to the time after which the removal of the primary stimulus no longer limits progression. The time courses shown are for the activation of transcription as indicated by levels of messenger RNA rather than protein expression. Note how gene activation occurs in waves. Thus *c-fos* is induced within minutes, the interleukin genes within a few hours, serine esterases after several days, and very late antigens (VLA) after about seven days. *(Reprinted by permission of the New England Journal of Medicine, Vol. 322, 515, 1991.)*

These VLA molecules belong to the integrin family of cell adherence molecules.

When CD28 binds its ligand B7-2 on the antigen-presenting cell, it also causes tyrosine phosphorylation. The tyrosine kinase acts on phospholipase C to promote calcium-dependent effects through calcineurin. As a result of CD28 signaling, the half-life of the mRNAs for TNF-α, IL-2, GM-CSF, and IFN-γ are all increased.

BIOLOGICAL CONSEQUENCES

It must not be assumed that the stimulation of a T cell by appropriately presented antigen is an all-or-nothing affair. In fact, it is becoming increasingly clear that T cells may become partially activated so that they express some, but not all, functions. The following sections should be read with this in mind.

Activation Antigens

When T cells are activated by appropriate exposure to antigen, they express a set of new surface proteins. These include CD25, CD26, CD30, CD69, CDw70, and CD71. CD25 is the α chain of the IL-2 receptor. The functions of CD26 are unknown. CD30 is structurally similar to the TNF receptor. CD69 and CDw70 are regulatory proteins associated in some way with protein kinase C. CD71 is the transferrin receptor. Cells require iron loading for entry into the S phase of the cell cycle. IL-2 induces the appearance of CD71 on the cell surface as early as 20 hours after stimulation. This protein binds transferrin with its associated iron. The complex is carried into the cytosol, where the iron is transferred to ferritin and the apotransferrin and CD21 are recycled to the cell surface. The iron is required for the proper functioning of cellular microtubules.

HELPER T-CELL SUBPOPULATIONS

In mice and humans there are two distinct subsets of helper T cells distinguishable by the mixture of lymphokines that they secrete. They are called helper 1 (Th1) and helper 2 (Th2) T cells (Fig. 11–16). These cells respond preferentially to antigen and co-stimulators presented by different antigen-presenting cell populations. Thus Th2 cells respond optimally to antigen presented by macrophages and less well to antigen presented by B cells. Th1 cells in contrast respond optimally to antigen presented by B cells. The relative proportions of these antigen-presenting cells may determine the nature of the helper T-cell response.

Th1 cells

Th1 cells characteristically secrete IL-2, IFN-γ, and lymphotoxin (TNF-β) within a few hours after stimulation by antigen. Although these lymphokines can promote B-cell proliferation and **immunoglobulin** secretion, especially IgM and IgG2a antibodies, they do not stimulate specific antibody formation. Th1 cells primarily act as helper cells for cell-mediated immune responses such as the delayed hypersensitivity reaction and macrophage activation, leading to increased resistance to intracellular parasites such as *Mycobacterium leprae* (the cause of leprosy) and *Leishmania.* The co-stimulator of Th1 cells is interleukin-12 secreted by B cells or macrophages. Th1 cells lack IL-1 receptors and will not therefore respond to IL-1.

Th2 cells

Th2 cells secrete IL-4, IL-5, IL-10, and IL-13 several days after exposure to antigen. This mixture of lymphokines stimulates B-cell proliferation and polyclonal immunoglobulin secretion but cannot act as helper cells in the delayed hypersensitivity or other

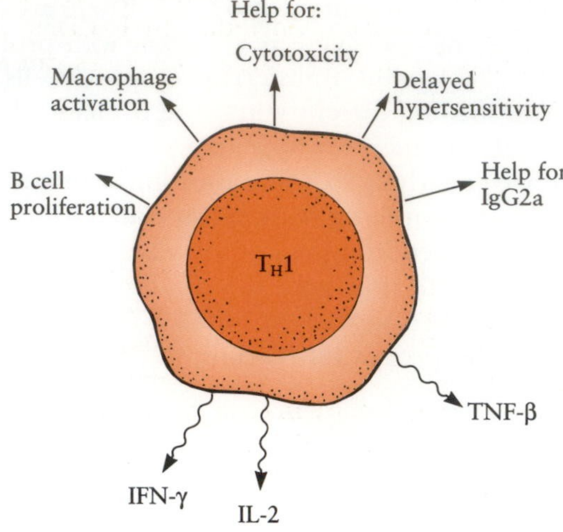

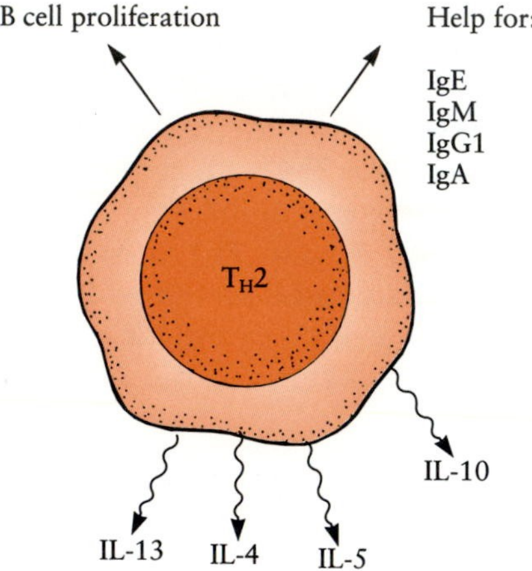

Figure 11–16 The differences between Th1 and Th2 cells. Th1 cells help different lymphocyte responses and secrete a different mixture of cytokines than Th2 cells.

cell-mediated reactions. Supernatant fluids from cultured Th2 cells enhance B-cell production of IgG1 and IgA up to 20-fold and production of IgE up to 1000-fold. A Th2 response is associated with enhanced immunity to some helminth parasites but with decreased resistance to leprosy and *Leishmania*. Th2 cells have IL-1 receptors and require IL-1 from macrophages or dendritic cells as a costimulator.

Th0 cells

A third helper T-cell population has also been described in mice that secretes a lymphokine mixture that is representative of both Th1 and Th2 cells. These cells, called Th0 cells, may be precursors of Th1 and Th2 or cells that are in transition between the two populations. They secrete IL-2, IL-4, IL-5, and IFN-γ. There is indeed evidence that some IL-2–secreting T cells may become IL-4–secreting cells after exposure to antigen, implying the existence of an interleukin switch and a change in phenotype from Th1 to Th2. The principal molecules that control this switch are IL-4 and IL-12. When cultured in the presence of IL-4, Th0 cells become Th2 cells. When cultured in the presence of IL-12, they become Th1 cells.

Although there are clearly three populations of helper T cells in mice, this is less clear in humans. A large proportion of human $CD4^+$ cells produce multiple lymphokines, including IL-2, IL-4, and IFN-γ and thus are Th0-like. However, human T cells that promote cell-mediated responses to mycobacterial antigens have a Th1-like phenotype (Fig. 11–17). In contrast, human T cells that are specific for allergens or helminth antigens are Th2-like.

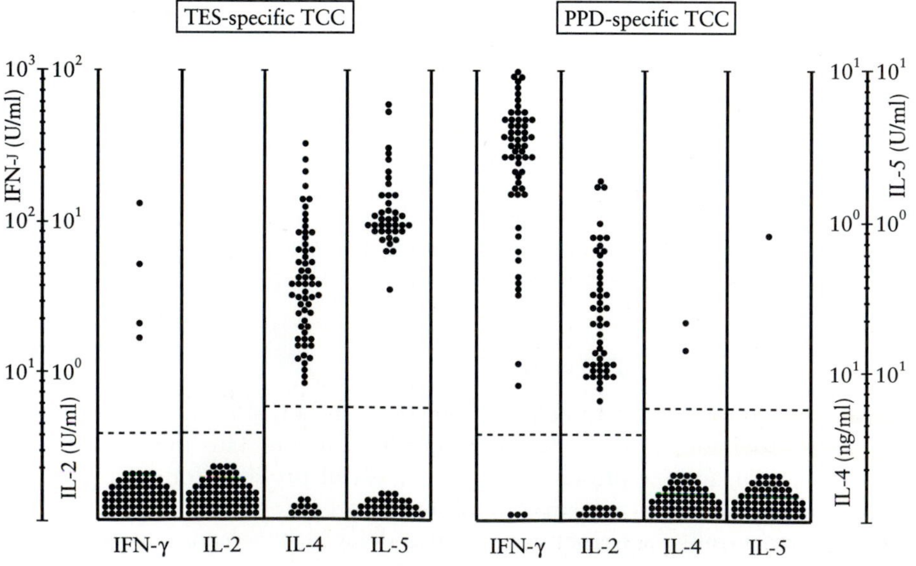

Figure 11–17 Cytokine production by T cells responding to either a parasite antigen (*Toxocara canis* secretory antigen—TES) or to a Mycobacterial antigen (purified protein derivative—PPD). Cytokines were measured in the supernatant fluids of antigen stimulated T cells. The response to TES is a Th2 response characterized by secretion of IL-4 and IL-5. The response to PPD is a Th1 response characterized by secretion of IFN-γ and IL-2. *(From Del Prete, G., et al. J. Clin. Invest., 88:346–350, 1991. With permission.)*

ALTERNATIVE PATHWAYS

In addition to the major activation pathway described previously, there are several alternative pathways for T-cell activation. The most important of these pathways is mediated through CD2 and its ligand CD58. Thus antibodies directed against CD2 can readily provoke T cells to divide. Since CD2 appears early in **ontogeny,** its stimulation may activate cells that have not yet developed their TCR–CD3 complex. CD2 stimulation may also potentiate the response of cells responding by the classical TCR-mediated pathway. Activation of human T cells through the CD2 pathway leads to tyrosine kinase activation and eventually to expression of the IL-2 receptor and synthesis of IL-2 and IFN-γ. Other cell surface proteins that may play a role in T-cell stimulation include Thy-1, CD59, and CD28. Antibodies to any of these molecules can promote T-cell division.

THE DEVELOPMENT OF T CELLS

Mouse T cells arise from stem cells originating within the fetal liver and bone marrow, where they eventually develop into prothymocytes carrying small amounts of Thy-1 and TL antigens (Fig. 11–18). When these cells colonize the thymus early in fetal life, they divide and differentiate. On leaving the thymus they give rise to T cells that populate different organs. In mice, T cells

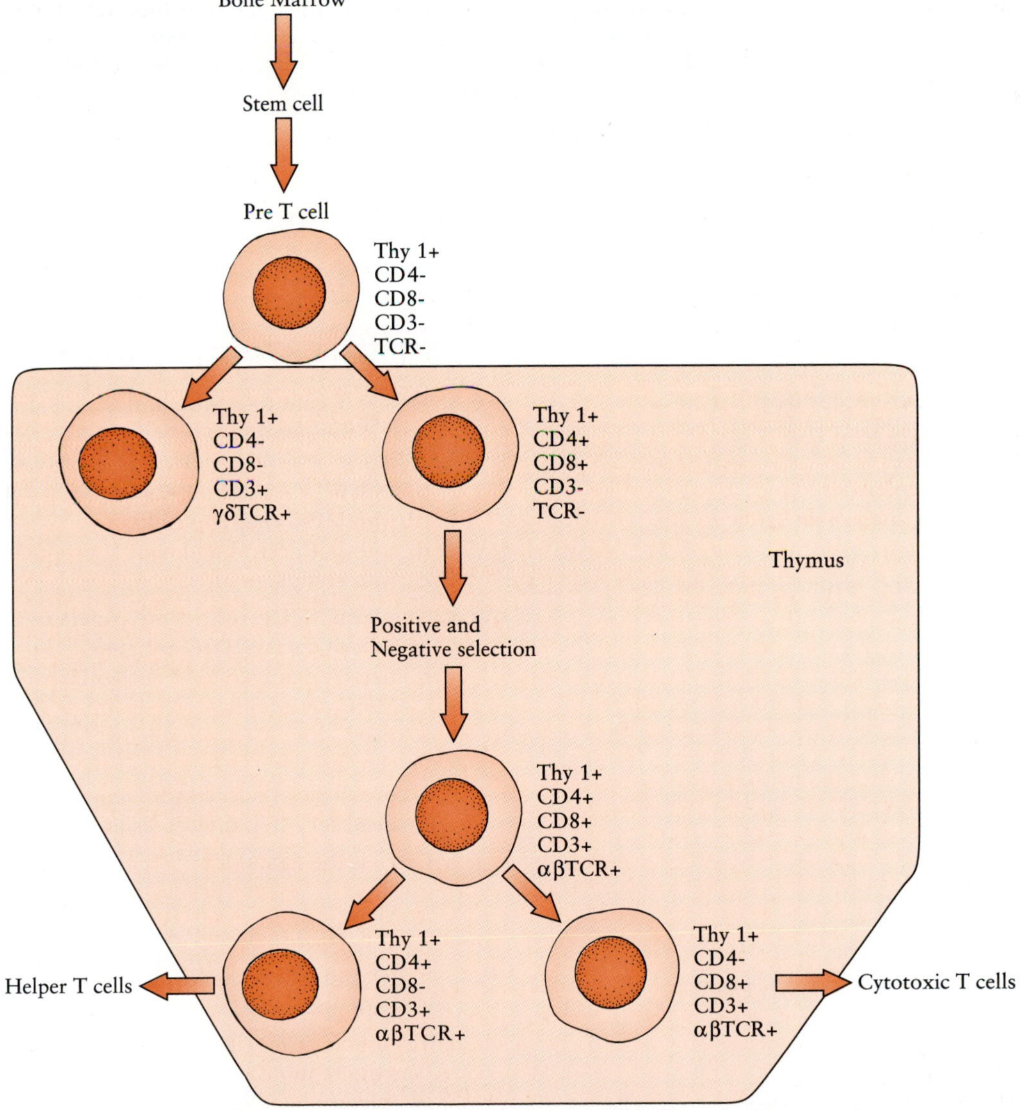

Figure 11–18 A simplified view of the development of T cells.

expressing TCR γ/δ leave the thymus first and emigrate to the skin. A second wave of γ/δ T cells then moves to the lining of the female reproductive organs and in tongue epithelium. A third wave colonizes the spleen, and a fourth wave colonizes the lining of the gastrointestinal tract. Later in development there is a switch to cells with α/β TCR. These cells predominate in the thymus and lymphoid tissues, although γ/δ cells may be found in epithelial tissues. A similar maturation process is seen in humans. The order in which waves of stem cells develop matches the order in which the genes for the different gamma-chain isoforms are arranged in the TCR gene. Developing T cells express different amounts of CD4 and CD8. Thus double-negative ($CD4^{-}8^{-}$) TCR T cells develop through a complicated series of intermediate steps, including a transient double-positive period ($CD4^{+}8^{+}$) to emerge as single-positive subsets that possess a selected TCR repertoire. When T cells leave the mouse thymus, they carry a persistent residue of Thy-1 but differentiate into the major subpopulations on the basis of whether they possess CD8 or CD4.

KEY WORDS

Activation antigen p. 149
α/β TCR p. 141
Alternative pathway p. 151
Antigen-presenting cell p. 140
B7-2 p. 145
Calcineurin p. 147
CD2 p. 151
CD3 p. 143
CD4 p. 144
CD8 p. 144
CD28 p. 145
CD40 p. 146
CD40-ligand p. 146
CD45 (Leukocyte common antigen) p. 147
fos p. 148
jun p. 145
Complementarity-determining region p. 142
CTLA-4 p. 146
Diacylglycerol p. 147
γ/δ T cells p. 142
G-protein p. 147
Helper T cell p. 140
Inositol trisphosphate p. 147
Interleukin-1 p. 145
lck tyrosine kinase p. 147
MHC restriction p. 140
ras p. 147
Signal transduction p. 146
T cell antigen receptor (TCR) p. 140
Th0 cells p. 150
Th1 cells p. 149
Th2 cells p. 149
Tyrosine kinase p. 147
Variable region p. 141
ζ and η chains p. 144

QUESTIONS

1. The T-cell antigen receptor may contain
   a. alpha chains and beta chains
   b. CD3 molecules
   c. CD4 or CD8 molecules
   d. gamma chains and delta chains
   e. all of the above
2. Helper T cells are distinguished by having which marker?
   a. CD2
   b. CD4
   c. CD3
   d. IL-2 receptor
   e. MHC class II
3. The major cytokines secreted by Th1 cells are
   a. IL-1 and IL-6
   b. IL-4 and IL-5
   c. IL-2 and IFN-γ
   d. IL-10 and IL-12
   e. TNF–α and GM-CSF
4. The helper cell population that promotes IgE production is composed of
   a. Th1 cells
   b. Th2 cells
   c. Th0 cells
   d. Tdh cells
   e. Te cells
5. Helper T cells recognize antigen on antigen-presenting cells as antigen
   a. alone
   b. with MHC class I product
   c. with MHC class II product
   d. with both class I and class II products
   e. with complement
6. The term "MHC restriction" refers to the
   a. inheritance of MHC antigens
   b. need for antigen to be recognized in association with MHC molecules
   c. need for MHC genes to control complement activity
   d. problems associated with allograft rejection
   e. need for MHC molecules in order to reject grafts
7. CD28 is the ligand for
   a. IL-1
   b. MHC class II
   c. antigen
   d. B7 and related molecules
   e. endotoxin
8. T cells with γ/δ TCRs are commonly found in high numbers in the mouse
   a. spleen
   b. lungs
   c. lymph nodes

d. liver
e. intestinal wall

9. The most important peptide chain involved in signal transduction through CD3 is
   a. γ chain
   b. δ chain
   c. ϵ chain
   d. ζ chain
   e. η chain
10. To induce an immune response, antigen-presenting cells must also secrete
    a. IL-6
    b. IL-2
    c. TNF-α
    d. IL-1
    e. IFN-γ
11. What are the advantages of possessing helper T cells? Would it not be more efficient if B cells could respond to antigen without requiring T-cell help?
12. Speculate on the possible functions of the γ/δ TCR present on some immature T cells. Why might two types of TCR have evolved?
13. What is the biological basis of MHC restriction between antigen-presenting cells and helper T cells?
14. Outline the ontogeny of human T cells. Why does it have to be so complex?
15. What function do helper T cells play in an antibody response?
16. Discuss helper T-cell heterogeneity. How many T-cell subsets can you identify? Draw up a table showing their identifying features and probable functions.
17. Outline the biochemical consequences of triggering a TCR with appropriately presented antigen. How do these consequences differ from those triggered by other stimuli such as interleukin-1?

Answers: 1e, 2b, 3c, 4b, 5c, 6b, 7d, 8e, 9d, 10d

SOURCES OF ADDITIONAL INFORMATION

Abehsira-Amar, O., et al. IL-4 plays a dominant role in the differential development of Th0 into Th1 and Th2 cells. J. Immunol., 148:3820–3829, 1992.

Abraham, R.T., et al. Signal transduction through the T-cell antigen receptor. TIBS, 17:434–438, 1992.

Acuto, O., and Reinherz, E.L. The human T cell receptor, structure and function. N. Engl. J. Med., 312:1100–1111, 1985.

Alexander, D.R., and Cantrell, D.A. Kinases and phosphatases in T-cell activation. Immunol. Today, 10:200–205, 1989.

Altman, A., Coggeshall, K.M., and Mustelin, T. Molecular events mediating T cell activation. Adv. Immunol., 48:227–360, 1990.

Armitage, R.J., et al. CD40L: A multi-functional ligand. Semin. Immunol., 5:401–412, 1993.

Bottomly, K. A functional dichotomy in CD4+ T lymphocytes. Immunol. Today, 9:268–273, 1988.

Boussiotis, V.A., et al. Activated human B lymphocytes express three CTLA-4 counterreceptors that costimulate T-cell activation. Proc. Natl. Acad. Sci. U. S. A., 90:11059–11063, 1993.

Cayabyab, M., Phillips, J.M., and Lanier, L.L. CD40 preferentially costimulates activation of $CD4^+$ T lymphocytes. J. Immunol., 152:1523–1528, 1994.

Cerottini, J.C., and MacDonald, H.R. The cellular basis of T cell memory. Annu. Rev. Immunol., 7:77–89, 1989.

Coffman, R.L. T-helper heterogeneity and immune response patterns. Hosp. Pract. [off.], 24:101–133, 1989.

Crabtree, G.R. Contingent genetic regulatory events in T lymphocyte activation. Science, 243:355–361, 1989.

Fiorentino, D.F., Bond, M.W., and Mossmann, T.R. Two types of mouse helper T cell: IV. Th2 clones secrete a factor that inhibits cytokine production by Th1 clones. J. Exp. Med., 170:2081–2095, 1989.

Grey, H.M., Sette, A., and Buus, S. How T cells see antigen. Sci. Am., 261:56–64, 1989.

Harding, C.V., and Unanue, E.R. Quantitation of antigen-presenting cell MHC class II/peptide complexes necessary for T cell stimulation. Nature, 346:574–576, 1990.

Hodgkin, P.D., and Kehry, M.R. The mechanism of T and B cell collaboration. Immunol. Cell Biol., 70:153–158, 1992.

Janeway, C.A. How the immune system recognizes invaders. Sci. Am., 269:73–79, 1993.

Kabelitz, D. Do CD2 and CD3-TCR T-cell activation pathways function independently? Immunol. Today, 11:44–46, 1990.

Kolesnick, R., and Golde, D. W. The sphinogomyelin pathway in tumor necrosis factor and interleukin-1 signalling. Cell, 77:325–328, 1994.

Kopf, M., et al. Disruption of the murine IL-4 gene blocks Th2 cytokine responses. Nature, 362:245–247, 1993.

Kourilsky, P., and Claverie, J-M. MHC-antigen interaction: What does the T cell receptor see? Adv. Immunol., 45:107–193, 1989.

Linsley, P.S., and Ledbetter, J.A. The role of the CD28 receptor during T cell responses to antigen. Annu. Rev. Immunol., 11:191–212, 1993.

Makgoba, M.W., Sanders, M.E., and Shaw, S. The CD2-LFA-3 and LFA-1-ICAM pathways: Relevance to T-cell recognition. Immunol. Today, 10:417–422, 1989.

Matis, L.A. The molecular basis of T cell specificity. Annu. Rev. Immunol., 8:65–82, 1990.

Mosmann, T.R., and Coffman, R.L. Heterogeneity of cytokine secretion patterns and functions of T helper cells. Adv. Immunol., 46:111–147, 1989.

Muñoz, E., et al. IL-1 signal transduction pathways I. Two functional IL-1 receptors are expressed in T cells. J. Immunol., 146:136–143, 1991.

Nesig, A., et al. Assembly of the T-cell antigen receptor. Participation of the CD3ω chain. J. Immunol., 151:870–879, 1993.

Ødum, N., et al. MHC class II molecules deliver costimulatory signals in human T cells through a functional linkage with IL-2-receptors. J. Immunol., 150:5289–5298, 1993.

Podolin, P.L. T cell activation. Clin Biotechnology, 3:137–150, 1991.

Quian, D., et al. The γ chain of the high-affinity receptor for IgE is a major functional subunit of the T-cell antigen receptor complex in γδ T lymphocytes. Proc. Natl. Acad. Sci. U. S. A., 90:11875–11879, 1993.

Robertson, M. T cell receptor—the present state of recognition. Nature, 317:768–771, 1985.

Schild, H., et al. The nature of major histocompatibility complex recognition by γδ T cells. Cells 76:29–37, 1994.

Schwartz, R.H. Costimulation of T lymphocytes: The role of CD28, CTLA-4, and B7/BB1 in interleukin-2 production and immunotherapy. Cell, 71:1065–1068, 1992.

Trinchieri, G. Interleukin 12 and its role in the generation of Th1 cells. Immunol. Today, 14:335–339, 1993.

Weaver, C.T, and Unanue, E.R. The costimulatory function of antigen-presenting cells. Immunol. Today, 11:49–55, 1990.

CHAPTER 12

Lymphokines and Cytokines

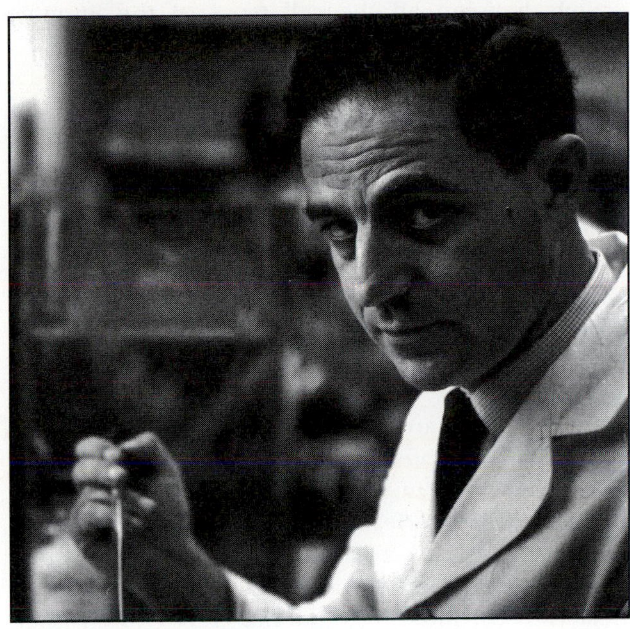

Alic Isaacs *Born in Britain in 1921, Alic Isaacs together with Jean Lindenmann discovered the antiviral substance that they called interferon in 1957. They were working on the phenomenon of viral interference. This is the process by which infection of cells by one virus prevents invasion by a second virus. They showed that when they treated a cell culture with a killed virus, the cell free fluid had the ability to make other cells resistant to virus infection. They isolated a protein from the culture fluid that had this antiviral property and called it interferon. This was the first cytokine to be identified.*

CHAPTER OUTLINE

CHAPTER CONCEPTS

1. T cells secrete many proteins and glycoproteins in response to appropriate stimulation.
2. The proteins secreted by lymphocytes are called lymphokines. The most important of these lymphokines are the interleukins.
3. The interleukins either enhance interactions between lymphocytes or act as cell growth stimulants. The most important of these are interleukin-2, secreted by Th1 cells, and interleukin-1, secreted by macrophages.
4. Th1 and Th2 cells secrete two distinct patterns of cytokines that result in occurrence of a very different immune response.
5. The interferons are antiviral or regulatory proteins. Interferon-γ is secreted by T cells. It has complex effects, including blockage of viral replication, activation of macrophages, and inhibition of tumor cell growth.
6. Lymphotoxins and tumor necrosis factors are proteins from macrophages or T cells that have the ability to kill some tumor cells or other abnormal cells.
7. Lymphocytes also secrete a mixture of growth factors that promote development of many types of cells in the bone marrow.

This chapter is devoted to a review of the properties of the proteins secreted by the cells of the immune system called cytokines or lymphokines. We review only the most important of these since there is an enormous number of them. Although each protein may be secreted by many different cell types, we examine them in relation to their major cell source. Thus we look again briefly at the major proteins produced by macrophages. Then we describe the proteins secreted by each of the two major populations of helper T cells. We also look at proteins secreted by cytotoxic T cells. We briefly review those proteins, derived from a variety of cells, that act as growth factors for cells of the immune system. Finally, the receptors for cytokines are examined and their classification described.

As a result of appropriate stimulation, cells of the immune system secrete a bewildering variety of proteins that mediate signaling between cells and so control the immune responses. The generic term for these regulatory proteins secreted by cells is **cytokine.** Cytokines secreted by lymphocytes are called **lymphokines.** At least 90 lymphokine-mediated activities have been recognized. In general, lymphokines are neither antigen-binding nor antigen-specific. Cytokines differ from conventional hormones in three important respects. First, they have effects on many different cell types, unlike classical hormones that tend to affect a single target organ. Second, cells rarely secrete only one cytokine at a time. Thus macrophages secrete at least four: IL-1, IL-6, IL-12, and TNF-α. Likewise, T cells usually secrete a complex mixture of lymphokines. Third, they appear to be "redundant" in their biological activities in that several different cytokines may have similar activities. For example, IL-1, TNF-α, TNF-β, IL-6, and MIP (macrophage inflammatory protein) all cause a fever. This has given rise to the concept of a cytokine network, a complex web of interactions between all the cell types of the immune system mediated by many different proteins.

THE MAIN CLASSES OF CYTOKINE

Interleukins are cytokines that regulate the interactions between lymphocytes and other leukocytes (Table 12–1). They are numbered sequentially in the order of their discovery. Because their definition is so broad, the interleukins are a heterogeneous mixture of proteins with very little in common except their name.

Interferons are glycoproteins synthesized in response to virus infection, to immune stimulation, or to a variety of chemical stimulators. They inhibit virus replication by interfering with viral RNA and protein synthesis (Table 12–2). Four major types of interferon are recognized: interferon alpha (IFN-α), interferon beta (IFN-β), interferon gamma (IFN-γ), and interferon omega (IFN-ω). Only IFN-γ is considered a true cytokine.

Tumor necrosis factors are two related cytokines, one derived from macrophages and one from T cells. As their name suggests, they both have the ability to kill tumor cells (Table 12–3).

Many cytokines act as **growth factors.** Thus they stimulate many cell populations to grow. They are

Table 12–1 The Interleukins

Name	MW(kDa)	Mainly Produced By	Major Targets	Receptor MW
IL-1α	17	Macrophages	Th2 cells, B cells	85
IL-1β	17	Macrophages	Th2 cells, B cells	85
IL-2	15	Th1 cells, NK cells	T cells, B cells	55/70
IL-3	25	T cells	Hematopoietic cells, stem cells	65/130
IL-4	20	Th2 cells	T cells, B cells, mast cells	140
IL-5	18	Th2 cells	B cells, T cells	60/130
IL-6	26	Fibroblasts, T cells	B cells	80/130
IL-7	25	Stromal cells	Immature lymphocytes	70
IL-8	8	Macrophages	T cells, neutrophils	67 + 59
IL-9	39	Th2 cells	T cells	64
IL-10	19	Th2 cells, B cells	Th1 cells	100
IL-11	24	Stromal cells	B cells	151
IL-12	75	Macrophages, B cells	Th1 cells, NK cells	110
IL-13	10	Th2 cells	B cells	?
IL-14	53	B cells	B cells	?
IL-15	15	Mononuclear cells	T cells	?

MW = molecular weight.

Table 12–2 The Interferons

Name	MW(kDa)	Mainly Produced By	Major Targets	Receptor MW
IFN-α	16–25	Macrophages	Many	120
IFN-β	16–25	Fibroblasts	Many	120
IFN-γ	20–25	Th1 cells, NK cells	Macrophages, many	90
IFN-ω	16–25	Trophoblast cells	Uterine cells	120

MW = molecular weight.

therefore very important in ensuring that the body is supplied with sufficient cells to defend itself.

Chemokines are a family of small proteins that play an important role in inflammatory reactions. A typical example of a chemokine is interleukin-8. Chemokines are discussed in Chapter 28.

CYTOKINES PRODUCED MAINLY BY MACROPHAGES

As described in Chapter 6, macrophages produce four major cytokines, namely IL-1, IL-6, IL-12, and TNF-α.

Interleukin-1

The properties of IL-1 are described in Chapters 11 and 28. IL-1 is the major costimulator of Th2 cells and is thus required to induce the production of their lymphokine mixture.

Interleukin-6

IL-6 has many properties in common with IL-1. It can act on many cell types (see Fig. 6–16). Thus it acts on B cells as a cofactor with IL-1 in IgM synthesis and IL-5 in IgA synthesis. Acting with IL-1, IL-6 promotes IL-2 and IL-2R production and differentiation of T cells. It induces T cells to move from G_0 to G_1, where they become more responsive to IL-2. IL-6 and IL-1 are both needed for optimal CD4 cell proliferation, while IL-6 alone is sufficient for CD8 cells. IL-6, like IL-1, plays an important role in the systemic inflammatory response. It acts with IL-1 to stimulate the production of acute-phase proteins by hepatocytes. It causes a fever, although it is much less potent than IL-1 or TNF-α. IL-6 stimulates formation of hematopoietic colonies from the bone marrow. IL-6 has weak antiviral activity and was once classified as an interferon.

The IL-6 receptor (CD126/CDw130) is a dimeric glycoprotein found on normal resting T cells or freshly activated B cells. One chain acts as the receptor, and the other chain is a signal transducer. T cells downregulate their IL-6 receptors on activation. B cells acquire IL-6 receptors when fully mature.

Interleukin-12

IL-12, produced by all the antigen-processing cells (macrophages, dendritic cells, and B cells), acts on Th1 cells and NK cells (see Fig. 6–17). It promotes Th1 cell differentiation from the Th0 stage. It is the costimulator of Th1 activity, inducing those cells to secrete IL-2 and IFN-γ and express IL-2R. IL-12 also inhibits some Th2 cell functions such as IgE formation by IL-4–stimulated B cells. A positive feedback loop exists since IFN-γ enhances IL-12 production. IL-12 enhances T and NK cell proliferation, enhances the cytotoxic activities of T cells in association with IL-2, and induces lymphokine-activated killer (LAK) cell production in association with TNF-α.

Tumor Necrosis Factor–α

TNF-α is a 17-kDa nonglycosylated protein produced by macrophages that acts as an inflammatory mediator and as a growth factor (see Fig. 6–18). Its production is triggered mainly by gram-negative bacteria, but it can also be produced in response to viruses, tumor cells, bacterial toxins, C5a, fungi, parasites, and IL-2.

Table 12–3 The Tumor Necrosis Factors

Name	MW(kDa)	Mainly Produced By	Major Targets	Receptor MW
TNF-β/LT	25	T cells	Tumors	55 + 75 + 95
TNF-α	17	Macrophages	Tumors	55 + 75 + 95

MW = molecular weight.

TNF-α can constitute as much as 1% of the secretory product of activated macrophages. IFN-γ enhances TNF-α secretion, whereas steroids downregulate its production. TNF-α may also be an integral membrane protein on activated T cells and macrophages and permit them to express cytotoxic activities when in contact with targets.

Two TNF receptors of 55 kDa (CD120a or TNFR-I) and 75 kDa (CD120b or TNFR-II) have been identified. They are found on virtually all cell types and bind both TNF-α and TNF-β. More recently, a third receptor that binds TNF-α but not TNF-β has been identified in the liver. Soluble forms of each receptor are found in serum and urine. These soluble receptors may bind free TNF and so modulate its activity. Each receptor may mediate distinct TNF activities. Thus cytotoxicity, antiviral activity, and fibroblast proliferation are mediated by TNFR-I, and thymocyte proliferation is mediated by TNFR-II. Signal transduction by these receptors is mediated by a sphingomyelin signaling pathway.

TNF-α stimulates the IL-1 and IL-6 genes and enhances the expression of CD11b/CD18. It inhibits a type I collagen gene and stimulates collagenase gene transcription. It also stimulates production of transcription factors such as *fos, myc,* and *jun.* As a result, it acts as a growth stimulator for fibroblasts. It has direct antiviral activity and synergizes with interferons. The tumor necrosis factors can, as their name implies, destroy tumor cells. Human recombinant TNF-α can also induce MHC class II expression on target cells. The TNFs probably have a physiological role in regulating the growth of normal cells.

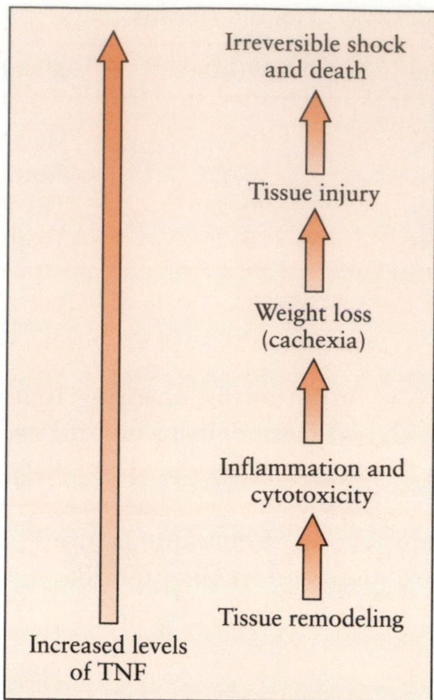

Figure 12–1 The effect of increasing doses of tumor necrosis factor. High doses may cause death as a result of septic shock.

Septic Shock. Individuals infected with gram-negative bacteria, such as *Escherichia coli* or *Salmonella typhi,* develop a characteristic syndrome. This includes fevers, rigors, myalgia, headache, and nausea. More severe infections cause acidosis, fever, hypotension, lactate release in tissues, elevation of plasma catecholamines, disseminated intravascular coagulation, and renal, hepatic, and lung injury. They induce enhanced endothelial procoagulant activity, promoting intravascular coagulation and capillary thrombosis. Eventually these infections induce a lethal condition called **septic shock** (Fig. 12–1). All these effects are mediated by the secretion of TNF-α and IL-1 by endotoxin-stimulated macrophages. If an animal is made resistant to endotoxin, its cells do not secrete TNF-α, implying that the toxic activities of gram-negative bacteria are almost entirely attributable to TNF-α.

Weight Loss. Animals exposed to chronic, sublethal doses of TNF-α lose weight and become anemic and depleted of protein. The weight loss occurs because TNF-α inhibits synthesis of enzymes necessary for the uptake of lipids by preadipocytes and causes mature adipocytes to lose stored lipids. TNF-α also acts on muscle cells and hepatocytes to stimulate their catabolism. As a result, it is responsible for the severe wasting seen in individuals suffering from cancer or chronic parasitic and bacterial diseases.

Immunomodulation. TNF-α can enhance the in vivo immune response at doses much lower than those that cause weight loss or tissue toxicity. It enhances proliferation of B and T cells and promotes the generation of cytotoxic T cells. In addition, it enhances IL-2–induced immunoglobulin production and augments IL-2–stimulated NK activity and proliferation of monocytes.

Inflammation. TNF-α is secreted during inflammation and stimulates the production of granulocyte-macrophage colony stimulating factor (GM-CSF), macrophage colony stimulating factor (M-CSF), and IL-1. TNF-α acts directly on the temperature-regulating center of the hypothalamus to cause a fever. It causes hepatocytes to secrete acute-phase proteins, enhances macrophage and neutrophil chemotaxis, and increases their phagocytic and cytotoxic activities. In addition, it enhances macrophage IL-1 secretion and neutrophil superoxide production. TNF-α stimulates fibroblast proliferation and induces growth factor secretion. IL-1 and TNF-α act on vascular endothelial cells to stimulate them to increase expression of leukocyte adhesion molecules, procoagulants, and MHC class I molecules.

Cytotoxicity. TNF-α kills some human tumor cell lines and causes necrosis in the center of some implantable tumors in vivo. The mechanism of TNF-induced cell lysis is not well understood, but the binding of TNF-α to its receptor is interpreted by some cells as a signal to initiate **apoptosis.** Cells that are resistant to TNF-α may still possess receptors but do not receive the apoptosis command. IL-1 and IFN-γ enhance the cytotoxicity of TNF-α. TNF-α may also have a toxic effect on normal cells since it is responsible for the epithelial, skin, and intestinal lesions that occur in graft-versus-host disease (Chapter 19).

LYMPHOKINES PRODUCED MAINLY BY T CELLS

T cells produce many different lymphokines that form two major groups—those produced by Th1 cells and those produced by Th2 cells. In general, the Th1-derived lymphokines tend to have biological activities that counteract the activities of the Th2-derived lymphokines. Thus Th1 and Th2 generation is regulated by the balance between these two groups of lymphokines.

LYMPHOKINES PRODUCED MAINLY BY Th1 CELLS

IL-12, produced by macrophages, dendritic cells, and B cells, specifically stimulates Th1 cell activity. Two key lymphokines are produced by Th1 cells, IL-2, and IFN-γ. Th1 cells also secrete TNF-β, which is discussed here as well.

Interleukin-2

Human IL-2 is a 15-kDa glycoprotein produced by Th1 cells in response to appropriately presented antigen and IL-12. It is produced in several glycoforms, but all have identical biological properties. IL-2 acts as an activation factor for both helper and cytotoxic T cells, B cells, and NK cells. Once IL-2 receptor expression is induced by IFN-γ, it also enhances macrophage function (Fig. 12–2).

To be responsive to IL-2, a T cell must first be activated by antigen and IL-12. This results in secretion of IL-2 and expression of IL-2R. When the IL-2 binds to its receptor, it is rapidly internalized and triggers the T cell to proliferate. Because IL-2 induces proliferation of activated T cells, it is a key component of the immune response. IL-2 also acts on B cells, promoting their growth and stimulating limited immunoglobulin synthesis (IgG2a). Acting on Th1 cells and NK (natural killer) cells, IL-2 stimulates proliferation and cytotoxicity and so generates LAK (**lymphokine-**

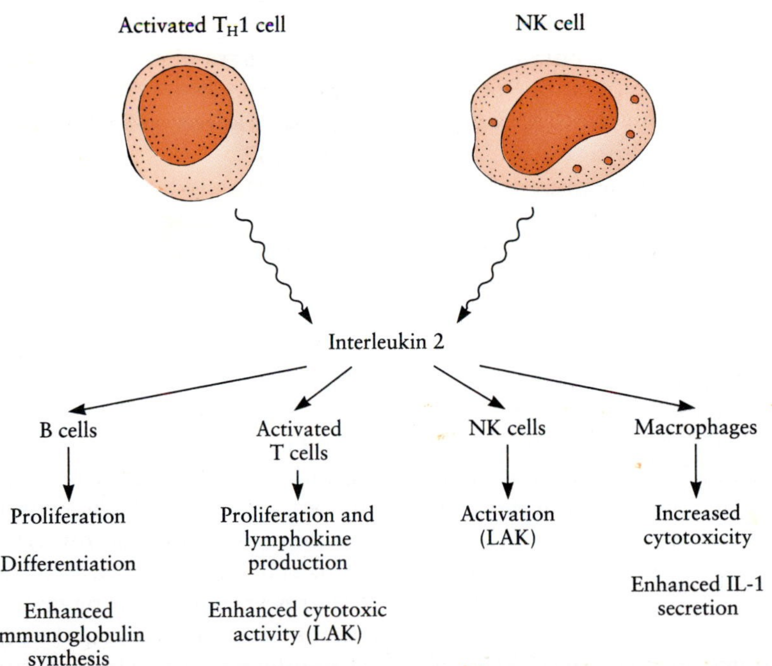

Figure 12–2 The origins and targets of interleukin-2.

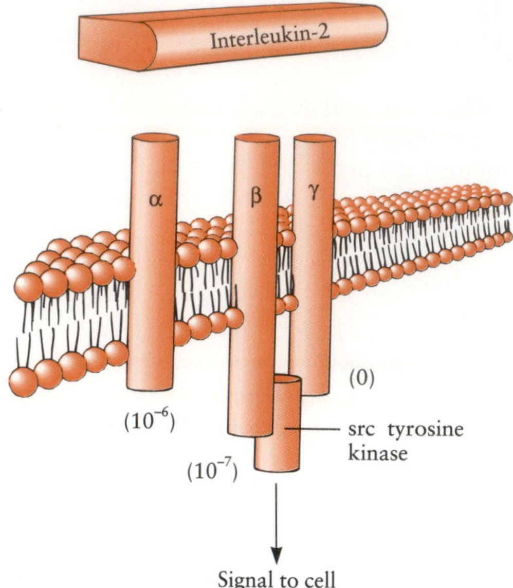

Figure 12–3 The three chains of interleukin-2 receptor. All have different affinities for IL-2 as shown in parentheses. However, when all three chains bind IL-2 the combined affinity is 10^{-11}. Both β and γ chains are needed for signal transduction through a tyrosine kinase of the *src* family.

activated killer) cells (Chapter 20). IL-2 also induces production of interferon-γ and IL-5 and regulates TNF-α receptor expression. IL-2 stimulates monocytes to secrete IL-1β, IL-6, and TNF-α and promotes their cytotoxicity.

IL-2 receptors are found on T and B lymphocytes, NK cells, thymocytes, and monocytes. These receptors are heterotrimers of α (CD25), β (CD122), and γ chains. (The γ chain is also a component of the IL-4 and IL-7 receptors.) Different combinations of these chains give rise to various forms of the IL-2 receptor (Fig. 12–3). A combination of β and γ chains is required for signal transduction, but the presence of the α chain is required for high-affinity binding of IL-2. Thus a receptor consisting only of β and γ chains has only low affinity (Kd = 10^{-9}M), whereas a receptor consisting of all three chains binds IL-2 with high affinity (Kd = 10^{-11}M). B cells have about 1000 high-affinity receptors and 10,000 low-affinity receptors. T cells have only high-affinity receptors, which they may shed on activation. This soluble IL-2 receptor (sIL-2R) may be measured as a marker of T-cell activation. Signal transduction by the β chain involves activation of a nonreceptor tyrosine kinase.

Interferon-γ

Human IFN-γ is a glycoprotein of 20 to 25 kDa produced by $CD8^+$ T cells, Th1 cells, and NK cells in response to IL-2. It occurs in two glycoforms that have identical biological activity. In the mouse, IFN-γ is a 38- to 80-kDa glycoprotein. The high-affinity receptor for IFN-γ (CDw119) is a single-chain integral membrane glycoprotein of 90 kDa found on many cell types.

IFN-γ has a selective effect on B-cell function (Fig. 12–4). For example, it stimulates B-cell production of Ig2a and lowers production of IgG3, IgG1, IgG2b, and IgE in mice. (IL-4, in contrast, stimulates production of IgG1 and IgE and lowers production of IgM, IgG3, IgG2a, and IgG2b.)

IFN-γ has a complex effect on T-cell functions. It enhances T-cell production of MHC class I but not MHC class II molecules. It induces Th1 cells to produce both IL-2 and IL-2R. IL-2 stimulates IFN-γ production by lymphocytes, and IFN-γ helps IL-2 initiate cytotoxic cell proliferation. IFN-γ also enhances the resistance of target cells to cell-mediated lysis. Thus, paradoxically, interferons may be immunosuppressive at the same time as they increase host resistance to tumors and to viruses. IFN-γ acts on Th2 cells to inhibit the production of IL-4 and, as a result, blocks IgE production in vitro. It can depress the mixed lymphocyte reaction but enhance allograft rejection. Depending on dose and timing, it can enhance or suppress the delayed hypersensitivity reaction.

Interferon-γ enhances the activities of **natural killer (NK) cells.** NK cells respond to an antigen by producing IFN-γ, so that a positive loop exists whereby activation of some NK cells results in interferon secretion and activation of other NK cells.

IFN-γ activates macrophages and greatly increases their ability to destroy ingested microorganisms (Chapter 18). It enhances the expression of FcγRI on macrophages and neutrophils and therefore promotes antibody-mediated phagocytosis as well as **antibody-dependent cell-mediated cytotoxicity** (ADCC) reactions.

Interferons have both enhancing and inhibitory effects on other cell functions. Their most important inhibitory effect is a slowing of the growth of normal and neoplastic cells. Interferons inhibit cell multiplication by synchronizing dividing cells into one phase of the cell cycle and then increasing the duration of the G_1 and S + G_2 phases. They also enhance expression of receptors for TNF-α. IFN-γ increases MHC class I expression on tumor cell lines, although it does not increase all gene products identically. It induces the appearance of MHC class II molecules on endothelial cells, keratinocytes, myeloid cells, some dendritic cells, and fibroblasts as well as on macrophages. As a result of this effect, interferons may upregulate the expression of MHC class I and II molecules on virus-infected cells. This in turn may lead to enhanced antigen recognition and cell destruction. When T cells secrete interferon during graft rejection, MHC class II

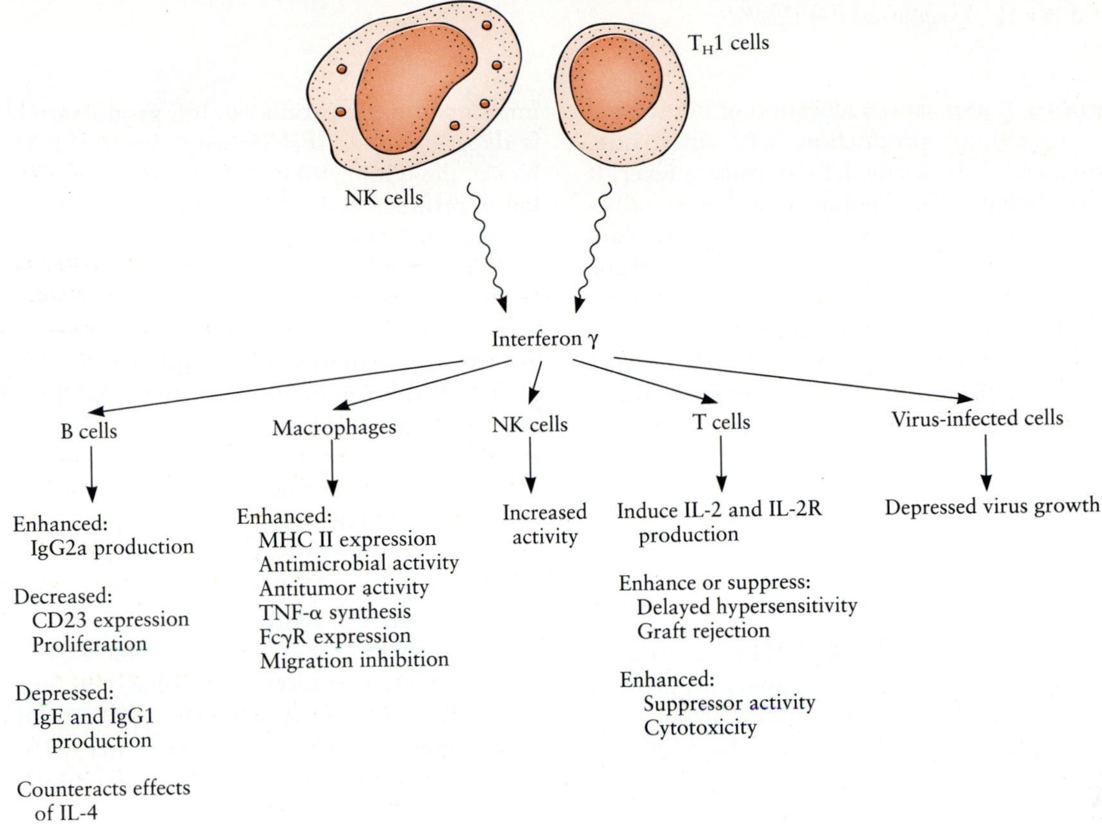

Figure 12–4 The origins and targets of interferon-γ.

molecules are induced on the endothelial cells of the graft as well as on the host's macrophages (Chapter 19). This enhances the intensity of graft rejection. Interferons increase MHC class III product synthesis by monocytes. Interferons usually downregulate hematopoietic cell proliferation. However, human IFN-γ can stimulate monocytes to secrete GM-CSF and induces T cells to secrete GM-CSF and IL-3. IFN-γ enhances the expression of cellular receptors for TNF and of **secretory component** on intestinal epithelial cells.

Lymphotoxin (Tumor Necrosis Factor–β)

Lymphotoxin (LT), or tumor necrosis factor–β (TNF-β), is a 25-kDa protein produced by activated $CD8^+$ and $CD4^+$ T cells. The $CD4^+$ producers are of the Th1 type. It has significant sequence homology with TNF-α, and both genes are located close together within the MHC class III region. Although normal B cells do not produce LT, some cultured B-cell lines do. Lymphotoxin may aggregate to form oligomers similar to TNF-α. It is found in both soluble and cell membrane–bound forms.

LT can induce apoptosis in some cells (Fig. 12–5). Malignant transformed cells are more susceptible than normal cells to LT killing. LT may act synergistically with IFN-γ since the interferon seems to prime the target for LT killing. LT may enhance the growth of some cell lines and the expression of some antigens

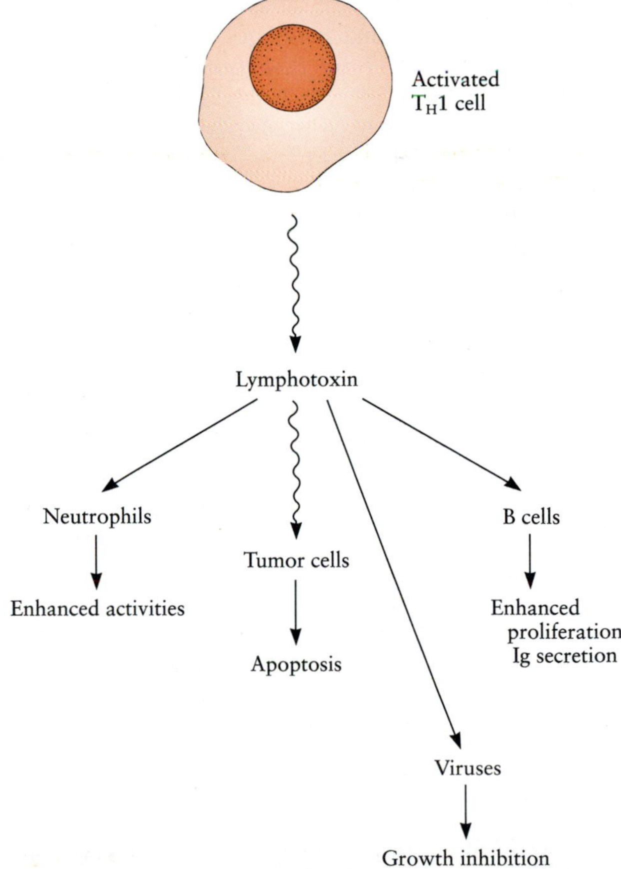

Figure 12–5 The origins and targets of lymphotoxin/TNF-β.

on cell surfaces. It also causes activation of neutrophil functions, including production of superoxide, phagocytosis, and cytotoxicity. LT can cause a fever. It acts on endothelial cells to enhance leukocyte adhesion and procoagulant activity. LT has some direct antiviral activity and exhibits synergy with interferons since it can induce the appearance of 2′–5′A synthase (Chapter 25). It can confer antiviral resistance on uninfected cells and kill virus-infected cells. It enhances proliferation of activated B cells and augments B-cell proliferation and immunoglobulin secretion induced by IL-2.

LYMPHOKINES PRODUCED MAINLY BY Th2 CELLS

Th2 cells secrete a mixture of lymphokines, including IL-4, IL-5, IL-6, IL-10, and IL-13, and generally provide helper activity for B-cell immunoglobulin production. They are costimulated to produce these lymphokines by interleukin-1.

Interleukin-4

IL-4 is a 20-kDa glycoprotein secreted by activated Th2 cells. It acts on a very wide range of cell types, including B cells, T cells, macrophages, endothelial cells, and mast cells (Fig. 12–6). IL-4 stimulates the growth and differentiation of B cells, promoting them to move from the G_1 phase of the cell cycle to the S phase, and enhances their expression of MHC class II molecules, Thy-1, FcγR, and CD23. It promotes the synthesis of IgG1, IgA, and IgE but suppresses IgG3 and IgG2b production, so that it regulates the type of immunoglobulin produced. IgE production by B cells is dependent on IL-4. Disruption of the IL-4 gene blocks the Th2 cytokine response so that these cells fail to produce IL-5, IL-9, and IL-10.

IL-4 enhances the development of cytotoxic T cells from resting T cells (Chapter 18). It can make helper T cells grow in the absence of IL-2 and stimulates fetal thymocyte, granulocyte, and mast cell growth. Interleukin-4 has complex effects on macrophages. It downregulates IL-1, IL-8, and TNF-α secretion by human monocytes. On the other hand, it increases their MHC class II expression, their antigen-presenting ability, and their cytotoxic activity. IL-4 induces giant-cell formation and can act as a growth stimulator for some macrophage cell lines.

The actions of IL-4 are neutralized by IFN-γ. Thus IFN-γ inhibits IgE synthesis, B-cell proliferation, and Thy-1 expression on T cells, all of which are enhanced by IL-4. IFN-γ enhances macrophage secretion of IL-1, TNF-α, and IL-6, all of which are inhibited by IL-4. It is interesting to note, however, that MHC class II expression in macrophages is enhanced by both IL-4 and IFN-γ. The IL-4 high-affinity receptor contains two chains—a ligand-binding protein of 140 kDa (CDw124) and a signal-transducing γ chain that is also a component of the IL-2 and IL-7 receptors. IL-4R expression is induced by IFN-γ.

Interleukin-5

Interleukin-5 is a disulfide-linked homodimeric glycoprotein of 45 kDa secreted by activated Th2 cells. In mice, IL-5 increases the growth and differentiation of activated B cells (Fig. 12–7). As a result, it stimulates IgG, IgA, and IgM production and enhances IL-4–in-

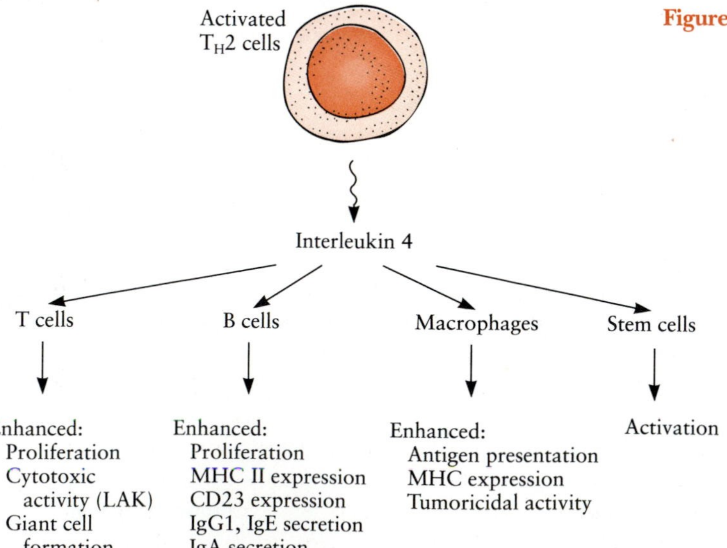

Figure 12–6 The origin and targets of interleukin-4.

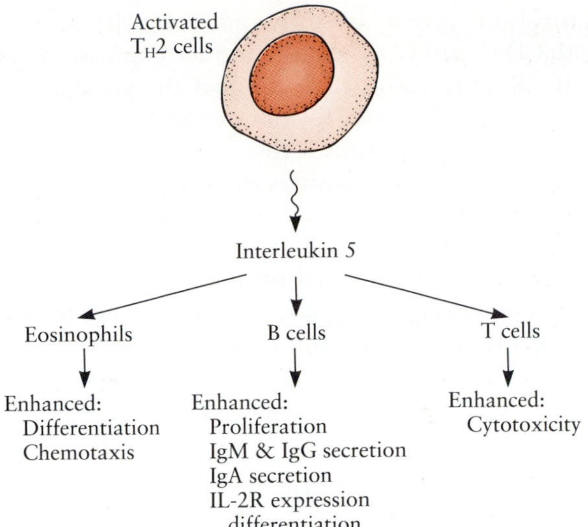

Figure 12–7 The origin and targets of interleukin-5.

duced IgE synthesis and FcεRII (CD23) expression. It enhances IgA production by B cells precommitted to IgA synthesis and makes them differentiate into IgA-producing plasma cells. It also induces the appearance of the homing receptor CD44 on the surface of activated B cells, suggesting that it has a role in regulating lymphocyte trafficking. In conjunction with IL-2, IL-5 induces cytotoxic T-cell activity, probably by inducing the expression of high-affinity IL-2R. IL-5 also stimulates growth and differentiation of eosinophil precursors in the bone marrow and is an eosinophil chemotactic factor. The IL-5 receptor is a heterodimer of 60 and 130 kDa. One of the receptor peptides, the 130-kDa β chain, is also employed in the IL-3 and GM-CSF receptors.

Interleukin-9

IL-9 is a 32- to 39-kDa single-chain glycoprotein secreted by activated Th2 cells. It induces antigen-independent growth of certain helper T-cell clones (both Th1 and Th2) but has no effect on cytotoxic T-cell clones. It enhances the proliferative response of bone marrow–derived mast cells to IL-3. It also potentiates the effect of IL-4 on B-cell IgE production.

Interleukin-10

In mice, IL-10 is an immunosuppressive glycoprotein of 19 to 21 kDa that is secreted by Th2 cells, by some B cells, and by activated macrophages (Fig. 12–8). IL-10 was originally called cytokine synthesis inhibiting factor (CSIF) since it appeared to downregulate cytokine production by Th1 cells in mice. It is now clear that IL-10 primarily acts on activated macrophages to suppress their secretion of IL-1, IL-12, TNF-α, and reactive oxygen radicals. It also reduces their antigen-presenting ability by downregulating MHC class II molecule expression and stimulating production of IL-1RA. The inhibitory effect of IL-10 on the production of IFN-γ and IL-2 by Th1 cells is a secondary effect, resulting from this suppression of macrophage or dendritic cell activity, most likely due to suppression of macrophage production of IL-12. IL-10 has no effect on Th2-cell cytokine synthesis. IL-10 supports mast cell proliferation, being synergistic with IL-3 and IL-4. In contrast to its suppressive effects on Th1 cells and macrophages, IL-10 is a potent stimulator of activated B cells, inducing MHC class II expression and promoting growth.

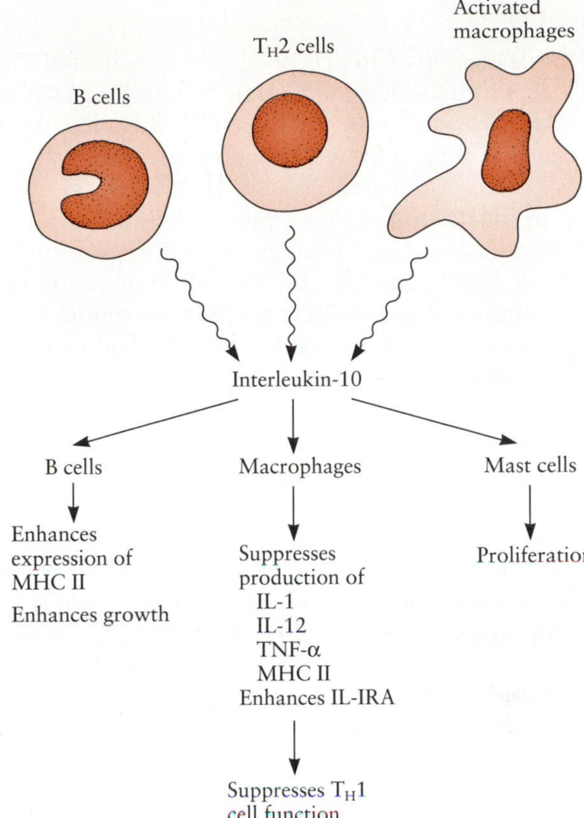

Figure 12–8 The origins and targets of interleukin-10.

The role of IL-10 in humans is less clear. IL-10 is produced by both Th1 and Th2 cells and downregulates the function of both cell types, probably by inhibiting macrophage accessory cell function. As a result, IL-10 in humans inhibits IL-4–induced IgE synthesis. Because of its suppressive effects on the secretion of the proinflammatory and immunomodulatory cytokines, IL-10 has many potential therapeutic applications. The IL-10 receptor is a peptide of 100 kDa that is structurally related to the IFN-γ receptor.

Interleukin-13

IL-13 is a protein of 10 kDa produced by activated Th2 cells. Its production is induced by ligation of CD28 on the T-cell surface, with a B7 or related molecule on the surface of an activated antigen-presenting cell. IL-13 has effects similar to those of IL-4. Thus on exposure to IL-13, monocytes upregulate MHC II and CD23 expression but downregulate FcR expression. In addition, IL-13 stimulates B-cell proliferation and differentiation increasing IgG4 and IgE secretion. It synergizes with IL-2 in regulating NK cell production of IFN-γ. Unlike IL-4, IL-13 has no effect on T cells.

LYMPHOCYTE-DERIVED GROWTH FACTORS

A large group of glycoproteins control the proliferation and maturation of the major blood cells. Each of these glycoproteins is produced by multiple cell types and in many cases affects several cellular targets. There are several well-characterized growth factors that influence the development of bone marrow precursor cells (Table 12–4).

Interleukin-3

IL-3 is a 15- to 25-kDa glycoprotein produced by activated T cells, NK cells, and mast cells. It stimulates the growth and maturation of bone marrow stem cells (Fig. 12–9). Injection of IL-3 causes a rise in blood eosinophils, neutrophils, and monocytes. IL-3 stimulates mast cell and basophil differentiation and activation. It activates eosinophils and promotes macrophage cytotoxicity and phagocytosis. IL-3 also promotes immunoglobulin secretion by B cells. The IL-3 receptor is a heterodimer whose β chain is shared by the receptors for IL-5 and GM-CSF.

Interleukin-7

IL-7 is a 25-kDa glycoprotein derived from bone marrow and thymic stromal cells. It induces proliferation of pre–B cells, thymocytes, and T cells and probably controls the activity of lymphoid stem cells. It acts on both $CD4^+$ and $CD8^+$ cells to induce expression of IL-2, IL-2R, and CD71. IL-7 supports the generation of cytotoxic T cells from thymocytes and so can induce the formation of LAK cells in mice and humans (Chapter 20). The activities of IL-7 are inhibited by TGF-β. There are two IL-7 receptors—a high-affinity one (CDw127) of 75 kDa expressed on T cells and a low-affinity one expressed on B cells and monocytes. The high-affinity receptor contains a γ chain identical to that found in IL-2R and IL-4R.

Interleukin-11

IL-11 is a protein of 19 kDa secreted by bone marrow stromal cells (fibroblasts). It acts as a growth factor on certain B-cell lines in association with IL-6. It also stimulates megakaryocyte colony formation in association with IL-3 and so may have a role in stimulating platelet production. Like IL-6, it can stimulate the production of hepatic acute-phase proteins.

Interleukin-14

A B-cell growth factor produced by T cells and some malignant B cells has been called interleukin-14. It is a secreted protein of 53 kDa that induces B-cell proliferation, inhibits immunoglobulin secretion, and selectively expands some B-cell subpopulations.

Interleukin-15

IL-15 is a 14-15 kDa protein produced by a wide variety of cells, especially peripheral blood mononuclear cells and epithelial and fibroblast cell lines. It has biological activities that are similar to IL-2 and acts as a T cell growth factor. It enhances proliferation of both helper (CD4) and cytotoxic (CD8) T cell populations. It can generate lymphokine-activated killer (LAK) cells from peripheral blood mononuclear cells. One reason for its similar biological properties to IL-2 is that it functions by binding to the β-chain of the IL-2 receptor.

Table 12–4 The Colony-Stimulating Factors

Name	MW(kDa)	Mainly Produced By	Major Targets	Receptor MW
G-CSF	23	Fibroblasts, macrophages	Hematopoietic cells	180
M-CSF	40/80	Fibroblasts, macrophages	Monocytes, stem cells	165
GM-CSF	18/22	T cells, other	Hematopoietic cells	70 + 130
IL-3	25	T cells	Hematopoietic cells	65 + 130
IL-7	25	Stromal cells	Immature lymphocytes	70
IL-11	24	Stromal cells	B cells	151

MW = molecular weight.

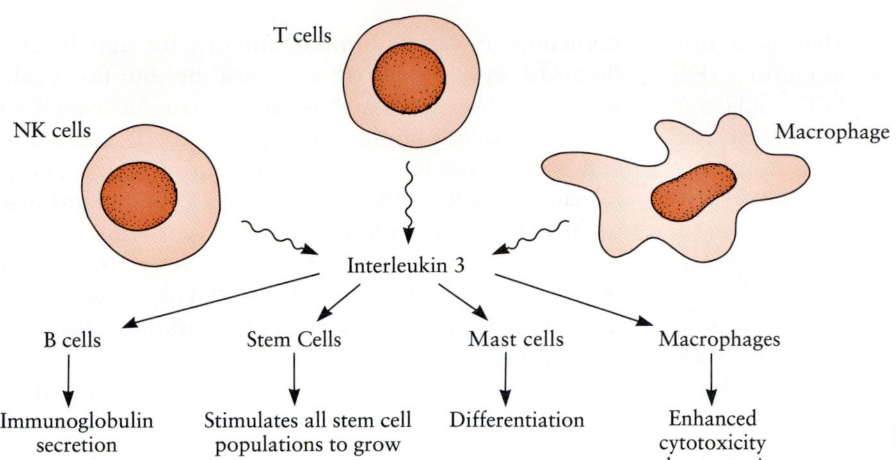

Figure 12–9 The origin and targets of interleukin-3.

Granulocyte Colony-Stimulating Factor (G-CSF)

IL-1 and TNF-α induce G-CSF production by endothelial cells and fibroblasts. G-CSF is an 18-kDa glycoprotein that stimulates the maturation of granulocyte progenitors into mature neutrophils and stimulates the production of superoxide by mature neutrophils. It also enhances neutrophil ADCC. (The term "colony-stimulating factor" refers to the ability of this molecule to promote the growth of clusters or colonies of granulocytes in tissue culture.)

Macrophage Colony-Stimulating Factor (M-CSF)

The M-CSFs are derived from fibroblasts and macrophages as well as from tissues as diverse as the yolk sac and mouse uterus. Their production is stimulated by IL-1 and TNF-α. They are two related homodimeric glycoproteins of 80 kDa and 40 kDa derived from a single precursor that is differentially spliced to generate the two mature proteins. M-CSFs induce the proliferation and differentiation of monocytes and macrophages and promote macrophage cytotoxicity.

Granulocyte-Macrophage Colony-Stimulating Factor (GM-CSF)

The main sources of GM-CSF in vivo are macrophages, especially if exposed to bacterial lipopolysaccharide or fibronectin. It may also be produced by bone marrow stromal cells. GM-CSF is an 18- to 22-kDa glycoprotein that acts as a growth factor for myeloid progenitor cells. It has a more limited spectrum of activity than IL-3 since its stimulating effect is limited to granulocyte, macrophage, and eosinophil progenitors. GM-CSF induces phagocytosis, superoxide production, and ADCC by neutrophils. It also stimulates neutrophil cytotoxicity, adhesion, aggregation, mobility, and accumulation at inflammatory sites. It increases surface expression of FcαR and CD11b/CD18. GM-CSF acts on eosinophils to enhance superoxide production and leukotriene C_4 synthesis. It activates macrophages, enhancing superoxide production, phagocytosis, tumoricidal activity, and expression of MHC class II molecules. GM-CSF also induces the synthesis of IL-1, IFN-γ, IFN-α, and prostaglandin E2. GM-CSF can also provide a signal to macrophages that provokes them to become "suppressor" macrophages and secrete IL-1RA.

Transforming Growth Factor–β (TGF-β)

TGF-β is a 25-kDa disulfide-linked glycoprotein homodimer that plays a central role in embryonic development, tumorigenesis, wound healing, fibrosis, and immunoregulation (Fig. 12–10). It was first identified

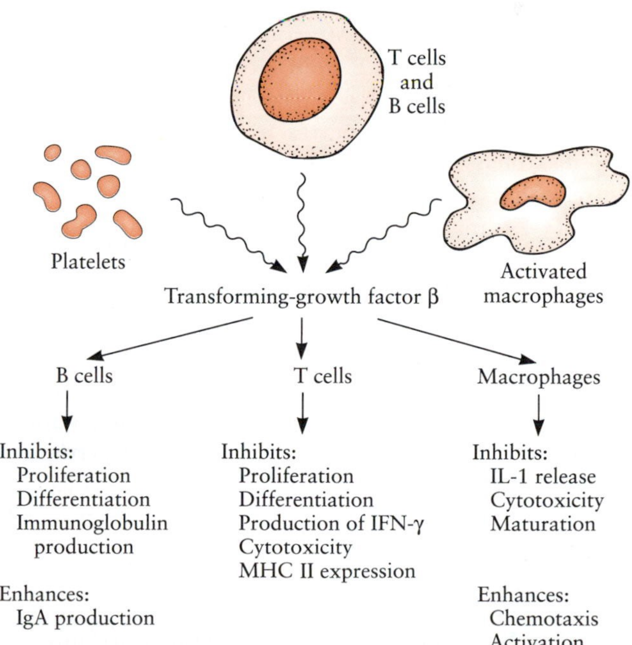

Figure 12–10 The origins and targets transforming growth factor-β.

as a product of viral-transformed cells that made normal cells act as if they were cancerous in culture. (For example, it could promote the growth of fibroblasts in suspension.) TGF-β is produced by platelets, activated macrophages, B cells, and T cells. Several isoforms are found: TGF-β1, β2, and β3 in mammals. (TGF-β4 and β5 have been described in chickens and *Xenopus* toads.) A heterodimer (TGF-β1.2) consisting of one of each type of chain has also been described.

TGF-β is produced in an inactive form and is activated by proteases before it can trigger its receptor. Thus monocytes bind pre-TGF-β and, when stimulated, secrete proteases that activate it. There are at least five TGF-β receptors, and different isoforms have different affinities for each receptor. Thus the biological potency of each isoform varies considerably.

TGF-β acts as both an inhibitory and a stimulatory molecule. Its stimulatory activities include fibroblast proliferation; fibroblast, monocyte, and neutrophil chemotaxis; stimulation of monocyte IL-1 and TNF-α secretion; and enhancement of helper T-cell function. Its inhibitory effects include inhibition of IL-1R expression, of the respiratory burst, and of the cytotoxic effects of activated macrophages. TGF-β also inhibits monocyte differentiation. TGF-β is a potent inhibitor of T- and B-cell proliferation. It can thus inhibit immune function in chronic inflammatory reactions. TGF-β inhibits production of IFN-γ by T cells as well as *myc* and GM-CSF gene transcription, IL-2R expression, MHC class II expression, cytotoxic T cell development, LAK production, NK cell production, and TNF-α production.

TGF-β inhibits proliferation and differentiation of B cells stimulated by interleukin-2. It thus blocks light-chain and IgG1 and IgM production as well as terminal differentiation of B cells. However, its effects on B cells are complex since it also causes lipopolysaccharide-stimulated lymphocytes to express IgA and enhances heavy-chain switching to IgA. IL-5 can synergize with TGF-β1 to promote an IgA response.

TGF-β may protect the host from chronic inflammation and promote healing since it promotes the ordered accumulation of inflammatory cell recruitment and vascular growth following subcutaneous inoculation. It is a chemoattractant for fibroblasts, regulates their growth, and upregulates the production of collagens as well as fibronectin. At the same time, it downregulates the secretion of collagenases. It acts on neutrophils to inhibit adherence to endothelial cells.

CYTOKINE RECEPTORS

All membrane-bound cytokine receptors are integral membrane proteins with both intracellular and extracellular domains. They also have two functional subunits, one for ligand binding and one for signal transduction, which may or may not be on the same peptide. These cytokine receptors can be classified into three major families based on their structure and activities—namely, those for hematopoietic growth factors (HGFRs), those for TNF (TNFR), and the members of the immunoglobulin superfamily (Chapter 7). A few receptors such as IL-2Rα (CD25) chain and the receptors for IL-8, TGF-β, and IFN-γ (CDw119) have a unique structure and do not belong to any obvious group.

The HGFRs include the receptors for IL-2, IL-3, IL-4, IL-5, IL-7, IL-9, G-CSF, and GM-CSF. They also include the common β chain of IL-3, IL-5, and GM-CSF receptors and the common γ chain of IL-2, IL-4, and IL-7 receptors. They have several large extracellular domains containing a number of conserved cysteine residues and a unique tryptophan-serine-X-tryptophan-serine sequence, where X is any amino acid. They also have a single membrane-spanning domain, and a large cytoplasmic domain without tyrosine ki-

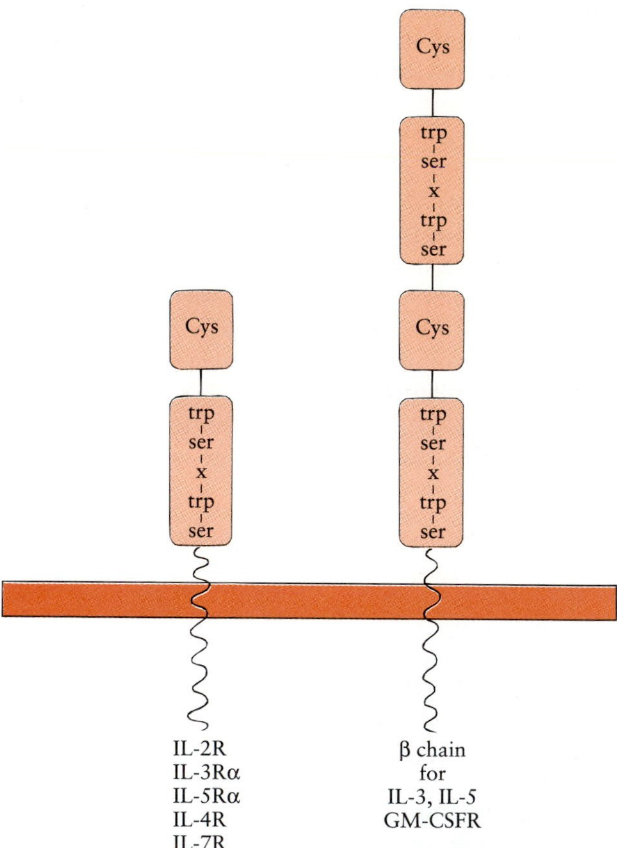

Figure 12–11 The structure of the cytokine receptors of the hematopoietic growth factor family. They are characterized by having one or two domains with a trp-ser-x-trp-ser sequence and one or two domains with multiple conserved cysteine residues.

nase or other enzymatic activity (Fig. 12–11). These receptors usually bind only one ligand. Many exist in high- and low-affinity forms.

The TNFR family are type II transmembrane proteins that include multiple peptide chains and multiple ligands. Besides the two TNFRs (CD120a and b), the family includes the B-cell protein CD40, the T-cell proteins CD27 and CD30 and the apoptosis signaling molecule fas. This family is identified by its unique repeats of a cysteine-rich domain. Both TNFs are trimers that probably aggregate their receptors on binding. The ligands for these receptors have significant sequence similarity.

The immunoglobulin superfamily receptors consist of immunoglobulin-like domains, and many but not all have intrinsic tyrosine kinase activity. Several of these receptors have been identified as proto-oncogene products such as *fms* and *kit.* They include the receptors for IL-1 (CDw121) and IL-6 (CD126). Some cytokine receptors have immunoglobulin domains in addition to the characteristic sequence of the HGFR.

KEY WORDS

QUESTIONS

1. Interferon-γ enhances
   a. ADCC
   b. NK-cell activity
   c. Macrophage activity
   d. T-cell–mediated cytotoxicity
   e. all of the above
2. Three major cytokines secreted by macrophages are
   a. IL-1, IL-2, and IL-3
   b. IL-2, IFN-γ, and IL-12
   c. IL-1, IL-12, and TNF–α
   d. IL-1, IFN-γ, and TNF–α
   e. TGF–β, IL-6, and TNF–α
3. Which of these is *not* a lymphokine?
   a. interferon
   b. histamine
   c. lymphotoxin
   d. tumor necrosis factor
   e. transforming growth factor
4. Interleukin-2 is produced by
   a. B cells
   b. Th1 cells
   c. macrophages
   d. neutrophils
   e. Th2 cells
5. Which of the following is a chemokine?
   a. interleukin-2
   b. interleukin-4
   c. interleukin-6
   d. interleukin-8
   e. interleukin-10
6. Septic shock is the toxic manifestation of the release of large amounts of
   a. interleukin-2
   b. interferon-γ
   c. transforming growth factor–β
   d. interleukin-8
   e. tumor necrosis factor–α
7. The key cytokines produced by Th1 cells are
   a. IL-4 and IL-10
   b. IL-1 and TNF-α
   c. IL-2 and IFN-γ
   d. IL-2 and TNF-α
   e. GM-CSF and IL-4
8. A major role of IL-4 is to promote the
   a. cell-mediated immune responses
   b. IgE responses
   c. macrophage activation
   d. IgG responses
   e. IgM responses
9. Interleukin-10 acts to
   a. enhance T-cell responses
   b. activate macrophages

c. suppress cytokine production
d. suppress antibody production
e. enhance IgE production

10. Transforming growth factor–β
    a. activates macrophages
    b. enhances T-cell functions
    c. enhances B-cell functions
    d. inhibits B-cell functions
    e. activates fibroblasts
11. Compare the biological activities of interleukin-1 with those of interleukin-12.
12. Given the availability of isolated, pure cytokines, which do you think might be of use in the treatment of human cancer?
13. Which of the following cytokines do you think is most essential for the proper functioning of the immune system: IL-1, IL-2, IFN-γ, TNF–β? Why?
14. Will interferons be of any use in treating human disease? In what diseases do you think they might be useful?
15. There is a growing tendency to develop a numerical classification for cytokines as shown by the long list of interleukins. Is this a better classification scheme than simply providing each cytokine with the name of its biological activity?
16. To what extent is the malaise associated with infections attributable to the systemic secretion of cytokines?
17. What are the advantages to an animal if the cytokines secreted during an immune response also stimulate bone marrow function?

Answers: 1a, 2c, 3b, 4b, 5d, 6e, 7c, 8b, 9c, 10e

SOURCES OF ADDITIONAL INFORMATION

Akira, S., et al. Biology of multifunctional cytokines: IL-6 and related molecules (IL-1 and TNF). FASEB J., 4:2860–2867, 1990.

Ambrus, J.L., et al. Identification of a cDNA for a human high molecular weight B cell growth factor. Proc. Natl. Acad. Sci. U. S. A., 90:6330–6334, 1993.

Beutler, B., and Cerami, A. The biology of cachectin/TNF—a primary mediator of the host response. Annu. Rev. Immunol., 7:625–655, 1989.

Beutler, B., and van Huffel, C. Unravelling function in the TNF ligand and receptor families. Science 264:667–668, 1994.

Cosman, D., et al. A new cytokine receptor superfamily. TIBS, 15:265–270, 1990.

Dalton, D.K., et al. Multiple defects of immune cell function in mice with disrupted interferon-γ genes. Science, 259: 1739–1742, 1993.

Davatelis, G., et al. Macrophage inflammatory protein-1: A prostaglandin-independent endogenous pyrogen. Science, 243:1066–1068, 1990.

Dinarello, C.A. Interleukin-1 and its biologically related cytokines. Adv. Immunol., 44:153–205, 1989.

Foxwell, B.M.J., Barrett, K., and Feldmann, M. Cytokine receptors: Structure and signal transduction. Clin. Exp. Immunol., 90:161–169, 1992.

Glauser, M.P., et al. Septic shock: pathogenesis. Lancer, 338: 732–736, 1991.

Grabstein, K. H., et al. Cloning of a T cell growth factor that interacts with the β chain of the Interleukin-2 receptor. Science, 264:965–968, 1994.

Groopman, J.E., Molina, J-M., and Scadden, D.T. Hematopoietic growth factors: Biology and clinical application. N. Engl. J. Med., 321:1449–1459, 1989.

Henderson, B., and Blake, S. Therapeutic potential of cytokine manipulation. TIPS, 13:145–152, 1992.

Henney, C.S. Interleukin 7: Effects on early events in lymphopoiesis. Immunol. Today, 10:170–173, 1989.

Howard, M., and O'Garra, A. Biological properties of interleukin 10. Immunol. Today, 13:198–200, 1992.

Kelso, A., and Metcalf, D. T-lymphocyte-derived colony-stimulating factors. Adv. Immunol., 48:69–105, 1990.

Kitamura, T., Ogorochi, T., and Miyajima, A. Multimeric cytokine receptors. TEM 5:8–14, 1994.

Lieschke, G.J., and Burgess, A.W. Granulocyte colony stimulating factor and granulocyte-macrophage colony-stimulating factor. N. Engl. J. Med., 327:28–35, 99–106, 1992.

MacNeil, I.A., et al. IL-10, a novel growth cofactor for mature and immature T cells. J. Immunol., 145:4167–4173, 1990.

McCartney-Francis, N.L., and Wahl, S.M. Transforming growth factor β: a matter of life and death. J. Leukocyte Biol. 55:401–409, 1994.

Malkovsky, M., et al. The interleukins in acquired disease. Clin. Exp. Immunol., 74:151–161, 1988.

Malefyt, R.W., et al. Effects of IL-13 on phenotype, cytokine production, and cytotoxic function of human monocytes. J. Immunol., 151:6370–6381, 1993.

Matias, S., et al. Activation of the sphingomyelin signalling pathway in intact EL-4 cells and in a cell-free system by IL-1β. Science, 259:519–522, 1993.

Metcalf, D. The hemopoietic regulators—an embarrassment of riches. Bioessays, 14:799–805, 1992.

Mizel, S.B. The interleukins. FASEB J., 3:2379–2388, 1989.

Old, L.J. Tumor necrosis factor (TNF). Science, 230:630–632, 1985.

Russell, S.M., et al. Interleukin-2 receptor γ chain: A functional component of the interleukin-4 receptor. Science, 262:1880–1883, 1993.

Sherry, B., and Cerami, A. Cachectin/tumor necrosis factor

exerts endocrine, paracrine, and autocrine control of inflammatory responses. J. Cell Biol., 107:1269–1277, 1988.

Smith, C.A., Farrah, T. and Goodwin, R.G. The TNF receptor superfamily of cellular and viral proteins: activation, costimulation, and death. Cell, 76:959–962, 1994.

Steward, W.P. Granulocyte and granulocyte-macrophage colony-stimulating factors. Lancet, 342:153–157, 1993.

Taga, T., and Kishimoto, T. Cytokine receptors and signal transduction. FASEB J., 7:3387–3396, 1993.

Tominaga, A., et al. Role of the interleukin 5 receptor system in hematopoiesis: Molecular basis for overlapping function of cytokines. Bioessays, 14:527–533, 1992.

Tracey, K.J., Vlassara, H., and Cerami, A. Cachectin/tumor necrosis factor. Lancet, i:1122–1125, 1989.

Uyttenhove, C., Simpson, R.J., and Van Snick, J. Functional and structural characterization of P40, a mouse glycoprotein with T-cell growth factor activity. Proc. Natl. Acad. Sci. U. S. A., 85:6934–6938, 1988.

Van Snick, J. Interleukin-6, an overview. Annu. Rev. Immunol., 8:253–278, 1990.

Wahl, S.M., McCartney-Francis, N., and Mergenhagen, S.E. Inflammatory and immunomodulatory roles of TGF–β. Immunol. Today, 10:258–261, 1989.

Zurawski, G., and de Vries, J.E. Interleukin 13, an Interleukin 4-like cytokine that acts on monocytes and B cells, but not on T cells. Immunol. Today, 15:19–26, 1994.

CHAPTER 13

The Nature of Antibodies

Rodney Porter *Rodney Porter was born in England in 1917 and died in 1985. He was awarded the Nobel Prize for his work on the structure of antibodies in 1972. Porter studied the structure of antibodies at a time when our knowledge of protein structure was in its infancy. Nevertheless, he was the first to show that enzymic digestion of immunoglobulins could break the molecule into fragments and on this basis devised the first plausible model of the IgG molecule. He was the first to suggest that the immunoglobulin molecule was Y shaped. Porter subsequently studied the activation of C1q and complement genes.*

CHAPTER OUTLINE

CHAPTER CONCEPTS

1. Antibodies belong to the class of proteins called immunoglobulins.
2. Several forms of antibody exist in each individual, and as a result, immunoglobulins can act against a great variety of invaders in many ways. These different forms of antibody are called classes. The five immunoglobulin classes are called IgG, IgM, IgA, IgD, and IgE.
3. All immunoglobulins are Y-shaped proteins. The "arms" of the Y bind antigen. The "tail" of the Y is responsible for the biological activity of the molecule, such as activating complement or binding to cells.
4. The ability of an immunoglobulin to bind antigen is determined by the sequence of amino acids in its variable regions. This sequence determines the shape of the antigen-binding site.
5. IgG is the major serum immunoglobulin. IgM is a very large immunoglobulin molecule that predominates in the primary immune response. IgA is responsible for immunity on body surfaces. IgD is a cell-bound immunoglobulin. IgE mediates allergies.
6. Immunoglobulins show significant variations in structure both between individuals and within an individual.
7. The combination of antigen and antibody is a reversible reaction. Thus the strength of binding of these components can be measured and gives useful information on the properties of antibodies.

This chapter begins with a brief background of the antibodies found in serum and how they are classified as proteins known as immunoglobulins. Then the structure of the most important immunoglobulin, IgG, is examined. This includes the structural features of each of its peptide chains and domains. To understand how this structure was arrived at, we spend a little time describing the tumors of plasma cells called myelomas, since these were of immense importance in providing material for study. The use of immunoglobulin domains in the family of proteins called the immunoglobulin superfamily is also described. Then we examine each major class of immunoglobulin, in turn describing its structure and key features. We briefly review the variations in antibody structure found in different individuals as well as the structural variations in the immunoglobulins within an individual. The last part of the chapter deals with a slightly different topic, the strength of binding of antigen and antibody. This binding strength, or affinity, has important effects on diagnostic tests as well as on resistance to disease, and its measurement and significance are discussed here.

The protective factors produced by the immune responses were found in serum and called **antibodies** by Emil von Behring in 1890. In 1930, Heidelberger showed that antibodies were proteins by demonstrating that pneumococcal polysaccharide antigen, when added to antiserum so that it bound antibody, formed a precipitate that contained large amounts of protein.

PROPERTIES OF ANTIBODIES

Antibodies can be detected in many body fluids, for example, tears, respiratory tract mucus, saliva, intestinal contents, urine, and milk. They are present in highest concentrations and most easily obtained in large quantities from blood **serum.** It is important to point out, however, that antibodies also act as the antigen-binding receptors of B cells (BCR). This dual function of antibodies, as cell receptors and as soluble serum proteins, is in marked contrast to the TCRs, which are found only in the cell-bound form.

Because antibodies can be isolated in very large quantities from serum, it has been possible to study their structure in great detail. Indeed, unlike TCRs, antibodies can be readily sampled and their levels measured. For example, antibodies, like other serum proteins, can be characterized by their electrophoretic mobility.

Electrophoretic Mobility

Since the overall charge on any protein depends on its amino acid composition, a mixture of proteins may be fractionated by subjecting it to an electrical potential at a standard pH. This technique is known as **electrophoresis.** The positively charged protein molecules are attracted toward the cathode, the neutral molecules remain stationary, and the negatively charged molecules are attracted toward the anode. Each protein moves at a rate dependent on its net charge.

When serum is electrophoresed, it consistently separates into four major fractions. The most negatively charged fraction consists of a single protein called serum **albumin.** The other three major fractions are classified into α, β, and γ **globulins** according to their electrophoretic mobility. Antibody activity is mainly found in the fraction nearest the cathode, the γ globulins, although some antibodies are also found among the β globulins. Because antibody proteins are all globulins, they belong to a class of glycoproteins called **immunoglobulins.**

Overall Structure

Five distinct immunoglobulin **classes** (or **isotypes**) have been identified (Table 13–1). The major immunoglobulin in serum is immunoglobulin G (usually abbreviated to IgG), a molecule of 160 kDa. The second major immunoglobulin in serum is a large molecule of 900 kDa called immunoglobulin M (IgM). The third immunoglobulin class is immunoglobulin A (IgA). IgA, a molecule of 360 kDa, is found in such

Table 13–1 The Major Immunoglobulin Classes of Humans

	IgG	IgM	IgA	IgE	IgD
Molecular weight (kDa)	160	900	360	200	160
% Carbohydrate	3	12	7	12	12
Electrophoretic mobility	γ	β	β − γ	β − γ	γ
Heavy chain	γ	μ	α	ϵ	δ
Heavy chain domains	4	5	4	5	4
Subclasses	γ1, γ2, γ3, γ4	None	α1, α2	None	None
Half-life (days)	21	5	6	2	3

METHODOLOGY

Serum Electrophoresis

Electrophoresis of serum is usually performed on a solid support that permits fluid to flow without allowing excessive diffusion of proteins. Typical supports include paper, agar, starch gels, and cellulose acetate. The electrophoresis support is saturated with buffer, and each end is connected to a power supply. This connection may be made simply by connecting the support to a buffer bath with wicks (see the figure).

A small volume of serum is placed on the support and a potential voltage applied. The voltage and duration of the process are adjusted according to the size of the support. When the electrophoresis is completed, the current is disconnected, and the support is removed and stained with a protein stain such as amido black, acid fuchsin, or ponceau S. After the mixture has been washed to remove unbound dye, each component is revealed as a stained band (see the figure).

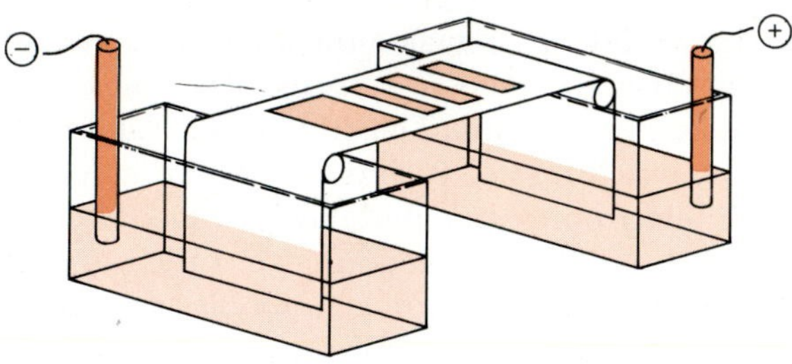

Electrophoresis of serum on a strip of paper or cellulose connecting two buffer baths.

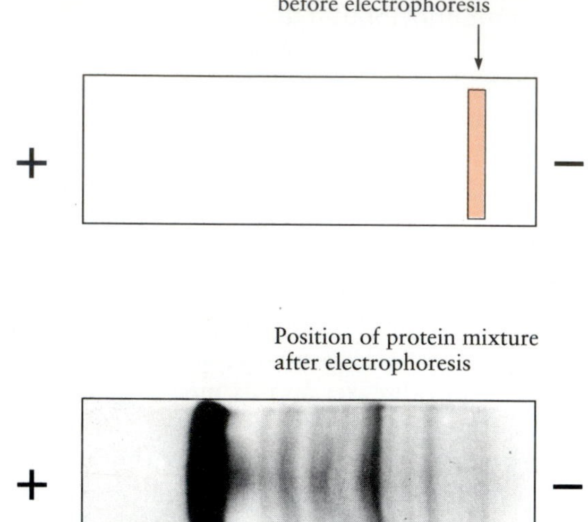

Electrophoresis of a serum sample on a strip of cellulose acetate. It can be resolved into about ten distinct bands in this way. *(Courtesy of Dr. S.H. An.)*

body secretions as saliva, milk, and intestinal fluid. Immunoglobulin D (IgD) (180 kDa) is mainly restricted to B-cell membranes but is present in very low concentrations in plasma. The fifth class, called immunoglobulin E (IgE), consists of molecules of 200 kDa that mediate allergic reactions.

STRUCTURE OF IMMUNOGLOBULINS

Immunoglobulin G

IgG, a glycoprotein of 160 kDa, is the predominant immunoglobulin class in serum. Its structure can serve as a model for the other immunoglobulins. IgG is a dimer of two disulfide-linked heterodimers. Each heterodimer contains a chain of 50 to 60 kDa called a heavy chain, disulfide-linked to a much smaller chain of about 25 kDa called a light chain (Fig. 13–1). The two heterodimers that are joined to form each IgG molecule are identical. Thus in total, an IgG molecule consists of four peptide chains—two heavy and two light.

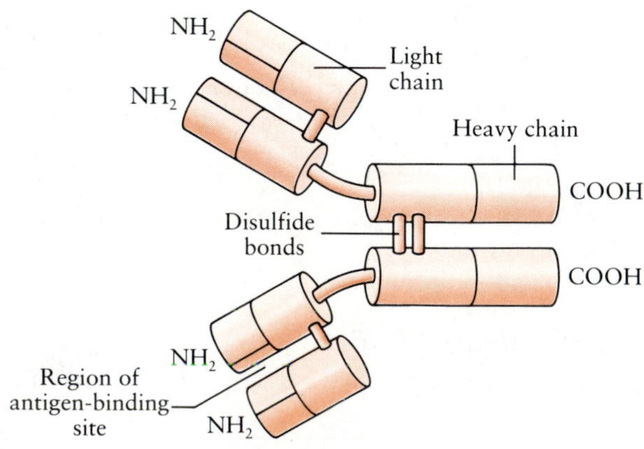

Figure 13–1 A simple model of an IgG molecule.

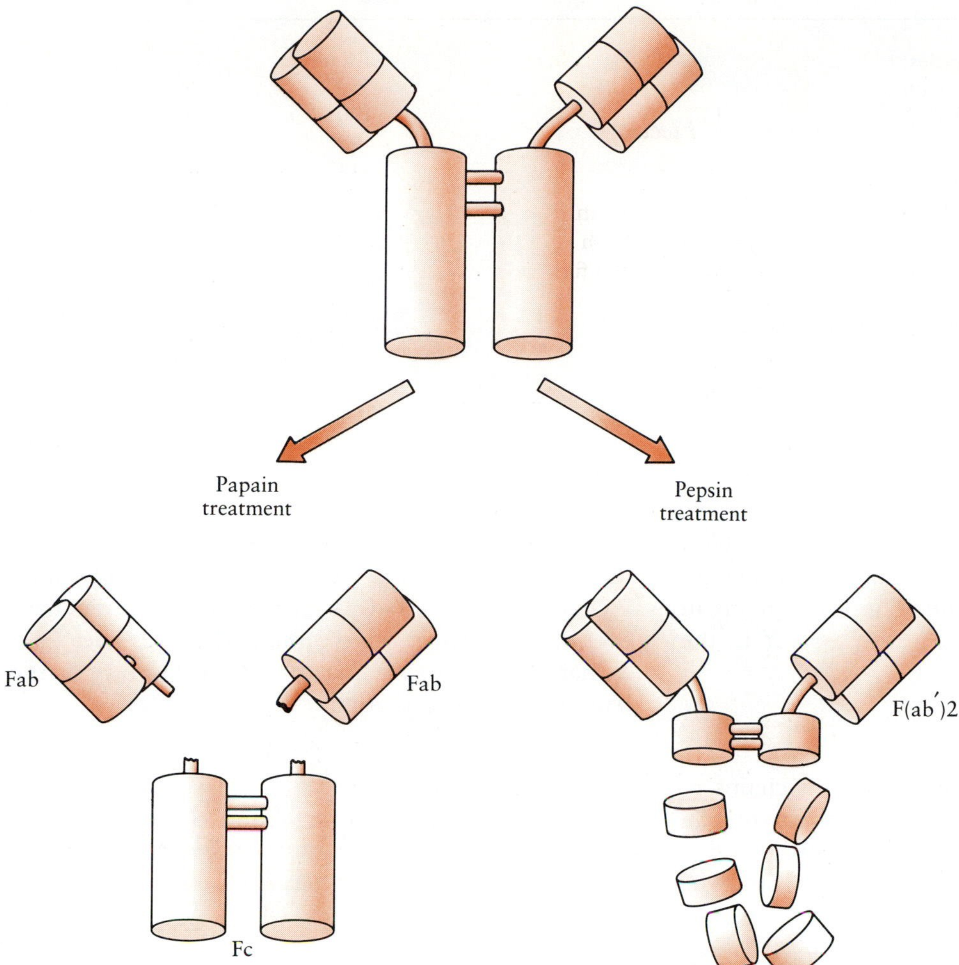

Figure 13–2 The effects of the proteolytic enzymes papain and pepsin on immunoglobulin G and the named fragments formed by this process.

If pure IgG is digested with the proteolytic enzyme papain, it splits into three approximately equal-sized fragments (Fig. 13–2). Two of these fragments retain the ability to bind antigen and therefore are called antigen-binding fragments, or **Fab fragments.** The third fragment obtained by papain treatment cannot bind antigen but is sometimes crystallizable. It is therefore called the **Fc fragment.** Other proteases such as pepsin may leave two Fab fragments joined together to produce a structure called F(ab′) 2.

When viewed with an electron microscope, IgG is seen to be Y-shaped (Fig. 13–3). These observations together with many others have shown that the complete IgG molecule has two binding sites for antigen, each located on one of the "arms" of the Y between a heavy and a light chain. (These are called the **Fab**

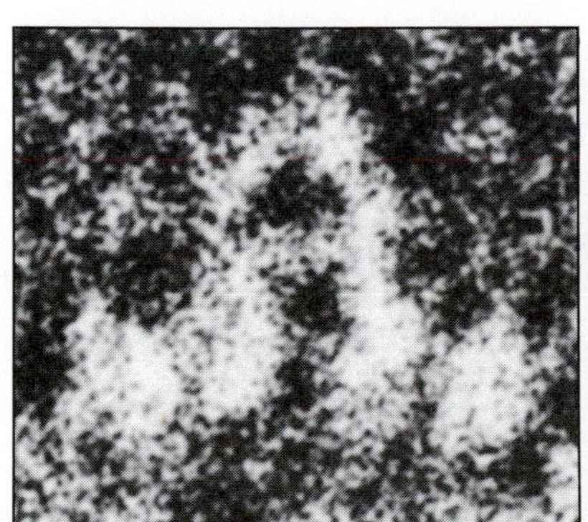

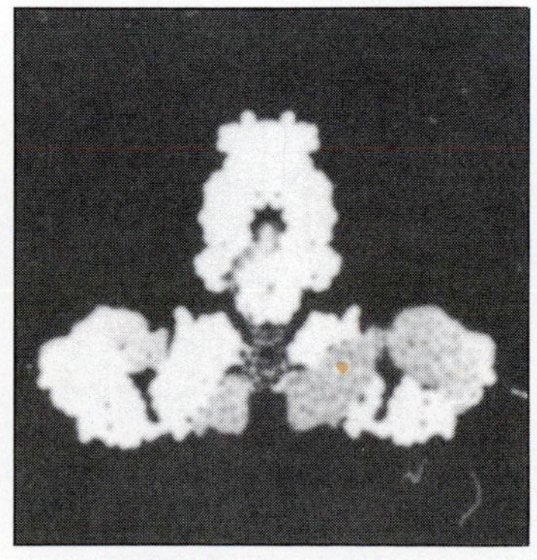

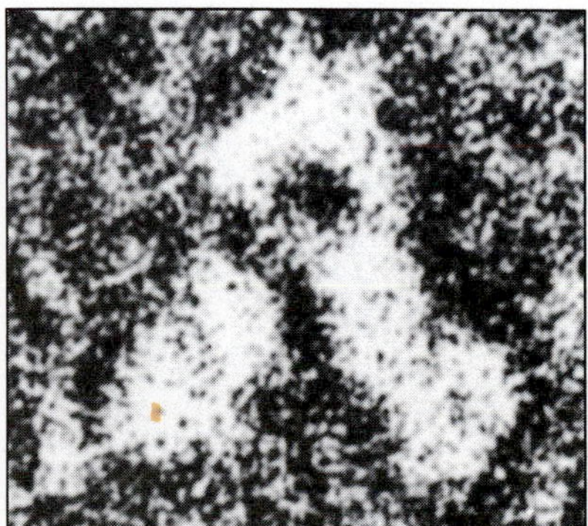

Figure 13–3 Electron micrographs of an IgG molecule. Compare the two electron micrographs with the computer-generated model in the center. This is rabbit IgG (original magnification ×2,042,500). *(From Roux, K.H., and Metzger, D.W. J. Exp. Med., 129:2548, 1982. With permission.)*

Table 13–2 The Subclasses of Human IgG

	IgG1	IgG2	IgG3	IgG4
Chemical Properties				
% of total Ig in serum	65	24	7	4
Molecular weight (kDa)	146	146	170	146
Heavy chain	γ1	γ2	γ3	γ4
Inter-heavy-chain bonds	2	4	13	2
Biological Properties				
Complement activation	++++	++	++++	(+)
Half-life (Days)	23	23	8	23
FcR binding				
Neutrophils	+	+	+	+
Macrophages	+	−	+	−
NK cells	+	−	+	−
Placental passage	+	−	+	+
Skin sensitization	+	−	+	+

regions.) The "tail" of the Y forms a third region called the **Fc region.** The paired heavy chains extend the full length of the molecule. The light chains in contrast are found only in the arms of the Y in the Fab regions (see Fig. 13–2).

Types of Light Chains

Each immunoglobulin molecule has light chains of one of two types regardless of their class or antigen-binding ability. They are called κ (kappa) and λ (lambda) and are only about 35% homologous. The ratio of κ to λ chains varies between species. In humans, about 65% of immunoglobulin molecules have κ chains, and 35% have λ chains. Mice and rats have over 95% κ chains, while cattle and horses have 98% λ chains. Monkeys such as the rhesus monkey or the baboon have 50% of each.

Subclasses of IgG

Within the major immunoglobulin classes, there are structural variants known as **subclasses** that result from differences between heavy chains. Their amino

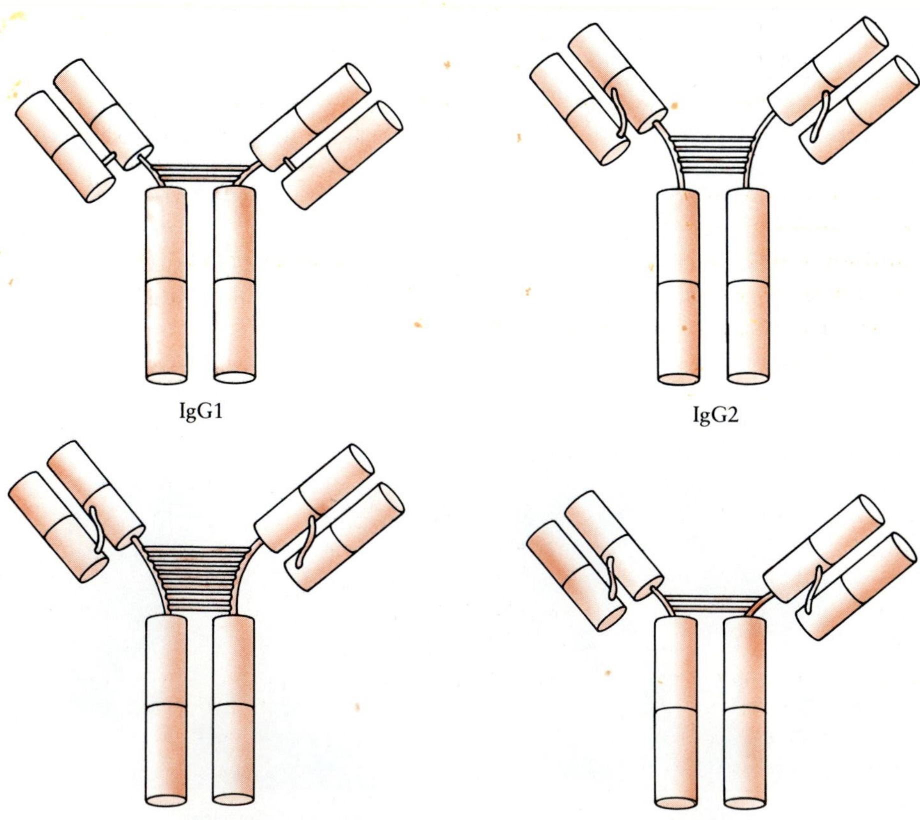

Figure 13–4 The basic structure of each of the human immunoglobulin G subclasses.

acid sequences and their antigenic differences are not great enough to constitute full classes, and they show a close overall relationship. Thus the four human IgG subclasses show greater than 90% sequence homology. Subclasses differ both in structure and in biological activities (Table 13–2). The four different IgG heavy chains in humans are designated γ1, γ2, γ3, and γ4. All normal persons possess all four subclasses. IgG molecules with γ1 heavy chains are known as IgG1, those with γ2 heavy chains are IgG2, and so on (Fig. 13–4). IgG3 has the shortest half-life, the lowest synthetic and highest catabolic rate, and an unusually high number of interchain disulfide bonds. Autoantibodies to clotting factors tend to belong to the IgG4 subclass, and autoantibodies to DNA tend to belong to the IgG1 and IgG3 subclasses. IgG4 is functionally monovalent and so will not precipitate or agglutinate antigen.

PRIMARY STRUCTURE OF IMMUNOGLOBULINS

The immunoglobulins found in serum are a mixture of molecules with antibody activity against a wide spectrum of epitopes. They represent a sample of the antibodies produced by that individual in response to a multitude of different antigenic stimuli. Because of this heterogeneity, it is not possible to use serum antibodies to analyze immunoglobulin structure in more than the general terms described previously.

However, immunoglobulins are secreted by **plasma cells.** Occasionally, a single plasma cell may become neoplastic, resulting in the growth of a clone of cancerous plasma cells that produces a single, absolutely homogeneous (**monoclonal**) immunoglobulin. As the tumor grows, large quantities of this monoclonal immunoglobulin are secreted into the bloodstream of affected individuals. Since a plasma cell tumor is called a **myeloma,** its monoclonal immunoglobulin product is called a myeloma protein (see page 208). Myeloma proteins may be purified and their structure analyzed in detail. One of the first tasks undertaken when myeloma proteins were recognized as monoclonal was to determine the sequence of amino acids in their light and heavy chains.

Light-Chain Sequences

Light chains each contain about 214 amino acids in two domains. If light chains from several different myelomas are studied, their amino acid sequences show important differences. In the C-terminal domain, the amino acid sequences are found to be almost identical. In contrast, the sequences in the N-terminal domain are very different in each light chain examined. For this reason, these two domains of a light chain are referred to as the **constant** (C_L) and **variable** (V_L) **regions,** respectively. This is, of course, a similar structure to that seen in TCR α and β chains where a variable domain is also linked to a constant domain.

Heavy-Chain Sequences

The heavy chains of IgG each consist of about 445 amino acids in four domains. In the center of the chain is an extended flexible hinge region. The sequence of amino acids in the N-terminal domain is different in each myeloma protein examined and thus constitutes a variable (V_H) region. The sequence in the other three domains located toward the C-terminus shows very few differences between different myeloma proteins and therefore forms a constant (C_H) region.

Variable Regions

When the amino acid sequences of many V regions from both light and heavy chains are examined in detail, two features emerge. First, the variation in amino acid sequence is largely restricted to three smaller regions within the entire variable region. These regions are therefore hypervariable. Between these **hypervariable regions** are the regions where the sequence is relatively constant. These are called **framework regions** (Fig. 13–5). We now know that the hypervariable regions are the regions that bind to an epitope. Their amino acid sequence determines the shape of the antigen-binding site (sometimes called the **paratope**), and they determine just which epitopes an immunoglobulin binds to (Fig. 13–6). They are therefore called **complementarity-determining regions** (CDRs). Each CDR is relatively short, consisting of six to ten amino acids. Thus the CDRs of light chains include residues 24 to 34 (CDR1), 50 to 56 (CDR2), and 89 to 97 (CDR3). The CDRs on heavy chains include residues 31 to 35 (CDR1), 50 to 65 (CDR2), and 95 to 102 (CDR3) (Fig. 13–7).

The amino acid sequences of framework regions are not absolutely constant in that some minor variations in sequence do occur. Analysis of the variability within the framework regions reveals that variable regions can be classified into subgroups. In humans, there are four subgroups of V_κ regions, six subgroups of V_λ regions, and four subgroups of V_H regions. In the mouse, there are at least 27 and perhaps as many as 100 V_κ subgroups but only two V_λ subgroups and seven V_H subgroups. (The precise number of subgroups depends, of course, on the degree of homology used to define the subgroup.)

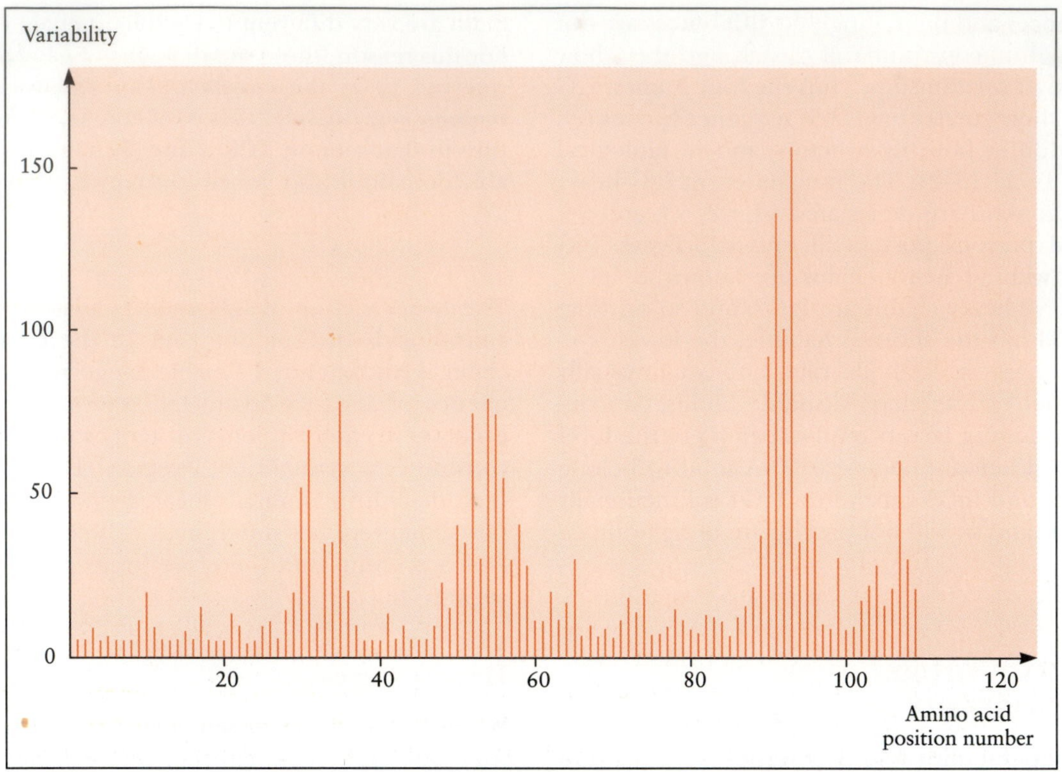

Figure 13–5 The variability of each amino acid position in an immunoglobulin light chain as shown by a Kabat-Wu plot. Note that there are three hypervariable regions, one of which is located close to the constant region around residues 95 to 100. This may be compared with a similar plot for the V region of the T cell antigen receptor (see Figure 11–4).

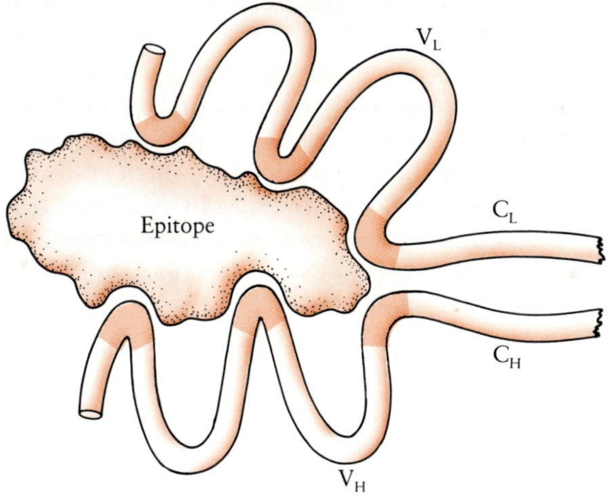

Figure 13–6 A schematic diagram showing the way in which light and heavy chains are folded so that the three complementarity determining regions in each chain form the walls of the antigen-binding groove.

Constant Regions

The C_H region is made up of three almost identical domains, each of 110 amino acids. They are structurally similar to the C_L domain. It is believed that these domains result from the repeated duplication of a single primordial gene coding for a basic domain of 110 amino acids. Presumably, this gene duplicated at an early evolutionary stage to form a linked constant and

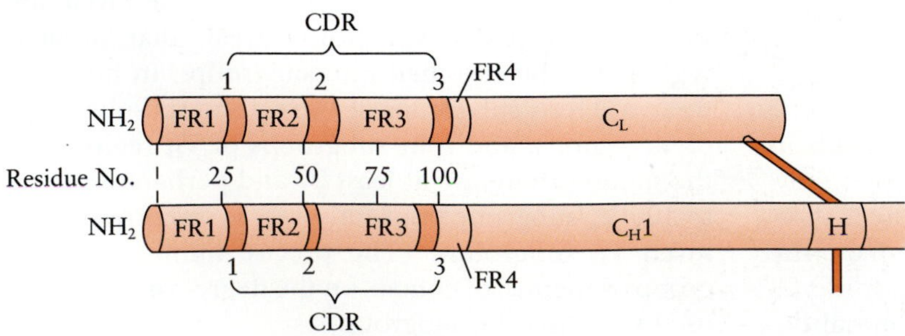

Figure 13–7 The location of the complementarity-determining, and framework regions in immunoglobulin light and heavy chains.

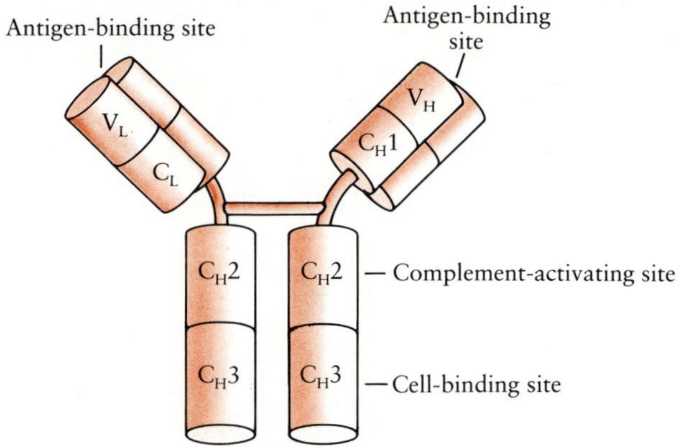

Figure 13–8 A model of an IgG molecule showing the domains and their biological activities.

variable domain and thus a primitive light chain (V_L and C_L). The constant domain was subsequently duplicated at least twice to form most of a heavy chain. One variable domain, one hinge region, and three constant domains make up a complete γ heavy chain. They are labeled from the N-terminal end, V_H, C_H1, H, C_H2, and C_H3. A similar arrangement is found in the α chains of IgA. In IgM and IgE heavy chains, an additional domain known as C_H4 is present. As a result, the Fc regions of IgM and IgE are larger than those of IgG or IgA.

Since heavy and light chains as well as both heavy chains are covalently linked, domains come together to form specialized regions within the immunoglobulin molecule. These paired domains provide a structure by which immunoglobulin molecules can exert their biological functions. Thus V_H and V_L form a paired domain that binds antigen, and C_H1 and C_L together act to stabilize the antigen-binding site. The paired C_H2 domains of IgG contain a site for the activation of the complement cascade (see Chapter 16) and a site that binds to receptors on phagocytic cells (Fig. 13–8). These domains also influence the catabolic rate of IgG. The structure of the heavy chain also regulates the transfer of IgG across the placenta to the fetus and antibody-mediated cellular cytotoxicity (Chapter 18), although these are probably due to the combined activities of several domains.

Hinge Region

On electron microscopy of IgG, it can be seen that the Fab regions are mobile and can swing freely around the center of the molecule as if they are hinged (Fig. 13–9). This hinge consists of about 12 amino acids located between the C_H1 and C_H2 domains. There is no homology between the hinge and other heavy-chain domains, and the sequence is unique for each immunoglobulin class and subclass. The hinge region contains many hydrophilic and proline residues (Fig. 13–10). The hydrophilic residues make the peptide chain unfold and thus make this region readily accessible to proteolytic enzymes. The proline residues

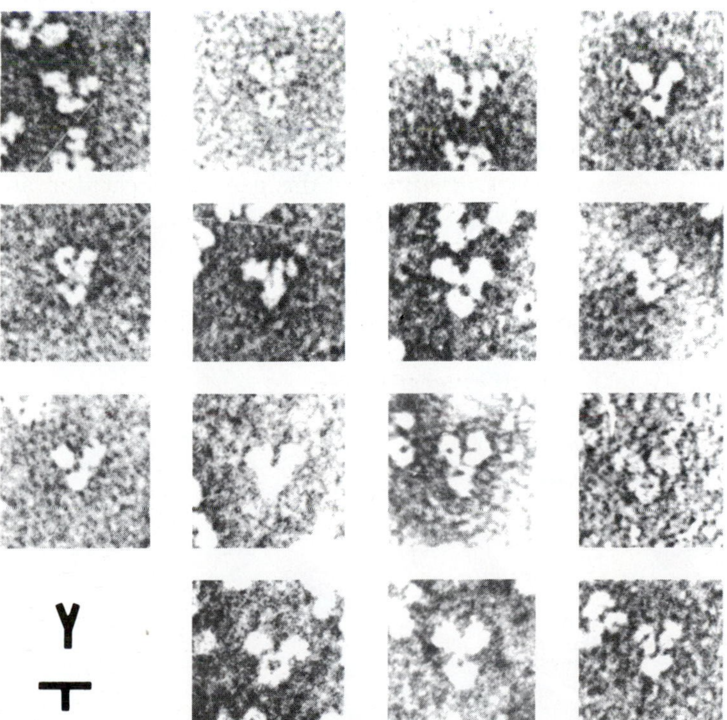

Figure 13–9 A series of electron micrographs of rabbit IgG showing how the Fab regions can move as a result of the flexibility of the hinge region, and a diagram showing the two extremes of the basic "Y" configuration. *(From Roux, K.H., and Metzger, D.W. J. Exp. Med., 129:2548, 1982. With permission.)*

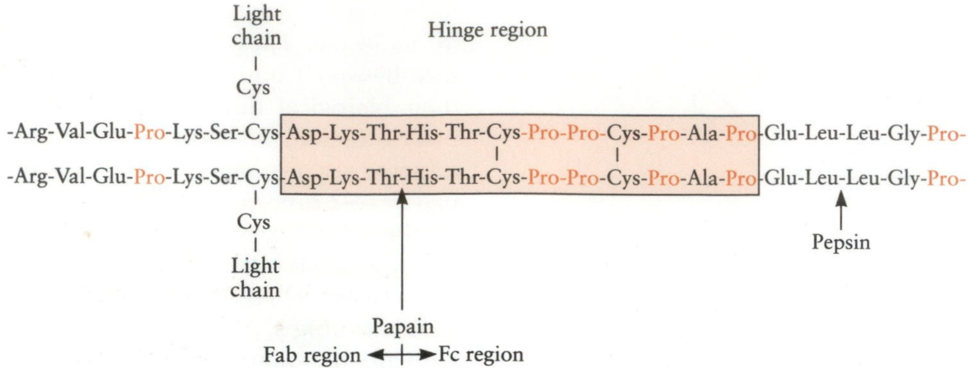

Figure 13–10 The amino-acid sequence within the hinge regions of the human IgG2 subclass. Note the large numbers of proline residues that confer flexibility and the cysteine residues that form interchain disulfide bridges. The diagram also shows the sites of pepsin and papain action.

make the chain flexible. This is the region where pepsin and papain act. This region also contains all the interchain disulfide bonds.

THE IMMUNOGLOBULIN SUPERFAMILY

Not only is the immunoglobulin domain the major structural unit of immunoglobulin constant regions, but individual immunoglobulin-like domains are key components of many other proteins involved in cellular interactions. These molecules form an **immunoglobulin superfamily.** A superfamily is a set of proteins that arise from a common ancestor and, as a result, share significant sequence homology.

An immunoglobulin domain consists of about 110 amino acids. Their sequence gives it a characteristic shape of two β-pleated sheets stabilized by an intrachain disulfide bond. The folding varies slightly between variable and constant domains. There are three β strands in one sheet and four in the other in a constant domain (Figs. 13–11 and 13–12). The β strands

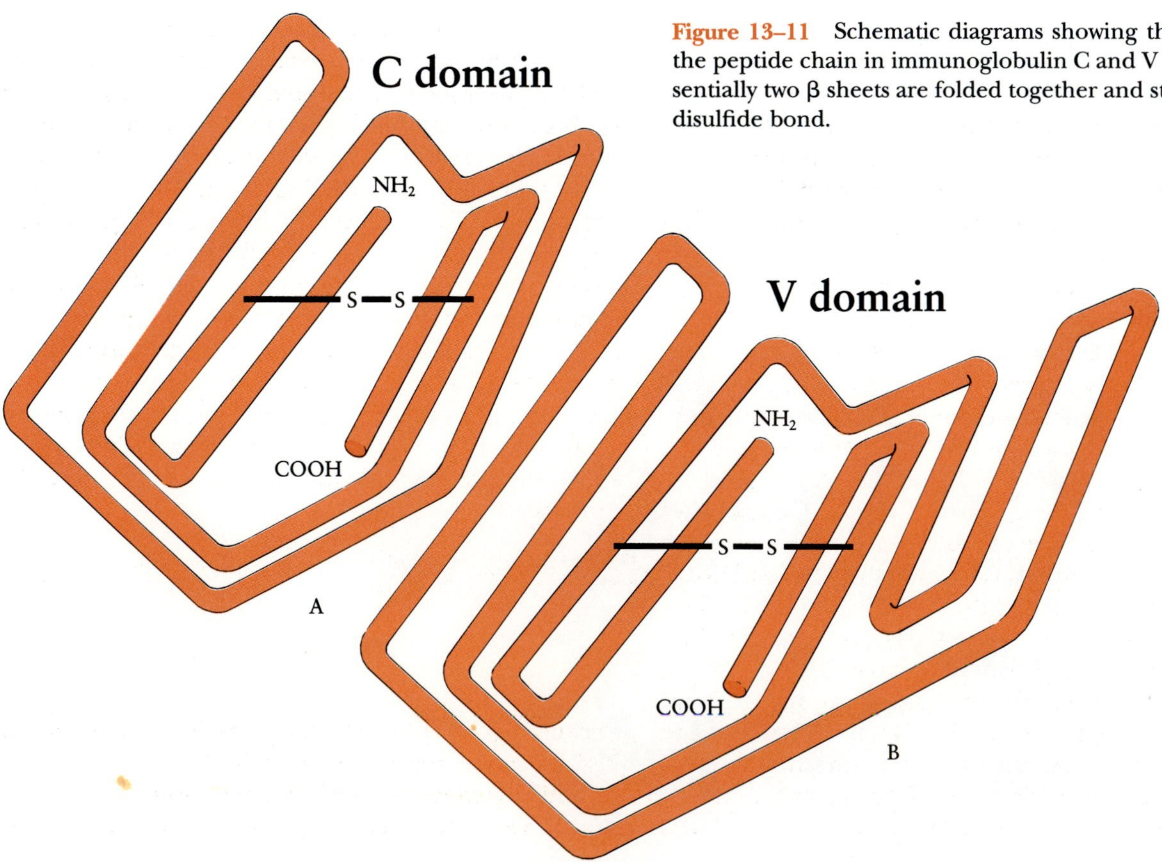

Figure 13–11 Schematic diagrams showing the folding of the peptide chain in immunoglobulin C and V domains. Essentially two β sheets are folded together and stabilized by a disulfide bond.

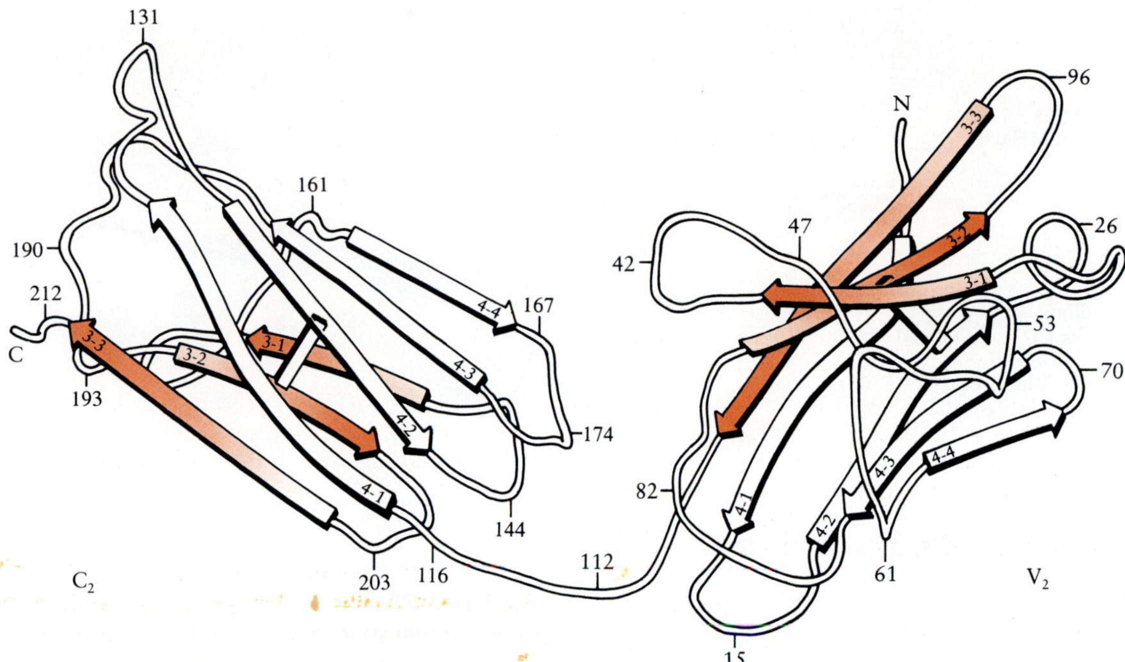

Figure 13–12 A schematic diagram showing the structure of two linked immunoglobulin domains—a light chain. Directional arrows are superimposed on segments participating in antiparallel β-pleated sheets. Three-chain layers are indicated by numbered striated arrows and four-chain layers by white arrows. Positions of representative residues are numbered. *(Reprinted with permission from A.B. Edmundson, K.R. Ely, E.E. Abola, M. Schiffer and N. Panagiotopoulos, Rotational allomerism and divergent evolution of domains in immunoglobulin light chains, Biochemistry, 14: 3953–3961 (1975) 1975. American Chemical Society.)*

consist of alternating hydrophobic and hydrophilic amino acid residues. The hydrophobic side chains are orientated toward the interior of the molecule and so stabilize the interaction between the two sheets. The hydrophilic chains point toward the exterior of the molecule. The intrachain disulfide bond stabilizes the structure and compacts it so that it is relatively insensitive to proteolytic enzymes.

The immunoglobulin superfamily consists of proteins that contain at least one immunoglobulin domain. The superfamily includes at least eight families of glycoproteins with multiple domains and 12 glycoproteins with only a single domain. The families with multiple domains include not only the immunoglobulin light and heavy chains, but also the α, β, γ, and δ chains of the T-cell receptors and the MHC class I and II molecules. **Secretory component,** the receptor that is used to carry polymeric immunoglobulins across mucosal surfaces, is a member of the multiple domain group. The amino acid sequence of β2-microglobulin, a member of the single domain group, is so similar to that of an immunoglobulin constant domain that it can activate complement and bind to the **Fc receptors** of macrophages. Members that contain only one domain include CD1, CD2, CD8, B7, as well as the CD3γ, δ, and ϵ chains of the TCR. Other members include brain and thymus antigens such as Thy-1 and OX-2, which are found on neurons and lymphocytes; neural cell surface glycoproteins such as N-CAM, found only on brain cells; and Po glycoprotein found in peripheral myelin (Fig. 13–13). Growth factor receptors such as those for M-CSF belong to this family. Some non-surface-associated molecules belong to this family. These include α1 B-glycoprotein found in serum and basement membrane link protein. In looking for the common features of the members of this gene family, it appears that all are involved in binding to another molecule. Most are found on cell surfaces, and none has enzymatic activity. In many cases, intercellular interaction is mediated by two different members of the superfamily as, for example, between TCR and MHC molecules.

IMMUNOGLOBULIN CLASSES

Immunoglobulin G

Because IgG is the immunoglobulin class found in highest concentration in blood, it plays the major role in antibody-mediated defense mechanisms (Fig. 13–14). It has a molecular weight of about 160 kDa and

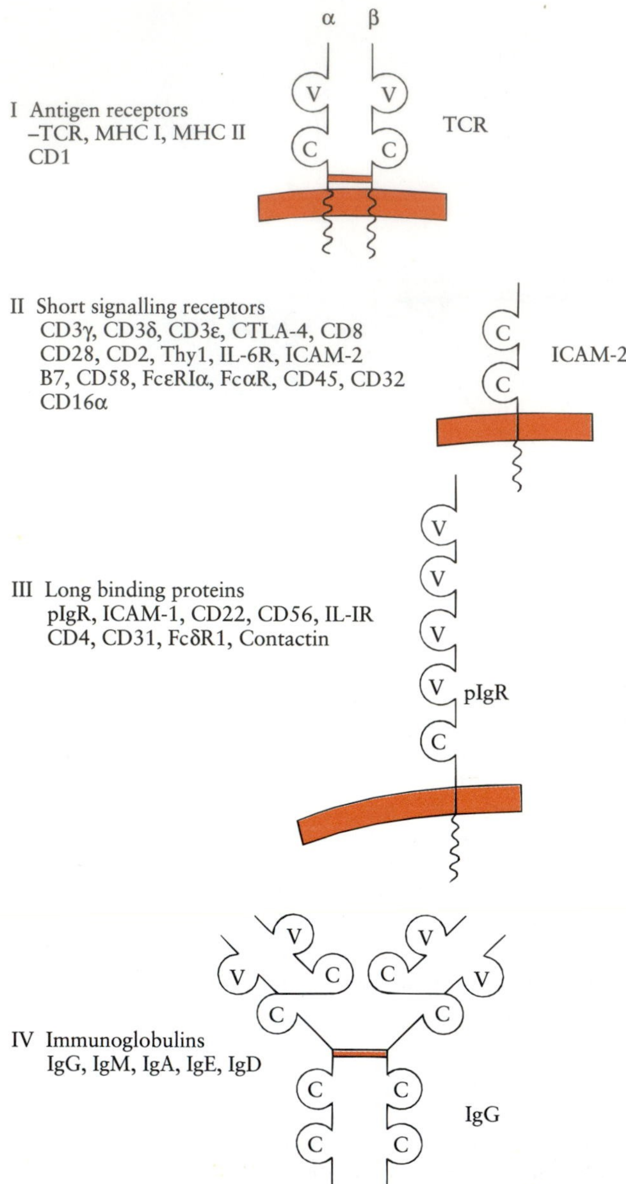

Figure 13–13 The members of the immunoglobulin superfamily. These fall into four major groups. (**a**) The antigen receptors, (**b**) the small receptors that contain one or two domains, (**c**) the large receptors that contain three or more domains, and (**d**) the immunoglobulins.

γ heavy chains. Because of its relatively small size, IgG can escape from blood vessels more easily than can the other immunoglobulin molecules. Therefore, it participates in the defense of tissue spaces and body surfaces. IgG can opsonize, agglutinate, and precipitate antigen (Chapter 17), but it can only activate the **classical complement pathway** when multiple IgG molecules have accumulated in a correct configuration on the antigen surface (Chapter 16).

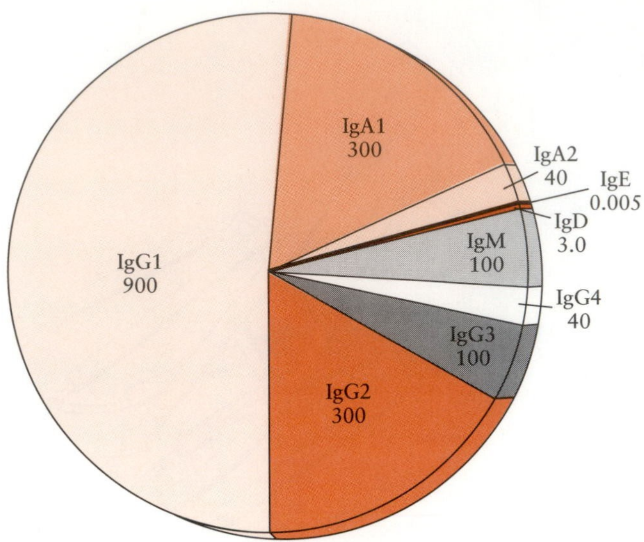

Figure 13–14 The immunoglobulins found in normal human serum. The numbers beside each class and subclass are average serum concentrations in mg/dl.

Immunoglobulin M

IgM is found in the second highest concentration after IgG in most mammalian serum; in humans, however, the IgA concentration in serum is slightly greater than that of IgM. Structurally, IgM is a polymer formed by five (rarely six) 180-kDa subunits (Fig. 13–15), so that its molecular weight is 900 kDa. Each subunit consists of two κ or two λ light chains and two μ heavy chains. Since each subunit possesses two antigen-binding sites, it might be anticipated that the valency of IgM for antigen would be 10. In practice, this valency is more commonly found to be 5 as a result of steric hindrance between antigen molecules. The μ chain has an additional fourth constant domain (C_H4) as well as an additional 20-amino-acid segment on its C-terminus, but it does not contain a hinge region. The site for complement activation by IgM is located on this C_H4 domain.

IgM **monomers** are linked by disulfide bonds in a circular fashion. A small cysteine-rich polypeptide called the **J chain** (15 kDa), coded for by a separate gene, binds two of the units to complete the circle. Since IgM molecules are normally secreted intact by plasma cells, the J chain must be considered an integral part of the molecule.

IgM is the major immunoglobulin class produced in a primary immune response. It is also produced during a secondary response, but this is commonly masked by the production of much larger quantities of IgG. In humans, for example, about 32 mg/kg of IgG is produced daily as opposed to 2 mg/kg of IgM. Although produced in relatively small quantities, when considered on a molar basis, IgM is more effi-

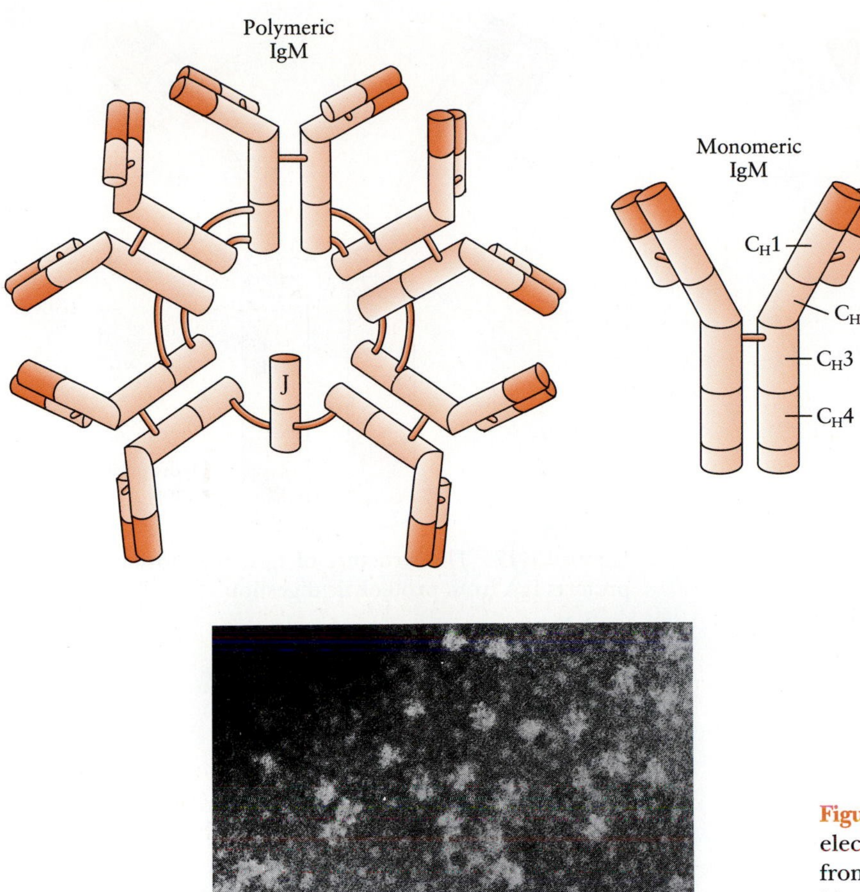

Figure 13–15 The structure of IgM and an electron micrograph of this immunoglobulin from bovine serum. Although IgM lacks a hinge region, the μ heavy chain is flexible between the C_H2 and C_H3 regions (approximate magnification ×240,000). *(Courtesy of Drs. K. Nielsen and B. Stemshorn.)*

cient than IgG at activation of the complement cascade, at opsonization, at virus neutralization, and at agglutination. Because of their large size, IgM immunoglobulins are confined to the blood and are therefore of little importance in conferring protection in tissue fluids or body secretions. When IgM acts as an antigen receptor on B-cell membranes (BCR), it is found only in the monomeric form (Chapter 14). This membrane-bound IgM also differs from the secreted form in that the C-terminus of the C_H4 domain is longer and contains a hydrophobic sequence that enables it to act as an integral membrane protein (Fig. 13–16).

Immunoglobulin A

Monomeric IgA has a molecular weight of 150 kDa. Each monomer has a typical four-chain structure consisting of paired κ or λ light chains and two α heavy chains. IgA, however, occurs naturally as a dimer. The two monomers are joined by a **J chain,** which links a Cα2 of one unit to the Cα3 of the other unit (Fig. 13–17). Higher polymers of IgA are occasionally found in serum. A membrane-bound form of IgA acts as a receptor on B cells. It differs from the secreted form in having a hydrophobic membrane-binding sequence at its C-terminus.

In many mammals, including humans, subclasses of IgA are recognized. In humans, differences in heavy-chain structure give rise to IgA1 and IgA2. There are also two variants of IgA2. One variant lacks disulfide bonds between the heavy and light chains, so that the molecule is held together only by noncovalent forces. Interchain bonds are present in the second IgA2 variant and in all IgA1 molecules. IgA is synthesized largely by plasma cells located on body surfaces. The IgA produced by the cells in the intestinal wall may either pass through epithelial cells into the intes-

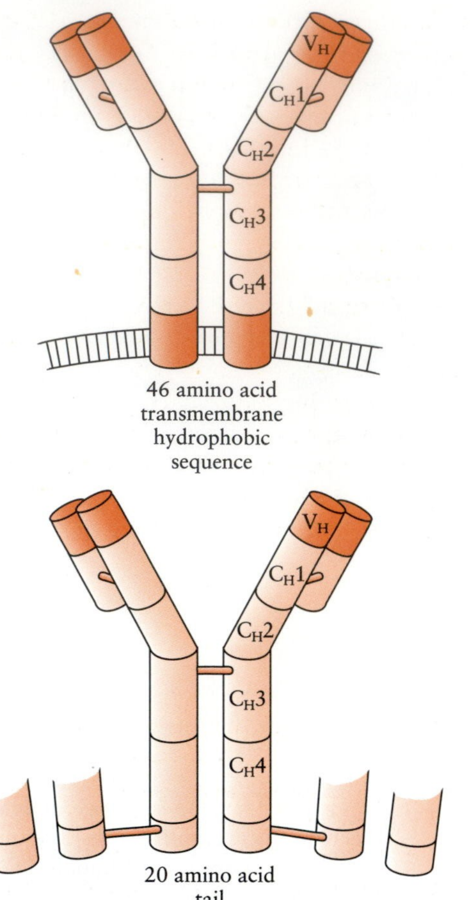

Figure 13–16 The differences in the heavy chains of cell-bound, and polymeric immunoglobulin M.

tinal lumen or, depending on the species, diffuse into the bloodstream. IgA within the bloodstream binds to hepatocytes and is carried through them into the bile. As the IgA is transported through intestinal epithelial cells or through hepatocytes, it is bound to a glycoprotein of 71 kDa known as **secretory component.** Secretory component binds covalently to IgA dimers to form a complex molecule called secretory IgA (SIgA). Secretory component protects IgA from digestion by intestinal proteolytic enzymes. Secretory IgA is of critical importance in protecting body surfaces against invading microorganisms since it is the major immunoglobulin in the intestinal, respiratory, and urogenital tracts; in milk; and in tears.

Immunoglobulin D

Immunoglobulin D molecules consist of two δ heavy chains and two κ or λ light chains. The molecular weight of IgD is about 170 kDa. IgD has only two domains in its heavy chains since it lacks a C_H2 domain. These two heavy-chain domains are separated by a long exposed hinge region (Fig. 13–18). IgD has no interchain disulfide bonds between its heavy chains and, as a result, is unusually susceptible to proteolysis. Since proteases are generated when blood clots, IgD cannot be found in serum but is present in low concentrations in plasma. Like IgE, IgD is readily denatured by mild heat treatment. IgD antibodies with activity against thyroid tissue, insulin, penicillin, nuclear antigens, and diphtheria toxoid have been described. Plasma IgD levels are twice as high in smokers than in

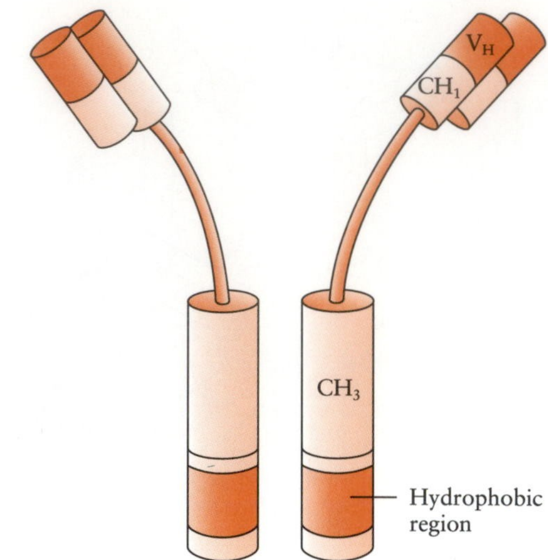

Figure 13–17 The structure of IgA. Secretory component protects IgA from proteolytic digestion.

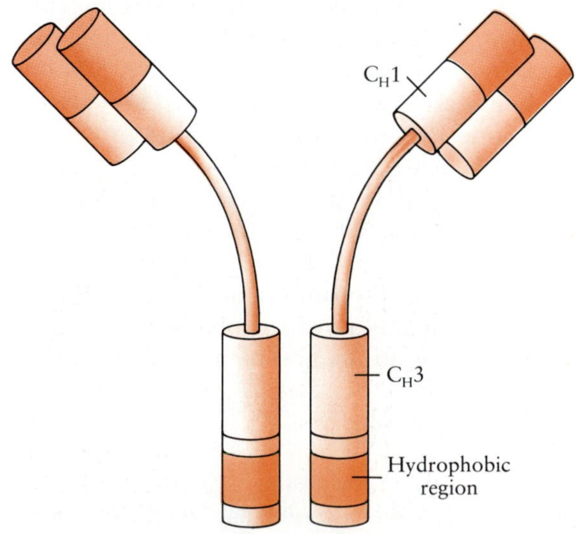

Figure 13–18 The structure of IgD showing the very long exposed hinge region that accounts for the extreme susceptibility of this molecule to proteolytic digestion.

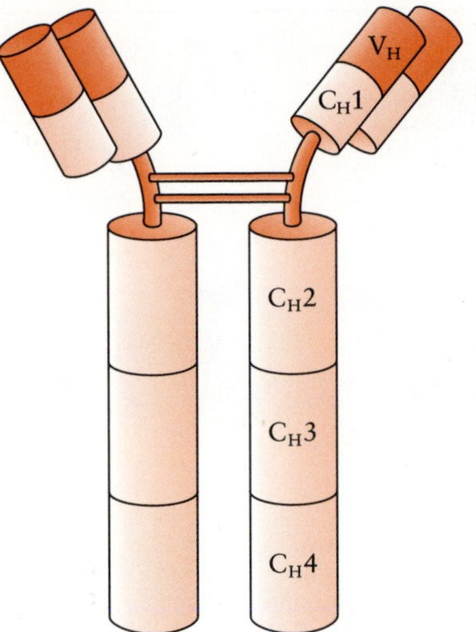

Figure 13–19 The structure of IgE. Note the presence of the additional domainthat permits the IgE to bind to Fc receptors on mast cells.

nonsmokers. Nevertheless, extensive IgD production is not a common feature of conventional immune responses. IgD is primarily a B-cell antigen receptor (Chapter 14).

Immunoglobulin E

An IgE molecule contains two κ or λ light chains and two ε heavy chains (Fig. 13–19). The hinge region is replaced by a constant domain so that each heavy chain contains four constant domains, and as a result, IgE has a molecular weight of 190 kDa. IgE is found in extraordinarily low concentrations in the serum of unparasitized individuals, varying from 20 to 500 ng/ml. This is approximately 1/40,000 of the concentration of IgG. IgE therefore does not neutralize antigens directly. Nevertheless, IgE is important since its biological activities are greatly amplified by binding to receptors on mast cells and basophils. As a result, it mediates **allergies** (type I hypersensitivity reactions, Chapter 29) and is largely responsible for immunity to invading parasitic worms. The Fc region of IgE binds strongly to high-affinity receptors on mast cells and basophils and, together with antigen, mediates the release of inflammatory agents from these cells. The receptor-binding site is formed by the 12 N-terminal amino acids of the C_H3 domains.

SECONDARY STRUCTURE OF IMMUNOGLOBULINS

The three-dimensional structure of immunoglobulin molecules is relatively constant. Nevertheless, the presence of hypervariable regions within the variable regions ensures that there are significant differences in molecular shape at the antigen-binding site. A monomeric immunoglobulin molecule consists of three globular regions (two Fab regions and one Fc region) linked by a flexible hinge (Fig. 13–20). Each of these globular regions is made up of paired domains. Thus the Fab regions each consist of two interacting do-

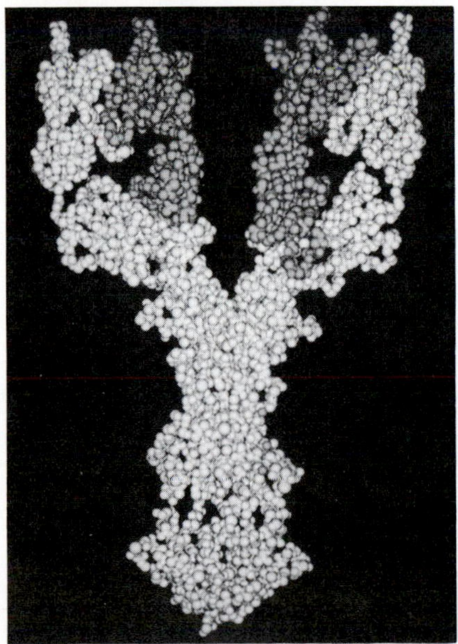

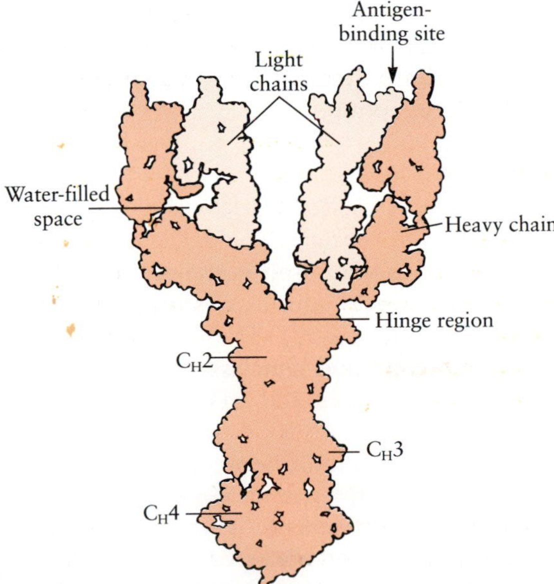

Figure 13–20 A molecular model of an IgE molecule showing the peptide backbone. The light chains are a darker shade than the heavy chains. The spaces within the molecule are normally filled with water. *(Courtesy of Dr. Scott Linthicum.)*

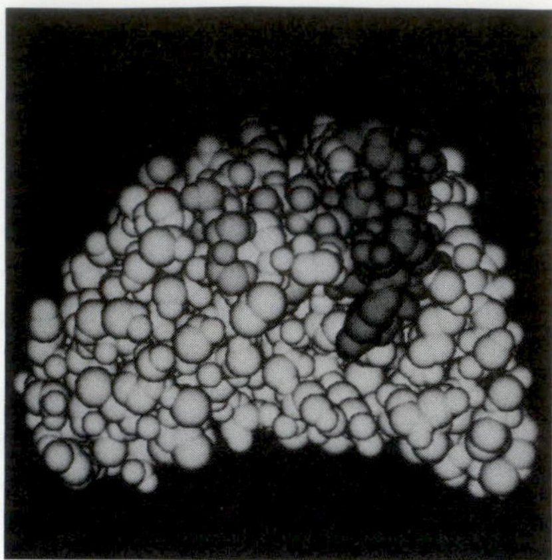

Figure 13–21 A side view of the antigen binding site of an immunoglobulin. The dark shaded atoms denote the location of one of the complementarity-determining regions. The antigen binding groove is at the top of the model. *(Courtesy of Dr. Scott Linthicum.)*

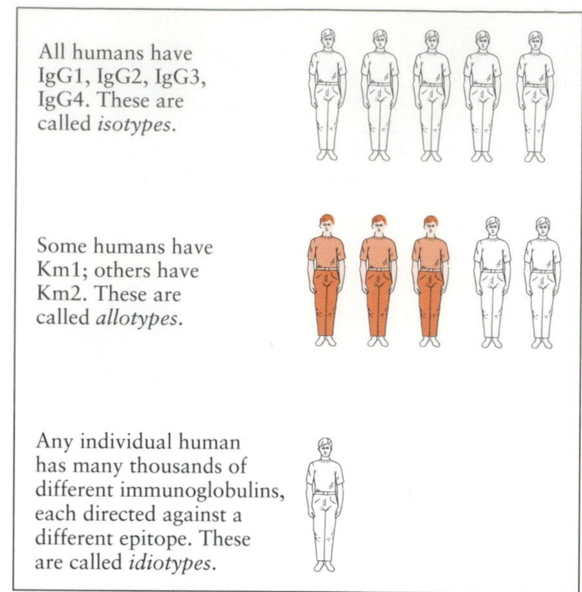

All humans have IgG1, IgG2, IgG3, IgG4. These are called *isotypes.*
Some humans have Km1; others have Km2. These are called *allotypes.*
Any individual human has many thousands of different immunoglobulins, each directed against a different epitope. These are called *idiotypes.*

Figure 13–22 The differences between isotypes, allotypes, and idiotypes.

mains (V_H–V_L and C_H1–C_L), and the Fc regions contain either two or three paired domains depending on the immunoglobulin class (i.e., C_H2–C_H2, C_H3–C_H3, and, in IgE or IgM, C_H4–C_H4). The polypeptide chains within each domain are closely intertwined.

In the Fab regions, the molecule's surface has a depression between the two variable domains, V_H and V_L. The peptides of the CDRs line this depression (Fig. 13–21). As a result, the surface of the depression has a highly variable shape. This depression forms the antigen-binding site. Because immunoglobulins are bilaterally identical, the CDRs on each of the Fab globular regions are also identical. As a result, the molecule has two identical antigen-binding sites and can bind two identical epitopes.

IMMUNOGLOBULIN VARIATIONS

The structural variations that result in the production of the different immunoglobulin classes occur in all animals of a species (Fig. 13–22). These variants are called classes or **isotypes** and subclasses. Thus IgG, IgM, IgA, and so on are classes, and IgG1, IgG2, IgG3, and so on are subclasses.

Another source of variation between proteins is the existence of minor sequence variations between the proteins of different individuals. These variations, which are inherited, are called **allotypes.** Provided the sequence variations are minor and involve replacement of similar amino acids, functional differences may not result. Immunoglobulin allotypes have been identified in many species, including humans, rabbits, and mice (Table 13–3). For example, three Km allotypes are found on κ light-chain constant regions. Km1 has a valine at position 153 and a leucine at position 191, Km1, 2 has an alanine at position 153 and a leucine at position 191, and Km3 has an alanine at position 153 and a valine at position 191. These three behave like alleles of a single gene.

The third group of structural variations in proteins is unique to the antigen-binding proteins such as immunoglobulins and the TCRs. These are formed by the variable regions on light and heavy chains. These variations are individually called **idiotopes** and collectively known as **idiotypes.** Some idiotopes may be located within the antigen-binding site. Others are located close by on non-antigen-binding areas of the V region. Idiotypes, like isotypes, are found in all individuals. However, because of the randomly generated sequence variations in this region, idiotypes are much more complex than isotypes.

Table 13–3 **Human Immunoglobulin Allotypes**

Location	Allotype
κ light chains	Km1, 2, 3
V_H region	Hv1
γ1 heavy chain	G1m1, 2, 3, 4, 17
γ2 heavy chain	G2m23
γ3 heavy chain	G3m5, 6, 11, 13, 14, 15, 16, 21, b, c
γ4 heavy chain	G4m, 4a, 4b
μ heavy chain	Mm1
α1 heavy chain	A2m1, 2

It is possible, on occasion, to provoke an animal to make antibodies against its own idiotypes. Thus an animal such as a rabbit can be immunized against a hapten. These antibodies can be isolated, purified, and stored. If injected with these stored antibodies after at least a year, the original donor rabbits make anti-idiotype antibodies.

ANTIGEN–ANTIBODY BINDING

When an antigen and its specific antibody combine, they interact through the chemical groups on the surface of the **epitope** and the CDRs of the immunoglobulin molecule. In classical chemical reactions, molecules are assembled through the establishment of firm covalent bonds. These bonds can be broken only by the input of a significant amount of energy—energy that is not readily available in the body. By contrast, the formation of noncovalent bonds provides a rapid and reversible way of forming complexes and permits reuse of molecules in a way that covalent bonding would not allow. Noncovalent bonds, however, act over short intermolecular distances and, as a result, form only when two molecules approach each other very closely. The binding of an epitope to the CDRs of an immunoglobulin molecule is exclusively noncovalent, so the strongest binding occurs when the two components are closely approximated and when the shape of the epitope and the shape of the CDRs conform to each other. This requirement for a close conformational fit has been likened to the specificity of a key for its lock.

Strength of Binding

The binding of antibody to its antigen is mediated by several types of noncovalent bonds such as hydrophobic and hydrogen bonding (Chapter 7). Indeed, hydrophobic bonding is probably of greatest importance. Each type of noncovalent bond is relatively weak in itself, but collectively the bonds may have a significant binding strength. All of these bonds act only across short distances and weaken rapidly as that distance increases. Electrostatic bond and hydrogen bond strengths are inversely proportional to the square of the distance between the interacting molecules; the Van der Waals forces and hydrophobic forces are inversely proportional to the seventh power of that distance. Clearly, therefore, the strongest binding between an epitope and immunoglobulin CDRs occurs only if their shapes match perfectly so that multiple noncovalent bonds can form. Nevertheless, antigens can bind to antibodies when they fit less than perfectly, although the number of bonds established and the strength of binding will clearly be less.

Measurement of Binding Strength

Because the combination of antigen and antibody is a reversible reaction

$$\mathrm{Ag} + \mathrm{Ab} \rightleftharpoons \mathrm{AgAb},$$

the strength of binding of the reactants can be defined through an affinity constant (K) by the mass action equation:

$$\mathrm{K} = \frac{[\mathrm{AbAg}]}{[\mathrm{Ab}]\ [\mathrm{Ag}]}$$

where [Ab] is the concentration of antibody-binding sites. [Ag] is the concentration of free epitopes, and [AbAg] is the concentration of immune complexes *at equilibrium.*

At the free epitope concentration $[\mathrm{Ag_c}]$, where only half the immunoglobulin-binding sites are filled, the concentration of antigen bound to antibody is equal to the concentration of free antibody, that is, [AbAg] = [Ab]. The equation may then be simplified to:

$$\mathrm{K} = \frac{1}{[\mathrm{Ag_c}]}$$

In other words, K is equal to the reciprocal of the concentration of free epitopes *when half the immunoglobulin-binding sites are occupied.*

If an antibody has a high **affinity** for its antigen, very little antigen is required to half saturate the antibody-binding sites. For example, antibody affinity constants can reach values as high as 10^{11} liters per mole. (This means that as small an amount as 10^{-11} moles per liter of free epitope (hapten) is required to occupy half the binding sites on a mole of this immunoglobulin.) Normally, antibody affinities fall within the range of 10^5 to 10^{10} liters per mole.

Affinity Maturation

Following the administration of antigen, the average affinity of the IgG antibodies produced gradually increases as an immune response progresses. In contrast, the affinity of IgM antibodies remains unchanged. The affinity of IgG antibodies also tends to rise as the dose of immunizing antigen is lowered. Baruj Benacerraf suggested that these two phenomena were linked. He suggested that a high initial dose of antigen can provoke many clones of antibody-producing cells with high, medium, or low affinity for antigen. The average affinity of the antibodies produced is relatively low (see Fig. 13–23). As an immune response progresses, however, the amount of available antigen is gradually reduced, which tends to stimulate only

METHODOLOGY

How to Measure Immunoglobulin Affinity

Immunoglobulin affinity can be measured by means of a method known as equilibrium dialysis. This technique involves placing a mixture of univalent **hapten** [H], for example, dinitrochlorobenzene, and antibody [Ab] in a dialysis sac (see the figure). The antibody cannot pass through the dialysis membrane, but the hapten does so freely. When equilibrium is reached, any hapten bound to the antibody [AbH] is retained within the sac. On the other hand, the unbound hapten, being free to diffuse through the membrane, is at an equal concentration both inside and outside the sac. If the total amount of hapten inside and outside the sac is measured, the amount of hapten bound to the antibody is easily measured by subtraction. The process can be repeated using different hapten concentrations.

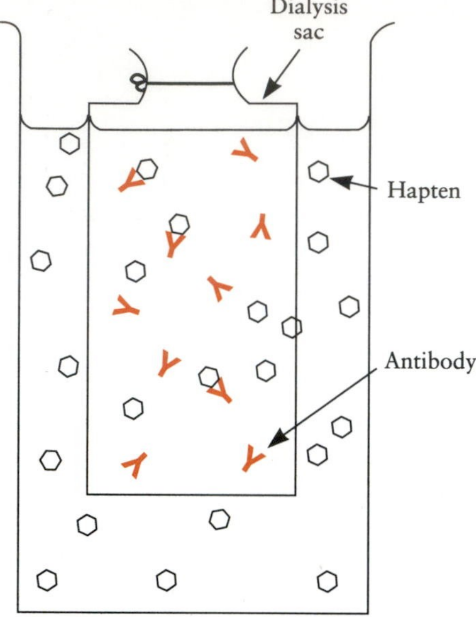

At equilibrium, therefore,

$$K = \frac{[AbH]}{[Ab]\ [H]} \qquad (1)$$

This can be simplified to:

$$K = \frac{b}{A_F \cdot c} \qquad (2)$$

where b is the concentration of bound hapten, A_F is the concentration of uncomplexed antibody-binding sites, and c is the concentration of free hapten at equilibrium. A_F can also be expressed as

$$(nA_T - b) \qquad (3)$$

where n is the number of binding sites on an antibody molecule (its valency) and A_T is the total concentration of antibody in the system. Thus equation 2 can be expressed as:

$$K = \frac{b}{(nA_T - b)c} \qquad (4)$$

If we divide this by A_T, we get:

$$K = \frac{b/A_T}{cnA_T/A_T - cb/A_T} \qquad (5)$$

Now b/A_T is the ratio of the moles of hapten bound per mole of antibody in the inner chamber; we can call this ratio r. Equation 4 may therefore be simplified:

$$K = \frac{r}{cn - cr} \qquad (6)$$

which can be rearranged to:

$$r/c = -Kr + Kn$$

In this system both r and c are variables, and K and n are constants. Equation 6 can therefore be considered a form of the straight-line equation ($y = ax + b$), where a is the slope and b the intercept of the vertical axis. Thus, if r/c is plotted against r, the slope will be −K, the intercept on the vertical axis will be Kn, and the intercept on the horizontal axis will be n, the valency of the antibody (see the figure). (This is known as a Scatchard plot.)

those cells with a high affinity for the antigen. The antibodies produced late in an immune response will therefore have a relatively high average affinity. This increase in affinity is also mediated through somatic mutation in the CDRs of IgG B-cell receptors. This is described in Chapter 15.

Biological Significance of Antibody Affinity

The affinity of a population of antibodies markedly affects their immunological properties, perhaps the most important of which is specificity. In general, high-affinity antibodies tend to be less specific for an-

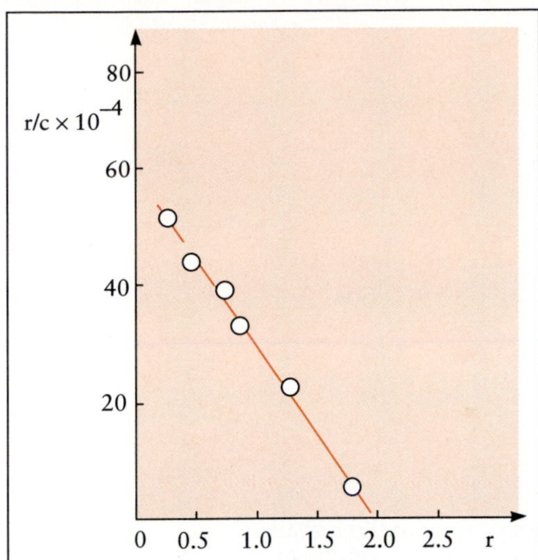

A plot of r/c against r (a Scatchard plot) for an absolutely pure immunoglobulin, for example, a monoclonal antibody. The intercept on the horizontal axis is 2.0 and indicates the valency of the immunoglobulin. The affinity of the antibody for the antigen is read from the vertical axis where r = 1 and is approximately 3×10^5 liters/mole.

When equilibrium dialysis is performed, it is usual to place a constant concentration of antibody (A_T) in the dialysis sac and permit it to react with a series of different hapten concentrations. The hapten concentration outside the sac is c, and r is the concentration of bound hapten in the dialysis sac divided by the total antibody concentration.

If a sample of pure **monoclonal antibody** reacts with a simple hapten and r/c is plotted against r, the points obtained fall in a straight line.

If, however, a serum sample reacts with the hapten, the line is not straight but curved (see the figure). The reason is that even a single epitope provokes the production of many immunoglobulins in a

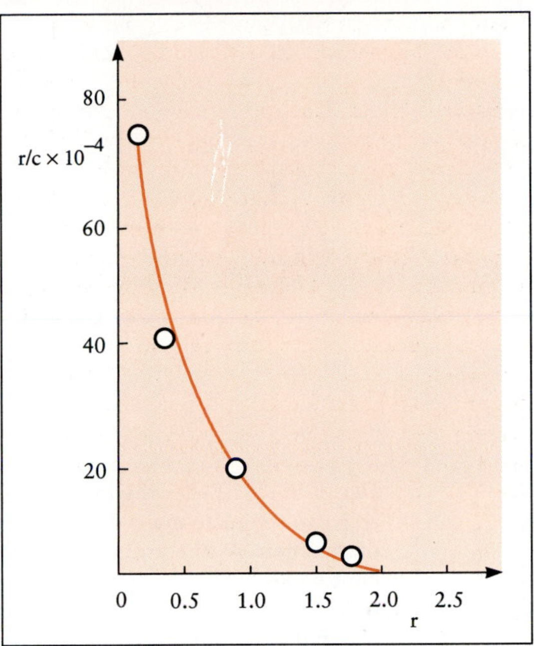

A Scatchard plot for an IgG fraction from serum. The intercept on the horizontal axis is 2.0 indicating the average valency of these antibodies. The line is curved since a mixture of different antibody molecules is present. Compare this to the previous figure in which a monoclonal antibody is examined. The average affinity of the mixture can be read from the vertical axis when r = 1 and is approximately 1.5×10^5 liters/mole.

serum, each having a different affinity for the epitope. Thus when an antigen induces antibody formation, not only are antibodies made against different epitopes, but even the antibodies made against a single epitope or hapten are very heterogeneous. Since affinity is calculated as the equilibrium constant when half the antibody-binding sites are occupied, it is usual to read the slope at a point where r = 1 and thus measure the average affinity of the antibody populations.

tigen than low-affinity antibodies. High-affinity antibodies combine strongly not only with their inducing epitope, but also with related epitopes, for which their affinity may be much lower. By contrast, low-affinity antibodies react with their inducing epitope but are unlikely to react detectably with any other epitope; they thus give the appearance of high specificity.

The strength of binding between antigen and antibody also influences the biological properties of the

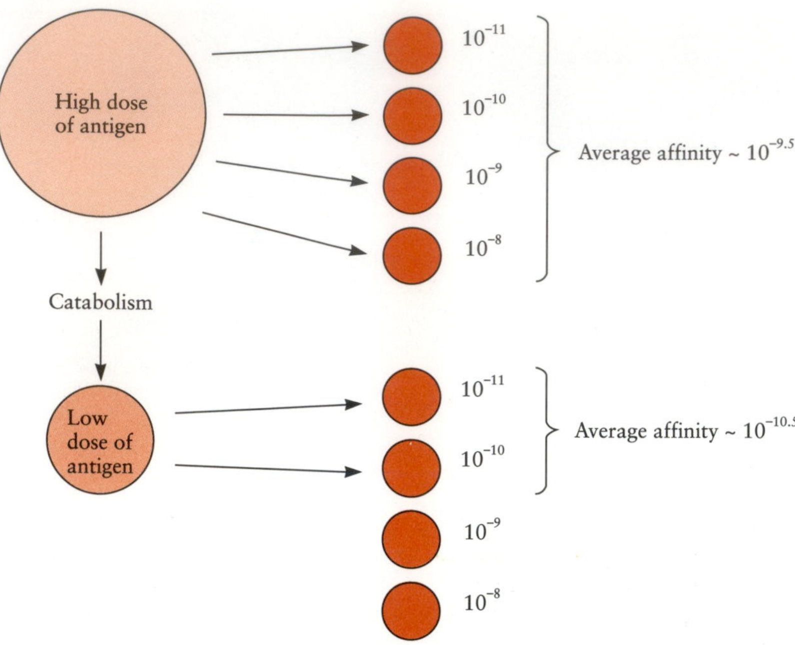

Figure 13–23 Following injection of antigen, the high dose of antigen available can stimulate many different clones including some with low affinity for the antigen. As the antigen is catabolized the amount of antigen becomes limiting and will therefore stimulate only high affinity B cell clones. This is reflected in differences in the affinity of the antibodies produced early and late in the immune response. It must not be forgotten that somatic mutation also contributes to this change in affinity.

antibody. Thus, high-affinity antibodies are superior to low-affinity antibodies at virus and toxin **neutralization.** The firm binding of antibody to these antigens apparently reduces the possibility of dissociation and so prevents the release of free virions or toxin from the complex. High-affinity antibody is also more effective at removal of foreign antigen from the bloodstream and at complement activation and red blood cell destruction through hemolysis.

BIOLOGICAL CONSEQUENCES OF ANTIGEN–ANTIBODY INTERACTION

The combination of antigen and antibody may block the toxic sites on toxins or prevent viruses from binding to cells, but it will not, by itself, induce the elimination or destruction of antigen. Thus the primary role of antibodies is to mark or identify the target antigen. It is only through reactions mediated by the constant regions of antigen-bound immunoglobulins that the major protective mechanisms are triggered. The variety of structures in each domain of all of the immunoglobulin classes and subclasses constitutes a system that provides a wide choice of biological consequences of **immune complex** formation. Antibody molecules therefore function as transducers that convert the initial antigen–antibody interaction into a protective response. After combining with antigen, antibody acquires the ability to activate the complement cascade, to bind effectively to phagocytic cells, or to provoke the degranulation of mast cells and basophils. (Activation of the complement cascade and the serologic tests that utilize this phenomenon are discussed in Chapter 16.) The most significant of these effects is the triggering of the antibody response itself, when antigen binds to cell membrane immunoglobulins on B lymphocytes (Chapter 14).

One suggestion on how these reactions are initiated is that a conformational change occurs in the immunoglobulin Fc region as a result of antigen combining with the Fab region, and the result is the generation of new, biologically active sites. Alternatively, and probably more likely, the active sites preexist hidden in the uncomplexed antibody molecule. When antigen binds to the antibody, the resulting conformational changes result in exposure of these hidden sites (Fig. 13–24).

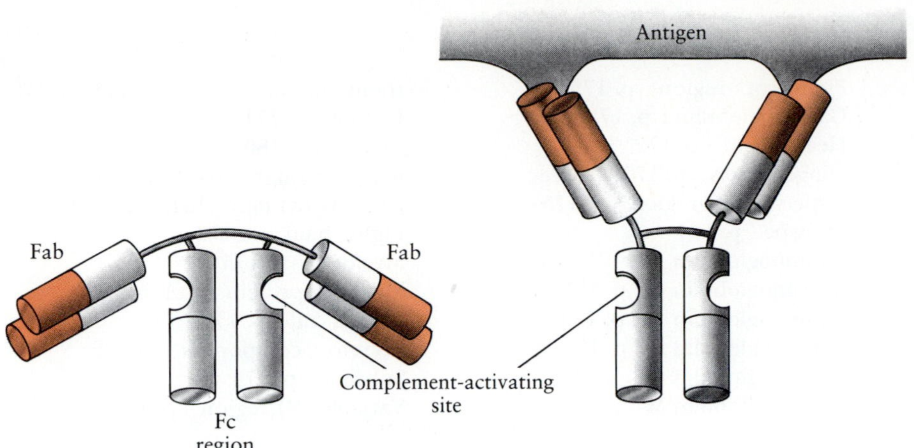

Figure 13–24 The complement activating site is located on the C_H2 area of the Fc region. The site is normally obstructed in uncomplexed immunoglobulin molecules. Combination of antigen with immunoglobulin exposes the activating site and therefore permits activation of the classical complement pathway.

Another type of cellular response to antigen–antibody–cell membrane interaction is degranulation. This occurs in mast cells, platelets, and, occasionally, neutrophils (Fig. 13–25). For example, if an antigen binds and links two mast cell–fixed IgE molecules, then a pathway is activated that makes the cell release the contents of its granules into the extracellular fluid (Chapter 29). Platelets that bind immune complexes release vasoactive factors and procoagulants. On occasion, neutrophils respond to immune complexes by releasing the contents of their lysosomes into extracellular fluid.

Cytotoxic activity may also be generated in response to immune complex formation (Chapter 18). Cells that can participate in this process (known as **antibody-dependent cellular cytotoxicity,** or ADCC) and destroy foreign cells on contact include macrophages, neutrophils, and some lymphocyte subpopulations. These cells exert their cytotoxicity regardless of whether they first bind to antibody-coated target cells or they bind antibody first and then bind to antigens on target-cell surfaces. A similar toxic effect may be mediated by eosinophils that bind to **helminths,** destroying them through IgG or IgE bound to the helminth surface (Chapter 28). Most serum immunoglobulins cannot bind to cell Fc receptors unless they are first complexed with antigen. Nevertheless, some antibodies, known as **cytophilic** or cytotropic antibodies, may bind to FcγRI and FcγRII even when uncomplexed. An example of a cytophilic antibody is IgE, which binds to FcεR on mast cells. (You can read more about Fc receptors and their activities on page 131.)

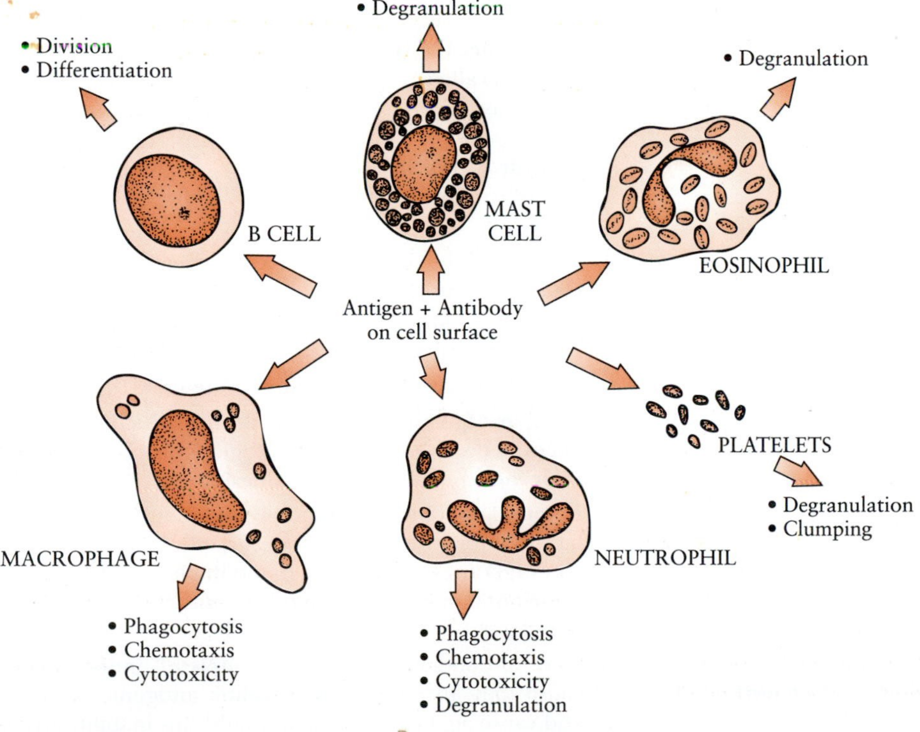

Figure 13–25 Some of the biological consequences of the combination of antigen and antibody on the surface of cells.

KEY WORDS

Affinity p. 185
Allotype p. 184
Antibody p. 171
Class p. 171
Complementarity-determining region p. 175
Constant (C) region p. 176
Degranulation p. 189
Electrophoresis p. 171
Equilibrium dialysis p. 186
Fab fragment (region) p. 173
Fc fragment (region) p. 173
Framework region p. 175
Gamma globulin p. 171
Heavy chain p. 172
Hinge region p. 177
Hypervariable region p. 175
Idiotype p. 184
Immunoglobulin p. 171
Immunoglobulin A p. 181
Immunoglobulin D p. 182
Immunoglobulin E p. 183
Immunoglobulin G p. 172
Immunoglobulin M p. 180
Immunoglobulin superfamily p. 178
Isotype p. 171
J chain p. 180
Kappa (κ) light chain p. 174
Lambda (λ) light chain p. 174
Light chain p. 172
Myeloma p. 175
Myeloma protein p. 175
Plasma cell p. 175
Secretory component p. 182
Subclass p. 174
Variable (V) region p. 175

QUESTIONS

1. How many constant domains does the heavy chain of IgE contain?
   a. two
   b. three
   c. four
   d. five
   e. six
2. The antigen-combining site of an antibody molecule determines its
   a. isotype
   b. anti-idiotype
   c. allotype
   d. hypertype
   e. idiotype
3. Which class of immunoglobulin activates complement upon binding with antigen?
   a. IgM
   b. IgA
   c. IgD
   d. IgE
   e. all of the above
4. Serum IgM molecules usually have
   a. 10 light chains
   b. 10 heavy chains
   c. a pentameric structure
   d. a J chain
   e. all of the above
5. Molecules in the immunoglobulin superfamily share
   a. antigen-binding sites
   b. domains
   c. variable regions
   d. amino groups
   e. proline residues
6. The major forces linking antigen to antibody are
   a. hydrogen bonds
   b. coulombic bonds
   c. covalent links
   d. hydrophobic bonds
   e. ionic bonds
7. Immunoglobulin binding to receptors on effector cells (e.g., phagocytes, mast cells) is due to which portion of the immunoglobulin molecule?
   a. Fab
   b. Fc
   c. Fd
   d. light chain
   e. V region
8. The hinge region of an immunoglobulin is flexible because it contains a large amount of which amino acid?
   a. serine
   b. cysteine
   c. threonine
   d. tyrosine
   e. proline
9. The complementarity-determining regions of immunoglobulins
   a. activate complement
   b. bind to cells
   c. mediate opsonization
   d. bind to antigen
   e. make the molecule flexible
10. The strength of binding between an antigen and its antibody is called
    a. avidity
    b. valency
    c. hydrophobicity
    d. electrophoretic mobility
    e. affinity
11. Why does the body need so many immunoglobulin classes and subclasses?
12. What is the biological significance of immunoglobulin domains? Why do you think that similar structures are found in many other unrelated proteins?
13. Compare IgA with IgM. How do the structural differences between the two immunoglobulin classes reflect their differing biological activities?
14. Immunoglobulin D is found predominantly on the surface of immature B cells. However, on occasion, low levels of IgD antibodies can be found in plasma. Why might these antibodies appear, and what might their function be?
15. Some antibodies are called "incomplete" antibodies because they are unable to cross-link antigenic particles and cause agglutination. How might this inability arise?

16. Draw a table outlining the chemical properties of the major human immunoglobulin classes and subclasses.
17. Which epitopes induce antibody formation when human IgG is injected into a rabbit? Which epitopes induce antibody formation when rabbit IgG is injected into a rabbit?
18. What are the differences between isotypes, allotypes, and idiotypes? What is the genetic basis of each of these differences?

Answers: 1c, 2e, 3a, 4e, 5b, 6d, 7b, 8e, 9d, 10e

SOURCES OF ADDITIONAL INFORMATION

Blattner, F.R., and Tucker, P.W. The molecular biology of immunoglobulin D. Nature, 307:417–422, 1984.

Burton, D.R. Antibody: The flexible adaptor molecule. TIBS, 15:64–69, 1990.

Capra, J.D., and Kehoe, J.M. Variable region sequence of five human immunoglobulin heavy chains of the V_{HIII} subgroup, definitive identification of four heavy chain hypervariable regions. Proc. Natl. Acad. Sci. U. S. A., 71:845–848, 1974.

Davies, D.R., Padlan, E.A., and Segal, D.M. Three-dimensional structure of immunoglobulins. Annu. Rev. Biochem., 44:629–667, 1975.

Edelman, G.M. The structure and function of antibodies. Sci. Am., 223:34–42, 1970.

Feinstein, A. Immunoglobulins and histocompatibility antigens. Nature, 282:230, 1979.

Greenspan, N.S., and Bona, C.A. Idiotypes: Structure and immunogenicity. FASEB J., 7:437–444, 1993.

Lin, L-C., and Putnam, F.W. Primary structure of the Fc region of human immunoglobulin D: Implications for evolutionary origin and biological function. Proc. Natl. Acad. Sci. U. S. A., 78:504–508, 1981.

Milstein, C. Monoclonal antibodies. Sci. Am., 243:66–74, 1980.

Natvig, J.B., and Kunkel, H.G. Human immunoglobulin, isotypes, subisotypes, genetic variants and idiotypes. Adv. Immunol., 16:1–59, 1978.

Nisonoff, A., Hopper, J.R., and Spring, S.B. The Antibody Molecule. Academic Press, New York, 1975.

Oi, V., et al. Lymphocyte membrane IgG and secreted IgG are structurally and allotypically distinct. J. Exp. Med., 151: 1260–1274, 1980.

Osler, A.G. On the precedence of 19S antibodies in the early immune response. Immunochemistry, 15:714–720, 1978.

Uzgiris, E.E., and Kornberg, R.D. Two-dimensional crystallization technique for imaging macromolecules, with application to antigen-antibody-complement complexes. Nature, 301:125–129, 1983.

Williams, A.F. The immunoglobulin superfamily takes shape. Nature, 308:12–13, 1984.

Wu, T.T., and Kabat, E.A. An analysis of the sequences of the variable regions of Bence-Jones proteins and myeloma light chains and their implications for antibody complementarity. J. Exp. Med., 132:211–250, 1970.

CHAPTER 14

The Response of B Cells to Antigen

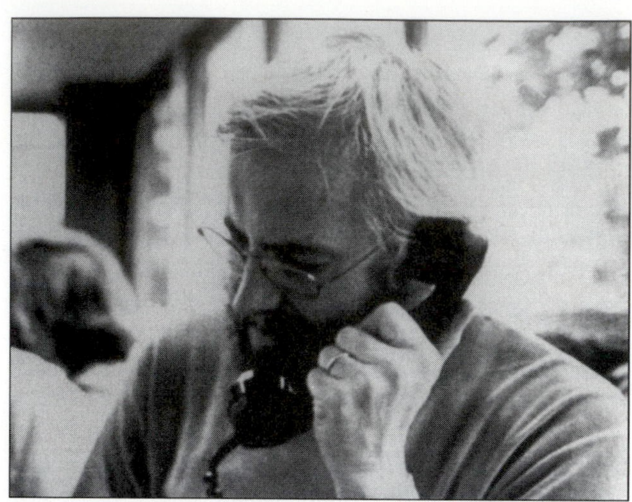

Georges Kohler *Georges Kohler was born in Germany in 1946. In 1984, he was awarded the Nobel prize together with his colleague Cesar Milstein for devising the process to generate hybridomas and the production of monoclonal antibodies. This was a remarkable methodological achievement that was all the more significant because Kohler was a post-doctoral student at the time of the discovery. He was seeking to devise a process by which he could combine the specificity of an antibody with the quantity of immunoglobulins produced by myeloma cells. One area of immunology in which the use of these antibodies has been critical is in the identification of cell-surface proteins through flow cytometry.*

CHAPTER OUTLINE

CHAPTER CONCEPTS

1. Each B cell carries many identical immunoglobulins on its surface specific for a single epitope. B cells are generated randomly from the bone marrow, each with a different specificity for antigen.
2. When an antigen binds to a B cell under the correct circumstances, it stimulates the B cell to divide and differentiate into plasma cells and memory cells.
3. To generate a B-cell response, antigen must be presented to a helper T cell. The helper T cell releases interleukins that together with antigen provoke a B cell to divide and differentiate.
4. Plasma cells are short-lived antibody-producing cells. The antibodies that they produce are specific for the inducing epitope.
5. Memory cells respond to antigen by division and differentiation in a manner similar to that of the original B cell.
6. B cells recognize free, unprocessed antigen, and antibodies, as a result, are directed against undenatured epitopes.
7. B cells can process antigens into peptides that can be recognized by T cells. Thus B cells can activate helper T cells.
8. A plasma cell tumor is called a myeloma. Cancerous plasma cells produce a single immunoglobulin product.
9. A myeloma cell may be fused with a plasma cell in the laboratory to form a hybridoma. Hybridomas produce monoclonal antibodies that are extremely pure and specific for a single epitope.

In this chapter we describe the properties of B cells and how they respond to antigen. We start with a brief historical note on the original theory on how cells responded to antigen. Then we describe in detail the structure of the B-cell antigen receptor and some of the costimulator molecules involved in turning on a B cell. B cells require stimulation from helper T cells, so we discuss how that happens and the role that cytokines play in the process. We also describe how B cells can act as antigen-processing cells. Then we follow the response of B cells to antigen from their initial stimulation until the B cells become plasma cells and memory cells. We also describe how this happens within germinal centers. The main section of the chapter ends with a brief review of the way in which B cells develop.

The last part of the chapter describes the tumors of B cells, called myelomas. These myelomas play a key role in the production of monoclonal antibodies by hybridomas. Since monoclonal antibodies have become an indispensable tool in immunology, a section is devoted to describing the technical details of their production.

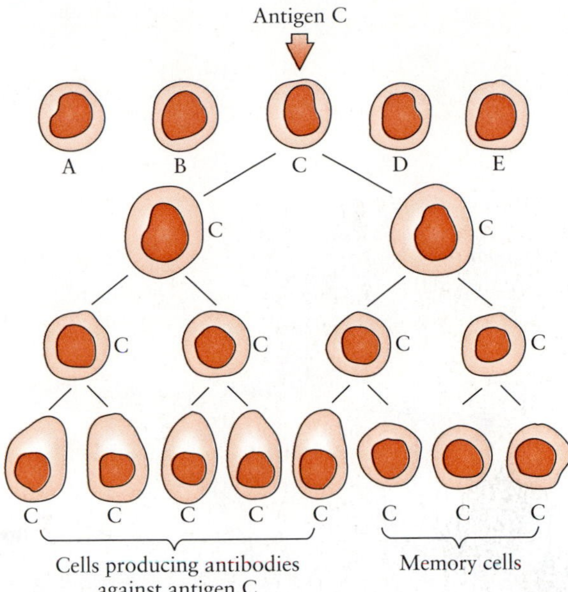

Figure 14–1 The clonal selection theory. Antigen promotes the growth and differentiation of a B cell clone (in this case clone C) responding only to that antigen. As a result antibody-producing cells, as well as memory cells specific for that antigen are produced.

CLONAL SELECTION

The general features of the cellular responses to antigen were first encompassed in what is known as the **clonal selection theory.** The basic postulates of this theory were developed by Sir Frank Macfarlane Burnet, who received the Nobel Prize for this work. They are as follows:

1. Lymphoid stem cells differentiate randomly to produce **clones** of lymphocytes, each of which is committed to respond to a single epitope (Fig. 14–1).
2. Antigen binding to lymphocyte receptors triggers them to proliferate and differentiate into antibody-producing cells, effector cells, and memory cells.
3. The specificity of the antibodies produced by a lymphocyte is identical to that of its antigen receptors.
4. Tolerance results when a clone of antigen-binding cells is destroyed or suppressed.

The clonal selection theory has been amply confirmed since it was first put forward, and in effect, it is the paradigm that defines modern immunology. (A paradigm (*par-a-dym*) is a generally accepted central concept.) Antigen-sensitive cells do exist as two distinct populations: the **B cells,** which produce antibodies, and the **T cells,** which mediate the cell-mediated immune reactions. Antigen binding to these cells induces them to proliferate and is the initiating event that triggers an immune response.

B CELLS

About 10^7 B cells are produced daily by the bone marrow in humans and they survive for several months. They are found in the **cortex** of lymph nodes, in the marginal zone in the spleen, and in Peyer's patches in the intestine. They do not circulate in the bloodstream in large numbers. Each B cell carries many identical receptors specific for a single epitope. These receptors are generated at random during B-cell development in a process described in Chapter 15. If a B cell encounters an epitope that will bind to its receptors, then under appropriate circumstances, it will respond by making antibodies.

The B Cell Antigen Receptor

Each B cell surface is covered with about 200,000 to 500,000 identical antigen receptors (BCRs). In addition to binding antigen and signaling the B cell to respond, the BCR triggers endocytosis of the antigen, leading to its proteolysis, processing, and expression of peptides associated with MHC class II molecules. B cells can thus capture, concentrate, and effectively present antigen to Th cells. Like the TCR, the BCR is a complex of several glycoproteins (Fig. 14–2). The specific antigen-binding structure is a cell membrane immunoglobulin. In immature B cells, this is a form of IgM whose C_H4 domain has a hydrophobic sequence that enables it to insert into the cell membrane

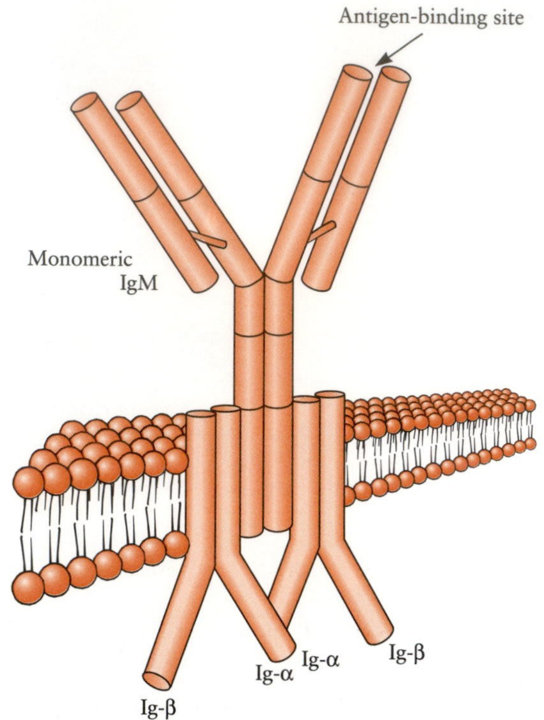

Figure 14–2 The structure of the B cell antigen receptor. The cell-surface immunoglobulin is linked to two heterodimers with extensive cytoplasmic domains. These serve as signal transducing molecules.

(Fig. 13–16). The receptor immunoglobulin is oriented so that its Fab regions are exposed and free to bind antigen, and its Fc region is noncovalently associated with proteins involved in signal transduction.

B-cell receptor immunoglobulins cannot transduce signals since their cytoplasmic domains contain only three amino acids. However, their C_H4 and transmembrane domains are linked to paired heterodimers (Ig–α/Ig–β), which have extensive cytoplasmic domains. In humans, Ig–α and Ig–β are 47- and 37-kDa glycoproteins. In mice, they are a little smaller (34 kDa and 39 kDa). Another glycoprotein called Ig–γ may also be associated with Ig–α. Ig–γ is a 34-kDa product of the β locus but is nonphosphorylated, underglycosylated, and truncated. (It contains 30 fewer amino acids.) The Ig–β chain appears to be identical irrespective of the immunoglobulin receptor class involved. There are, however, differences in glycosylation of the Ig–α chains associated with IgM, IgG, and IgD that could reflect distinct signaling pathways. IgD can also be found unlinked to Ig–α/β and attached to the cell membrane by a glycosyl-phosphatidylinositol linker.

Ig–α and Ig–β are both members of the immunoglobulin superfamily. They have an extracellular C-terminal domain that binds to the N-terminal domain on the immunoglobulin heavy chain. The cytoplasmic domains of both Ig–α and Ig–β contain a conserved sequence of 22 amino acids that is also found in the TCR CD3 and ζ chains as well as on CD23. This sequence is probably noncovalently associated with several src tyrosine kinases, including lyn, blk, lck, and fyn. The latter shares homology with the ZAP tyrosine kinase of the TCR.

Binding of antigen to the membrane immunoglobulin, especially if two receptors are cross-linked, triggers activation of the src kinases and results in phosphorylation and activation of several proteins, including two phospholipases C and possibly a G-protein (Fig. 14–3). As with the TCR, these reactions are dynamically regulated by CD45 tyrosine phosphatases. Subsequent hydrolysis of phosphatidylinositol and calcium mobilization lead to activation of a protein kinase C and transcriptional activation of several genes, including MHC class II molecules, the transferrin receptor (CD71), and transcription factors such as fos and myc. Clearly more than one signal is required for B cells to be induced to proliferate since cross-linking of antigen receptors alone is not sufficient. Second signals such as binding by CD19, CD40, IL-2, or IL-4 as well as helper T cell contact are required to move lymphocytes to a point in activation or to drive lymphocytes from the G_1 phase to the S phase of the cell cycle. It is suggested that the antigen provides a "competence" signal that activates resting B cells to enter G_1, and a "progression" signal such as IL-2 or IL-4 induces activated cells to proceed to S phase and divide.

An immature B cell usually carries only receptor IgM on its surface. As B cells mature, their receptor immunoglobulins change so that they go through a stage in which they have two types of receptors, IgM and IgD, on their surface. At this stage the B cells become able to respond to antigen. Following exposure to antigen, B cells switch their BCRs again to IgM and subsequently to IgG, IgA, or IgE.

The CD19 Complex

Although the BCR must bind antigen in order for a B cell to be stimulated, there is also a requirement for additional cross-linking. This may be mediated by a very large B cell surface protein complex containing CD19 (Fig. 14–4). CD19 occurs in this multimolecular complex with CR2 (a complement receptor, also called CD21), a protein called TAPA-1 (CD81), and the lyn tyrosine kinase. CD19 is a 95-kDa transmembrane glycoprotein of the Ig superfamily, and CR2 (CD21) is a 140-kDa molecule similar to the selectins.

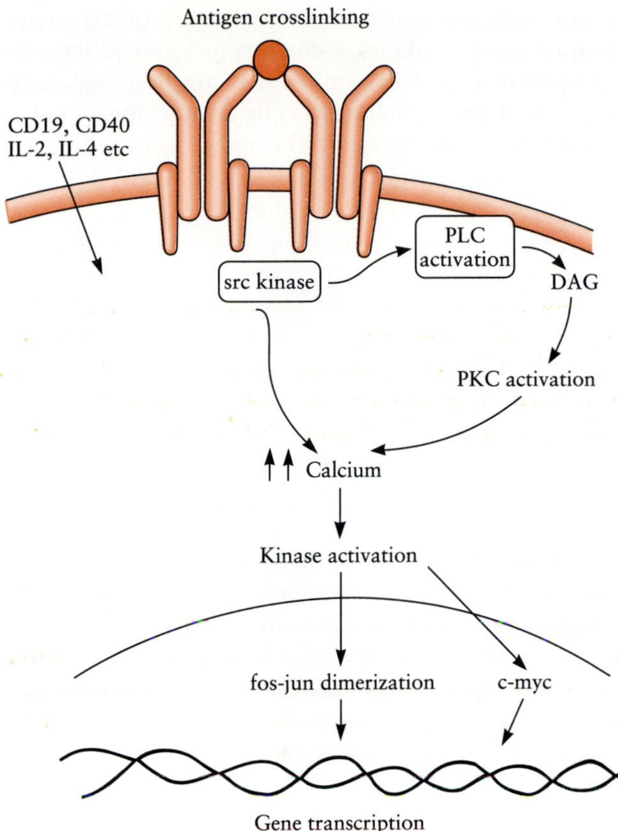

Figure 14–3 Signal transduction by the BCR occurs when two immunoglobulins are crosslinked. The resulting tyrosine kinase activation leads eventually to gene transcription and cell activation. (PLC, phospholipase C; DAG, diacylglycerol; PKC, protein kinase C)

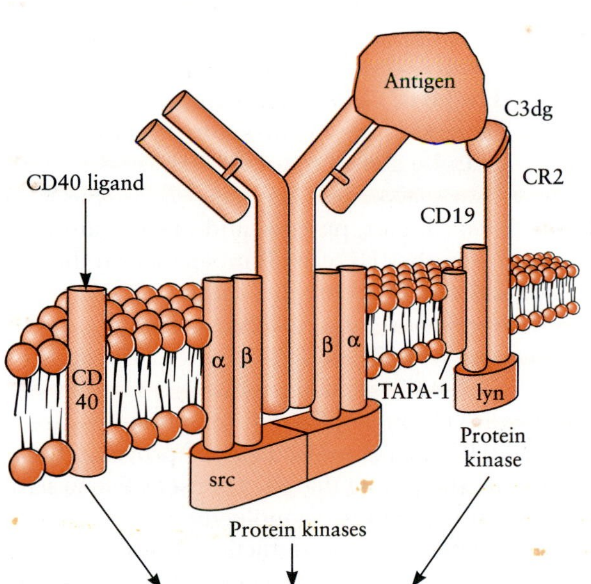

Figure 14–4 In order for a B cell to respond to antigen it must be costimulated by several other molecules including CD40 and CR2. These costimulators enhance the signals sent to the B cell.

It is suggested that the CR2 may bind to antigen linked to the BCR. The CR2–CD19 complex then associates with the BCR, and its tyrosine kinase promotes signal transduction, augmenting phospholipase C production. The dual binding of BCR and the CD19 complex may greatly improve the stability of the interaction and lowers the threshold for antigen receptor stimulation by about 100-fold. Thus B lymphocytes can be stimulated to divide when approximately 100 receptors per cell are co-ligated with the CD19 complex.

The function of the BCR thus parallels in many respects the function of the TCR. For example, the CD19 complex has similar coreceptor functions to CD4 and CD8. All are members of the immunoglobulin superfamily, each is physically associated with a tyrosine kinase, and each decreases the threshold for antigen receptor stimulation when co-ligated with the antigen receptor. It is also of interest to note that independent ligation of CD19, CD4, or CD8 suppresses signaling through the antigen receptors.

It is appropriate to emphasize at this stage that the B-cell receptor for antigen has the same antigen-binding ability as do antibody molecules. Thus they can bind to free, intact antigenic molecules in solution. This is very different from the TCR that can bind and respond only to processed antigen fragments presented on a MHC class II molecule. This difference in the antigen-recognizing ability of B and T cells is significant in that B cells are thus capable of responding to a much greater variety of antigenic stimuli than are T cells. Likewise, antibodies are directed, not against breakdown products of antigens, but against the intact antigen molecules. As a result, antigen–antibody interactions are usually dependent on maintenance of the three-dimensional conformation of the antigen. A good example of this is seen with tetanus toxoid. Antibodies raised against the intact molecule bind only to the intact molecule and cannot bind to proteolytic fragments such as would be produced by macrophage processing.

HELPER T CELLS AND B CELLS

As noted previously, removal of the neonatal thymus prevents the development of a cell-mediated response and significantly reduces antibody production. The thymus must therefore be necessary for successful antibody production. Studies have shown that the presence of some Th cells is necessary for an optimal B-cell response (Table 14–1). Some $CD4^+$ T cells respond to exogenous peptides bound to MHC class II molecules on antigen-presenting cells. These $CD4^+$ cells then activate antigen-specific B cells. The helper effect is mediated in two distinct stages. In the first stage, cell–B cell contact, mediated by antigen MHC class II antigens and CD19, is required. This is a mandatory stage since even in the presence of high concentrations of cytokines, some B cell–T cell adherence is required to provide sufficient activating signal. In the second stage, helper T cells release interleukins that stimulate the activated B cells to differentiate.

Table 14–1 Demonstration That B and T Cells Collaborate in Antibody Formation

Mice were lethally irradiated to destroy all their lymphocytes. They were then reconstituted using syngeneic bone marrow cells (B cells), thoracic duct lymphocytes (T cells), or both. The mice were then inoculated with sheep erythrocytes and their immune response monitored using the Jerne plaque assay (page 203) with the following results.

Reconstituted with	Anti-Sheep RBCs Plaque-Forming Cells
10^7 B cells + 10^8 sheep RBCs	310 ± 45
10^9 T cells + 10^8 sheep RBCs	0
10^7 B cells + 10^9 T cells + 10^8 sheep RBCs	2103 + 61

Source: Mitchell, G.F., and Miller, J.F.A.P. Immunological significance of the thymus and thoracic duct lymphocytes. Proc. Nat. Acad. Sci. U. S. A., 59:296–303, 1968. With permission.

THE RESPONSE OF B CELLS TO ANTIGEN

STAGE ONE—CELL–CELL INTERACTIONS

Helper T-Cell–B-Cell Interaction

Unprimed B cells require direct physical contact with an activated helper T cell to initiate their response to antigen. The T cells are activated by antigen-presenting cells some of which may also be B cells. This physical contact occurs through MHC class II molecules and antigen and through the binding of several key receptors on the cell surfaces. These receptors are of two types. Some of the receptors are nonspecific cell adhesion molecules such as integrins and selectins. These are not restricted to lymphocytes. Other receptors are found only on B and T cells. They must play a specific signaling role. These key receptor ligand interactions include those between CD40 and its ligand CD40-L, between B7-2 (CD80) and its ligands CD28 and CTLA-4, and between the CD19 complex and its ligand.

Successful cooperation between Th and B cells is restricted by MHC class II molecules expressed on the B cells. Using a simple in vitro culture system, for example, it can be shown that cooperation occurs only between Th and B cells of the same MHC class II haplotype. B cells ingest, process, and present antigen to Th cells through MHC class II molecules on the B cell surface and the TCR complex on the Th cell surface. B cells therefore send a **MHC-restricted** antigen-presenting signal to Th cells. This interaction not only triggers a Th cell into activity but may activate the B cell as well. Treatment of B cells by antibodies directed against MHC class II molecules can provoke them to proliferate, suggesting that they are sensitive to signals transmitted through these molecules.

MHC class II-TCR interaction alone cannot account for all the consequences of T-B cell interaction. T cells and B cells can recognize different epitopes on the same antigenic molecule. Helper T cells recognize processed epitopes and B cells recognize intact molecules or haptens. Thus the delivery of T-cell help to a B cell requires **linked recognition** of at least two epitopes on the antigen molecule. Once B cells are activated, their secondary responses can occur in the absence of linked recognition.

The CD40 molecule is expressed on B cells. Its ligand CD40-L is found on activated T cells. When cross-linked, CD40 stimulates B cell proliferation, prevents B cell apoptosis, and permits immunoglobulin class switching. If the CD40–CD40-L interaction is blocked, B cells are unable to respond to T cells. Molecules of the B7-2 (CD80) family are found on all antigen-presenting cells including B cells. Of their two ligands, CD28 is found on all T cells while CTLA-4 is found only on activated T cells. CD28 and CTLA-4 must bind their ligands if a T cell is going to respond to an antigen-presenting B cell. Signaling through B7-2-CD28 is required for the T cell to become fully activated. Thus when T and B cells come together they signal to each other not only through antigen but also through a dialogue between these two receptor systems. The signals delivered through CD40 and CD28 are both required to promote cell proliferation following antigen binding. Likewise both signals prevent apoptosis of the responding cells.

Antigen Processing by B Cells

B cells can be very effective antigen-presenting cells. Antigen binds to the immunoglobulin in the BCR. B cells then endocytose and fragment this **exogenous antigen** prior to linking it to MHC class II molecules and presenting it to the helper T cells while at the same time expressing B7-2 and secreting IL-1 and IL-12 (Fig. 14–5). Each B cell will, of course, bind only one type of antigen. This makes the B cells much more efficient than macrophages that must ingest any foreign material that comes their way. This is especially true in a primed animal in which many B cells may be available to bind a specific antigen. B cells can thus activate Th cells with 1/1000 of the antigen concentration of nonspecific antigen processing cells such as macrophages. They may be especially effective in presenting antigen to memory T cells. During the antigen presentation process, B and Th cells come into contact over a large area of apposition. It takes up to 6 to 8 hours for the B cells to activate T cells, and they can retain this activity for up to 72 hours after ingesting antigen. This presentation by B cells seems to be especially important in lymph nodes. Cytokines such as interferon-γ

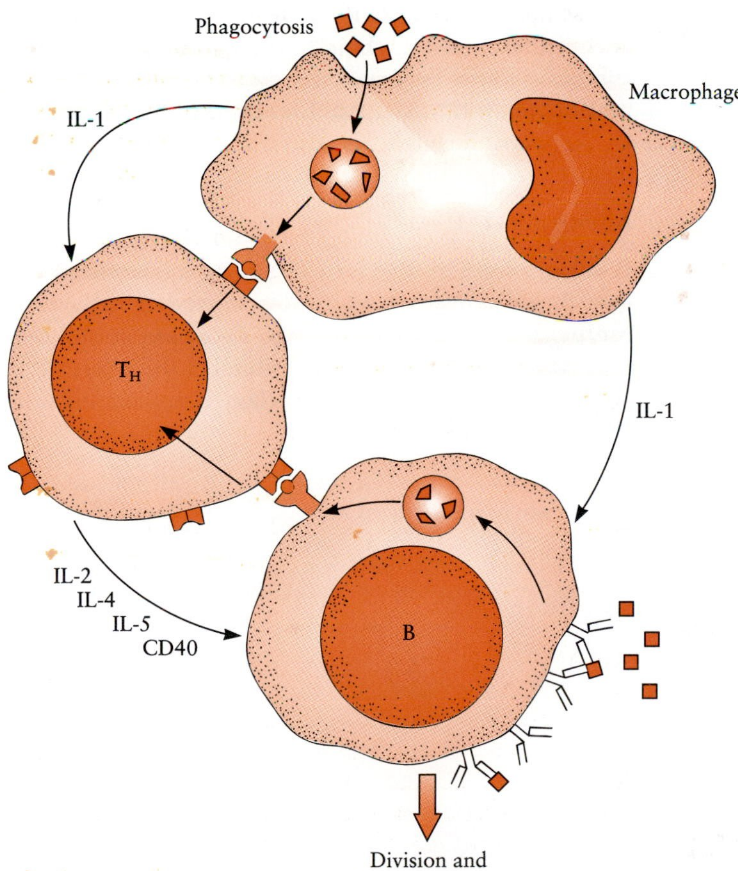

Figure 14–5 An outline of the cell interactions that occur when a B cell responds to antigen. Macrophages present antigen to a helper T cell and costimulate it with interleukin-1. The B cell is stimulated by free antigen and responds to it under the influence of IL-2, IL-4, and IL-5. B cells may also present antigen to T cells.

(IFN-γ) that promote MHC class II expression enhance antigen presentation by B cells.

STAGE TWO—ACTIVATION BY CYTOKINES

Helper Cell Subpopulations

Normal mouse and human helper T cells are divided into two major subpopulations with distinctly different functions called Th1 and Th2. These two subpopulations have different growth characteristics, respond in different ways to cytokines, and, most important, produce different mixtures of cytokines. It is probable that different types of antigen-presenting cells (e.g., macrophages, dendritic cells) activate different helper cell subpopulations and so induce different types of immune response (Table 14–2). Thus Th2 cells are optimized to help antibody formation by B cells. Th1 cells in contrast are optimized to help cell-mediated immune responses.

Th1 cells secrete IL-2 and IFN-γ that do not stimulate specific antibody formation even though they promote B-cell proliferation and polyclonal immunoglobulin secretion. Th2 cells in contrast secrete IL-4, IL-5, IL-6, IL-10, and IL-13. Consequently, supernatant fluids from cultured Th2 cells that contain these lymphokines enhance B-cell production of IgG1 and IgA up to 20-fold and production of IgE up to 1000-fold.

Table 14–2 The Amount of Immunoglobulin in Each Class Produced by B Cells in the Presence of Either Th1 or Th2 Cells

Class	Th2 cell (ng/ml)	Th1 cell (ng/ml)
IgG_1	21,600	<8
IgG_{2a}	39	14
IgG_{2b}	189	<8
IgG_3	354	<8
IgM	98,000	248
IgA	484	<1
IgE	187	<1

Adapted from Coffmann, R.L., et al. Immunol. Rev., 102:5, 1988. With permission from the author and Munksgaard International Publishers Inc.

Interleukins That Stimulate B Cells

Interleukin-2

IL-2 is secreted by Th1 cells in response to appropriately presented antigen in association with IL-12. It stimulates the appearance of IL-2 receptors (IL-2R) on B cells. Thus, stimulated B cells develop about 1000 high-affinity receptors and 10,000 low-affinity receptors for IL-2. The responsiveness of a B cell to IL-2 is determined by the level of expression of these receptors. Resting B cells do not express IL-2R until acted on by IL-1.

IL-2 acts with antigen and T-cell adherence to trigger a B-cell response. These three stimuli together make the B cells enlarge and proceed into the S phase of the cell division cycle (Fig. 14–6). This activates the B cell but does not in itself trigger antibody production. To complete the activation process, these responding B cells must then be acted on by IL-4 and IL-5. These molecules bind to the activated B cells and stimulate them further, ensuring their division and making them differentiate into immunoglobulin-producing plasma cells.

Interleukin-4 (IL-4)

IL-4 is secreted by activated Th2 cells. It stimulates the growth and differentiation of activated B cells, promoting them to move from the G_1 phase of the cell cycle to the S phase. IL-4 also enhances the expression by activated B cells of MHC class II molecules and

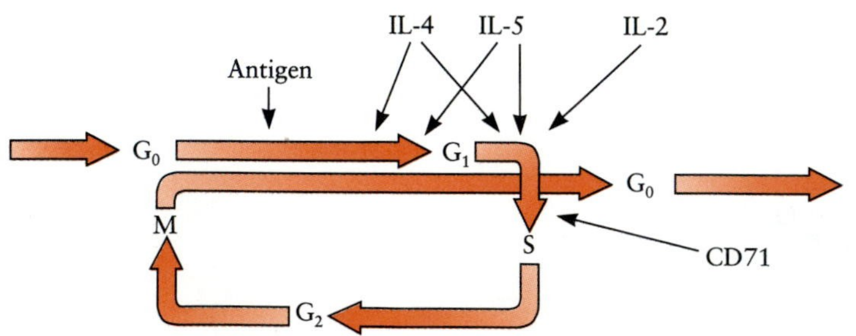

Figure 14–6 The sites in the B cell mitotic cycle where interleukins act.

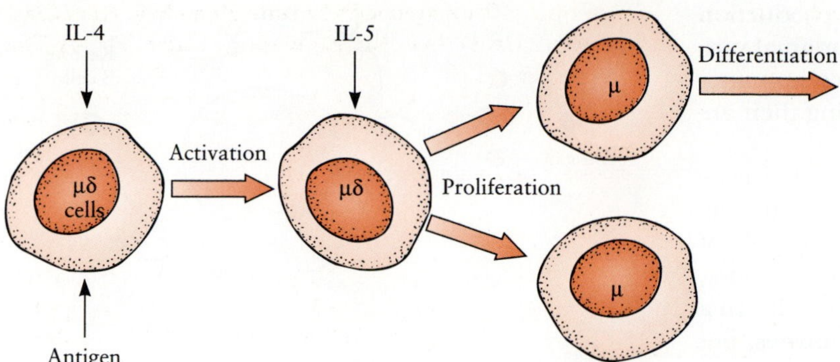

Figure 14–7 Interleukins 4 and 5 act at different stages during the response of B cells to antigen. In this simplified diagram, IL-4 activates the cells and IL-5 stimulates their proliferation.

other surface proteins such as Thy-1 and FcγR. It also induces mouse B cells to switch classes to IgG1 or IgE production while suppressing IgG3 and IgG2b production.

The actions of IL-4 are inhibited by IFN-γ, which has opposite biological effects. For this reason, Th1 cells, which secrete IFN-γ, may inhibit IgE synthesis. The proliferative effect of IL-4 on B cells is enhanced by IL-1. There are approximately 300 IL-4 receptors on a B cell.

Interleukin-5 (IL-5)

Interleukin-5 is also secreted by Th2 cells. IL-5 is a differentiation factor that acts on activated B cells to enhance their differentiation into plasma cells. It increases IL-2R expression (Fig. 14–7), stimulates IgG and IgM production, and enhances IL-4–induced IgE production. IL-5 selectively enhances IgA production in Peyer's patch B cells from the intestine. IL-6 promotes this B cell response to IL-5.

STAGE THREE—THE B CELL RESPONSE

When a mature B cell encounters antigen that binds to its surface receptors, in association with the correct stimuli from helper T cells, it responds by division and differentiation (Fig. 14–8). The responding B cells lose their cell membrane IgD and secrete IgM specific for the inducing antigen. As their response to antigen continues, these immunoglobulin-producing B cells

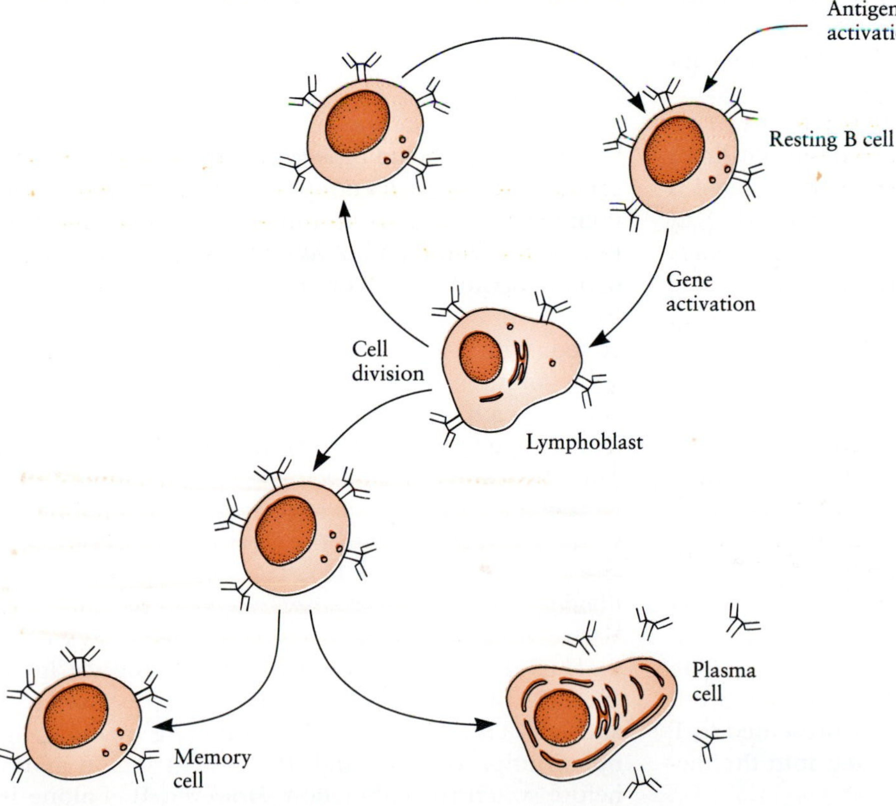

Figure 14–8 Activation of B cells leads to cell division and differentiation into memory cells and plasma cells.

switch their heavy-chain constant-region production from μ to γ, ϵ, or α while retaining their original variable regions. As a result, they can produce immunoglobulins of different classes without altering their antigen specificity.

When antigen enters the body, it can bind directly to a B cell with appropriate receptors. The term **clonotype** is used to describe a clone of B cells capable of responding to a single epitope. In a normal mouse, there exist at any one time about 3×10^8 B cells. In a newborn mouse, there are about 10^4 clonotypes, but their numbers increase as the mouse matures and may reach 10^8. Presumably, this increase reflects increased use of alternative sets of V genes and somatic mutation in these V genes (Chapter 15). In an adult mouse, the number of cells within a given clonotype varies as a result of exposure to different antigens over the animal's lifetime. For some antigens, there may be as few as 10 responsive cells in the spleen or bone marrow; for others, there may be as many as 10^4 cells.

T and B cells must interact closely to ensure that an antibody response occurs. Since each clonotype is present at a low average frequency, there is a low probability for effective contact. Indeed, as discussed in Chapter 9, T and B cells have different migration patterns and tend to segregate in different locations. It is unclear, therefore, just how these cells encounter each other. It is assumed that activated T cells circulate until they encounter the "correct" B cell. This seems to be a very inefficient process, especially in young animals in which clonotype frequency is low. As an individual ages, however, the most "used" clonotypes expand and the process increases in efficiency.

Each resting B cell carries both IgM and IgD receptor molecules on its surface with about ten times as many IgD molecules as IgM. These unstimulated B cells may secrete a small quantity of monomeric IgM into the medium spontaneously. When antigen binds to BCR immunoglobulins in the presence of helper T cells, IL-2, or IL-4, a signal is transmitted into the cell. This signal eventually leads to enhanced expression of B-cell MHC class II antigens and receptors for transferrin, IL-2, IL-4, IL-5, IL-6, TNF-α, and TGF-β and starts the process that leads to cell division.

At the same time, the B-cell membrane immunoglobulins, which are normally free to diffuse across the plane of the lipid membrane, aggregate in patches on the cell surface. Eventually these patches aggregate to form a single "cap" (Figs. 14–9 and 14–10). **Capping** is not energy-dependent but is associated with the contractile elements of the cell, especially myosin. The capped immune complexes are taken into the B cell through endocytosis and destroyed or presented to T cells. (In some cases, they may be shed into the medium.)

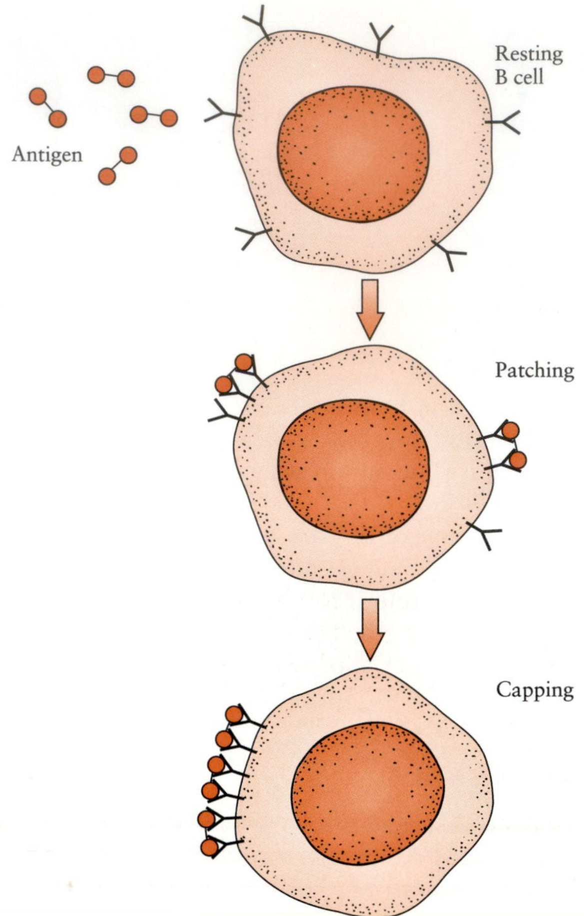

Figure 14–9 A schematic diagram showing the process of patching and capping in response to antigen.

An appropriately stimulated B cell enlarges and commences to divide repeatedly. Some of its progeny cells develop a rough endoplasmic reticulum, increase their rate of synthesis of immunoglobulins, and start to secrete immunoglobulins (Fig. 14–11). The half-life of immunoglobulins in or on the cells eventually drops from 20 to 30 hours to 2 to 4 hours as they are secreted. Within a few days, the cell switches to synthesizing another immunoglobulin class. This switch occurs while a B cell is in the germinal center and results in a change from IgM to IgG, IgA, or IgE production. This class switch occurs as a result of deletion of unwanted heavy-chain gene segments and the joining of variable-region genes to the next available constant-region genes (Chapter 15). The specificity of the antibody produced remains unchanged. Class switching is controlled by cytokines such as IL-4, IFN-γ, and TGF-β. Thus IL-4 directs mouse B cells to produce IgG1 and IgE while it directs human B cells to produce IgG4 and IgE. The IL-4 directs this switching by inducing transcription of the ϵ and γ1 constant region genes before switch recombination. However, IL-4 alone is

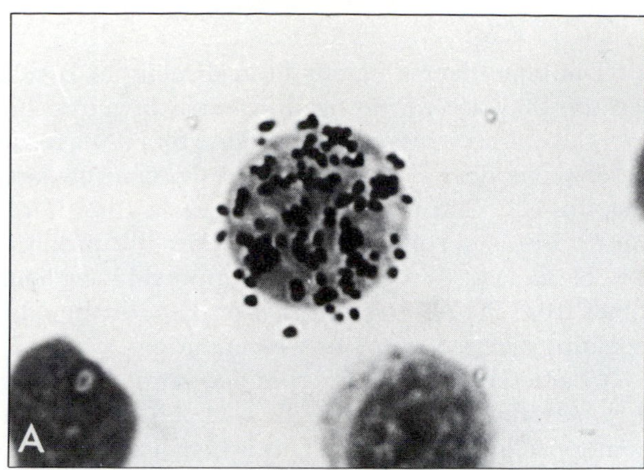

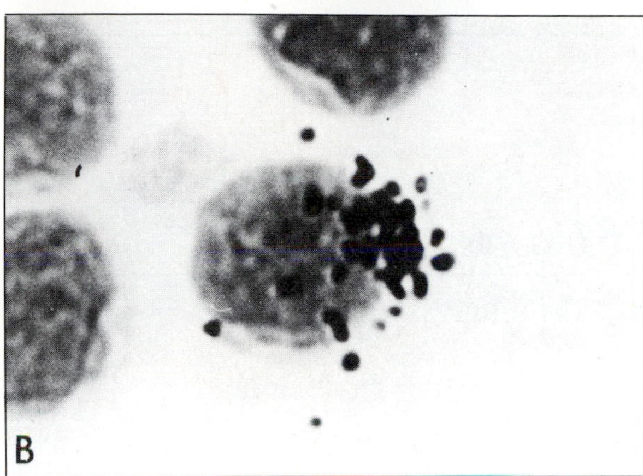

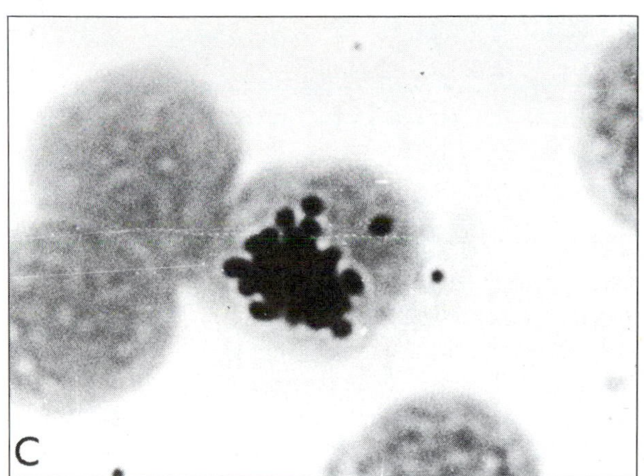

Figure 14–10 The capping phenomenon: Autoradiographs of lymph-node cells from an immunized mouse incubated with bound, tritiated, polymerized flagellin. (**A**) Uniform distribution of antigen after incubation at 0°C for 30 minutes. (**B** and **C**) Aggregation of antigen at a polar region after incubation at 37°C for 15 minutes. (A, B, and C original magnification ×2400.) *(From Diener, E., and Paetkau, W.H. Proc. Natl. Acad. Sci., 69:2364, 1972. With permission. Courtesy of Dr. Diener.)*

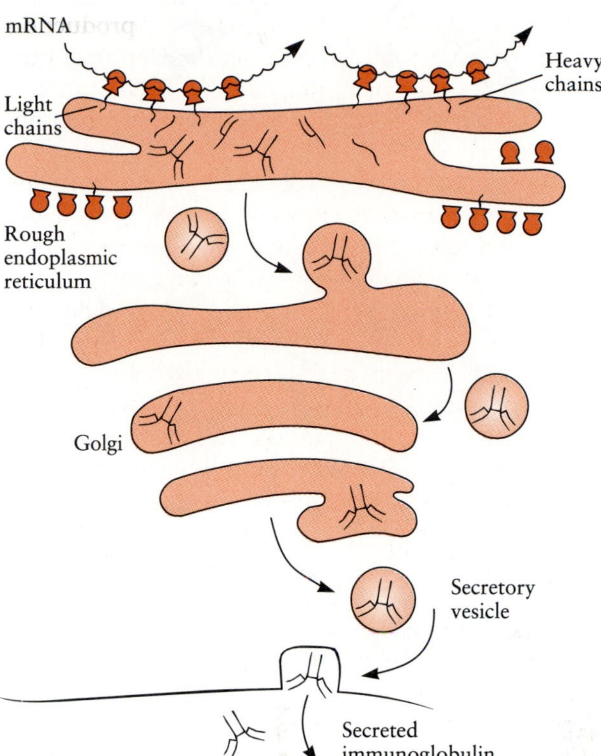

Figure 14–11 A schematic diagram depicting the synthesis of an immunoglobulin peptide chain by a plasma cell. The genes for this protein are translated from DNA in the nucleus into messenger RNA through a number of intermediate steps. The messenger RNA, acting through transfer RNA, transcribes this information into a polypeptide chain. The assembled immunoglobulin molecules are secreted through reversed pinocytosis.

insufficient for class switching and additional signals are required to complete the process. In humans, the additional signal can be provided by physical contact with T cells through CD40 and its ligand (page 197). In mice the second signal may be IL-5. IFN-γ stimulates a switch to IgG2a and IgG3 in mouse B cells and effectively suppresses the effects of IL-4. TGF-β may also be an IgA switch factor in both mouse and human cells.

As was described in Chapter 13, as an immune response progresses, there is a gradual increase in antibody affinity for antigen. This increase is probably not due to changes in individual B cells but reflects progressive somatic mutation and selection occurring within responding B-cell populations.

Thymus-Independent Antigens

Certain antigens can provoke antibody formation in the absence of helper T cells. These so-called **T-independent antigens** are usually simple repeating polymers, such as *Escherichia coli* lipopolysaccharide, polymerized salmonella flagellin, pneumococcal polysaccharide, dextrans, levans, and polyglutamic acid.

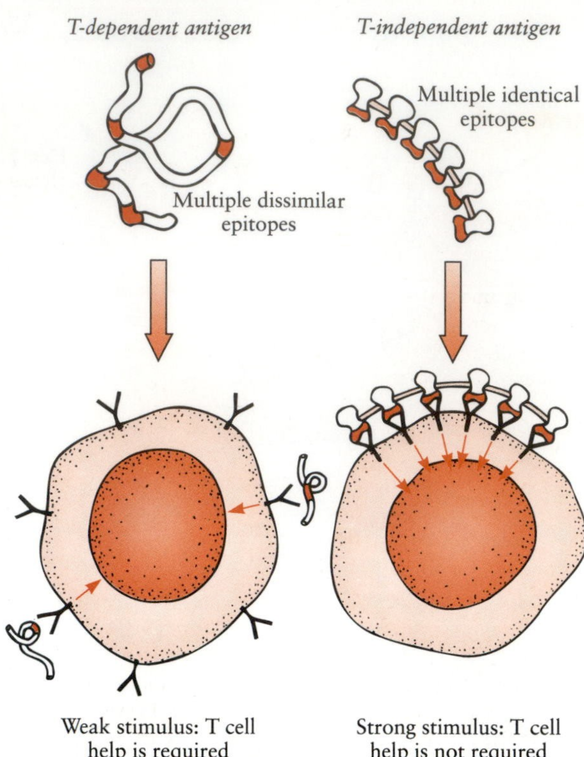

Figure 14–12 The differences between T-dependent and T-independent antigens in stimulating a B cell response.

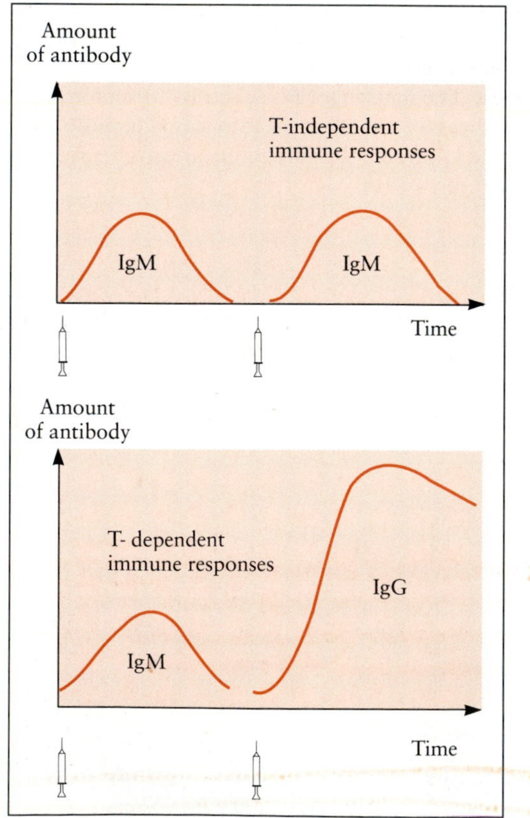

Figure 14–13 The different character of the antibody responses induced by T-independent and T-dependent antigens. The switch from IgM to IgG production does not occur with T-independent antigens.

T-independent antigens bind directly to B cells. Because they are repeating polymers, they cross-link several B-cell antigen receptors at one time. As a result, the effective dose of these epitopes is relatively large (Fig. 14–12). The matrix of epitopes in these molecules provides a sufficient stimulus for the proliferation of at least some B cells. Supporting evidence comes from the observation that haptens arranged in repeating arrays become less T-dependent.

Characteristically, T-independent antigens trigger only IgM responses in B cells and fail to generate memory cells (Fig. 14–13). This is because they fail to

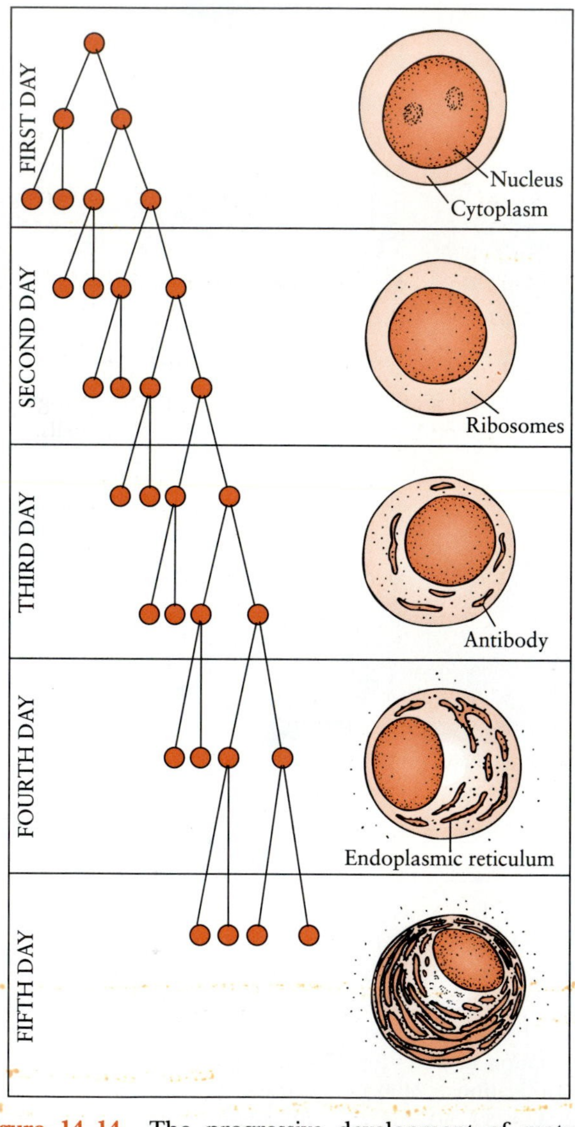

Figure 14–14 The progressive development of mature plasma cells after lymphocyte triggering by antigen: This process takes about five days and eight generations. The mature plasma cell has a small nucleus and extensive rough endoplasmic reticulum. It is almost totally devoted to immunoglobulin production. *(From How Cells Make Antibodies by G.J.V. Nossal. © 1964 by Scientific American Inc. All rights reserved.)*

METHODOLOGY

How to Detect Individual Antibody-Forming Cells

It is a relatively simple matter to identify individual cells producing antibody against sheep red blood cells. The test (known as a Jerne plaque assay, after its originator Dr. Niels Jerne) can be performed by mixing a suspension of antibody-producing cells (usually spleen cells from an inoculated animal) with sheep red blood cells and stabilizing the mixture in an agarose gel (agarose is a purified form of agar). When the mixture is incubated at 37° C, antibodies released from the producing cells diffuse into the agarose and combine with nearby red blood cells. If hemolytic complement is incorporated in the agarose and if the antibody being produced is IgM, then antibody-coated red blood cells will be lysed. As a result, there appears a clear zone or plaque owing to the local lysis of red blood cells around each antibody-producing cell (see figure). These are known as **plaque-forming cells** (PFCs). The test may be modified to detect cells producing antibodies of other immunoglobulin classes by incorporating specific antiglobulins in the agarose. Thus, anti-IgA will reveal IgA-producing cells, and so on. The test may also be employed to detect antibodies to soluble antigens, if these antigens are first chemically linked to the red blood cells.

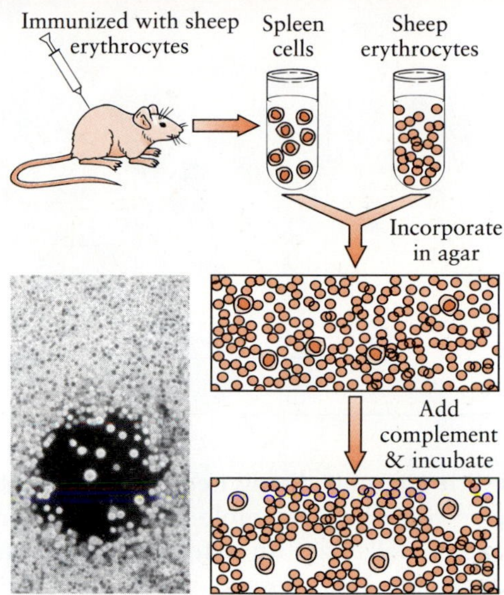

The Jerne plaque technique: Antibodies released by lymphocytes cause red cell hemolysis in the presence of complement. A clear zone or plaque therefore surrounds each antibody producing cell. In the photomicrograph, the large cell closest to the center of the plaque is probably the antibody-producing cell.

induce the formation of interleukins from helper T cells and so cannot trigger the switch from IgM to IgG production. In addition, if T-dependent and T-independent forms of the same antigen are given simultaneously, the response is additive, implying that different B-cell populations are being triggered.

PLASMA CELLS

Plasma cells develop from B cells in response to antigen. As a result, a series of cells that are intermediate in morphology between lymphocytes and plasma cells can be identified (Fig. 14–14). These "plasmablasts" develop in secondary lymphoid tissues in the areas where T-cell and B-cell cooperation occurs. Plasmablasts are found, therefore, at the margin between the lymph node cortex and paracortex and in the mantle zone in the spleen. Once developed, fully mature plasma cells usually migrate away from these areas and may eventually be found distributed throughout the body. Plasma cells are found in greatest numbers in the spleen, the medulla of lymph nodes, and in the bone marrow (Chapter 9).

Plasma cells are ovoid cells, 8 to 9 μm in diameter (Fig. 14–15). They have a round, eccentrically placed

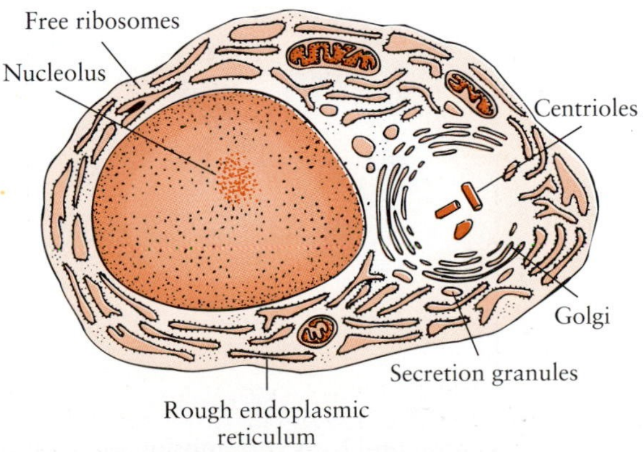

Figure 14–15 The major structural features of a plasma cell.

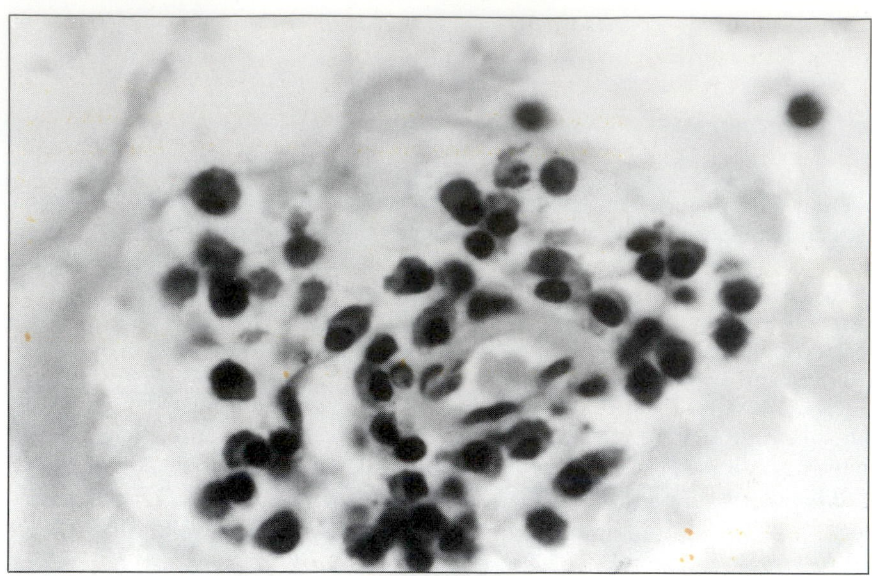

Figure 14–16 Plasma cells in the medulla of a lymph node (original magnification ×450). *(Courtesy of Dr. S. Yamashiro.)*

nucleus with unevenly distributed chromatin. As a result, the nucleus may resemble a clock face or cart wheel. Plasma cells have an extensive cytoplasm that is rich in rough endoplasmic reticulum and stains strongly with basic dyes and pyronin as well as a large pale-staining Golgi apparatus (Figs. 14–16, 14–17). They can synthesize up to a million molecules of immunoglobulin per hour. On occasion, the production of immunoglobulins may be so rapid that they accumulate within cells in vesicles known as Russell bodies. Normally, however, immunoglobulin molecules are secreted by exocytosis soon after they are formed. The immunoglobulin produced by a plasma cell is of identical antigen-binding specificity to the original antigen receptor on the parent B cell. One population of plasma cells, found in the lymph node medulla and

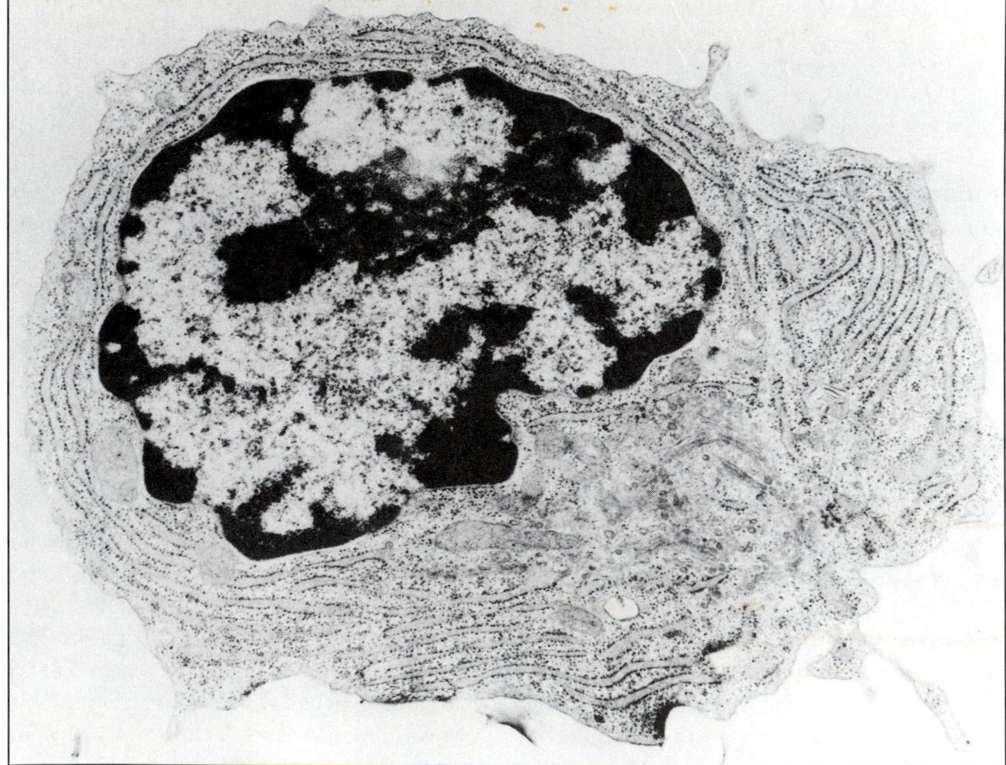

Figure 14–17 A transmission electron micrograph of a plasma cell. Its major structural features can be identified by reference to Figure 14–15. *(Courtesy of Dr. Scott Linthicum.)*

the splenic red pulp, has a short life span of approximately three days; another population has a life span of three to four weeks. Loss of plasma cells does not immediately result in decreased serum antibody levels, since the immunoglobulins once secreted decline slowly through catabolism.

MEMORY CELLS

One reason why the primary immune response terminates is because many responding B cells and plasma cells are simply removed by apoptosis. If all these cells died, however, immunological memory could not develop. Clearly some B cells must survive to become **memory cells.** The survival of memory cells is under the control of a gene called *bcl*-2. *Bcl*-2 is expressed in memory cells but is absent in the short-lived B cells and plasma cells. Activation of the *bcl*-2 gene and expression of its product permits a cell to avoid death by apoptosis and to differentiate into a memory cell. Thus, *bcl*-2 is a "survival gene" that ensures the prolonged survival of these cells by preventing apoptosis. (*Bcl*-2 is a proto-oncogene whose overexpression causes B cell lymphomas since the cells fail to die.)

Memory cells form a reserve of antigen-sensitive cells to be called upon on subsequent exposure to an antigen (Fig. 14–18). Evidence suggests that there are probably two types of memory cell. One population consists of small long-lived resting cells that produce IgG. These cells, unlike plasma cells, do not have a characteristic morphology but resemble other small lymphocytes. On exposure to antigen, they proliferate and differentiate into plasma cells without undergoing further hypermutation. The antibodies produced by these cells have a high affinity for the antigen. Any further mutation may not be beneficial. These cells are not found within germinal centers. Many of these memory cells or their precursors migrate from the spleen and lymph nodes to colonize the bone marrow.

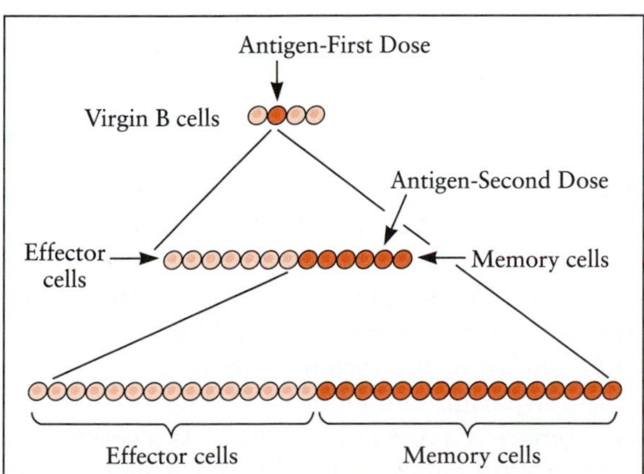

Figure 14–18 A schematic diagram showing the response of B cells to antigen leading to a great increase in the size of the available responding B cell pool—these are memory cells.

There is probably another type of memory cell population that consists of large dividing IgM-producing cells. These cells persist within germinal centers, their continued survival being dependent on the presence of a source of IL-2 as well as antigen on follicular dendritic cells. When restimulated by antigen, these cells undergo **somatic mutation.** As a result of this, antibody affinity may continue to improve even in a memory response.

If a second dose of antigen is given to a primed animal, it is met by many memory B cells, which respond in the manner described previously for antigen-sensitive B cells (Fig. 14–19). As a result, a secondary immune response is much greater than a primary immune response. The lag period is shorter since more antibody is produced and it is detected earlier. IgG is also produced in preference to the IgM that is characteristic of the primary response.

GERMINAL CENTERS

As described in Chapter 9, the development of **germinal centers** in the secondary lymphoid tissues such as lymph nodes and spleen parallels the development of memory B cells. These germinal centers are now known to be the location where many critical events in the life of a B cell occur (Fig. 14–20). Thus they are the sites of antigen-driven cell proliferation, **somatic mutation,** and positive and negative selection of B-cell populations. Thus B cells stimulated by appropriate presentation of antigen and helper T cells elsewhere in a lymphoid organ migrate to a germinal center about six days after the response begins. They then divide rapidly at the middle of the germinal center. These large dividing cells give a germinal center its pale appearance in histological sections. During this stage of rapid B-cell division, their V-region genes mutate at a rate of about one mutation per division. This results in the generation of many B cells whose immunoglobulin receptors differ in many respects from those of the parent cell. Once these cells have been clonally expanded, a process that takes 10 to 20 days, they then migrate to the periphery of the germinal center, where they encounter the antigen on **dendritic cells.** Antigen is trapped by dendritic follicular cells in germinal centers that hold it as antigen–antibody complexes close to B cells. Because of somatic mutation, some of the germinal center B cells bind antigen with greater affinity than the dendritic cells, but many, per-

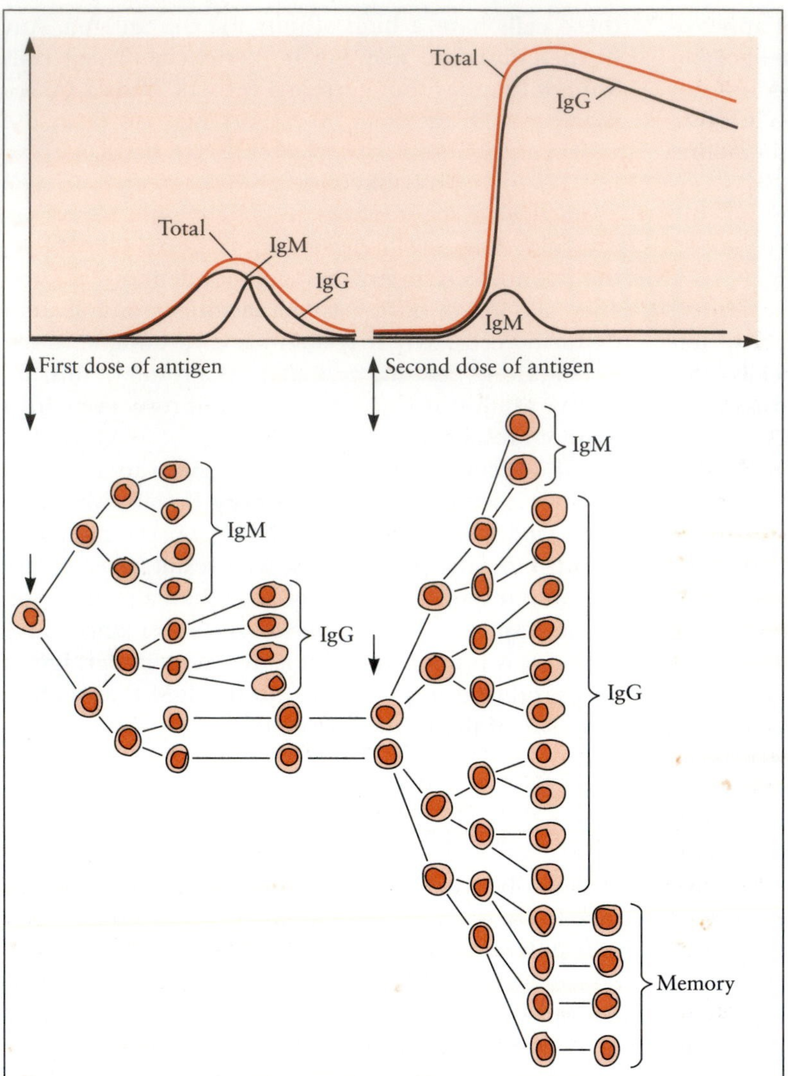

Figure 14–19 The response of B cells to antigen and the relationship between this response and serum antibody levels. *(With permission from Bellanti, J.A.,* Immunology, *W.B. Saunders, 1979. Courtesy of Dr. Bellanti.)*

haps the great majority, bind the antigen less strongly. If somatic mutation has resulted in greater affinity for the antigen, then those B cells will divide further and leave the center to form either plasma cells or memory cells. Most mutated B cells, however, do not exhibit enhanced antigen binding. These cells undergo rapid **apoptosis** and their debris is removed by macrophages. Thus the B cells that emerge from a germinal center are very different from the population of cells that entered it.

In addition to hypermutation, B cells undergo an immunoglobulin **class switch** within germinal centers. Thus it is here that B cells switch from IgM to other classes. This switch occurs around the time the B cells are undergoing somatic hypermutation and is mediated by IL-4, IL-5, IFN-γ, and TGF-β. Germinal center B cells carry a glycoprotein of 50 kDa called CD40 on their surface. CD40 is a receptor whose ligand (CD40-L) is a 33-kDa glycoprotein found on activated T cells. Binding of CD40-L to CD40 mediates B-cell proliferation in the absence of other stimuli and promotes IgE and IgG4 production in the presence of IL-4. Monoclonal antibodies to CD40 stimulate the B cell so that proteins are phosphorylated, the cells proliferate, and apoptosis of germinal center cells is prevented. Under some circumstances, they may also promote a change to the memory B-cell phenotype. The importance of CD40 is shown in the rare immunodeficiency disease X-linked hyperimmunoglobulin M syndrome. In these patients there is a defect in CD40-L. As a result of failure of CD40 to bind its ligand, the B cell does not make the switch from IgM production to other classes. Consequently, high IgM levels are produced but very little of the other classes.

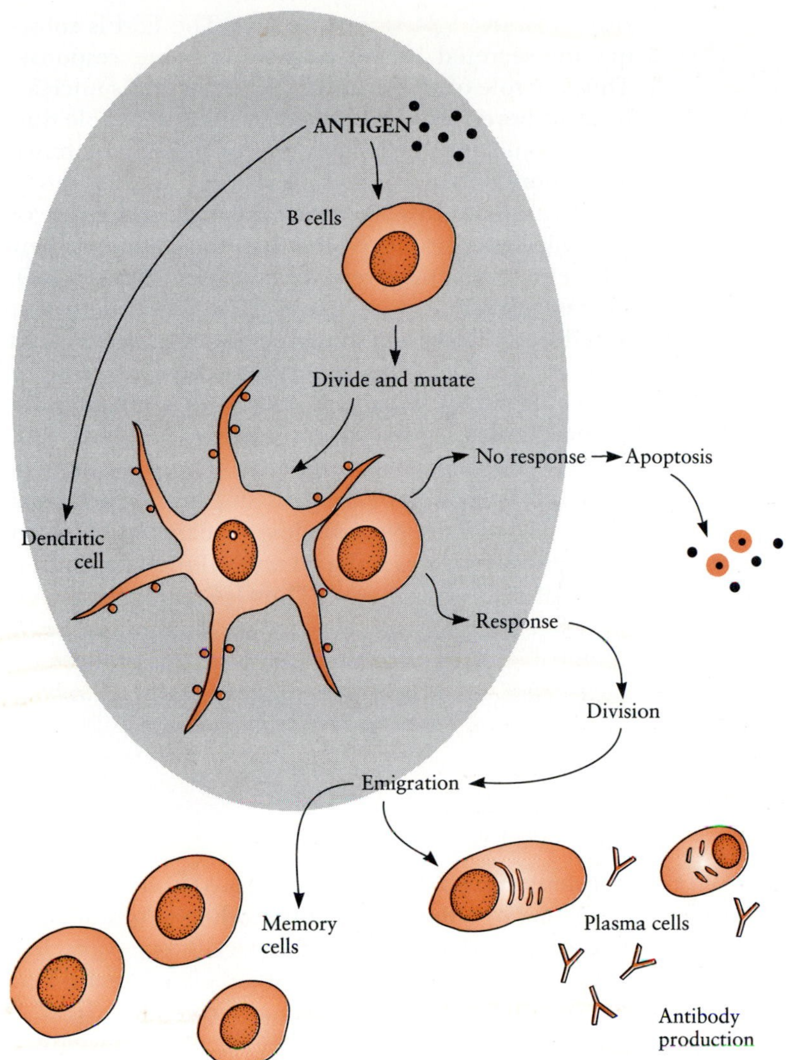

Figure 14–20 The events that occur in a germinal center. B cells divide and mutate in the middle of the germinal center. At the periphery the B cells that can respond to antigen are preferentially stimulated to divide while the unresponsive cells undergo apoptosis.

B-CELL ONTOGENY

B cells arise from pluripotent **stem cells** in the bone marrow, as may be shown by the fact that only bone marrow cells can reconstitute a lethally irradiated animal. These pluripotent stem cells develop first into progenitor B cells under the influence of IL-7 secreted by nearby stromal cells. As they develop, the progenitor B cells begin the process of gene rearrangement that culminates in the production of an immunoglobulin molecule. They first develop into pre-B cells, which can be identified because they possess FcR and contain the μ heavy chains of IgM within their cytoplasm (Fig. 14–21). These pre-B cells also synthesize the other components of the BCR, Ig–α and Ig–β. Somatic mutation also occurs in the pre-B cell using the enzyme terminal deoxynucleotidyl transferase (TdT). TdT is a DNA polymerase that can add nucleotides to a DNA segment in the absence of a template. It is found only in the nucleus of pre-B cells. It is estimated that at least half of the developing B cells die at the pre-B stage. B-cell survival depends on the formation of a receptor on the cell surface that can receive a survival signal. Failure to generate this receptor results in the cell's death. Once a functioning antigen receptor is formed, the cells are considered to be immature B cells. Immature B cells begin to synthesize immunoglobulin light chains, and eventually complete IgM and IgD molecules are transported to the cell surface. Only when they reach this stage of development are B cells ready to encounter antigen and mediate an immune response.

One subpopulation of B cells carries the surface antigen CD5. CD5 is a 67-kDa glycoprotein that binds specifically to another B cell–specific molecule CD72. These $CD5^+$ cells are also called B-1 cells. They have somewhat different properties from the conventional B cells in that they home to the peritoneal and pleural

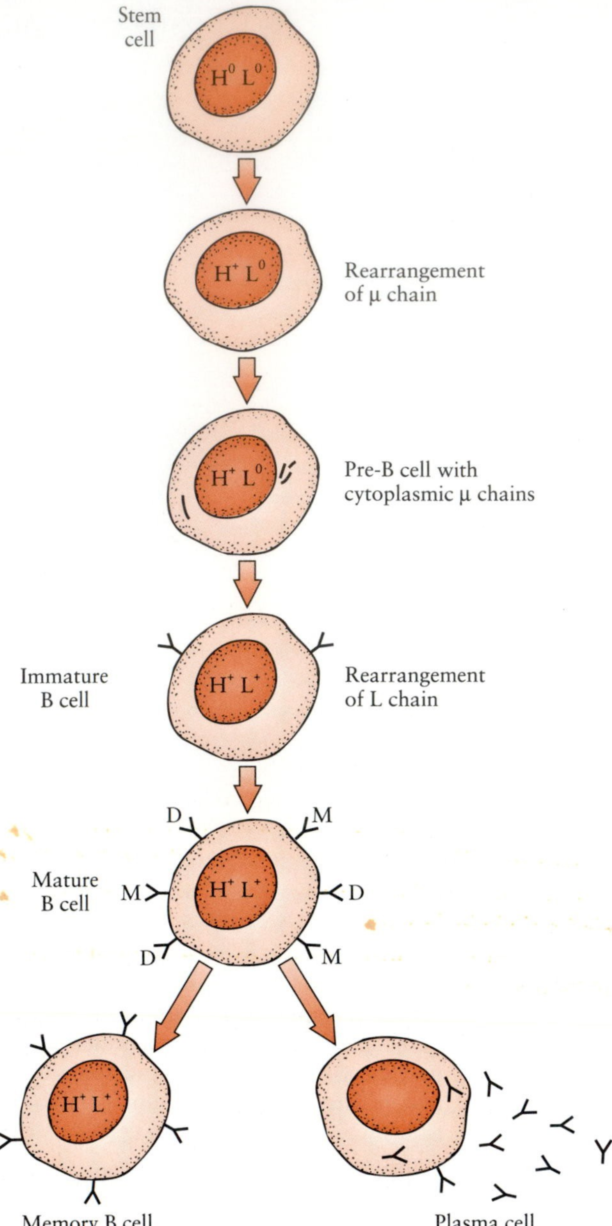

Figure 14–21 The progressive development of B cells from a bone marrow stem cell.

cavities and have the potential to renew themselves. B-1 cells may be a distinct B-cell lineage originating from precursors in the fetal liver or omentum rather than the bone marrow. Alternatively, they may be in a long-lived, self-renewing state as a result of a particular contact with antigen. B-1 cells may play an important role in some autoimmune diseases such as rheumatoid arthritis.

The Role of IgD

The vast majority of mature unstimulated B cells express both IgM and IgD receptors on their surface, and there is about ten times as much IgD as IgM on these cells. When B cells are activated, IgD is downregulated and IgM is upregulated. The IgM is subsequently secreted in the primary immune response. Thus the role of IgD as an antigen receptor is unclear. It was believed that IgD played an important role during the switch of B cells from susceptibility to tolerance to responsiveness, but contradictory results raised questions about this. To get a clear view of the role of IgD, mice were produced that had a mutation in their IgD gene so that they were selectively deficient in IgD. In these deficient mice, mature B cells were present in reduced numbers, although immature (pre-B) cells appeared normal. They had two to three times more surface IgM than normal B cells. Other classes were normal, indicating that class switching was unaffected. The IgD-deficient mice responded well to T-independent and T-dependent antigens and normal germinal centers developed. Nevertheless, somatic mutation in these antibodies was reduced and affinity maturation of the B cells was delayed. Thus IgD seems to be responsible in some way for the recruitment of high-affinity B cells in the early phase of the primary response, perhaps simply by enhancing antigen binding.

MYELOMAS

If a single B cell becomes cancerous, it may give rise to a clone of immunoglobulin-producing tumor cells. The structural features of these cells may vary, but some are recognizable as plasma cells (Fig. 14–22). Plasma cell tumors are known as **myelomas.** Because tumors arise from a single precursor cell, myelomas secrete a single homogeneous immunoglobulin product known as a myeloma protein.

The term **gammopathy** or hypergammaglobulinemia is used to denote any condition in which a pathological increase in immunoglobulin levels occurs. In general, the gammopathies are of two types. **Monoclonal gammopathies,** as are found in myeloma patients, are characterized by a rise in a single molecular type of immunoglobulin. This is seen as a narrow, sharp peak on an electrophoretic scan (Fig. 14–23). **Polyclonal gammopathies** are characterized by an overall rise in gamma globulin levels, which may also be identified on **electrophoresis.** Because there is a rise in all of the proteins in the globulin region, a broad peak is produced on the electrophoretic scan.

Myeloma proteins may belong to any immunoglobulin class. In general, the prevalence of myeloma proteins of the various immunoglobulin class correlates well with their relative quantities in normal serum, implying that the tumor arises as a result of a random mutation of a single plasma cell. Light-chain disease is a condition in which light chains alone are

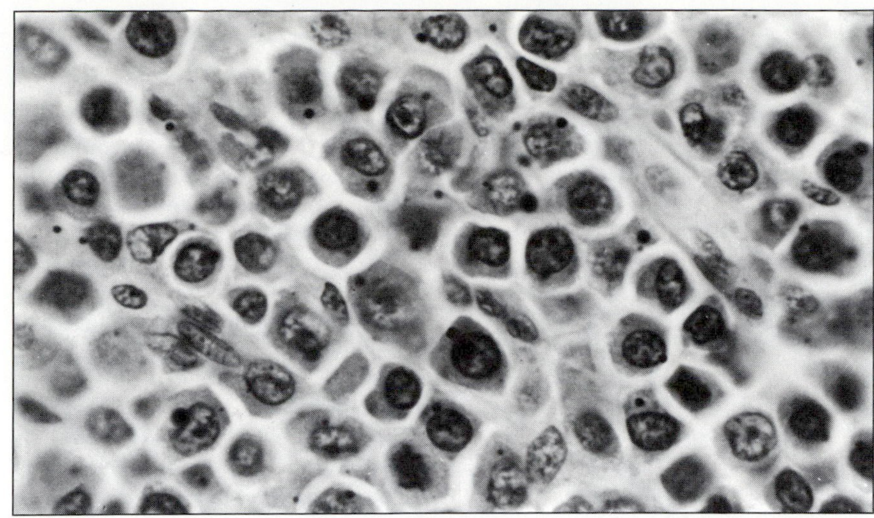

Figure 14–22 A photomicrograph of a section of a myeloma: Compare the cells in this section with those in Figure 10–8 (original magnification ×900). *(Courtesy of Dr. R.G. Thompson.)*

produced or the production of light chains is greatly in excess of the production of heavy chains. Similarly, there are rare variants of this condition in which abnormal heavy chains are produced. These are termed heavy-chain diseases.

Myelomas have been reported to occur in humans, mice, dogs, cats, horses, cows, pigs, and rabbits (Fig. 14–24). Because myeloma cells also break down bone, the presence of myelomas growing within bones may lead to multiple fractures. The increased viscosity of the blood as a result of its very high immunoglobulin content can lead to heart failure. Because light chains are relatively small (25 kDa), they pass through the glomeruli and are excreted in the urine. Unfortunately, they are toxic for renal tubular cells, and as a result, myelomas cause kidney failure. These light chains may be detected by electrophoresis of concentrated urine or, in some cases, by heating the urine. Light chains precipitate when heated to 60° C but redissolve as the temperature is raised to 80° C. Proteins possessing this curious property are known as **Bence-Jones proteins** (Dr. Bence-Jones first described them), and their presence in urine is suggestive of a myeloma.

Individuals with myelomas are profoundly immunosuppressed because of the overwhelming commitment of the body's immune resources to the production of neoplastic plasma cells. In addition, the replacement of normal bone marrow tissue by tumor cells and the suppression of B-cell responses induced

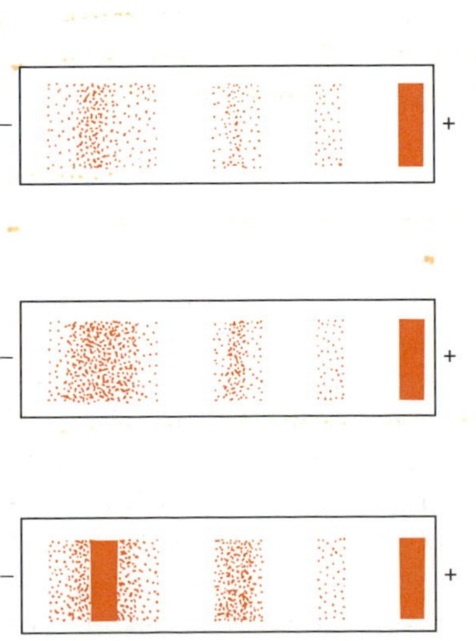

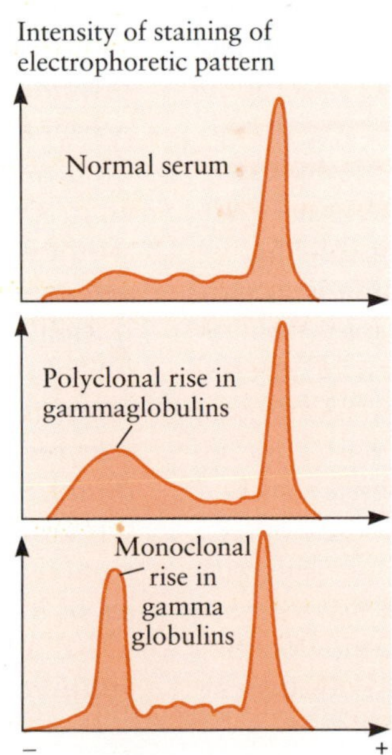

Figure 14–23 Serum electrophoretic patterns showing the normal pattern and the differences observed between monoclonal and polyclonal gammopathies.

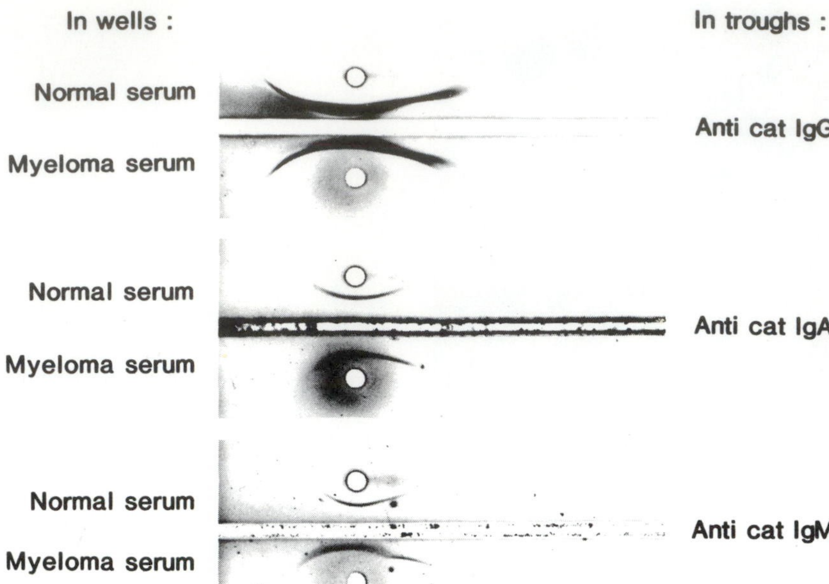

Figure 14–24 Immunoelectrophoresis is a technique that demonstrates the presence of immunoglobulins of each specific class. For details see Figure 17–16. This illustration shows three separate immunoelectrophoresis patterns obtained using anti IgG, IgA, and IgM. The use of anti-IgG and anti-IgA shows little difference between the normal serum and the myeloma serum. The presence of an IgM myeloma protein in the bottom well causes the arc of precipitate to be distorted in a characteristic fashion. The line is much thicker than the control and it forms two distinct arcs. *(Courtesy of Dr. G. Elissalde.)*

by elevated serum immunoglobulin also contribute to the suppression. As a result, these patients commonly suffer from severe bacterial infections.

HYBRIDOMAS

The plasma cells in myelomas can synthesize immunoglobulins in the usual manner. Unfortunately, since plasma cells become neoplastic in an entirely random manner, the immunoglobulins that they produce are not usually directed against any antigen of practical importance. Nevertheless, myeloma cells can be grown in tissue culture, where they can survive indefinitely. From a practical point of view, it would be highly desirable to set up a system to obtain large quantities of absolutely pure, specific immunoglobulin directed against an antigen of interest. This can be done by fusing a normal plasma cell making the antibody of interest, with a myeloma cell with the capacity for prolonged growth in tissue culture. The resulting mixed cell is called a **hybridoma** (Fig. 14–25).

The first stage in making a hybridoma is to generate antibody-producing plasma cells. This is done by immunizing a mouse against the antigen of interest and repeating the process several times to ensure that it mounts a good response. Two to four days after administration of antigen, the mouse's spleen is removed and broken up to form a cell suspension. These spleen cells are suspended in culture medium together with a special mouse myeloma cell line. It is usual to use myeloma cells that do not secrete immunoglobulins since this simplifies purification later on. Polyethylene glycol is added to the mixture. This compound induces many of the cells to fuse (although it takes about 200,000 spleen cells on average to form a viable hybrid with one myeloma cell). If the fused cell mixture is cultured for several days, any unfused spleen cells will die. The myeloma cells would normally survive, but they are eliminated by a simple trick.

There are two biosynthetic pathways by which cells can produce nucleotides and hence nucleic acids (Fig. 14–26). The myeloma cells are selected so that they lack the enzyme hypoxanthine phosphoribosyl transferase and as a result cannot utilize hypoxanthine in the culture medium to produce inosine, a pyrimidine precursor. They are obliged to utilize an alternative biosynthetic pathway involving thymidine. The fused cell mixture is therefore grown in a culture containing three compounds: hypoxanthine, aminopterin, and thymidine (HAT medium). Aminopterin is a drug that prevents cells from making their own thymidine. Since the myeloma cells cannot use hypoxanthine and the aminopterin stops them from using the alternative synthetic pathway, they cannot make nucleic acids and will soon die. Hybrids made from a myeloma and a normal cell will grow; they possess hypoxanthine phosphoribosyl transferase and can therefore use the hypoxanthine and thymidine in the culture medium and survive. The hybridomas divide rapidly in the HAT medium, doubling their numbers every 24 to 48 hours. On average, about 300 to 500 hybrids can be isolated from a mouse spleen, although not all make antibodies of interest.

If a mixture of cells from a fusion experiment is cultured in wells on a plate with about 5×10^4 my-

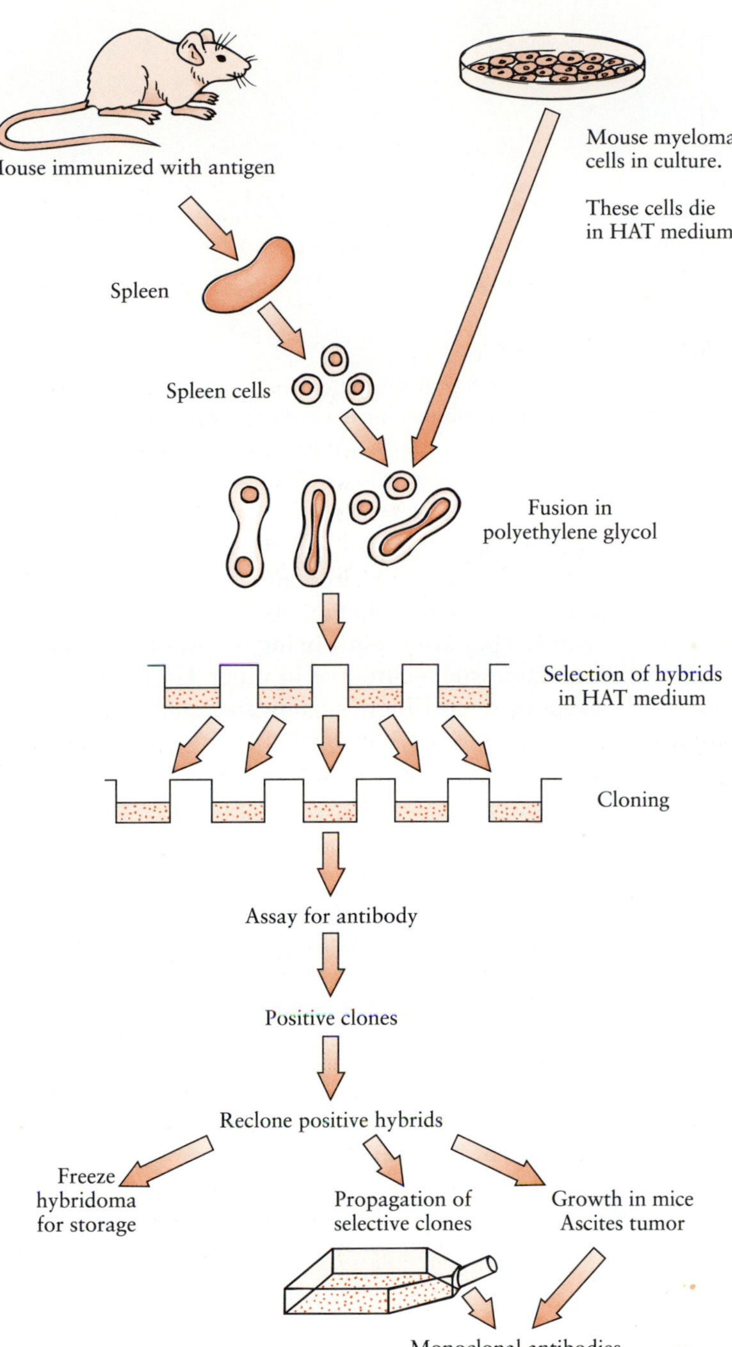

Figure 14–25 The production of monoclonal antibodies from cloned hybridoma cells. The hybrids are generated by fusing spleen cells with myeloma cells in the presence of polyethylene glycol.

eloma cells per well, it is usual to obtain about one hybrid in every three wells. After culturing for two to four weeks, the growing cells can be seen and the supernatant fluid can be screened for the presence of antibodies. It is essential to use a good sensitive assay at this time. Radioimmunoassays or ELISAs are preferred (Chapter 17). Clones that produce the desired antibody are grown in mass culture and recloned to eliminate non-antibody-producing hybrids.

Unfortunately, antibody-producing clones tend to lose this ability after being cultured for several months. Thus it is usual to make large stocks of hybridoma cells and store them frozen in small aliquots. These can then be thawed as required and grown in bulk culture. Alternatively, the hybridoma cells can be injected intraperitoneally into mice. Since they are tumor cells, the hybridomas grow rapidly and provoke the effusion of a large volume of fluid into the mouse

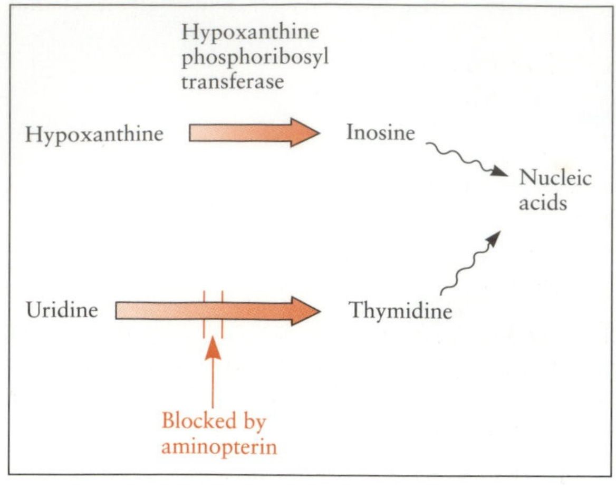

Figure 14–26 A highly simplified diagram showing the two pathways involved in nucleic acid synthesis. Myeloma cells used in hybridoma formation lack hypoxanthine phosphoribosyl transferase.

peritoneal cavity. This fluid is rich in monoclonal antibody and can be readily harvested.

Within recent years, **monoclonal antibodies** produced by hybridomas have become the preferred source of antibodies for much immunological research. They are absolutely specific for single epitopes and are available in large amounts. Because of their purity, they can function as standard chemical reagents. They are rapidly being incorporated into clinical diagnostic techniques in which large quantities of antibody of consistent quality are required.

KEY WORDS

Antigen processing p. 197
Antigen-sensitive cell p. 193
B cell p. 193
B-cell antigen receptor (BCR) p. 193
Bence-Jones proteins p. 209
Capping p. 200
CD5 p. 207
CD19 p. 194
CD40 p. 197
CR2 (CD21) p. 194
Class switch p. 206
Clonal selection theory p. 193
Clone p. 193
Clonotype p. 200
Dendritic cells p. 205
Germinal center p. 210
Hybridoma p. 210
Ig–α p. 194
Ig–β p. 194
Immunoglobulin D p. 208
Immunoglobulin M p. 193
Interleukin-2 p. 198
Interleukin-4 p. 198
Interleukin-5 p. 199
Linked recognition p. 197
Memory cell p. 205
Monoclonal antibody p. 212
Monoclonal gammopathy p. 208
Myeloma p. 208
Myeloma protein p. 208
Plasma cell p. 203
Polyclonal gammopathy p. 208
Somatic mutation p. 205
Th1 cells p. 198
Th2 cells p. 198
Thymus-independent antigens p. 201
Tyrosine kinase p. 194

QUESTIONS

1. A hybridoma is a cell formed by fusion of a
   a. plasma cell with a plasma cell of another species
   b. T cell with a myeloma cell
   c. macrophage with a myeloma cell
   d. T cell with a B cell
   e. plasma cell with a myeloma cell
2. Mature antibody-producing cells are called
   a. immunoblasts
   b. histiocytes
   c. T cells
   d. neutrophils
   e. plasma cells
3. Which T-cell–derived cytokine triggers the class switch from IgM to IgE production?
   a. IL-2
   b. IFN-γ
   c. TGF-β
   d. IL-4
   e. IL-5
4. What is the major activity occurring within germinal centers?
   a. antibody production
   b. antigen trapping
   c. somatic mutation

   d. antigen presentation
   e. plasma cell formation

5. The central paradigm of the immune response is called
   a. somatic mutation
   b. antigen presentation
   c. MHC restriction
   d. clonal selection
   e. antigen recognition

6. The B-cell antigen receptor typically consists of
   a. CD3 and receptor immunoglobulin
   b. ζ chain and receptor immunoglobulin
   c. Ig–α/β and CD3
   d. Ig–α/β and receptor immunoglobulin
   e. ζ chain and CD3

7. To optimize a B-cell response to antigen, which two additional molecules are required to link to antigen?
   a. CD23 and CR2
   b. CR2 and CD19
   c. CD40 and CR2
   d. CD19 and CD4
   e. CD19 and CD3

8. The need for a Th cell and a B cell to recognize two different epitopes on an antigen is called
   a. linked recognition
   b. MHC restriction
   c. covalent bonding
   d. signal transduction
   e. antigenic variation

9. When IL-2 acts on B cells, it stimulates the expression of
   a. IL-3
   b. TGF-β
   c. IL-4
   d. IL-1R
   e. IL-2R

10. A common feature of T-independent antigens is that they are
    a. globular proteins
    b. simple repeating polymers
    c. very large molecules
    d. carbohydrate
    e. bacterial in origin

11. What are the advantages and disadvantages of monoclonal antibodies over polyclonal antibodies in experimental and diagnostic immunology?

12. Some investigators suggest that macrophages are minor antigen-presenting cells and that dendritic cells and B cells may be more significant. What evidence is available to support this argument?

13. Where are plasma cells found in the body? Are these locations of biological significance, or is it immaterial just where plasma cells secrete antibodies?

14. How would you diagnose a myeloma? How might you differentiate a gammopathy developing as a result of a myeloma from one that has benign origins?

15. Suggest experiments that might be used to determine the life span of memory B cells.

16. Can a B cell present antigen to a helper T cell? Can a helper T cell stimulate a B cell? How might these two cells communicate in a bidirectional fashion?

17. What is the physical basis of the interaction between a B cell and a helper T cell? Rank the interacting cell surface molecules in order of importance.

18. List the cytokines that are required in order for optimal T-cell–B-cell interaction to occur. What is the role of each cytokine?

Answers: 1e, 2e, 3d, 4c, 5d, 6d, 7b, 8a, 9e, 10b

SOURCES OF ADDITIONAL INFORMATION

Armitage, R.J., et al. CD40L: A multifunctional ligand. Semin. Immunol., 5:401–412, 1993.

Berek, C. The development of B cells and the B-cell repertoire in the microenvironment of the germinal center. Immunol. Rev., 126:5–19, 1992.

Berek, C., and Ziegner, M. The maturation of the immune response. Immunol. Today, 14:400–404, 1993.

Burnet, F.M., and Fenner, F. The Production of Antibodies. MacMillan, New York, 1949.

Cambier, J.C., and Campbell, K.S. Membrane immunoglobulin and its accomplices: New lessons from an old receptor. FASEB J., 6:3207–3217, 1992.

Carter, R.H., and Fearon, D.T. CD19: Lowering the threshold for antigen receptor stimulation of B lymphocytes. Science, 256:105–107, 1992.

Clark, E.A., and Ledbetter, J.A. How B and T cells talk to each other. Nature, 367:425–428, 1994.

Clark, M.R., Campbell, K.S., Kazlauskas, A., et al. The B cell antigen receptor complex: Association of Ig-α and Ig-β with distinct cytoplasmic effectors. Science, 258:123–125, 1992.

Esser, C., and Radbruch, A. Immunoglobulin class switching: Molecular and cellular analysis. Annu. Rev. Immunol., 8: 717–735, 1990.

Feldman, M., and Basten, A. Cell interactions in the immune response: Specific collaboration across a cell impermeable membrane. J. Exp. Med., 136:49–67, 1972.

Finkelman, F.D., et al. Lymphokine control of *in vivo* immunoglobulin isotype selection. Annu. Rev. Immunol., 8: 303–333, 1990.

Finkelman, F.D., Lees, A., and Morris, S.C. Antigen presentation by B lymphocytes to $CD4^+$ T lymphocytes in vivo: Importance for B lymphocyte and T lymphocyte activation. Semin. Immunol., 4:247–255, 1992.

Hodgkin, P.D., et al. Separation of events mediating B cell

proliferation and Ig production by using T cell membranes and lymphokines. J. Immunol., 145:2025–2034, 1990.

Kincade, P.W., et al. Cells and molecules that regulate B lymphopoiesis in bone marrow. Annu. Rev. Immunol., 7:111–143, 1989.

Kipps, T.J. The CD5 B cell. Adv. Immunol., 47:117–185, 1989.

Kohler, G., and Milstein, C. Continuous cultures of fused cells secreting antibody of predefined specificity. Nature, 256:495–497, 1975.

Korsmeyer, S.J. Bcl-2: a repressor of lymphocyte death. Immunol Today, 13:285–288, 1992.

Miedema, F., and Melief, C.J.M. T cell regulation of human B cell activation. Immunol. Today, 6:258–259, 1983.

Mitchell, G.F., and Miller, J.F.A.P. Cell-to-cell interaction in the immune response II. The thoracic duct lymphocytes. J. Exp. Med., 128:821–837, 1968.

Mitchell, G.F., and Miller, J.F.A.P. Immunological significance of the thymus and thoracic duct lymphocytes. Proc. Natl. Acad. Sci. U. S. A., 59:296–303, 1968.

Myers, C.D. Role of B cell antigen processing and presentation in the humoral immune response. FASEB J. 5:2547–2553, 1991.

Nitschke, L., et al. Immunoglobulin D–deficient mice can mount normal immune responses to thymus-independent and -dependent antigens. Proc. Natl. Acad. Sci. U. S. A., 90: 1887–1891, 1993.

Nossal, G.J.V. The molecular and cellular basis of affinity maturation in the antibody response. Cell, 68:1–2, 1992.

Paige, C.J., and Wu, G.E. The B cell repertoire. FASEB J., 3: 1818–1824, 1989.

Potash, M.J. B lymphocyte stimulation. Cell, 23:7–8, 1981.

Reth, M., et al. Structure and signalling function of B cell antigen receptors of different classes. *In* Molecular Mechanisms of Immunological Self-Recognition, pp. 69–75. Academic Press, New York, 1993.

Sprent, J.T and B Memory cells. Cell, 76:315–322, 1994.

Unanue, E.R. Cooperation between mononuclear phagocytes and lymphocytes in immunity. N. Engl. J. Med., 303: 977–985, 1980.

Venkitaraman, A.R., et al. The B-cell antigen receptor of the five immunoglobulin classes. Nature, 352:772–781, 1991.

Vitetta, E.S., et al. Cellular interactions in the humoral immune response. Adv. Immunol., 45:1–105, 1989.

Waldman, T.A., and Broder, S. Polyclonal B-cell activators in the study of the regulation of immunoglobulin synthesis in the human system. Adv. Immunol., 32:1–63, 1982.

CHAPTER 15

The Genetic Basis of Antigen Recognition

Susumu Tonegawa *Susumu Tonegawa was born in Japan in 1939 but has worked for most of his career in the United States. He was awarded the 1987 Nobel Prize for developing the techniques that showed that antibody diversity was generated by recombination of multiple genes. He pursued the first evidence that gene rearrangements were involved in creating antibody diversity. By using restriction enzymes and recombinant DNA techniques he showed how B cells generated different antibody molecules by recombination between multiple gene segments and has also shown that TCR do exactly the same.*

CHAPTER OUTLINE

Immunoglobulin Diversity
Immunoglobulin Gene Structure
Generation of V Region Diversity
Potential Immunoglobulin Diversity
Constant Region Diversity

TCR Diversity
TCR Gene Structure
Generation of TCR V Region Diversity
Where Does This Happen?
Potential TCR Diversity

CHAPTER CONCEPTS

1. Both immunoglobulins and the T-cell antigen receptor proteins are constructed through the use of multiple gene segments.
2. Each peptide chain is coded for by three or four gene segments. These are a variable gene segment (V), a joining gene segment (J), and a constant gene segment (C). Some peptide chains also employ a fourth gene segment called D. The V, D, and J gene segments are selected at random from a large population of these segments in the genome.
3. Amino acid sequence diversity is generated by random splicing of these multiple gene segments.
4. Additional sequence diversity may be generated through N-region insertion and variability at splice sites.
5. Diversity at CDR 1 and CDR 2 is generated in immunoglobulins through somatic mutation. Somatic mutation is not employed in the TCRs because of the risk of generating autoimmune responses.
6. The immunoglobulin class produced by a B cell is determined by selection of one of a family of heavy-chain genes.
7. These genetic processes ensure that both immunoglobulins and TCRs have the diversity needed for the successful recognition of the vast majority of potential antigens.

The receptors for antigen found on B cells and T cells must be able to bind an enormous variety of molecules. In this chapter we describe the mechanisms that generate the receptor molecules that can bind specifically to all these antigens. The structure of the three gene families that code for immunoglobulins is described, followed by how their structural diversity is generated and the magnitude of the diversity is possible. Since there are five classes of immunoglobulins depending on the heavy chains used to construct them, this section concludes with a description of the mechanisms by which this heavy-chain switching occurs. The second part of the chapter covers TCR diversity. Thus the structure of the four gene families that code for TCR peptide chains is discussed, along with the mechanisms by which TCR diversity is generated and the magnitude of possible diversity.

One of the central problems in understanding the immune system is how its cells can recognize the great variety of infectious agents that can invade the body. Given that microorganisms can also evolve and change rapidly, the immune system must be able to respond not only to existing organisms, but also, within reason, to newly evolved organisms. Nevertheless, the original studies on the specificity of the antibody responses to haptens (Chapter 4) suggested that almost any organic molecule could be used as a hapten and provoke production of specific antibody. Any theory of antibody production must account for this ability to produce an enormous variety of highly specific antibody molecules. Similar considerations apply to the cell-mediated immune responses. The range of specific immune responses implies the existence of a huge number of different antigen-specific T-cell receptors. The earliest theories about antibody formation assumed that the variability of epitopes was so great that antibodies directed against all of them could not be preformed but must be synthesized on demand following exposure to an antigen. These theories, formulated by Linus Pauling, suggested that an epitope would function as a template around which its specific antibody could be constructed. Once built, the newly formed antibody, it was postulated, would be stabilized by interchain bonding and then released into the bloodstream. Although attractive, theories of this type (known as instructive theories since the epitope "instructed" the antibody) were never supported by experimental evidence. For example, it could never be shown that antigen was present when antibodies were being synthesized. More important, instructive theories could never be reconciled with the realities of protein synthesis and the central dogma of molecular biology, namely, that information flows only from DNA through RNA to protein, not the reverse.

An alternative theory (known as the selective theory) was put forward by Niels Jerne. This held that the information needed to make all possible antibodies is already stored in an animal's genome, and all that is required for an antibody to be produced is that the necessary gene be "selected" by exposure to specific epitope. Consequently, the specific immunoglobulin could be synthesized through use of appropriate messenger RNA. This selective theory is compatible not only with our knowledge of molecular biological mechanisms, but also with our knowledge of the factors that govern protein configuration. Thus all the information for the structure of antibodies and T-cell antigen receptors must be encoded in our genes.

The shape of a protein molecule and hence its ability to bind antigen are in no way governed by a physical template. The shape depends on the folding of its peptide chains, which is governed in turn by the amino acid sequence. Each amino acid in a peptide chain exerts an influence on its neighboring amino acids, which determines their relative orientation. The final shape of a peptide chain, therefore, represents the total effect of all amino acids in the chain as the peptide assumes its most energetically favorable conformation. The shape of a protein is determined only by its amino acid sequence, and that sequence is determined by the sequence of bases in the DNA coding for that molecule.

IMMUNOGLOBULIN DIVERSITY

In 1965 Dreyer and Bennett suggested that each immunoglobulin peptide chain was coded for by two distinct genes, one for the variable region and one for the constant region (as opposed to a single gene coding for an entire heavy or an entire light chain). In this way, they suggested, a single constant region gene could be combined with any one of many variable region genes. Subsequent studies of the genetics, serology, and structure of immunoglobulin molecules by Susumu Tonegawa and others confirmed the correctness of this suggestion.

This finding, that immunoglobulin chains were coded for by multiple genes, immediately simplified the major problem of all selective theories of antibody formation. Instead of having to store information about all possible immunoglobulin peptide chains, it is necessary to store only the information (genes) about all the variable regions and to match these, when required, with a small number of constant re-

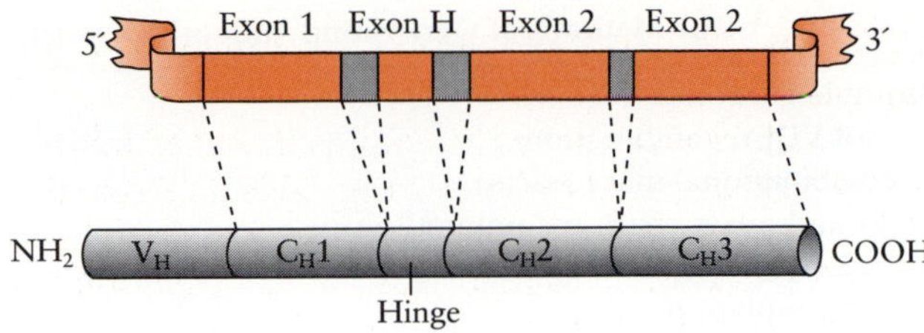

Figure 15–1 Immunoglobulin heavy chain constant regions are coded for by four exons; one for each constant domain and one for the hinge region.

gion genes in order to produce a complete range of immunoglobulin molecules.

Immunoglobulin Gene Structure

Susumu Tonegawa demonstrated that three gene families code for immunoglobulin peptide chains and they are found on different chromosomes. One codes for κ light chains, one for λ light chains, and one for heavy chains. Each of these three gene families contains a large number of **gene segments** coding for variable regions (V genes), one or more gene segments coding for constant regions (C genes) (Fig. 15–1), and a variable number of very small gene segments known as **J (joining)** or D (diversity) segments. Antibody diversity arises as a result of recombination between these gene segments.

Lambda Light-Chain Genes

Lambda light chains are coded for by a gene family found on chromosome 22 in humans and 16 in mice. The V_λ and C_λ regions of the light chain are coded for by separate gene segments. The V_λ gene segment codes for most of the V_λ region up to the amino acid located at position 95 from the N-terminus. The C_λ gene segment codes for the C-terminal domain starting at position 110. The intervening 15 amino acids in the V region are coded for by a short gene segment known as J_λ. The human λ light-chain family consists of two different V_λ segments, four different J_λ segments (one of which is defective), and one C_λ gene segment (Fig. 15–2).

Kappa Light-Chain Genes

Kappa light chains are coded for by a gene family found on chromosome 2 in humans and 6 in mice. This also consists of three types of gene segment—V, J, and C. In humans it contains 75 to 80 V_κ segments, five J_κ segments, and one C_κ segment.

Immunoglobulin Heavy-Chain Genes

The immunoglobulin heavy-chain gene family is located on chromosome 14 in humans and 12 in mice. In humans, this contains between 100 and 300 V_H gene segments. These V_H segments are divided into six distinct families based on their sequence homology. (V gene segment families are defined as V gene segments having more than 50% sequence homology at the protein level.) In mice there may be as many as 1000 to 1500 V_H segments divided into nine families. The size of these V_H families varies among mouse strains. Up to 40% of these V_H genes may have crippling mutations that turn them into nonfunctional **pseudogenes.** These pseudogenes are not entirely useless, however, since they may serve as reservoirs for

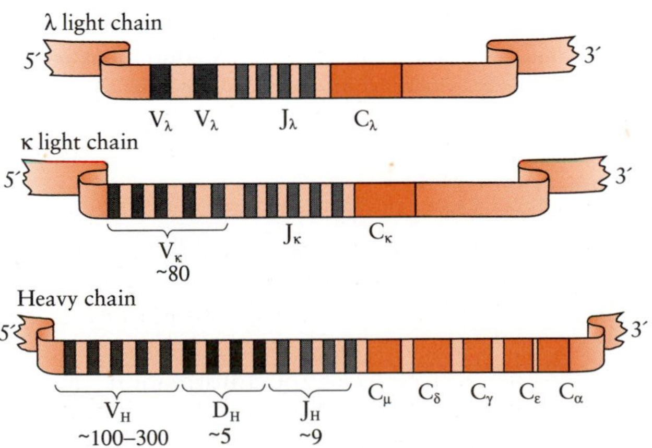

Figure 15–2 The structure of the genes from lambda light chains, kappa light chains, and heavy chains. Each is located on a different chromosome.

short segments of DNA that can be used to increase diversity in other V_H segments. The heavy-chain genes also include several J_H segments (four in mice, nine in humans) situated 3′ to the V_H segments. In addition, several additional short gene segments called D segments (D for diversity) are located between the V_H and J_H segments (see Fig. 15–2). In mice there are about 12 D segments containing from 6 to 17 bases. In humans there are at least five D gene segments.

The rest of the heavy-chain genes are separated from the J gene segments by a large intron. They consist of a series of constant region genes, one gene for each immunoglobulin class. They are arranged in the order -Cμ-Cδ-Cγ-Cε-Cα- along the chromosome.

Generation of V Region Diversity

To generate a great variety of antibodies with different specificity for antigen, it is necessary to generate great diversity in the amino acid sequence in the variable regions of both light and heavy chains (Table 15–1). Since the amino acid sequence is determined by the nucleotide sequence in the genes coding for the variable regions, mechanisms must exist for generating this sequence diversity. As might be expected, several such mechanisms exist.

Gene Rearrangement

The most obvious mechanism of generating V region diversity is to randomly select one V gene segment from the available pool and join it to a randomly selected J gene segment. Since there are multiple V and J gene segments available, the number of possible combinations can be very large. For example, if there are 100 V gene segments and 10 J gene segments, then $100 \times 10 = 1000$ different V regions can be made.

Table 15–1 The Methods of Generating Antibody Diversity

Multiple germ line V genes
VJ and VDJ recombination
Recombinational inaccuracies
Light and heavy chain recombination
N-region addition
Somatic mutation

The V, J, and C gene segments are normally separated by introns in the DNA. When B cells develop and make light chains, these **introns** must be removed so that the gene segments can be joined to form a continuous sequence (Fig. 15–3). This occurs in two stages. Randomly selected V and J segments are first joined to form a complete V region gene. Once assembled, this V region gene is then joined to the C gene segment to form a complete light chain. The V and J gene segments are joined (or **spliced**) by **looping out** and then excising the intervening DNA. The V and J gene segments have special sites at each end that enable these intervening sequences to be deleted and splicing to occur.

The reader is advised to read this paragraph in conjunction with viewing Figure 15–4 since the concept is difficult to describe in words. At the 3′ end of a V gene segment is a conserved set of seven bases (a heptamer) separated by 23 bases from a conserved set of nine bases (a nonamer) (CACAGTG—23 bases—ACAAAAACC). At the 5′ end of a J gene is a hep-

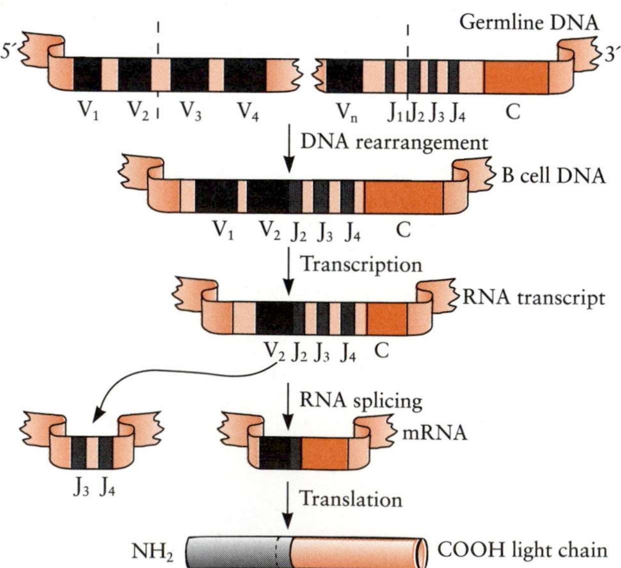

Figure 15–3 The production of a κ light chain. The germline cell possesses a complete set of V and J genes. As a B cell differentiates, V and J genes are excised at random, so that any one V gene may be joined to any one of the J genes. When the B-cell DNA is transcribed to mRNA, the V-J complex is spliced to the C_k gene to form a continuous mRNA segment. This is translated into a complete light chain. L is a leader segment.

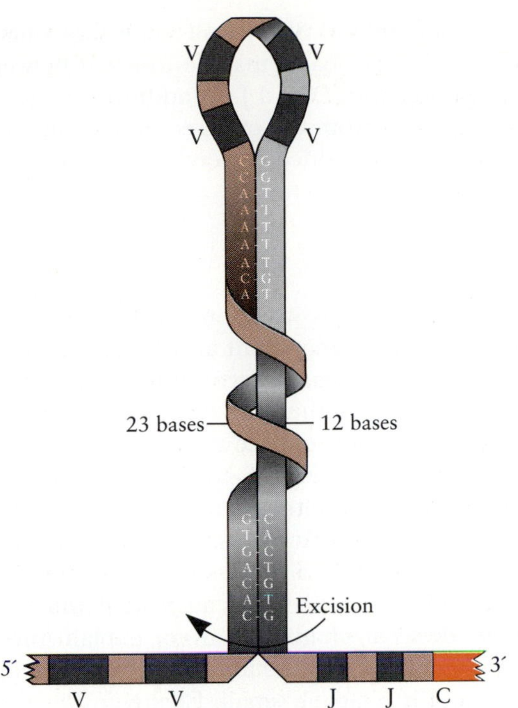

Figure 15–4 The mechanism of V–J joining: a portion of the nucleotide sequence is looped out so that the seven nucleotide segment and the nine nucleotide segment come together prior to removal of the loop.

tamer separated by 12 bases from a nonamer (GGTTTTTGT—12 bases—CACTGTG). When the intervening sequence is looped out, the heptamers CACAGTG and CACTGTG pair up and both nonamers also pair so that a loop of DNA is formed. The spacers of 12 or 23 bases that separate the heptamers and nonamers correspond to one or two turns in the DNA helix, respectively. The looped out segment is then chopped off by an enzyme called a recombinase, leaving the V gene joined directly to the J gene. The joined V–J gene segments remain separated from the C gene segment until they are transcribed into messenger RNA, the V–J transcript is then spliced to the C transcript, and the complete light chain is translated from this mRNA.

Light-chain assembly requires splicing of one V, one J, and one C gene segment. When an immunoglobulin heavy chain is assembled, the situation is made more complex by the presence of the D gene segment between the V and J gene segments. Thus construction of the V region of an immunoglobulin heavy chain requires the splicing of V_H, D, and J_H gene segments. This use of three randomly selected gene segments enormously increases the amount of variability possible. For example, if one combines segments from a pool of 100 V gene segments, 10 J gene segments, and 10 D gene segments, then 100 × 10 × 10 = 10,000 V different regions are possible.

Although the random selection of gene segments from two or three pools generates many V region genes, not all combinations of gene segments necessarily provide usable peptides. Some combinations may result in a nucleotide sequence that cannot be translated into protein. For example, nucleotides are read as triplets that each codes for a specific amino acid. If the bases are to be read correctly, then the sequence must be in the correct reading frame. If additional bases are inserted so that the reading frame is changed, the resulting gene may code for a totally different amino acid sequence. If this happens as a result of inappropriate splicing, out-of-phase rearrangements are transcribed into full-length mRNA but translation of these is prematurely terminated. It is probable that nonproductive arrangements are produced two out of three times during B-cell development. If this happens, the B cell has several additional opportunities to produce a functional antibody. For example, pre–B cells initially use one of the κ chain genes (Fig. 15–5). If this fails to produce a functional light chain, the pre–B cells switch to the other κ allele

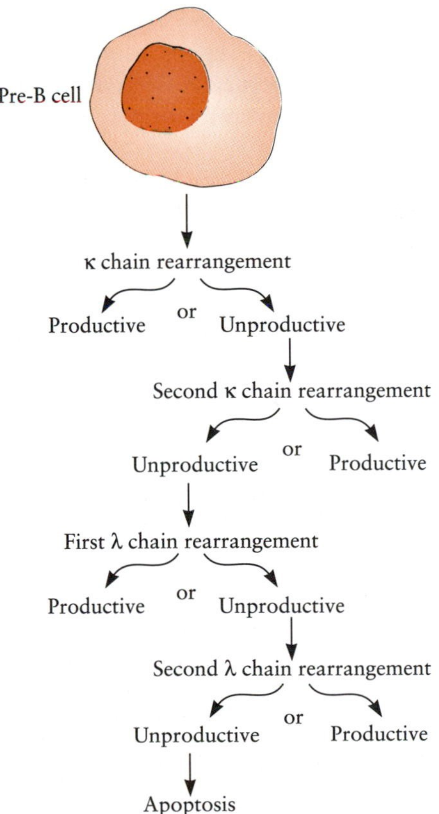

Figure 15–5 The progressive series of gene rearrangements that occur as B cells develop.

for a second attempt. If both κ alleles are unsuccessfully rearranged or deleted, the B cell then uses one of the λ alleles, and if this fails, the second λ allele represents the last resort. If all these efforts fail to produce a functional light chain, a B cell develops that cannot make a functional immunoglobulin. This cell cannot be stimulated by antigen and eventually undergoes **apoptosis** without participating in an immune response.

Joining Position

Although random joining of gene segments generates significant V region diversity, additional mechanisms play a role in increasing this diversity still further. For example, the actual base at which V and J gene segments join can vary, and changes in the base sequence occur at the splice site around position 96. This variability in V–J joining may be important in generating different amino acid sequences in this region of the light chain.

Base Insertion

In immunoglobulin heavy-chain gene processing, yet another mechanism can come into play that generates considerable sequence diversity. Random bases may be inserted or deleted at the V–D and D–J splice sites. This process is called N-region addition and is mediated by the enzyme terminal deoxynucleotidyl transferase. Terminal deoxynucleotidyl transferase adds bases to the exposed 3′ ends of gene segments. Between one and ten bases may be inserted between V and D and between D and J. In addition, some bases may be removed from the exposed ends. Whether bases are removed or added depends on the orientation of the base chains.

Somatic Mutation

Although tremendous variation can be generated by gene segment rearrangement and N-region addition, these mechanisms cannot account for all the variability known to occur within immunoglobulin V regions. For example, there are three areas of hypervariability within V regions (see Fig. 13–5). One of these, CDR3, is found around position 96 and can be readily explained by the mechanisms described previously. However, CDR1 and CDR2 are located some distance from the V–J or V–D–J splice sites associated with CDR3. Likewise, these mechanisms cannot explain the variability of human λ light chains where the size of the V gene pool is relatively small. Clearly, other mechanisms of generating antibody variability must be operating (Fig. 15–6).

If antibodies are isolated at intervals after immunization with a hapten and the amino acid sequences of their V regions are studied, it can be shown that there is a progressive change in the amino acid sequence as the immune response progresses. These changes cannot be accounted for by gene segment rearrangement but result from a process called **somatic**

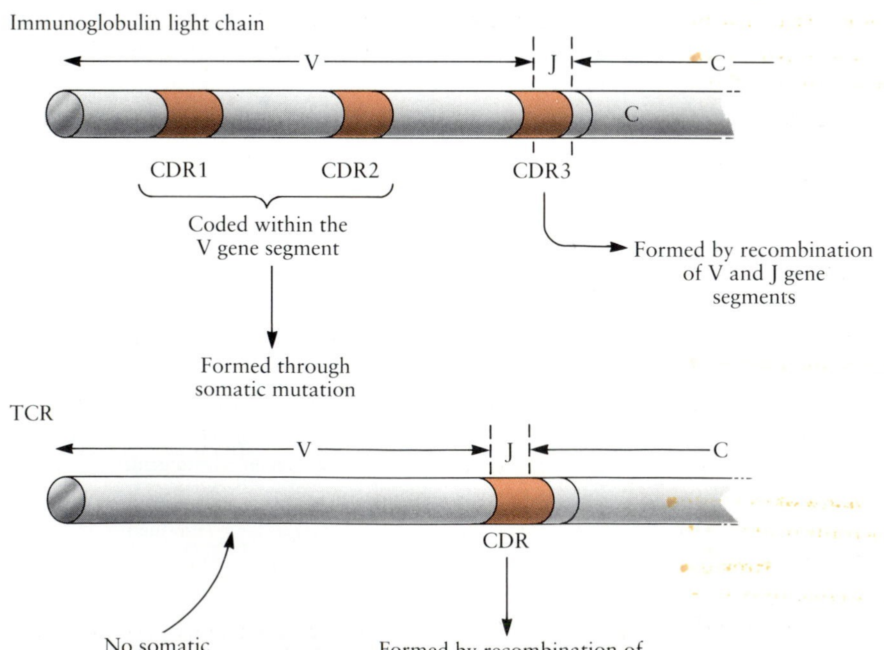

Figure 15–6 Although CDR3 is readily accounted for by recombination of V and J gene segments, CDR1 and CDR2 must be generated in a very different fashion in immunoglobulins, namely by somatic mutation. Somatic mutation is not permissible in the T cell receptor (TCR) so that TCRs have only one CDR. You should compare this diagram with the Kabat-Wu plots of immunoglobulins (Fig. 13–5) and TCR (Fig. 11–4).

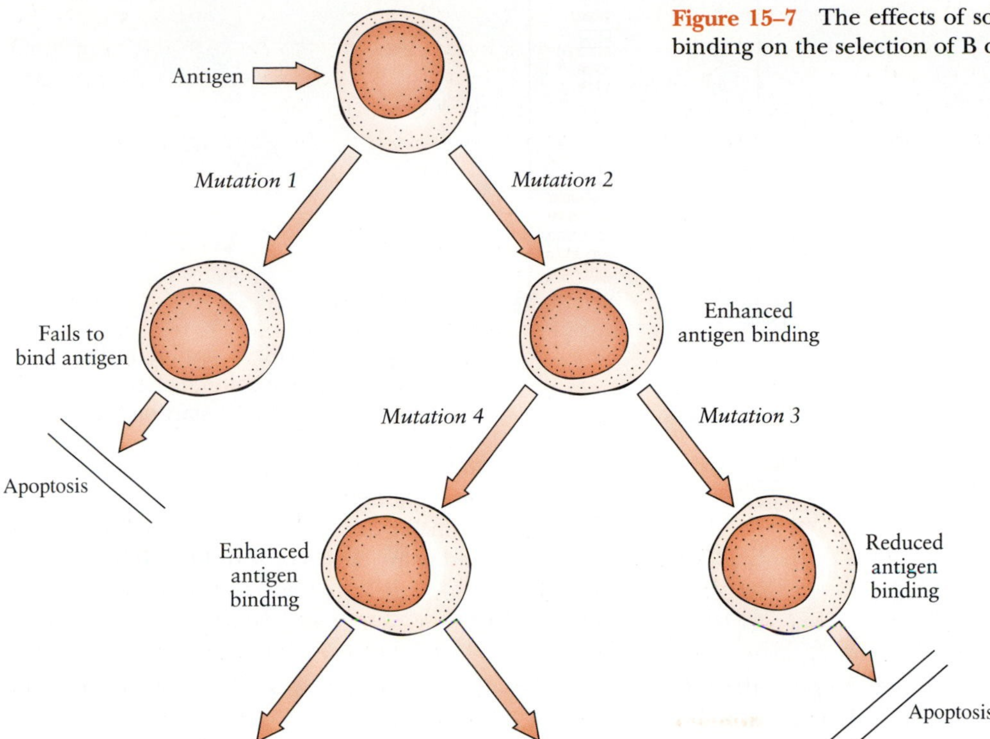

Figure 15–7 The effects of somatic mutation and antigen binding on the selection of B cell populations.

mutation. For example, animals can be immunized against the hapten phosphoryl choline and monoclonal antibodies obtained. When the amino acid sequences of the V_H region of 19 related monoclonal antiphosphoryl choline antibodies were studied, 10 had identical sequences and the remaining 9 had numerous amino acid substitutions. These sequence changes resulted from changes in single bases within the V_H gene segment. The changes were tightly clustered at the sites of CDR1 and CDR2 and fell off sharply on both sides. This localization is probably due to selection of the responding B cells. That is, V region genes of responding B cells probably mutate at random in many positions. However, B cells whose immunoglobulin receptors fail to bind antigen are not stimulated to multiply. They do not produce immunoglobulins and die off. In contrast, those B cells that make the "correct" mutation are selectively stimulated by antigen and proliferate (Fig. 15–7). The "correct" mutation is one that modifies the immunoglobulin binding site so that its affinity for the antigen is increased. The better the fit (i.e., affinity), the greater is the stimulus to divide (Fig. 15–8). This somatic mutation occurs in germinal centers of lymph nodes and the spleen and is described in Chapter 14. Somatic mutation occurs most frequently in B cells that have undergone heavy-chain class switching. These somatic mutations accumulate in immunoglobulin genes over successive B-cell generations and occur at a frequency of 10^{-3} to 10^{-4} per base pair per cell per generation. Practically, this means that one mutation occurs each time a B cell divides. This is a very high mutation rate and suggests that these mutations are generated by a specific mutagenic mechanism.

When the anti–phosphoryl choline immunoglobulins described previously were analyzed, only IgM contained the initial amino acid sequence. Those immunoglobulins that had altered sequences were either IgG

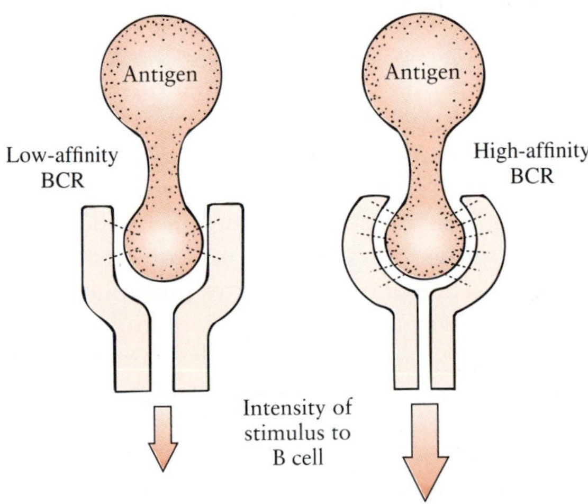

Figure 15–8 The intensity of B cell stimulation is related directly to the affinity of binding between an antigen and the B cell immunoglobulin receptor.

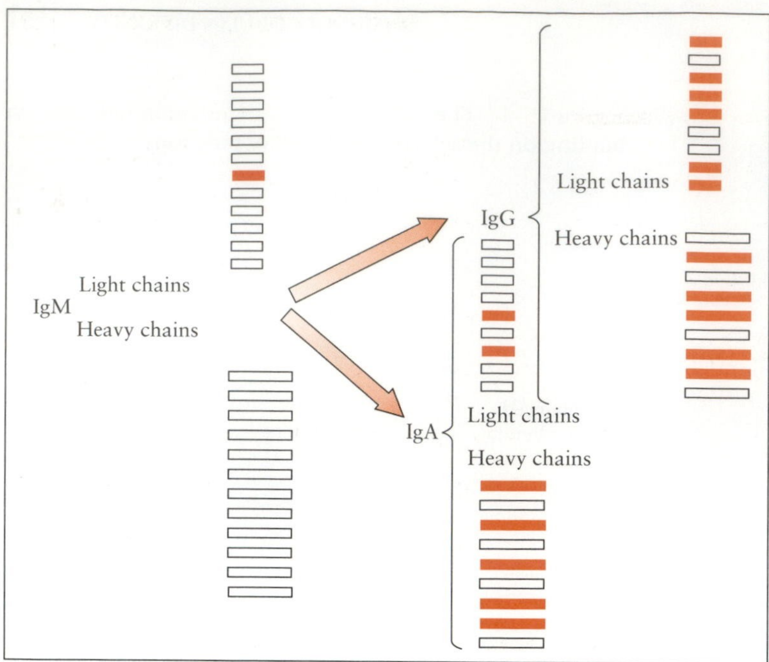

Figure 15–9 Somatic mutation during an immune response. The colored chains represent those that differ from the starting V region sequence at the initiation of the response to a defined hapten (phosphoryl choline). Only one IgM light chain was different as opposed to major changes that occurred in the light and heavy chains of both IgA and IgG.

or IgA (Fig. 15–9). Since IgM production precedes IgG and IgA production, it is suggested that the somatic mutation mechanism was not activated until after a responding cell had become committed to utilizing a specificV region segment. This may also account for the fact that the affinity of IgM antibodies does not increase as the immune response proceeds whereas the affinity of IgG antibodies does (Chapter 17).

Potential Immunoglobulin Diversity

The splicing process described previously can generate diversity in CDR3 and hence antigenic specificity in three ways. First, as occurs in the case of the V_κ family, only one of a possible 80 V gene segments is selected for transcription as is only one of the five J gene segments. This random joining of segments provides a source of variability in the base sequence since there are 400 possible V_κ–J_κ combinations. The use of a third gene segment randomly selected from a pool clearly increases the diversity that is generated as a result of combinational joining. Thus with 300 V_H, 5 D, and 9 J_H gene segments there may be 13,500 ($300 \times 5 \times 9$) V_H sequences generated in humans. In addition, the presence of two splice sites multiplies the potential for diversity arising as a result of imprecise joining of the DNA. Junctional site diversity is also increased by N-region addition, although, as pointed out, many of the combinations so formed may be of little functional use.

Since both light- and heavy-chain CDRs contribute to the antibody-binding site, the number of potential combinations and hence binding specificities is about 1.8×10^{16}. This estimate does not, of course, take into account somatic mutation. This figure may be compared with the estimate of 1×10^7 epitopes that the immune system may recognize (Table 15–2).

Constant Region Diversity

The constant region genes for immunoglobulin heavy chains are selected without regard to V region diversity. As a result, any individual V region may be associated with any one of the five immunoglobulin class constant regions.

For each class, a set of four or five C_H exons each codes for a single constant domain as well as for the hinge region and transmembrane domains. For example, an IgM constant region gene consists of five C_μ exons, and an IgA constant region gene contains four C_α exons. During an antibody response, the immunoglobulin classes are synthesized in a fairly constant sequence. Thus a B cell first uses the $C\mu$ gene to make IgM. At an appropriate time it also begins to transcribe the $C\delta$ gene and then makes both IgM and IgD. The IgM is firmly bound to the cell membrane in the immature B cell. Once the B cell matures, the IgM is secreted. Eventually, a B cell switches to using either $C\gamma$, $C\alpha$, or $C\epsilon$ and becomes committed to synthesizing immunoglobulins of one of the other major classes, IgG, IgA, or IgE. All these changes require that the cell transcribe the appropriate C genes and switch to the genes for another class, as required.

Class Switching

The genes required for all immunoglobulin heavy chains are found on chromosome 14 in humans and chromosome 12 in mice (Fig. 15–10). The selection of the required VDJ segments is an early event in the

Table 15–2 Potential Immunoglobulin and TCR Diversity in Humans

	Immunoglobulin		TCR αβ		TCR γδ	
	IgH	L(κ)	α	β	γ	δ
V	300	80	50	57	8	3
D	5	—	—	2	—	3
J	9	5	75	13	5	3
1. Possible combinations of V, D, and J $A = (V \times D \times J)$	13,500	400	3750	1482	40	27[1]
Number of splice sites (n)	2	1	1	2	1	3
2. Five base variation at each splice site $B = A(\times 5)^n$	$=3 \times 10^5$	=2000	=18,750	=37,050	=200	=3375
3. N-region insertion at each splice site[2] $C = B(\times 5461)^n$	$=9 \times 10^{13}$	$=2 \times 10^3$*	$=10^8$	$=10^{12}$	$=10^6$	$=5 \times 10^{14}$
4. Combination of two chains[3]	$H \times L = 1.8 \times 10^{16}$		$\alpha \times \beta = 10^{20}$		$\delta \times \delta = 5 \times 10^{20}$	

Note: 1. In δ chains both D segments may be employed, i.e., V × D × D × J.
2. Up to six bases may be inserted at each splice site. This can give 5461 different possible combinations of four bases, i.e., $4^6 + 4^5 + 4^3 + 4^2 + 4^1 + 4^0$.
3. This figure is an upper theoretical limit. The actual figure will be very much lower than this because of the need to read sequences in frame and because of major constraints on the amino acids that can be used to make a functional V region. *N-region addition has not been recorded in light chains.

development of a B cell, and once made, the selection cannot be changed (Fig. 15–11). In contrast, the selection of genes for a specific immunoglobulin class is a relatively late event and switches as the immune response progresses. Thus, VDJ segments remain separated from the C genes until just before transcription into mRNA. The unwanted C genes are excised, and the required C gene is spliced to the VDJ segments. If IgM is to be synthesized, the VDJ genes are spliced directly to the C_μ gene (Fig. 15–12). If IgA is to be synthesized, the genes coding for C_μ to C_ϵ inclusive are excised and the VDJ genes are spliced to the C_α gene (Fig. 15–13). There are several possible ways in which the intervening genes can be deleted (Fig. 15–14). The simplest is looping-out deletion. In this situation the two gene segments come together by looping out and then excising the intervening material in a manner similar to that seen in V–J joining. An alternative mechanism is called sister chromatid exchange. This occurs when the chromosome duplicates during mitosis. An unequal crossing over occurs between genes on sister chromatids, thereby joining the genes.

Heavy-chain class switching operates in a manner very different from V gene segment splicing. On the 5′ side of each C_H gene (except Cδ) is a switch (S) region. Recombination occurs within these switch regions. Each switch region can recognize and splice to the switch region associated with another heavy-

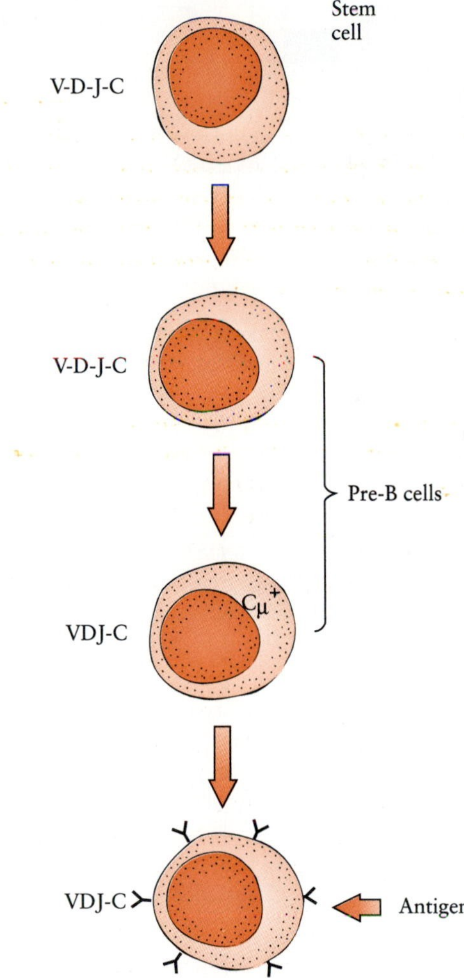

Figure 15–11 The time course of immunoglobulin gene rearrangement in the developing B cell. D-J joining is an early event in pre-B cells. VDJ-C joining occurs only when the B cell responds to antigen and RNA splicing occurs.

(Text continues on p. 225)

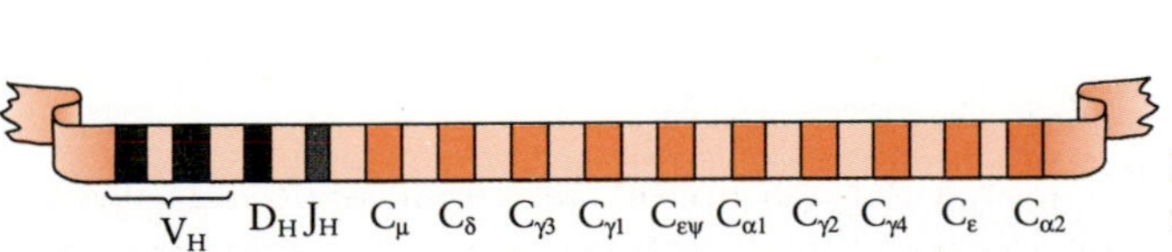

Figure 15–10 The structure of a complete immunoglobulin heavy chain gene.

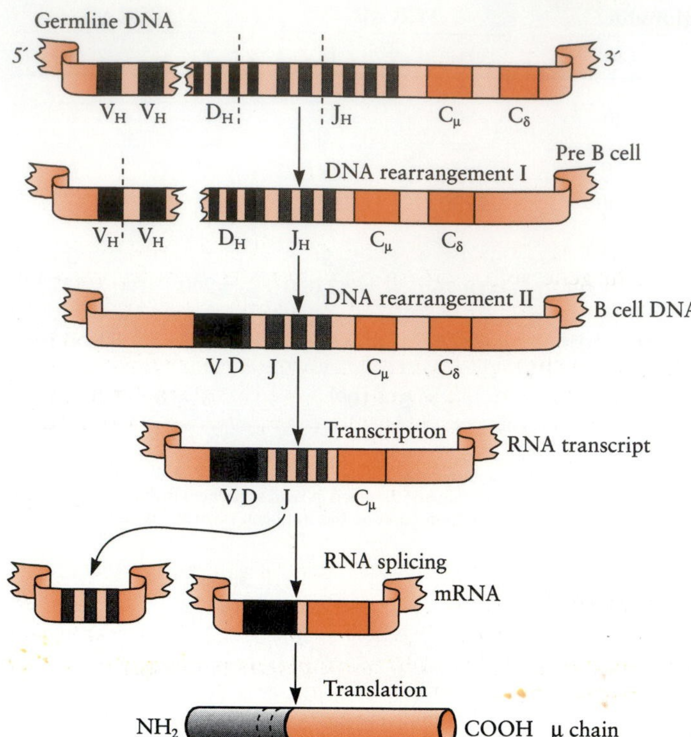

Figure 15–12 The production of an IgM heavy chain. The complete gene sequence is present in germ-line and embryonic cells. As B cells differentiate, they randomly delete some V_H genes, some D genes, and some J genes and link V_HD and J genes together. Following transcription and RNA splicing, a continuous mRNA sequence is generated and translated into a complete IgM heavy chain.

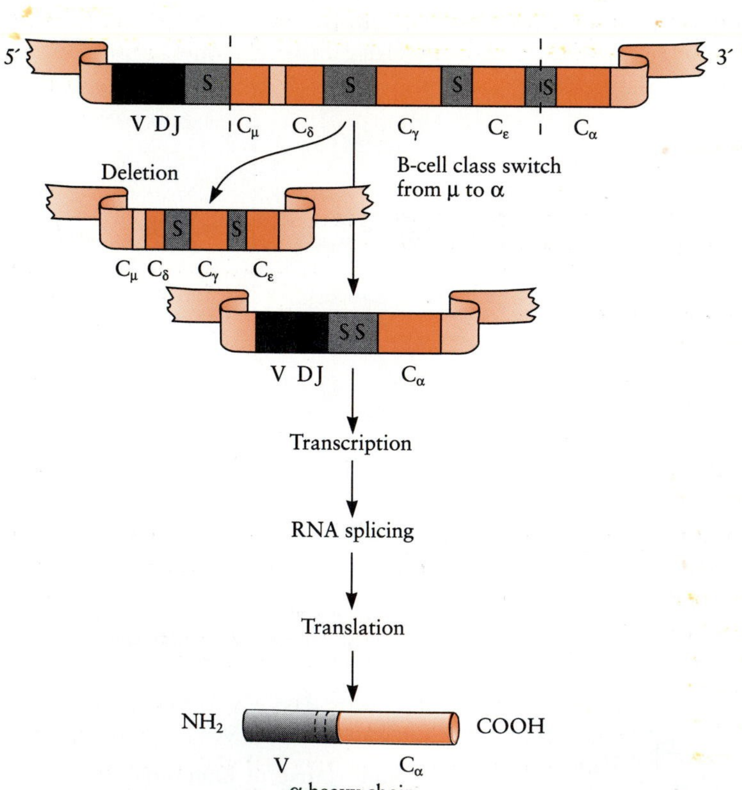

Figure 15–13 The production of an IgA heavy chain. An unstimulated B cell is committed to generate IgM antibodies. At some time during the response of the B cell to antigen, the V-D-J gene is disconnected from the Cμ gene and connected to the Cα gene. As a result, the B cell DNA codes for an IgA heavy chain.

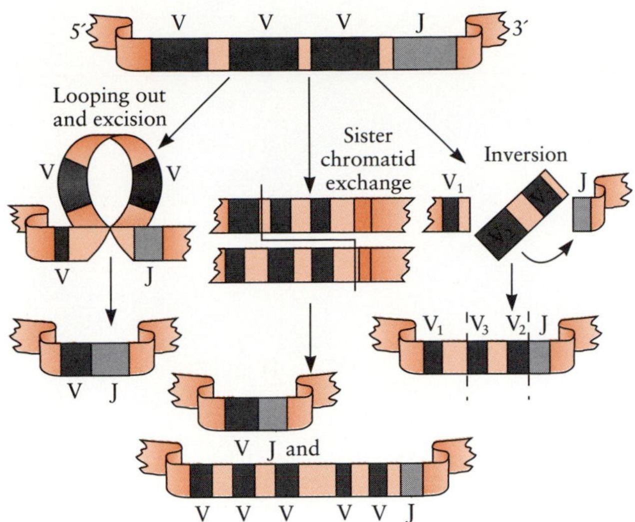

Figure 15–14 Several different mechanisms of immunoglobulin gene splicing. In looping out, the gene segments are joined while the intervening sequences are looped out and destroyed. In sister chromatid exchange, the gene segment on one chromatid is directly linked to a gene segment on the sister chromatid. In inversion, a segment of chromatid is excised, rotated, and replaced so that it has reversed polarity. This process is seen in some TCR genes.

chain gene. For example, for IgA to be produced, the Sα region binds to the Sμ region, looping out and deleting all the intervening DNA and directly joining VDJ to Cα (Fig. 15–13). Each S region may be designed for a different gene rearrangement. For example, there may be a distinct site for joining Sμ to Sγ and another for joining Sμ to Sα, and so on. Class switching in B cells is regulated by several cytokines, most notably IL-4, and by physical contact through CD40 and T cells. The IL-4 regulates the activity of DNA binding factors that control transcriptional activity and may target specific switch regions.

How Immunoglobulin Gene Transcription Is Regulated

B cells do not make complete immunoglobulin chains until after V–J or V–D–J rearrangements are completed. As with other genes, each V gene segment has a promoter sequence located on the 5′ side of the site where transcription is initiated. A promoter contains the start site for RNA transcription and signals where this should begin. When bound to the promoter, the RNA polymerase begins to transcribe the V gene DNA.

However, the binding of RNA polymerase to the V gene promoter is very weak and V gene transcription will not occur spontaneously, prior to V–D–J rearrangement (Fig. 15–15). Once the gene segments are rearranged, the process of transcription speeds up considerably as a result of the activities of enhancers. Enhancers are sequences that increase the utilization of promoters. In immunoglobulin genes, there is an enhancer sequence located between the J and C gene segments. In the unrearranged genes this enhancer is too far away to have an effect on the promoter region. Once the unused V gene segments are excised, however, the enhancer and promoter are moved much closer together. The enhancer is then able to activate transcription through the promoter. As a result, the enhancer may increase V gene transcription up to 10,000-fold.

IgM and Cell Membranes

IgM can exist either as a monomeric membrane-bound antigen receptor or as a secreted pentameric antibody. The heavy chain of the membrane-bound form has a transmembrane C-terminal domain con-

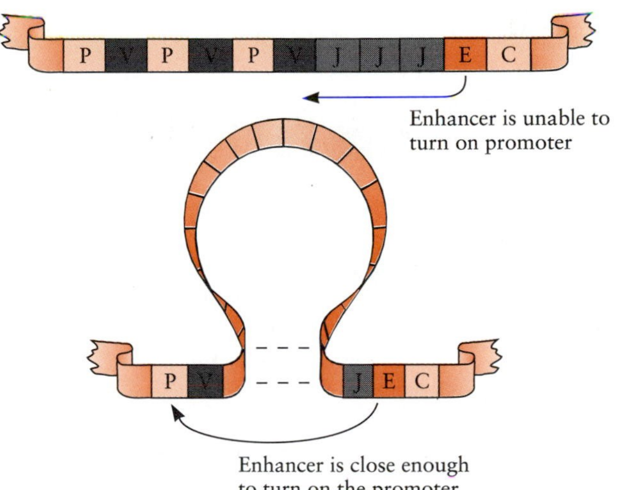

Figure 15–15 Enhancer sequences cannot activate promoter sequences unless the two are located close together. Thus in unrearranged immunoglobulin genes, gene transcription cannot occur. Once the intervening gene segments are looped out, the enhancer is placed close to the promoter and gene transcription is turned on.

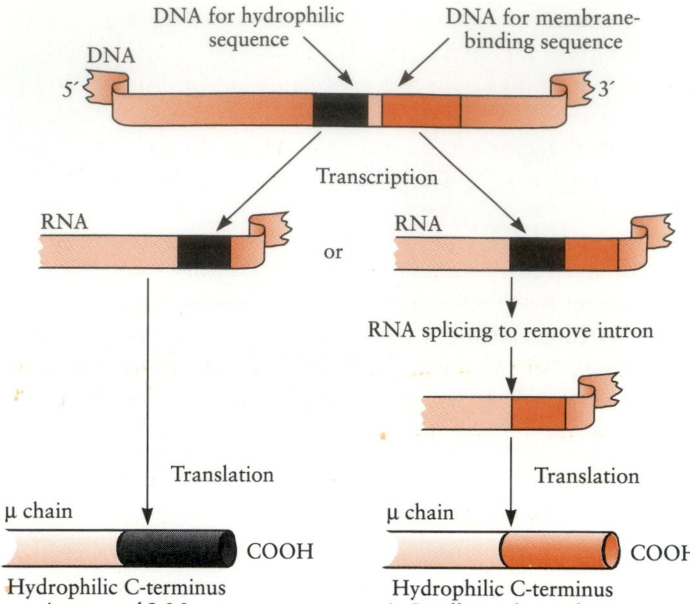

Figure 15–16 The way in which a B cell can produce both the membrane-bound form and the secreted form of IgM. The difference arises as a result of alternative splicing of precursor mRNA.

sisting of hydrophobic amino acids. This domain is absent from the secreted antibody. The switch between the two forms depends on the differential splicing of different domains. Thus in the DNA coding for IgM, there are two short exons, $C_\mu S$ and $C_\mu M$, located 5′ to $C_\mu 4$ (Fig. 15–16). When IgM is to be synthesized, all the Cμ exons are transcribed. To produce membrane IgM, the mRNA is cleaved so that the $C_\mu S$ gene segment is deleted and the $C_\mu 4$ exon is spliced directly to the $C_\mu M$ exon. $C_\mu M$ codes for the hydrophobic domain that is inserted into the lymphocyte membrane. To produce secreted IgM, the exon coding for the $C_\mu M$ exon is deleted and translation is stopped after $C_\mu 4$ and $C_\mu S$ are read.

IgM and IgD Production

During B-cell maturation, the cells go through a period during which they can synthesize IgM and IgD simultaneously. This is also accomplished by a special

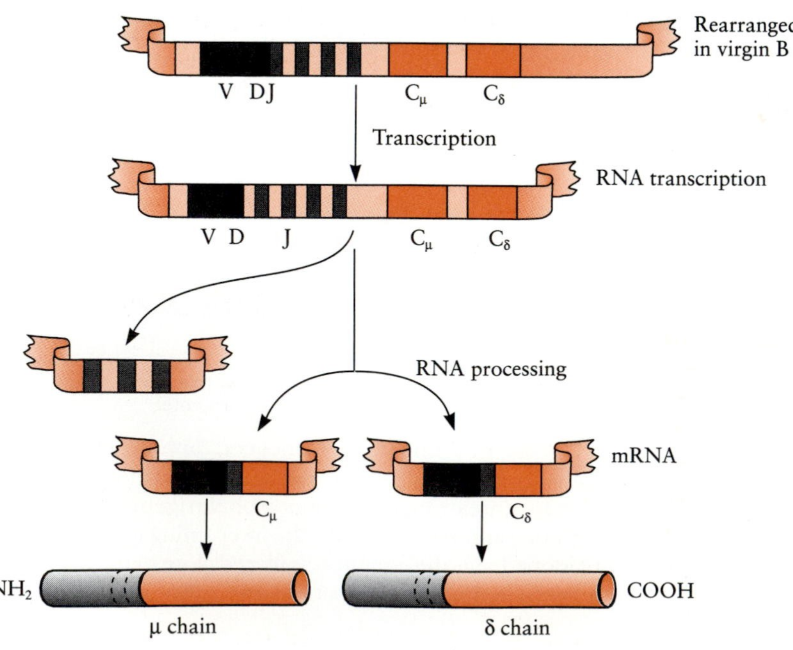

Figure 15–17 The way in which a B cell can express both IgM and IgD on its surface at the same time as a result of differential RNA processing.

process. The Cμ and Cδ genes are located close together under control of a single switch region, and as a result, both genes are transcribed into mRNA at the same time. The mRNA is then spliced in one of two ways. The VDJ segment may be joined to the Cμ genes to produce μ heavy chains or to the Cδ genes to produce δ heavy chains (Fig. 15–17).

TCR DIVERSITY

Like immunoglobulins, the four peptide chains that make up the two types of TCR can bind specific epitopes. They do this because each consists of a variable region attached to a constant region. The diversity of the TCR V region is generated in much the same way as in the immunoglobulins but with some interesting and significant differences. T-cell receptor and immunoglobulin gene rearrangements are specific in that immunoglobulin genes are not rearranged in T cells and TCR genes are not rearranged in B cells.

TCR Gene Structure

The four T-cell receptor peptide chains are coded for by four separate gene families (α, β, γ, and δ) that resemble the immunoglobulin light- and heavy-chain gene families. They differ from immunoglobulins, however, in that two of the families (α and δ) are very closely linked. All four families contain V, J, and C gene segments, and β and δ gene families also contain D gene segments (Fig. 15–18).

Another difference from the immunoglobulin gene families is that each TCR gene family contains two constant region genes. In the α/δ family these two C region genes are functionally and structurally different so that one codes for the α constant region (Cα) and the other codes for the δ constant region (Cδ). The two Cγ genes are functionally similar but structurally different. The two Cβ genes, in contrast, are indistinguishable, although Cβ2 is most commonly used. All helper and cytotoxic T cells rearrange and express TCR α and β genes, while a small subpopulation uses TCRγ and δ.

The Alpha and Delta Chains

The human TCR α and δ gene families are found on chromosome 14. They are unique in that the set of δ chain genes are embedded within the α chain genes. The entire complex is about one megabase in size. When T cells are very immature, they use δ chains for their receptors. As T cells mature, however, they become committed to the use of the α/β receptor, and the δ chain gene segments are then looped out and deleted. The α chain family contains V, J, and C gene segments that are separated into two regions by the δ chain genes. On the 5′ side of the δ chain genes, about 50 Vα gene segments are grouped into 12 families. The rest of the α gene family is on the 3′ side of the

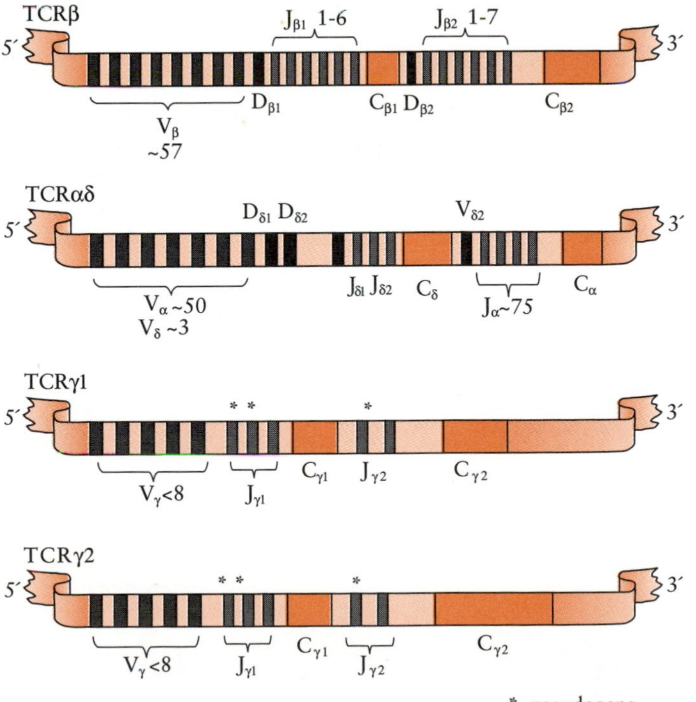

Figure 15–18 The gene structure of the TCR peptide chains. The α and δ chains are coded for by intermingled sets of genes. There are two different γ chain genes with differences in their Cγ2 region.

δ chain gene segments. It consists of about 75 Jα gene segments followed by a single Cα gene. Obviously, many more J gene segments are here than are found in immunoglobulin loci. The δ gene family contains V, D, J, and C gene segments. There are about three Vδ segments, divided into those that are used in adults and those used in the fetus, as well as three Dδ gene segments and three Jδ segments located upstream of one Cδ gene. The Dδ gene segments are short and very dissimilar. Both Dδ gene segments may contribute to the final product so that a complete Vδ gene may consist of V, D1, D2, and J gene segments.

The Beta Chain

The TCR β gene family is located on chromosome 7 in humans and 6 in mice. It contains about 23 Vβ gene segments in mice and 57 in humans, located upstream of two nearly identical D–J–C clusters (see Fig. 15–17). There are about six functional J gene segments in each cluster. Any of the Vβ gene segments may be joined to either of the two D–J–C gene clusters. An additional Vβ gene segment has been identified about 10 kilobases downstream (3′) of the second Cβ gene. This V gene segment may be used to form a functional β chain gene by chromosome inversion (see Fig. 15–14). The other Vβ segments are located 5′ to the D–J–C clusters and rearrange by conventional mechanisms (either by looping out and deletion of intervening DNA or by sister chromatid exchange). The two D gene segments of the TCRβ family are very similar in sequence and length, and their use is optional. As a result, a Vβ segment may join to a Dβ or Jβ segment. In both mice and humans there are two Cβ genes, each divided into four exons.

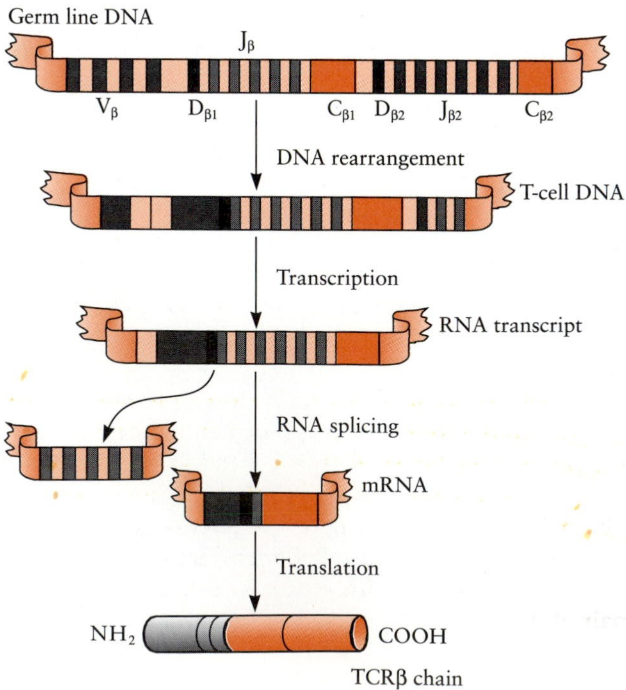

Figure 15–19 The formation of a TCR β chain from its genes. The processes (and possibly the enzymes) involved are identical to those involved in immunoglobulin synthesis.

Table 15–3 The Generation of TCR Diversity

Multiple germ-line genes
VJ and VDJ recombination
Combinational inaccuracies
Chain combinations
N-region addition

The Gamma Chain

The TCR γ chain gene family is found on chromosome 7 in humans and chromosome 14 in mice. There are 14 Vγ gene segments in humans, of which eight are functional, as well as five Jγ segments and two Cγ genes. The Cγ genes are composed of four exons. There is no Dγ region, so Vγ segments combine directly with Jγ segments.

Generation of TCR V Region Diversity

The diversity in amino acid sequences seen in the TCR chains is generated by the recombination of V, J, and D gene segments (Table 15-3). Diversity is generated by mechanisms similar to those that operate in the immunoglobulin system. Thus the various gene segments that are separate in the germ line are brought together by DNA rearrangement and deletion as the T cells differentiate (Fig. 15–19).

Gene Rearrangement

TCR α and γ chains use V and J segments to form their V region genes. TCR β and δ chains use V, D, and J segments to form their V region genes. Delta chains have the ability to use both of their D segments, and as a result, V–D–D–J constructs can be formed. In TCR β and δ chains the reading frame of the D segments is commonly changed and can yield productive rearrangements. This is a rare event in immunoglobulins. Looping out and deletion account for more than 75% of the T-cell receptor rearrangements. The remainder of the rearrangements are due either to unequal sister chromatid exchange or to inversion—moving an inverted segment of gene into a position beside a segment in the opposite orientation (see Fig. 15–14). The looping out and deletion of TCR gene segments is mediated by signals identical to those of immunoglobulins. The heptamer/nonamer signal sequences are indistinguishable from those found in immunoglobulin genes. Thus the same joining enzyme (a recombinase) probably acts on both immunoglobulin and TCR genes. Several V segments show diversity in the position of their joining points.

Base Insertion

Random bases may be inserted at the V, D, and J junctions as a result of the activities of terminal deoxynucleotidyl transferase. As many as five bases between V and D and four between D and J can be added by this mechanism. In humans, the TCR-δ chains are composed predominantly of VDDJ joined by very large N regions. N-region addition is probably the most significant component of TCR junctional diversity.

Somatic Mutation

There is one major difference between diversity-generating mechanisms in immunoglobulins and the TCR. Somatic mutation does not occur in TCR V region genes. Although T cells must be able to recognize a foreign epitope in association with the presenting MHC molecule, it is essential that they recognize self MHC molecules while not responding to self-antigens. If random somatic mutation were to occur, it would carry the unacceptable risk of altering MHC restriction and so rendering the foreign epitope unrecognizable. It might also lead to generation of T cells reactive with self-antigens and thus to autoimmunity.

Where Does This Happen?

The TCR genes are rearranged and first expressed in the developing thymus. It is believed that the developing T cell first attempts to rearrange the genes for γ/δ and, if this is not productive, proceeds to make α/β. Alternatively, it has been suggested that γ/δ and α/β belong to two different cell lines and express these receptors independently. The rearrangements and expression of TCRs occur in sequential and ordered fashion. Thus in mice there are three waves of T-cell development in the thymus. First, a population of γ/δ-bearing T cells using certain Vγ gene segments transiently appears. γ/δ T cells in the epidermis carry similar γ/δ heterodimers and may originate from this early population. Then a second population of γ/δ T cells appears using different Vγ gene segments. This may give rise to the cells found in the intestinal mucosa. Finally, T cells bearing α/β TCR appear. These are the cells that are found in peripheral blood.

Rearrangement in the TCR δ locus appears to be the earliest event in T-cell development. Because of the geometry of the TCR δ/α loci, joining of a Vα segment to a Jα segment invariably deletes the Dδ, Jδ, and Cδ genes on that allele (see Fig. 15–17). Thus α chain rearrangements eliminate any possibility of δ chain expression.

Although TCR genes bear close resemblance to the immunoglobulin genes, there are some interesting differences between them. Thus TCR genes usually use many fewer V segments than do immunoglobulins. TCR γ uses only eight V segments as compared to the 250 to 1000 employed by immunoglobulin heavy chains. This difference is accentuated when paired V regions (one from each chain) are com-

A CLOSER LOOK

One Cell–One Receptor?

It has been generally accepted that each T cell (and B cell) carries antigen receptors specific for a single epitope. This is the "one cell–one receptor" rule. However, it has also been recognized that each cell actually has the potential to produce two receptor chains of two types. This is because in diploid cells there are two copies of each gene. Thus, for example, any T cell is theoretically capable of producing two different Tα chains. This does not normally occur because of a process called allelic exclusion that permits only one chain to be expressed. Thus the presence of one rearranged Vβ chain prevents rearrangement of the other Vβ chain.

In contrast, both Vα alleles rearrange simultaneously. Recent studies by Elisabetta Padovan and her colleagues have now shown that up to one third of TCR αβ T cells can express two different Vα chains simultaneously. These cells may therefore be specific for more epitopes than cells with only a single receptor and may play an important role in immunity.

Padovan, E., et al. Expression of two T cell receptor α chains: Dual receptor T cells. Science, 262:422–424, 1993.

pared. On the other hand, TCRs use many more J region genes than do immunoglobulins. If the diversity in immunoglobulin CDR3 is compared with the single CDR in the TCR, the diversity of the VDJ junction in TCRs is much greater than that seen in immunoglobulins.

Potential TCR Diversity

In the human TCR α gene there are 75 Jα segments, 50 Vα segments, some N addition diversity, and some junctional diversity (i.e., $75 \times 500 \times 5461 \times 5 = 1 \times 10^8$) (see Table 15–2). After correction for codon redundancy and correct reading frame, the number of TCR α chains is about 10^6. In the Tβ chain in humans there are about 57 Vβ segments. They have two D segments and 13 J segments. Thus, in humans, there are $57 \times 2 \times 13 = 1482$ possible combinations of Vβ. In addition, there is junctional diversity where regions join at the VDJ junction. Rearrangements in some cases may occur at up to seven positions on the 5′ side of Jβ. Likewise, the use of Dβ elements is flexible, and many portions of Dβ can be used. Thus, there are about 60 ways to use the 13 bases of Dβ1 and about 50 ways to use the 11 bases of Dβ2 (some combinations cannot be used). Thus, in total, there are perhaps 110 possible Dβ combinations. As a result of all this, in the human TCRβ there are about 10^{12} possible sequences. This factor can be reduced threefold because of the necessity to preserve the VJ reading frame. Likewise, because of redundancy in the genetic code (two or more codons can code for one amino acid), we can reduce this by a factor of 30, leaving us with about 5×10^9 possible VDJβ sequences. (A somewhat similar figure can be arrived at for the mouse. Given that a mouse has 5×10^7 T cells, that is much more potential diversity than a mouse would ever be able to use.)

In the Vδ chain, a combination of V region diversity, two D regions, three sites where N-addition can occur, and diversity in V-J joining position can generate about 10^{14} possible amino acid sequences. While in the Vγ, 10^6 combinations are possible. There is no difficulty, therefore, in accounting for the enormous diversity seen in the TCRs.

KEY WORDS

C gene segment p. 217
Class switching p. 222
D gene segment p. 217
Enhancer element p. 225
Gene family p. 217
Gene rearrangement p. 218
Gene segment p. 217
Heavy-chain gene p. 217
Immunoglobulin diversity p. 216
Immunoglobulin genes p. 217
Instructive theory p. 216
J gene segment p. 217
Joining position p. 220
Light-chain gene p. 217
Looping out p. 218
Nonproductive rearrangement p. 219
N-region addition p. 220
Pseudogene p. 217
Reading frame p. 219
Selective theory p. 216
Sister chromatid exchange p. 223
Somatic mutation p. 220
Splicing p. 218
Switch region p. 223
TCR gene p. 227
V gene segment p. 217

QUESTIONS

1. Which genes code for an immunoglobulin light chain?
   a. one V gene, one C_L gene
   b. one V gene, one J gene, one C_L gene
   c. one light-chain gene
   d. one V gene, one J gene, one D gene, one C_L gene
   e. one V gene, one J gene, one C_H gene
2. One characteristic that distinguishes the construction of immunoglobulin heavy chains from that of light chains is
   a. somatic mutation
   b. the use of V gene segments
   c. looping out of gene segments
   d. the use of a D gene segment
   e. the use of a J gene segment
3. The insertion of one or more additional bases at a gene splice site is called
   a. N-region addition
   b. somatic mutation
   c. antigenic variation
   d. sister chromatid exchange
   e. looping out
4. A key difference between the generation of diversity in TCR and immunoglobulin variable regions is the use of
   a. sister chromatid exchange
   b. class switching
   c. looping out
   d. N-region addition
   e. somatic mutation
5. The κ and λ gene families code for
   a. immunoglobulin heavy chains
   b. immunoglobulin light chains
   c. TCR heavy chains
   d. immunoglobulin constant regions
   e. immunoglobulin V regions

6. When one or two bases are added to a DNA molecule, this results in a(n)
   a. frame shift
   b. gene deletion
   c. class switch
   d. heavy-chain switch
   e. class switch
7. The immunoglobulin class switch results from
   a. deletion of D gene segments
   b. somatic mutation
   c. deletion of unwanted C gene segments
   d. deletion of unwanted J gene segments
   e. N-region addition
8. The only two immunoglobulin classes that can be expressed simultaneously on a B-cell surface are
   a. IgM and IgG
   b. IgA and IgG
   c. IgG and IgD
   d. IgM and IgD
   e. IgD and IgE
9. How many constant region genes are found in each TCR gene family?
   a. one
   b. two
   c. three
   d. four
   e. five
10. Which TCR gene product can express two diversity (D) gene segments?
   a. α chain
   b. β chain
   c. γ chain
   d. δ chain
   e. μ chain
11. Explain the phenomenon of affinity maturation, both by the cellular selection theory and on the basis of somatic mutation and natural selection of responding cells.
12. How is each of the three hypervariable regions in the variable region of a kappa light chain produced?
13. Explain how a B cell can first produce IgM alone, then IgM and IgD, and finally IgM alone again.
14. What is the maximum number of antibody-binding sites that can be generated in a mouse by combinatorial diversity in both light and heavy chains?
15. Speculate as to why two types of light chains are required by the body.
16. What is somatic mutation? How might it come about? What is its significance? Why does it not occur in the TCR?
17. What mechanisms might be involved in inducing the switch in immunoglobulin class from IgM to IgA production in the wall of the intestine?
18. Compare the generation of TCR diversity with the generation of immunoglobulin diversity. Are the differences between the two processes significant, and do they have any implications for the nature of the antigen recognized by B or T lymphocytes?

Answers: 1b, 2d, 3a, 4e, 5b, 6a, 7c, 8d, 9b, 10d

SOURCES OF ADDITIONAL INFORMATION

Baltimore, D. Somatic mutation gains its place among the generators of diversity. Cell, 26:295–296, 1981.

Boss, M.A. Enhancer elements in immunoglobulin genes. Nature, 303:281–282, 1983.

Brack, C., et al. A complete Ig gene is created by somatic recombination. Cell, 15:1–14, 1978.

Calame, K.L. Immunoglobulin gene transcription: Molecular mechanisms. Trends Genet., 5:395–399, 1989.

Crews, S., et al. A single VH gene segment encodes the immune response to phosphorylcholine: Somatic mutation is correlated with the class of antibody. Cell, 25:59–66, 1981.

Davis, M.M. Molecular genetics of T-cell antigen receptors. Hosp. Pract. [off.], 23:157–170, 1988.

Davis, M.M., and Bjorkman, P.J. T-cell antigen receptor genes and T-cell recognition. Nature, 334:395–402, 1988.

Dreyer, W.J., and Bennett, J.C. The molecular basis of antibody formation: A paradox. Proc. Natl. Acad. Sci. U.S.A., 54: 964–969, 1965.

Early, P., and Hood, L. Allelic exclusion and non-productive immunoglobulin gene rearrangements. Cell, 24:1–3, 1981.

Eckhardt, L.A. Immunoglobulin gene expression only in the right cells at the right time. FASEB J., 6:2553–2560, 1992.

Gearhart, P.J., and Bogenhagen, D.F. Clusters of point mutations are found exclusively around rearranged antibody variable genes. Proc. Natl. Acad. Sci. U.S.A., 80:3439–3443, 1983.

Gershenfield, H.K., et al. Somatic diversification is required to generate the Vκ genes of MOPC 511 and MOPC 167 myeloma proteins. Proc. Natl. Acad. Sci. U.S.A., 78:7674–7678, 1981.

Gottlieb, P.D. Immunoglobulin genes. Mol. Immunol., 17:1423–1435, 1980.

Kallenbach, S., et al. Three lymphoid-specific factors account for all junctional diversity characteristic of somatic assembly of T-cell receptor and immunoglobulin genes. Proc. Natl. Acad. Sci. U.S.A., 89:2799–2803, 1992.

Lai, E., Wilson, R.K., and Hood, L.E. Physical maps of the mouse and human immunoglobulin-like loci. Adv. Immunol., 46:1–59, 1989.

Leder, P. Genetic control of immunoglobulin production. Hosp. Pract. [off.] 18:73–82, 1983.

Marcin, K.B., and Cooper, M.D. New views of the immunoglobulin heavy-chain switch. Nature, 298:327–328, 1982.

Marx, J.L. Immunoglobulin genes have enhancers. Science, 221:735–737, 1983.

Moss, P.A.M., Rosenberg, W.M.C., Bell, J.I. The human T cell receptor in health and disease. Ann. Rev. Immunol. 10:71–96, 1992.

Purkerson, J., and Isakson, P. A two-signal model for regulation of immunoglobulin isotype switching. FASEB J., 6:3245–3252, 1992.

Robertson, M. Chopping and changing in immunoglobulin genes. Nature, 287:390–392, 1980.

Siebenkotten, G., et al. The murine IgG1/IgE class switch program. Eur. J. Immunol., 22:1827–1834, 1992.

Tonegawa, S. Somatic generation of antibody diversity. Nature, 302:575–581, 1983.

CHAPTER 16

The Complement System

Hans Müller-Eberhard *Hans Müller-Eberhard was born in Germany in 1927. Working in the United States, Dr. Müller-Eberhardt was the first to systematically analyze the components of the complement system using modern biochemical techniques. In this way he worked out the detailed biochemistry of the complement activation process. Thus he was the first to isolate and define C3 and its breakdown products. He identified the enzymes that activated it including factor B. He identified the metastable binding site of C3b and demonstrated the initiating factors for the alternate complement pathway.*

CHAPTER OUTLINE

CHAPTER CONCEPTS

1. The complement system is a set of proteins that act together to destroy invading microorganisms.
2. There are three separate enzyme pathways within the complement system: the classical pathway initiated by antibody bound to antigen; the alternative pathway initiated by foreign surfaces; and the terminal pathway that results in the destruction of microorganisms.
3. Activation of the complement system results in inflammation, chemotaxis, opsonization, and cell lysis. It also regulates the immune system.
4. The complement system is carefully regulated to ensure that it does not act in an uncontrolled fashion.
5. Deficiencies of some complement components may result in an individual becoming susceptible to infections or autoimmune disease.
6. A serologic test, called the complement fixation test, uses complement to measure antibody levels in serum. It is easy to read but somewhat complex to perform.

In this chapter we describe the properties of a family of proteins called the **complement** system. This complement system plays a critical role in the defense of the body in association with antibodies and the cells of the immune system. After discussing the complement proteins, we describe several ways by which the complement system can be turned on and how it can lead to the destruction of invading microorganisms. We also explain how such a potent system has to be regulated. Once you understand the basic features of the complement system, you will learn its role in many aspects of defense against infection such as chemotaxis, opsonization, and inflammation. We describe briefly how the complement proteins are inherited and the effects of not having a functioning complement system. We then review the relationship of the complement system to other important body systems. Finally, we will briefly describe how complement can be used in two very useful diagnostic tests.

The properties of antibody molecules bound to antigen are very different from those of free antibody. For example, antigen-bound antibody can bind to Fc receptors on phagocytic cells and thus functions as an **opsonin.** New epitopes appear on antigen-bound antibody molecules. These epitopes may be regarded as foreign and therefore provoke the formation of autoantibodies called **rheumatoid factors** (Chapter 31). The development of these new epitopes and biological activities is a result of conformational changes in the immunoglobulin **Fc region.** Normally, the Fc region is masked by the Fab regions. When antibody binds antigen through the Fab regions, the shape of the molecule changes and the Fc region is exposed. The active sites on the Fc region thus become available and free to exert their biological functions.

A second mechanism that accounts for the increased activity of antigen-bound antibodies is the number of available active sites. Thus a single immunoglobulin molecule may be unable to initiate reactions by itself. However, when several immunoglobulin molecules bind closely together on an antigenic surface and multiple Fc regions are exposed, the combined stimulus may be sufficient to initiate subsequent reactions.

COMPLEMENT

There are several essential biochemical systems that, if activated in the absence of effective control mechanisms, could lead to disease or death. Examples of such systems include the blood clotting system, the **fibrinolytic** system, and the **kinin** system. Uncontrolled activation of these could cause uncontrollable hemorrhage, extensive intravascular thrombosis, and severe disturbances in vascular permeability, respectively. Nevertheless, these systems must be readily available and able to be rapidly activated when required. Thus, if the clotting system was activated slowly, a person might bleed to death before his or her blood clotted. To ensure that these systems are rapidly activated yet controllable, linked enzyme reactions are used. The general principle of these reactions is that the products of one reaction catalyze a second reaction, whose products then catalyze a third reaction, and so on (Fig. 16–1). This results in an expanding "cascade" of reactions. Provided that many of the intermediate products either are present in limiting quantities, have a very short half-life, or are easily inhibited, it is possible to ensure that the reactions proceed very rapidly yet in a controlled fashion.

Enzymatic chain reactions of this type are known as **cascade reactions** and usually require a "trigger" to initiate the reaction chain. For example, in the case of the blood clotting system, activation of Hageman factor by altered surfaces is the initiating event that sets the system in motion.

Complement is a chain of enzymes whose activation eventually results in the disruption of cell membranes and the destruction of cells or invading microorganisms. Complement is an essential part of the body's defenses, but it must be carefully regulated since uncontrolled activation can lead to massive cell

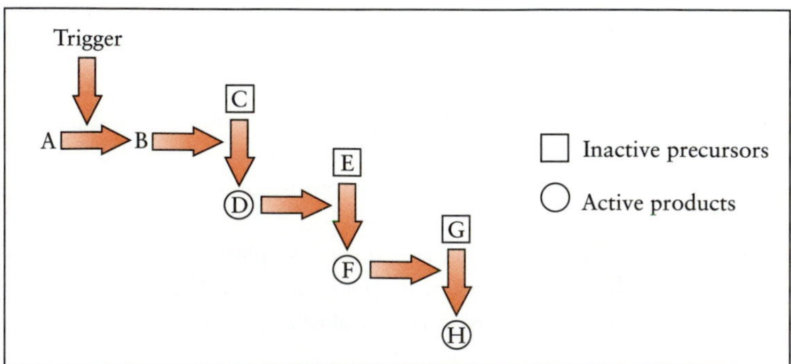

Figure 16–1 Cascade reactions: The process is started by a trigger that initiates the conversion of an inactive proenzyme A to an active enzyme B. Enzyme B acts to convert proenzyme C to active enzyme D. Enzyme D in turn converts proenzyme E to active enzyme F, and so forth. The net effect is a very rapid acceleration in these reactions.

destruction. The complement system was discovered in 1893 by the Belgian scientist Jules Bordet, who called it "alexine." However, Paul Ehrlich later introduced the rival term "complement."

COMPLEMENT COMPONENTS

The proteins that form the complement system are either labeled numerically with the prefix C—for example, C1, C2, C3—or designated by letters of the alphabet—B, D, P, and so forth. There are at least 19 of these components; they are all serum proteins and together they make up about 10% of the globulin fraction of serum. The molecular weights of the complement components vary between 24 kDa for factor D and 460 kDa for C1q. Serum concentrations in humans vary between 20 μg/ml of C2 and 1300 μg/ml of C3 (Table 16–1). Complement components are synthesized at various sites throughout the body. Most C3, C6, C8, and B are made in the liver, and C2, C3, C4, C5, B, D, P, and I are synthesized by macrophages. As a result, these components are readily available for defense at sites of inflammation where macrophages accumulate. B, C4, and C2 are coded for by genes located in the MHC class III region (Chapter 8).

Table 16–1 Complement Components

Name	MW (kDa)	Serum Concentration (μg/ml)
Classical pathway		
C1q	460	80
C1r	83	50
C1s	83	50
C4	200	600
C2	102	20
C3	185	1300
Alternative pathway		
D	24	1
B	90	210
Terminal components		
C5	204	70
C6	120	65
C7	120	55
C8	160	55
C9	70	60
Control proteins		
C1-INH	105	200
C4-bp	550	250
H	150	480
I	88	35
P	4 × 56	20
Vitronectin	83	500

Classical Complement Pathway

The first component of the classical complement pathway (Fig. 16–2) consists of three separate proteins, C1q, C1r, and C1s, held together by calcium-dependent bonds. In the presence of calcium-chelating agents (such as EDTA), the C1 trimolecular complex falls apart. C1q looks like a six-stranded whip when viewed by electron microscopy (Fig. 16–3). It is composed of 18 polypeptide chains of 25 kDa (six each of A, B, and C). The three chains combine to form a triple helix, and each triple helix constitutes a single

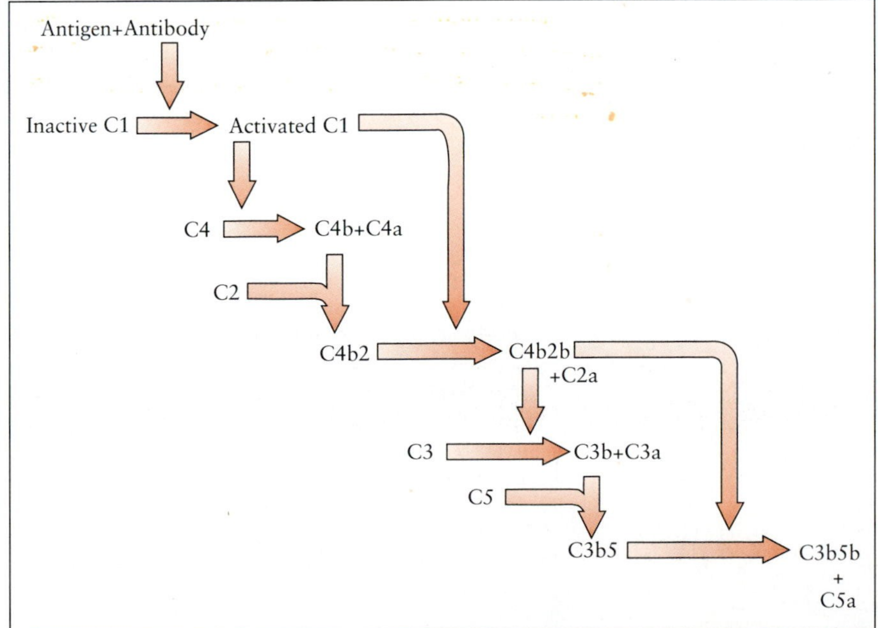

Figure 16–2 The classical complement pathway. The pathway is activated by antigen binding to a single IgM molecule or two closely spaced IgG molecules. C4b2a is the classical C3 convertase.

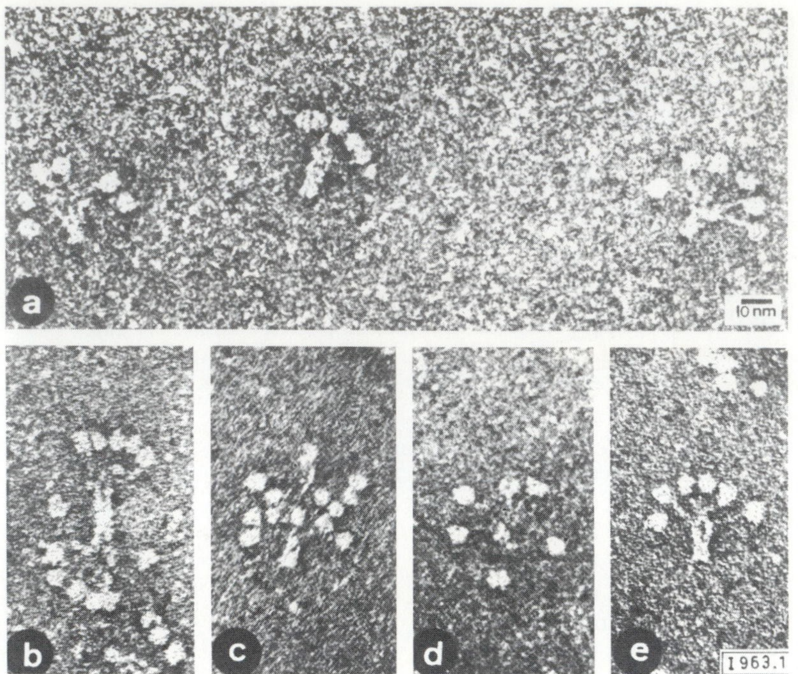

Figure 16–3 Different aspects of the C1q molecule as shown by electron microscopy (original magnification ×500,000). *(From Knobel, H.R., Villiger, W., and Isliker, H. Eur. J. Immunol., 5:78–82, 1975. With permission.)*

strand of the molecule. The tips of each strand can bind to an immunoglobulin attached to antigen but not to free immunoglobulin molecules. The site on an IgG molecule that binds C1q is located on its C_H2 domain (Fig. 16–4). In IgM the complement activating site is on the C_H3 domain. This complement-activating site is present on unbound immunoglobulin molecules but is probably masked by the Fab regions. Only when an immunoglobulin binds to the antigen is the site exposed to C1q.

C1r and C1s form a complex composed of two molecules of each component. The complex assumes a figure-eight structure situated between the C1q strands. Activation of C1 in normal serum is normally

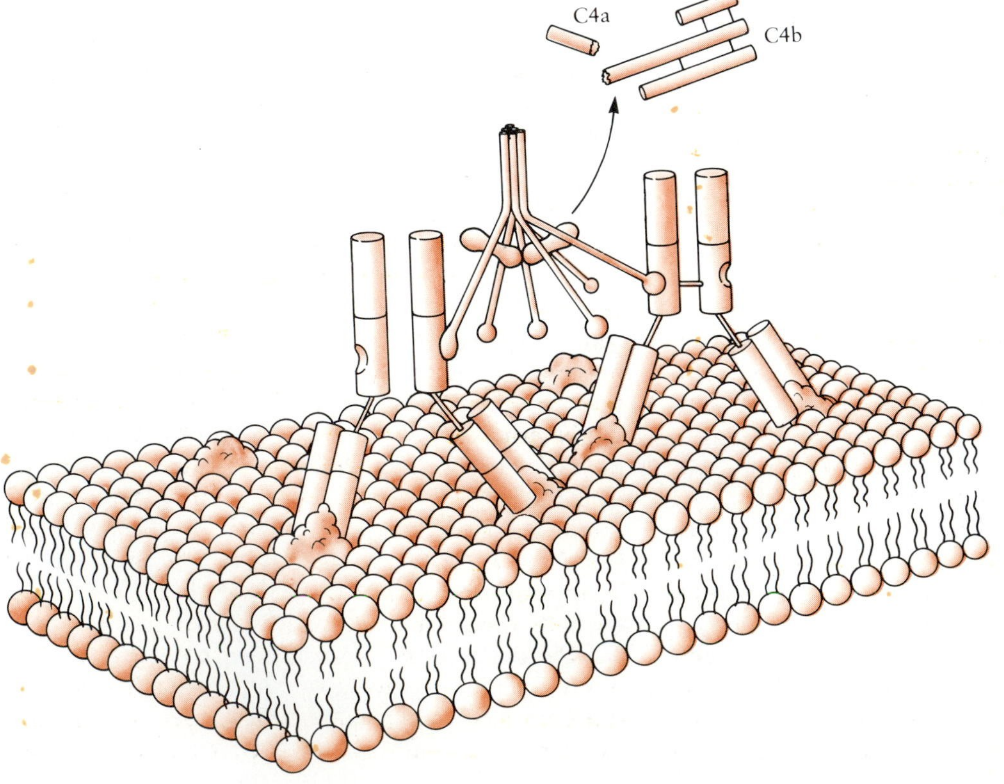

Figure 16–4 The activation of C4 by the C1 complex. The C1 is activated by two membrane bound IgG molecules. C1r and C1s are located between the C1q strands.

METHODOLOGY

How to Detect and Measure Complement

The proteins of the complement system are found in all normal sera. If antibody-coated red blood cells (sheep red blood cells are commonly used) are added to normal serum, the complement system is activated and, as a result, the red blood cells are lysed. The hemolytic activity of serum varies between species; guinea pig complement is highly hemolytic and horse serum is nonhemolytic. As a result, guinea pig serum is commonly used as a source of complement. The level of complement activity in a serum can be measured by using CH_{50} units. One CH_{50} is the amount of complement required to lyse 50% of a standard preparation of antibody-coated sheep red blood cells.

Hemolysis of antibody-coated sheep red blood cells detects the activity of the **classical complement pathway.** In the presence of calcium-chelating agents such as EDTA, the classical pathway is blocked. The alternative complement pathway can then be detected by measuring hemolysis of foreign red blood cells after activating the cascade with endotoxin or aggregated immunoglobulins.

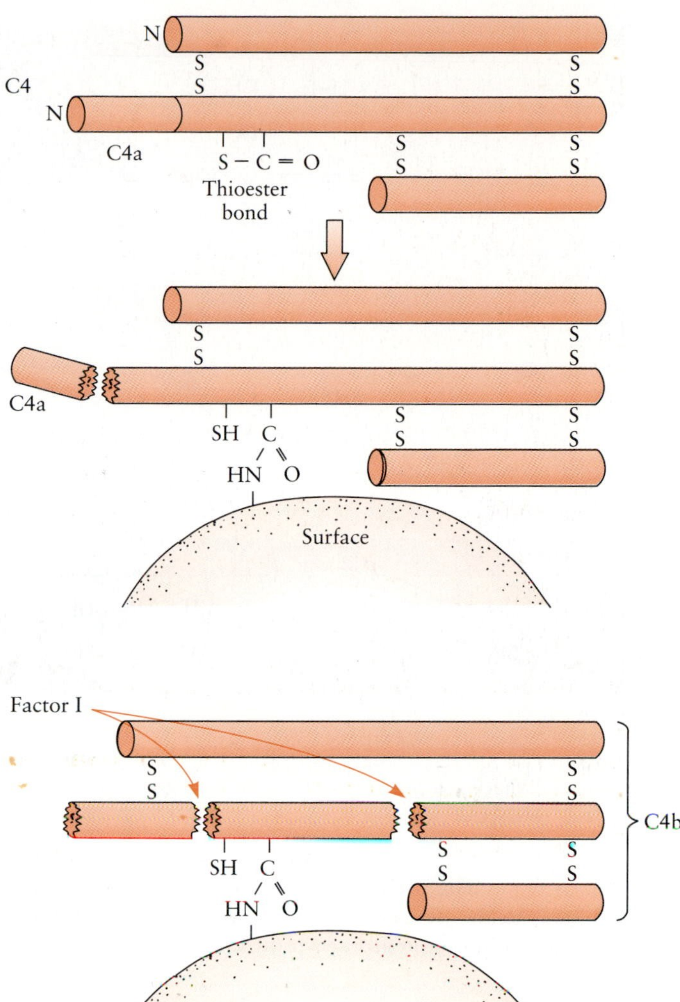

Figure 16–5 The structure of C4: C1s activates C4 by cleaving off C4a leaving C4b. At the same time a thioester bond is broken permitting the C4b to bind to nearby surfaces. Factor I inactivates C4b by cleaving the α chain to form C4bi.

prevented by a protein called C1 inhibitor (C1-INH). C1-INH binds to active C1r and C1s, preventing their activation and removing active C1r and C1s from the complex. This inhibitory effect is overcome by complement activators such as immune complexes. The binding of two or more C1q strands to an immunoglobulin results in a conformational change in the C1q that is transmitted to C1r. As a result, C1r changes to reveal an active proteolytic site, which cleaves a peptide bond in C1s to convert that molecule to an enzymatically active form.

The stimuli that can activate C1 include single antigen-bound molecules of IgM or paired antigen-bound molecules of IgG. The polymeric IgM structure readily provides two closely spaced complement-activating sites. Two IgG molecules must be located very close together to have the same effect. As a result, IgG is much less active than IgM in activating complement. C1 may also be activated directly by the surface proteins of some viruses or by bacteria such as *Escherichia coli* and *Klebsiella pneumoniae.*

Although the complement components are numbered, they do not act in numerical order. Thus, activated C1 acts on C4 and C2. C4 and C2 in turn act on C3. C4, the fourth component of complement, is a three-chain glycoprotein, containing an α chain of 93 kDa, a β chain of 78 kDa, and a γ chain of 33 kDa (Fig. 16–5). C4 is synthesized as a single-chain precursor molecule (pro-C4), which is converted to multichain C4 by the action of plasmin. (C3 and C5 are processed in a similar manner.) Activated C1s acts on the α chain of C4 to cleave off a small fragment from the N-terminus, called C4a (9 kDa), and to leave the major fragment C4b. Removal of C4a causes a change in the shape of the C4b, and as a result, an exposed thioester bond between a cysteine and a glutamine residue is broken (Fig. 16–6). This break in the thioester bond generates a reactive carbonyl group that enables the C4b to bind to any nearby cell membrane.

C2, the second component of complement, is a single-chain glycoprotein of 102 kDa. A C2 molecule can bind to C4b in a magnesium-dependent fashion to form a complex C4bC2. Activated C1s then splits

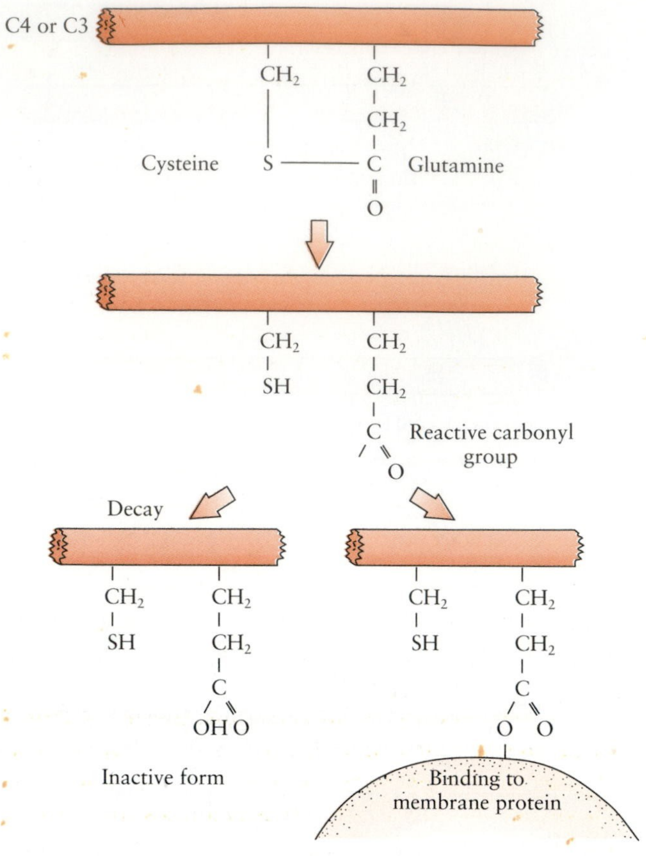

Figure 16–6 The breakage of the thioester bond on C4 or C3 leads to the development of a reactive carbonyl group which can bind to membrane proteins.

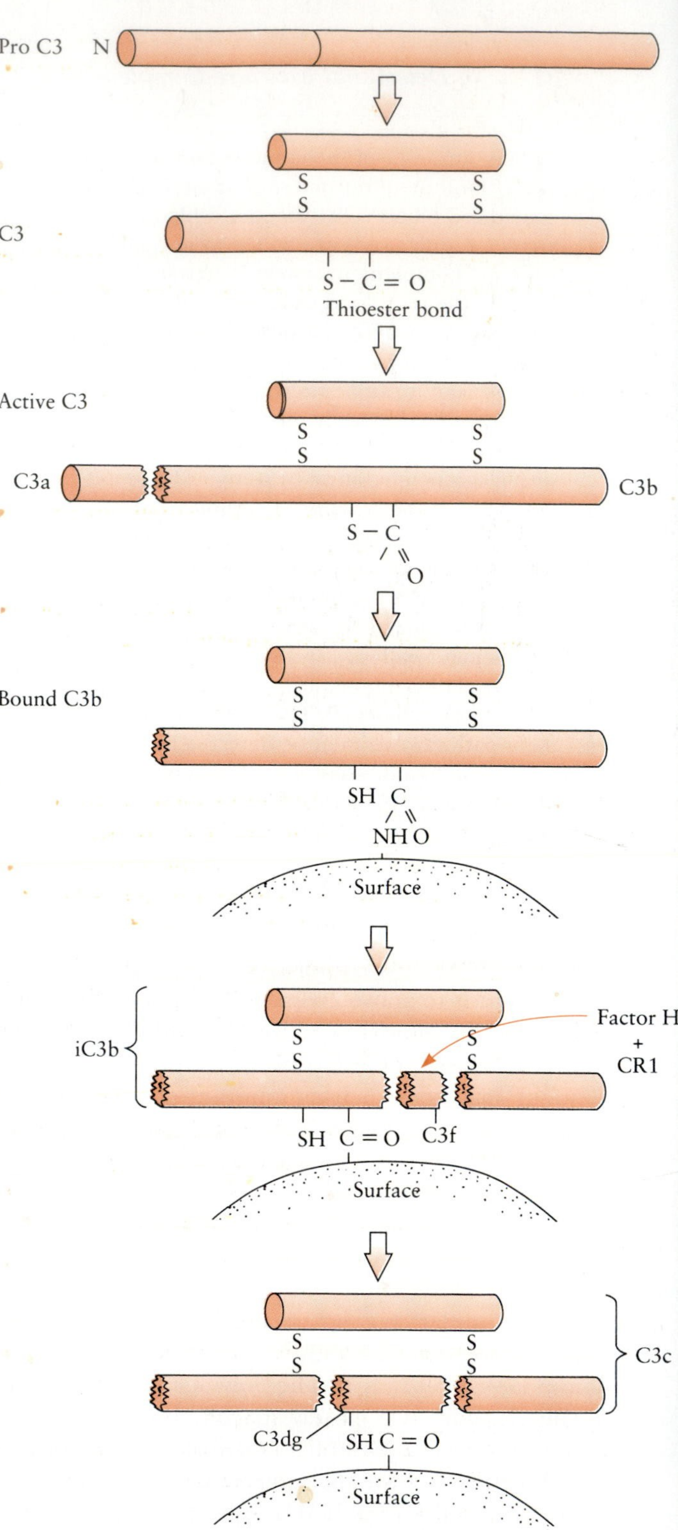

Figure 16–7 The activation and breakdown of C3: C3a has anaphylatoxic activity. C3b is the biologically active form of C3 that initiates the terminal portion of the complement pathway.

the bound C2 into C2a (30 kDa) and C2b (70 kDa) to generate a molecular complex called C4b2b. C1s cannot act on free C2 in solution. The C2 must be bound to the C4 before it can be cleaved. This **substrate modulation** occurs at several points in the complement cascade. It presumably confines the activities of proteases such as C1s to the targets under attack. The C2b component of the C4b2b complex is a protease, and it can activate C3. C4b2b is therefore called classical **C3 convertase.** C4b2b breaks down spontaneously as the C2b decays to C2bi, although the remaining C4b can reform the complex if additional C2b is provided. C4b can also interact with C1 to permit more efficient activation of C2.

Third Component of Complement

By far the most important of the complement components is C3. C3 is a heterodimeric β globulin of 190 kDa containing an α chain of 120 kDa and a β chain of 70 kDa linked by two disulfide bonds (Fig. 16–7). C3 is synthesized by liver cells and macrophages and is the complement component of highest concentration in serum. For the complement cascade to proceed, C3 must be activated by proteolytic cleavage. C4b2b, the classical C3 convertase, acts on the N-terminus of the α chain of C3 to split off a small 6-kDa fragment called C3a. The remaining portion of the molecule is called C3b. As in the activation of C4, C3 has an exposed thioester bond that is broken when C3a is cleaved off. The carbonyl group exposed in this way enables C3b to bind to cell surfaces. The activation of C3b by C4b2b is a major step since each C4b2b complex can activate as many as 200 C3 molecules.

Alternative Complement Pathway

Although C3 can be activated by the classical pathway as described previously, it can also be activated by a second route called the **alternative complement pathway** (Fig. 16–8).

Native C3 breaks down slowly but spontaneously in plasma, and as a result, small amounts of C3b are continuously generated in normal serum. The thioester bond on C3b enables it to bind to nearby cell surfaces. This cell-bound C3b either may be inactivated or may initiate the alternative pathway of complement activation. Which event occurs depends on the affinity of C3b for a protein called factor H, and this in turn depends on the nature of the surface.

When factor H interacts with sialic acid and other neutral or anionic polysaccharides on cell surfaces, its binding to surface-bound C3b is enhanced. Sialic acid increases the affinity of factor H for C3b up to 100-fold. On the other hand, on surfaces deficient in sialic acid, the binding of factor H to C3b is depressed. Most mammalian cell surface glycoproteins are heavily glycosylated and thus do not trigger the alternative complement pathway. In contrast, activating surfaces that inhibit factor H and thus permit activation of C3b include cell walls from many bacteria, polyanions such as bacterial lipopolysaccharides, and nucleic acids, small polysaccharides, viruses, aggregated immunoglobulins, foreign particles such as asbestos, and cells that have had their sialic acid artificially removed by means of neuraminidases. Antibodies may permit activation of the alternative pathway on cell membranes by masking sialic acid residues. In the presence of activating surfaces, the binding of factor H is reduced and the C3b is free to bind factor B.

Some bacteria such as the K1 strain of *E. coli* can evade destruction by the alternative pathway because they are covered by sialic acid and so resemble normal mammalian cells. These organisms cause severe diarrhea in newborn infants since infants cannot use the classical pathway until they can produce appropriate immunoglobulins. Adults and older children can activate complement through the classical pathway and are not susceptible to this strain.

In the absence of an activating surface, factor H binds to C3b. This alters the α chain of C3b in such a way that it becomes susceptible to an enzyme called factor I. As a result, factor H accelerates the decay of C3b. Factor H also regulates the C5 convertase activity of C3b by competing with C5 for binding to C3b. Factor I is a serine esterase that cleaves the α chain of C3b, producing inactive iC3b. Subsequent attack by another protease splits the iC3b into a large fragment called C3c and a small fragment called C3dg. In a normal, stable situation, factors H and I destroy C3b as fast as it is generated, and as a result, the alternative pathway for C3-convertase activation remains inactive.

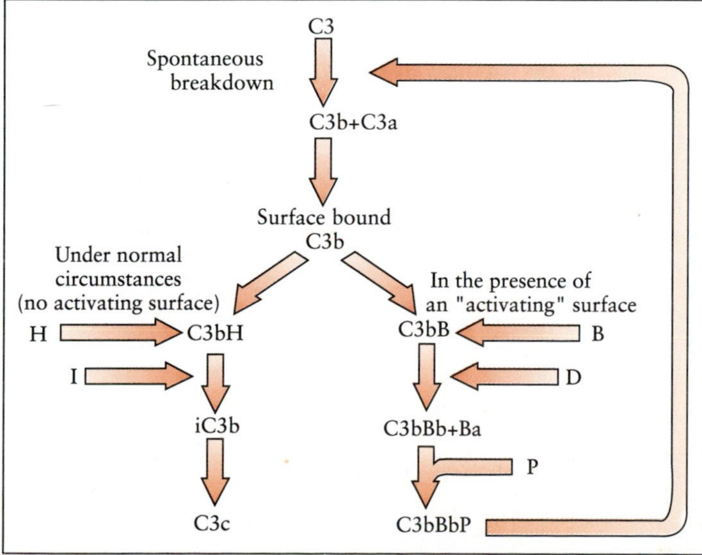

Figure 16–8 The alternate pathway of complement activation. C3b may bind either factor B or factor H. Binding of factor B leads to the progressive activation of the alternate pathway. Binding to factor H leads to inactivation of C3b.

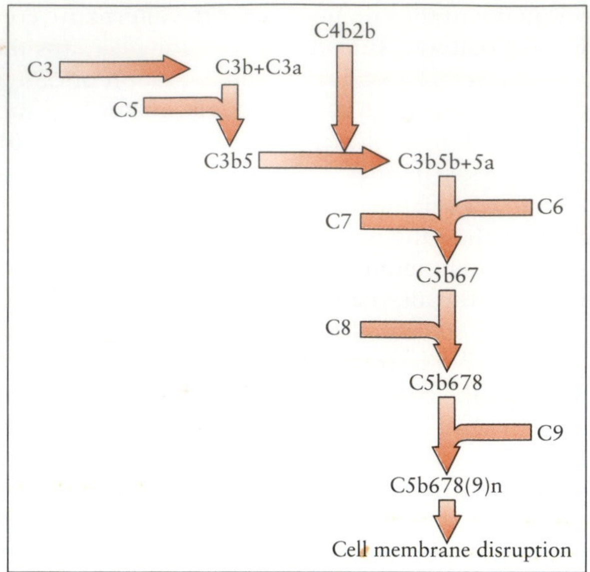

Figure 16–9 The terminal complement pathway. Initiated by C3b, it terminates in the production of polymerized C9 and cell-membrane lysis.

Factor I also cleaves the α chain of C4b, causing loss of biological activity, especially its C3 convertase activity.

In the presence of an activating surface, the ability of factor H to bind C3b is greatly reduced and the C3b can bind a protein called factor B. Factor B is a 90-kDa protein that binds to C3b through a magnesium-dependent bond. This complex, C3bB, has weak C3-convertase activity. However, bound factor B is cleaved by the enzyme known as factor D into a soluble fragment of 30 kDa called Ba and a bound 60-kDa fragment called Bb. The complex C3bBb has potent C3-convertase activity. Factor B is mainly synthesized in the liver, but some is made in macrophages. Its synthesis is enhanced in macrophages by IFN-γ.

Factor D is a 24-kDa serine esterase that cleaves factor B into Ba and Bb. Factor D, however, acts only on B after it is bound to C3b, another example of substrate modulation. As a result, its overall activity is regulated by the supply of C3b. Factor D is synthesized by monocytes.

The alternate C3 convertase, C3bBb, is capable of splitting C3 and so generating more C3b, although it has a half-life of only 5 minutes. If factor P (also called properdin) binds to the complex, the half-life of C3bBbP is extended to 30 minutes. Factor P consists of cyclic polymers of variable sizes (e.g., dimers, trimers, tetramers) of identical 56-kDa subunits. Its avidity for C3bBb increases with the size of the polymer.

Other Methods of Generating C3b

C3b may be generated through the activities of many proteolytic enzymes on C3. Examples of such enzymes include those from activated phagocytic cells or those generated by thrombin on platelets. As a result, C3b is generated in sites of inflammation or in areas of thrombosis.

Terminal Complement Pathway

Both C4b2b, the classical C3 convertase, and C3bBb, the alternate C3 convertase, act on C3 to cleave off C3a and generate surface-bound C3b. C3b acts as the initiator of the terminal complement pathway (Fig. 16–9) by binding to C5 and rendering the C5 susceptible to either C4b2b or C3bBb. (Thus these enzymes are also C5 convertases.) It was assumed that C3b bound to these convertases and altered their specificity so that they attacked C5. It is more likely, however, that substrate modulation occurs. Thus C5 cannot be cleaved by either C4b2b or C3bBb unless it is first bound to C3b (Fig. 16–10).

C5 is a heterodimer of 204 kDa consisting of an α chain (141 kDa) and a β chain (63 kDa) linked by disulfide bonds. Once bound to C3b, C4b2b or C3bBb split the α chain of C5 to remove an 11-kDa peptide

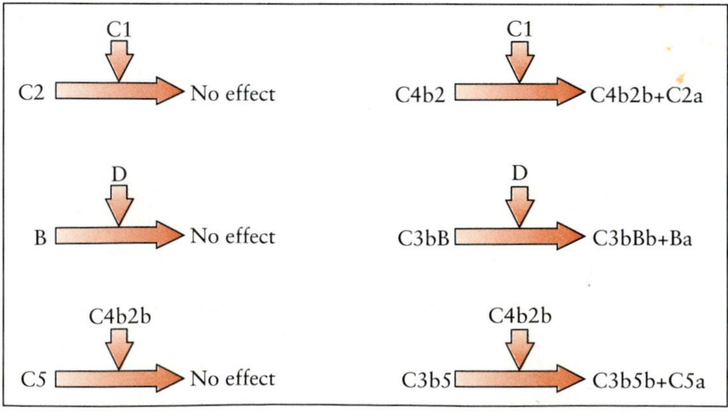

Figure 16–10 The phenomenon of substrate modulation. The enzymes C1, D, and C4b2b cannot act on their substrates unless the substrates are first bound to another molecule.

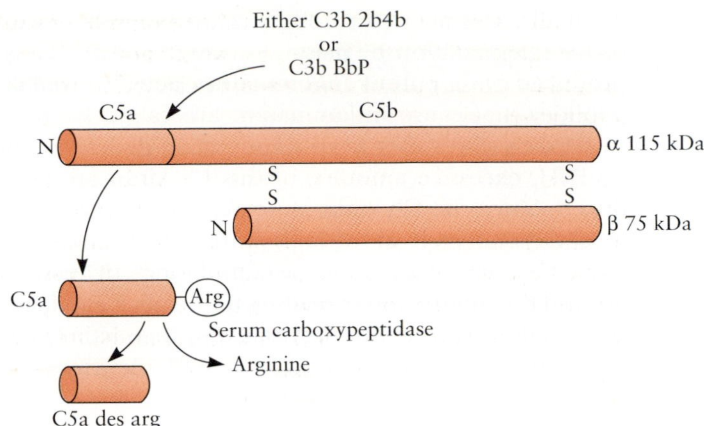

Figure 16–11 The structure, activation, and breakdown of C5. C5 convertase (C3b) cleaves off C5a leaving enzymically active C5b. C5a is cleaved by a carboxypeptidase (anaphylatoxin inactivator) which removes the terminal arginine to leave C5a des arg.

known as C5a. The remainder of the molecule is termed C5b (Fig. 16–11). C5b then binds to C6 to form a complex C5b6. In the unbound state, C5b decays rapidly. C6 is a single polypeptide chain of 120 kDa. It is not cleaved on binding to C5b.

C7, a 120-kDa molecule, can bind to the C5b6 complex either in solution or bound to the cell membranes. Unbound C5b67 rapidly decays to C5i67, a complex that is cytolytically inert but has chemotactic activity. When C5b67 is formed, it generates a metastable membrane-binding site. If it does not bind to a membrane, the complex self-aggregates and is inactivated.

C8 is a three-chain molecule of 163 kDa. It consists of a disulfide-linked α–γ dimer and a noncovalently associated β chain. C8 binds to active cell membrane–bound C5b67 complex, undergoes a conformational change, and penetrates the lipid bilayer of the cell.

The entire C5b678 complex acts as an initiator to induce the polymerization of C9. C9 is a single-chain protein of 71 kDa. C9 molecules come together to form a tubular structure consisting of poly C9 and called the **membrane attack complex** (MAC). MAC has a composition of C5b678(9n), where n lies between 1 and 18. The minimal and maximum molecular weights of the MAC are 660 and 1850 kDa. The MAC forms a large doughnut-shaped structure that can insert itself into a cell membrane (Fig. 16–12). As a result, the central "hole" of the MAC forms a transmembrane channel in the membrane and osmotic lysis of the target cell occurs (Fig. 16–13).

Mannose-Binding Proteins

Recently, a third way of activating the complement system has been identified. It involves production of mannose-binding proteins. When macrophages ingest bacteria or other foreign material, they are stimulated to secrete IL-1, IL-6, and TNF-α. These three proteins, and especially IL-6, act on hepatocytes, stimulating them to secrete several new proteins. One of the proteins secreted by liver cells in response to IL-6 is a mannose-binding protein. Mannose is a major component of bacterial cell walls and is thus found on many bacterial surfaces. The mannose-binding protein thus at-

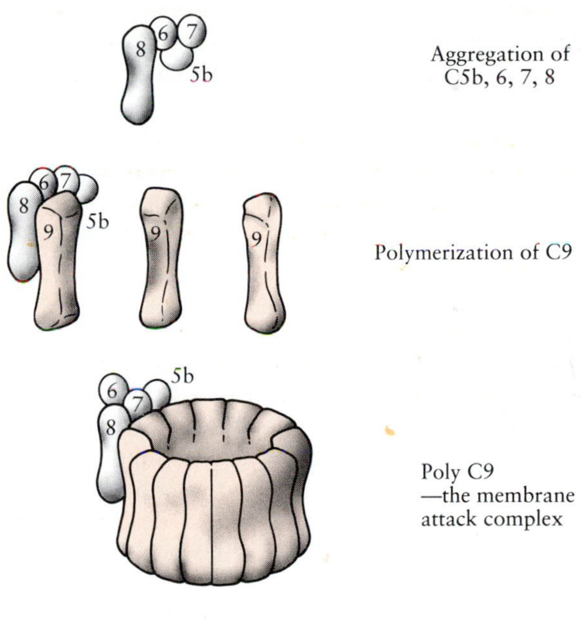

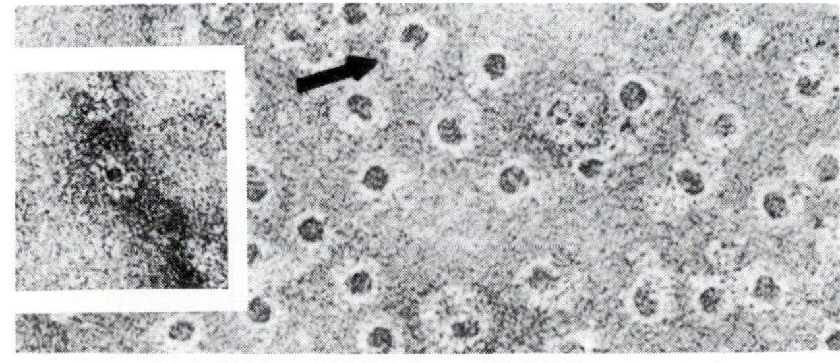

Figure 16–12 The structure of poly C9, the membrane attack complex together with a photomicrograph of human (and rabbit) poly C9 on a red cell surface. *(From Podack, E.R. and Dennert, G. 1983. Nature 307: 442. Used with permission.)*

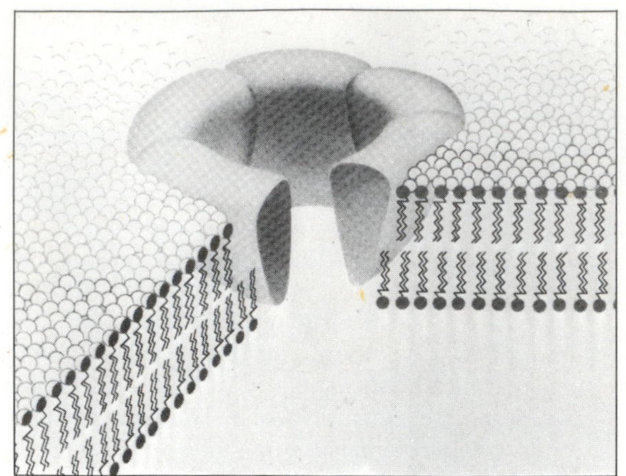

Figure 16–13 Artist's conception of a membrane attack complex created by the polymerization of C9. *(From The Complement System, by M. Mayer. © 1973 by Scientific American, Inc. All rights reserved.)*

taches to any bacteria in the bloodstream. When it binds to mannose residues on a bacterium, the mannose-binding protein acts as a potent opsonin. It also significantly enhances complement deposition by the alternative pathway. Despite its structural similarities to C1q, mannose-binding protein alone cannot activate the classical complement pathway. However, in serum it is associated with an 83-kDa serine protease, and together both molecules can activate the classical pathway.

REGULATION OF THE COMPLEMENT SYSTEM

The consequences of complement activation are so significant and potentially dangerous that the system must be very carefully regulated. This regulation is accomplished through the activities of control proteins.

The most important regulator of the classical pathway is C1 inhibitor. C1 inhibitor (C1-INH) is a 105-kDa glycoprotein that controls the assembly of C4b2b (classical C3 convertase) by binding and blocking the activities of active C1r and C1s (Fig. 16–14). Normally, C1s acts on C4b2 to cleave off C2a. After further degradation by plasmin, a fragment of C2a is produced with potent inflammatory activity. (Small peptides that cause inflammation are called **kinins.**) In persons who suffer from a congenital deficiency of C1-INH, excessive amounts of this C2 kinin are produced. Since the C2 kinin increases blood vessel permeability, affected individuals suffer from attacks of tissue edema, with fluid accumulating in tissues around the mouth and spreading to the neck and face. If this edema makes the larynx swell, the victim may suffocate. The disease caused by C1-INH deficiency is known as hereditary angioedema.

Six glycoproteins have been described that regulate the activities of C3b and C4b. These are factor H and C4-bp in plasma and complement receptor 1 (CR1, CD35), CD55 (decay-accelerating factor, DAF), CD59 (protectin), and CD46 (membrane cofactor protein) on cell membranes. The genes for complement receptor 1, complement receptor 2 (CR2, CD21), C4-bp, and CD55 as well as factor H and CD46 are found on human chromosome 1.

C4-binding protein (C4-bp) is a heat-stable protein that binds C4b, inhibits formation of the C4b2b convertase, and accelerates its decay by displacing C2b from C4b. It also accelerates the decay and dissociation of C2b and destroys its activity. It may act in a similar manner to factor H of the alternate pathway since it acts with factor I to cleave the α chain of C4b into two inactive fragments, C4d and C4c (Fig. 16–15).

CD55, or decay-accelerating factor, is a glycoprotein of 70 kDa attached by a GPI anchor to red blood cells, neutrophils, lymphocytes, monocytes, platelets, and endothelial cells. Its function is believed to protect normal cell membranes from complement attack. It can bind to either the classical (C4b2b) or the alternate (C3bBb) convertases and accelerate their decay when these molecules bind to a cell surface. It can also bind to newly formed C3b or C4b to impede the binding of C2b or Bb. In the disease called paroxysmal nocturnal hemoglobinemia, a deficiency of CD55 results in the accumulation of C3b on red cell surfaces. As a result, these red cells are unusually susceptible to

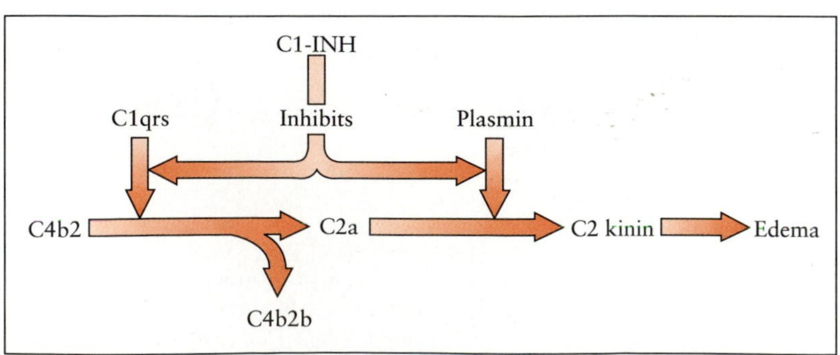

Figure 16–14 The site of action of C1 inhibitor and the pathogenesis of hereditary angioedema.

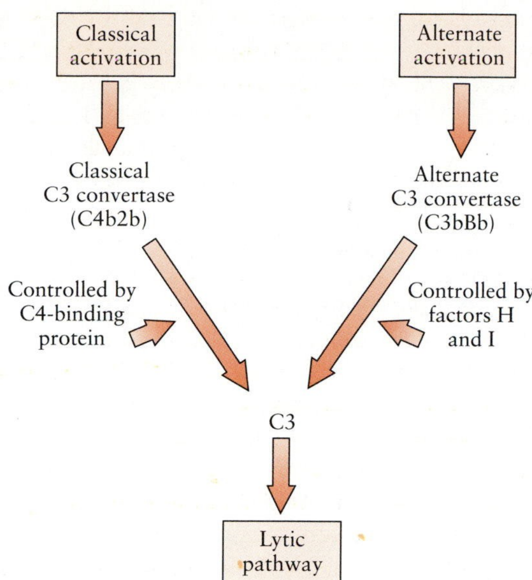

Figure 16–15 The basic control mechanisms of the complement cascade.

complement-mediated lysis and affected individuals suffer from periodic episodes of severe red cell destruction.

CD59, or protectin, is an 18 to 20-kDa GPI-linked glycoprotein that prevents lysis of human red cells, endothelial cells, and leukocytes by homologous complement. It acts at the final stages of MAC formation by competing with C9 for a binding site on C8. It blocks C5b678 insertion into the cell membrane and so inhibits C9 polymerization.

CD46, or membrane cofactor protein, is a heterodimeric glycoprotein with 56 and 66 kDa peptides. It functions as a receptor for iC3b or C3b. It is not found on red cells but is widely distributed on endothelial cells. CD46 controls C3 turnover on cells by acting as a cofactor for the cleavage of C3b or C4b by factor I.

Control of the C56789 complex is mediated by two adhesive glycoproteins called vitronectin and clusterin. Up to three molecules of vitronectin and one of clusterin can bind to C5b67 complex, make it water soluble, and stop it from binding to the cell membrane.

Finally, target cells such as neutrophils can resist lysis by removal of the membrane attack complexes by endocytosis or exocytosis. These cells form membrane vesicles that contain the membrane attack complexes. These vesicles are then either shed into the extracellular fluid or ingested and destroyed.

Complement Receptors

There are four major cellular receptors for fragments of C3 (Table 16–2). These are called CR1 (CD35), CR2 (CD21), CR3 (CD11a/CD18), and CR4 (CD11c/CD18).

CR1 binds C3b, iC3b, and C4b; it is a glycoprotein existing in four allotypic forms ranging from 160 to 250 kDa. CR1 is present on human red cells, neutrophils, eosinophils, monocytes, macrophages, B cells, and some T cells, although it may be shed and readsorbed to cells fairly freely. Red cells carry about 500 to 600 CR1 molecules per cell, and resting neutrophils and monocytes carry about 5000. Mature B cells carry about 20,000 to 40,000 CR1 molecules per cell. Red cell CR1 accounts for 90% of all the CR1 in the blood. CR1 probably removes immune complexes from the circulation. (**Immune complexes** can bind to CR1 on red cells, and the coated red cells are then removed in the liver and spleen.) Patients with the autoimmune disease systemic lupus erythematosus (SLE) may have a CR1 deficiency and are thus unable to remove these immune complexes effectively. CR1 binds C3b strongly and C4b weakly. Particles coated with C3b

Table 16–2 Complement Receptors

Molecule	MW (kDa)	Binds	Cells
CR1 (CD35)	250 220 190 160	C3b, C4b, iC3b	Red cells, macrophages, neutrophils, B cells, some T cells, dendritic cells
CR2 (CD21)	145	C3d, C3dg, iC3b, C3u	B cells
CD11a/CD18 (CR3)	165α 95β	iC3b	Macrophages, neutrophils, NK cells
CD11c/CD18 (CR4)	150α 95β	iC3b	Macrophages, monocytes, neutrophils
CD46 (MCP)	56/66	C3a, C3b	Platelets, macrophages, T cells, B cells, endothelial cells
CD55	70	C4b2b, C3bBb	Platelets, red cells, other leukocytes
C3a/C4aR	83–114	C3a, C4a	Mast cells, granulocytes
C5aR	50	C5a, C5a des arg	Mast cells, granulocytes, macrophages, platelets
C5aR (CD88)	42	C5a	Mast cells, neutrophils, macrophages

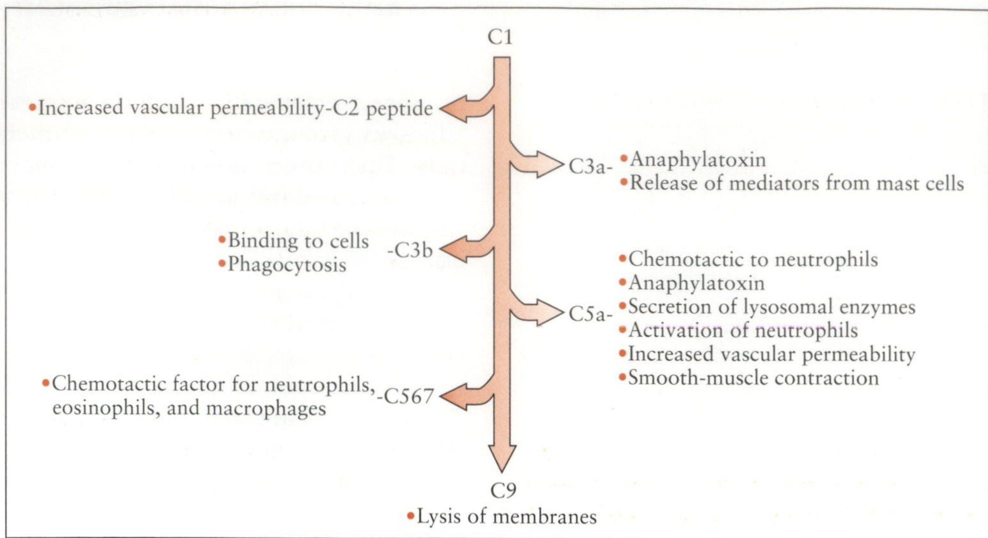

Figure 16–16 The biological consequences of complement activation.

bind to cells through these receptors in a process called immune adherence.

CR2 is a 145-kDa glycoprotein found on most B lymphocytes but not on macrophages or neutrophils. It binds particles bearing the C3d fragment of C3. It is part of the CD19 complex that is believed to act as a coreceptor for antigen. It also serves as the receptor for Epstein-Barr virus and for the IgE receptor CD23 (also called FcεRII). The genes for CR1 and CR2 are so closely linked that some investigators have suggested that they are coded for by the same gene.

CR3 is an integrin that binds iC3b. It is found on macrophages, neutrophils, and NK cells. A genetic deficiency of CR3 (leukocyte adherence deficiency, or LAD) has been described in which affected individuals suffer from severe recurrent infections (Chapter 26). CR3 differs from the other complement receptors since it binds iC3b in a calcium-dependent fashion.

Neutrophils, T cells, NK cells, macrophages, and some platelets carry CR4. It binds iC3b as well as a fragment of C3 called C3dg.

Monocytes, T cells, B cells, granulocytes, platelets, fibroblasts, and endothelial cells have a C1q receptor. This is implicated in the binding of immune complexes, inhibition of IL-1 production, opsonization, and cytotoxicity. Since C1q also binds to extracellular matrix proteins such as collagen or fibronectin, these may have a role in mediating the adherence of cells such as fibroblasts and platelets to the extracellular matrix.

CONSEQUENCES OF COMPLEMENT ACTIVATION

The complement system plays a major role in host defense both through destruction of invading microorganisms and through mediation of inflammation. In addition, the complement system may cause tissue injury as a result of excessive inflammatory responses (Fig. 16–16).

Complement-Mediated Opsonization

Phagocytic cells possess both FcR and CR1. As a result, both antibody and complement-coated particles bind to these cells and can be phagocytosed. If, for some reason, these particles cannot be ingested, then neutrophils may be induced to secrete their lysosomal enzymes. These enzymes, once free in the tissues, cause inflammation and tissue damage and may activate C3 or C5.

Removal of Immune Complexes

One of complement's normal functions is to protect against excessive immune complex formation. By binding to immune complexes and promoting their removal through red cell CR1 and phagocytosis, it acts to clear immune complexes from the blood. Deficiencies of some complement components or some complement receptors allow circulating immune complexes to accumulate in inappropriate sites such as the kidney and cause tissue damage.

Complement-Mediated Immune Regulation

In recent years, it has become difficult to escape the conclusion that the complement system plays a significant role in regulating the immune system. As might be expected, C3 plays a dominant role. B cells but not T cells possess CR1 and can thus bind C3b. C3b can block B-cell differentiation. C3a is immunosuppressive, blocking helper T-cell activity and inhibiting NK cell activity. In contrast, C5a stimulates both T- and B-cell responses by enhancing IL-1 production by mac-

Table 16–3 Complement-Derived Chemotactic Factors

Factor	Target
C5a	Neutrophils, eosinophils, macrophages
C5a des arg	Neutrophils, eosinophils, macrophages
Cb567	Neutrophils, eosinophils
Bb	Neutrophils
C3e	Promotion of leukocytosis

rophages. Brief stimulation of cells with C5a through its receptor (CD88) results in increased expression of CR1, CD11a/CD18, CD11c/CD18, CD16, and CD45.

Complement-Mediated Cytolysis

If the complement sequence proceeds to completion, **membrane-attack complexes** are formed. These complexes can be seen by electron microscopy. They are ring-shaped structures with a central electron-dense area surrounded by a lighter ring of poly C9 (see Fig. 16–13). MACs are incorporated into the lipid bilayer of cells, and as a result, channels are formed through which the cell is lysed. It is not clear whether the channel is through the center of the complex or around its periphery. Nevertheless, the lysis of cells by complement is compatible with the formation of ''holes'' in the cell membranes. C9 belongs to a class of proteins called **perforins** that are implicated in cell lysis mediated not only by complement but also by cytotoxic T cells, NK cells, and eosinophils. T-cell perforin, C6, C7, C9, C8α, and C8β are structurally related (Chapter 18).

For gram-negative bacteria, a single ''hit'' by poly C9 is insufficient for bactericidal activity, since this only disrupts the outer bacterial cell membrane. To complete the lytic process by rupturing the inner cell membrane, the enzyme **lysozyme** is required. Thus bacterial destruction by complement is a ''two-hit'' process (Chapter 25).

Complement-Mediated Chemotaxis

During activation of the complement cascades, several chemotactic peptides are generated, including C5a, C5b67, C3e, and Bb (Table 16–3). There appear to be subtle differences in the nature of their targets. C5b67 is chemotactic for neutrophils and eosinophils only, whereas C5a attracts not only neutrophils and eosinophils but also macrophages and basophils. When C5a attracts neutrophils, it stimulates their metabolism, especially their respiratory burst, aggregates them, and causes them to release thromboplastins and promotes upregulation of CR1, CD11a/CD18, and CD11c/CD18. C3e, a fragment derived from the α chain of C3c, causes a leukocytosis on intravenous administration since it mobilizes leukocytes from bone marrow.

Complement-Mediated Inflammation

C3a and C5a function as **anaphylatoxins;** that is, they cause smooth muscle such as that found in the bronchi or intestinal wall to contract (Fig. 16–17). This reaction is independent of histamine but may be due to **leukotrienes** since it is inhibited by lipoxygenase inhibitors. C3a and C5a also provoke degranulation of mast cells at low concentrations (10^{-12} and 10^{-15} M)

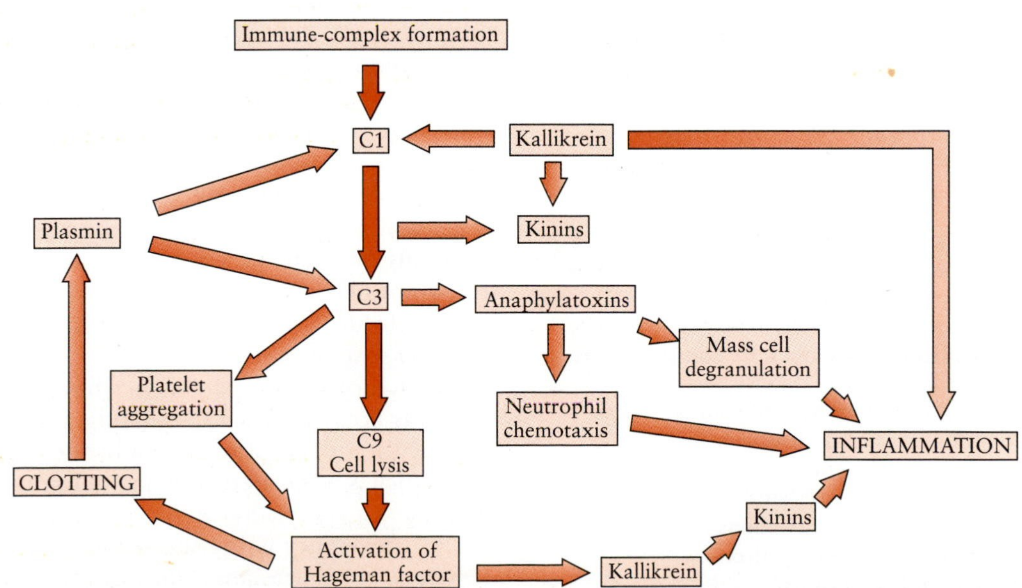

Figure 16–17 Some of the more important interactions between the clotting, complement, and inflammatory systems.

and act on platelets to promote their release of histamine and serotonin. They increase vascular permeability and cause lysosomal enzyme release from neutrophils and thromboxane release from macrophages. The name "anaphylatoxin" was introduced to describe the toxic activities of immune complexes containing C3a and C5a. If these peptides are treated with carboxypeptidases or a naturally occurring serum inhibitor (anaphylatoxin inactivator) to remove the terminal arginine, their toxic activities are inhibited, although they retain the ability to block the activities of the intact peptides, and treated C5a (C5a des arg) retains its chemotactic activity.

COMPLEMENT GENETICS

C4, C2, and factor B are coded for by genes located within the MHC class III region. The C3 locus is found close to the MHC in mice, but in humans it is on a different chromosome. Genes for C4-bp, CD55, CR1, CR2, CD46, and factor H in humans are found on the long arm of chromosome 1 (band q32), and all consist of a single, repeated 60-amino-acid domain. This is called the RCA (regulation of complement activation) gene cluster. The cluster thus codes for all the C3 receptors.

Table 16–4 Disorders Associated with Complement Deficiencies

Deficient Component	Clinical Problem
C1q	Systemic lupus erythematosus (SLE), bacterial infections, hypogammaglobulinemia, glomerulonephritis
C1r	Renal disease, SLE, rheumatoid disease, recurrent infections
C1s	SLE
C4	SLE
C2	Rheumatoid arthritis, SLE, glomerulonephritis, recurrent infections
C3	Severe repeated bacterial infections, glomerulonephritis
C5	SLE, recurrent infections
C6	Recurrent neisserial infections, (gonorrhea or meningococcal disease)
C7	Recurrent neisserial infections
C8	SLE, recurrent neisserial infections
C9	Neisserial infections
C1-INH	Hereditary angioedema
I	Severe repeated infections
H	Hemolytic-uremic syndrome
P	Neisserial infections

In humans, C4 is located in two tandem loci, so that most individuals have two forms of C4—C4A and C4B. (These are not alleles; they are isotypes.) Six amino acid residues account for the difference between C4A and C4B. Although total C4 deficiency is uncommon, deficiency of either C4A or C4B is not. The two forms are functionally different since they tend to bind to different groups on their substrate. C4A binds to amino groups and C4B to hydroxyl groups. When C4A binds to cells, it gives rise to the Rogers blood group. When C4B binds to red blood cells, it gives rise to the Chido blood group. Anti-Rogers (Rg) and anti-Chido (Ch) are IgG antibodies found in the serum of patients who have received transfusions and lack either C4A or C4B. Anti-Rg detects epitopes on the C4A γ chain, and anti-Ch detects an epitope on the C4B γ chain. In mice, a nonhemolytic variant of the C4 molecule is called Slp (sex-linked protein). Slp is homologous with C4 but is hemolytically inactive since only C4 is cleaved and activated by C1s. The synthesis of Slp is regulated by testosterone in some mice. The gene for factor B, known as Bf, is located close to the C2 gene within the MHC class III region but is under independent control.

COMPLEMENT DEFICIENCIES

In general, congenital deficiencies of complement components result in an increased susceptibility to infections and immune complex disease (Table 16–4). The most severe conditions occur in individuals deficient in C3. They suffer from severe recurrent bacterial infections, especially from staphylococci or streptococci, in a manner similar to individuals lacking antibodies (Chapter 26). In contrast to the severe effects of a C3 deficiency, congenital deficiencies of other complement components are not necessarily lethal. Thus individuals with C6 or C7 deficiencies have been described who are quite healthy. This lack of discernible effect suggests that the terminal portion of the complement pathway leading to lysis may not be biologically essential. Notwithstanding this, other individuals deficient in C5 or C6 tend to suffer from recurrent neisserial meningitis, as do properdin-deficient individuals. In C5-deficient mice, phagocytosis of *Neisseria gonorrhoeae* is delayed.

Two forms of C1q deficiency have been reported. One has nonfunctional C1q (antigenic activity only), and in the other form the molecule is absent. This deficiency is associated with the autoimmune disease SLE.

C2 deficiency is the most common complement deficiency in caucasians. Many of the C2-deficient individuals are healthy. Almost half of C2-deficient individuals develop SLE-like diseases. C9 deficiency is especially common in Japanese, where it predisposes individuals to meningococceal meningitis.

The most common predisposing defect for SLE is a null allele at the C4A locus. The C4A null allele occurs in about 35% of all lupus patients compared with 10% to 15% of the normal population. About 90% of patients with null alleles at both C4A and C4B loci develop SLE. In the absence of these complement components, C3 does not bind to immune complexes, and as a result, the complexes do not attach to CR1. When red cells bind immune complexes using CR1, these are subsequently destroyed in the spleen and liver and it is likely that this is a major function of CR1. Patients with SLE may have lower numbers of CR1 on their red cells and thus show reduced clearance of immune complexes.

Deficiencies of the complement control proteins have also been described. The development of hereditary angioedema as a result of a deficiency of the C1 inactivator has been described in an earlier section (page 242). A congenital deficiency of factor H or factor I leads to a secondary C3 deficiency as a result of uncontrolled activation of the alternate pathway, and patients suffer from red cell destruction and kidney failure (hemolytic-uremic syndrome).

COMPLEMENT FIXATION TEST

The activation of the complement system by immune complexes results in the generation of the MAC, which can disrupt cell membranes. If the immune complexes are generated on red blood cell surfaces, then the red blood cell membranes are lysed and hemolysis occurs. It is possible to use this reaction to measure serum antibody levels. The most important test of this type is the hemolytic complement fixation test (CFT) (Fig. 16–18). The CFT is one of the most widely applicable of all immunological techniques. Once the required reagents are prepared and standardized, the CFT may be used to detect many immune interactions. The test is easily read and does not depend on the availability of purified suspensions of antigens. It is therefore commonly used in the diagnosis of viral diseases. The most important disadvantage of the CFT is its complexity, particularly with regard to the standardization and preparation of the reagents required.

The hemolytic CFT is performed in two parts. First, antigen and the serum under test (deprived of its complement by heating at 56° C) are incubated in the presence of normal guinea pig serum, which provides a source of complement. (Guinea pig serum is

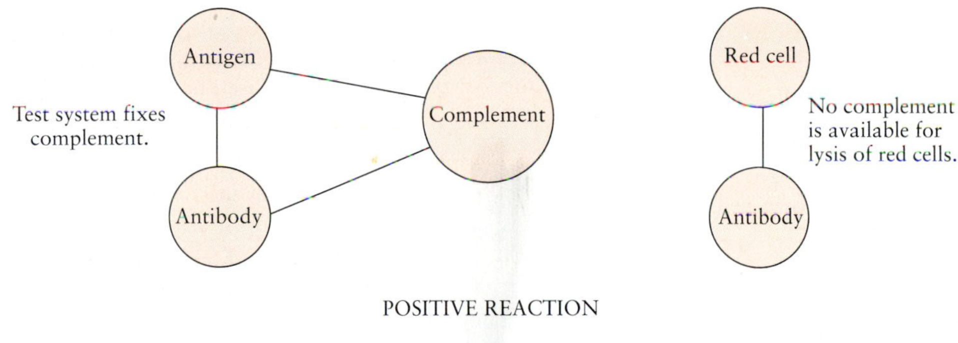

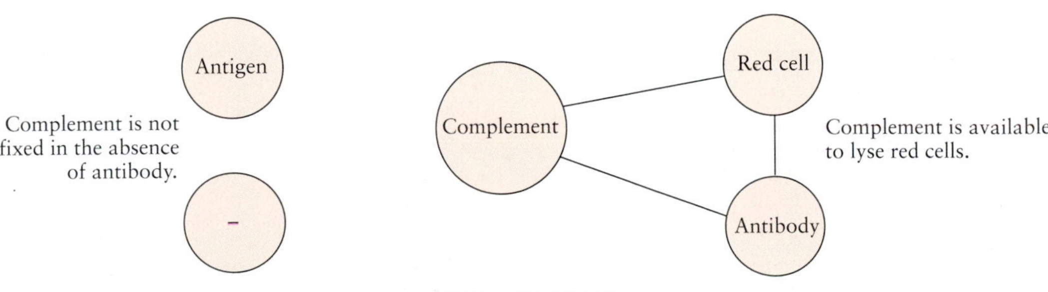

Figure 16–18 The principle of the complement fixation test. Complement, if fixed (activated) by antigen and antibody, is unavailable to lyse the indicator system. In the absence of antibody, the complement will remain unfixed and available for lysis of the indicator system. *(After Roitt, I., Essential Immunology. Blackwell Scientific Publications, Oxford, 1971.)*

most commonly used because its complement has a high hemolytic activity; that is, it lyses red blood cells well.) After the antigen–antibody–complement mixture is allowed to react for a short period, the amount of free complement remaining is measured by adding an "indicator system" consisting of antibody-coated sheep red blood cells. Lysis of these red blood cells, seen as the development of a transparent red solution, is a negative result, since it indicates that complement was not activated or "fixed" and antibody was therefore absent from the serum under test. Absence of lysis (seen as a cloudy red blood cell suspension) is a positive result. It is usual to titrate the serum being tested so that if antibodies are present in that serum, then the reaction in each tube will depend on the dilution of serum (Chapter 17). The titer may be considered to be the reciprocal of the highest dilution of serum in which no more than 50% of the red blood cells are lysed.

Before a hemolytic CFT is performed, all reagents to be used—antigen, complement, sheep red blood cells, and antibody against these red blood cells (**hemolysin**)—must be carefully standardized. For example, addition of the correct amount of complement is critical, since too little complement results in incomplete lysis, whereas excessive complement is not completely activated by immune complexes and therefore leads to false-negative results. Excessive antigen interferes with complement fixation, and insufficient antigen may fail to fix complement in demonstrable amounts. The hemolysin should be heated at 56° C for 30 minutes before being used in order to destroy its content of complement.

Cytotoxicity Tests

Complement can cause membrane damage not only to red blood cells but also to nucleated cells and protozoans. Antibodies against cell surface antigens may therefore be measured by mixing a cell suspension with complement and the serum under test. If antibodies are present in the serum, they bind to the cells, complement is activated, and the cells are lysed. The presence of dead cells can be measured by adding a dye such as trypan blue or eosin Y to the cell suspension. Living cells do not take up these dyes, and dead ones stain intensely. This form of test is employed in the identification of MHC class I molecules (Chapter 5) and in the diagnosis of toxoplasmosis.

KEY WORDS

Activating surfaces p. 239
Alternative complement pathway p. 239
Chemotaxis p. 245
Chido antigen p. 246
Classical complement pathway p. 235
Complement p. 234
Complement components p. 235
Complement deficiencies p. 246
Complement fixation test p. 247
Complement receptors p. 243
Complement regulatory proteins p. 242
C1 inhibitor p. 237
Convertase p. 238
Cytolysis p. 245
Cytotoxicity tests p. 248
Hereditary angioedema p. 242
Immune complex p. 244
Inflammation p. 245
Mannose-binding proteins p. 241
Membrane attack complex p. 241
MHC class III genes p. 246
Opsonization p. 244
Perforins p. 245
RCA gene cluster p. 246
Rogers antigen p. 246
Substrate modulation p. 238
Terminal complement pathway p. 240

QUESTIONS

1. Which complement component produces an anaphylatoxin?
   a. C1
   b. C2
   c. C3
   d. properdin
   e. immunoconglutinin
2. Complement used for serological tests is usually derived from
   a. horses
   b. guinea pigs
   c. sheep
   d. cattle
   e. rabbits
3. Which complement component has chemotactic properties?
   a. C1a
   b. C2a
   c. C4a
   d. C5a
   e. C6a
4. What is the major ligand for CR1?
   a. C1
   b. C2
   c. C3b
   d. C5
   e. C7

5. Activation of the classical complement cascade requires interaction of antigen with
   a. at least two IgG molecules
   b. five IgG molecules
   c. an IgA molecule
   d. a single IgG molecule
   e. at least two IgA molecules
6. Which autoimmune disease appears to be provoked by a deficiency of C4?
   a. thyroiditis
   b. systemic lupus erythematosus
   c. rheumatoid arthritis
   d. myasthenia gravis
   e. multiple sclerosis
7. Which complement component has the highest concentration in serum?
   a. C1q
   b. C1r
   c. C3
   d. factor B
   e. C9
8. The alternative complement pathway is inhibited by
   a. factor H
   b. factor B
   c. factor D
   d. factor P
   e. factor F
9. C9 is a member of which family of proteins?
   a. immunoglobulins
   b. acute-phase proteins
   c. cytokines
   d. addressins
   e. perforins
10. Which of the following is coded for by MHC class III genes?
   a. C3
   b. properdin
   c. C9
   d. factor B
   e. C5a
11. Horse complement is not hemolytic. If healthy horses do not need to kill foreign cells with complement, is it important in immunity?
12. Compare the lytic activities of poly C9 and the perforins released by cytotoxic T cells.
13. Outline the alternative complement pathway. Predict what would happen to individuals suffering from a congenital deficiency of properdin.
14. What might happen if the body lost its ability to regulate the amount of C3b formed?
15. What role does the complement system play in generating acute inflammation?
16. What are the functions of the cell membrane complement receptors? What might be the result of a deficiency of CR1?
17. Speculate on why the class III genes are located within the major histocompatibility complex.
18. Why do many complement component deficiencies result in the deposition of immune complexes in body organs?

Answers: 1c, 2b, 3d, 4c, 5a, 6b, 7c, 8a, 9e, 10d

SOURCES OF ADDITIONAL INFORMATION

Chan, A.C., et al. Identification and partial characterization of the secreted form of the fourth component of human complement: Evidence that it is different from the major plasma form. Proc. Natl. Acad. Sci. U. S. A., 80:268–272, 1983.

Chenoweth, D.E., and Hugli, T.E. Demonstration of a specific C5a receptor on intact human polymorphonuclear leukocytes. Proc. Natl. Acad. Sci. U. S. A., 75:3943–3947, 1978.

Colten, M.R., Alper, D.A., and Rosen, F.S. Genetics and biosynthesis of complement proteins. N. Engl. J. Med., 304:653–656, 1981.

DiScipio, R. The relationship between polymerization of complememt component C9 and membrane channel formation. J. Immunol. 147:4239–4247, 1991.

Fearon, D.T., and Austin, K.F. Current concepts in immunology. The alternative pathway of complement. A system for host resistance to microbial infection. N. Engl. J. Med., 303:259–263, 1980.

Hourcade, D., Holers, V.M., and Atkinson, J.P. The regulators of complement activation (RCA) gene cluster. Adv. Immunol., 45:381–416, 1989.

Ishii, Y., et al. Structure of the human C2 gene. J. Immunol., 151:170–174, 1993.

Johnson, C.A., et al. Molecular heterogeneity of C2 deficiency. N. Engl. J. Med., 326:871–874, 1992.

Kaufmann, T., et al. Genetic basis of human complement C8β deficiency. J. Immunol., 150:4943–4947, 1993.

Lublin, D.M., and Atkinson, J.P. Decay-accelerating factor: Biochemistry, molecular biology, and functions. Annu. Rev. Immunol., 7:35–58, 1989.

Matsushita, M., and Fujita, T. Activation of the classical complement pathway by mannose-binding protein in association with a novel C1s-like serine protease. J. Exp. Med., 176:1497–1502, 1992.

Molina, H., et al. A molecular and immunological characterization of mouse CR2. Evidence for a single gene model of mouse complement receptors 1 and 2. J. Immunol., 145:2974–2983, 1990.

Morgan, B.P. Complement. Clinical Aspects and Relevance to Disease. Academic Press, London, 1990.

Morgan, B.P., Dankert, J.R., and Esser, A.F. Recovery of human neutrophils from complement attack: Removal of the membrane attack complex by endocytosis and exocytosis. J. Immunol., 138:246–253, 1987.

Müller-Eberhard, H.J. The membrane attack complex of complement. Annu. Rev. Immunol., 4:503–528, 1986.

Osler, A.G. Complement Mechanisms and Functions. Foundations of Immunology Series. Prentice-Hall, Englewood Cliffs, N.J., 1976.

Pangburn, M.K., et al. Activation of the alternative complement pathway: Recognition of surface structures on activators by bound C3b. J. Immunol., 124:977–982, 1980.

Podack, E.R., et al. Membrane attack complex of complement: A structural analysis. J. Exp. Med., 151:301–313, 1980.

Porter, R.R., and Reid, K.B.M. The biochemistry of complement. Nature, 275:699–704, 1978.

Reid, K.B.M., and Day, A.J. Structure-function relationships of the complement components. Immunol. Today, 10:177–180, 1989.

Rittner, C., and Schneider, P.M. Complexity of MHC class III genes and complement polymorphism. Immunol. Today, 10:401–403, 1989.

Roos, M.H., Atkinson, J.P., and Shreffler, D.C. Molecular characterization of the Ss and Slp (C4) proteins of the mouse H-2 complex: Subunit composition, chain size polymorphism and an intracellular (Pro-Ss) precursor. J. Immunol., 121:1106–1115, 1978.

CHAPTER 17

The Measurement of Antigen and Antibody Combination

Rosalyn S. Yalow *Rosalyn S. Yalow was born in New York in 1921. She was awarded the Nobel Prize for the development of the technique of radioimmunoassay in 1977. Dr. Yalow was not an immunologist but an endocrinologist who was studying diabetes mellitus. She required a method of measuring the very low levels of antibodies against insulin found in the serum of diabetics and developed the idea of labeling the insulin with a radioisotope. This technique was eventually refined so that exquisitely small levels of molecules such as hormones could be readily measured. The technique of radioimmunoassay is now used in many fields other than endocrinology.*

CHAPTER OUTLINE

CHAPTER CONCEPTS

1. Serology is the science of measuring antibody or antigen in body fluids. Serologic techniques are widely employed as diagnostic tests for infectious diseases.
2. Primary binding tests such as radioimmunoassays, immunoperoxidase techniques, immunofluorescence techniques, and ELISAs are very sensitive although sometimes complex assays.
3. Secondary binding tests that involve secondary reactions such as agglutination, or precipitation, are usually simple but relatively insensitive assays.

In this chapter you will read about the many techniques used to measure antibodies or antigens. These are used extensively as diagnostic tests as well as research tools. These techniques can be classified into major categories. We first describe the methods that directly measure the binding of antigen and antibody. These include such techniques as radioimmunoassays, ELISA tests, and western blotting. We then discuss some of the methods that measure antigen–antibody binding by their secondary effects. These tests include precipitation and agglutination assays.

Antibody molecules combine reversibly with antigens to form immune complexes. The detection and measurement of these reactions form the basis of **serology,** a subdiscipline of immunology.

Three groups of techniques are used to detect and measure antigen–antibody combination. The most sensitive technique (in terms of the amount of antigen or antibody detectable) is to directly measure or visualize the immune complexes formed in an in vitro system. These tests are known as **primary binding tests.**

An alternative approach, however, is to detect and measure the consequences of antigen–antibody interaction. These consequences include precipitation of soluble antigens; clumping (**agglutination**) of particulate antigens; **neutralization** of bacteria, viruses, or toxins; and activation of the complement system. These tests are known as **secondary binding tests.** They are usually less sensitive than primary binding tests but may be easier to perform.

The third type of test, the **tertiary binding test,** measures the consequences of immune responses in vivo. Measurement of the protective effects of antibody falls into this category. This type of test is much more complex than primary or secondary binding tests, but the results reflect the practical significance of the immune response.

REAGENTS EMPLOYED IN SEROLOGICAL TESTS

Serum. The most common source of antibody is **serum,** obtained by allowing a blood sample to clot and the clot to retract. Serum can be stored frozen and used when desired. If necessary, the serum can be depleted of complement activity by heating to 56° C for 30 minutes.

Complement. **Complement** is a normal constituent of all fresh serum, but the complement in fresh, unheated guinea pig serum is the most efficient in hemolytic tests. Serum used as a source of complement for serological applications should be stored frozen in small volumes and used promptly once thawed. Complement should not be repeatedly frozen and thawed since some components are unstable.

Antiglobulins. Because immunoglobulins are proteins, they can function as antigens when injected into an animal of a different species (Chapter 13). For example, purified human immunoglobulins can be injected into rabbits, which respond by making specific antibodies known as **antiglobulins.** Depending on the purity of the injected immunoglobulin, it is possible to make nonspecific antiglobulins against immunoglobulins of all classes or to make specific antiglobulins directed against single classes, allotypes, or idiotypes (Chapter 13). Antiglobulins are essential reagents in many immunological tests.

Monoclonal Antibodies. Myeloma cells can be fused with normal plasma cells that are actively producing specific antibody. The resulting hybridomas can be selected to combine the most desirable qualities of both parent cells and produce homogeneous, specific antibody when cultured (for details see Chapter 14). These hybridoma-derived **monoclonal antibodies** are pure and specific, can be used as standard chemical reagents, and can be obtained in almost unlimited amounts. Monoclonal antibodies derived from hybridomas are frequently used to replace conventional **antiserum** in immunodiagnostic tests.

PRIMARY BINDING TESTS

Primary binding tests are performed by allowing antigen and antibody to combine and then measuring or visualizing the amount of immune complex formed. It is usual to use radioisotopes, fluorescent dyes, or enzymes as labels to identify one of the reactants.

Radioimmunoassays

Radioimmunoassays for Antibody. One widely employed primary binding test for antibody is called the RAST (radioallergosorbent test). In this technique, antigen-impregnated cellulose disks are immersed in test serum so that specific antibody binds to the antigen. After washing, the disk is immersed in a solution containing radiolabeled antiglobulin. The antiglobulin binds to the disk only if antibodies have first bound to the antigen. The level of antibody activity in the test serum is measured by counting the radioactivity associated with the disk. If antiglobulins specific for a single immunoglobulin class are used, it is possible to determine the activity of specific antibodies of that

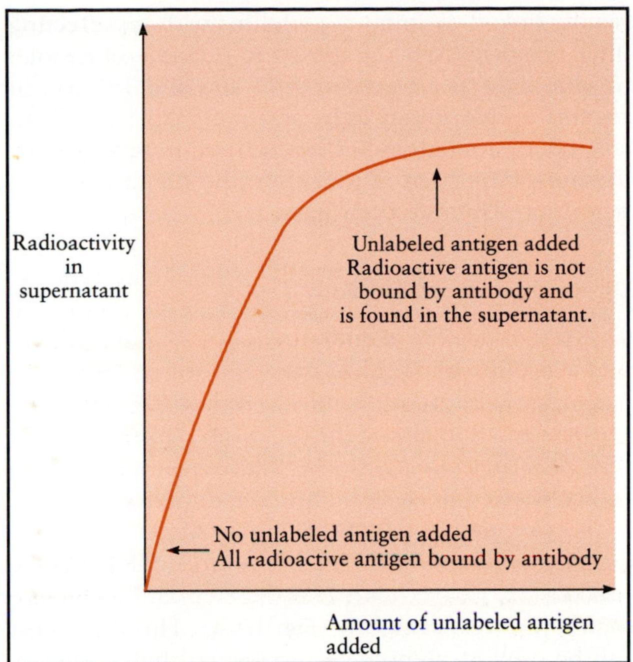

Figure 17–1 The principle of competitive radioimmunoassay. The proportion of isotope-labeled antigen released from immune complexes into the supernatant fluid is related to the amount of unlabeled antigen added to the mixture.

class in a serum. The RAST is used to measure IgE antibody levels in allergic individuals.

Radioimmunoassays for Antigen. Competitive immunoassays can detect very small amounts of antigen. They are based on the principle that unlabeled antigen may displace radiolabeled antigen from immune complexes. The amount of labeled antigen displaced is directly related to the amount of unlabeled antigen added (Fig. 17–1). Assays of this type may differ in the way that the antigen is labeled and measured. The most common method is the competitive radioimmunoassay in which antigen is labeled with an isotope such as tritium (H^3), $carbon^{14}$, or $iodine^{125}$. When radiolabeled antigen is mixed with its specific antibody, the two combine to form immune complexes that may be precipitated out of solution with ammonium sulfate. The radioactivity of the supernatant fluid provides a measure of the amount of unbound labeled antigen (Fig. 17–2). If unlabeled antigen is added to a mixture of labeled antigen and unbound antibody, the unlabeled antigen competes with the labeled antigen for antibody-binding sites. As a result, some labeled antigen is unable to bind antibody, and the amount of radioactivity in the supernatant increases. If a standard curve is first constructed by using known

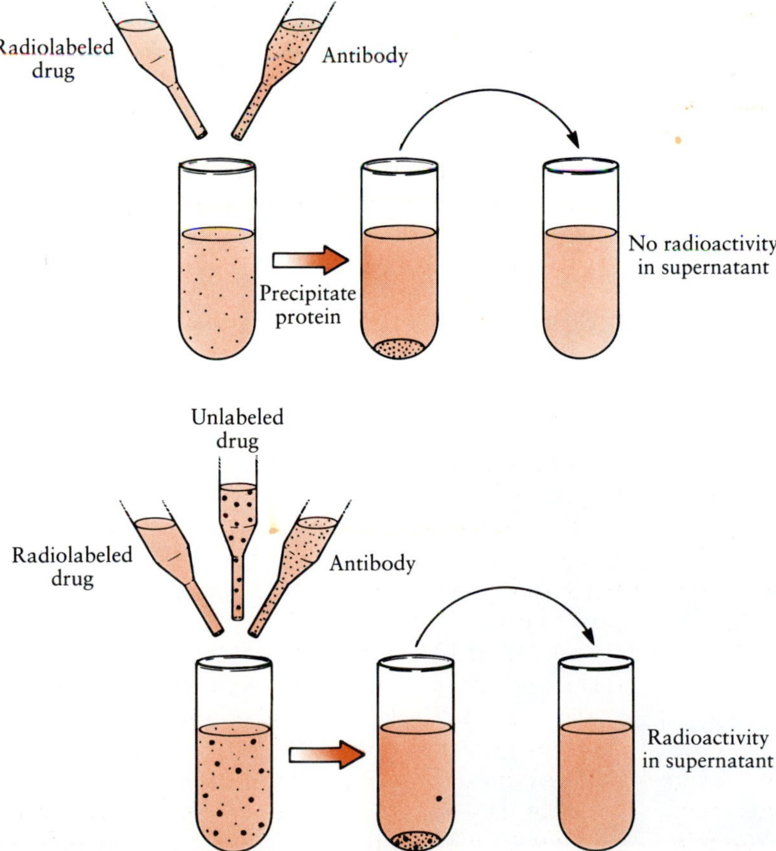

Figure 17–2 A radioimmunoassay. Addition of unlabeled antigen displaces labeled antigen from immune complexes. If the antibodies are precipitated free labeled antigen remains in the supernatant.

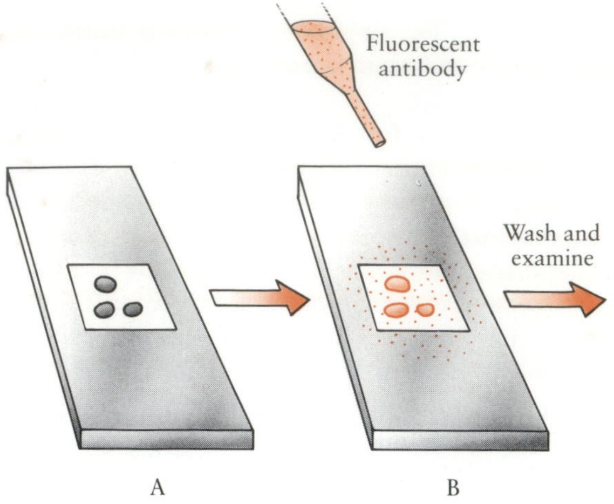

Figure 17–3 The direct fluorescent antibody test is employed to detect antigen by using FITC-labeled antibody directed against that antigen. Antigen in a frozen section, smear, or culture will bind FITC-labeled antibody.

amounts of unlabeled antigen, then the amount of antigen in a test sample may be measured by reference to this standard curve. This type of test is extremely sensitive and commonly used for detecting trace amounts of drugs. For example, morphine can be measured in urine at concentrations of 10^{-8} to 10^{-10} M by means of competitive radioimmunoassay.

Immunofluorescence Assays

Fluorescent dyes are commonly employed as labels in primary binding tests, the most important being fluorescein isothiocyanate (FITC), although other fluorescent dyes such as rhodamine or phycoerythrin can also be employed. FITC is a yellow compound that is readily linked to immunoglobulins without affecting their reactivity. When exposed to invisible ultraviolet or blue light at a wavelength of 290 and 145 μm, it reemits visible green light at about 525 μm. FITC-labeled immunoglobulins can be used in several ways, the most important of which are the direct and indirect fluorescent antibody tests.

Direct Fluorescent Antibody Test

This test is used to identify the presence of antigen. Antibody, directed against a specific antigen such as a bacterium or virus, is chemically linked to FITC. A tissue or smear containing the organism is fixed to a glass slide and incubated with the **fluorescent antibody.** It is then washed to remove unbound antibody (Fig. 17–3). If examined by dark-field illumination under a microscope with an ultraviolet light source, the antigenic particles that have bound the labeled antibodies are seen to fluoresce brightly (Fig. 17–4). This direct test can be used to identify bacteria when their numbers are very low, as when examining the sputum of persons suspected of shedding *Mycobacterium tuberculosis* or when examining smears from lesions for the presence of the clostridial organisms. It may also be employed to detect viruses growing in tissue culture or tissues from infected animals, such as rabies virus in the brains of infected animals or antigens of human immunodeficiency virus on the surface of infected cells.

Indirect Fluorescent Antibody Test

The indirect fluorescent antibody test can be used for the detection of antibodies in serum or the demonstration and identification of antigens in tissues or cell

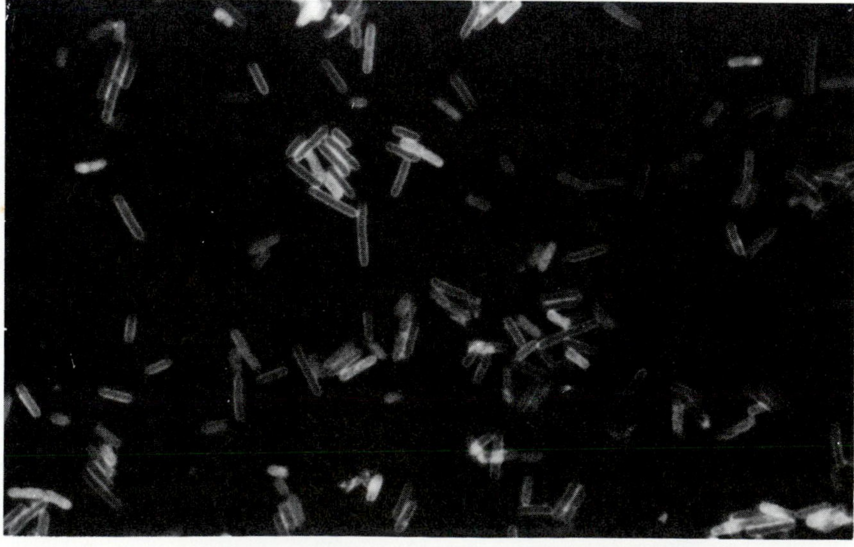

Figure 17–4 Direct immunofluorescence of a smear of *Clostridium chauveoi. (Courtesy of Dr. C.L. Gyles.)*

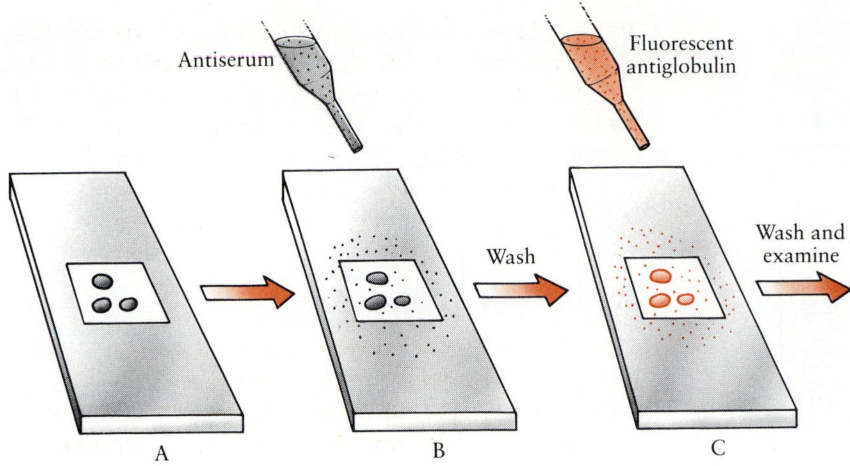

Figure 17–5 The indirect fluorescent antibody test. Antiserum under test is first bound to antigen on a slide. After washing, the bound antibody is labeled with a FITC-antiglobulin. The slide is examined after washing to remove unbound antiglobulin.

cultures (Fig. 17–5). During testing for antibody, antigen is employed in the form of a tissue smear, frozen section, or cell culture on a slide or coverslip. This is incubated in a serum suspected of containing antibodies to that antigen, and the serum is then washed off, leaving only specific antibodies bound to the antigen.

These bound antibodies may be visualized after incubating the smear in FITC-labeled antiglobulin serum. When the unbound antiglobulin is removed by washing and the slide is examined, fluorescence indicates that antibody was present in the test serum. The quantity of antibody in the test serum may be estimated by examining increasing dilutions of serum on different antigen preparations.

The indirect fluorescent antibody test has several advantages over the direct technique. Since each antibody molecule binding to antigen itself binds several labeled antiglobulin molecules, the fluorescence is considerably brighter than in the direct test. Similarly, use of antiglobulin sera specific for each immunoglobulin class allows the class of the specific antibody present in the serum to be determined. All tests of this nature must be accompanied by the use of appropriate controls.

Particle Fluorescence Immunoassay

Immunofluorescence assays can be automated and quantitated by means of particle immunoassays. In these assays antigen is bound to polystyrene particles and mixed with test serum. After incubation, the particles are filtered out and exposed to fluorescent antiglobulin. After the suspension is filtered again and washed to remove unbound reagents, the particle suspension is placed in a fluorometer and the intensity of fluorescence measured. Several variations of this technique can be employed, including inhibition assays in which the ability of unlabeled test serum to inhibit the binding of fluorescent antibodies to the antigen on the particles is measured.

Flow Cytometry

This technique is described in Chapter 10. It involves labeling a cell suspension with a fluorescent monoclonal antibody to a cell surface antigen. The cell suspension is then passed across a laser beam and the characteristics of the labeled and unlabeled cells are determined. It provides a rapid quantitative technique for the analysis of complex cell mixtures.

Immunoenzyme Assays

The most important of the immunoenzyme assays are the **enzyme-linked immunosorbent assays (ELISAs).** As with other primary binding tests, they may be used to detect and measure either antibody or antigen.

ELISA Tests

In the indirect ELISA for antibody detection, polystyrene tubes or wells in polystyrene plates are first incubated with the antigen solution (Fig. 17–6). Protein antigens bind firmly to polystyrene so that after unbound antigen is removed by vigorous washing, the tubes remain coated with antigen. The coated tubes can be stored until required. The serum under test is added to the tubes so that specific antibodies in the serum bind to the antigen on the tube wall. After incubation and washing to remove unbound antibody, antiglobulin chemically linked to an enzyme is added to the tubes. This complex binds to the bound antibody, and after incubation and washing, the bound antiglobulin can be detected and measured by addition of the enzyme substrate. The enzyme and substrate are selected so that a colored product develops in the tube. The intensity of the color that develops is

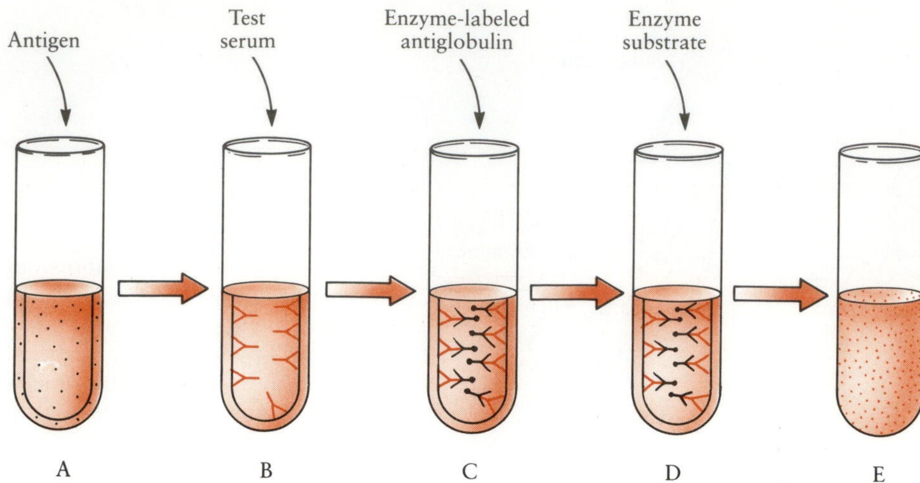

Figure 17–6 A schematic diagram showing the procedures for the indirect ELISA technique.

proportional to the amount of enzyme-linked antiglobulin that is bound, which in turn is proportional to the amount of bound antibody. The color intensity can be estimated visually or, preferably, determined spectrophotometrically.

A modification of this technique, the antibody sandwich ELISA, is a highly sensitive method of detecting antigen. It involves first coating polystyrene tubes with specific antibody (capture antibody). The antigen solution is then added so that the antigen is bound by the capture antibody. This is followed (after washing) by specific antibody, enzyme-labeled antiglobulin, and substrate, as described for the indirect technique. In this test, the intensity of the color reaction is related directly to the amount of bound antigen. Because these tests involve the formation of antibody–antigen–antibody layers, they are called sandwich ELISAs (Fig. 17–7). Sandwich ELISAs are used to detect antigen when the amount in the sample is too dilute to be detectable when directly adsorbed to the surface. Thus this is the test employed to detect circulating virus in blood from patients with AIDS. Because of their simplicity, ELISA tests are widely used for the immunodiagnosis of many bacterial, viral, and parasitic infections.

Immunoperoxidase Techniques

Enzymes conjugated to immunoglobulins or antiglobulins can also be used to identify specific antigens in tissue sections. Horseradish peroxidase is the most widely employed enzyme label. The tests are performed in a manner similar to the immunofluorescence tests. In the direct immunoperoxidase test, the tissue section is treated with the enzyme-labeled antibody. After washing, the tissue is incubated in the appropriate enzyme substrate. Bound antibody is detected by the presence of a brown deposit at the site of antibody binding. In the indirect test, bound antibody is detected by means of labeled antiglobulin. Immunoperoxidase techniques have a significant advantage over the immunofluorescence techniques; the tissue can be examined by conventional light microscopy and stained so that structural relationships are easier to see.

Western Blotting

One solution to the problem of identifying protein antigens in a complex mixture is use of a technique called western blotting (Figs. 17–8 and 17–9), a three-

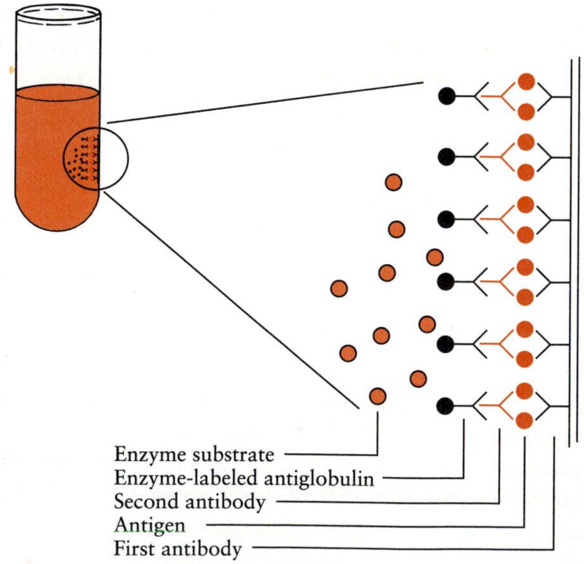

Figure 17–7 The principle of the "sandwich" ELISA. The basic procedure involves sequential addition and removal of the reagents in order from polystyrene tubes.

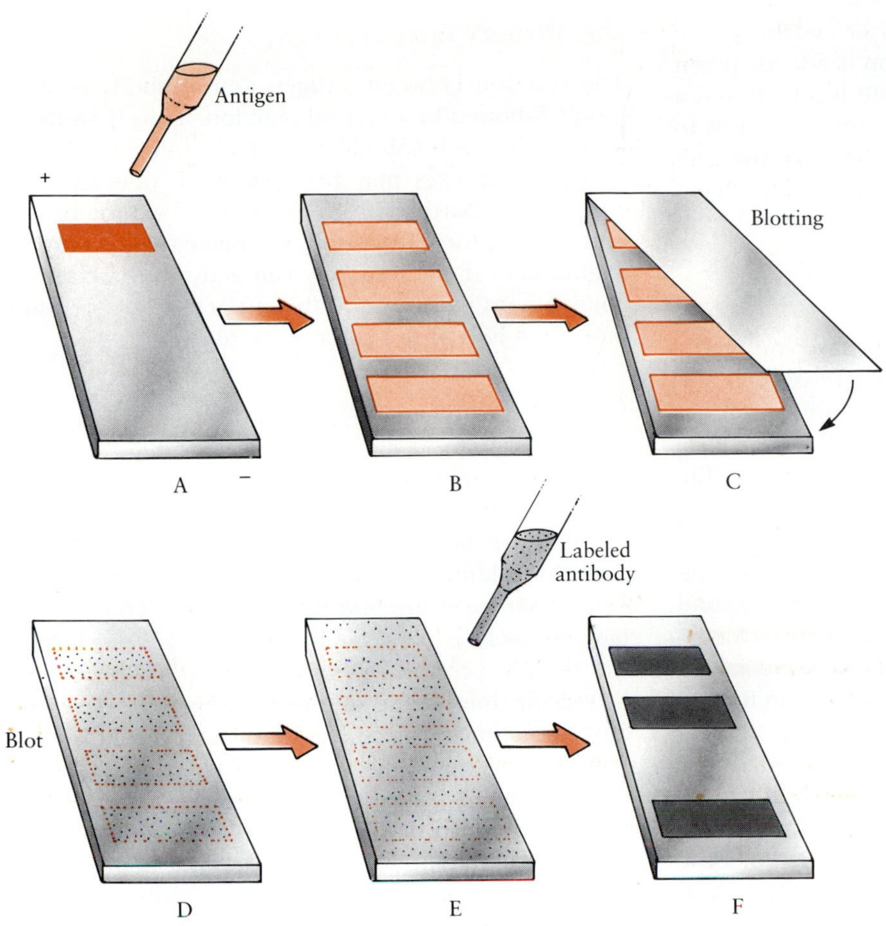

Figure 17–8 The technique of Western blotting. An antigen mixture is electrophoresed, blotted and the blot then developed using a labeled antibody.

stage primary binding test. Stage 1 involves electrophoresis of the protein mixture on gels so that each component is resolved into a single band. Stage 2 involves blotting or transfer of the electrophoresed protein to an immobilizing nitrocellulose membrane. This membrane is placed on top of the gel to be transferred, and the two are sandwiched between sponges saturated with buffer. The sandwich is supported between rigid plastic sheets and placed in a buffer reservoir. An electrical current is passed between the sponges. The protein bands are rapidly transferred from the gel to the nitrocellulose membrane without loss of resolution. Stage 3 involves visualization of transferred antigens by means of an enzyme immunoassay or radioimmunoassay. When an enzyme immunoassay is employed, the membrane is first incu-

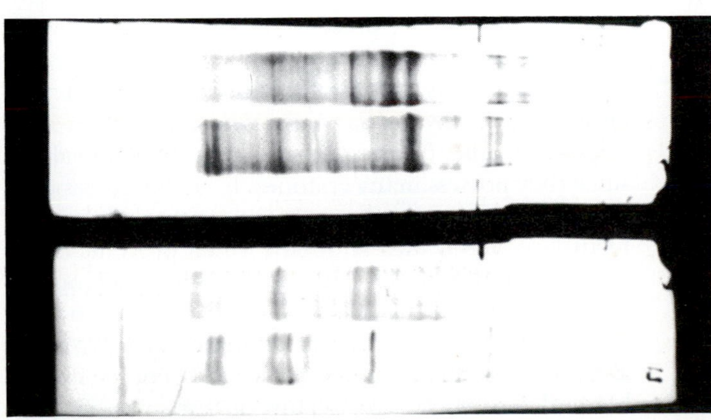

Figure 17–9 Two Western blots of bovine serum directed against antigens from the liver fluke *Fasciola hepatica*. *(Courtesy of Dr. C. Hicks.)*

bated in specific antiserum. After washing, an enzyme-labeled antiglobulin solution is added. When this is removed by washing, substrate is added and a color develops in the bands where antibody binds to antigen. When isotope-labeled antiglobulin is used, an autoradiograph must be made and the labeled band identified by darkening of a photographic emulsion.

Avidin-Biotin Immunoassay

Although radioisotopes and enzymes have commonly been used as labels for primary binding tests, each possesses certain disadvantages. For example, isotopes may have a short half-life, are potentially hazardous, and may require expensive detection devices. Enzymes, although stable and relatively cheap, are large molecules that may inhibit antibody activity or lose enzymatic activity in the process of conjugating with antiglobulin. One alternative to the use of enzymes and radioisotopes as labels is to use the small molecule biotin and its specific binding protein called avidin. Biotin has a molecular weight of 244 Da and can be very easily and gently bound to proteins without affecting their biological activity. Avidin is a small protein found in egg white that binds strongly and specifically to biotin. Avidin can be conjugated with an enzyme such as horseradish peroxidase. This avidin-peroxidase complex rapidly and specifically binds to biotin-labeled proteins and thus forms an effective and sensitive label.

Other Labels Used in Primary Binding Tests

Although about 20 enzymes have been used in enzyme immunoassays, the most popular choices include alkaline phosphatase, horseradish peroxidase, and β-galactosidase. These enzymes are inexpensive and readily assayed. Enzyme assays involving the production of luminescent or fluorescent products, for example, luciferase, may be many times more sensitive than conventional enzyme assays, but they require sophisticated instruments to detect and measure the luminescence produced. Several other labels have been used as alternatives to radioisotopes, enzymes, or fluorescent dyes in primary binding tests. For example, reagents linked to the iron-containing protein ferritin may be used to identify the location of antigens on cell surfaces examined by electron microscopy. The ferritin molecule is a protein of 700 kDa containing 23% iron as ferric hydroxide or phosphate. The iron is concentrated within the molecule and on electron microscopy can be detected as a characteristic electron-dense spot. Therefore, if ferritin is linked to an immunoglobulin, the location of antigen may be readily observed on electron micrographs.

SECONDARY BINDING TESTS

The reaction between antigen and antibody is commonly followed by a second reaction. Thus if antibodies combine with soluble antigens in solution, the resulting complexes may precipitate. If, however, the antigens are particulate (for example, bacteria or red blood cells), then antibodies will make them clump or agglutinate. If the antibody can activate the classical complement pathway and the antigen is on a cell surface, then lysis of the cell may result.

Precipitation

If a suitable amount of a clear solution of soluble antigen is mixed with its antisera and incubated at 37° C, the mixture becomes cloudy within a few minutes, then flocculent, and within an hour or so a white precipitate settles to the bottom of the tube. This process may be analyzed by examining the effect of altering the relative proportions of antigen and antibody. If increasing amounts of soluble antigen are mixed with a constant amount of antibody, the amount of precipitate formed depends on the relative proportions of the reactants. No precipitate is formed at very low antigen concentrations. As the amount of antigen added increases, a precipitate forms and increases in amount until it reaches a maximum. With the addition of even more antigen, the amount of precipitate begins to diminish, until eventually none is observed in tubes containing a large excess of antigen over antibody (Fig. 17–10).

In the first of these reactions where antibody is in excess, only a little antigen is bound to antibody, free antibody may be found in the supernatant, and little precipitate is deposited. In contrast, when maximum precipitation occurs, both antigen and antibody are completely complexed, and neither can be detected in the supernatant. This is known as the equivalence zone, and the ratio of antibody to antigen is said to be in optimal proportions. When antigen is added to excess, little precipitate is formed, although soluble immune complexes are present and free antigen may be found in the supernatant.

These results can be explained by the fact that antibodies are bivalent and can cross-link only two epitopes at a time, but protein antigens are multivalent, since they possess many epitopes. In the mixtures containing excess antibody, each antigen molecule is effectively covered with antibody, which prevents cross-linkage and thus precipitation. When the reactants are in optimal proportions, the ratio of antigen to antibody is such that extensive cross-linking occurs. As the antigen–antibody complex grows, it becomes insoluble and eventually precipitates out of solution (Fig.

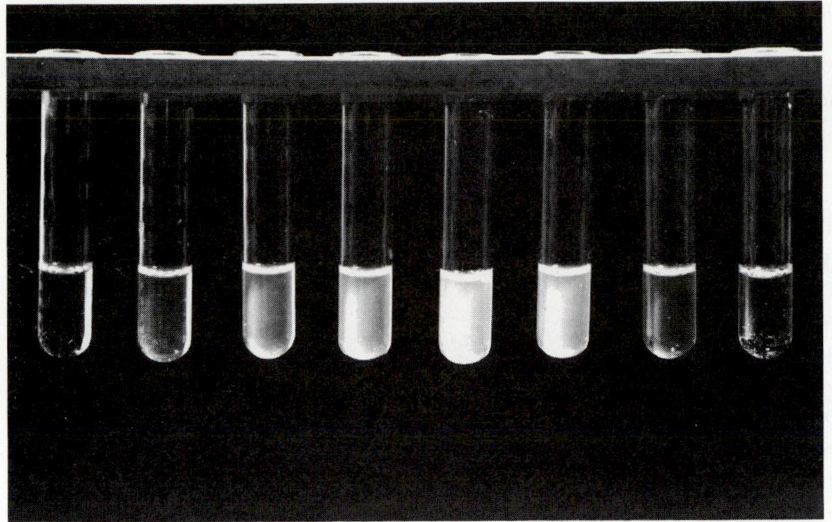

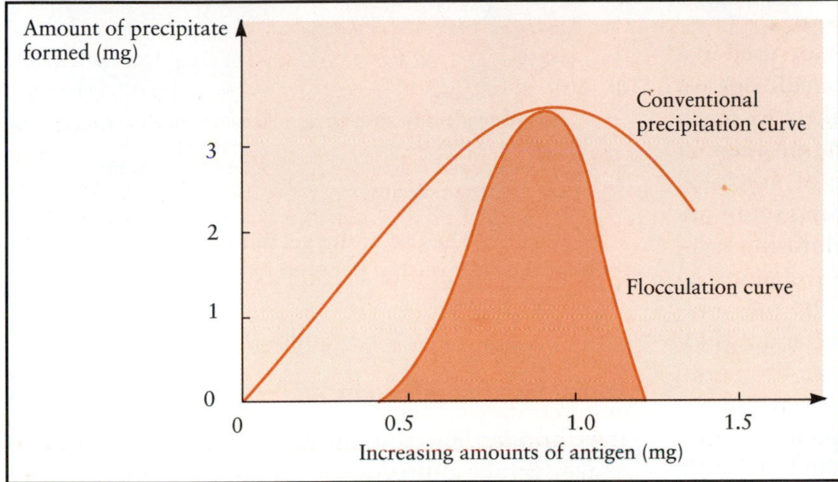

Figure 17–10 The effect of mixing increasing amounts of bovine serum albumin with a constant amount of rabbit antiserum to that antigen. The tube with the greatest amount of precipitate is the one in which the proportion of antigen to antibody is optimal. A quantitative precipitation curve of this test graphically shows the effect of adding increasing amounts of antigen to a constant amount of antibody.

17–11). In mixtures in which antigen is in excess, each antibody molecule is bound to a pair of antigen molecules. Further cross-linkage is impossible in this case, and since these complexes are small and soluble, no precipitation occurs.

The cells of the mononuclear-phagocytic system are most efficient at binding and removing from the bloodstream complexes formed at optimal proportions and in antibody excess. Immune complexes formed in antigen excess are poorly removed by phagocytic cells in vivo but are deposited within vessel walls and in the glomeruli of the kidneys, where they cause an acute inflammatory reaction classified as a type III hypersensitivity reaction (Chapter 30).

Laser Nephelometry

Nephelometry is the measurement of light scattering by particles in suspension. If optically clear solutions of antigen and antibody are mixed, the resulting precipitate will make the mixture appear cloudy. The quantity of immune complexes formed can be accurately measured by shining the beam of a helium-neon

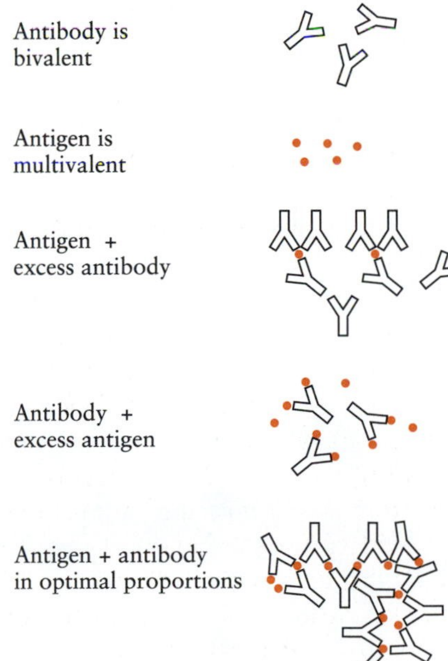

Figure 17–11 The mechanism of immune precipitation: In antibody excess and in antigen excess only small, soluble immune complexes are produced. At optimal proportions, however, large insoluble complexes are generated.

laser through the solution. This technique can be used for a rapid antibody assay.

Immunodiffusion

A simple method of detecting the precipitation reaction is to cut two round wells in a layer of agar in a Petri dish. One well is filled with antigen and the other with antiserum; the reactants diffuse out radially. A circular concentration gradient therefore is established for each reactant, and these eventually overlap. Optimal proportions for the occurrence of precipitation occur in one zone of the superimposed gradients, and an opaque white line of precipitate develops between the wells (Fig. 17–12). This technique is known as the **gel-diffusion** or double-diffusion test.

If several antigen–antibody mixtures are used in a gel-diffusion test (e.g., if antibodies are made against whole serum), each component is unlikely to reach optimal proportions in exactly the same position in the agar. Consequently, a separate line of precipitation appears in the gel for each interacting set of antigens and antibodies present. Immunodiffusion techniques may also be used to determine the relationship between antigens. If two antigen wells and one antibody well are set up as in Figures 17–12 and 17–13, then lines of precipitate will form between each antigen well and the antibody well. If the two antigens are identical, then the two lines will be completely confluent. If the two antigens are totally unrelated, then the lines will not interact but will cross each other. If the two antigens possess epitopes in common, then the lines will merge with spur formation. The antigen that forms the line that continues as a spur possesses epitopes not present in the other.

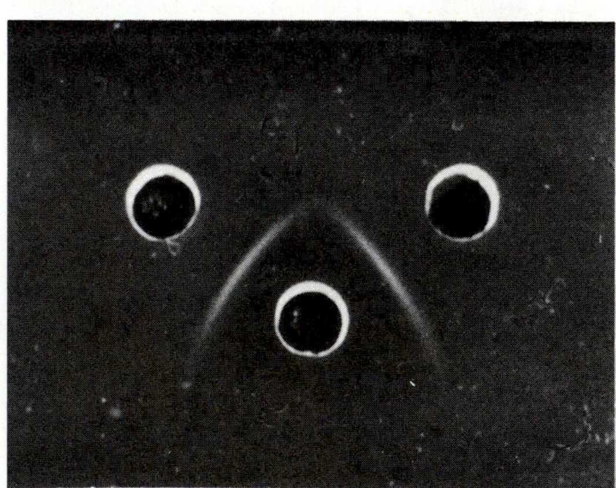

Figure 17–12 Precipitation in agar gel: Antigen and antibody diffusing from their respective wells precipitate in a region where optimal proportions are achieved. In this example, the two top wells each contain identical antigens. The antibody is in the bottom well. The precipitation lines fuse as a result of the identity of the antigen.

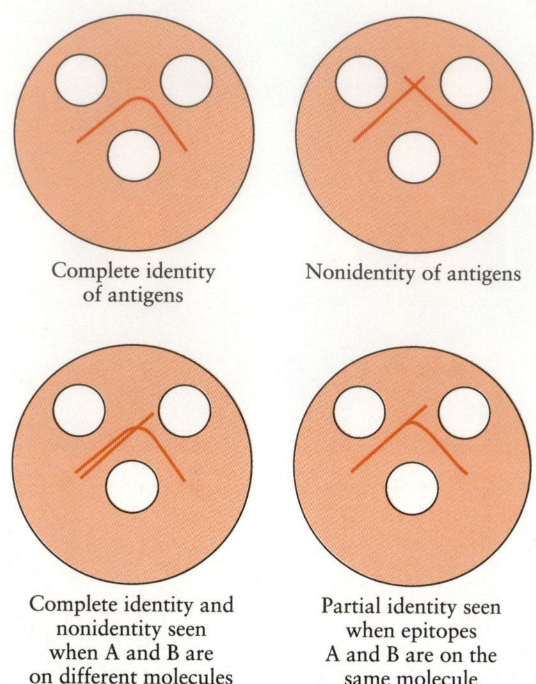

Figure 17–13 The use of the gel-diffusion technique to determine the relationship between two antigens.

Radial Immunodiffusion

If antigen is allowed to diffuse from a well into agar in which specific antiserum is already incorporated, then a ring of precipitate indicating the zone of optimal proportions will form around the well. The area within this ring of precipitate is directly proportional to the amount of antigen placed in the well. A standard curve may therefore be constructed using known amounts of antigen. Unknown solutions of antigen can then be accurately assayed by comparing the diameters from unknowns with the standard curve (Fig. 17–14).

If, instead of being permitted to diffuse into agar-containing antiserum as in the radial immunodiffusion technique, the antigen is driven into the antiserum agar by electrophoresis, then the ring of precipitation around each well becomes deformed into a "rocket." The length of the rocket is proportional to the amount of antigen placed in each well. This technique, known as electroimmunodiffusion, can also be employed to measure antigen (Fig. 17–15).

Immunoelectrophoresis and Related Techniques

Although double-diffusion techniques give a separate precipitation line for each antigen–antibody system in a mixture, it is often difficult to resolve all the components in a complex mixture in this way. One technique that can be used to improve the resolution of the system is to separate the antigen mixture by elec-

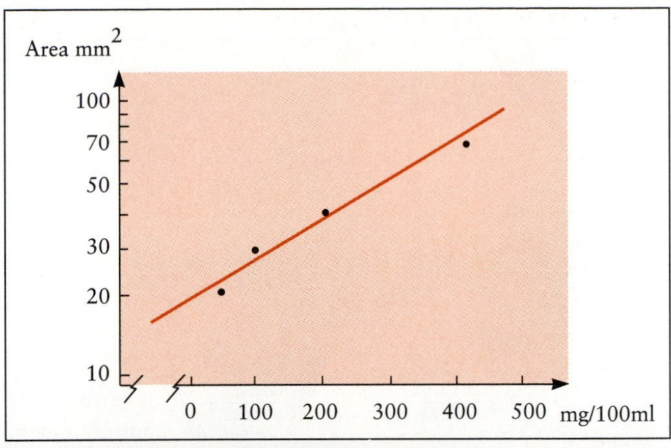

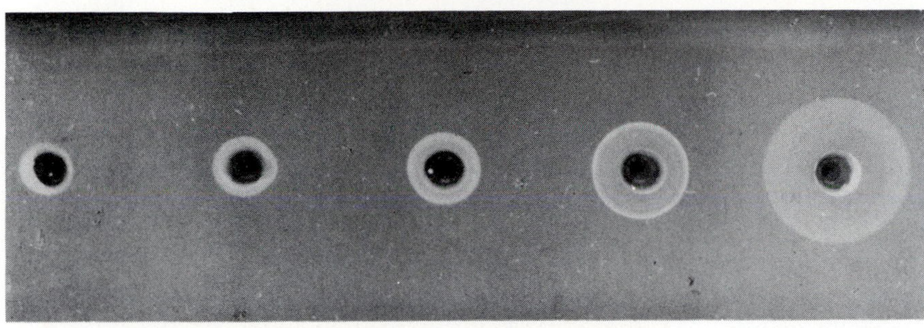

Figure 17–14 Radial immunodiffusion: In this example, antiserum to IgA is incorporated in agar and is used to measure serum IgA levels.

trophoresis before undertaking immunodiffusion. This technique, known as **immunoelectrophoresis,** is usually employed to identify proteins in body fluids.

Immunoelectrophoresis involves the electrophoresis of the antigen mixture in agar gel in one direction. A trough is then cut in the agar just to one side of and parallel to this line of separated proteins. Antiserum is placed in this trough and allowed to diffuse laterally. When the diffusing antibodies encounter antigen, curved lines of precipitate are formed. One arc of precipitation forms for each of the constituents in the antigen mixture (Fig. 17–16). This technique can be used to resolve the proteins of normal serum into 25 to 40 distinct precipitation bands. The exact number depends on the strength and **specificity** of the antiserum employed (Fig. 17–17). By means of this technique, it is possible to readily identify the absence of a normal serum protein such as occurs in individuals with a congenital deficiency of some complement components. It is also possible to detect the presence of excessive amounts of a serum protein, for example, the raised immunoglobulin level found in persons with a plasma cell tumor (myeloma).

Two-Dimensional Immunoelectrophoresis. Two-dimensional immunoelectrophoresis is a combination of electrophoresis and electroimmunodiffusion. The first stage involves the electrophoresis of a mixture

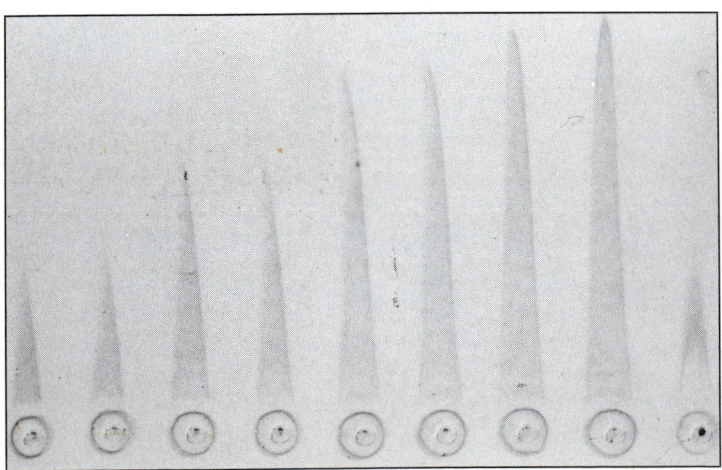

Figure 17–15 Electroimmunodiffusion: In this example, the level of the iron-binding protein lactoferrin from milk is being estimated. The height of each "rocket" is proportional to the amount of antigen placed in each well. *(Courtesy of Dr. R. Harmon.)*

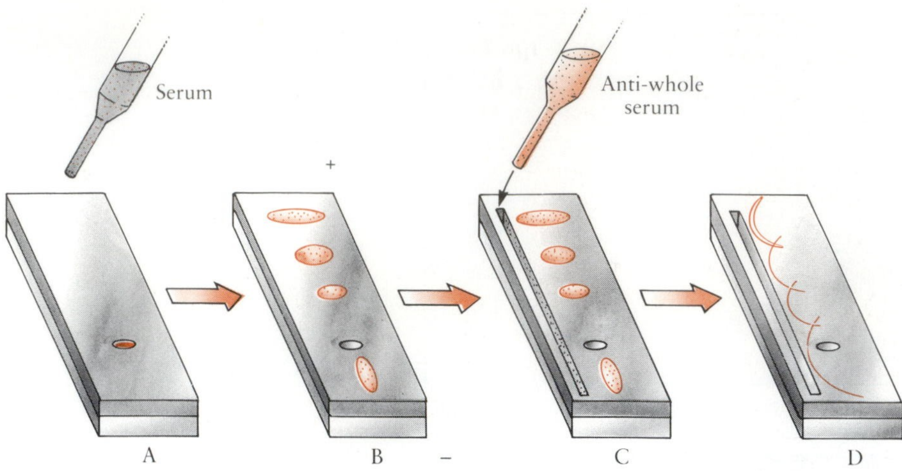

Figure 17–16 The principle of immunoelectrophoresis. For details see the text.

such as serum in one direction. Following this first stage, the current is reset at a right angle to the first electrophoretic run and the serum is reelectrophoresed into agar containing polyspecific antiserum. As a result, a series of large precipitate arcs is formed in the agar (Fig. 17–18). These arcs may be quantitated by measuring their surface area. Individual components can be identified either by means of specific antisera or by looking for a line of identity with a known antigen.

Counterimmunoelectrophoresis. When subjected to electrophoresis in agar gel, some antibodies move toward the cathode because of a flow of buffer through the agar in that direction. This phenomenon is called electroendosmosis. If an antigen is strongly negatively charged so that it moves toward the anode in spite of this flow, then it is possible by a suitable arrangement of wells in an agar plate to drive antigen and antibody together by electrophoresis. A precipitate can be produced within a few minutes. This technique, known as counterimmunoelectrophoresis, can be used for the rapid identification of bacteria and mycoplasma and for the diagnosis of some viral diseases.

Titration of Antibodies

Although the detection of antibodies or antigen is sufficient for many tests, it is usually desirable to arrive at some estimate of the amount of reactants present. The amount of specific antibody is often determined by **titration.** Titration is a procedure in which the serum under test is made up into a series of increasing dilutions (Fig. 17–19). Each dilution is then tested for activity in the test system. The reciprocal of the highest dilution giving a positive reaction is known as the **titer,** or titre (depending on geographical location), and provides a measure of the amount of antibody in that serum.

Agglutination

In the same way that bivalent antibodies may link soluble antigens to form an insoluble complex that then precipitates, antibody may cross-link particles, resulting in their clumping or **agglutination** (Fig. 17–20). Agglutination may be produced by mixing a suspension of antigenic particles, such as bacteria or red cells, with antiserum (Fig. 17–21). Antibody combines rapidly with the particles, the primary interaction, but ag-

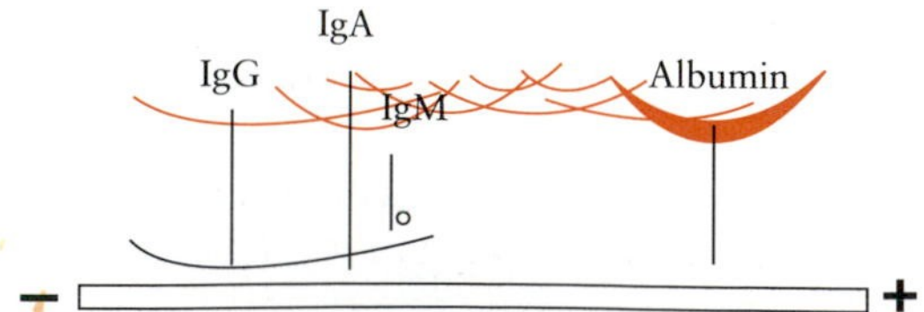

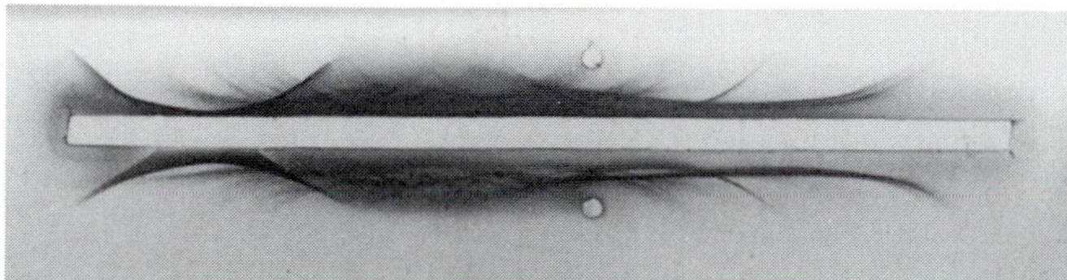

Figure 17–17 Immunoelectrophoretic patterns obtained by first electrophoresing human serum and then developing the pattern with rabbit antihuman serum.

Figure 17–18 Two-dimensional immunoelectrophoresis of normal catfish serum. The anode is on the right. The point of application is on the left. *(Courtesy of Dr. D.H. Lewis.)*

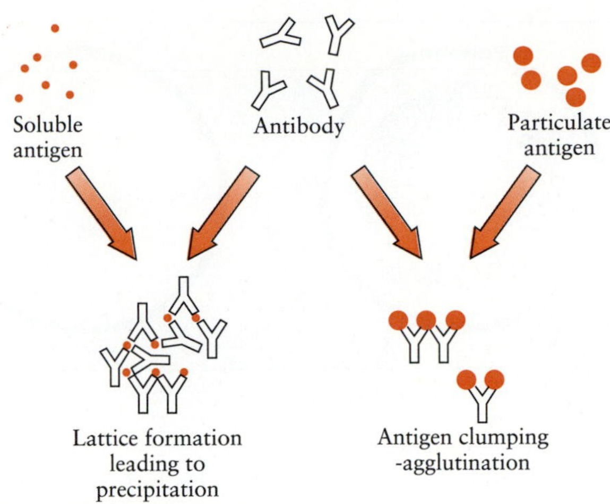

Figure 17–20 The essential similarity of precipitation and agglutination reactions.

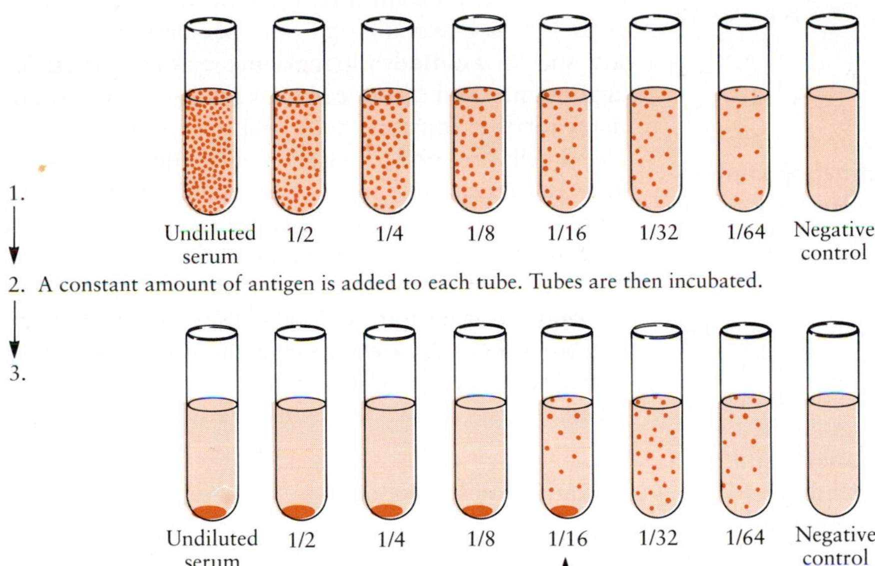

Figure 17–19 The principle of antibody titration: Serum is first serially diluted in a series of tubes. A constant amount of antigen is then added to each tube and the mixture incubated. At the end of the incubation period in this example, agglutination has occurred in all tubes up to a serum dilution of 1/16. The titer of the serum is therefore 1/16.

glutination is a much slower process, since adherence between particles occurs only when they touch each other. Normally, the suspensions of antigen particles are stable, since the particles are prevented from clumping by a negative charge on their surface. However, immunoglobulins are positively charged, and therefore, on coating particles, they reduce their surface charge. In addition, some immunoglobulins, especially IgM, can bridge charged particles by extending out beyond the effective range of this potential. As a result, the particles can approach closely, bind, and agglutinate. Antibodies differ in their ability to promote agglutination. IgM antibodies are considerably more efficient than IgG or IgA antibodies in producing this form of reaction (Table 17–1).

If excess antibody is added to a suspension of antigenic particles, then just as in the precipitation reaction, each particle may be so heavily coated by antibody that agglutination is inhibited. This lack of reactivity seen at high antibody concentrations is termed a **prozone.** Another cause of prozone formation is the presence in a serum of antibodies that do

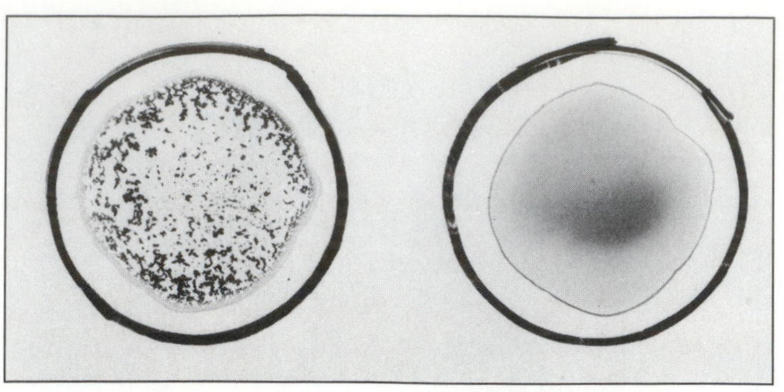

Figure 17–21 Bacterial agglutination (on the left) compared to a nonagglutinated bacterial suspension on the right.

not cause agglutination even when bound to the particles. These nonagglutinating antibodies are also known as **incomplete antibodies.** Their lack of agglutinating activity is probably a result of restricted movement in their hinge region (Chapter 13), causing them to be functionally monovalent. This is a feature of antibodies of the IgG4 subclass.

Antiglobulin Tests

To test for the presence of nonagglutinating antibodies on the surface of particles such as bacteria or red blood cells, a direct antiglobulin test can be used (Fig. 17–22). The particles are first washed to remove any unbound antibody. The washed particles are then mixed with an antiglobulin serum, and if nonagglutinating antibodies are present, agglutination will occur.

To test for the presence of nonagglutinating antibodies in a serum, an indirect antiglobulin test is used. In this technique, the serum under test is first incubated with antigen particles, which adsorb the antibodies. After washing to remove unbound antibody, the coated particles are mixed with an antiglobulin serum. On reacting with bound antibody, the antiglobulin cross-links the particles and causes agglutination.

Table 17–1 Role of Specific Immunoglobulins in Diagnostic Tests

Property	IgG	IgM	IgA
Agglutination	+	+++	+
Complement fixation	+	+++	−
Precipitation	+++	+	+−
Virus neutralization	+	++	+
Time of appearance after exposure to antigen (in days)	3–7	2–5	3–7
Time to reach peak titer (in days)	7–21	5–14	7–21

Passive Agglutination

Since agglutination is a more sensitive technique than precipitation, it is sometimes desirable to convert a precipitating system to an agglutinating one. This can be done by chemically linking soluble antigen to inert particles, such as red blood cells, bacteria, or latex, so that specific antibody will make the sensitized particles agglutinate. Red blood cells are among the best particles for this purpose: tests that employ coated red blood cells are termed passive **hemagglutination** tests. Some antigens, such as the bacterial lipopolysaccharides, adsorb naturally to red blood cells, and it is possible to use this phenomenon to advantage in diagnostic tests. Unfortunately, these lipopolysaccharides are also adsorbed to red blood cells in vivo so that the red blood cells are destroyed by the antibacterial immune responses; as a result, anemia is a feature of many diseases caused by gram-negative organisms.

Agglutination Inhibition

If free soluble antigen is added to an agglutinating system, then the agglutination is inhibited. A practical example of this is used for the rapid diagnosis of pregnancy. A suspension of latex particles or sheep red cells is coated with human chorionic gonadotropin (HCG) and mixed with a solution containing anti-HCG that is just able to agglutinate the particles. HCG is a hormone secreted by the placenta, and detectable quantities are found in urine just a few days after the blastocyst has implanted itself in the wall of the uterus. If urine containing HCG is added to the particle-antiserum mixture, the free antigen competes with the particles for the antibody and prevents agglutination. Thus a woman is probably pregnant if the agglutination is inhibited and not pregnant if agglutination occurs. Agglutination inhibition procedures are employed as immunoassays for certain clotting components, for certain viral antigens, and as methods for rapidly monitoring antibody responses in some virus infections such as influenza.

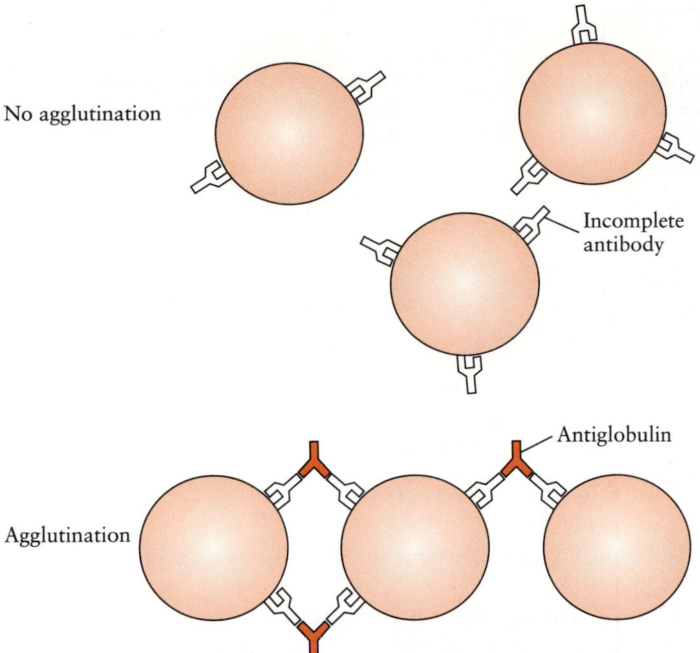

Figure 17–22 The principle of the direct antiglobulin test. The presence of an antiglobulin antibody is required to agglutinate erythrocytes coated with nonagglutinating antibody.

DIAGNOSTIC APPLICATIONS OF IMMUNOLOGICAL TESTS

The immune responses can be used in two general ways in the diagnostic laboratory. First, specific antibody can be used to detect or identify an antigen. Second, by detecting specific antibody in serum, it is possible to determine whether an individual has been exposed to a specific antigen. This can assist in establishing a diagnosis or determining the degree of exposure of the population to an infectious agent.

One feature that must be considered in the interpretation of serological tests is the possibility of errors. Technical errors are usually prevented by incorporation of appropriate controls into the test system. Other errors, however, are largely unavoidable. These may be of two types: false-positive results and false-negative results. A test in which a large proportion of the positive results are false is considered nonspecific, whereas one with a very high proportion of false-negative results is considered insensitive. In general, the level of these errors is set by the criteria used to differentiate positive from negative reactions. If these criteria are adjusted so that the number of false-positive results is reduced, then an increasing proportion of false-negative results will be encountered, and vice versa. Thus, highly sensitive tests tend to be relatively nonspecific, and highly specific tests are generally insensitive. The establishment of criteria in reading tests and, from this, the sensitivity and specificity of a test are determined both by the requirements of the test procedure and by the results of false-positive and false-negative reactions. In ideal tests, it would be desirable for the criteria used in interpreting the test results to be so obvious and absolute that each test would be absolutely sensitive and specific. Unfortunately, such ideal tests are uncommon. The use of two tests in tandem may overcome this problem. Thus in diagnosing AIDS, two tests are employed. A relatively sensitive, nonspecific, and inexpensive ELISA is first used to screen out the great majority of individuals with true-negative results. The few samples that are positive by ELISA can then be tested by the sensitive, specific, and expensive western blot method. The error rate of this technique is very low indeed.

As has been evident from the discussions earlier in this chapter, the advantages and disadvantages of each immunodiagnostic test vary according to the specific requirements of the investigator; the nature of the antigen employed; and the complexity, sensitivity, and specificity of each method. In general, the selection of a diagnostic test represents a compromise between its sensitivity, its specificity, and its complexity—that is, the number of steps involved, the technical expertise required, the cost of the test, and the nature of the equipment needed to conduct the test. Although precise guidelines cannot be drawn up, it is usually most appropriate to use the most sensitive and specific test that can be satisfactorily performed with the available technical assistance and equipment.

KEY WORDS

Agglutination p. 262
Antiglobulin p. 252
Antiglobulin tests p. 264
Avidin-biotin immunoassay p. 258
Direct fluorescent antibody tests p. 254
ELISA tests p. 255
Equivalence zone p. 258
Fluorescent antibody test p. 254
Gel-diffusion test p. 260
Hemagglutination p. 264
Immune complex p. 262
Immunodiffusion p. 260
Immunoelectrophoresis p. 260
Immunoenzyme assay p. 255
Immunofluorescence assay p. 254
Immunoperoxidase techniques p. 256
Indirect fluorescent antibody tests p. 254
Incomplete antibody p. 264
Laser nephelometry p. 259
Monoclonal antibody p. 252
Optimal proportions p. 258
Particle fluorescence immunoassay p. 255
Passive agglutination p. 264
Precipitation p. 258
Primary binding tests p. 252
Radial immunodiffusion p. 260
Radioimmunoassay p. 252
Secondary binding tests p. 258
Serology p. 252
Serum p. 252
Tertiary binding test p. 252
Titration p. 263
Western blotting p. 256

QUESTIONS

1. Which of the following is a quantitative precipitation technique?
   a. counterimmunoelectrophoresis
   b. gel diffusion
   c. immunoelectrophoresis
   d. radial immunodiffusion
   e. western blot
2. Which of these antibody assays is a primary binding test?
   a. complement fixation test
   b. immunoelectrophoresis
   c. agglutination test
   d. fluorescent antibody test
   e. agglutination inhibition
3. Serum IgM levels may be measured by means of
   a. a direct agglutination test
   b. an indirect fluorescent antibody test
   c. radial immunodiffusion
   d. passive hemagglutination
   e. immunoelectrophoresis
4. The gel-diffusion technique is
   a. a primary binding test
   b. a secondary binding test
   c. a tertiary binding test
   d. a test for cell-mediated immunity
   e. none of the above
5. The most sensitive immunological test in terms of the amount of antibody detectable is
   a. radioimmunoassay
   b. complement fixation test
   c. direct agglutination test
   d. gel-precipitation test
   e. viral hemagglutination test
6. Incomplete antibodies are antibodies that
   a. lack a Fc region
   b. lack a Fab region
   c. cannot agglutinate antigen
   d. cannot bind antigen
   e. cannot bind to Fc receptors
7. The ligand for the small protein avidin is
   a. immunoglobulin
   b. biotin
   c. fluorescein
   d. ferritin
   e. antiglobulin
8. Which of these is a secondary binding test?
   a. ELISA
   b. immunoelectrophoresis
   c. western blot test
   d. competitive radioimmunoassay
   e. mouse protection test
9. Immune complex precipitates are formed in
   a. antigen excess
   b. antibody excess
   c. the zone of equivalence of antigen and antibody
   d. the presence of concentrated acid
   e. the absence of electrolyte
10. Antiglobulins are
    a. incomplete antibodies
    b. complement-fixing antibodies
    c. antibodies against immunoglobulins
    d. immunofluorescent antibodies
    e. agglutinating antibodies
11. What advantages do primary binding tests have over other immunological diagnostic tests? What disadvantages do they have?
12. Why is the ELISA test rapidly becoming a favored test for many immunological procedures? What advantages does the ELISA hold over other immunological tests?
13. Outline the procedures for the direct and indirect fluorescent antibody tests. What advantages do these tests have over other primary binding tests?
14. What problems or errors may arise in performing an agglutination test? How may the test be modified to overcome these problems or errors?

15. Which immunological tests could be used to identify the species of origin of a piece of meat? How are these tests performed?
16. What is immunoelectrophoresis used for in diagnostic laboratories? How reliable is this test?
17. What are the advantages and disadvantages of the western blotting technique?

Answers: 1d, 2d, 3c, 4b, 5a, 6c, 7b, 8b, 9c, 10c

SOURCES OF ADDITIONAL INFORMATION

Delaat, A.N.C. Primer of Serology. Harper & Row, New York, 1976.

Diamond, B.A., Yelton, D.E., and Scharff, M.D. Monoclonal antibodies: A new technology for producing serologic reagents. N. Engl. J. Med., 304:1344–1349, 1981.

Friedman, H., Linna, T.J., and Prior, J.E., eds. Immunoserology in the Diagnosis of Infectious Diseases. University Park Press, Baltimore, 1979.

Haber, E., and Krause, R.M., eds. Antibodies in Human Diagnosis and Therapy. Raven Press, New York, 1977.

Hill, H.R., and Matsen, J.M. Enzyme-linked immunosorbent assay and radioimmunoassay in the serologic diagnosis of infectious diseases. J. Infect. Dis., 147:258–263, 1983.

Journal of Immunological Methods. Elsevier North Holland (Biomedical Press), Amsterdam.

Macy, E., Kemeny, M., and Saxon, A. Enhanced ELISA: How to measure less than 10 picograms of a specific protein (immunoglobulin) in less than 8 hours. FASEB J., 2:3003–3009, 1988.

Rose, N.R., and Friedman, H., eds. Manual of Clinical Immunology, 3rd ed. American Society for Microbiology, Washington, D.C., 1986.

Spiegelberg, H.L. Biological activities of immunoglobulins of different classes and subclasses. Adv. Immunol., 19:259–294, 1974.

Wictor, T.T., Flamand, A., and Koprowski, H. Use of monoclonal antibodies in diagnosis of rabies virus infection and differentiation of rabies and rabies-related viruses. J. Virol. Methods, 1:33–46, 1980.

Wilchek, M., and Bayer, E.A. The avidin-biotin complex in immunology. Immunol. Today, 5:39–43, 1984.

Yalow, R.S. Radioimmunoassay: A probe for fine structure of biologic systems. Science, 200:1231–1245, 1978.

CHAPTER 18

Effector T-Cell Function

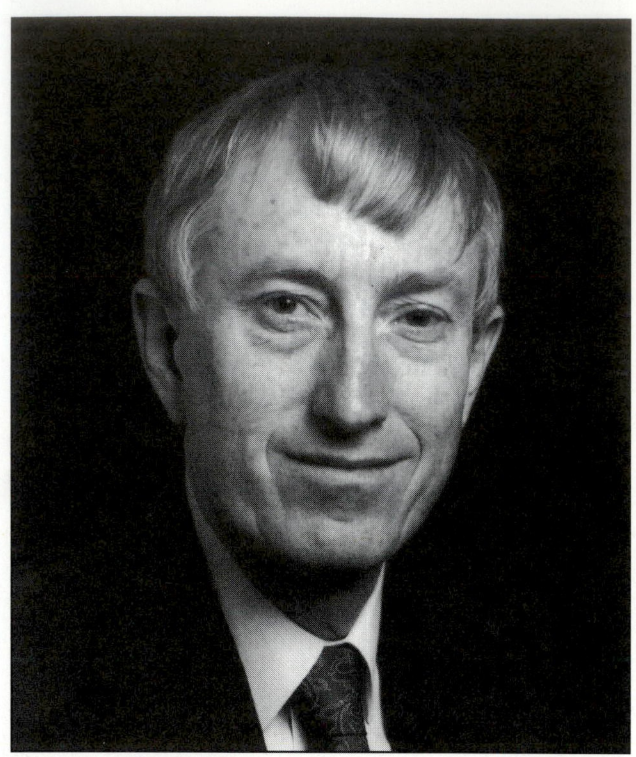

Peter Doherty *Peter Doherty is an Australian pathologist who was studying the response of mice to the lymphocytic choriomeningitis virus. He was interested in the role of T cells and T cell cytotoxicity in this disease. He and his colleague Rolf Zinkernagel showed that cytotoxic T cells from infected mice could only destroy virus-infected cells if the targets had the same MHC molecules on their surface as the cytotoxic cells. This is the phenomenon of MHC restriction. This finding that MHC molecules could play a role in antigen recognition eventually gave rise to our modern understanding of the role that MHC molecules play in antigen presentation.*

CHAPTER OUTLINE

CHAPTER CONCEPTS

1. Cytotoxic effector T cells recognize antigen fragments that have been appropriately processed and presented on a class I MHC molecule.
2. Cytotoxic T cells adhere to and destroy target cells by inducing apoptosis through the release of cytotoxic factors such as perforins, granzymes, and lymphotoxins.
3. Cytotoxic T-cell targets include virus-infected cells, chemically modified cells, and tumor cells.
4. Effector T cells can release macrophage-activating factors that enhance immunity to intracellular bacteria.
5. Effector T cells participate in a characteristic skin inflammatory reaction called the delayed hypersensitivity reaction.

In this chapter the key mechanisms of the cell-mediated immune responses are described. First, the antigens that stimulate this type of immune response are discussed, followed by the properties of cytotoxic T cells and how these cells bind to target cells and deliver a lethal hit to them. We discuss the cytotoxic factors that T cells are able to inject into target cells and consider the regulation of this process. For the sake of completeness, macrophage-mediated and other forms of cell-mediated cytotoxicity are discussed here.

In addition to the actions of cytotoxic T cells, the cell-mediated immune responses encompass the activation of macrophages by T cells. A related cell-mediated effect is the inflammatory reaction known as the delayed hypersensitivity reaction. Following a description of T-cell memory, the chapter concludes with some of the ways in which cell-mediated immune responses can be measured.

In Chapter 11 we discussed the role of helper T cells. However, another form of T cell, the cytotoxic T cell, actively defends the body by destroying foreign or altered cells. These cells play a major role in removing virus-infected cells, cancer cells, and foreign grafts. T cells also play a role in protection against microorganisms that can live within cells. These intracellular organisms are protected against destruction by antibodies and complement and can be eliminated only by cell-mediated immune responses.

ANTIGENS THAT PROVOKE CYTOTOXIC T-CELL RESPONSES

Endogenous Antigens

As described in Chapter 8, cytotoxic T cells respond specifically to **endogenous antigens.** These are antigens synthesized within cells and associated with MHC class I molecules on the cell surface. Clearly, these antigens differ in many important respects from those recognized by B cells. For example, the T cells respond to fragments of antigen molecules, while B cells in contrast can bind and respond to intact antigen molecules. As a result, antibodies formed against proteins are directed against conformational epitopes (i.e., against epitopes with defined shapes). On the other hand, the delayed hypersensitivity skin response, a T-cell–mediated reaction, can be shown to be directed against sequential epitopes (i.e., epitopes with the same sequence of amino acids but not necessarily the same shape as the original molecule). These sequential epitopes are generated by proteolysis and insertion of peptides in the groove on the MHC molecule. Since endogenous peptides must be folded in order to fit into the groove on the MHC class I molecule, it is clear that their precise conformation cannot be important in their recognition.

The two major populations of T cells, those that have γ/δ TCR and those that have α/β TCR, also appear to recognize different antigens. T cells with α/β TCR recognize peptides presented on polymorphic MHC class Ia molecules. In contrast, T cells with γ/δ TCR probably recognize peptides presented on nonpolymorphic MHC class Ib molecules. γ/δ T cells also preferentially recognize conserved characteristic structures from bacteria. These include heat shock proteins and N-formylated peptides. Heat shock proteins are major antigens in Q fever, syphilis, tuberculosis, and leprosy (Chapter 25).

Superantigens

When T cells are exposed to a foreign antigen, usually only a very small proportion of the population, perhaps less than 1 in 10,000 cells, is able to respond. However, some antigens are exceptionally potent and stimulate a high proportion of the T-cell population to divide, perhaps as high as one in five cells. It was

CLARIFICATION

Minor Lymphocyte-Stimulating (mls) Antigens

Mls antigens are cell surface proteins that function as superantigens in mice. As described previously, they can stimulate T-cell division by linking certain TCR Vβ regions to a MHC class II molecule. These mls antigens are the products of genes derived from endogenous retroviruses. The mouse genome, unlike the human genome, may contain over 1000 of these endogenous retroviral sequences. Since these superantigens are inherited, they are present on the T-cells of very young animals and are treated as self-antigens. As a result, T cells that can react with them are eliminated in the thymus through negative selection. The biological significance of this is unknown. Perhaps the stimulation of T cells by mls superantigens may make them more susceptible to retrovirus infection; alternatively, the elimination of reactive T cells by the thymus may protect the mouse from infection by eliminating susceptible cells.

A CLOSER LOOK

Superantigens Are Bad for You!

Toxic shock syndrome is a disease that usually affects menstruating women. It is characterized by development of a fever, low blood pressure leading to vascular collapse, the development of a skin rash resembling scarlet fever, conjunctivitis, and damage to organs such as the kidney, liver, and intestine.

The disease is associated with the use of vaginal tampons that permit overgrowth of certain strains of *Staphylococcus aureus.* These bacteria grow in the tampon and produce a toxin called TSST-1. TSST-1 is a potent superantigen binding specifically to the $V_{\beta}2$ regions on the TCR. As a result it acts as a T cell mitogen and as many as 20% of normal T cells in human blood may respond to it. In order to act as a superantigen and stimulate these T cells, TSST must first bind to MHC class II molecules on antigen-presenting cells such as macrophages. The toxin–MHC class II complex then binds to the $V_{\beta}2$ positive T cell and stimulates it to respond.

The clinical signs of toxic shock syndrome are therefore a result (at least in part) of the massive release of cytokines, especially TNF-α and IL-1, by stimulated T cells or by macrophages activated by these T cells.

originally thought that these molecules acted as nonspecific mitogens. Analysis showed, however, that they specifically stimulated cells whose TCR contained specific Vβ regions. These powerful antigens have been called **superantigens.** Superantigens come from bacteria such as streptococci, staphylococci, and mycoplasma and from viruses such as mouse mammary tumor virus, rabies virus, and possibly human immunodeficiency virus (HIV). In mice, some cell surface molecules called minor lymphocyte-stimulating (mls) antigens (Mls-1[a] and Mls-2[a]) can also act as superantigens. These mls antigens are now known to be the products of retroviral genes. The responses to superantigens are not MHC restricted (i.e., they don't depend on specific MHC haplotypes), but the presence of MHC antigens is required for an effective response. Superantigens bind to TCRs bearing certain Vβ regions, such as $V_{\beta}3$, $V_{\beta}8$, $V_{\beta}5$, $V_{\beta}11$, $V_{\beta}12$, and $V_{\beta}17a$. (As described in Chapter 15, the TCR antigen-binding site is formed by domains called V regions. The many different Vβ regions fall naturally into groups based on their amino acid sequences.) Superantigens bind to both a TCR Vβ region and to an MHC class II molecule on the antigen-presenting cell. As a result, they tightly cross-link the T cell and the antigen-presenting cell. Superantigens do not bind to the antigen groove of the MHC class II molecule but attach elsewhere on its surface (Fig. 18–1). Because of their strong binding, a powerful T-cell response is triggered.

APOPTOSIS

Any multicellular organism must ensure that the production of new cells is approximately equal to the destruction of old cells in order to maintain a constant size. Likewise, an animal must be able to remove old, damaged, or abnormal cells that would otherwise in-

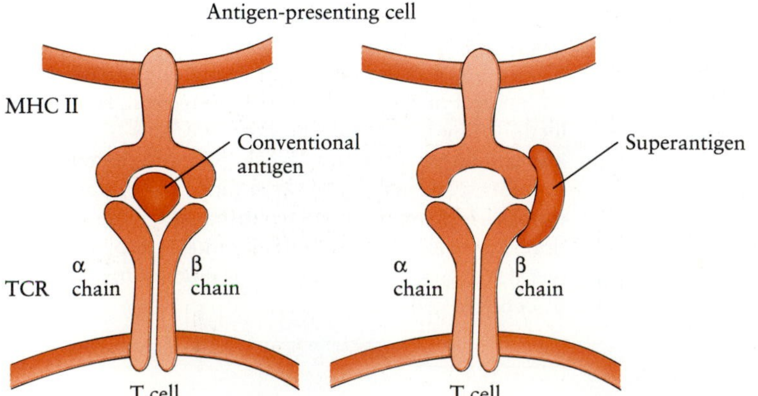

Figure 18–1 The stimulus transmitted to a T cell by superantigen binding to a TCR is significantly greater than that transmitted by a conventional antigen.

terfere with normal tissue functions. Thus cell death is a physiological process that is necessary for normal body function. This physiological cell death is not random but is in fact a well-defined active process known as **apoptosis.** (''Apoptosis'' is a Greek word describing the falling away of petals from flowers or the leaves from trees.) It has distinctive morphological and structural features that are very different from those of pathological cell death or necrosis. For example, in cells dying by apoptosis, their nuclear chromatin condenses against the nuclear membrane (Fig. 18–2). The cells shrink as a result of water loss. They detach from the surrounding cells. Eventually nuclear fragmentation and cytoplasmic budding occur to form characteristic fragments called apoptotic bodies (Fig. 18–3). These bodies are rapidly phagocytosed and destroyed by nearby macrophages. The fragmentation of the cells does not lead to the release of cellular contents, and the phagocytosis of the apoptotic bodies does not lead to inflammation. As a result, cell death can occur without damage to adjacent cells or tissues.

The first detectable event of cells dying through apoptosis is an increase in intracellular calcium. This in turn leads to activation of a number of enzymes that initiate the structural changes. These include a calcium-dependent endonuclease that disrupts the cell's DNA. A characteristic feature of apoptotic cells is frag-

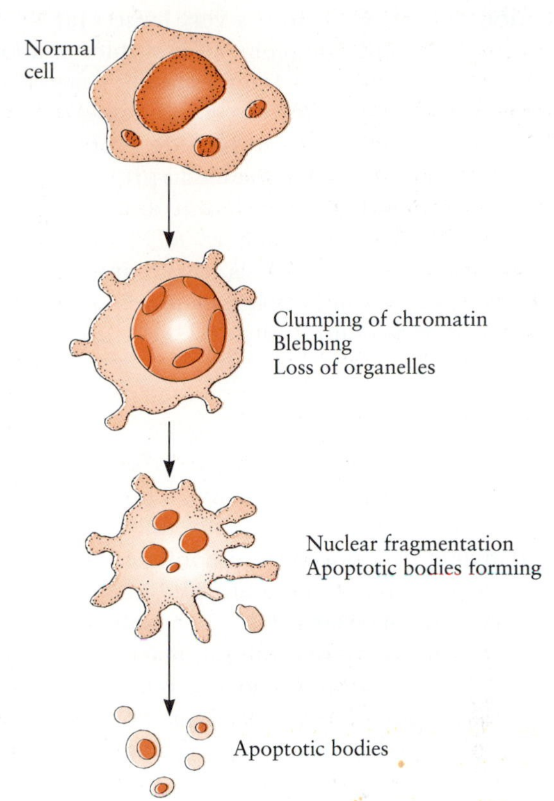

Figure 18–2 The key structural features of cell death by apoptosis.

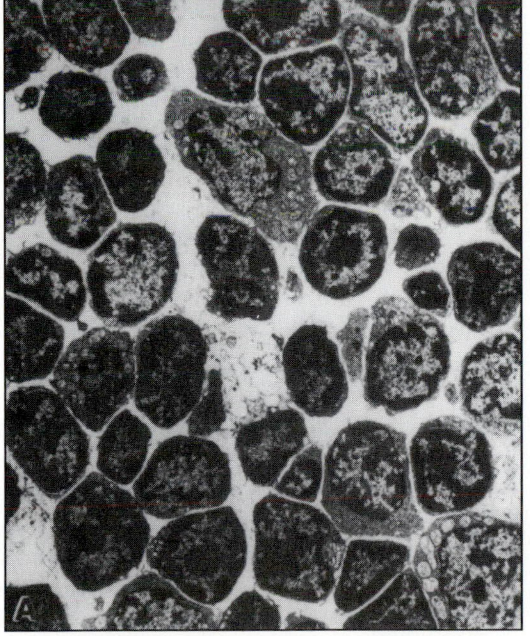

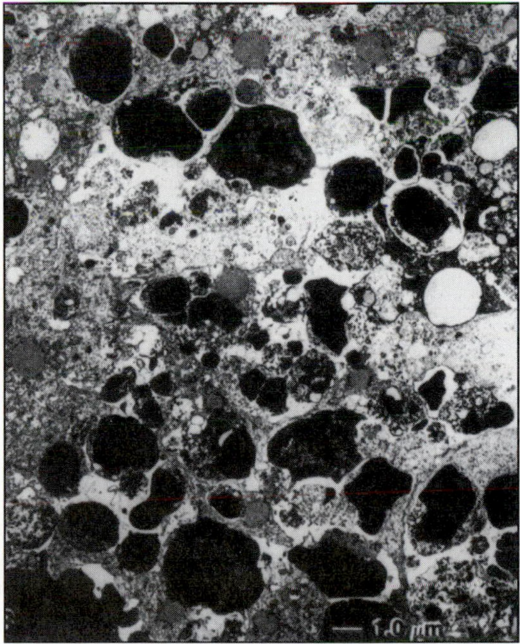

Figure 18–3 The ultrastructural features of apoptosis in the thymic cortex. Thymuses from a control animal (left) or from an animal injected with anti-CD3 antibody to induce thymocyte apoptosis (right). The apoptotic thymocytes show the characteristic nuclear condensation, cell shrinkage, and formation of apoptotic bodies. *(Courtesy Y. Shi, M. Szalay, D. R. Green, Sem. Immunol, 4:335, 1992. With permission.)*

mentation of their DNA into a very large number of low molecular weight fragments in multiples of about 200 base pairs. This DNA fragmentation may be responsible for the characteristic chromatin condensation. Other enzymes activated include proteases that can degrade cytoplasmic and skeletal proteins. Apoptosis is an energy-dependent event and usually requires an energy source such as ATP. Agents that block RNA or protein synthesis usually inhibit apoptosis in lymphocytes, although they may increase it in other cells such as neutrophils.

Apoptosis in Inflammation

Neutrophils usually die during an inflammatory response. If they release their contents, especially large amounts of lysosomal enzymes, they may cause severe tissue damage. As a normal acute inflammatory reaction proceeds, apoptotic neutrophils are phagocytosed intact by macrophages. As a result, their contents are not released. In addition, macrophages that consume these neutrophils do not release cytokines or vasoactive lipids. The macrophages recognize the apoptotic neutrophils as a result of changes in their surface lipids. Clearly, phagocytosis of apoptotic cells is an efficient way of removing unwanted cells without releasing toxic cell contents.

Apoptosis in Immunity

Apoptosis is a critical process in the normal functioning of the immune system. Thus the destruction of self-reactive thymocytes in the developing animal and the cytotoxic effects of $CD8^+$ T cells, lymphokines, and tumor necrosis factors on target cells are all mediated by apoptosis. A cell surface glycoprotein called fas appears to play a key role in this process. Fas (CD95) is a transmembrane protein of 45 kDa that belongs to the TNF receptor/CD40 family. It is widely distributed on cells, especially thymocytes. Some monoclonal antibodies against fas can induce apoptosis on cells expressing fas on their surface. This suggests that fas is a receptor for a ligand that induces apoptosis.

The fas ligand is a 40 kDa protein related to TNF that can induce apoptosis in target cells that express fas. Fas ligand is expressed in activated spleen cells and thymocytes, suggesting that it may play a role in cell-mediated cytotoxicity as well as in the destruction (**negative selection**) of self-reactive T cells.

If the gene that codes for fas is defective, it gives rise to the *lpr* (lymphoproliferation) defect in homozygous mice. These mice do not express fas on their thymocytes. As a result, the thymocytes do not undergo apoptosis and are released into the secondary lymphoid organs. Here they proliferate excessively, resulting in lymphadenopathy. Because self-reactive cells are not destroyed, many of these cells are reactive to self-antigens. As a result, *lpr* mice develop autoimmune disease similar to systemic lupus erythematosus.

CYTOTOXIC T-CELL RESPONSES

Two or possibly three interacting cells are required for an effective cytotoxic T-cell response to occur: an infected cell as a source of endogenous antigen, an effector cell (the cell that will actually mediate the response), and, to a much lesser extent, a Th1 cell.

The conventional view of these responses has been that endogenous antigen binding to the TCR stimulates these T cells to progress from rest into G_1. These cells will then express few IL-2 receptors and their expression of TCR is reduced. IL-2 secreted by Th1 cells binds to these receptors, triggering them to proceed from the G1 into the S phase of the cell cycle and make large numbers of IL-2 receptors. Once a critical number of IL-2 receptors have bound IL-2, then DNA synthesis and mitosis occur and the effector cells begin to divide (Fig. 18–4). If the antigenic stimulus persists, the continued presence of IL-2 is sufficient to stimulate repeated division of the effector T cells (Fig. 18–5). T-cell activation is thus a two-stage process. Antigen drives the effector cell from rest into G_1. IL-2 drives it from G_1 to S and maintains it in a stimulated state. Stimulated cells can exert a cytotoxic effect on target cells bearing appropriate cell surface antigens. IL-2 stimulates the proliferation of the cell that produces it (autocrine effect) as well as that of nearby cells (paracrine effect). In the absence of sufficient antigen, reexpression of TCR occurs and the number of IL-2 receptors drops.

This conventional view has now been questioned by studies on IL-2–deficient mice. Surprisingly, these mice, which cannot produce any IL-2, seem to have almost normal T-cell responses. T cells develop normally, their distribution is normal, and there is only a minor reduction in their cytotoxic T-cell responses. These mice do, however, show an increase in Th2-cell activity and may produce significantly increased levels of IgG1 and IgE as a result of excessive IL-4 production.

Irrespective of the precise mechanism, processed antigen binds to the TCR–CD8 receptor complex of these cells and sends a signal causing the T cells to respond. Responding T cells divide after 24 to 48 hours and then differentiate as genes are sequentially activated. As they divide, the T cells eventually differentiate into memory-cell and effector-cell populations. Effector cells are larger than unstimulated lympho-

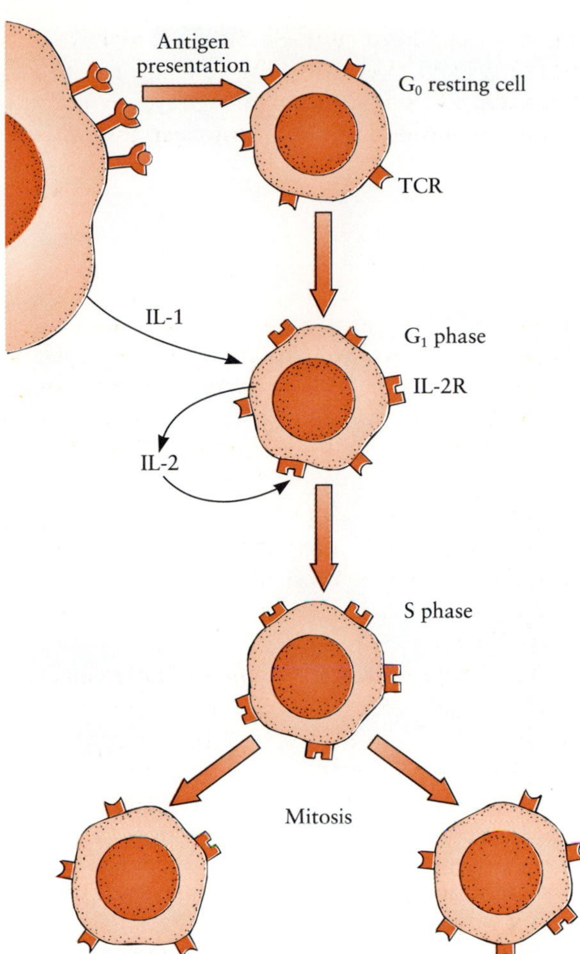

Figure 18–4 The changes that occur in T cells as they respond to antigen and to costimulation by IL-1. The G_1 phase T cell is driven to gain IL-2 receptors while at the same time losing its antigen receptors. If these IL-2 receptors are filled by IL-2 then the cell is driven to divide. If IL-2 is not present, the cell will return to the resting stage, lose IL-2 receptors, and reexpress its antigen receptors.

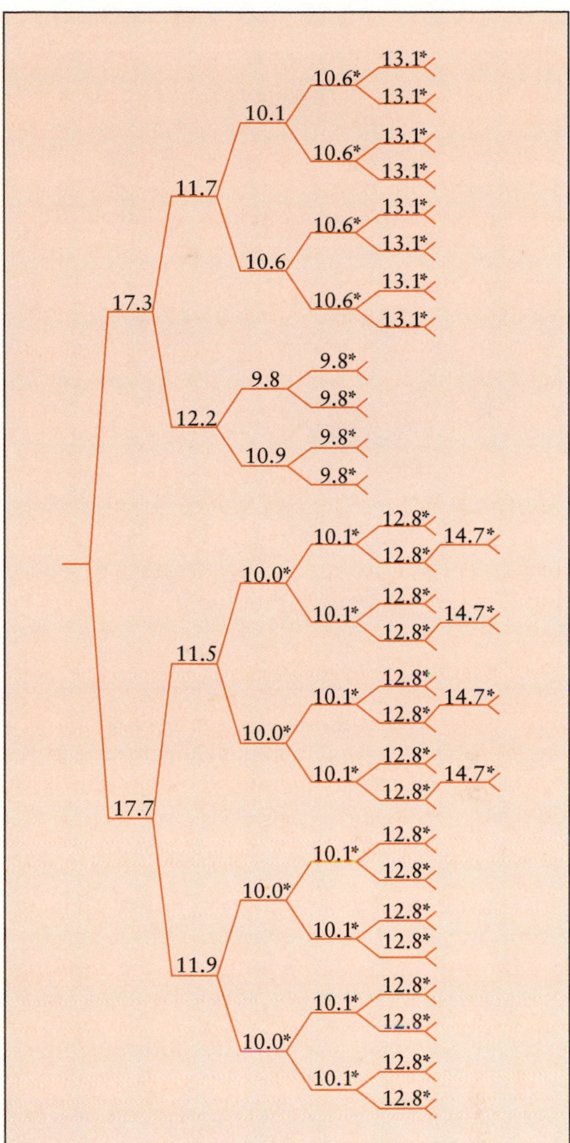

Figure 18–5 A reconstruction of the fate of a single responding human T lymphocyte for 64 hours after exposure to antigen (Tuberculin). Generation times are shown in hours over appropriate lines: those marked with an asterisk are average times in a group of cells where it was impossible to follow the fate of a single cell, but where the mitoses within the group were clearly seen. *(From Marshall, W.H., Valentine, F.T., and Lawrence, H.S. J. Exp. Med., 131:327, 1969. With permission.)*

cytes, and their cytoplasm may become **pyroninophilic,** reflecting their ability to synthesize proteins (Figs. 18–6 and 18–7).

The process through which cytotoxic T cells destroy target cells carrying foreign epitopes on their surface can be arbitrarily divided into three phases: adhesion, lethal hit, and dissociation and target death.

The Adhesion Phase

Adhesion between a cytotoxic T cell and its target is mediated by the same set of adherence molecules that is involved in the interaction of helper T cells with antigen-presenting cells. Cytotoxic T cells possess CD8 molecules as part of their TCR complex. When the TCR binds an epitope bound to a MHC class I molecule, the two terminal domains of the CD8 heterodimer bind to the α3 domain of the MHC molecule. This binding of CD8 to the MHC molecule is usually necessary for a successful cytotoxic T-cell response. Experimentally, the presence of antibodies directed against CD8 can block cytotoxic target recognition. The blocking effect of this anti-CD8 can be overcome

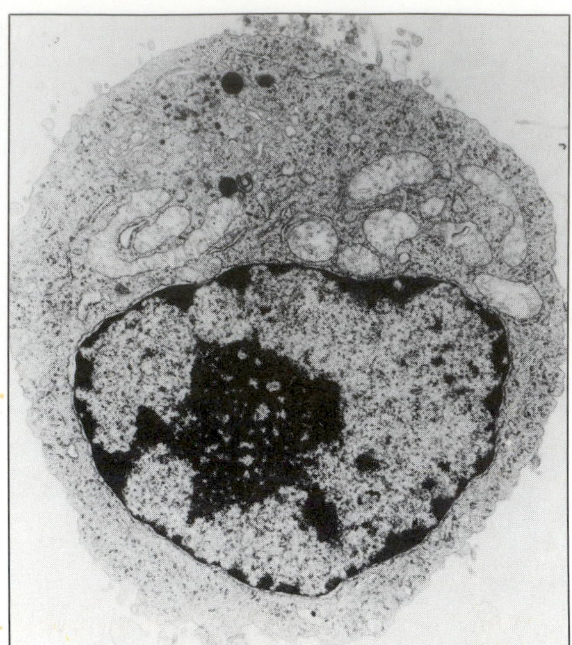

Figure 18–6 A transmission electron micrograph of a responding T cell—a lymphoblast. Compare this to Figure 7–2. Note the extensive cytoplasm, ribosomes, and large mitochondria. *(Courtesy of Dr. S. Linthicum.)*

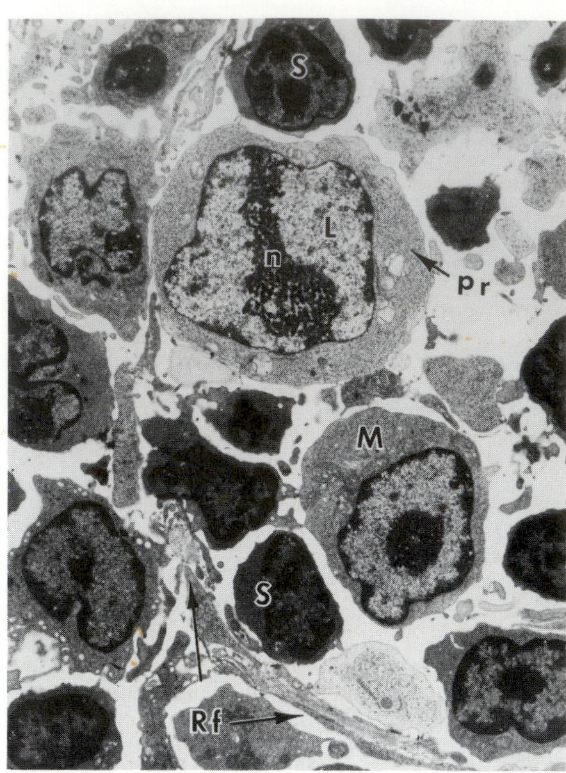

Figure 18–7 A transmission electron micrograph of the paracortex of a lymph node responding to antigen. The large cell is a lymphoblast (L) containing a large nucleolus (n) and numerous polyribosomes (pr); medium sized lymphocytes (M) and small lymphocytes (S) are also seen as well as some reticular fibrils (Rf). *(From Cellular Functions in Immunity and Inflammation, by J.J. Oppenheim, D.L. Rosenstreich, and M. Potter, eds. Copyright 1981 by Elsevier Science Publishing Co., Inc. Courtesy of Dr. A.O. Anderson.)*

by increasing the concentration of either MHC or antigen on the surface of the target. Thus it may be that the CD8 molecules simply enhance the binding between a T cell and its target. If the T-cell receptor has a very high affinity for the target, the CD8 molecule might not be required at all. In this case the TCR alone could confer both antigen and self-MHC specificity on the cell expressing it.

In addition to the effects of antigen, TCR, and CD8, optimal cytotoxic T-cell responses require that a second signal is transmitted between the T cell and its target. As with CD4$^+$ helper cells, CD8$^+$ cytotoxic cells are rapidly activated when their CD28 surface receptors bind to their ligand, the B7-2 (CD80) molecule on the target cell. The costimulation mediated by CD28/B7-2 may permit cytotoxic T cells to kill target cells in the absence of help from CD4$^+$ helper cells. Clearly, tumors or virus-infected cells that express B7-2 are much more sensitive to killing by cytotoxic T cells than target cells that fail to express this molecule. Peter Linsley and his colleagues have shown that tumor cells expressing B7-2 are very rapidly destroyed by cytotoxic T cells. Additional adhesion between cytotoxic T cells and their target is mediated by bonding between CD2 and CD58 and between CD11a/CD18 and ICAM-1.

CLARIFICATION

B7, B70, or B7-2?

The ligand for the T cell receptor molecule CD28 has had various names in the literature. Originally called B7, it was subsequently found that the biological activity resided in only one of its isoforms. The active isoform was called by different investigators B70 or B7-2. (I tend to prefer B7-2 since it simply denotes a different isoform from the original molecule B7-1.) To compound the confusion, B7 has also been designated CD80. (I decided not to use the CD80 designation in this book since the designation is very recent.)

This confusion over the nomenclature of an important molecule is an excellent example of the difficulties encountered by the discovery of new molecules in a fast moving field such as immunology.

Target Cell Perforation

Cytotoxic T-cell granules contain the components necessary for the lytic process (Fig. 18–8). These granules are of two types. Type I are electron-dense granules, and type II granules are vesicular. Each of the type II granules consists of a central electron-dense core of variable size surrounded by 30- to 70-nm vesicles. The active cytotoxic molecules, **granzymes** and **perforins,** are probably complexed to proteoglycans within type I granules.

Perforins. Perforins are proteins found in the type I granules of cytotoxic T cells, LAK cells, or NK cells but not in macrophages or granulocytes. Two perforins have been identified. Perforin P1 is a protein of 65 to 70 kDa that polymerizes in the target cell membrane to form a polymer of 10^3 kDa. This polymer of polyperforin is a tubular complex that produces large (160Å) lesions in target cell membranes. Perforin P2 is found in NK cells and produces small (50 to 70Å) lesions when it polymerizes. The perforins thus act in a similar manner to C9, the lytic molecule of the complement cascade, although C9 produces lesions that are a slightly different size (100Å). The perforins belong to the same protein family as C7, C8α, C8β, and C9. All share the central portion of their peptide chain that contains 280 amino acids or about two thirds of the molecule. This conserved central portion may be the domain that interacts with the target cell membrane. The importance of perforins in the cytotoxic process has been disputed. Thus some T cells do not express perforins yet are fully cytotoxic. Likewise some target cell lysis can occur in the absence of Ca^{++}, an ion that is required for perforin assembly. Nevertheless the significance of perforins is demonstrated by the results of disrupting the perforin genes and so producing a mouse that is unable to produce perforins. Examination of these mice shows that they are unable to overcome infections with viruses such as vaccinia or lymphocytic choriomeningitis because of a failure to develop cytotoxic T cells.

It is believed that the perforins may either lyse target cells directly, or additionally permit the entry of cytotoxins, especially granzymes, into the target cells so triggering apoptosis.

Cytotoxic Factors

Granules of mouse cytotoxic T cells contain eight serine esterases called granzymes A to F. In humans, three granzymes have been isolated. (In NK cells, related enzymes called fragmentins occur.) The granzymes represent about 90% of the total type I granule contents. Granzymes belong to a family of granule enzymes found also in mast cells and neutrophils. All are proteases, many with trypsin-like activities.

Cytotoxins. Lymphotoxin (LT or TNF-β) secreted by antigen-stimulated T cells kills nearby sensitive tumor cells. The LT probably binds to fas (or a related pro-

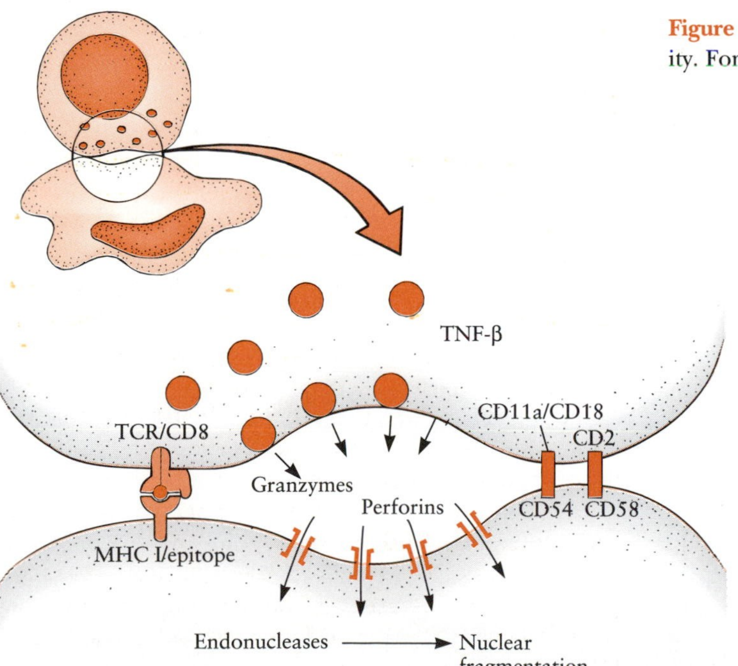

Figure 18–8 The mechanism of T-cell–mediated cytotoxicity. For details, see text.

METHODOLOGY

How to Measure Cell-Mediated Cytotoxicity

Viable cells have the ability to take up sodium chromate. If the cell dies or is disrupted in any way, however, the chromium is released into the extracellular fluid. Radioactive ^{51}Cr may be used in this way to measure cell death (see figure).

Lymphocytes from an immune animal are mixed in an appropriate ratio with ^{51}Cr-labeled target cells. The mixture is centrifuged gently to ensure that all the cells are in contact. It is then incubated for between 4 and 24 hours. At the end of this time, the cell suspension is centrifuged and the presence of ^{51}Cr in the supernatant measured. The amount of chromium released is related directly to the amount of cytotoxicity occurring. The chromium released in the absence of cytotoxic cells must also be measured and subtracted from that released in the presence of cytotoxic cells in order to get a true reading.

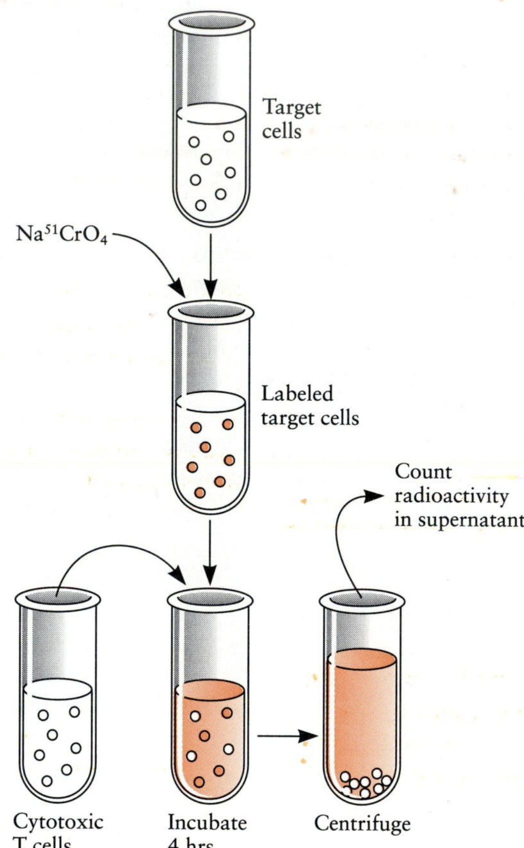

The measurement of cell-mediated cytotoxicity in vitro. The degree of target cell lysis is determined by measuring the release of radioactive chromium from these cells.

tein) on the target and induces apoptosis. Mitochondria are the first organelles to show damage, usually after about 2 to 3 hours. By 3 to 4 hours the nuclear membrane shows signs of damage, and nuclear fragmentation occurs after 4 hours of exposure to LT. By 16 hours more than 90% of target cells are dead.

Another probable cytotoxin is fas-ligand. As mentioned, this protein, a member of the TNF family, can bind to fas on target cells and induce apoptosis. Fas-ligand is expressed on activated thymocytes and splenocytes. Evidence suggests that expression of fas on target cells is an essential prerequisite for the destruction of some targets by cytotoxic T cells.

Target Cell Apoptosis

The actual killing of a target cell occurs by apoptosis. This can be a very rapid process. Within seconds after contact between a T cell and its target, the organelles and the nucleus of the target cell disrupt simultaneously. The T cell can then disengage itself and move on to find another target. The process can be divided into three discrete stages.

Cytoplasmic Reorganization. T-cell activation by target cells is accomplished by a cytoplasmic rearrangement as the T-cell microtubule organizing center, the cytoplasmic granules, and the Golgi complex reorient toward the contact site. This interaction requires Mg^{2+}, and cellular microfilaments and microtubules are involved in the interaction. The redistribution of organelles toward the target cell occurs within a few minutes of binding to the target.

Attack on the Target Membrane. After the T cell comes into contact with its target and the appropriate signal is given, the T cell's Golgi complex and microtubules align themselves so that the granules are released at the region of cell contact. The granules move into a space that forms between the cytotoxic T cell and its target. In this extracellular environment, the pH and salt concentration cause the perforins and granzymes to be released from the granule proteoglycan. Once the granules contact the target, the perforin molecules polymerize in a calcium-dependent reaction to form polyperforins (Fig. 18–9). Polyperforins insert themselves into the target membranes so that they form transmembrane channels. Although C9 inserts into membranes only after activation by C5b-8, T-cell perforin does not require such initiation. It can bind to the cell membrane, insert into the lipid bilayer, and polymerize. Between 12 and 18 monomers come together to form a closed cylinder, although incomplete rings are also seen.

It is not clear just how the cytotoxic T cell is not damaged during this process. It is possible that the perforin is carefully oriented toward the target or that the granule contents are directed toward the target by the TCR. When two cytotoxic cells meet each other, only one is triggered, perhaps the one with highest affinity for antigen.

Effects on the Target. When they kill targets, cytotoxic T cells cause a rapid increase in intracellular Ca^{++} and fragmentation of target-cell DNA into nucleosome-sized fragments (Fig. 18–10). These events are typical of cells undergoing apoptosis. This nuclear DNA fragmentation occurs very early in the cytotoxic process and is due to activation of an endonuclease in the target cell. Lymphotoxin may be responsible for this. It is not mediated by a T-cell–derived endonuclease but granzyme A is a potent mediator of DNA breakdown in permeabilized target cells.

Regulation of Cytotoxicity

When a T cell encounters a target, one prerequisite is that target cell and cytotoxic T cell must have MHC class I molecules in common. This is an example of

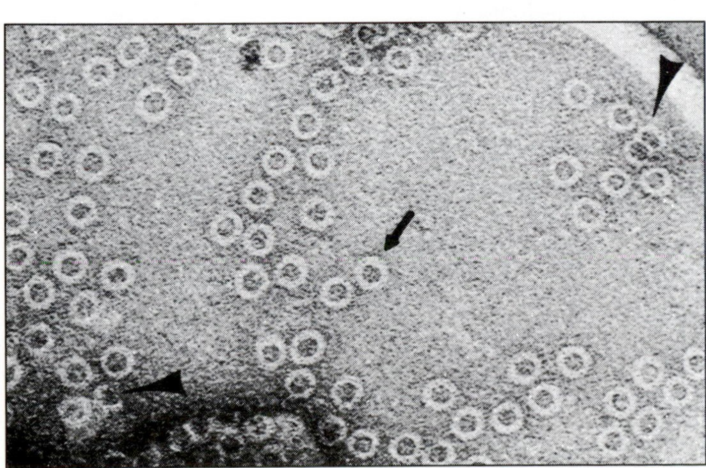

Figure 18–9 Polyperforins from human NK cells on the surface of a rabbit erythrocyte target. The arrowheads point to incomplete rings and double rings. *(From Podack, E.R., and Dennert, G. 1983. Nature 301:442. Used with permission.)*

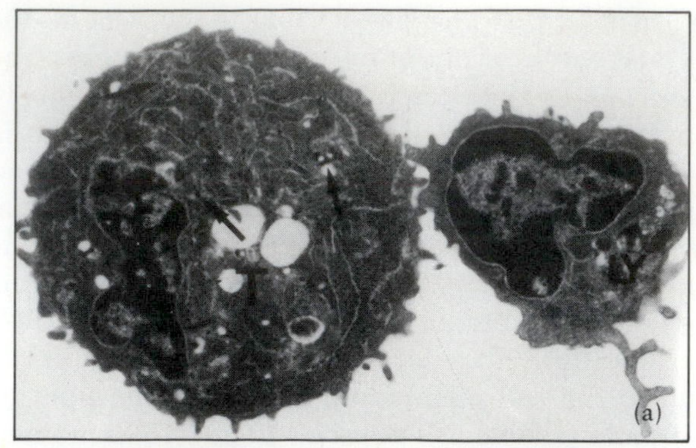

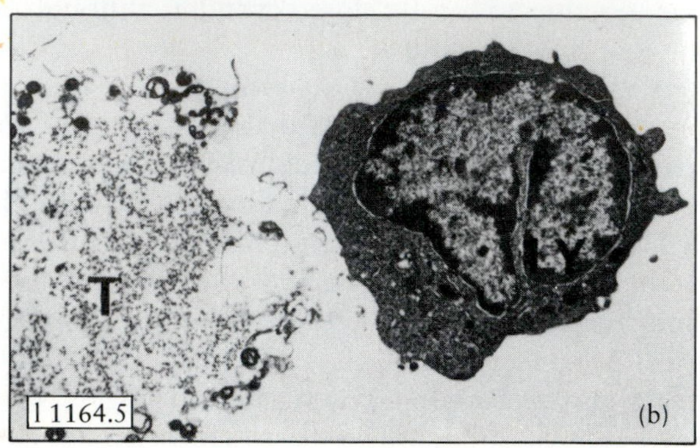

Figure 18–10 Destruction of EL4 target cells by cytotoxic T cells. (A) Conjugation between a peritoneal exudate lymphocyte and a target cell. Note the lysosome-like bodies (LY) and the nuclear fragmentation of the target cell (T). (B) A lymphocyte with a lysed target cell. *(From Zagury, D., et al. Eur. J. Immunol., 5:1975, p. 881. Used with permission.)*

MHC restriction and simply reflects the need for the T cell to recognize the target-cell antigens associated with a MHC class I molecule. Perforin activity in cytotoxic T cells is enhanced 2- to 14-fold by IL-2, IL-3, IL-4, IL-6, and to a lesser degree TNF and IFN-γ. Perforin-mediated cytolysis is regulated by some plasma proteins, especially lipoproteins, and by vitronectin.

MACROPHAGE CYTOTOXICITY

Under some circumstances, macrophages destroy target cells or organisms without ingesting them. This may be either a slow antibody-independent process or a rapid antibody-dependent process (ADCC, see later). For example, on phagocytosis of bacteria or parasites, macrophages release nitric oxide and TNF-α. The nitric oxide is directly toxic to nearby bacteria and cancer cells. The TNF-α is, of course, cytotoxic for some tumor cells. Some nonspecific macrophage cytotoxicity may be attributable to the release of C3b, cytotoxic peroxides, or proteases.

OTHER MECHANISMS OF CYTOTOXICITY

One other mechanism of cell-mediated toxicity, although not mediated by T cells, is mentioned here for comparison (Fig. 18–11, Table 18–1). Cells that possess FcγRI or FcγRII may bind to foreign target cells or bacteria by means of specific antibody and then become cytotoxic. These cells include monocytes, eosinophils, neutrophils, B cells, and NK cells (see page 309). The mechanism of this **antibody-dependent cell-mediated cytotoxicity (ADCC)** is unknown. However, neutrophils and eosinophils probably act through a respiratory-burst mechanism, which causes destruc-

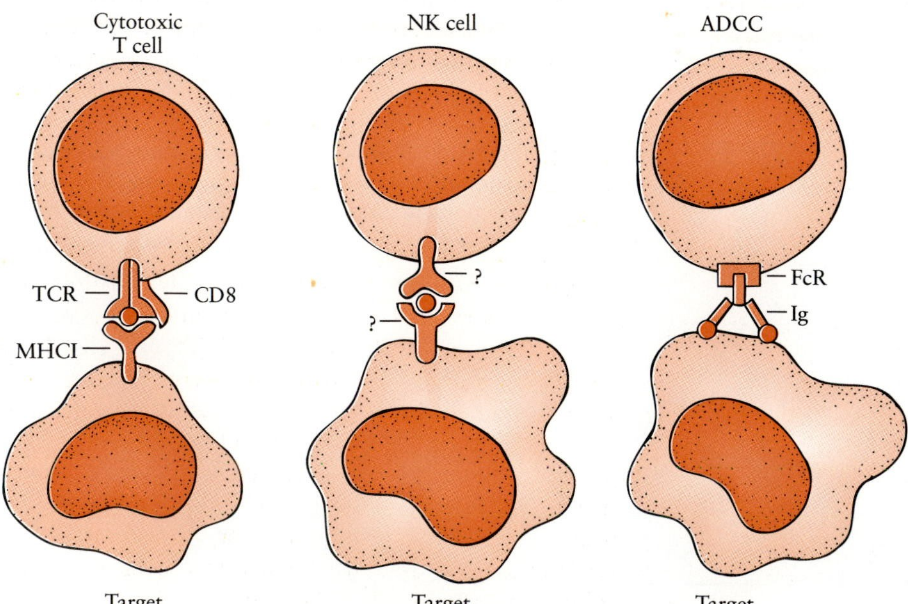

Figure 18–11 The three mechanisms whereby cytotoxic cells can adhere to their targets. The mechanisms of binding and of cytotoxicity of NK cells are unknown.

Table 18–1 A Comparison of Three Assays for Cell-Mediated Cytotoxicity Using ^{51}Cr-Labeled Cells as Targets

Cytotoxic Cells	Time (hours)	Mechanism	MHC Restricted	Antigen Specific
Normal lymphocytes	24	NK activity	−	−
Normal lymphocytes + Specific antibody	6	ADCC activity	−	+
T cells from an immunized donor	1	T-cell–mediated cytotoxicity	+	+

tion of target cell membranes (Chapter 5). ADCC is slower and less efficient than direct T-cell–mediated cytotoxicity, taking at least 6 to 18 hours to occur. Whether a macrophage participates in an ADCC reaction depends on the level of expression of its FcR and on its degree of activation. Thus IFN-γ or GM-CSF promotes macrophage ADCC but not IL-2, IL-3, IL-4, IL-6, or TNF-α.

OTHER CELL-MEDIATED IMMUNE REACTIONS

Immunity to Intracellular Bacteria

Bacteria such as *Listeria monocytogenes, Mycobacterium tuberculosis, Brucella abortus,* and *Legionella pneumophila* and protozoa such as the *Leishmania* and *Toxoplasma gondii* are readily engulfed by macrophages but are resistant to intracellular destruction. They can survive and multiply within normal macrophages (Table 18–2). This is an environment free of antibody, and as a result, the humoral immune response is ineffective against these organisms. Passively transferred antiserum does not confer protection, although passively transferred lymphocytes do (see the table in the Methodology box on page 281).

Protection against this type of infection is afforded by activated macrophages (Fig. 18–12). Although macrophages from unimmunized animals are normally incapable of destroying these organisms, this ability is acquired by the macrophages of infected animals about ten days after onset of infection. This activation is brought about by Th1- and NK-cell–derived IFN-γ supplemented by exposure to microbial products. These products include endotoxin, muramyl dipeptide, and cell wall carbohydrates (glucans and mannans). In many cases, activation is a multistage process. For example, inflammatory macrophages may first be primed by interferon. In a second step, these primed macrophages are activated by bacterial products.

The process of macrophage activation is modulated by Th1 and NK cells. For example, macrophage-derived cytokines can stimulate NK cells to secrete IFN-γ. This NK-derived IFN-γ then activates the macrophages. At the same time, antigen presented to Th1 cells by the macrophages stimulates the lymphocytes to secrete more IFN-γ and IL-2. The IL-2 can activate the NK cells still further, enhancing their output of IFN-γ and activating more macrophages. The process is downregulated by IL-10 secreted by Th2 cells.

Activated macrophages secrete increased amounts of many proteins. Thus they release TNF-α as well as large quantities of proteinases, which can act locally to activate complement components. They secrete interferons and interleukin-1 as well as thromboplastin, prostaglandins, fibronectins, plasminogen activator, and the complement components C2 and B. Activated macrophages express increased quantities of MHC class II molecules on their surface and thus have an enhanced ability to process antigen for the immune response.

Activated macrophages are enlarged and have increased membrane activity, especially ruffling, increased formation of pseudopodia, and increased pinocytosis (uptake of fluid droplets) (Fig. 18–13). They move more rapidly in response to chemotactic stimuli. They contain elevated levels of lysosomal enzymes and respiratory-burst metabolites, and they are more avidly phagocytic than normal cells. Activated macrophages have an enhanced ability to kill intracellular organisms or tumor cells by generating nitric oxide. The nitric oxide can destroy nearby tumor cells and intracellular parasites such as *Leishmania major.* Interferon-

Table 18–2 Some Facultative Intracellular Pathogens

Protozoa	*Toxoplasma gondii* *Leishmania donovani* *Trypanosoma cruzi*
Bacteria	*Mycobacterium tuberculosis* *Legionella pneumophila* *Brucella abortus* *Salmonella typhimurium* *Listeria monocytogenes* *Corynebacterium ovis* *Francisella tularensis* *Nocardia asteroides*
Fungi	*Candida albicans* *Histoplasma capsulatum* *Cryptococcus neoformans*

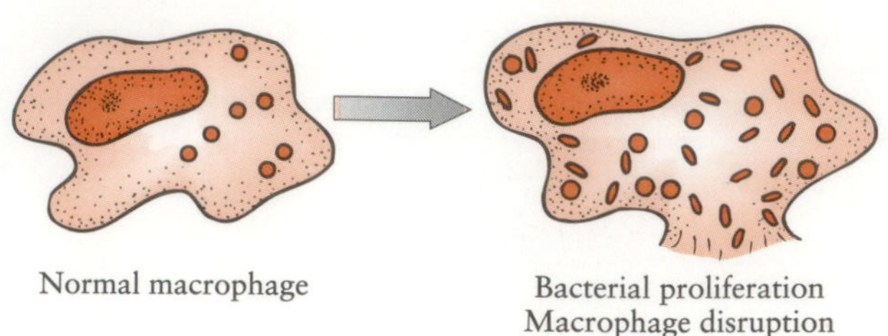

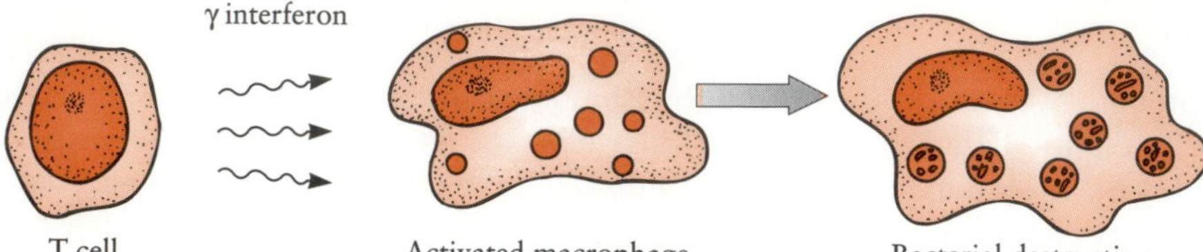

Figure 18–12 Normal macrophages are killed by proliferating intracellular bacteria. Interferon released by T cells can activate macrophages and so enable them to kill otherwise resistant intracellular bacteria.

γ–activated macrophages can also inhibit the growth of some intracellular bacteria such as *Legionella pneumophila* by limiting the availability of iron. They do this by downregulating their transferrin receptors so that less transferrin is endocytosed and by reducing the concentration of intracellular ferritin, the major iron storage protein in macrophages.

Urokinase is a ubiquitous serine protease that converts plasminogen to plasmin. The plasmin activates procollagenase, which degrades connective tissues. Macrophages have a receptor for urokinase that is regulated by IFN-γ and TNF-α. Thus when macrophages are activated, they bind urokinase to their surface, which makes them more invasive for tissues.

All these changes result in an increased ability of the cells to kill microorganisms (Fig. 18–14). The changes themselves are a reflection of a form of **acquired cell-mediated immunity.** The response of these macrophages tends to be nonspecific, particularly in *Listeria* infections, and activated macrophages are capable of destroying a wide range of normally resistant bacteria. Thus an animal recovering from an infection with *Listeria monocytogenes* develops increased resistance to infection by *M. tuberculosis.* The development of these activated macrophages often coincides with the appearance of delayed (type IV) hypersensitivity to intradermally administered antigen.

In humans and mice, a large proportion of γ/δ T cells recognize mycobacteria. *M. tuberculosis*–activated γ/δ T cells express IL-2R, secrete IL-2, and can lyse target cells infected with *M. tuberculosis.* They clearly have a major role to play in defense against these organisms.

Cell-Mediated Resistance to Viruses

Although serum antibodies and complement can neutralize free virus and destroy virus-infected cells, it is the cell-mediated immune mechanisms that are most important in controlling virus diseases. This is readily seen in immunodeficient humans (Chapter 26). Those with a defect in their ability to mount an anti-

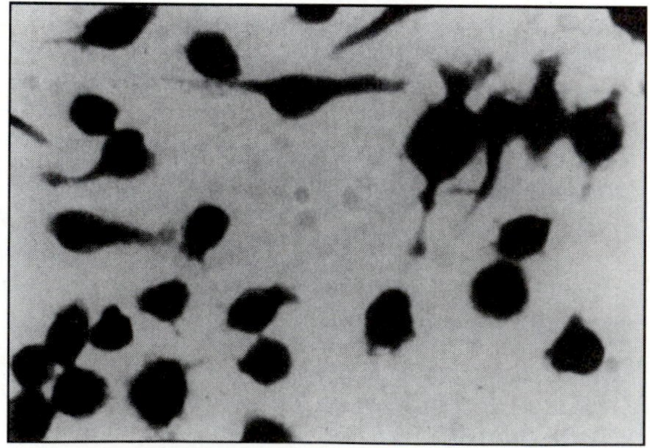

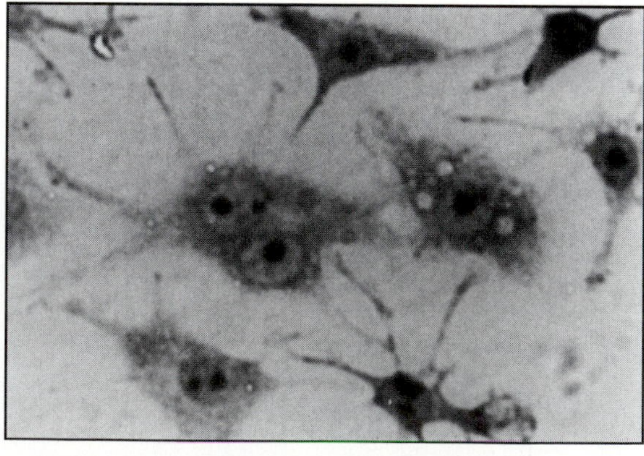

Figure 18–13 Stained culture of mouse macrophages. Left: Normal unstimulated macrophages. Right: Macrophages activated by exposure to IFN-γ and acemannan. Note the cytoplasmic spreading of the activated cells. These cells are much more phagocytic and secrete large amounts of cytokines and nitric oxide. *(Original magnification × 400. Courtesy of Dr. Linna Zhang.)*

METHODOLOGY

How It Was Shown That Lymphocytes Mediated Immunity to Intracellular Bacteria

Mice were infected with a sublethal dose of *Listeria monocytogenes.* Once they developed immunity, they were killed and their spleens removed. Normal, syngeneic mice were then inoculated with either normal or immune live mouse spleen cells (equivalent to the cells from four spleens), normal or immune dead (frozen) spleen cells, or normal or immune mouse serum (0.2 ml) and challenged with virulent *L. monocytogenes* at the same time (see figure). The results were as follows.

Response to *L. monocytogenes*

Treatment	Deaths/Challenged
Control (no cells or serum)	9/10
Live normal cells	6/10
Live immune cells	0/10
Frozen normal cells	9/10
Frozen immune cells	10/10
Immune serum	19/19
Normal serum	19/20

The only effective way of conferring adoptive immunity is through the transfer of live immune cells. Dead cells or antibodies are ineffective.

From Miki, K., and Mackaness, G.B. The passive transfer of acquired resistance to *Listeria monocytogenes.* J. Exp. Med., 120:93–103, 1964.

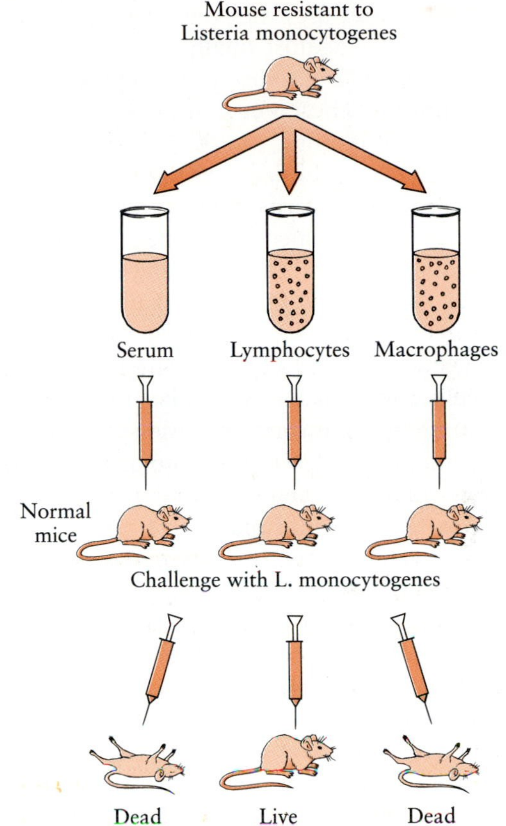

The experiment of G.B. Mackaness that demonstrated that macrophage activation and resistance to intracellular bacteria (in this case *Listeria monocytogenes*), was mediated by lymphocytes. Only lymphocytes could adoptively transfer immunity although the macrophages of the donor mouse were activated.

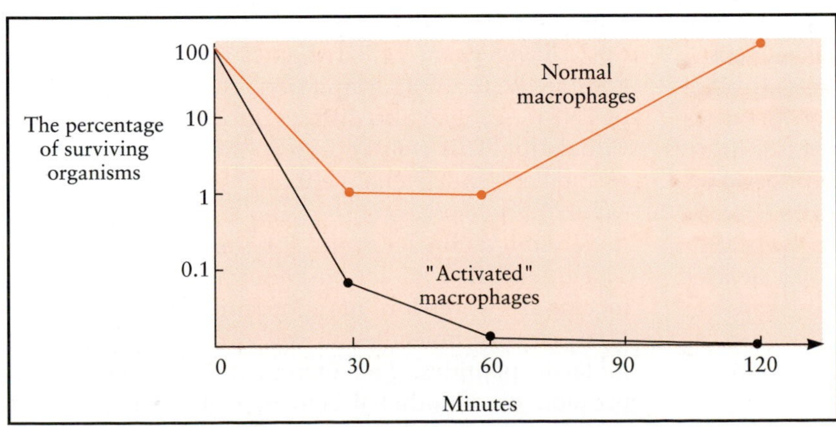

Figure 18–14 The destruction of *Listeria monocytogenes* when mixed in vitro with cultures of normal macrophages and "activated" macrophages from Listeria-infected mice.

body-mediated response suffer severely from recurrent bacterial infections but tend to respond normally to smallpox vaccination and to recover from mumps, measles, chickenpox, poliomyelitis, and influenza. In contrast, humans who have a congenital deficiency in their cell-mediated immune response are commonly resistant to bacterial infection but highly susceptible to viral diseases. Thus, they may die from generalized vaccinia if vaccinated against smallpox.

Virus-infected cells develop new antigens on their surface membranes. These new antigens may arise as a result of viruses budding from the cell surface or as a result of the cell synthesizing virus-coded protein. For example, oncogenic viruses may provoke infected cells to produce tumor-specific antigens as they become neoplastic. Virus-infected cells may therefore be recognized as foreign and thus provoke immunological attack.

In studies of specific T-cell–mediated cytotoxicity for tumor cells and virus-infected cells, it was noticed that lymphocytes from normal unsensitized donors exerted a considerable degree of cytotoxicity on their own. This spontaneous cell-mediated cytotoxicity is largely due to NK cells and may be of major importance in antiviral immunity. It is markedly stimulated by interferon, and as a result, NK activity is rapidly enhanced soon after onset of virus infection. Spontaneous cell-mediated cytotoxicity therefore provides protection before the development of specific cell-mediated cytotoxicity.

Macrophages also exert significant antiviral activity. Viruses are readily taken up by macrophages and are usually destroyed. If the viruses are noncytopathic but able to grow within macrophages, then a persistent infection will result. Under these circumstances, the macrophages must be activated to eliminate the virus. This activation is readily accomplished by IFN-γ.

DELAYED HYPERSENSITIVITY REACTIONS

When certain antigens are injected into the skin of a sensitized animal, an inflammatory response, taking many hours to develop, may occur at the injection site. Since this **delayed hypersensitivity** reaction cannot be transferred from sensitized to normal animals by serum, but only by adoptive transfer of live lymphocytes, it is apparently cell mediated. Delayed hypersensitivity reactions of this sort are classified as type IV hypersensitivity reactions and occur as a result of the interaction between the injected antigen and sensitized T lymphocytes. An important example of a delayed hypersensitivity reaction is the tuberculin response, the skin reaction that occurs in an individual with tuberculosis following an intradermal injection of tuberculin.

Tuberculin Reaction

Tuberculin is the name given to extracts of *M. tuberculosis* (or the closely related *M. bovis* or *M. avium*), which are employed as antigens in the skin testing of humans or animals to identify those with tuberculosis. The most important extract is **purified-protein-derivative (PPD) tuberculin.** Its major antigenic component is probably HSP 65.

When PPD tuberculin is injected into the skin of a normal individual, there is no significant inflammation. On the other hand, if it is injected into a person sensitized by infection with the tubercle bacillus, a delayed hypersensitivity response occurs. Following injection of tuberculin into the skin of such an individual, very few changes are detectable for several hours. By 24 hours, however, local vasodilation and increased vascular permeability occur, which result in redness and the development of a firm lump at the injection site. On histological examination, the reaction differs from a conventional acute inflammatory response in that the infiltrating cells are **mononuclear cells,** that is, macrophages and lymphocytes (Fig. 18–15). A few neutrophils are found in the early stages of the reaction. The inflammation reaches its greatest intensity by 24 to 72 hours and may persist for several weeks before gradually fading. In very severe reactions, tissue destruction may occur at the injection site.

The tuberculin reaction is an immunologically specific inflammatory reaction mediated by T cells. When an individual develops tuberculosis, mycobacterial antigen is taken up by antigen-presenting cells and presented to $CD4^+$ T cells. Some of these T cells respond by developing into a functional subgroup of T cells that can be called Tdh cells (dh-delayed hypersensitivity). Not all $CD4^+$ T cells become Tdh cells. Indeed, the T-cell types involved are clearly heterogeneous: some are MHC class II^+, and others are MHC class II^-. Once generated, Tdh cells can enter the circulation and respond to antigen entering the body by any route. Since delayed hypersensitivity can be elicited many years after exposure to antigen, it is reasonable to suggest that some Tdh cells are very long lived. When antigen is injected intradermally, some is taken up by Langerhans cells (Chapter 8), which then migrate to the draining lymph node under the influence of TNF-α. Antigen also diffuses into nearby capillaries. Circulating Tdh cells recognize this antigen, adhere to capillary endothelial cells, and emigrate from the capillaries to encounter the antigen (Fig. 18–16). This initial emigration takes many hours to occur. IFN-γ, TNF-α, and IL-1 all act on endothelial cells to cause increased expression of adherence molecules and MHC class II molecules as well as the release of chemotactic peptides. The increased MHC class II expression by endothelial cells may also cause a local

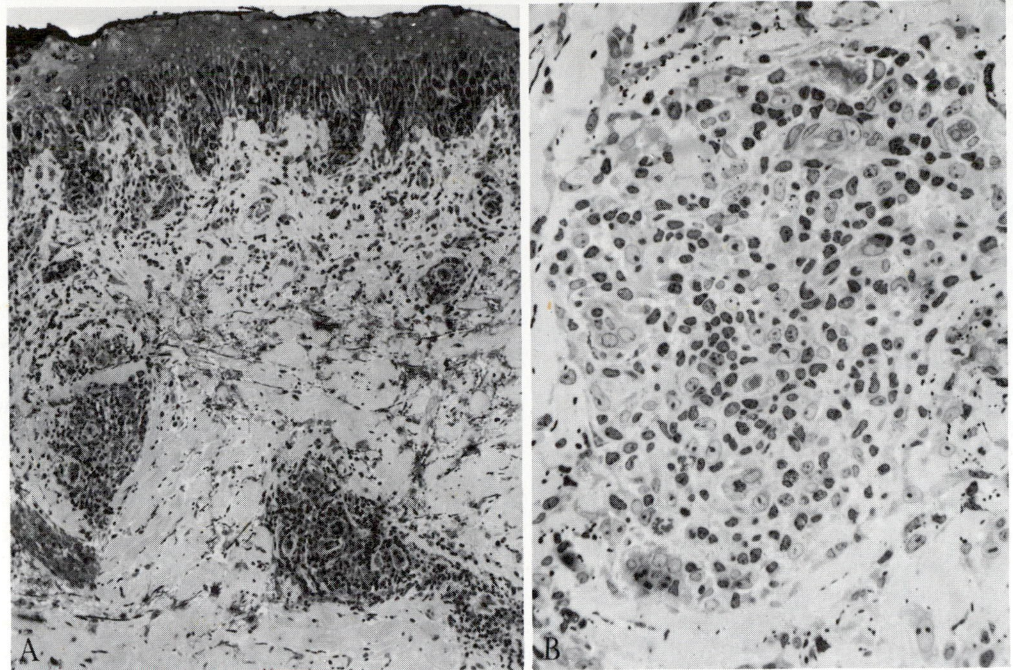

Figure 18–15 Forty-eight hour tuberculin reaction elicited in a PPD-sensitive individual. Massive perivascular mononuclear aggregates are seen in the deep dermis, and a more diffuse infiltrate in the superficial dermis. Epon thin section (1 μm) stained with Giemsa (original magnification: A × 110, B × 400). *(From Parker, C.W., ed. Clinical Immunology. W.B. Saunders, Philadelphia, 1980. With permission.)*

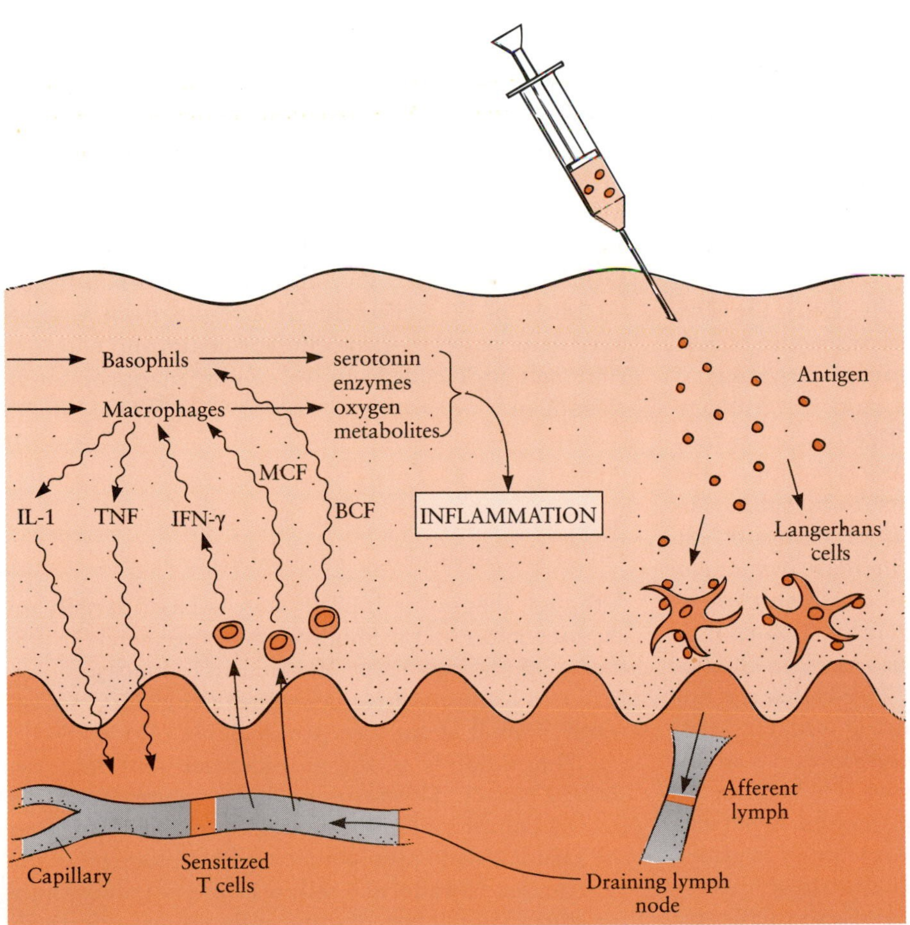

Figure 18–16 The pathogenesis of the delayed hypersensitivity reaction. Intradermally inoculated antigen binds to Langerhans cells in the epidermis and is carried by them to the draining lymph node where T cells are sensitized. The sensitized T cells return to the skin where they release chemotactic and activating factors for macrophages and basophils. These cells in turn release cytokines and inflammatory factors that mediate the delayed inflammatory reaction within the epidermis.

CLARIFICATION

The Origin of Cells in a Delayed Hypersensitivity Reaction

Delayed hypersensitivity against PPD tuberculin can be readily induced in guinea pigs by infecting the animals with BCG, an attenuated strain of *M. bovis.* This delayed hypersensitivity can be adoptively transferred from a sensitized to a nonsensitized syngeneic animal by means of lymphocytes, confirming that it is indeed a cell-mediated immune reaction. The transferred cells can be labeled by administering tritiated thymidine together with antigen to the donor a few days previously so that the active cells will take up the thymidine. If, after transfer, the recipients are injected intradermally with PPD and the resulting DTH reactions examined by autoradiography, it can be shown that no more than 5% of the lymphocytes in the lesion are radioactively labeled. Likewise, if the reverse experiment is conducted and unlabeled lymphocytes are transferred to a recipient who has received tritiated thymidine for several days, then only about 10% to 15% of the lymphocytes in the lesion will be unlabeled. Thus only a very small proportion of the lymphocytes seen in a delayed hypersensitivity reaction are actually specific for the antigen. The vast majority are recruited nonspecifically.

increase in antigen-presenting ability. Once the Tdh cells encounter antigen, they release vasoactive factors such as serotonin as well as macrophage-activating cytokines, macrophage chemokines, and macrophage chemotactic cytokines. The initial T-cell response may also generate a lymphokine that attracts basophils. Basophil-derived serotonin (in rodents) or histamine (in humans) enhances the migration of mononuclear cells into the lesion. The resulting increase in vascular permeability and the opening up of gaps between endothelial cells in capillaries permit more Tdh cells to migrate from the blood into the tissues. The chemotactic factors cause immigration of T cells that are not specifically sensitized for the inducing antigen. Macrophages also accumulate in the lesion and may be activated as a result of the release of IFN-γ. Some of the tissue damage seen in intense delayed hypersensitivity reactions may be due to the release of enzymes and reactive oxygen metabolites from these macrophages. The macrophages ingest and eventually destroy the injected antigen. This permits the tissues to return to normal.

Cells other than lymphocytes and macrophages also participate in the delayed hypersensitivity reaction. Thus neutrophils are found in the early stages of the reaction. In some species, basophils are prominent in the inflammatory reaction (see Fig. 30–16). This type of reaction, called **cutaneous basophil hypersensitivity** (CBH), can be transferred between animals with antibody, with purified B cells, or with T cells. CBH is therefore a heterogeneous phenomenon. CBH is observed in chickens in response to intradermal Rous sarcoma virus, in rabbits in response to schistosomes, and in humans in allergic contact dermatitis and renal allograft rejection.

T-CELL MEMORY

When T cells respond to antigen by the production of effector cells, memory T cells are also generated. Memory T cells can be distinguished from virgin T cells by expressing different cell surface antigens, by secreting a different mixture of cytokines, and by having different activation requirements. Most memory T cells are $CD8^+$. In addition, they express increased levels of CD2, CD11a/CD18, and CD44. Differentiation of T cells from unstimulated to activated and memory cells also involves a change in CD45, the lymphocyte common antigen, from high-molecular-weight to low-molecular-weight forms. Memory T cells produce more IL-4 and IFN-γ and seem to respond more strongly to stimulation of their TCR complex than virgin T cells. However, they secrete less IL-2 and proliferate less since they appear to be locked in the G1 phase of the cell cycle. This results from a resistance to increases in intracellular calcium and may prevent memory T cells from responding inappropriately to antigen.

MEASUREMENT OF CELL-MEDIATED IMMUNITY

Although diagnostic immunology is based largely on the detection of antibodies, measurement of cell-mediated immune responsiveness in humans and animals may be desirable under some circumstances. Currently, several techniques are available (Table 18–3).

The simplest is the intradermal skin test, described earlier. The resulting inflammatory response may be considered cell mediated, provided that it has the characteristic time course and histology of a type IV reaction. Intradermal skin tests are not always convenient, and injection of some antigens may effectively

Table 18–3 Methods of Evaluating Cell-Mediated Immunity

Descriptive assays
1. Count blood lymphocyte numbers
2. Count T cell numbers by flow cytometry using the CD3 marker
3. Measure the CD4:CD8 ratio by flow cytometry

In vitro functional assays
1. Lymphocyte responses to mitogens such as Concanavalin A
2. Measure cytokine release from mitogen or antigen-stimulated cells.

In vivo functional assays
1. Elicit contact dermatitis to an allergen such as dinitrochlorobenzene, fluorescein etc.
2. Intradermal skin test to either microbial antigens such as candida or PPD or mitogens such as phytohemagglutinin.
3. Ability to reject a small skin allograft.

cause sensitization. For these reasons, in vitro tests may be more appropriate. The in vitro tests are designed to measure either the proliferation of T lymphocytes in response to antigen or their production of lymphokines.

To measure T-cell proliferation in response to antigen, a suspension of purified peripheral blood lymphocytes from the individual to be tested is mixed with antigen and cultured for 48 to 96 hours. Twelve hours before harvesting, thymidine labeled with the radioactive isotope tritium is added to the cultures. Normal, nondividing lymphocytes do not take up thymidine but dividing cells do, because they are actively synthesizing DNA. If the T cells are proliferating, they will take up the tritiated thymidine and the radioactivity of the washed cells will provide a measure of the degree of proliferation. The greater the response of the cells to antigen, the greater the radioactivity. The ratio of the radioactivity in the stimulated cultures to the radioactivity in the controls is called the stimulation index.

The measurement of lymphokine release by T cells is a much more complicated procedure. One of the most common techniques involves incubating a purified lymphocyte suspension with antigen. After 24 to 48 hours, the supernatant fluid of the culture is removed and assayed for individual cytokines by ELISAs or radioimmunoassays (Chapter 17).

It is sometimes useful to measure an animal's ability to mount cell-mediated immune responses in general. One way to do this is to apply a skin graft and measure its survival time. A much simpler technique is to paint the animal's skin with a sensitizing hapten such as dinitrochlorobenzene. The intensity of the resulting dermatitis provides a rough estimate of the animal's ability to mount a cell-mediated immune response.

An alternative in vitro technique is to measure the response of lymphocytes to mitogenic lectins such as **phytohemagglutinin,** or **concanavalin A** (Chapter 10). The intensity of the lymphocyte proliferative response, as measured by tritiated thymidine uptake, provides an estimate of the reactivity of an animal's lymphocytes. In addition, if phytohemagglutinin is injected intradermally, it provokes a reaction with many features of a delayed hypersensitivity response. This is a convenient and rapid method of assessing an animal's ability to mount a cell-mediated response without the need for first sensitizing the animal to an antigen. The response to phytohemagglutinin is nonspecific, however, and its interpretation may be difficult.

None of the currently available techniques to measure cell-mediated immunity, with the possible exception of intradermal testing and cytokine ELISAs, lends itself readily for use by anyone but investigators in well-equipped laboratories. The measurement of cell-mediated immunity has become an increasingly important feature of the analysis of immune reactivity, however, and refined and simpler techniques are expected to become available in the future.

KEY WORDS

Acquired cell-mediated immunity p. 280
Activated macrophage p. 279
Antibody-dependent cellular cytotoxicity p. 278
Apoptosis p. 270
Autocrine effect p. 272
B7-2 (CD80) p. 274
CD28 p. 274
Cell-mediated cytotoxicity p. 276
Cytotoxic macrophage p. 278
Cytotoxic T cell p. 272
Delayed hypersensitivity p. 282
Effector cell p. 272
Endogenous antigen p. 269
Facultative intracellular bacteria p. 279
Granzyme p. 275
Interferon-γ p. 279
Interleukin-2 p. 272
Intracellular organism p. 279
Intradermal skin test p. 284
Lectin p. 285
Lymphoblast p. 274
Lymphotoxin p. 275
Memory T cell p. 284
NK cell p. 282

Paracrine effects p. 272
Perforin p. 275
Phytohemagglutinin p. 285
Stimulation index p. 285
Superantigen p. 269
Tritiated thymidine p. 285
Tuberculin reaction p. 282
Type IV hypersensitivity p. 282
Urokinase p. 280

QUESTIONS

1. *M. tuberculosis* infection is diagnosed by a(n)
   a. skin test
   b. complement fixation
   c. indirect fluorescent antibody test
   d. passive hemagglutination test
   e. antiglobulin test
2. The major mechanism of immunity to *Listeria monocytogenes* involves
   a. generation of IgE
   b. complement-mediated lysis of the organism
   c. inhibition of intracellular multiplication by antibodies
   d. destruction by activated macrophages
   e. destruction by antibody-mediated cytotoxicity
3. Macrophages can be cytotoxic for other target cells as a result of the release of
   a. superantigens
   b. perforins
   c. complement
   d. immunoglobulins
   e. nitric oxide
4. A typical delayed hypersensitivity reaction reaches maximum size in
   a. seconds
   b. minutes
   c. 6 to 10 hours
   d. 24 to 48 hours.
   e. 7 to 10 days
5. The enzyme urokinase is found on the surface of macrophages and
   a. activates complement
   b. kills intracellular bacteria
   c. helps them move through tissues
   d. helps them present antigen
   e. mediates delayed hypersensitivity
6. The process of physiologically mediated cell death is called
   a. necrosis
   b. lysis
   c. hemolysis
   d. apoptosis
   e. osmosis
7. At least one of the functions of CD8 on the surface of lymphocytes interacting with another cell is to
   a. bind antigen
   b. bind to MHC class I molecules
   c. bind to MHC class II molecules
   d. bind to Fc receptors
   e. bind surface immunoglobulin
8. Among the key molecules that mediate cell-mediated cytotoxicity are
   a. perforins
   b. immunoglobulins
   c. complement
   d. integrins
   e. selectins
9. Superantigens are molecules that
   a. are nonspecific T-cell mitogens
   b. bind to specific TCR Vβ regions
   c. bind to immunoglobulin Fc regions
   d. bind to specific antigens
   e. bind to complement receptors
10. Which of the following techniques is a method of measuring cell-mediated immunity?
    a. radial immunodiffusion
    b. response to intradermally injected antigen
    c. passive cutaneous anaphylaxis
    d. passive hemagglutination
    e. complement fixation
11. Comment on the suggestion that *M. tuberculosis* is a fairly nonpathogenic organism and that, in fact, it is the immune response to tuberculosis that is responsible for the development of lesions in this disease.
12. What is a superantigen? Outline their probable mode of action and biological significance.
13. Explain the physical basis of MHC restriction. What are the biological consequences of this restriction? Why might such a process have evolved?
14. Outline the mechanism by which endogenous antigen is processed within cells. How is this processing pathway kept separate from the processing of exogenous antigen? Speculate on the consequences if the two pathways were joined.
15. What is the relative importance of granzymes, perforins, and lymphotoxins in T-cell cytotoxicity? Why might three distinct mechanisms of cytotoxicity have evolved in a single cell type?
16. Give examples of facultative intracellular parasites. How have they succeeded in evading the immune response? Why is this sort of organism not much more common?
17. Many diseases such as tuberculosis and AIDS are associated with a significant loss of body weight. How does this come about? Can it be of any possible benefit to the body?
18. Some investigators have suggested that a special population of T cells, called Tdh cells, is involved in mediating a delayed hypersensitivity reaction. What evidence is there for the existence of these cells as a separate population?

Answers: 1a, 2d, 3e, 4d, 5c, 6d, 7b, 8a, 9b, 10b

SOURCES OF ADDITIONAL INFORMATION

Barr, P.J., and Tomei, L.D. Apoptosis and its role in human disease. Biotechnology, 12:487–494, 1994.

Berke, G. Lymphocyte-triggered internal target disintegration. Immunol. Today 12:396–399, 1991.

Bierer, B.E., et al. The biologic role of CD2, CD4 and CD8 in T cell activation. Annu. Rev. Immunol., 7:579–599, 1989.

Born, W., et al. Recognition of heat shock proteins and γδ cell function. Immunol. Today, 11:40–41, 1990.

Brandtzaeg, P., et al. Epithelial homing of γδ T cells. Nature, 341:113, 1989.

Chen, L., et al. Costimulation of antitumor immunity by the B7 counterreceptor or the T lymphocyte molecules CD28 and CTLA-4. Cell, 71:1093–1102, 1992.

Cohen, J.J. Apoptosis. Immunol. Today, 14:126–130, 1993.

Davodeau, F., et al. Surface expression of two distinct functional antigen receptors on human γδ T cells. Science, 260: 1800–1802, 1993.

Doherty, P.C., Blanden, R.B., and Zinkernagel, R.M. Specificity of virus-immune effector T cells for H-2K or H-2D compatible interactions: Implications for H-antigen diversity. Transplant. Rev., 29:89–124, 1976.

Green, D.R., and Cotter, T.G. Apoptosis in the immune system. Sem. Immunol. 4:355–362, 1992.

Kagi, D., et al. Cytotoxicity mediated by T cells and natural killer cells is greatly impaired in perforin-deficient mice. Nature 369:31–37, 1994.

Kagi, D., et al. Fas and perforin pathways as major mechanisms of T cell-mediated cytotoxicity. Science, 265:528–530, 1994.

Krahenbuhl, O., and Tschopp, J. Perforin-induced pore formation. Immunol. Today, 12:399–402, 1991.

Kündig, T.M., et al. Immune responses in interleukin-2-deficient mice. Science, 262:1059–1061, 1993.

Larrick, J.W., and Wright, S.C. Cytotoxic mechanism of tumor necrosis factor-α. FASEB J., 4:3215–3223, 1990.

Lowin, B., et al. Cytolytic T-cell cytotoxicity is mediated through perforin and Fas lytic pathways. Nature, 370:650–652, 1994.

Lu, P., et al. Perforin expression in human peripheral blood mononuclear cells. J. Immunol., 148:3354–3360, 1992.

Nossal, G.J.V. Life, death and the immune system. Sci. Am., 269:53–62, 1993.

O'Rourke, A.M., and Mescher, M.F. The roles of CD8 in cytotoxic T lymphocyte function. Immunol. Today, 14:183–187, 1993.

Rouvier, E., Luciani, M.F., and Golstein, P. Fas involvement in Ca (2+)-independent T cell–mediated cytotoxicity. J. Exp. Med., 177:195–200, 1993.

Scott, P. IL-12: Initiation cytokine for cell-mediated immunity. Science, 260:496–497, 1993.

Smyth, M.J., and Ortaldo, J.R. Mechanisms of cytotoxicity used by human peripheral blood $CD4^+$ and $CD8^+$ T cell subsets. J. Immunol., 151:740–747, 1993.

Sprent, J. T and B memory cells. Cell, 76:315–322, 1994.

Stanley, K., and Luzio, P. Perforin: A family of killer proteins. Nature, 334:475–476, 1988.

Stout, R.D., and Suttles, J. T cell–macrophage cognate interaction in the activation of macrophage effector function by Th2 cells. J. Immunol., 150:5330–5337, 1993.

Suda, T., et al. Molecular cloning and expression of the fas ligand, a novel member of the tumor necrosis factor family. Cell, 75:1169–1178, 1993.

Takahashi, T., et al. Generalized lymphoproliferative disease in mice, caused by a point mutation in the fas ligand. Cell, 76:969–976, 1994.

Taniguchi, T., and Minami, Y. The IL-2/IL-2-receptor system: A current overview. Cell, 79:5–8, 1993.

Townsend, A., and Bodmer, H. Antigen recognition by class I-restricted T lymphocytes. Annu. Rev. Immunol., 7:601–624, 1989.

Tschopp, J., and Nabholz, M. Perforin-mediated target cell lysis by cytolytic T lymphocytes. Annu. Rev. Immunol., 8:279–302, 1990.

CHAPTER 19

Organ Transplantation

Donnal Thomas *Donnal Thomas is a surgeon who was born in Texas in 1920. He was the first to use bone marrow allografts to treat leukemia patients. He demonstrated how total body irradiation followed by bone marrow reconstitution could eradicate leukemia cells and save the patient. He performed the first marrow allograft in 1969. He developed methods using immunosuppressive drugs and radiation to reduce graft-versus-host disease and in effect took bone marrow transplantation from an experimental procedure to practical therapy. Thomas was awarded the Nobel Prize in 1990.*

CHAPTER OUTLINE

CHAPTER CONCEPTS

1. Foreign organ grafts are rapidly rejected by untreated recipients as a result of the recognition of MHC class I and class II molecules on the graft by host T cells.
2. T cells invade the graft and cause its destruction by direct cytotoxicity. Antibodies and complement may also cause graft rejection, especially in second-set reactions and in xenografts.
3. Drugs that inhibit the immune responses such as cyclosporine permit prolonged allograft survival and have made transplantation of kidneys, heart, skin, and liver practical and effective procedures.
4. During pregnancy, the fetus protects itself against maternal immune responses by means of local immunosuppressive mechanisms active in the placenta and uterus.

In this chapter the process of rejection of foreign organ grafts is discussed. This is, of course, a cell-mediated immune response, although antibodies also play a role in rejection. The chapter opens with a brief explanation of why grafts are rejected. A description of the pathology of graft rejection and the mechanisms involved in the rejection process follows. Some special features of heart and liver grafts are described. Bone marrow grafts give rise to a condition called graft-versus-host disease, which is also discussed.

Some grafts, most notably the fetus, are not rejected by the body. The ways in which the fetus and some other organs resist rejection are described. Organ transplants, which have become an integral part of modern medicine as a result of the use of successful immunosuppressive drugs, are described as well as the paradoxical immunosuppressive effects of blood transfusions. Organ grafting has also raised some complex ethical issues, which are considered at the end of the chapter.

Although the immune responses first attracted the attention of scientists by virtue of the body's ability to recognize and eliminate invading microorganisms, the observation that animals also reject foreign skin grafts led to a broader view of the function of the immune system as a whole. In a complex multicellular organism, cells must interact. If a cell becomes abnormal and, as a result, its surface structure becomes modified, then the change may be recognized by other cells. The cell-mediated immune system is designed to identify and eliminate cells with modified surfaces. Among the cell surface structures that serve as the recognition units are the MHC class I and II molecules. These molecules are characteristic of an individual animal rather than of specific organs or cell types, although their distribution is not uniform throughout the body. The cells that recognize and respond to foreign MHC molecules include T cells and natural killer (NK) cells.

Tissue may be surgically grafted between different parts of an animal's body. This type of graft is called an **autograft.** Good examples of autografts include the transplanting of skin from an unburned to a burned area in plastic surgery, and the use of a leg vein to bypass blocked cardiac arteries. Autografts present no immunological problems whatsoever.

Isografts are tissues transplanted between genetically identical individuals (Fig. 19–1). Thus a graft between identical (monozygotic) twins is an isograft. Similarly, grafts between inbred mouse strains are isografts and present no immunological difficulties since the immune system of the recipient cannot differentiate between the graft and normal "self" tissue.

Allografts are grafts transplanted between genetically different members of the same species. Thus most of the grafts performed on humans are of this type as organs are transplanted from a donor who is usually unrelated to the graft recipient. Allografts induce a rejection reaction that must be controlled if the grafted tissue is to survive. Organs that are now routinely allografted in human patients include kidney, heart, lung, liver, bone marrow, bone, and pancreas.

Figure 19–1 Three major forms of graft and their approximate survival times in untreated recipients.

Xenografts are grafts between animals of different species. Thus the transplant of a baboon heart into a human infant is a xenograft. Xenografted organs are very different from the organs of the host and provoke an intense immune response that is very difficult to suppress.

WHY ALLOGRAFTS ARE REJECTED

The successful surgical transplantation of organs became feasible only about 100 years ago with the development of aseptic surgical techniques. Thus the process of allotransplantation is entirely artificial. It has no counterpart in nature and cannot have a physiological role. However, the identification and destruction of altered "self" cells—for example, virus-modified or chemically modified cells or tumor cells—are of great importance to an animal. Two hypotheses have been put forward to explain why foreign MHC molecules should trigger such an immune response. One is that the host T cells are really responding to the peptide epitopes that are invariably present in the groove of the MHC molecules of the grafted cells. It is almost impossible to physically separate a MHC molecule from the peptide located in its antigen-binding groove. As a result, a foreign cell inevitably carries many foreign epitopes on its MHC molecules. Alternatively, the host may be simply responding to the large quantity of foreign MHC protein present on the graft cells. Foreign MHC molecules are present on the cells of organ grafts in large amounts. This is in contrast to self MHC molecules linked to a processed epitope, which are present on the surface of antigen-presenting cells in much lower frequency. In other words, the response to cells bearing foreign MHC molecules may simply be a dose effect.

ALLOGRAFT REJECTION

If a kidney is transplanted between unrelated individuals, it will survive for about a week (Fig. 19–2). Other organs may last somewhat longer, but they are eventually rejected. The rejection process results in disruption of blood vessels and, in the kidney, decreased urine production and stoppage of function. If a second graft is transplanted from the same donor, it will be rejected by the recipient within one to two days without ever becoming functional. The accelerated reaction to a second graft is known as a **second-set reaction** and is a secondary immune response. The second-set reaction is specific for a graft from the original donor or from a donor genetically identical to the first. It is not restricted to any particular site or to any specific organ, since MHC class I molecules are present on most nucleated cells.

Pathology of Allograft Rejection

The events that take place during graft rejection vary between different types of grafts. For example, if a skin graft is placed on an animal, it takes several days for blood vessels and lymphatic connections to be established between the graft and the host. Only when these connections are made can host cells enter the graft and commence the rejection process. The first sign of rejection is a neutrophil accumulation around the blood vessels at the base of the graft. This is followed by an infiltration of **mononuclear cells** (lymphocytes and macrophages) that eventually extends throughout the grafted skin. The first signs of tissue damage are observed in the capillaries of the graft, whose endothelium is destroyed. As a result, the blood clots and stops flowing through these vessels and tissue death follows rapidly. In contrast, the blood supply to a transplanted kidney is established at the time of transplantation. The whole organ gradually becomes infiltrated with mononuclear cells that cause progressive damage

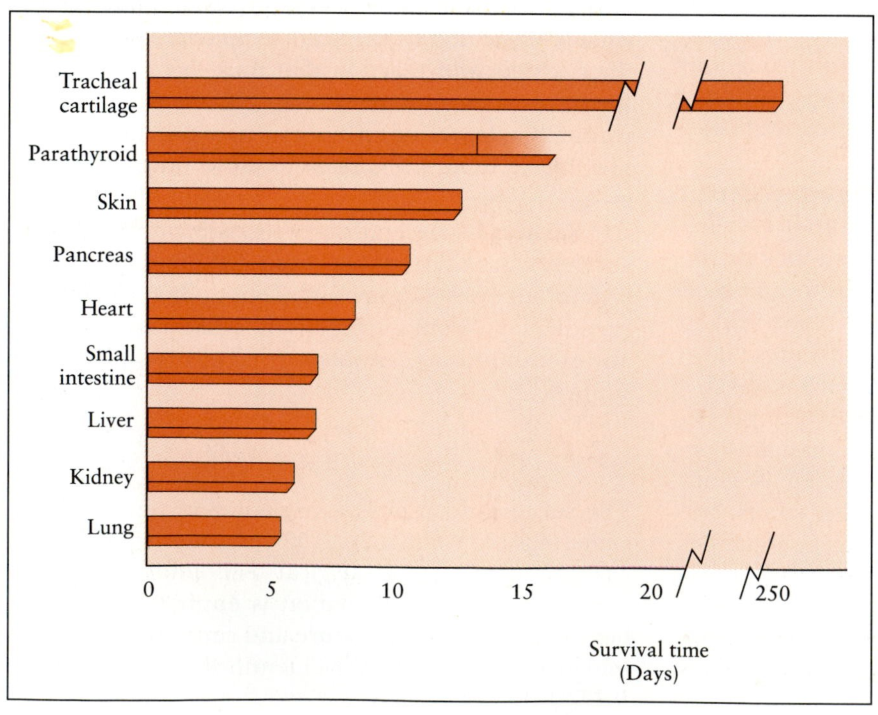

Figure 19–2 The average survival time of allografts in dogs that did not receive immunosuppressive treatment.

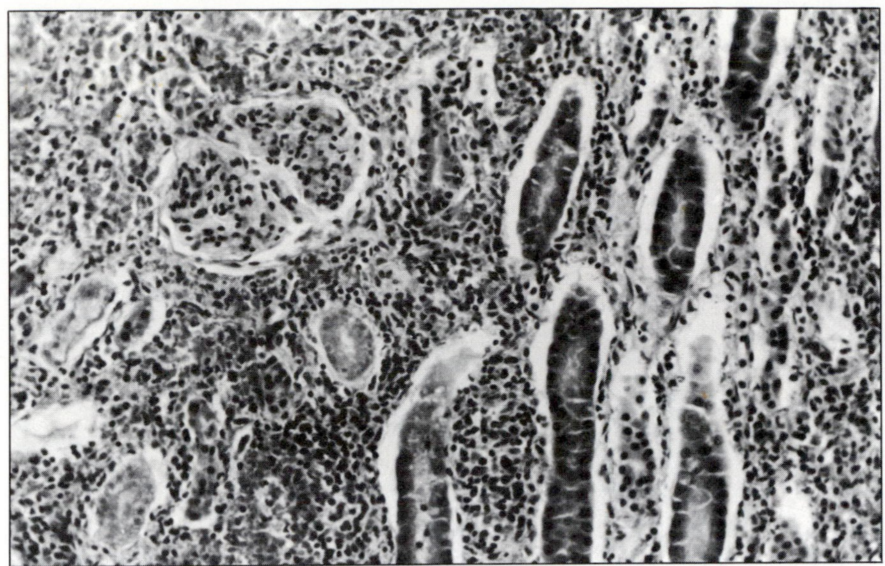

Figure 19–3 A section of canine kidney that had been acutely rejected after 14 days in an allogeneic recipient. The kidney is completely infiltrated with mononuclear cells. *(Original magnification × 350. Courtesy of Dr. R.G. Thomson.)*

to the endothelial lining of small blood vessels (Fig. 19–3). Tissue destruction, stoppage of blood flow, hemorrhage, and death of the grafted kidney follow thrombosis of these vessels. Similar mononuclear cell infiltrations are seen in heart and liver grafts (Figs. 19–4 and 19–5).

In a second-set reaction, blood vessels usually do not have time to grow into a skin graft, since an extensive and destructive mononuclear cell and neutrophil infiltration rapidly develops within the graft bed. Similarly, the blood vessels of second-set kidney grafts rapidly become blocked as a result of the action of antibodies and complement on the vascular endothelium. Xenografts are usually rejected extremely promptly. For example, pig kidney xenografts transplanted to dogs are rejected in 10 to 20 minutes because of the presence of natural antibodies to pig antigens in dog serum.

Mechanisms of Allograft Rejection

As described, the rejection of grafts is of two general types. In **first-set reactions,** grafts are rejected in a matter of days. In second-set reactions or xenografts, rejection is a much more rapid process. Both rejection processes result in damage to vascular endothelium and other cells in the graft that can be reached by the host's T lymphocytes. The first-set reaction can be divided into two stages. First, information about the antigens of the graft must reach antigen-sensitive cells,

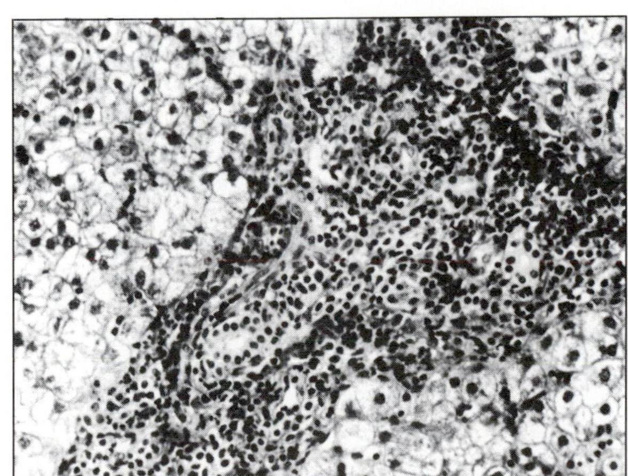

Figure 19–4 Acute liver allograft rejection. The liver is infiltrated in the portal region with a mixture of mononuclear cells. The portal vessels are hard to identify because of the endothelial damage. *(H&E. × 186. Courtesy of Dr. R. Jaffe.)*

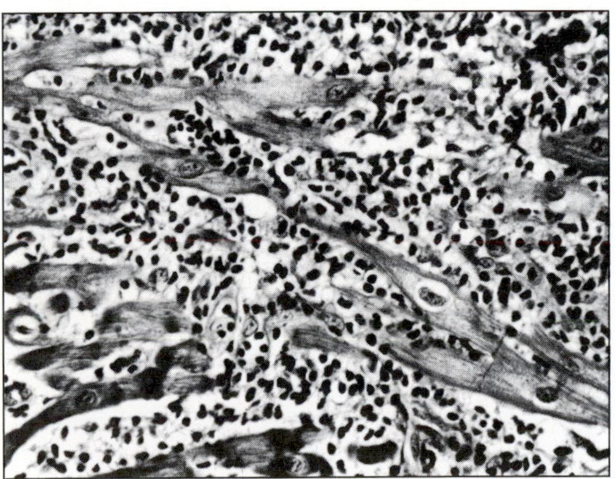

Figure 19–5 Severe cardiac allograft rejection is characterized by a mononuclear cell infiltrate with severe damage to the cardiac muscle cells. *(H&E. × 385. Courtesy of Dr. R. Jaffe.)*

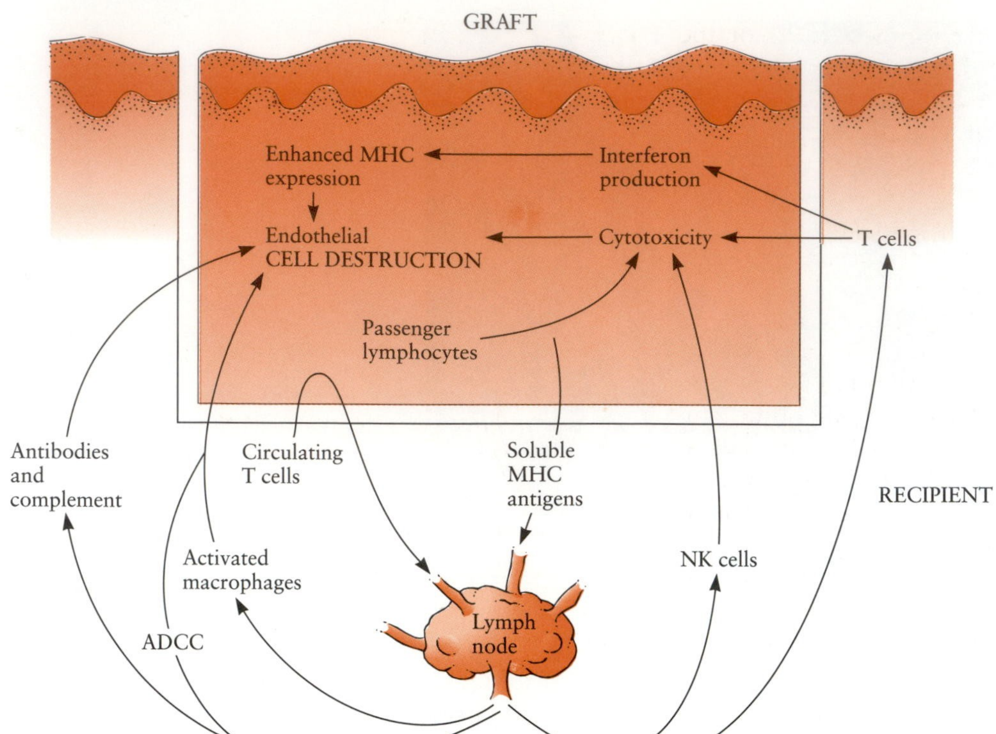

Figure 19–6 A schematic diagram showing the basic mechanisms involved in allograft rejection.

usually located in the draining lymph node. Second, effector cells from this lymph node must invade the graft and mediate its destruction.

Sensitization of the Recipient

Graft antigens may reach the antigen-sensitive cells of the recipient by several routes (Fig. 19–6). Thus graft cells may release soluble MHC molecules that enter the host, and graft dendritic cells may emigrate into host lymphoid tissues (Fig. 19–7). More important, circulating host T cells may, on passing through the vessels of the graft, encounter foreign MHC molecules either naturally on cell surfaces or on presentation by host dendritic cells that also invade the graft. The cells lining the blood vessels of the graft carry MHC class I molecules. If the grafted organ also contains donor macrophages, dendritic cells, or B cells, then the circulating $CD4^+$ T cells will also encounter foreign MHC class II molecules. The activated $CD4^+$ T cells may then

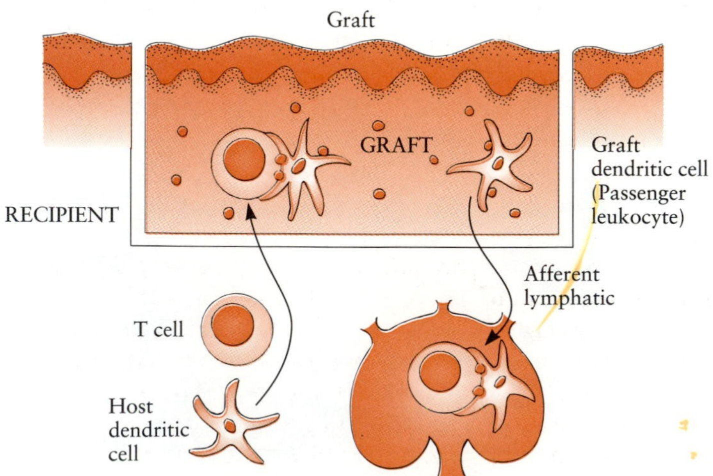

Figure 19–7 The role of dendritic cells in allograft rejection. Graft dendritic cells may leave the graft and present antigen to T cells in the host lymph node. In addition, host dendritic cells may enter the graft, trap graft antigen, and present it to host T cells.

pass from the graft in either the lymphatics or the blood vessels and lodge in the draining lymph node as a result of local triggering of the lymphocyte "trap" (Chapter 9). At least in skin grafts, the lymphatic route is probably of major importance, since rejection is considerably delayed if a graft is prevented from developing lymphatic connections with the host.

The CD4$^+$ cells that recognize the MHC class II molecules on the graft cells move to the draining lymph node and activate CD8$^+$ effector T cells within that node by secreting interleukin-2. The paracortical regions of lymph nodes draining a graft therefore contain increased numbers of large pyroninophilic lymphoblasts. The numbers of these cells are greatest about six days after grafting and decline rapidly once the graft has been rejected. In addition to these signs of an active T-cell–mediated immune response, it is usual to observe some germinal-center formation in the cortex and plasma-cell accumulation in the medulla, suggesting that antibody formation also occurs.

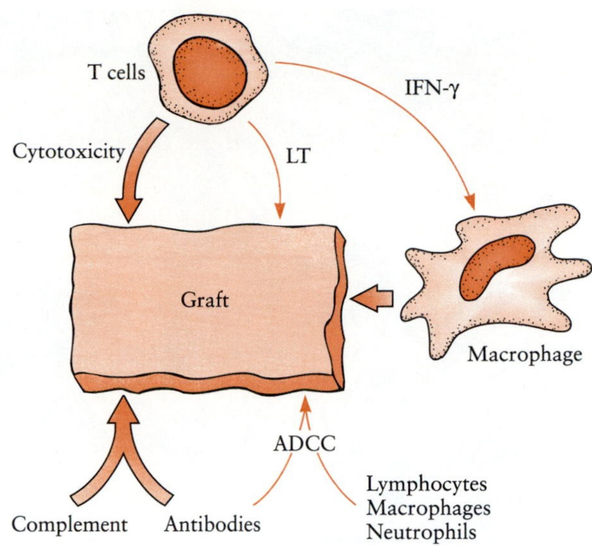

Figure 19–8 The relative importance (as denoted by the thickness of the arrows) of the different immune processes involved in graft rejection.

Destruction of the Graft

As a result of the events that occur in the draining lymph node, CD8$^+$ T cells leave the node in the efferent lymph and reach the graft through the blood. When these cells enter the graft, they recognize its MHC class I molecules (or possibly peptides associated with these class I molecules) and bind and destroy the vascular endothelium and other accessible cells by direct cytotoxicity. As a result, hemorrhage, platelet aggregation, thrombosis, and stoppage of blood flow occur. The grafted organ dies because of the failure of its blood supply. If biopsies of the grafts are performed and the lymphocytes within them examined, host MHC class I–specific T cells predominate early in the allograft response and MHC class II–specific T cells from the host predominate later. When T cells are stimulated by antigen, they produce, among other factors, interferon-γ. IFN-γ causes the cells of the graft to express increased quantities of MHC molecules. During allograft rejection, MHC class I and especially MHC class II expression is increased in transplanted tissues. As a result, the graft becomes an even more attractive target for the host's cytotoxic T cells.

Although cytotoxic T cells are most important in destroying foreign grafted tissues, antibodies also play a significant role in graft rejection (Fig. 19–8). Antibodies directed against graft MHC class I molecules act with complement and neutrophils, or through antibody-dependent cytotoxic cell activity, to cause vascular endothelial cell destruction (Chapter 18). Antibodies seem to be of greatest importance in second-set and xenograft reactions.

Cardiac Allografts

The most common form of rejection of cardiac allografts is acute cell-mediated rejection. It is associated with massive lymphocytic infiltration and cellular necrosis (see Fig. 19–5). In some cases, however, the rejection process may be slow, and as a result, the pathology of the rejected organ may be uniquely different. Endothelial cells are stimulated to proliferate in the walls of cardiac blood vessels, leading to a condition called allograft arteriosclerosis (Fig. 19–9). Ordinary arteriosclerosis is focal and eccentric, whereas allograft arteriosclerosis is generalized throughout the graft and is concentric. Cytokines released by invading T cells cause endothelial smooth muscle cells to grow. The resulting obliteration of the blood vessel lumen eventually results in heart failure. A similar lesion is sometimes seen in renal allografts undergoing chronic rejection.

Liver Allografts

Liver transplantation is a very successful and practical treatment for massive liver failure. Thus 85% of adults receiving a liver allograft return to full function with appropriate immunosuppressive treatment, although about 75% of these patients develop some form of infection. Massive hemorrhage is the greatest threat to survival. One-year survival is about 70%; two-year survival is more than 60%.

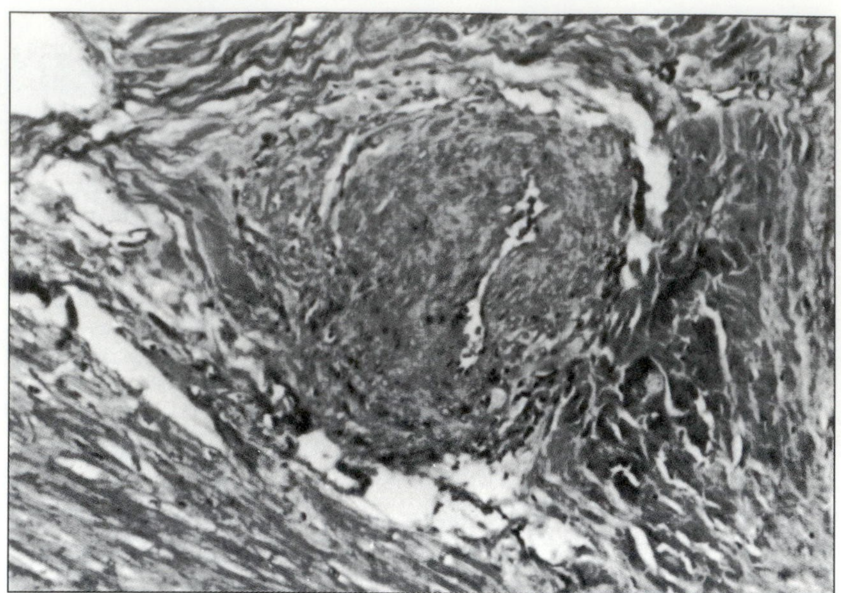

Figure 19–9 A section of coronary artery from a canine heart allograft showing allograft arteriosclerosis. Note how the artery is almost completely blocked as a result of proliferation of the cells lining the artery. *(Original magnification × 100. From Penn, O.C., et al. Transplantation, 22:313, 1976. With permission of Williams and Wilkins, Publishing Co.)*

GRAFT-VERSUS-HOST DISEASE

If healthy lymphocytes are injected into the skin of an allogeneic recipient, local inflammation occurs as a result of the lymphocytes attacking the host cells—the graft attacks the host. Provided the recipient has a functioning immune system, this **graft-versus-host (GVH) reaction** is not serious, since the recipient is able to destroy the foreign lymphocytes and thus terminate the reaction. If, however, the recipient cannot reject the grafted lymphocytes, then these cells may cause uncontrolled destruction of the host's tissues, disease, and eventually death.

An animal is unable to efficiently reject grafts from its parents since it possesses and is tolerant to some of their MHC molecules. If, on the other hand, parental lymphocytes are injected into their offspring, the cells will attack the recipient because they recognize its cells as foreign—it carries the antigens derived from the other parent (Fig. 19–10). GVH disease may also occur if the recipient of a lymphoid cell graft has been immunosuppressed or is immunodeficient. This is of major importance in individuals who receive bone marrow or thymus allografts to cure congenital immunodeficiencies or leukemia (Chapter 26). In order to treat some forms of leukemia, patients are made totally lymphopenic by whole-body radiation and are then given a healthy marrow transplant. Although the recipient cannot reject the allogeneic marrow cells, the marrow cells can attack the host cells and cause severe GVH disease.

The lesions generated in GVH disease depend on the MHC disparity between graft and host. When the difference is in the MHC class I molecules, then the

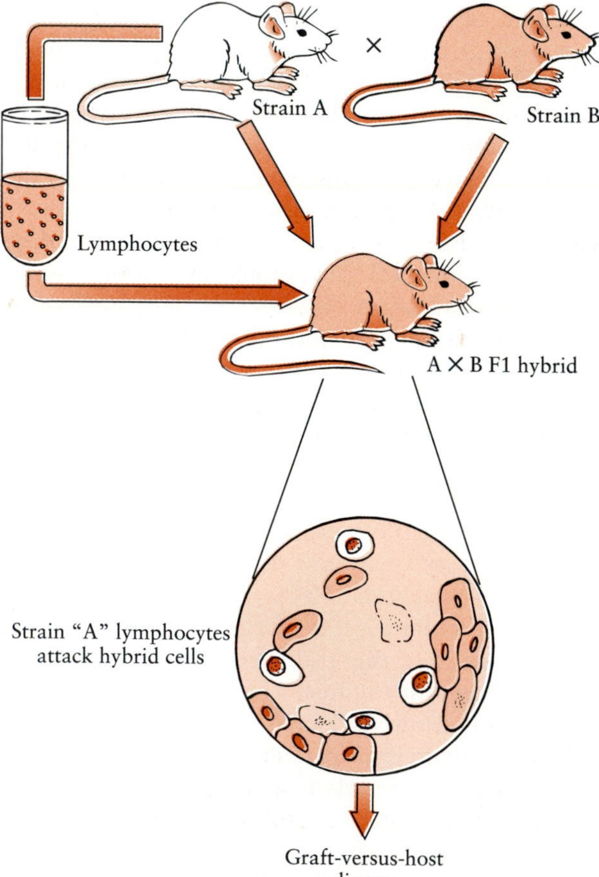

Figure 19–10 Graft versus host disease. If parental lymphocytes are injected into their offspring, the grafted cells will attack the recipient. The recipient animal will not attack the grafted cells since they possess surface antigens in common.

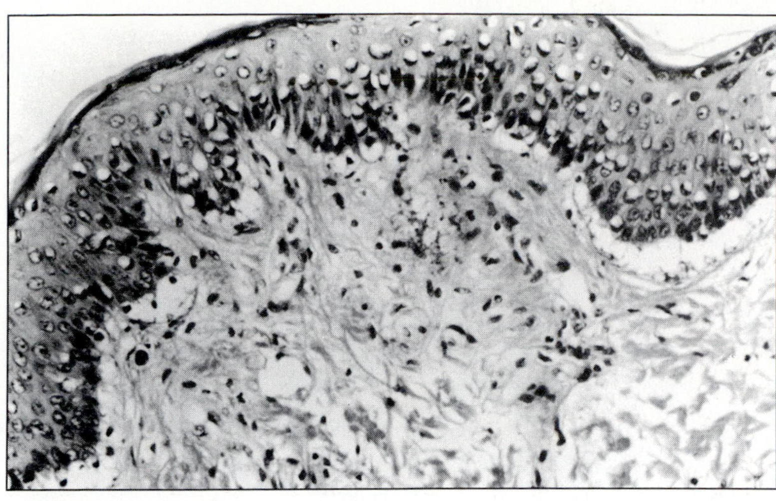

Figure 19–11 A section of skin from a patient who developed graft-versus-host disease following bone marrow transplantation. Note the disruption at the base of the epidermis leading to cleft formation and vesiculation. *(Courtesy Dr. Daniel Weisdorf.)*

invading lymphocytes attack all the nucleated cells of the host. The result is a wasting syndrome characterized by bone marrow destruction, leading to total loss of blood cells (pancytopenia) and aplastic anemia, loss of T and B cells, and hypogammaglobulinemia. Individuals suffering from this type of GVH disease develop a lymphocyte infiltration of the intestine, skin, and liver (Fig. 19–11). These lymphocytes secrete TNF-α, which causes mucosal destruction and diarrhea, skin and mouth ulcers, liver destruction, and jaundice. Affected persons stop growing and may die if the condition is untreated. If, in contrast, there is a MHC class II disparity between the graft and host, then the graft and host lymphocytes will interact. This may lead to immunostimulation, autoantibody formation, and clinical signs resembling those of systemic lupus erythematosus (Chapter 31) and polyarthritis. Treatment of GVH disease is similar to the immunosuppressive therapy used to prevent graft rejection.

GRAFTS THAT ARE NOT REJECTED

Privileged Sites

Certain areas of the body, such as the anterior chamber of the eye, the cornea, and the brain, lack effective lymphatic drainage. Although antigen derived from grafts made in these sites may reach lymphoid tissues, cytotoxic effector cells cannot reach the graft, and these grafts survive relatively well. For this reason, corneal allografting is a successful procedure.

Sperm

Allogeneic sperm, of course, can successfully and repeatedly penetrate the female reproductive tract without usually provoking a significant immune response. Seminal plasma is immunosuppressive, and sperm exposed to this fluid are nonimmunogenic, even after washing. Prostatic fluid, one of the immunosuppressive components of seminal plasma, also inhibits complement.

Pregnancy

Because the fetus possesses antigens derived from its father, it is an allograft within the mother. Nevertheless, the fetus is consistently successful in establishing and maintaining itself through pregnancy, in spite of great histocompatibility differences (Fig. 19–12). The uterus is not a privileged site, since grafts of other tissues, such as skin, made in the uterine wall are readily rejected. Under some circumstances, the mother may make antibodies against fetal blood group antigens, and these can destroy fetal red blood cells either in utero, as in humans, or following ingestion of colostrum, as occurs in other mammals (Chapter 30).

The immunological destruction of the fetus is prevented by the combined activities of several immunosuppressive mechanisms. First, MHC molecules are not expressed on preimplantation embryos or oocytes. Once the placenta forms, the fetus is protected from the mother's immune system by the trophoblast (that part of the placenta in closest contact with maternal tissue). Cells within the trophoblast do express MHC class I molecules, but these are not the highly polymorphic class Ia molecules. Instead the cells make HLA-G, a nonpolymorphic class Ib molecule. This molecule, found only in the trophoblast, fails to trigger a T cell response and protects cells against NK cell-mediated lysis. As might be expected, trophoblast cells do not express MHC class II molecules. Another major cell-membrane structure found on trophoblast cells is CD46, also known as membrane cofactor protein. This

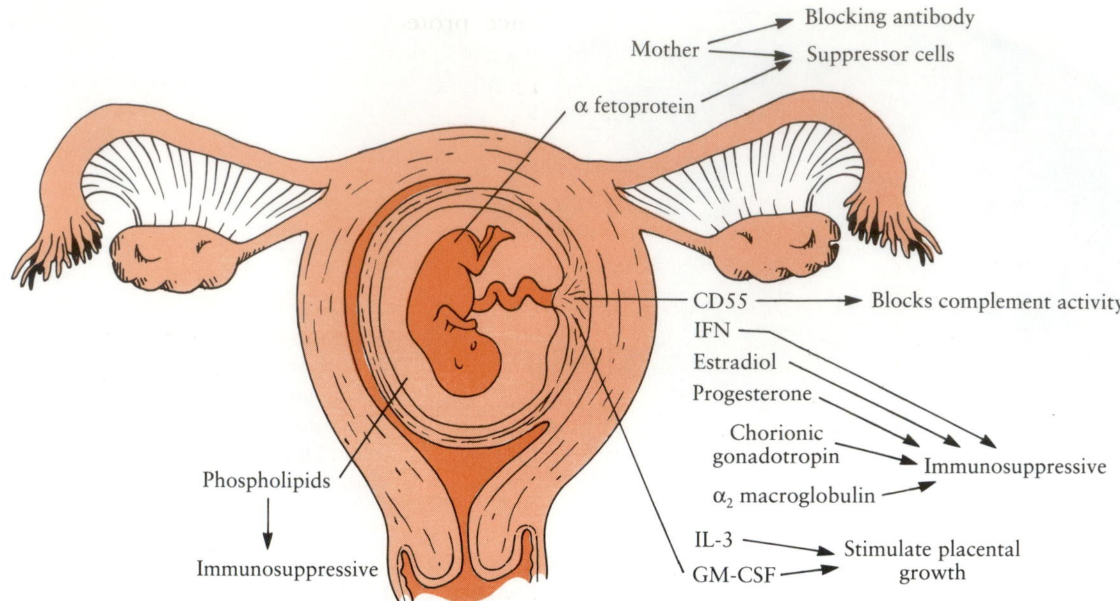

Figure 19–12 Some of the factors that interfere with the maternal immune response to the fetus and so permit a successful pregnancy.

molecule inhibits complement activation and so prevents complement-mediated cell lysis. In addition CD55 (decay accelerating factor—Chapter 16) is incorporated in the trophoblast at the fetomaternal interface and so protects it against complement attack. Cytokines such as IFN-γ, which usually enhance MHC expression, have no effect on trophoblast cells. Nevertheless, in some mouse strains, up to 95% of pregnant animals make antibodies against the fetal MHC. In other strains none of the mothers makes these antibodies. These antibodies develop only at the end of a second pregnancy and are not cytotoxic. Up to 40% of women make antibodies to fetal MHC molecules after giving birth. Notwithstanding this, the presence of these antibodies has no apparent effect on the course of the pregnancy.

Second, the fetus is a source of locally active immunosuppressive factors, including the hormones estradiol and progesterone and possibly also chorionic gonadotropin. The major protein in fetal serum, α-fetoprotein, may be immunosuppressive because of its ability to stimulate suppressor cell function. (Some isoforms of α-fetoprotein are immunoregulatory; others are not.) Some pregnancy-associated glycoproteins and a trophoblast-derived interferon (IFN-ω) have immunosuppressive properties. Amniotic fluid is rich in immunosuppressive phospholipids.

Third, blocking antibodies may be produced in response to fetal antigens. These coat placental cells, masking antigens and thus preventing their destruction by maternal T cells. These antibodies can be eluted from the placenta and shown to suppress other cell-mediated immune reactions against paternal antigens, such as graft rejection. Absence of this blocking antibody accounts for some cases of recurrent abortion in women. Nevertheless, it can be shown that totally immunodeficient mice can have successful pregnancies.

It must not be assumed from the foregoing list of immunosuppressive factors that the pregnant female is grossly immunosuppressed. In fact, the **immunosuppression** generated by the fetus is very local in nature. Pregnant animals have only minor deficiencies in cell-mediated immune reactivity to nonfetal antigens, showing, for example, a slight delay in the rejection of skin grafts or transient unreactivity to the tuberculin skin test (Chapter 18). NK cell activity is also suppressed during pregnancy. Nevertheless a local immune response to fetal antigens stimulates placental function. Thus hybrid placentas are larger than the placentas of inbred animals, and females tolerant to paternal antigens have smaller placentas than intolerant females. Other studies show that mothers sensitized to paternal MHC molecules have better fetal survival. This effect is due to the stimulatory effect of maternal IL-3 and GM-CSF on trophoblast growth.

Cultured or Stored Organs

If an organ or tissue is grown in tissue culture or stored frozen, its potential for successful transplantation is greatly enhanced (Fig. 19–13). Thus, if thyroid tissue is grown in an organ culture for about 25 days before transplantation, it will survive in a recipient as if it were

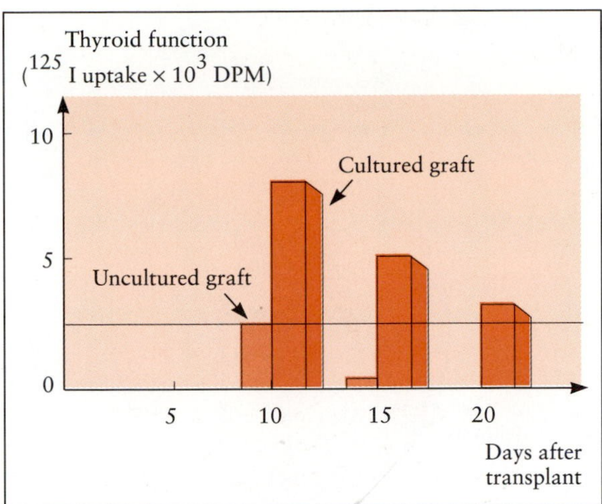

Figure 19–13 Thyroid allografts normally function for less than ten days before they are destroyed. (As measured by a thyroid function uptake of less than 2.5 × 10^3 d.p.m.) If cultured in vitro, for several weeks prior to grafting, they may survive for longer than 20 days. *(From Lafferty, K.J., et al. Science, 188:260, 1975. Copyright 1975 by the American Association for the Advancement of Science. With permission.)*

an isograft. Uncultured thyroids are rejected in about 10 days. If a recipient is first primed by exposure to an uncultured graft, it will then be able to reject a cultured one. Conversely, if an accepted cultured graft is returned to the donor, it will provoke acute rejection. This phenomenon is believed to be due to the loss of dendritic cells from the thyroid tissue during culture. If true, the rejection of a thyroid allograft may be largely a response of host lymphocytes to graft dendritic cells. This suggestion is supported by the observation that pretreatment of donors with cytotoxic drugs or antiserum against dendritic cells reduces the subsequent allograft rejection. Similarly, if a kidney is disrupted into its constituent cells and each cell population is tested for its ability to provoke allograft rejection, by far the most immunogenic components are the MHC class II positive ''passenger'' leukocytes (see Fig. 19–7).

Freeze-dried pig dermis is useful in covering extensive wounds, and there have been reports of successful bone and aortic allografts after storage of the grafts in liquid nitrogen. Treatment of pig heart valves with glutaraldehyde also renders them immunologically inert so that they can be employed to replace defective valves in humans.

Immunologically Favored Organs

On some occasions, individuals who have received a liver allograft do not mount a response against it, whereas others may mount a weak and easily suppressed response. Liver allografts may, by their presence, protect another allograft, such as a kidney from the same donor, resulting in its prolonged survival. If recipients are first sensitized and then given liver allografts, subsequent grafts are retained, the implication being that liver grafting also blocks immunological memory. These effects probably result from the release from the liver of immunosuppressive molecules, such as hepatic glycoferroprotein and bilirubin. In some individuals who have maintained functioning renal allografts for several years, immunosuppressive therapy may be gradually reduced and occasionally is totally discontinued as a graft acceptance becomes complete. This phenomenon may be an example of high-dose tolerance, that is, tolerance that results when the amount of antigen is in excess of the ability of the antigen-sensitive cells to respond to it (Chapter 21). In the case of graft acceptance following prolonged immunosuppression, the antigen-sensitive cells are gradually eliminated by the immunosuppressive agents. Once their numbers are sufficiently low, the mass of grafted tissue may be sufficient to establish and maintain tolerance. Alternatively, over time, the graft vascular epithelium may be replaced by cells of host origin. As a result of the formation of this barrier, the graft is protected against rejection.

SUPPRESSION OF THE ALLOGRAFT RESPONSE

The techniques available for inhibiting the allograft response (**immunosuppression**) may be classified into three general groups. First, the most widely employed techniques involve drugs or radiation that, by inhibiting cell division, reduces the multiplication of antigen-sensitive cells upon encountering antigen. This approach is crude and dangerous, since other rapidly proliferating cell populations, such as intestinal epithelium and bone marrow cells, may also be destroyed. Second, techniques that selectively eliminate T cells can be used. This can be done by means of specific anti–T cell serum, by monoclonal antibodies, by selective drugs, or by thoracic duct drainage. Third, the normal immune responses are well controlled by the body, and although not widely employed at present, manipulation of these natural control mechanisms is likely to provide useful techniques for the control of the allograft response in the future.

Radiation

X-radiation exerts its effect on cells by several mechanisms. The simplest is through ionizing rays hitting an essential, unique molecule within the cell. The most important of these molecules are the nucleic acids, particularly DNA. A loss of even one nucleotide entails a permanent mutation of a gene, with potentially lethal effects on the progeny of the affected cell. Another mechanism involves the formation of reactive free oxygen and hydroxyl radicals as a result of ionization in aqueous solutions. These free radicals can react with dissolved oxygen to form peroxides that have toxic effects on many cell processes. Although x-radiation is of some use in prolonging graft survival in many experimental animals, it is usually necessary to irradiate the recipient for several days prior to grafting. This is impractical in clinical situations since it is not possible to predict when an organ such as a kidney will become available.

Corticosteroids

Species may be classified as corticosteroid sensitive or corticosteroid resistant on the basis of the ease with which they can be depleted of lymphocytes. Most laboratory rodents are much more sensitive to the immunosuppressive effects of corticosteroids than humans, and care should therefore be taken not to extrapolate laboratory animal results directly to humans (Table 19–1).

Steroid receptors differ from membrane receptors in that they are found inside the cell and act by regulating DNA transcription. Thus steroids are absorbed directly into a cell, where they bind to receptors in the cytoplasm. The steroid-receptor complexes are then transported to the nucleus, where they bind to the chromatin protein and the DNA (Fig. 19–14).

Table 19–1 The Effects of Corticosteroids on the Immune System

Neutrophils
Neutrophilia
Depressed chemotaxis
Depressed margination
Depressed phagocytosis
Depressed ADCC
Depressed bactericidal activity
Macrophages
Depressed chemotaxis
Depressed phagocytosis
Depressed bactericidal activity
Depressed IL-1 production
Depressed antigen processing
Lymphocytes
Depressed proliferation
Depressed T-cell responses
Impaired T-cell–mediated cytotoxicity
Depressed IL-2 production
Depressed lymphokine production
Immunoglobulins
Minimal decrease
Complement
No effect

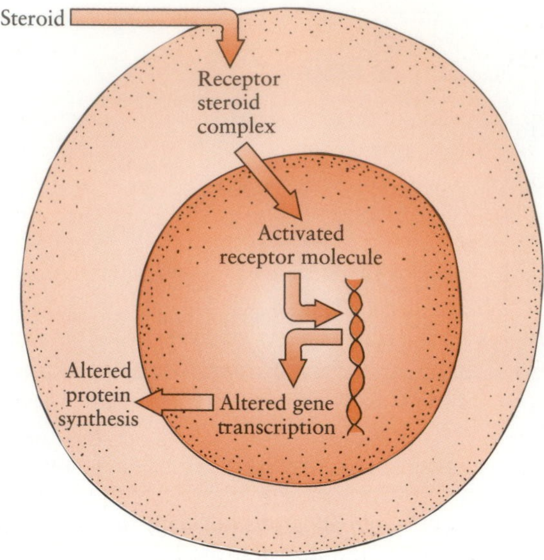

Figure 19–14 The mode of action of corticosteroids in altering protein synthesis by cells. Steroids bind to cytoplasmic receptors. The receptor-steroid complexes bind to the cell's DNA and trigger new protein synthesis.

As a result of derepression of RNA transcription, new mRNA is formed and initiates new protein synthesis. These newly synthesized proteins mediate the effects of corticosteroids. For example, a steroid-induced protein called macrocortin or lipomodulin blocks the activities of cell membrane phospholipase A2 and thus prevents the synthesis of prostanoids such as the leukotrienes and prostaglandins.

Corticosteroids act in four areas. They have effects on leukocyte circulation; they influence the immune effector mechanisms in lymphocytes; they modulate the activities of inflammatory mediators; and they modify protein, carbohydrate, and fat metabolism.

The effects of corticosteroids on leukocyte circulation vary between species. In humans, the number of circulating eosinophils, basophils, and lymphocytes declines abruptly when steroids are administered. The number of neutrophils, on the other hand, increases as a result of a decreased adherence to vascular endothelium (margination) and reduced emigration into inflamed tissues. Neutrophil, monocyte, and eosinophil chemotaxis is suppressed by corticosteroids, but neutrophil random migration is enhanced. Corticosteroids suppress the cytotoxic and phagocytic abilities of neutrophils. Macrophage production of interleukin-1 is also diminished.

Corticosteroids in high doses cause thymic atrophy and trigger T-cell apoptosis. In lower doses they depress the ability of T cells to produce IL-2 and thus interfere with their responses to mitogens such as concanavalin A and phytohemagglutinin. Lymphocyte proliferation in the mixed lymphocyte reaction is suppressed, suggesting that lymphocytes interfere with the recognition of MHC class II molecules. (The mixed lymphocyte reaction is the proliferation that occurs when two populations of allogeneic lymphocytes are cultured together.) NK and some ADCC reactions may be refractory to corticosteroid treatment, and in cattle corticosteroids may increase serum interferon levels. The effects on antibody responses during corticosteroid therapy are variable and depend on the time of administration and the dose given. In general, B cells tend to be corticosteroid resistant, and enormous doses are usually required to depress antibody synthesis.

The synthetic glucocorticoids suppress acute inflammation. They inhibit the increase in vascular permeability and vasodilation. As a result, they prevent edema formation and fibrin deposition. At the same time, corticosteroids block the emigration of leukocytes from the capillaries. This is accomplished by depressing metabolism in neutrophils and macrophages, thus inhibiting the release of lysosomal enzymes and impairing antigen processing by macrophages. In the later stages of inflammation, they inhibit capillary and fibroblast proliferation (perhaps by blocking interleukin-1 production) and enhance collagen breakdown. As a result, corticosteroids delay wound healing.

Cytotoxic Drugs

The major immunosuppressive drugs, having been designed to inhibit cell division, act on various stages of nucleic acid synthesis and activity. Three of these drugs, cyclophosphamide, azathioprine, and methotrexate, are highly toxic and fairly nonspecific in their actions. Their use in immunology is now largely confined to the aggressive treatment of autoimmune diseases. They have been replaced by two very specific and relatively nontoxic drugs, cyclosporine and FK506, in the treatment of allograft rejection.

The alkylating agents, the most important of which is cyclophosphamide, cross-link DNA helices, preventing their separation, and thus inhibit cell division and gene transcription (Fig. 19–15). As a result,

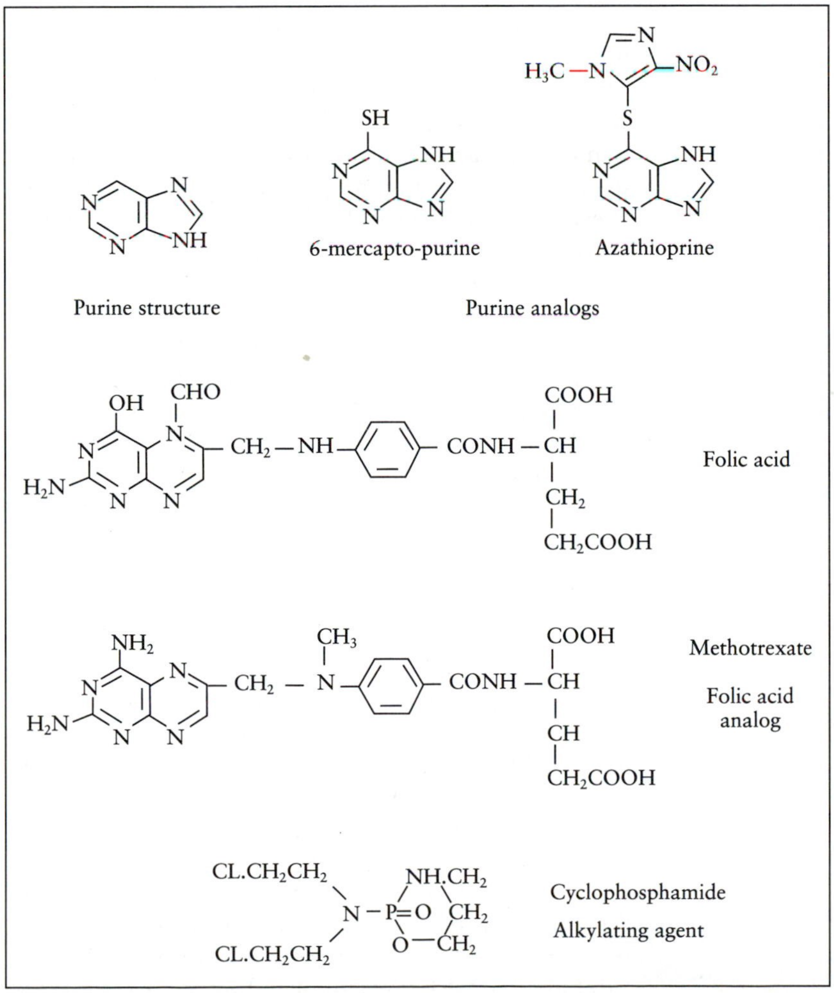

Figure 19–15 The structure of some important immunosuppressive agents.

they are toxic for both resting and dividing cells, especially for dividing immunocompetent cells. They impair both B-cell and T-cell responses, especially the primary immune response. They suppress macrophage function and therefore have an antiinflammatory effect. Their main toxic effect is bone marrow suppression, leading to loss of white cells (leukopenia) with predisposition to infection, thrombocytopenia, and anemia.

Unlike cyclophosphamide, azathioprine affects only proliferating, not resting, lymphocytes and is therefore a less potent immunosuppressant. It can suppress both primary and secondary antibody responses if given after antigen exposure. Like cyclophosphamide, its major toxic effect is bone marrow depression.

Methotrexate is a folic acid antagonist that blocks the synthesis of tetrahydrofolate, leading to failure to synthesize thymidine and purine nucleotides. It can suppress antibody formation, and its side effects are similar to those seen with cyclophosphamide and azathioprine.

Less commonly used drugs include the alkaloids vincristine and vinblastine, which bind to tubulin and prevent mitosis; hydroxyurea, which blocks DNA synthesis; and 5-fluorouracil, which is a pyrimidine analog. Some drugs used in the treatment of autoimmune disorders are also immunosuppressive. These include aspirin and the gold salts sodium aurothiomalate and aurothioglucose, which depress antigen-induced blastogenesis.

Cyclosporine

Perhaps the single most important step in the development of routine, successful organ allografting has been the development of potent but selective immunosuppressive agents. Of these, cyclosporine (Sandimmune) has been by far the most successful.

Cyclosporine is an immunosuppressive polypeptide derived from two species of fungi, *Tolypocladium inflatum* and *Cyclindrocarpon lucidum.* These fungi yield several natural forms of cyclosporin, including cyclosporins A, B, C, D, E, and H. Cyclosporine refers specifically to cyclosporin A (CyA). This molecule of 1203 Da consists of 11 amino acids arranged in a circular fashion (Fig. 19–16). Ten of the 11 amino acids are of conventional structure; the eleventh is unique. Be-

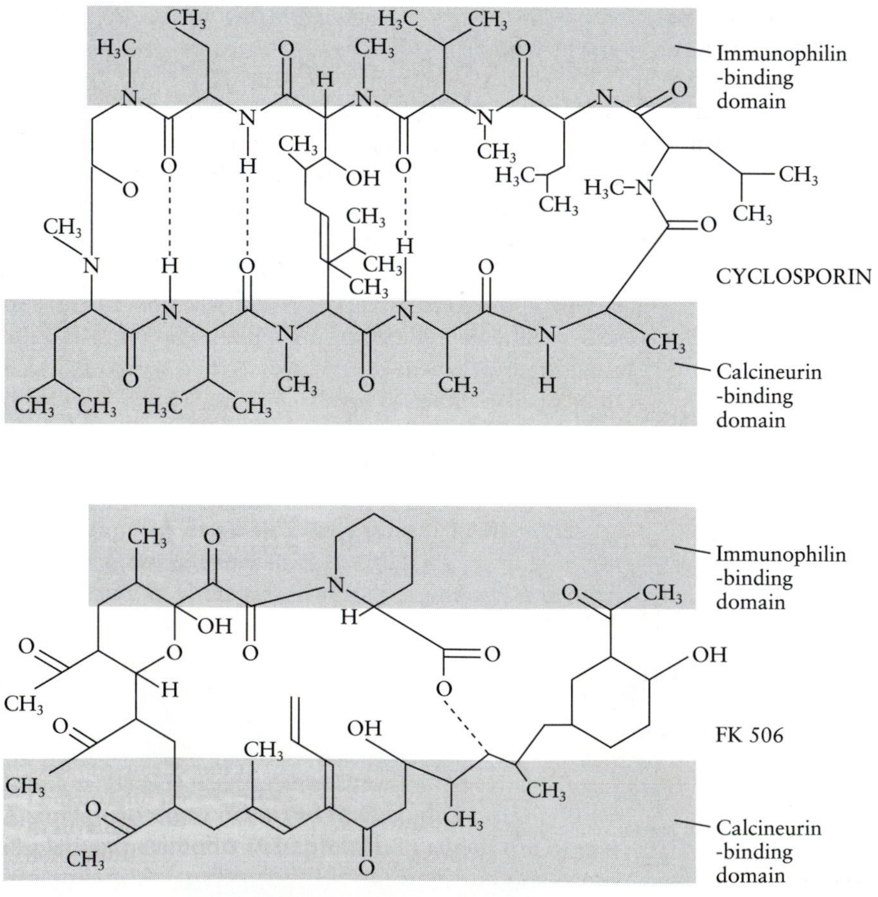

Figure 19–16 The structure of cyclosporin A and FK 506 denoting their two binding domains, one for immunophilins and one for calcineurin.

cause of this structure, CyA has two distinct surfaces that allow it to bind two proteins simultaneously. Thus CyA enters the cytoplasm and binds to intracellular receptor proteins called immunophilin. One immunophilin is an enzyme called cytophilin which is a rotamase. Rotamases can rotate the bonds between proline and other amino acids in a peptide chain and thus control protein folding. Since cyclosporine blocks this activity, it has been assumed that this was its major mechanism of action. It has been shown, however, that some derivatives of cyclosporine can bind rotamases but are not immunosuppressive. Recently, it has been shown that the cyclosporine–immunophilin complex also binds to the intracellular transmitter calcineurin and so interferes with signal transduction from the TCR by preventing activation of protein kinases C and the production of the DNA-binding protein NF-AT. As a result, cyclosporine-treated T cells fail to make IL-2 or IFN-γ and inhibit the expression of IL-2 receptors. The net effect of cyclosporine treatment is therefore the blocking of the helper T-cell response without affecting resting lymphocytes.

Cyclosporine inhibits the production of IFN-γ by helper T cells and hence prevents IFN-γ induction of MHC class I expression in grafts. Since corticosteroids have a similar effect, the combination of corticosteroids and cyclosporine is especially potent and can give essentially 100% survival of renal allografts. The side effects of cyclosporine include hypertension, tremors, and hyperesthesia, hypertrichosis (hair growth, even in nude mice!), gingivitis, fatigue, anemia, and renal failure. Patients receiving cyclosporine occasionally develop B-cell lymphomas. Epstein-Barr virus has the ability to transform B cells into immortal (tumor) cells in culture, but in vivo this is prevented by cytotoxic T cells (Chapter 25). Cyclosporine inhibits these cytotoxic T cells and hence permits the malignant B cells to grow in an uncontrolled fashion. Because of its specificity for cytotoxic and some helper T cells, cyclosporine suppresses graft rejection while at the same time leaving other immune functions intact. It therefore has a significant advantage over other older immunosuppressants. The use of cyclosporine has made tissue transplantation a routinely successful and safe procedure (Fig. 19–17).

FK506 is an experimental drug that is obtained from the soil bacterium *Streptomyces tsukubaensis.* It is an 882-kDa macrolide antibiotic containing several complex functional groups incorporated in a 23-membered ring. It has very similar biochemical and clinical effects to those of cyclosporine, although it is much more potent. It can rapidly terminate acute rejection episodes with few apparent side effects. FK506 binds to an immunophilin called macrophilin, a rotamase unrelated to cyclophilin and calcineurin that inhibits T-cell signal transduction. Rapamycin is a macrolide antibiotic that has many structural features in common with FK506, although its immunosuppressive profile is different. It binds to a rotamase, but its intracellular target is unknown.

Depletion of Lymphocytes

Because of the many adverse side effects of the nonspecific immunosuppressive agents (not the least important of which is an increased predisposition to infection), considerable effort has been made to find more specific alternative immunosuppressive procedures. One relatively simple technique that largely depletes T cells is to administer an antiserum specific for T lymphocytes. Antilymphocytic serum (ALS) suppresses the cell-mediated immune response and leaves

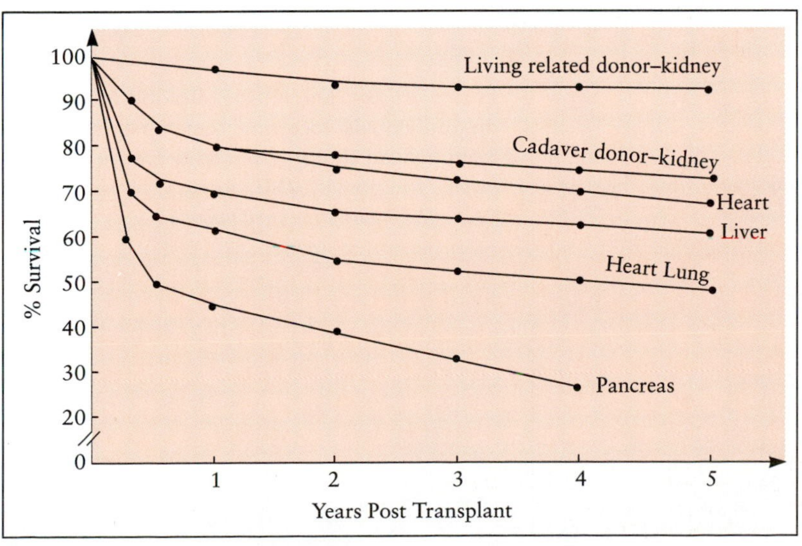

Figure 19–17 The survival of organ graft recipients in the years following transplantation. *(From Sheil, A.G.R. Transplant Proc., 19:2782–2786, 1987. Reprinted by permission of Grune and Stratton, Inc., and the author.)*

the humoral immune response relatively intact. In practice, ALS has proved to be of variable efficiency and causes severe side effects. In experiments in mice, ALS-treated animals have been shown to accept rat xenografts, whereas clinical use of ALS in humans has not been universally accepted as being useful. A much more specific antiserum with precise targeting is monoclonal anti-CD3. Anti-CD3 is directed only against T cells and appears to be very effective in reversing graft rejection in humans. An even more specific monoclonal antibody is anti-IL-2R. This attacks only activated lymphocytes. Anti-IL-2R helps prevent renal graft rejection and has fewer side effects than traditional antilymphocyte globulin.

Blood Transfusions

For many years, the administration of blood transfusions to potential renal allograft recipients prior to transplantation was discouraged on the grounds that they might sensitize the recipient and so hasten graft rejection. Experience has shown, however, that multiple transfusions given prior to transplantation may enhance graft survival in some individuals. In one series, transfusions given before renal allografts changed an 18% survival in controls to 70% survival at one year (Fig. 19–18). In another series, 20% of untransfused recipients showed accelerated rejection as compared with 10% of transfused recipients. The mechanism of this transfusion effect is becoming clearer. The transfused blood must contain leukocytes to be effective. The transfusion should be given well before grafting (up to two years) since transfusions administered immediately prior to transplantation are less effective. The best results are obtained when the blood donor and the recipient share at least one MHC class II molecule, so that one HLA-DR antigen is matched and one is mismatched. In these cases, graft survival is improved and alloantibody formation is decreased. Transfusion recipients show a significant decrease in cytotoxic T-cell precursors. This decrease is not seen if the transfused blood is totally DR-mismatched or if it is completely matched.

THE ETHICS OF ORGAN GRAFTING

Transplantation is a highly sophisticated and very expensive form of treatment. The cost of a liver graft in 1989 was $56,900. With complications, the amount could rise to as high as $151,200, and a medical evaluation beforehand costs up to $5200. Society must judge whether spending such enormous sums on treating one patient is justified since it is probable that similar sums spent on preventive medicine programs could yield much greater benefits for many patients.

Extensive organ replacement is now possible. For example, the lung, heart, liver, and kidney of a single donor have been transplanted into a recipient, and organ transplants are unquestionably life-saving (Fig. 19–19). Nevertheless, there are many more potential transplant recipients than there are donors of organs such as the heart and liver. In 1988, for example, between 15,000 and 20,000 people with terminal heart disease could have benefited from a heart transplant. New hearts were given to 1647 patients. A very similar ratio applies to liver and lung grafts. Who then will receive a graft? Should the young be favored over the aged on the grounds that they have more to contribute to society? Should the well educated be favored over the poorly educated? How important is emotional stability and the ability to withstand the stress of waiting many months for a graft that might never become available? Recipients must wait an average of nine months for a heart, five months for a liver, and eighteen months for lungs. How does the physician decide

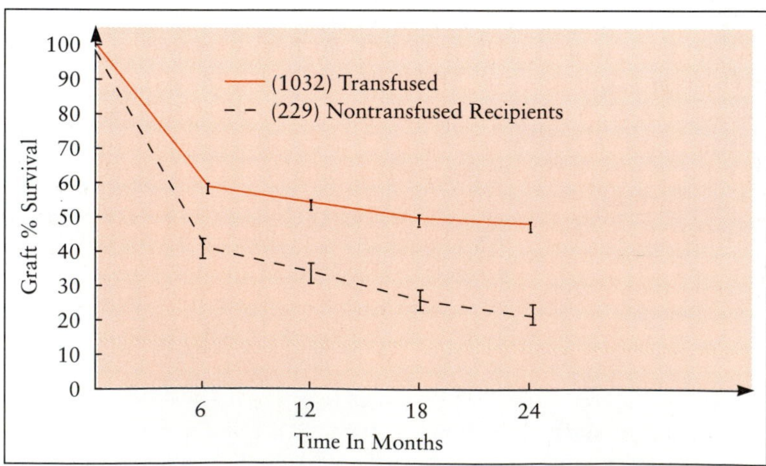

Figure 19–18 The effects of blood transfusion on individuals receiving their first transplant from a dead donor. *(From Spees, E.K., et al. Transplant. Proc., 13:155, 1981. Reprinted by permission of Grune and Stratton, Inc., and the author.)*

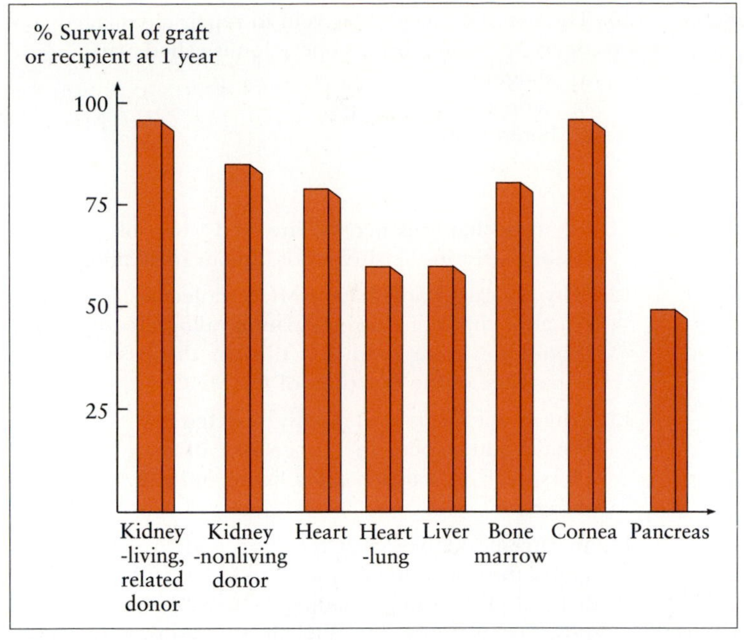

Figure 19–19 The percentage survival of human organ grafts at one year after transplantation.

who will receive a graft and live and who will be denied a graft and die? These are exceedingly difficult decisions and place enormous stress on the physicians who have to make them. The problems are compounded by the potential commercialization of transplantation. Thus the buying and selling of body organs has become a significant business in India and the Middle East. Although the immunological problems associated with organ transplantation are largely overcome, these and other ethical issues remain to be resolved.

KEY WORDS

Allograft p. 289
Allograft arteriosclerosis p. 293
Antilymphocyte serum p. 301
Autograft p. 289
Azathioprine p. 300
Blocking antibody p. 296
Blood transfusion p. 302
Bone marrow allograft p. 294
Cardiac allograft p. 293
Corticosteroid p. 298
Cultured allograft p. 296
Cyclosporine p. 300
Cytophilin p. 301
Cytotoxic drug p. 299
Cytotoxic T cell p. 293
Dendritic cell p. 292
α-Fetoprotein p. 296
First-set reaction p. 291
FK506 p. 301
Graft destruction p. 293
Graft rejection p. 290
Graft-versus-host reaction p. 294
Immunosuppression p. 297
Interferon-γ p. 293
Isograft p. 289
Liver allograft p. 293
Macrocortin p. 298
MHC class Ib molecule p. 295
Passenger leukocyte p. 295
Pregnancy p. 297
Privileged site p. 295
Second-set reaction p. 290
Sensitization p. 292
Trophoblast p. 295
Xenograft p. 289
X-radiation p. 297

QUESTIONS

1. Transplanted cells are mainly destroyed by
   a. B cells
   b. macrophages
   c. neutrophils
   d. T cells
   e. eosinophils

2. The fetus can be considered a(n)
   a. allograft
   b. xenograft
   c. heterograft
   d. isograft
   e. antigraft

3. An untreated primary kidney graft in a dog will not normally survive longer than
   a. 2 days
   b. 10 days
   c. 25 days
   d. 50 days
   e. 100 days
4. Corneal grafts are not rejected because they
   a. are resistant to lymphocytotoxic activity
   b. have no lymphatic drainage
   c. do not possess histocompatibility antigens
   d. are not exposed to antibodies
   e. are nonantigenic
5. Cyclosporine binds to which intracellular messenger?
   a. tyrosine kinase
   b. interleukin
   c. phospholipase
   d. cyclophosphamide
   e. calcineurin
6. One of the ways by which allograft rejection is prevented is through administration of
   a. antibodies to CD3
   b. anticomplementary antibodies
   c. immunoglobulin E
   d. anti-interferon antibodies
   e. anti-rhesus antibodies
7. Corticosteroids mainly suppress allograft rejection by suppressing
   a. macrophage function
   b. T-cell function
   c. neutrophil function
   d. antibody synthesis
   e. NK cell function
8. The major targets of cytotoxic T cells within a kidney allograft are
   a. neutrophils
   b. macrophages
   c. vascular endothelial cells
   d. glomerular cells
   e. proximal tubule cells
9. The major clinical problem associated with bone marrow allografts in humans is
   a. contact dermatitis
   b. aplastic anemia
   c. allograft rejection
   d. graft arteriosclerosis
   e. graft-versus-host disease
10. The use of a piece of leg vein to repair a damaged coronary artery in a bypass operation is called a(n)
    a. allograft
    b. xenograft
    c. heterograft
    d. autograft
    e. isograft
11. What mechanisms may be involved when blood transfusions assist in the survival of human renal allografts?
12. Why are class I and class II MHC molecules apparently equally important in regulating allograft rejection? Would you have predicted this on the basis of your knowledge of the functions of the MHC?
13. Why is the fetus not rapidly rejected by its mother? Would you expect the pregnancy of an interspecies cross, such as that between a horse and a donkey, to be different?
14. In chronic kidney graft rejection and in cardiac allografts, the rejection process stimulates vascular endothelial cell proliferation, leading to blockage of blood vessels. Speculate on the possible mechanisms by which immune attack might stimulate cell proliferation.
15. Xenografts can be very rapidly rejected. Is this inevitable, or is it possible that we can manipulate the immune system so that we will be able to use xenografts as easily as we now use allografts? Speculate on methods of overcoming the xenograft response.
16. Outline the mechanisms of graft-versus-host disease. How might this disease be overcome?
17. Why do women not normally make antibodies to sperm? Under what conditions might a woman become sensitized, and what might the consequences of producing antisperm antibodies be? Could this be used as an effective contraceptive?
18. To what extent is it appropriate that developed countries spend large sums on organ transplantation?
19. What might happen if an individual is given a kidney graft that is incompatible at the class I MHC locus but compatible at the class II locus? Might the rejection process differ if the graft was compatible at the class I locus but incompatible at the class II locus?

Answers: 1d, 2a, 3b, 4b, 5e, 6a, 7b, 8c, 9e, 10d

SOURCES OF ADDITIONAL INFORMATION

Anderson, D.J., and Tarter, T.H. Immunosuppressive effects of mouse seminal plasma components *in vivo* and *in vitro*. J. Immunol., 128:535–539, 1982.

Athanassakis, I., et al. The immunostimulatory effect of T cells and T cell products on murine fetally derived placental cells. J. Immunol., 138:37–44, 1987.

Bach, F.H., and Platt, J.L. Xenotransplantation: A view of issues. Transplant. Proc., 24(suppl. 2):49–52, 1992.

Baumann, G. Molecular mechanism of immunosuppressive agents. Transplant. Proc., 24(suppl 2):4–7, 1992.

Carpenter, C.B. Immunosuppression in organ transplantation. N. Engl. J. Med., 322:1224–1226, 1990.

DeWit, D., et al. Preferential activation of Th2 cells in chronic graft-versus-host reaction. J. Immunol., 150:361–366, 1993.

Harrison, R.W., and Lippman, S.S. How steroid hormones work. Hosp. Pract., 24:63–76, 1989.

Häyry, P., et al. Cellular and molecular mechanisms in allograft arteriosclerosis. Transplant. Proc., 24:2359–2361, 1992.

Hunt, J.S., and Orr, H.T. HLA and maternal-fetal recognition. FASEB J., 6:2344–2348, 1992.

Jagannath, S., Barlogie, B., and Tricot, G. Hematopoietic stem cell transplantation. Hosp. Pract., 28:79–86, 1993.

Johnson, P.M. Immunobiology of the human placental trophoblast. Exp. Clin. Immunogenet., 10:118–122, 1993.

Kahan, B.D. Cyclosporine. N. Engl. J. Med., 321:1725–1738, 1989.

Krensky, A.M., et al. T-lymphocyte-antigen interactions in transplant rejection. N. Engl. J. Med., 312:510–517, 1990.

Murray, J.E. Human organ transplantation: Background and consequences. Science, 256:1411–1415, 1992.

Pflügl, G., et al. X-ray structure of a decameric cyclophilin-cyclosporin crystal complex. Nature, 361:91–93, 1993.

Robertson, R.P. Pancreatic and islet transplantation for diabetes—cures or curiosities? N. Engl. J. Med., 327:1861–1868, 1992.

Rowlands, D.T., Hill, G.S., and Zmijewski, C.M. The pathology of renal homograft rejection: A review. Am. J. Pathol., 85:774–812, 1976.

Sargent, I.L. Maternal and fetal immune responses during pregnancy. Exp. Clin. Immunogenet., 10:85–102, 1993.

Schwartz, R.S. Therapeutic uses of immune suppression and enhancement. Hosp. Pract., 16:93–101, 1981.

Strom, T.B., and Carpenter, C.B. Transplantation: Immunogenetic and clinical aspects. Hosp. Pract., 18:135–150, 1983.

Stütz, A. Immunosuppressive macrolides. Trans. Proc., 24(suppl 2):22–25, 1992.

Tanio, J.W., and Eisen, H.J. Medical aspects of cardiac transplantation. Hosp. Pract., 61–74, 1993.

Thompson, A.W. Immunobiology of cyclosporin A: A review. Aust. J. Exp. Biol. Med. Sci., 61:147–172, 1983.

Thompson, A.W. FK-506 enters the clinic. Immunol. Today, 11:35–36, 1990.

Van Rood, J.J., and Claas, F.H.J. The influence of allogeneic cells on the human T and B cell repertoire. Science, 248: 1388–1393, 1990.

CHAPTER 20

Resistance to Tumors

William B. Coley *In 1892, the American surgeon William B. Coley noted that some of his cancer patients showed spontaneous cures when they became septicemic as a result of bacterial infections. He therefore pioneered the use of Coley's toxin, a killed suspension of common gram-positive and -negative bacteria for the treatment of cancer. When injected, the toxin caused very severe fevers and other side effects but a significant number of cancer patients showed tumor regression. Because of its toxicity, the use of Coley's toxin was eventually abandoned. With hindsight, Coley had devised a way of increasing TNF levels in his patients in a form of septic shock. The TNF helped destroy some tumors, especially sarcomas. This was apparently the first recorded form of tumor immunotherapy.*

CHAPTER OUTLINE

CHAPTER CONCEPTS

1. When cells become neoplastic, they may gain new antigens and provoke an immune response. This immune response may destroy the cancer cells.
2. Tumor antigens are typically weakly antigenic and do not stimulate a strong immune response.
3. It is not uncommon for cancer to immunosuppress an affected individual so that the immune response to tumor antigens is ineffective.
4. Immunity to most solid tumors is mediated by NK cells. Antibodies may be effective against circulating tumor cells. T cells and macrophages may also contribute to resistance to cancer.
5. Resistance to cancer may be enhanced by immunostimulating agents such as mycobacteria. Some cytokines such as IFN-γ may have limited activity against certain tumors. Some vaccines are effective against virus-induced cancers of animals.
6. Activated lymphocytes called lymphokine-activated killer (LAK) cells may be of considerable assistance in treating some cancers.

Occasionally, a cell becomes altered in such a way that it grows uncontrollably and gives rise to a tumor. Some of these cancerous cells may be sufficiently abnormal that they can trigger an immune response. Sometimes that immune response can destroy the tumor. In this chapter the immune response to cancer cells is described. We describe how cancer cells differ from normal, how the immune system can detect these differences, and how cancer cells can be attacked, especially by natural killer cells, a specialized population of lymphocytes. Other mechanisms of immunological destruction of tumor cells are also explained, followed by a discussion of why the immune system does not destroy cancer cells all the time. The ways in which these cells can evade the immune system are described. The methods used to stimulate the immune system to provoke tumor-cell destruction are reviewed. This includes some of the most recent advances in cancer therapy and the possible use of antitumor vaccines. The chapter concludes with a somewhat different topic, tumors of the immune system. These tumors are not uncommon, and they can provide us with very interesting information on immune mechanisms.

Normal body processes depend on careful regulation of cellular activity. One such activity is cell division. If cells are permitted to multiply, it is essential that they do so only as and when required. Unfortunately, as a result of exposure to certain chemicals, virus infection, or mutation, a cell may break free of the constraints that normally regulate cell division. At the same time it may develop the ability to invade sites that are normally occupied by other cells. A cell that is proliferating in an uncontrolled fashion gives rise to a growing clone of cells that eventually develops into a tumor, or **neoplasm.** If these cells remain clustered at a single site, the tumor is said to be **benign.** Benign tumors can usually be removed through surgery. In some cases, however, tumor cells may break off from the main tumor mass and be carried by the blood or lymph to distant sites, where they lodge and continue to grow. This form of tumor is said to be **malignant.** The secondary tumors that arise in these distant sites are called metastases. Malignant tumors are very difficult to treat since it may be impossible to surgically remove all metastases. Malignant tumors are subdivided according to their tissue of origin. Tumors arising from epithelial cells are called **carcinomas,** and those arising from mesenchymal cells such as muscle, lymphoid, or connective tissue cells are called **sarcomas.** A **leukemia** is a tumor derived from hematopoietic cells.

Although the essential difference between a normal cell and a tumor cell is a loss of control of cell division or of apoptosis, tumor cells may also show other abnormalities. For example, the proteins on the surface of a tumor cell may be different from those found on normal cells. These new or altered proteins may be recognized by the cells of the immune system and so trigger immunological attack.

TUMORS AS ALLOGRAFTS

When organ transplantation became a common and widespread procedure it was observed that patients with prolonged graft survival as a result of immunosuppressive therapy were two to six times more likely to develop cancer than were nonimmunosuppressed individuals (Fig. 20–1). It has also been observed that patients with congenital or acquired immunodeficiency syndromes show an increased tendency to develop malignant tumors. (For example, AIDS patients may develop Kaposi's sarcoma.) It was therefore suggested that the immune system was responsible for the prevention of cancer. From this suggestion the concept of a surveillance function for the cell-mediated immune system was developed, providing a stimulus for a growing interest in the role of the immune responses in preventing tumor development.

Evidence has shown, however, that this view of T cells forming as a system that identifies and destroys tumor cells alone is no longer tenable. One of the most important pieces of evidence to discredit this theory is the observation that nude (*nu/nu*) mice (Chapter 26), although deficient in T cells, are no more susceptible than normal mice to chemically induced or spontaneous tumors (although they are more susceptible to tumors induced by the polyoma virus). The natural resistance of nude mice, and indeed of all normal animals, to tumors is probably dependent on NK cells. Mice deficient in NK cells (*bg/bg* mice) do show an increased susceptibility to spontaneous tumors.

Although the original surveillance hypothesis has therefore had to be greatly modified, there is good evidence that antitumor defense mechanisms do exist and may be enhanced in order to protect an individual against cancer. However, there is a great difference between the strong and effective cell-mediated immune response triggered by allografts and the quantitatively much poorer responses to the very weak antigens associated with tumor cells.

Tumor Antigens

Tumor cells that are functionally different from their normal precursors may also be antigenically different in that they gain or lose cell membrane proteins (Fig. 20–2). They may, for example, lose histocompatibility

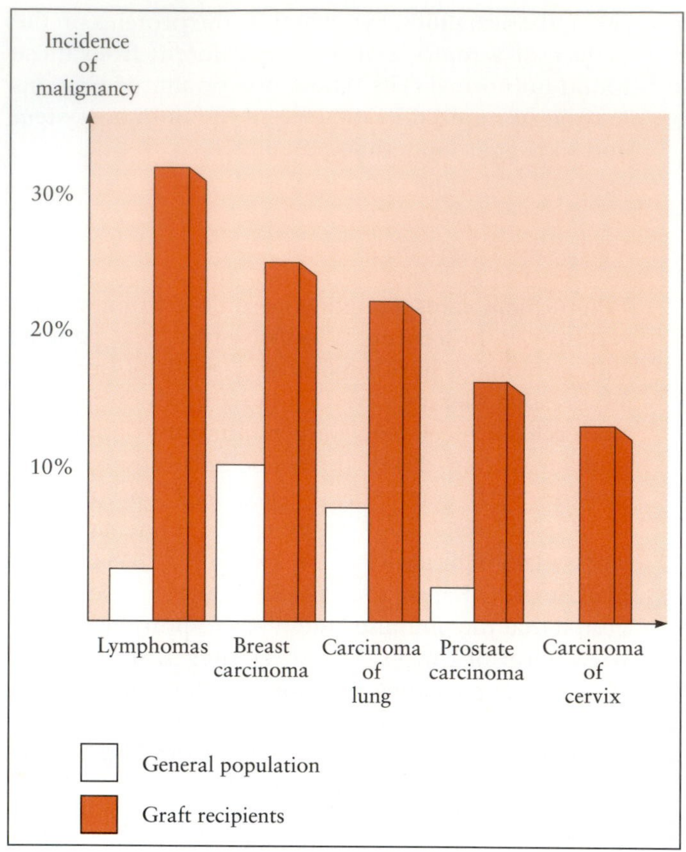

Figure 20–1 The prevalence of malignancies in human allograft recipients. Data derived from Penn, I. Tumor incidence in human allograft recipients. *(Transplant. Proc., 11:1047, 1979.)*

or blood group molecules. Some tumors of the intestine, such as colon carcinomas, may lose the ability to produce mucus. More commonly, tumor cells gain proteins, and some naturally occurring tumors of adult humans are characterized by the production of antigenic proteins normally found only in the fetus. For example, tumors of the gastrointestinal tract may produce a glycoprotein known as carcinoembryonic antigen (CEA) (CD66e), which is usually found only in the fetal intestine. The appearance of CEA in serum may therefore indicate the presence of a colon or rectal adenocarcinoma. Other examples of the generation of fetal proteins by tumor cells (oncofetal antigens) include production of α-fetoprotein by hepatoma cells (α-fetoprotein is normally found only in the fetal liver), and squamous cell carcinoma cells may possess antigens found in normal fetal liver and skin. These oncofetal antigens are poor immunogens and do not provoke protective immunity perhaps because they are not recognized by both $CD4^+$ and $CD8^+$ cells. Melanomas carry characteristic gangliosides while many carcinomas have antigenic mucins. Nevertheless even when tumor cells develop completely new antigens, the body may not be able to recognize or respond to them.

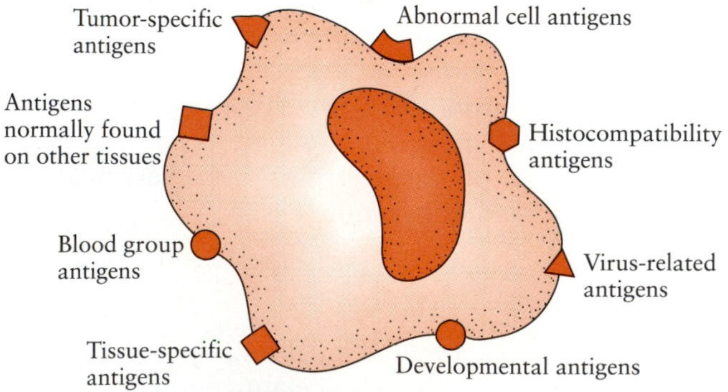

Figure 20–2 Some of the antigens that can develop on the surface of cancer cells and provoke an immune response.

Tumors induced by oncogenic viruses tend to gain new antigens characteristic of the inducing virus. These antigens, although coded for by the viral genome, are not part of the virus particle. An example of this type of antigen is the FOCMA antigen found on the neoplastic lymphoid cells of cats infected with feline leukemia virus. Antibodies against FOCMA will prevent tumor formation in infected cats.

Chemically induced tumors differ from the virus-induced variety by developing new surface antigens unique to the tumor and not to the inducing chemical (Fig. 20–3). Tumors induced by a single chemical in

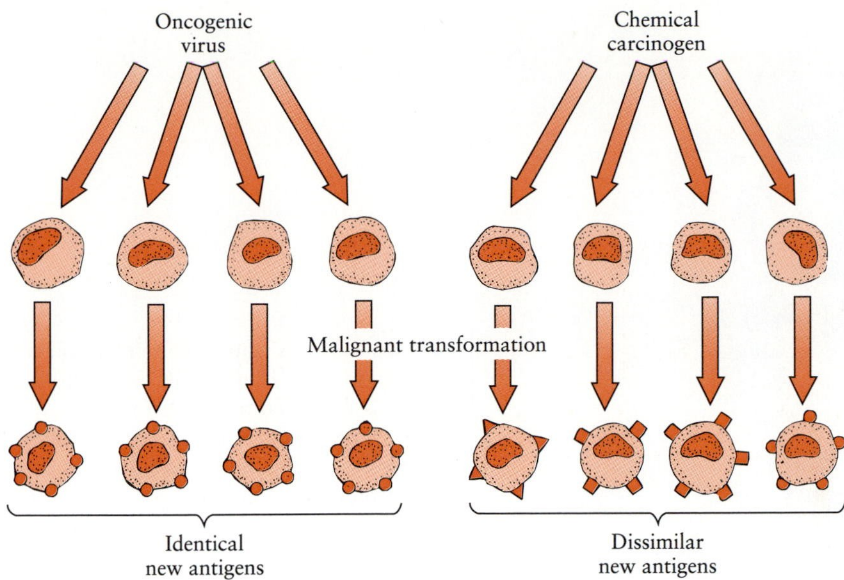

Figure 20–3 The development of new antigens on tumor cells induced by viruses or chemical carcinogens.

different animals of the same species may be antigenically quite unrelated; even within a single chemically induced tumor mass, antigenically distinct subpopulations of cells exist. As a result, resistance to one chemically induced tumor does not prevent growth of a second tumor induced by the same chemical.

IMMUNE RESPONSES TO TUMOR ANTIGENS

If tumor cells are sufficiently different from normal, they will be regarded as foreign and stimulate an immune response. The major mechanisms of tumor-cell destruction involve NK cells, although cytotoxic T cells, activated macrophages, and antibodies also participate in this process (Fig. 20–4).

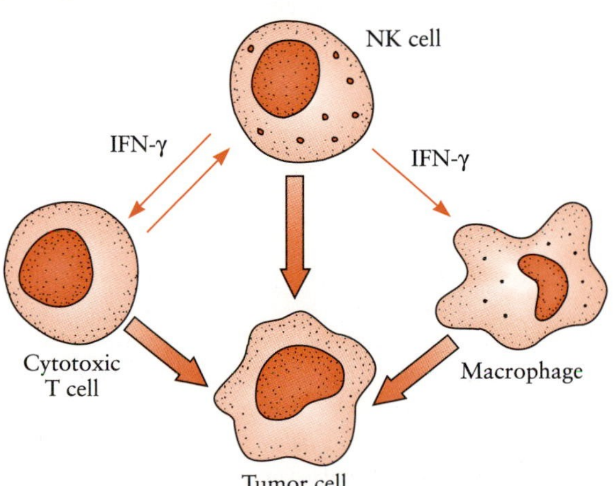

Figure 20–4 The three key cells that may kill tumor cells and the central role of interferon in regulating this response.

NATURAL KILLER CELLS

Basic Features

When peripheral blood lymphocytes are examined for B-cell and T-cell markers, some cells do not fall into these two categories but represent a distinct third population of lymphocytes. These non-T, non-B cells are called natural killer (NK) cells. NK cells are large lymphocytes with obvious granules in their cytoplasm (Fig. 20–5). They originate in the bone marrow, and their production is promoted by IL-2, IL-12, and IFN-γ. NK cells are closely related to T cells in that both develop from a common precursor. These cells, if they mature in the thymus, differentiate into T cells. If, however, they differentiate outside the thymus, they become NK cells. In humans, NK cells represent 10% to 15% of the mononuclear cells or 5% to 8% percent of all white cells in blood. NK cells do not recirculate like T cells and are not found in the thoracic duct lymph. They are mainly found in the secondary lymphoid organs, a smaller number is found in the bone marrow, and none is found in the normal thymus. (NK cells are, however, found in the thymus of combined immunodeficient mice [*scid*] and in nude mice.) They have been identified in mice, rats, humans, dogs, cats, swine, horses, cattle, and chickens.

Several NK cell subpopulations exist. These differ not only in their surface antigens but also in their biological activities. Some, for example, act on circulating tumor cells (NK cells), and others are more effective against solid tumors. Some T cells that express the γ/δ T-cell antigen receptor show a cytotoxic activity that resembles that of NK cells. These are, however, not true NK cells but are NK-like T cells.

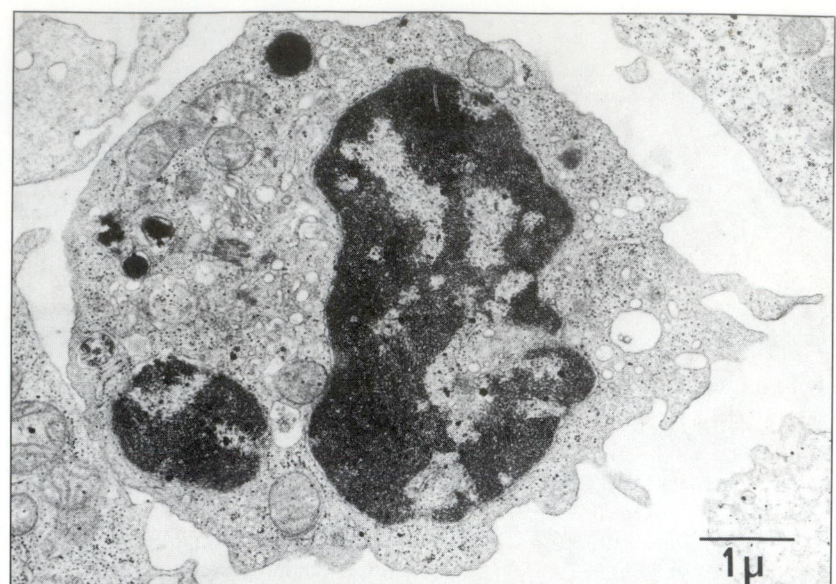

Figure 20–5 A transmission electron micrograph of a human NK cell: The nucleus is indented and rich in chromatin. The cytoplasm is abundant and contains many granules. Numerous mitochondria, centrioles, and a Golgi apparatus are visible. (Original magnification ×17,000.) *(From Carpen, O., Virtanen, I., and Saksela, E. J. Immunol., 128:2691, 1982. With permission.)*

Surface Markers

NK cells do not express CD4, immunoglobulins, or the T-cell antigen receptor (TCR). They do, however, express two major surface antigens: CD56 (NKH-1), a 175- to 185-kDa glycoprotein, and CD11a/CD18 (LFA-1) (Fig. 20–6). CD56 is a type I integral cell membrane glycoprotein member of the immunoglobulin superfamily that serves as an adhesion molecule. NK cells also possess CD2, CD11b, and CD38 as well as the low-affinity receptors for antibody, CD16 (FcγRIII). A small percentage of NK cells express CD8 and low levels of Thy-1.

NK cells do not express either the TCR or CD3 receptor chains and thus cannot recognize antigen through the TCR. Nor is the cytotoxic activity of NK cells dependent on the expression of MHC class I or class II molecules on their targets. However, they do have two other possible antigen-recognizing complexes. One such complex consists of CD16 together with the ζ chain and the tyrosine kinases lck and ZAP 70. The other consists of CD2 in association with the ζ chain as occurs in T cells. CD28 is also found on NK cells and plays a role in their proliferation and cytokine production.

CD16. CD16 is a protein of 50 to 70 kDa that binds IgG with low affinity (FcγRIII). It is expressed on NK cells, granulocytes, and macrophages. The forms found on granulocytes and NK cells differ in several functional respects. The form on granulocytes is a 30-kDa molecule attached to the cell membrane by a phosphatidyl-inositol linker. The form on macrophages and NK cells is a 36- to 38-kDa transmembrane protein. It is linked to either the γ chain of FcεRI (in macrophages) or to the ζ chain (in NK cells). Trig-

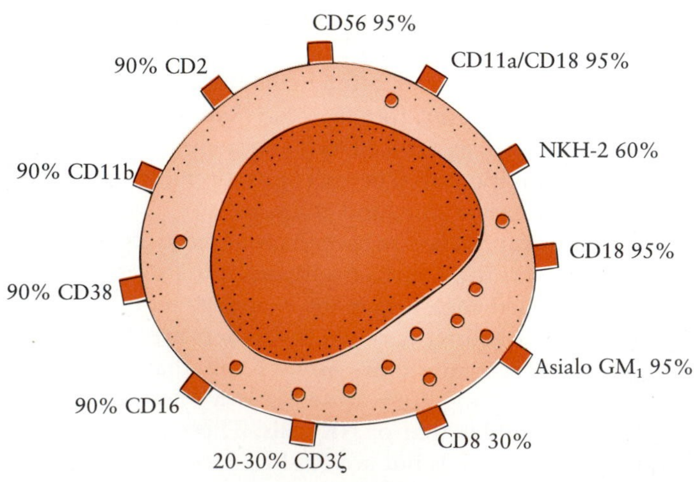

Figure 20–6 The receptors found on the surface of NK cells and the proportion of NK cells that carry them.

gering of NK cells through CD16 causes activation of the lck tyrosine kinase, a rise in intracellular Ca^{++}, and cellular activation, although it does not result in proliferation. The NK cells are triggered to produce IFN-γ, TNF-α, and CD25. Lymphocytes that express CD16 can mediate antibody-dependent cellular cytotoxicity (ADCC), whereas $CD16^+$ granulocytes cannot. CD16 may be spontaneously released by NK cells so that the NK cell may rapidly detach from the antibody-coated target after delivering the lethal hit.

Antibodies to the glycosphingolipid asialo-GM_1 together with complement completely eliminate mouse NK activity and partially abrogate cytotoxic T-cell responses. This molecule is apparently expressed on both NK cells and cytotoxic T cells, but the NK cells are apparently more sensitive to lysis. Asialo-GM_1 is also present on activated macrophages.

Functions

NK cells do not require prior sensitization or antigen presentation through MHC molecules to kill their targets. Nevertheless, they can bind to target cells and effectively cause their destruction. NK cells kill their targets by both antibody-dependent and independent processes. The antibody-dependent process is mediated through CD16 described previously. The antibody-independent process is mediated through perforins and cytotoxins although the nature of the target recognition molecules is unknown. Perforins and granzymes are constitutively expressed in NK cells. Their level of expression is increased by exposure to IL-2 and IL-12. The perforin (perforin P2) is a molecule of 70 to 72 kDa (slightly larger than that produced by T cells). It produces characteristic small (50–70A) lesions in target cell surfaces. Presumably cytotoxic granzymes are inserted into the target cells through the perforin channels. For example, NK cells secrete a 32 kDa protease called fragmentin that can induce DNA fragmentation and apoptosis in target cells. NK cells may also secrete soluble apoptosis-inducing molecules including TNF-α and LT. NK cell activity is regulated by exposure to IL-2 and IFN-γ. Thus production of TNF-α is augmented by IFN-γ. Proliferation of NK cells is stimulated by IL-2 and the IL-2R. Culture of NK cells in the presence of IL-2 induces *in vitro* differentiation of NK cells so that their cytotoxic activity is enhanced and they can lyse normally resistant tumor targets. These activated cells are called lymphokine activated killer (LAK) cells.

At first it was believed that NK cells functioned only as an antitumor surveillance system. It was later shown, however, that they were also active against a range of targets, including xenogeneic cells and virus-infected cells (Fig. 20–7). They also have antibacterial activity against organisms such as *Staphylococcus aureus* and *Salmonella typhimurium* as well as some fungi. Most evidence that supports a role for NK cells in immunity to tumors is derived from studies on tumor-cell lines grown in vitro. Thus NK cells destroy cultured tumor-cell lines, and a correlation exists between the level of NK activity measured in vitro and resistance to tumor cells in vivo. It is possible to increase resistance to tumor growth in vivo by passive transfer of NK cells. NK cells will destroy human leukemia, myeloma, and some sarcoma and carcinoma cells in vitro, and this activity is enhanced by IFN-γ. NK cells have also been shown to invade small primary mouse tumors. Patients with Chédiak-Higashi syndrome have a selective deficiency of NK cells and a high prevalence of lymphoproliferative diseases. Some carcinogenic agents, such as urethane, dimethylbenzanthracene, and low doses of radiation, can inhibit NK activity. It is also of interest to note that certain stressors such as surgery may depress NK activity and so promote tumor growth.

Control of NK Activity

NK cells are responsive to IL-3, IL-4, and IL-12 (Fig. 20–8). IL-4 stimulates NK function and enhances cytotoxicity, and IL-3 prevents the death of cultured NK cells. Although NK cells are active in the nonimmunized animal, virus infections or administration of in-

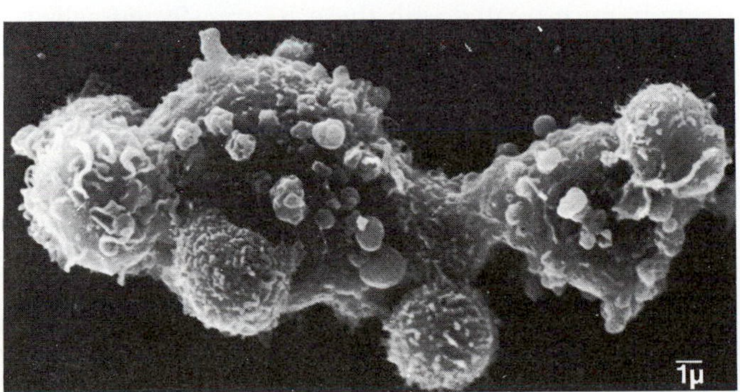

Figure 20–7 A cluster of NK cells and target cells after four-hour incubation in the presence of 1000 units/ml of IFN-γ. The NK cells at the bottom center are bound to two target cells. The round extrusions on the surface of the target cells indicate their impending death. The cell on the extreme left is a monocyte. *(From Carpen, O., Virtanen, I., and Saksela, E. J. Immunol., 128:2691, 1982. With permission.)*

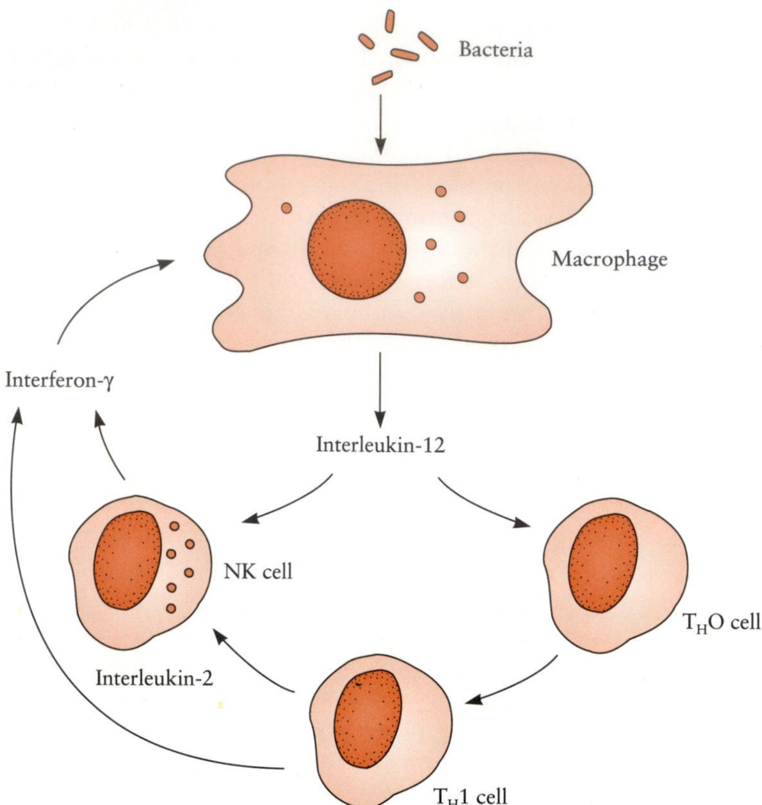

Figure 20–8 The interrelationship between NK function and other cells of the immune system. IL-12 secreted by macrophages stimulates NK cells to secrete IFN-γ. The IL-12 also stimulates Th1 cell production which also stimulates NK activity through IL-2 and IFN-γ.

terferon inducers stimulates NK activity above normal levels. In these diseases, macrophages phagocytose the invading organisms and as a result produce TNF-α and IL-12 within a few hours. These then induce IFN-γ production by NK cells. This IFN-γ enhances NK activity by promoting the rapid differentiation of pre-NK cells. It also activates the macrophages. The NK cell–interferon system probably plays a critical role in resistance to neoplasia. Neutralization of interferon by means of specific antisera enhances tumor growth in mice, presumably by depressing NK cell activity. Finally, IFN treatment may be useful in the treatment of some human cancers. For example, IFN-α is of significant benefit in the treatment of hairy-cell leukemia (a B-cell tumor) and Kaposi's sarcoma. Thus, collectively, NK cells acting under the influence of interferon may have an important role in **immune surveillance** against tumors.

OTHER CELLULAR DEFENSES

T-Cell–Mediated Immunity

It is occasionally possible to detect a cell-mediated response to tumor antigens either by skin testing or by use of an in vitro test such as a cytotoxicity assay (Fig. 20–9). Lymphocytes from some tumor-bearing animals may exert a cytotoxic effect on tumor cells cultured in vitro. Nevertheless, antitumor responses by T cells are probably significant only in the control of virus-induced tumors.

It may be possible, however, to use T cells to destroy cancer cells by manipulating the way in which they recognize these cells. You may recollect, for example, that cytotoxic T cells must be stimulated not only by antigen-MHC, but also by CTLA-4/CD28-B7 binding (Chapter 11). Thus it has been suggested that a cancer cell that lacks surface B7 (CD80) will not be able to deliver a triggering signal to the cytotoxic T cell. Peter Linsley and his colleagues have shown that transfecting B7 into cancer cells made them highly susceptible to T-cell killing. If these cells are injected into mice, they are rapidly destroyed, whereas unmodified cancer cells cause lethal tumors. In addition, mice treated with these B7 transfected cells become resistant to the unmodified cancer cells, suggesting that, once primed, the cytotoxic T cells no longer require B7.

Macrophage-Mediated Immunity

In some experimental tumor systems, macrophages may exert an antineoplastic function. This is especially true of macrophages activated by exposure to interferon. These activated macrophages may act in either

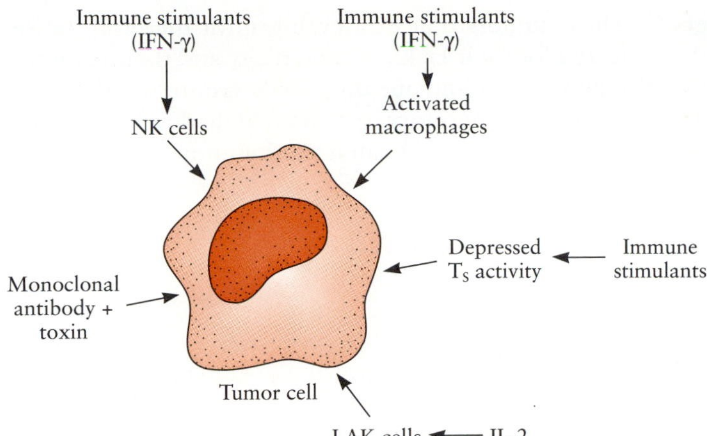

Figure 20–9 Some of the ways in which the immune system can be stimulated to mount a protective response to tumors.

a specific or nonspecific fashion by releasing nitric oxide and reactive oxygen metabolites that can kill cancer cells. Nonspecific activation of macrophages by bacillus Calmette-Guérin (BCG) (*Mycobacterium bovis*) or *Propionibacterium acnes* results in enhanced production of interleukin-1 or TNF-α and subsequent activation of helper T-cell activity. Interleukin-1 has a cytostatic effect on some tumors, and TNF-α may have potent antitumor activity. Unfortunately, many malignant tumors release a factor, which inhibits macrophage activation, and the macrophages of tumor-bearing individuals may show defective mobilization and chemotaxis. In animals in which tumor growth is inhibited, macrophages use nitric oxide synthase to metabolize arginine to nitric oxide. In animals with progressive tumor growth, the arginine metabolic pathway may switch to using arginase to produce ornithine, a molecule without cytotoxic effects.

Antibody-Mediated Immunity

Antibodies to tumor cells are commonly found in many tumor-bearing animals; for instance, about 50% of sera from dogs with lymphosarcomas contain precipitating antitumor antibodies. These antibodies may be of some protective significance, since in conjunction with complement, they may be capable of lysing free tumor cells. Antibodies do not appear to be effective in destroying the cells in solid neoplasia.

FAILURE OF IMMUNITY TO TUMOR CELLS

The fact that neoplasia are so readily induced in experimental animals and are so common in humans testifies to the inadequacies of the immunological protective mechanisms. Studies of tumor-bearing individuals have indicated the existence of several mechanisms by which immune systems fail to reject tumors.

Immunosuppression

Tumor-bearing animals are commonly immunosuppressed. This suppression is most clearly seen in animals bearing lymphoid tumors, in which tumors of B cells suppress antibody formation, whereas tumors of T-cell origin generally suppress the cell-mediated immune responses. The suppression observed in leukemia virus infections is possibly a reflection of a generalized disturbance in the lymphoid system brought about by these viruses. In contrast, immunosuppression observed in animals bearing chemically induced tumors appears to be due in some cases to the release of immunosuppressive factors, such as prostaglandins, or nitric oxide from the tumor cells or from tumor-associated macrophages. Finally, the presence of actively growing tumor cells represents a severe protein drain on an animal. This protein loss may impair the immune response.

Much of the immunosuppression seen in tumor-bearing individuals may, however, be attributed to the development of "suppressor" cells. These are $CD8^+$ cells that have the ability to turn off some immune responses. They may be cytotoxic T cells that are exerting a suppressive effect. Thus administration of antibodies against suppressor cells to tumor-bearing mice inhibits tumor growth as a result of inhibition of suppressor-cell function. Enhanced suppressor-cell activity can be detected in the sera of patients with osteogenic sarcomas, thymomas, myelomas, and Hodgkin's disease and in many tumor-bearing animals. In most cases, these suppressor cells are undoubtedly T cells generally bearing CD8. In others, however, the suppressor cell may be macrophage-like, having the ability to adhere to surfaces, and in still others, it may have B-cell characteristics.

An excellent example of the role of immunosuppression is seen in skin cancer induced by ultraviolet (UV) light in mice. Thus if UV radiation–induced skin cancers are transferred to the skin of normal syngeneic mice, they are rejected. If they are transferred to chronically UV-irradiated mice, they grow successfully. This is because a population of suppressor T cells develops in UV-irradiated skin before tumors appear. If adoptively transferred to normal mice, these suppressor cells prevent rejection of a subsequent tumor challenge. The suppressor cells must recognize a common antigen present on all UV-induced tumors since a UV-irradiated mouse will accept all tumors induced by UV light but will reject tumors induced by other agents. These suppressor cells are I-J$^+$, CD4$^-$, and CD8$^-$. However, they may be very heterogeneous. Thus in a study of suppressor clones isolated from UV-induced tumors, one clone expressed and rearranged immunoglobulin RNA and the other expressed TCR RNA. The simplest explanation is that these cells may originate from an uncommitted lymphoid stem cell.

It must also be pointed out that other immunosuppressive mechanisms are operative in UV-radiated skin. For example, irradiated keratinocytes release interleukin 10. IL-10 suppresses cytokine release and will therefore interfere with skin immunity.

Inhibitors of Immunity

Although tumor cells are antigenic and may stimulate a protective cell-mediated immune response, the humoral immune response commonly has an opposite effect. Thus, the serum of tumor-bearing animals, when given to other tumor-bearing animals, may cause the tumor of the second animal to grow even faster, a phenomenon known as **enhancement.** Serum of this type may also effectively inhibit the cytotoxicity of T cells for tumor cells in vitro (see box). Alternatively,

blocking antibodies may be produced. These are non-complement-fixing antitumor antibodies that mask tumor antigens and thus protect the tumor cells from attack by cytotoxic T cells. In general, the presence or absence of these blocking factors correlates well with the state of progression of a tumor.

Tumor-Cell Selection

There are two major selection mechanisms by which tumor cells can evade the host's immune response and so enhance their own survival. One is "sneaking through," the process by which a tumor may not provide sufficient stimulus to trigger an immune response until it has reached a size at which it cannot be controlled by the host. Remember that even a very small tumor may contain an enormous number of cells. For example, a 10-mm-diameter tumor contains about 10^9 cells. Second, tumor cells that are antigenically different from the host will induce a strong immune response and be eliminated without leading to disease. Those tumors that do develop must therefore be selected for their lack of antigenicity and their resulting inability to stimulate the host's immune system. To this extent, therefore, tumors that do develop have, by definition, already beaten the immune system.

TUMOR IMMUNOTHERAPY

Immune Stimulation

Two general approaches have been used in attempts to cure or slow tumor growth through immunotherapy (Table 20–1). The simplest is to nonspecifically stimulate the immune system (see Fig. 20–9). Obviously, any improvement in an animal's immune abilities will tend to enhance its resistance to tumors, although a cure may be expected only if the tumor mass is small or surgically excised. The most widely used immune stimulant is the attenuated strain of *M. bovis,* BCG. This organism, being a **facultative intracellular parasite,** activates macrophages and stimulates interleukin-1 release, so promoting T-cell activity. It may be given systemically, or more effectively, it may be injected directly into the tumor mass. Most information on the use of BCG has come from studies of patients with melanomas or bladder cancer. Direct injection of BCG into skin melanoma metastases may cause complete regression not only of the injected lesion, but also, occasionally, of uninjected skin metastases. However, visceral metastases usually remain unaffected. BCG enhances survival or remission length in some leukemias, and its direct application in bladder cancer gives complete or partial response rates of up to 70%. BCG, however, can cause severe lesions at the site of injection and, occasionally, hypersensitivity reactions.

Table 20–1 Some Approaches to Tumor Immunotherapy

Nonspecific immune stimulation
Microbial products
BCG, *P. acnes,* yeast glucans, Levamisole
Complex carbohydrates
Acemannan
Cytokines
Interferons, TNF, IL-2, IL-4
Cytokine-activated cells
LAK cells, TIL cells
Passive immunization
Monoclonal antibodies
Alone or conjugated to toxins
Active immunization
Crude tumor cell extracts
Isolated specific tumor cell antigens
Chemically modified tumor cells
Vaccination against oncogenic viruses
Feline leukemia

CLARIFICATION

How Blocking Antibodies Were Discovered

Blocking antibodies were first demonstrated by their ability to reduce the cytotoxic effects of specific immune T cells. A simple ^{51}chromium release assay was set up for which it was anticipated that the cytotoxic T cells would destroy the cancer cells. Sera from a variety of sources, including both normal and tumor-bearing individuals, were added to the cultures. The sera from patients with tumors significantly reduced the T-cell cytotoxicity (see figure).

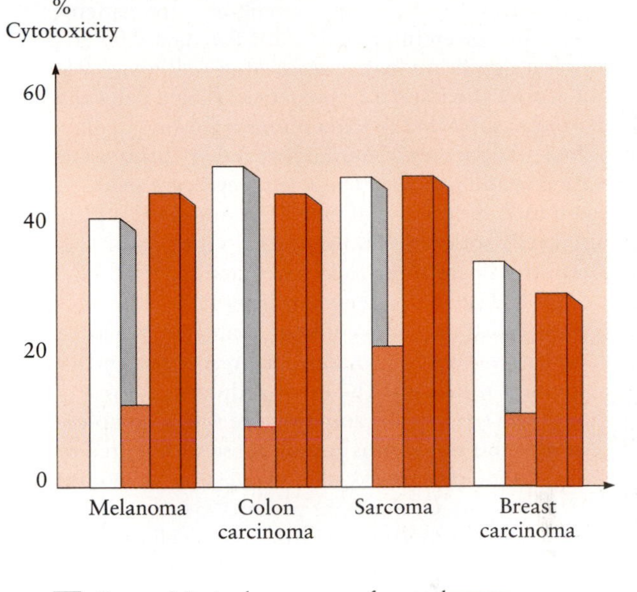

The effect of serums from different sources on the cytotoxic effects of T cells on various tumors. *(Data derived from Hellstrom, K.E., and Hellstrom, I. Adv. Immunol. 18:209–277, 1974.)*

Other immunostimulants that have been employed, generally with less success than BCG, include *P. acnes,* levamisole, and various mixed bacterial vaccines. Complex carbohydrates such as the mannose polymer acemannan have also been used with encouraging results. A related technique that has been used to treat human skin tumors is to paint them with the contact allergen dinitrochlorobenzene. The local hypersensitivity reaction provoked by this compound preferentially damages the tumor cells and so promotes tumor regression.

Active Immunization

Specific immunotherapy, the second major approach, has also been attempted by many investigators. This involves vaccinating the patient either with lysed whole tumor cells or with purified antigens. Some investigators have focused on using purified tumor-cell components such as gangliosides from melanomas or mucins from carcinomas as antigens. Other investigators have used very crude antigen mixtures such as lysed cells mixed with BCG vaccine or other immunomodulators and reinjected the mixture into the host. Because so many tumors can evade the immune response, it is necessary to treat the tumor cells to enhance their antigenicity. Thus, X-irradiated cells and neuraminidase- or glutaraldehyde-treated cells have been tested in tumor vaccines. The results obtained to date with tumor vaccines have been erratic and usually unsatisfactory.

Cytokine Therapy

Many attempts have been made to treat cancer patients with isolated cytokines but with limited success. Interferons, for example, are effective only against certain selected tumors. Thus 70% to 90% of patients

METHODOLOGY

LAK Therapy

To receive LAK therapy (see figure), the patient is first given interleukin-2 for five days. This IL-2 causes a significant leukocytosis. After waiting for 24 hours after the last dose, the patient is bled daily and the blood subjected to leukophoresis for one week. (The leukocytes are removed and the erythrocytes returned to the patient.) These leukocytes are incubated in IL-2 during that time. The unwashed cultured cells are then retransfused into the patient intravenously for three successive days.

Of 102 cancer patients receiving IL-2-activated LAK cells who could be evaluated, 8 had complete responses, 15 had partial responses, and 10 had minor responses. The median duration of response was ten months among those with a complete response and six months among those with partial responses. One patient was still in remission 22 months after treatment. Of the eight patients with a complete response, five were suffering from renal cell cancer with metastases to the lungs following radical nephrectomy. Two had melanomas with pulmonary metastases and one had a lymphoma. The tumor masses in the lungs shrank dramatically following LAK treatment.

Rosenberg, S.A., et al. N. Engl. J. Med., 80:889–897, 1987.

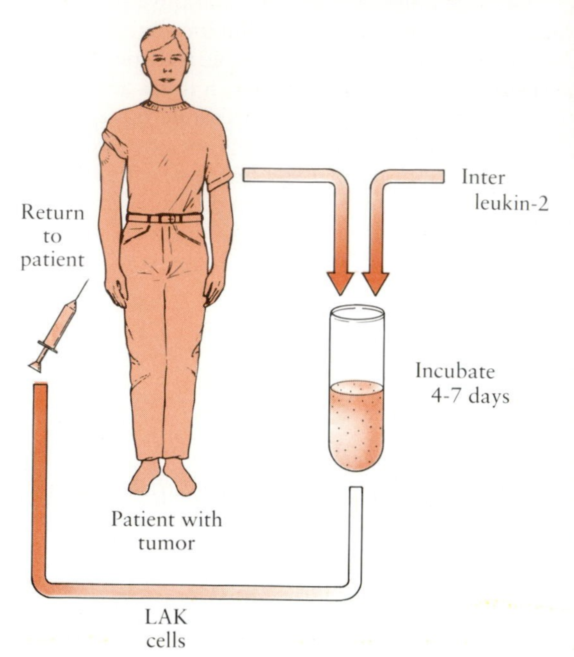

The production of LAK cells. For details see text.

with hairy-cell leukemia treated with IFN-α show complete or partial remission. The antitumor activities of TNF are synergistic with the interferons. Administration of interleukin-2 to patients with melanoma induces a positive response in 25% to 40% of cases. This may be due to induction of specific tumor-reactive T cells.

One major difficulty in cytokine therapy has been the toxicity shown by many of these agents. For example, TNF-α produces clinical signs similar to those induced by endotoxin. IL-2 is extremely toxic when administered alone. In low doses it induces fever, chills, nausea, and weight gain. Another common side effect is a "capillary leak syndrome," resulting in massive pulmonary edema as a result of increased capillary permeability in the lung. IL-2 also produces hematological abnormalities such as anemia, thrombocytopenia, and eosinophilia. Patients may develop a severe, very itchy rash, neuropsychiatric changes, and endocrine abnormalities. Thus IL-2 is a very hazardous protein and has limited usefulness when used alone. Preliminary trials of IL-4 have indicated that toxic effects, similar to those seen with IL-2, occur in treated patients.

LAK and TIL Cell Therapy

If NK cells or NK-like ($CD4^-8^-$) T cells are incubated in the presence of IL-2 for four days, they develop cytotoxic properties and are called **lymphokine-activated killer** cells. Injection of these LAK cells into mice with experimental lung tumors can lead to cancer remission. A combination of LAK cells and IL-2 has given encouraging results when administered to humans with cancer.

LAK cells are of two major types. About 40% are derived from NK cells (NK-LAK), and the remainder are T cell derived (T-LAK). They arise from two different precursors. NK-LAK cells are mainly $CD16^+$, $CD3^+$, and $CD56^+$. They have large granular lymphocyte morphology and act as NK cells. They release LAK-1, a 120-kDa cytotoxic protein as their effector molecule. IL-4 and IL-7 also induce LAK activity in mice, although IL-2 is five times better than IL-7 and 200 times better than IL-4 in this regard. Production of LAK cells may be enhanced by IL-1, IFN-α, and IFN-γ. IL-3 can inhibit LAK activity and block LAK activation by IL-2.

In an effort to obtain even better results, tumors have been surgically removed from cancer-bearing patients. The lymphocytes found within these tumors were removed and cultured in the presence of IL-2 for four to six weeks so that their numbers could grow significantly. These tumor-infiltrating lymphocytes (TIL cells) recognize and infiltrate only the tumors they come from. Given, together with IL-2, back to the donor patients, they produce very high remission rates. The most encouraging results have been obtained in patients with melanomas, colorectal cancer, and kidney cancer. TIL therapy, especially when used in conjunction with cytotoxic drugs, may be an important new treatment for cancer.

Antibody Therapy

Passive immunization using serum from immune animals is generally considered undesirable because of the risk of tumor enhancement through transfer of blocking antibodies. Monoclonal antibodies may be raised against tumor-specific antigens. These antibodies can be used to destroy tumors, either when given alone or when complexed to highly cytotoxic drugs or potent radioisotopes, which they carry directly to the tumor cells. Experiments using these immunotoxins have given encouraging results. Tumor immunotherapy using mouse monoclonal antibodies is hindered by the fact that the antibody is foreign for humans. As a result, it is immunogenic and rapidly cleared from the body. These problems may be alleviated by using chimeric immunoglobulins that consist of murine V regions fused to human C regions. These chimeric antibodies are less immunogenic than pure mouse antibodies when injected into humans. It is also possible to engineer antibody molecules for specific tissues by making them bifunctional. Thus a hybrid antibody can be constructed. One Fab region can be specific for a tumor antigen, and the other can be specific for a toxin, cytotoxic drug, or radioactive isotope. This antibody can be administered to a patient and allowed to bind to the target tumor before the therapeutic agent is given. The therapeutic agent will then bind specifically to the target via the artificial receptor formed by the bifunctional antibody. These bifunctional antibodies have been constructed chemically or by the use of modified hybridoma techniques.

Successful Antitumor Vaccines

In contrast to the techniques described previously, most of which have met with only partial success, there do exist established successful techniques for vaccination against tumor viruses. The most important of these are the vaccines against feline leukemia in cats. These vaccines usually contain high concentrations of the major viral antigens, and immunity is almost entirely directed against the glycoproteins of the feline leukemia virus. Success has also been achieved with inactivated wart vaccines in animals, probably because virus-specific antigens develop on tumor-cell membranes and function as immunogens. It is probable that future successes are more likely to be due to the development of vaccines against oncogenic viruses such as these rather than to more complex immunological approaches directed toward the destruction of tumor cells.

LYMPHOID TUMORS

The normal immune response, whether antibody or cell mediated, involves the induction of a burst of rapid proliferation in lymphocytes. This burst of proliferation must be carefully controlled. Although uncontrolled lymphoid cell function may induce autoimmunity, uncontrolled lymphocyte proliferation may result in the development of a lymphoma or lymphosarcoma. In the lpr mouse strain, mutation of the *fas* gene results in a failure of apoptosis. Affected mice, as a result, suffer from excessive lymphoproliferation and autoimmunity. It is no accident that individuals suffering from autoimmune disease are very much more likely than normal individuals to develop lymphoid cell tumors. Any cells of the immune system may become neoplastic. Tumors of B cells, T cells, and macrophages are well recognized. Plasma-cell tumors (myelomas) have made major contribution to our understanding of immunoglobulin structure. An interesting form of lymphoid tumor is Burkitt's lymphoma.

Burkitt's Lymphoma

Burkitt's lymphoma is a tumor of B cells that has a unique geographical distribution (Fig. 20–10). It is mainly found in countries with a high prevalence of malaria. It is also associated with infection by Epstein-Barr virus (EBV), a herpesvirus that is the cause of infectious mononucleosis. A combination of malaria and EBV infection is a significant stimulus for Burkitt's lymphoma. This tumor is also associated with a consistent chromosome aberration (Fig. 20–11); one part of chromosome 8 is invariably translocated to another

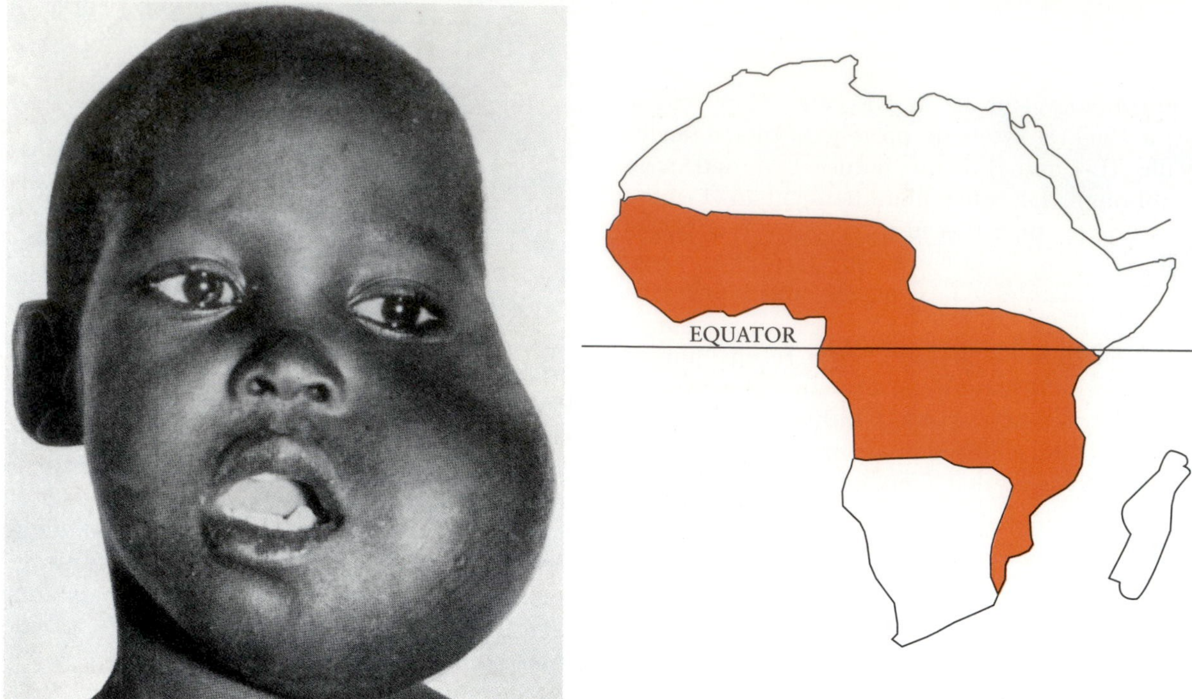

Figure 20–10 (A) A five-year-old boy with Burkitt's lymphoma. (B) The "lymphoma belt" across central Africa where Burkitt's lymphoma is endemic. There are tumor-free regions within this area and occasional cases have been reported north and south of it. *(From Burkitt, D.P., and Wright, D.H. Burkitt's Lymphoma, Livingstone, Edinburgh, 1970. With permission.)*

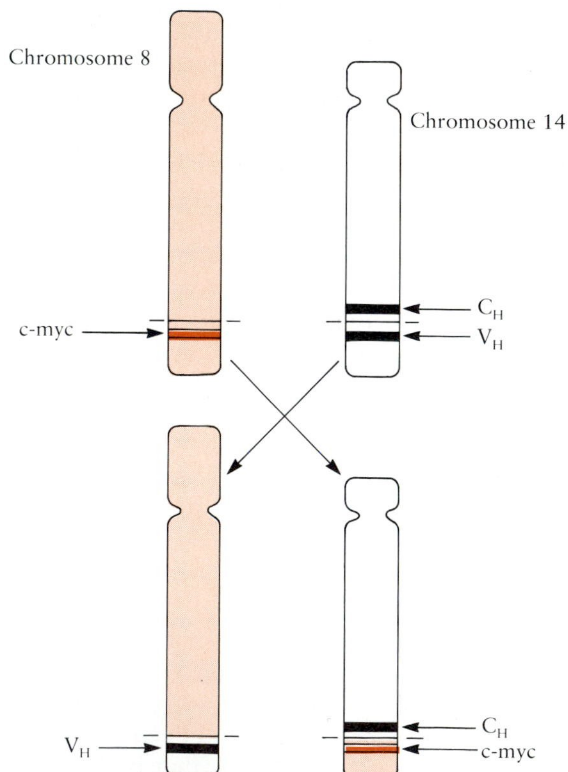

Figure 20–11 The chromosomal translocation that occurs in Burkitt's lymphoma. There is an exchange of portions of chromosomes 8 and 14 leading to the juxtaposition of the oncogene *c-myc* with the immunoglobulin heavy chain constant region gene enhancer.

chromosome, either 2, 14, or 22. In about 75% of cases, chromosome 8 is translocated to 14, in 15% of cases to 22, and in 10% of cases to chromosome 2. The break in chromosome 8 occurs at the location of the proto-oncogene called *c-myc*. The break in the other chromosome occurs between the V and C genes of an immunoglobulin chain (V_H on chromosome 14, V_λ on chromosome 22, and V_κ on chromosome 2). Similar translocations are seen in mouse myelomas between the *c-myc* gene on chromosome 15 and the V_H and V_κ genes on mouse chromosomes 12 and 6, respectively.

Excessive production of the *c-myc* product, a cellular growth stimulant, leads to cancer. In the case of Burkitt's lymphoma the translocation of chromosome 8 brings the *c-myc* gene under the influence of the immunoglobulin heavy-chain gene enhancer. Alternatively, the *c-myc* gene may be translocated without a controlling gene so that, in its new location, it acts in an unregulated fashion. The net result of all this is the development of a fast-growing tumor of B cells in affected individuals. A related effect is seen in follicular B-cell lymphomas. In these tumors, there is a reciprocal translocation between the immunoglobulin μ chain gene on chromosome 14 and an oncogene on chromosome 18 called *c-bcl-2*. *Bcl-2* is a cell membrane protein that inhibits apoptosis rather than affect cell proliferation. Many other lymphomas and leukemias have visible chromosomal abnormalities. These are commonly translocations involving cellular proto-oncogenes and immunoglobulin loci in B-cell tumors or the TCR loci in T-cell tumors.

Lymphocytes may become malignant at almost any stage of their development. As a result, they carry markers characteristic of that stage of differentiation. Chronic lymphatic leukemia (CLL) cells, for example, usually carry markers characteristic of mature B cells. Acute lymphoblastic leukemias (ALLs), on the other hand, usually consist of null cells; rarely can they be identified as T cells, and even more rarely are they B cells. Myelomas are clearly mature plasma cells.

By using markers such as surface immunoglobulins, MHC class II molecules, and specific antisera for CD molecules, lymphoid tumors can be classified according to the stage of development of the lymphocytes involved. The precise identification of the stage of development helps to determine the prognosis for treatment success. For example, T-ALL and B-ALL have a poor prognosis, but lymphomas arising from pre-B and stem cells (common-ALL) have a better prognosis. These tumors, commonly seen in children, carry CD10 (also called CALLA—common acute lymphoblastic leukemia antigen) and respond well to vincristine, prednisolone, and L-asparaginase. CALLA is a membrane-bound endopeptidase of 95 to 100 kDa found on many normal and neoplastic B cells.

CLL is a disease of older individuals, being rare in individuals under 50. It is usually benign unless the tumor is producing monoclonal immunoglobulins. These CLL cells express CD5 weakly, suggesting that they may originate as B-1 cells (Chapter 14).

KEY WORDS

Activated macrophage p. 312
Antitumor vaccine p. 317
BCG vaccine p. 314
Benign tumor p. 307
Blocking antibody p. 314
Burkitt's lymphoma p. 317
CALLA (CD10) p. 318
Carcinoembryonic antigen p. 308
Carcinoma p. 307
Chromosome translocation p. 317
Chronic lymphatic leukemia p. 318
Epstein-Barr virus p. 317
FcRIII (CD16) p. 310
α-Fetoprotein p. 308
Immune stimulant p. 314
Immune surveillance p. 307
Immunosuppression p. 313
Interferon-γ p. 309
Lymphoid tumor p. 317
Lymphokine-activated killer (LAK) cell p. 316
Malignant tumor p. 307
Neoplasm p. 307
Natural killer cells p. 309
Oncogenic virus p. 308
Sarcoma p. 307
Suppressor cell p. 313
Tumor p. 307
Tumor antigen p. 307
Tumor-cell selection p. 314
Tumor enhancement p. 314
Tumor-infiltrating lymphocyte p. 316
Tumor necrosis factor p. 313

QUESTIONS

1. Tumor cells may be killed by
   a. antibodies and complement
   b. activated T cells
   c. natural killer cells
   d. activated macrophages
   e. all of the above
2. One mechanism by which tumors evade immunological destruction is
   a. secretion of anticomplementary factors
   b. release of lymphotoxins
   c. production of immunosuppressive molecules
   d. destruction of neutrophils
   e. alternate pathway cytotoxicity
3. Carcinoembryonic antigen is characteristically secreted by tumors of the
   a. kidney
   b. blood
   c. lungs
   d. bone
   e. gastrointestinal tract
4. NK cells develop from which cell lineage?
   a. T cell
   b. B cell
   c. myeloid
   d. erythroid
   e. none of the above
5. Ligation of CD16 on NK cells results in
   a. antibody-dependent cellular cytotoxicity
   b. cell-mediated cytotoxicity
   c. LAK cell induction
   d. interferon production
   e. TNF production
6. Macrophage antitumor activity is mainly mediated by
   a. interleukin-6 and interleukin-1
   b. tumor necrosis factor and interleukin-1
   c. nitric oxide and interleukin-6
   d. nitric oxide and tumor necrosis factor
   e. interleukin-2 and interleukin-1

7. LAK (lymphokine-activated killer) cells are induced by
   a. interleukin-2
   b. tumor necrosis factor–α
   c. interferon-γ
   d. interleukin-1
   e. interleukin-7
8. Tumor enhancement is
   a. promotion of tumor growth by drugs
   b. promotion of tumor growth by antibody
   c. promotion of tumor growth by NK cells
   d. inhibition of tumor growth by NK cells
   e. inhibition of tumor growth by antibody
9. Immunity to some tumors may be stimulated by treatment of patients with
   a. *Streptococcus pyogenes*
   b. *Staphylococcus aureus*
   c. *Escherichia coli*
   d. *Listeria monocytogenes*
   e. *Mycobacterium bovis*
10. Burkitt's lymphoma
    a. occurs in high frequency in Central America
    b. is caused by Epstein-Barr virus
    c. involves chromosomal translocation between chromosomes 12 and 15
    d. is associated with trypanosome infection
    e. involves activation of the proto-oncogene *c-myb*
11. Outline a suggested regimen of immunotherapy for cancer. Why do you think such a regimen would work?
12. Can we take the concept of immune surveillance seriously in view of the high prevalence of cancer in the population?
13. Outline the types of changes that occur in the surface antigens of a cell when it becomes neoplastic?
14. Compare the process of allograft rejection with the possible rejection of a tumor. What are the similarities and differences between the two processes?
15. What is the biological significance of NK cells?
16. Outline the production of TIL and LAK cells. Why does this procedure appear to work but other forms of antitumor immunotherapy have been disappointing?
17. Discuss the unique features of Burkitt's lymphoma and Kaposi's sarcoma.
18. Discuss the problems and successes achieved with cytokine treatment of spontaneous tumors. How would you suggest that these problems can be overcome?
19. There is a growing interest in the use of immunotoxins—monoclonal antibodies conjugated to toxins and directed against antigens associated with tumor cells. Outline the advantages and disadvantages of this form of therapy.

Answers: 1e, 2c, 3e, 4e, 5a, 6d, 7a, 8b, 9e, 10b

SOURCES OF ADDITIONAL INFORMATION

Boehm, T., and Rabbits, T.H. The human T cell receptor genes are targets for chromosomal abnormalities in T cell tumors. FASEB J., 3:2344–2359, 1989.

Broder, S., and Waldmann, T.A. The suppressor cell network in cancer. N. Engl. J. Med., 299:1281–1284, 1983.

Brooks, C.G., et al. The majority of immature fetal thymocytes can be induced to proliferate to IL-2 and differentiate into cells indistinguishable from mature natural killer cells. J. Immunol., 151:6645–6656, 1993.

Bumol, T.F., et al. Monoclonal antibody and an antibody-toxin conjugated to a cell surface proteoglycan of melanoma cells suppress in vivo tumor growth. Proc. Natl. Acad. Sci. U. S. A., 80:529–533, 1983.

Herberman, R.B., and Ortaldo, J.R. Natural killer cells: Their role in defenses against disease. Science, 214:24–30, 1981.

Hercend, T., and Schmidt, R.E. Characteristics and uses of natural killer cells. Immunol. Today, 9:291–293, 1988.

Janeway, C.A. A primitive immune system. Nature, 341:108, 1989.

Karn, I., and Friedman, H. Immunosuppression and the role of suppressive factors in cancer. Adv. Cancer Res., 25:271–321, 1977.

Lopez, D.M., et al. Nuclear disintegration of target cells by killer B lymphocytes from tumor-bearing mice. FASEB J., 3: 37–43, 1989.

McConkey, D.J., et al. NK cell induced cytotoxicity is dependent on a Ca^{++} increase in the target. FASEB J., 4:2661–2664, 1990.

Mills, C.D, et al. Macrophage arginine metabolism and the inhibition or stimulation of cancer. J. Immunol., 149:2709–2714, 1992.

Ortaldo, J.R., and Longo, D.L. Human natural lymphocyte effector cells: Definition, analysis of activity and clinical effectiveness. J. Natl. Cancer Inst., 80:999–1010, 1988.

Ritz, J. The role of natural killer cells in immune surveillance. N. Engl. J. Med., 320:1748–1749, 1989.

Schwartz, R.S. Therapeutic uses of immune suppression and enhancement. Hosp. Pract. [Off.], 16:93–101, 1981.

Tao, M-H., and Levy, R. Idiotype/granulocyte-macrophage colony-stimulating factor fusion protein as a vaccine for B-cell lymphoma. Nature, 362:755–758, 1993.

Trinchierl, G. Biology of natural killer cells. Adv. Immunol., 47:187–376, 1989.

Vivier, E., et al. Tyrosine phosphorylation of the FcγRIII (CD16): ζ complex in human natural killer cells. Induction by antibody-dependent cytotoxicity but not by natural killing. J. Immunol., 146:206–210, 1991.

Westermann, J., and Pabst, R. Distribution of lymphocyte subsets and natural killer cells in the human body. Clin. Invest., 70:539–544, 1992.

CHAPTER 21

Tolerance

Peter Medawar *Peter Medawar was British although born in Brazil in 1915. His early studies on skin grafting led to elucidation of the genetic basis of the rejection process. For example, he had distinguished between rejection of the different forms of grafts such as grafts between monozygotic and dizygotic twins. Medawar was aware of Burnet's concepts of tolerance and devised experiments to test them. He received the Nobel Prize for his experiments that showed that mouse embryos inoculated with foreign tissues became tolerant to them. This was the first experimental demonstration of tolerance. He died in 1987.*

CHAPTER OUTLINE

CHAPTER CONCEPTS

1. Tolerance to self-antigens is an essential prerequisite for a functioning immune system.
2. Both B and T cells can be made tolerant, although T cells are considerably easier than B cells to make tolerant.
3. Self-reactive T cells undergo apoptosis in the thymus in a process called negative selection.
4. Self-reactive T cells that evade negative selection can be made tolerant by other mechanisms operating outside the thymus.
5. One common way by which T and B cells can be made tolerant is through apoptosis induced by an incomplete stimulating signal.
6. Suppressor T cells can also suppress inappropriate T-cell responses and give rise to functional tolerance.

One of the key features of the immune system is tolerance. **Tolerance** is the process by which the body ensures that immune responses are directed against foreign antigens while ensuring that the immune system does not react to normal self-antigens. Thus the body could not have a functioning immune system if it reacted with normal body components. In this chapter tolerance to self-antigens is described and its mechanisms analyzed. Since tolerance can occur in both T and B cells, we first discuss how tolerance is induced in T cells. This involves both the selection of T cells in the thymus whose receptors do not bind self-antigens. It also involves stimulation of T cells that can react with foreign antigens. This section also addresses other methods the body uses to suppress self-reactive T cells. The second part of the chapter describes the induction of tolerance in B cells. This too involves several mechanisms acting at different stages in B-cell development. The chapter ends with a discussion of incomplete tolerance and the duration of tolerance.

The interaction between T cells, B cells, and antigen-presenting cells forms a complex cellular network. Part of this complexity stems from the need to control the immune responses. All physiological processes are the subject of careful and rigorous control mechanisms. The immune system is by no means unique in its complexity, and the patterns of interaction that have been uncovered are only a reflection of the superb sophistication of biological systems in general.

Although the role of the immune system is to attack and destroy invading microorganisms, it must also be able to distinguish the foreign invaders from normal self-antigens. The immune system must be able to destroy a great variety of invading microorganisms yet not attack the body. The immune responses must be carefully regulated to ensure that they are appropriate with respect to both quality and quantity. More important than this, however, is the absolute need to prevent the immune system from attacking normal body tissues. The immune system must be constructed so that it attacks only foreign or abnormal tissues and is tolerant of normal, healthy tissues. As might be anticipated, several mechanisms ensure that self-tolerance is established.

Since both T and B lymphocytes generate antigen-binding receptors at random, it is clear that the production of self-reactive cells cannot be prevented. The diversity of antigen-binding receptors is generated by combination of gene segments, by N-region addition, and by somatic mutation. An animal cannot control the amino acid sequences and hence the antigen specificity of these receptors. Tolerance induction must therefore require the selective elimination of those lymphocytes that develop receptors that can bind self-antigens.

TOLERANCE

Tolerance is the specific immunological unresponsiveness of an individual to an antigen. Physiologically, this tolerance is directed against antigens from normal tissues. Experimentally, however, tolerance can be established against any antigen by the use of appropriate procedures. These procedures include administration of antigen to very immature animals, administration of very large or very small doses of antigen, or administration of antigen treated in such a way that it cannot be processed properly.

McFarlane Burnet and Frank Fenner suggested in 1948 that the development of self-tolerance was associated with the stage of development of antigen-sensitive cells when they first encounter antigen. Burnet suggested that lymphocytes were somehow trained not to respond to an antigen as a result of interaction with that antigen during fetal life.

Support for this suggestion came from observations on chimeric calves. In 1945 Ray Owen noted that when cows have dizygotic (nonidentical) twin calves, the blood vessels in their two placentas may fuse. As a result, the blood of the calves intermingles freely, and stem cells from one animal can colonize the other. Owen found that these calves each possessed a mixture of red blood cells, some of their own and some originating from the other calf. In spite of being genetically and thus antigenically dissimilar, the foreign blood cells persisted indefinitely. It was clear that each calf was fully tolerant to the presence of foreign red blood cells. Burnet and Fenner suggested that this could happen only because each calf was exposed to the foreign cells early in fetal life at a time when antigen-sensitive cells become tolerant on exposure to antigens. Cells from an unrelated calf would be rejected normally if administered after birth. Rupert Billingham, Leslie Brent, and Peter Medawar conducted a variation of this "natural" experiment by inoculating the spleen cells from mice of one inbred strain (CBA) into the newborns of a second inbred strain (A). When the A mice grew to maturity, they would accept skin grafts from strain CBA mice but not from unrelated strains (Fig. 21–1). This tolerance was very clearly demonstrated when they succeeded in having a skin graft from a black mouse grow on a white one. This experiment also supported Burnet and Fenner's hypothesis. It is of interest to note that Burnet had previously tried unsuccessfully to induce tolerance by

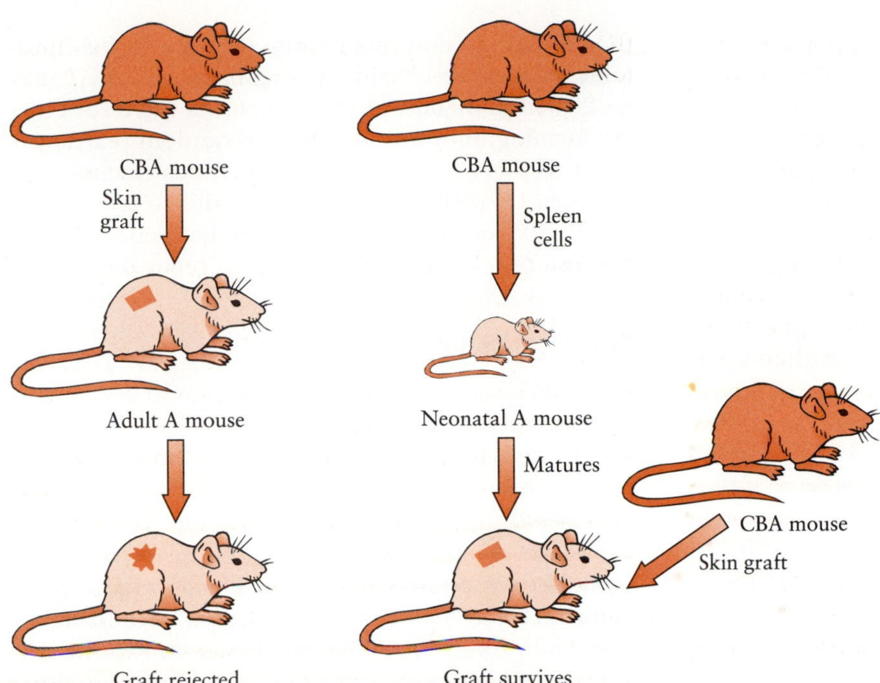

Figure 21–1 Medawar's tolerance experiment. He succeeded in making a CBA skin graft survive on a strain A mouse by injecting the recipient at birth with CBA strain spleen cells. Thus the strain A animal was fully tolerant to CBA cells.

injecting chick embryos with influenza vaccine. This difference in results was due to the persistence of antigen. The influenza vaccine survived for only a short period in the chicks, and the temporary tolerance induced could not be sustained. The spleen cells in Billingham's mice, in contrast, survived. Tolerance therefore persisted because the foreign antigen persisted.

Further evidence that prenatal exposure causes tolerance is seen in guinea pigs with a congenital deficiency of the fourth component of complement (C4). If injected with guinea pig C4, these deficient animals make anti-C4 antibodies. Thus guinea pigs have the ability to make antibodies to guinea pig C4 but not, however, in the presence of natural C4. The timing of the switch from tolerance induction to normal immune responsiveness varies between species. In laboratory rodents, it occurs soon after birth. In the other domestic mammals and humans, it occurs during the first third of gestation.

By reconstituting lethally irradiated mice with T or B cells derived from normal or tolerant syngeneic donors, tolerance can be shown to occur in both major lymphocyte populations. The susceptibility of T and B cells to tolerance induction, however, differs considerably. Thus T cells develop tolerance rapidly and easily, and they remain in that state for a long time (more than 100 days). B cells develop tolerance much more slowly (about 10 days as compared with 24 hours for T cells) and return to normal within 50 days. (This may simply be a result of ongoing somatic mutation in BCRs, resulting in the progressive generation of new, antigen-specific B cells.) Since helper T cells are required for both cell- and antibody-mediated immune responses, an animal may be functionally tolerant even though its B cells are fully responsive to an antigen (Fig. 21–2).

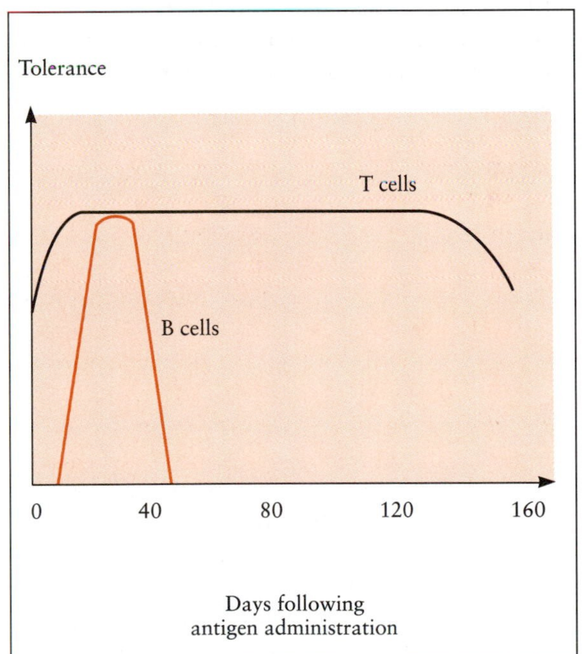

Figure 21–2 The duration of tolerance in T and B cells in mice. T cells develop tolerance rapidly and it lasts for a long period. In contrast, B cells can only be made tolerant for a short period. This probably results from the continuous production of B cells with specificity for the inducing antigen as a result of somatic mutation.

T-CELL TOLERANCE (Fig. 21–3)

Selection of the TCR

The ability of T cells to respond to an antigen depends on the availability of the correct antigen receptors. Tolerance to a specific antigen results if an animal lacks T cells with receptors specific for that antigen. Although theoretically the body may use any of the possible combinations of α/β or γ/δ T-cell receptors generated by the processes outlined in Chapter 15, far fewer are actually used by mature T cells than might be anticipated. Two selective processes are employed to limit the combinations employed (Fig. 21–4). First, there is positive selection of T cells carrying receptors that can recognize the animal's own MHC antigens. Second, there is negative selection (or clonal deletion) of those T cells bearing receptors for self-antigens and so potentially able to mount autoimmune responses that may be damaging to the host.

The range of available TCRs is determined within the thymus. When thymocyte progenitors arrive in the thymus from the bone marrow, they express no TCRs. On arriving in the thymus, γ/δ TCR expression precedes α/β expression by two days. The β-chain genes are first rearranged so that β chain can be expressed in the cytoplasm. This is followed by α-chain rearrangement and the formation of complete α/β receptors. Immature T cells possess only about 10% of the number of receptors seen in mature T cells and carry both CD4 and CD8 molecules as well. As the T cells develop within the thymus, the thymocytes lose either CD4 or CD8 so that they express just one of these proteins. TCR expression increases greatly until there may be 20,000 to 40,000 TCRs per cell. As the α/β TCRs are forming, they are subjected to random rearrangement of their gene segments, which, as discussed in Chapter 15, could potentially give rise to as many as 10^9 to 10^{12} receptors. This does not happen, however, because of selection of the receptor repertoire.

Positive Selection

Since all T cells must recognize antigen in association with self-MHC molecules, it is first necessary to select those T cells having receptors that can bind these molecules. This stimulation is called positive selection. Positive selection appears to be based on the ability of the thymocyte to respond to self-MHC molecules presented by thymic epithelial cells. Those cells that recognize self-MHC molecules are stimulated, and clonal expansion results. Positive selection occurs in the thymic cortex at the immature double-positive stage of thymic development. The binding of the TCR to either MHC class I or class II molecules on the thymic epithelial cells also directs differentiation of the selected cells into either $CD4^-8^+$ or $CD4^+8^-$ populations.

Negative Selection

In 1959 Joshua Lederberg suggested that self-tolerance resulted from the physical destruction of self-reactive lymphocytes during embryonic life, the

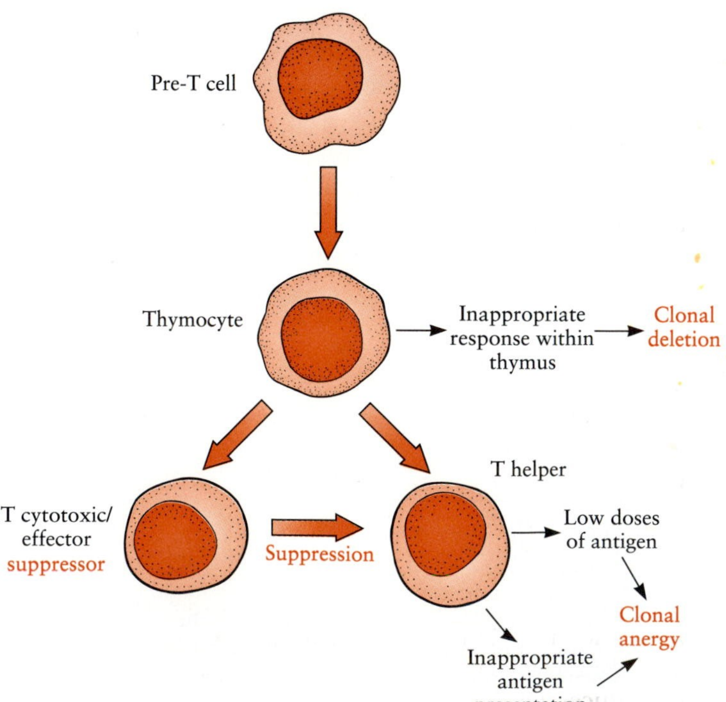

Figure 21–3 Some of the mechanisms of induction of tolerance in T cells.

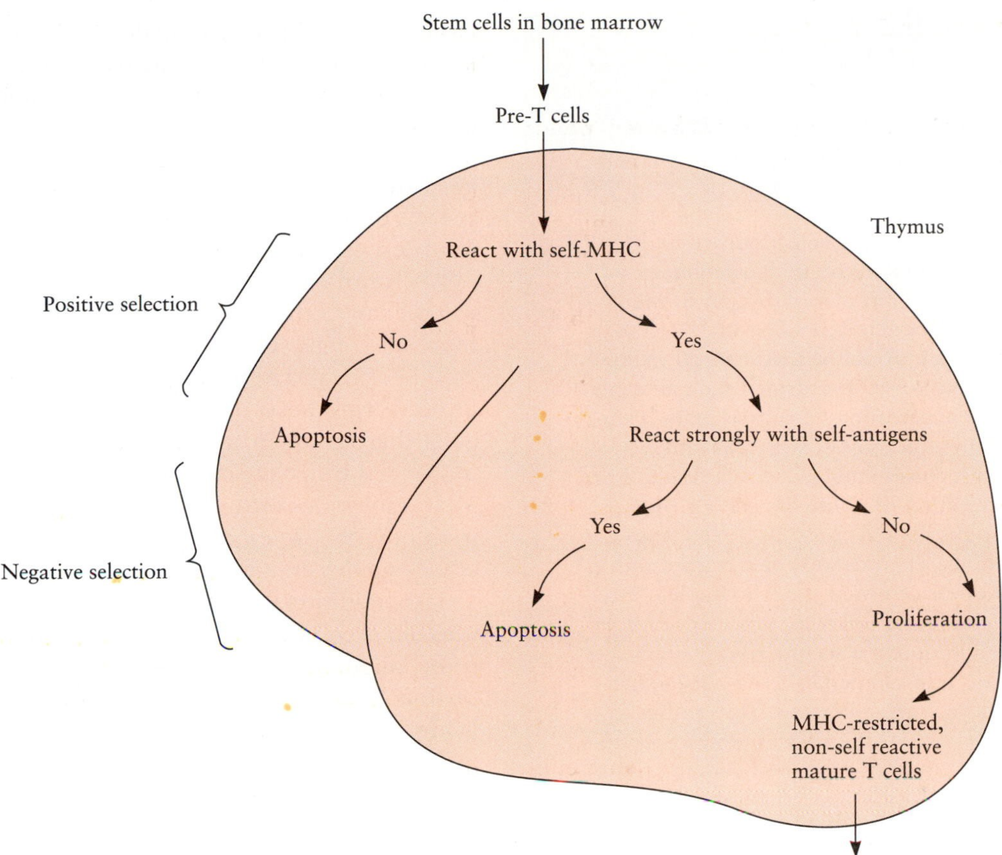

Figure 21–4 The T cell repertoire in the developing animal is determined by sequential positive and negative selection in the thymus. This ensures that only MHC-restricted, non-self reactive T cells reach maturity.

clonal deletion theory. Evidence has now accumulated to show that this theory is largely correct and that **clonal deletion,** the elimination of T cells capable of responding to self-antigens, also occurs in the thymus as T cells develop. The timing and extent of this deletion depend on the affinity of the TCR for self-antigen. T cells that bind self-antigens with high affinity are deleted earlier and more completely than low-affinity cells. Negative selection occurs when the TCR on a thymocyte binds either to ligands on antigen-presenting cells or thymic epithelial cells and triggers thymocyte apoptosis. Because T cells with anti-self activity are deleted by this negative selection process, this process leaves "holes" in the repertoire of immune responses to foreign antigens. This may adversely affect the defenses of the body. A good example of such a hole in the repertoire is seen in strain 13 guinea pigs, whose T cells cannot respond to the synthetic peptide poly-L-lysine. If a hapten such as DNP is attached to a poly-L-lysine carrier, these guinea pigs cannot respond to the DNP either, although they can respond to DNP attached to another carrier molecule such as bovine serum albumin.

Superantigens, as described in Chapter 18, bind to a specific Vβ domain of an α/β TCR. Because there are a limited number of different Vβ domains among the α/β TCR, many more T cells can bind a superantigen than can bind a conventional antigen peptide. For example, if there are about 20 Vβ domains, then one T cell in 20 will bind a superantigen. Some superantigens, the Mls antigens, for example, are produced by **retroviruses** (page 24). These retroviruses can infect sperm or egg cells and integrate themselves fully into an animal's genome. In fact, most mice and humans may carry several of these integrated viruses in their genome with no apparent ill effect. The proteins made by these integrated viruses are, in effect, self-antigens. Phillipa Marrack and John Kappler have examined the role of these retrovirus superantigens in tolerance. They have shown, for example, that a retrovirus called mammary tumor virus–7 (MTV-7) produces a superantigen that binds to Vβ8.1. They found

CLARIFICATION

Negative Selection and Intrathymic Transplantation

If, as a result of negative selection, self-reactive T cells are eliminated in the thymus, then it may be argued that a graft placed directly in the thymus would not be rejected since any T cell that chose to attack it would be automatically eliminated. Andrew Posselt performed such an experiment using pancreatic islet cells as grafts. Conventional attempts to graft islet cells into patients have consistently failed as a result of the allograft rejection response. Posselt's group, however, transplanted islet cells directly into the thymus of rats. The rats were also treated with antilymphocytic serum in order to deplete T-cell reserves. These islet cell allografts survived indefinitely. In addition, if a second donor strain islet allograft was transplanted to a site outside the thymus, this was also tolerated. Thus in this experiment, immature thymocytes were forced to mature in a thymic environment containing foreign MHC antigens. Those T cells capable of responding to the graft were probably deleted or inactivated. No other immunologic abnormalities were observed in the tolerant rats.

that in mice that do not carry MTV-7 about 8% of the T cells carry Vβ8.1 as part of their TCR. In contrast, mice that had MTV-7 integrated into their genome had no T cells with Vβ8.1 as part of their receptor. Similar results have been obtained using other retroviral superantigens. Thus the presence of a self-antigen, in this case the MTV-7 superantigen, led to the complete elimination of T cells carrying a receptor that could react with it. These T cells must have been destroyed at some stage in their development.

Additional evidence for the importance of negative selection is seen in *lpr* mice. These mice suffer from severe autoimmunity and lymphoproliferation involving T cells. The defect in these animals has been identified as a mutation in the fas protein. The fas protein, you may remember, is a key component of the normal apoptotic pathway (Chapter 18). Because of the *lpr* mutation, fas cannot transmit an apoptosis signal to T cells. As a result, negative selection cannot occur, self-reactive thymocytes survive, and autoimmunity develops.

The clonal deletion theory has one major problem, namely how tolerance can be induced to self-antigens that do not normally occur within the thymus. Evidence suggests that many of these antigens are in fact taken up by macrophages and B cells and carried to the thymus. Others such as those in the eye or brain may not be processed in this way, and tolerance to them may not fully develop. Some, such as sperm, do not appear until later in life and thus are not present during development of tolerance.

It is not clear how both positive and negative selection can both occur within the thymus and not remove all T cells. Thus in one case, self-reactive T cells are selected, and in the other, they are destroyed. One favored hypothesis to explain this contradiction suggests that although all cells that can bind self-MHC molecules are positively selected, only those that bind MHC molecules with very low or very high affinity are subsequently deleted. Thus the moderate-affinity clones survive and are able to present foreign antigen.

An additional factor that probably determines whether a cell will be positively or negatively selected is the amount of antigen presented to cells within the thymus. Thus if the concentration of a specific antigen is high (as one might anticipate from a self-antigen), multiple TCRs will be occupied on each thymocyte. This level of binding generates a signal that will trigger apoptosis and thus negative selection. In contrast, if only low concentrations of an antigen are available, these will occupy only a few TCRs on each thymocyte. This level of signal causes positive selection and thymocyte proliferation.

Transgenic Mice and Tolerance

If all T cells are made tolerant by negative selection within the thymus, then all possible "self"-antigens must be expressed within the thymus. If these antigens are not encountered in the thymus, then the T cells should never become tolerant to them. Studies using transgenic mice, however, have shown that the development of tolerance requires more than just negative selection within the thymus.

Transgenic mice are mice that carry foreign genes. When these foreign genes are expressed, their products should be treated as self-antigens. By using tissue-specific promoters, the expression of these "foreign" antigens can be restricted to specific cell types. For example, J.F.A.P. Miller and his colleagues have taken the gene for a MHC class I protein called K^b, linked it to an insulin promoter, and by microinjection inserted the whole DNA construct into fertilized allogeneic (H-2B) mouse eggs (Fig. 21–5). The offspring of these eggs developed into normal mice but expressed the foreign MHC protein on the β cells of their pancreatic islets. The T cells of these mice were

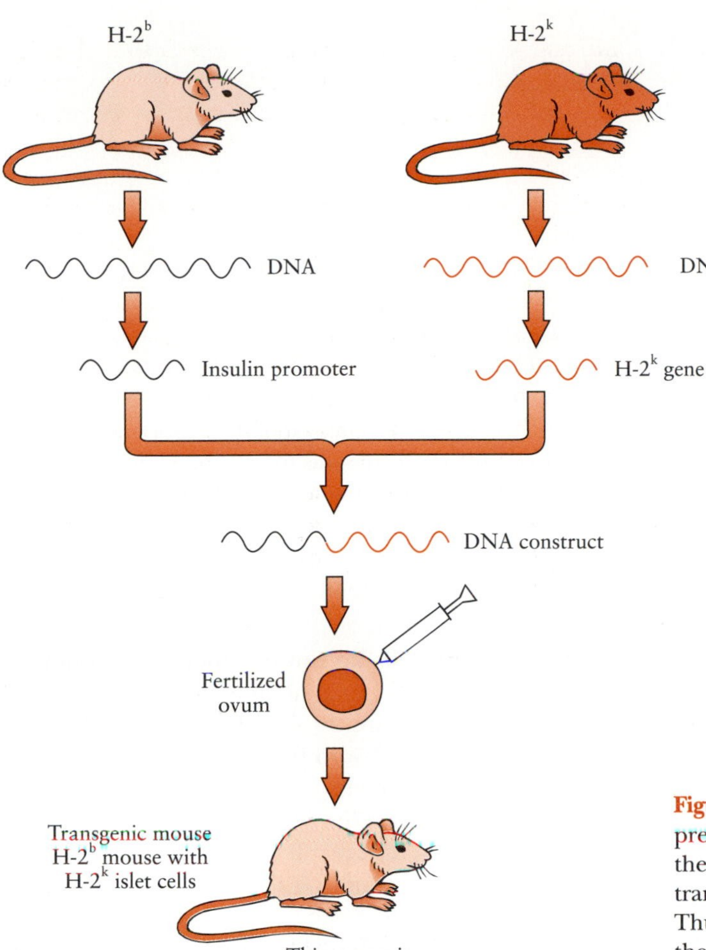

Figure 21–5 The construction of a transgenic mouse expressing a foreign MHC antigen on its islet cells. Although the foreign antigen has never penetrated the thymus of the transgenic animal, it does not induce an immune response. Thus negative selection by exposure to self-antigens within the thymus cannot be the only mechanism of tolerance induction.

tolerant to the foreign molecule despite the fact that it is not found in the thymus. This tolerance faded once all the islet cells died and the transgene was lost. Thus T cells specific for these extrathymic antigens are not always clonally deleted or permanently silenced. Similar tolerance to a transgene product has been observed when it is presented linked to a metallothionine promoter (restricted to the liver, kidney, and pancreas) or to a myelin basic protein promoter (restricted to the brain).

Clonal Anergy

Because self-tolerance is essential to survival as described previously, the body does not rely solely on negative selection of thymocytes to ensure that self-reactive T cells are suppressed. For example, not all self-antigens can enter the thymus and induce negative selection. The self-reactive T cells that escape from the thymus must be suppressed by other mechanisms. This functional suppression is called clonal **anergy.** Clonal anergy, the prolonged, antigen-specific suppression of a clone of T cells depends on the signals delivered to the T cell. As pointed out in Chapter 11, binding of antigen to a TCR alone is insufficient to trigger a T-cell response. Indeed, occupation of the TCR by a MHC–epitope complex in the absence of an additional costimulatory signal leads to clonal anergy.

Normally, any protein solution contains some spontaneously aggregated molecules. These aggregated molecules are readily taken up and processed and presented by macrophages and are thus highly immunogenic. About 30 years ago, David Dresser showed that if a solution of such a protein, for example, bovine gamma globulin, is ultracentrifuged so that all the aggregates are removed, then the aggregate-free solution will induce tolerance when injected into adult rabbits. This tolerance is specific, since it is directed only against the bovine gamma globulin. It occurs as a result of clonal anergy that developed because the antigen-sensitive helper T cells were exposed to free (not macrophage-bound) antigen. As a result, the helper cells failed to receive a second triggering signal. The second signal is normally provided by the

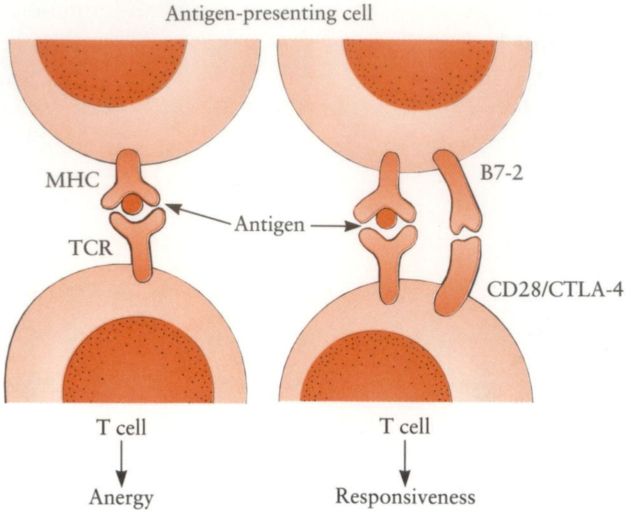

Figure 21–6 T cells will only respond to processed antigen when they also receive a stimulus from another receptor such as CD28 or CTLA-4. In the absence of this costimulus, anergy and tolerance result.

binding of B7-2 (CD80) on an antigen-presenting cell to CD28 on the T cell (Fig. 21–6).

Activation of the TCR by an MHC–epitope complex alone activates the tyrosine kinase and phospholipase C of the T cell and raises intracellular Ca^{++}. This in itself is not sufficient to activate the IL-2 gene. In fact, it triggers a repressor gene that makes the cell anergic. In this case the T cell may become incapable of producing IL-2 on restimulation. Only if IL-1 is present and CD28–B7 binding occurs will the induction of the repressor be blocked and the cell activated.

Anergic Th1 cells do not produce IL-2, although small amounts of IL-2 may be detected using antigen concentrations that are 30 to 100 times higher than normal. There is a 93% to 95% drop in the levels of IL-2 mRNA in these cells. IL-3 production is reduced by 87% and IFN-γ is reduced by 33%. Once induced, this anergy can last for several weeks. The anergic state may be reversed by exposing the cell to high concentrations of IL-2.

Very high doses of antigen induce a form of tolerance known as **immune paralysis** (Fig. 21–7). This is probably a form of clonal anergy. The high doses of

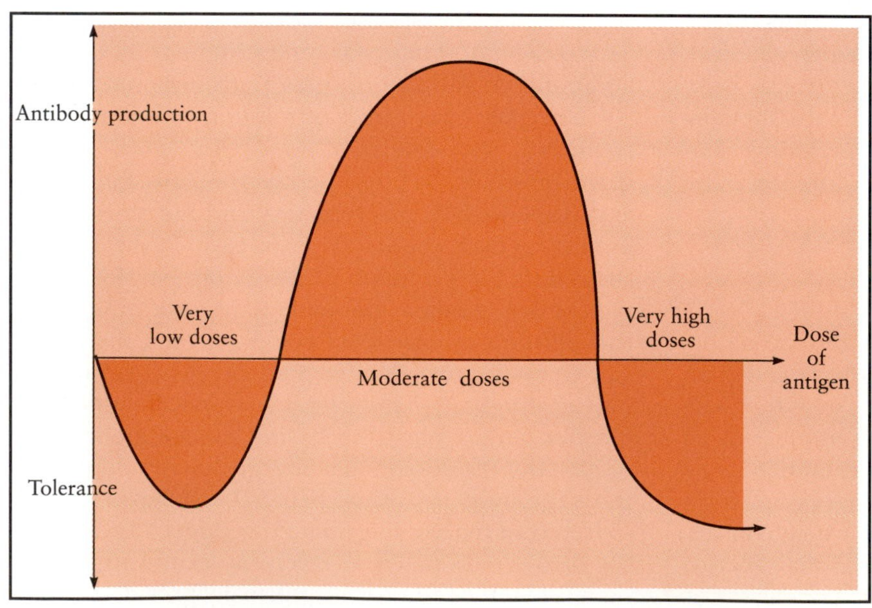

Figure 21–7 High and low zone tolerance. Both low and high doses of antigen will cause tolerance. Moderate doses induce a conventional immune response.

antigen probably bypass antigen-presenting cells, reach helper T cells directly, and make them anergic.

T Suppression

Although many investigators now doubt their existence, cytotoxic T cells may act as suppressor cells and play a role in the induction and maintenance of self-tolerance. If suppressor-cell function is depressed, as occurs in diseases such as systemic lupus erythematosus (Chapter 31) and in certain strains of mice (especially the New Zealand Black × New Zealand White hybrid), then autoimmune diseases tend to develop. Suppressor T cells can provoke a tolerance that is transmissible to other animals by adoptive transfer (Fig. 21–8). Suppressor cells appear to be involved in tolerance induced both by very low doses of antigen (low-zone tolerance) and by deaggregated antigen, since that tolerance may be transmitted to syngeneic recipients by lymphocytes from tolerant donors. Low-zone tolerance is maintained by the action of suppressor T cells, which are triggered by a dose of antigen much lower than that required by helper T cells.

B-CELL TOLERANCE

Unlike the T-cell–receptor repertoire, antibody diversity is generated within the B-cell repertoire in two phases. The first occurs by gene rearrangement in the primary lymphoid organs; the second phase involves somatic mutation in germinal centers in peripheral lymphoid organs. Thus B cells have several opportunities to become reactive to self-antigens, and self-reactive B cells can be readily found in lymphoid organs. These cells do not necessarily make autoanti-

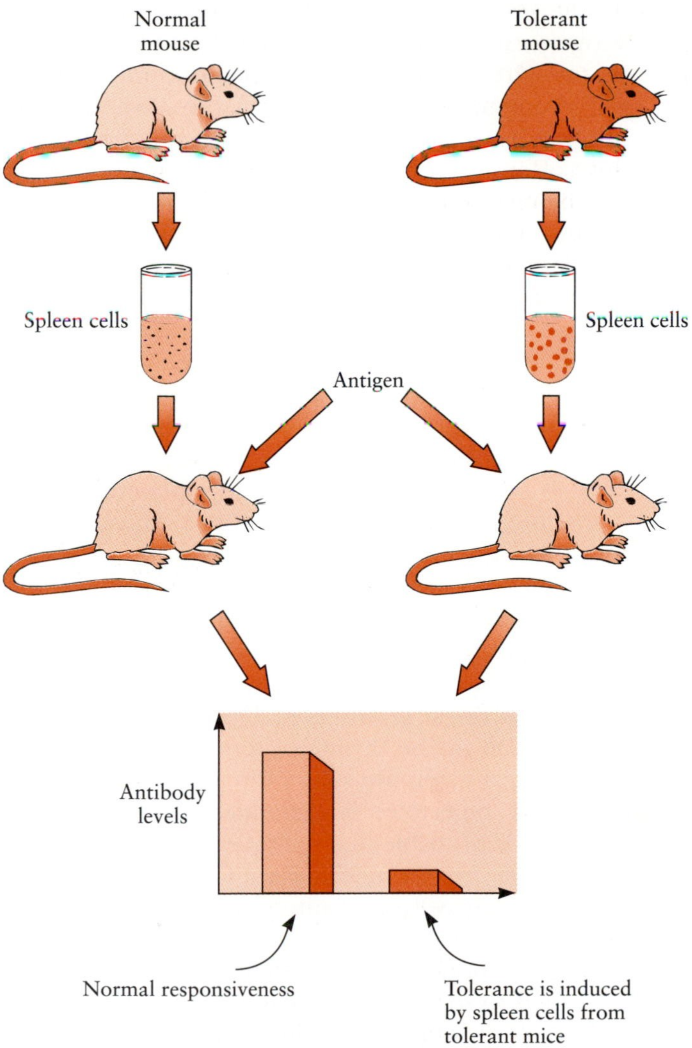

Figure 21–8 Evidence for the existence of suppressor T cells. Spleen cells from a tolerant mouse will produce a suppressed immune response when adoptively transferred to recipient animals. Treatment of the spleen cells from tolerant animals with anti-Thy-1 reversed the suppressive effect in recipient animals. It is, of course, possible that Th2 cells producing IL-10 could have a similar effect. *(From an experiment by Basten A., et al., J. Exp. Med. 140: 199, 1974.)*

bodies. If all the components necessary for a B-cell response to antigen are not present (i.e., antigen-presenting cells and helper T cells) or if suppressor T cells are active, then the B cell may be unable to respond effectively to antigen and so appear to be nonfunctional (Fig. 21–9). Since their response depends on helper T cells, their failure to produce anti-self antibodies could simply result from the absence of self-reactive helper T cells rather than any change in the B cells themselves. This is not, however, a foolproof method of ensuring no self-reactivity. In situations in which T cells but not B cells are tolerant, antibody production can be stimulated by using either cross-reacting epitopes or a new carrier molecule to stimulate nontolerant helper T cells. Thus tolerance to a hapten can be broken by administering it on a carrier that is unrelated to the carrier originally used to provoke tolerance. The new carrier molecule stimulates a new and unsuppressed population of helper T cells. Similarly, one can break tolerance to a large antigenic molecule by administering a cross-reactive antigen. The epitopes on the cross-reactive molecule will act as new carrier epitopes and recruit new helper cells. Thus other mechanisms must also be available to ensure B-cell tolerance. Four of these mechanisms of B-cell tolerance are clonal abortion, clonal exhaustion, functional deletion, and blockage of B-cell receptors.

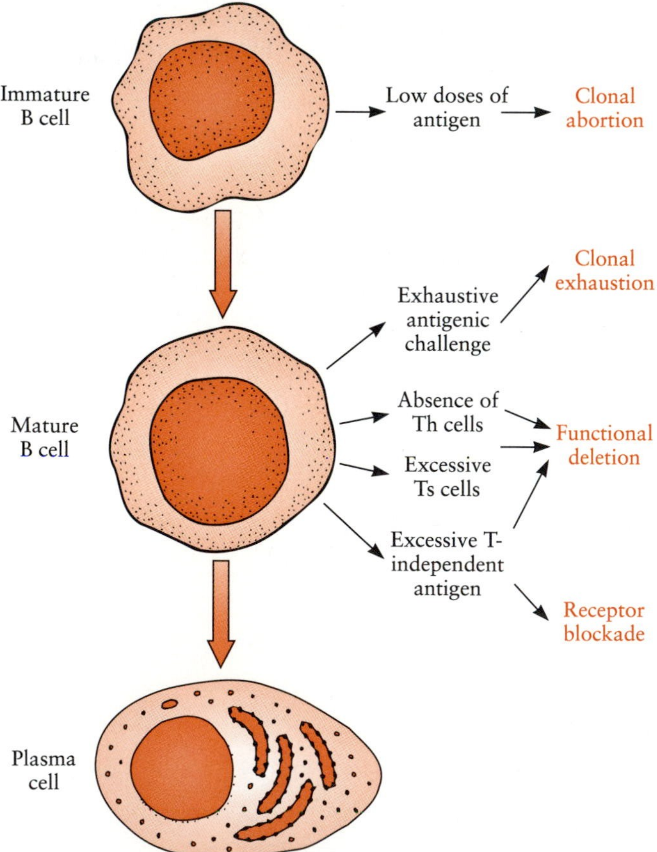

Figure 21–9 The mechanisms of induction of tolerance in B cells.

Clonal Abortion and Anergy

The original hypothesis of Burnet and Fenner, that all anti-self cells were turned off prior to birth, is not entirely true since, as mentioned previously, some anti-self B cells are present in adults. However, immature B cells can be made tolerant by exposure to self-antigen. These developing B cells can be made tolerant only when they have successfully arranged their V-region genes and are committed to express complete IgM molecules on their surface. (This is the distinguishing feature between pre–B cells and immature B cells.) After a few days these immature B cells begin to express additional cell surface molecules such as IgD, L-selectin, CR1, CR2, and CD23. It is at this early immature stage that self-reactive B cells can be eliminated. They encounter large amounts of self-antigens on the surface of nearby cells. If they bind this self-antigen strongly, then the BCR will transmit a signal that results in arrest of development and apoptosis within one to three days. An immature B-cell population can be rendered tolerant by one millionth of the dose of antigen required to render a mature B-cell population tolerant. One reason for this is that immature animals are deficient in helper T cells and IL-2. The ability of B cells to develop tolerance also is inversely related to their possession of IgD. The cells of very young animals are low in IgD, and lymphocytes readily develop tolerance following removal of IgD from the cell surface. In addition, immature B cells may be unable to regenerate cell surface immunoglobulins after capping, and macrophages of newborn animals present antigen poorly to lymphocytes.

Self-reactive B cells must bind to a critical threshold of self-antigen to be rendered tolerant. This results in selective silencing of high-affinity B cells. Presumably, the failure to make low-affinity anti-self B cells tolerant poses little threat of autoimmune disease because the low affinity of the antibodies produced does not permit tissue destruction. This tends to parallel the situation in the thymus, where T cells with high affinity for antigen are selectively destroyed. Because the B-cell repertoire is generated in two distinct phases, it is probable that self-tolerance is also generated by two distinct mechanisms. Clonal abortion occurs only if B cells encounter antigen early in their development within the bone marrow. A second mechanism of deleting self-reactive peripheral B cells is needed to suppress cells that acquire self-reactivity as a result of somatic mutation. This involves the development of clonal anergy.

Inappropriate Antigen Presentation

An experimental method of inducing B-cell tolerance by inappropriate antigen presentation is through the use of a hapten attached to a nonimmunogenic carrier. Thus, if a normally effective hapten, such as DNP, is linked to a nonimmunogenic carrier, such as a copolymer of D-glutamic acid and D-lysine, then it will provoke tolerance. In this case, the tolerant animal will specifically fail to respond to the same hapten bound to an immunogenic carrier. Thus the B cells must be specifically tolerant. This presumably results from a failure of the B cell to receive appropriate stimuli from helper T cells.

Clonal Exhaustion

B cells subjected to repeated, exhaustive antigenic stimulation may be stimulated to differentiate into short-lived plasma cells. If all B cells develop into plasma cells, then no memory B cells will remain to respond to subsequent doses of antigen. A related method of inducing tolerance involves giving an animal immunosuppressive drugs simultaneously with antigen. If, for example, cyclophosphamide or cyclosporine (drugs that destroy dividing lymphocytes) is given at the same time as antigen, then specific tolerance, rather than a normal immune response, may result. These drugs may either lower the threshold for tolerance induction or block the differentiation of antigen-stimulated cells. Cyclophosphamide, for example, prevents the B cell from renewing its antigen receptors.

Blockage of B-Cell Receptors

Some polymeric antigens such as pneumococcal polysaccharide can bind irreversibly to B-cell receptors. These antigens thus effectively freeze the B-cell membrane and block any further responses by these cells. The B cells will recover when the antigen is removed.

INCOMPLETE TOLERANCE

Tolerance need not be total, and it can be generated in selected segments of the immune system. For example, tolerance in the response involving a single immunoglobulin class can be produced and Bϵ cells are much more readily made tolerant than Bγ or Bμ. As a result, it is possible to decrease susceptibility to allergies by desensitizing injections of antigen (Chapter 26).

DURATION OF TOLERANCE

When an animal is made tolerant, the duration of tolerance depends on the persistence of antigen and the ability of the bone marrow to generate fresh antigen-sensitive cells. When the antigen is metabolized, the tolerance fades. If the antigen is persistent (such as occurs in calf chimeras or with an animal's own tissue antigens), then tolerance also persists. This is because, in the presence of antigen, newly formed antigen-sensitive cells are triggered to undergo apoptosis immediately after their receptors bind self-antigen. Treatment that promotes bone marrow activity, such as low doses of radiation, hastens the fading of tolerance, whereas immunosuppressive drug treatment has the opposite effect. It is possible to break tolerance to some self-antigens, such as thyroglobulin or myelin basic protein (Chapter 31), by administering the antigen with a potent adjuvant (such as Freund's complete adjuvant).

KEY WORDS

Clonal abortion p. 330
Clonal anergy p. 327
Clonal deletion p. 325
Clonal exhaustion p. 331
Immune paralysis p. 328
Incomplete tolerance p. 331
Negative selection p. 324
Positive selection p. 324
Receptor blockage p. 331
Superantigen p. 325
Suppressor cell p. 329
TCR selection p. 324
Thymus p. 324
Tolerance p. 322

QUESTIONS

1. What cell types can be made tolerant?
   a. T cells alone
   b. B cells alone
   c. NK cells alone
   d. T and B cells
   e. macrophages

2. Tolerance is most readily induced in
   a. older animals
   b. female animals
   c. immunologically stimulated animals
   d. young animals
   e. fetal animals

3. Where does the major step in inducing T-cell tolerance occur?
   a. spleen
   b. bone marrow
   c. bursa of Fabricius
   d. thymus
   e. Peyer's patches
4. The elimination of self-reactive T cells in the thymus is called
   a. negative selection
   b. positive selection
   c. clonal selection
   d. apoptosis
   e. clonal anergy
5. T cells of *lpr* mice fail to undergo apoptosis because of a defect in the
   a. *lck* gene
   b. *ras* gene
   c. *myc* gene
   d. *fos* gene
   e. *fas* gene
6. When a T cell is made tolerant by receiving insufficient or inappropriate costimulation, this is called
   a. negative selection
   b. clonal anergy
   c. clonal selection
   d. positive selection
   e. receptor blockade
7. Very high doses of antigen induce
   a. negative selection
   b. clonal selection
   c. immune paralysis
   d. receptor blockade
   e. T suppression
8. Clonal abortion is a mechanism of inducing tolerance in
   a. T cells
   b. B cells
   c. NK cells
   d. TIL cells
   e. LAK cells
9. Which antigen can cause B-cell tolerance by blocking receptors?
   a. interleukin
   b. endotoxin
   c. gamma globulin
   d. serum albumin
   e. pneumococcal polysaccharide
10. Which treatment may hasten the fading of tolerance?
    a. immunosuppression
    b. immunostimulation
    c. low-dose radiation
    d. high-dose radiation
    e. complement depletion
11. Classify the ways in which a fetal animal becomes tolerant to self-antigens. How can this form of tolerance be broken?
12. What are the risks of inducing tolerance to an antigen such as a vaccine? What steps might you take to ensure that tolerance is not provoked?
13. One possible treatment of severe allergies would be to provoke tolerance to the offending allergens. How might this be done, and why, if it's so easy, is this not done routinely?
14. Predict the results of loss of self-tolerance in an animal.
15. How might thymocytes be both positively and negatively selected at the same time?
16. How has the use of transgenic mice contributed to our understanding of the role of the thymus in the generation of tolerance?
17. What is incomplete tolerance? What is its significance?

Answers: 1d, 2e, 3d, 4a, 5e, 6b, 7c, 8b, 9e, 10c

SOURCES OF ADDITIONAL INFORMATION

Ashton-Rickhardt, P.G., et al. Evidence for a differential avidity model of T cell selection in the thymus. Cell, 76:651–663, 1994.

Cohen, J.J. Glucocorticoid-induced apoptosis in the thymus. Semin. Immunol. 4:363–369, 1992.

Ferrick, D.A., et al. Thymic ontogeny and selection of αβ T cells. Immunol. Today, 10:403–407, 1989.

Fuchs, E.J., and Matziger, P. B cells turn off virgin but not memory T cells. Science, 258:1156–1158, 1992.

Gimmi, C.D., et al. Human T-cell clonal anergy is induced by antigen presentation in the absence of B7 costimulation. Proc. Natl. Acad. Sci. U. S. A., 90:6586–6590, 1993.

Green, D.R., and Cotter, T.G. Apoptosis in the immune system. Semin. Immunol., 4:355–362, 1992.

Hartley, S.B., et al. Elimination of self-reactive B lymphocytes proceeds in two stages: Arrested development and cell death. Cell, 72:325–335, 1993.

Jekins, M.K., and Miller, R.A. Memory and anergy: Challenges to traditional models of T lymphocyte differentiation. FASEB J., 6:2428–2433, 1992.

Lorenz, R.G., and Allen, P.M. Thymic cortical cells can present self antigens in vivo. Nature, 337:560–562, 1989.

Marrack, P., and Kappler, J.W. The T cell repertoire for antigen and MHC. Immunol. Today, 9:308–314, 1988.

Marrack, P., and Kappler, J.W. How the immune system recognizes the body. Sci. Am., 269:81–89, 1993.

Marrack, P., McCormack, J., and Kappler, J.W. Presentation of antigen, foreign major histocompatibility complex proteins and self by thymus cortical epithelium. Nature, 338: 503–505, 1989.

Miller, J.F.A.P., et al. T-cell tolerance in transgenic mice expressing major histocompatibility class I molecules in defined tissues. Immunol. Rev., 107:109–123, 1989.

Posselt, A.M., et al. Induction of donor-specific unresponsiveness by intrathymic islet transplantation. Science, 249: 1293–1295, 1990

Pullen, A.M., Kappler, J.W., and Marrack, P. Tolerance to self antigens shapes the T-cell repertoire. Immunol. Rev., 107:125–139, 1989.

Ramsdell, F., and Fowlkes, B.J. Maintenance of in vivo tolerance by persistence of antigen. Science, 257:1130–1138, 1992.

Takahama, Y., and Singer, A. Post-transcriptional regulation of early T cell development by T cell receptor signals. Science, 258:1456–1462, 1992.

Zepp, F., and Staerz, U.D. Thymic selection process induced by hybrid antibodies. Nature, 336:473–476, 1988.

CHAPTER 22

Regulation of the Immune Response

Niels K. Jerne *Niels Jerne was born in England in 1911 but mainly worked in Switzerland. He was awarded the Nobel prize for his theoretical studies on the immune system in 1984. He is credited with the concept that the thymus is the site where lymphocytes multiply, differentiate, and become functional T cells. He developed the network theory that antibodies stimulate antiidiotypes and with antibodies form a functional network. He also was the first to expound the selective theory that explained how antibodies could be produced that respond to a great variety of antigens.*

CHAPTER OUTLINE

CHAPTER CONCEPTS

1. The immune response is antigen driven. Antigen starts the response. Once antigen is removed, the immune response ceases. Antigens that persist cause a prolonged immune response.
2. Antibody levels are controlled by negative-feedback. Thus, once antibodies are produced, they act to turn off excessive antibody production.
3. Epitopes differ in their ability to provoke B-cell, helper-T-cell, or suppressor-T-cell responses. This may be related to the way in which the antigen is processed and presented to T cells.
4. Suppressor T cells may play a role in regulating the immune response in many situations, although their existence is hotly debated.
5. Interacting networks involving both antibodies (idiotypes) and cells are also involved in regulating immune responsiveness.
6. The nervous system greatly influences the functioning of the immune system. Stress especially can be immunosuppressive.

In this chapter you will read about the regulation of the immune response. Clearly, any complex multicellular system requires careful regulation. The first part of the chapter describes how antigen and especially specific epitopes regulate the immune responses. The second section describes how antibody exerts feedback on the B-cell responses. In addition, it explains how idiotypes have the ability to stimulate their own immune responses and so regulate the immune system. Some of the genetic influences that control immunity are described, followed by a discussion of regulatory cells, including the contentious issue of suppressor cells. These suppressor cells can include not only T cells but also macrophages and natural suppressor cells. The chapter concludes with a description of a new, exciting area of immunology, namely the regulation of the immune system by the nervous system as well as the effects that the immune system may have on the nervous system.

The immune responses, though essential for the protection of the body, can cause severe damage if permitted to act in an uncontrolled fashion. The most damaging result of a lack of regulation is the development of immunity to self-antigens. This is normally controlled by the development of tolerance, as described in the previous chapter. Failure to develop tolerance may result in the production of autoantibodies and the development of autoimmune disease (Fig. 22–1). Tolerance, however, cannot be the only mechanism of immune regulation employed. An ability to regulate the magnitude of an immune response is also essential. An inadequate immune response may lead to immunodeficiency and increased susceptibility to infection. An excessive immune response may result in the development of a disease such as amyloidosis (Chapter 28). Failure to control the burst of lymphocyte proliferation that occurs during immune responses may lead to the development of lymphoid cell tumors. Failure to control the immune response to the fetus may lead to abortion (Chapter 19). The immune responses must therefore be carefully regulated to ensure that they are appropriate in quality and quantity. As might be anticipated, several control mechanisms exist to accomplish this regulation.

ANTIGEN REGULATES IMMUNE RESPONSES

Immune responses are antigen driven. They commence only on exposure to antigen, and once antigen is eliminated, the responses stop. If antigen persists, then the stimulus persists, and as a result, the immune response is prolonged. Thus a prolonged response occurs after immunization with poorly metabolized antigens such as the bacterial polysaccharides or with antigen incorporated in oil or insoluble adjuvants so that the antigen cannot be rapidly eliminated.

The quantity of antigen also influences the nature of the immune response. We have already discussed in Chapter 17 the phenomenon of **affinity maturation** in which high doses of antigen provoke production of low-affinity antibody and low doses induce high-affinity antibody. Antigen-sensitive lymphocytes respond to an epitope only if the antigen is presented to them at

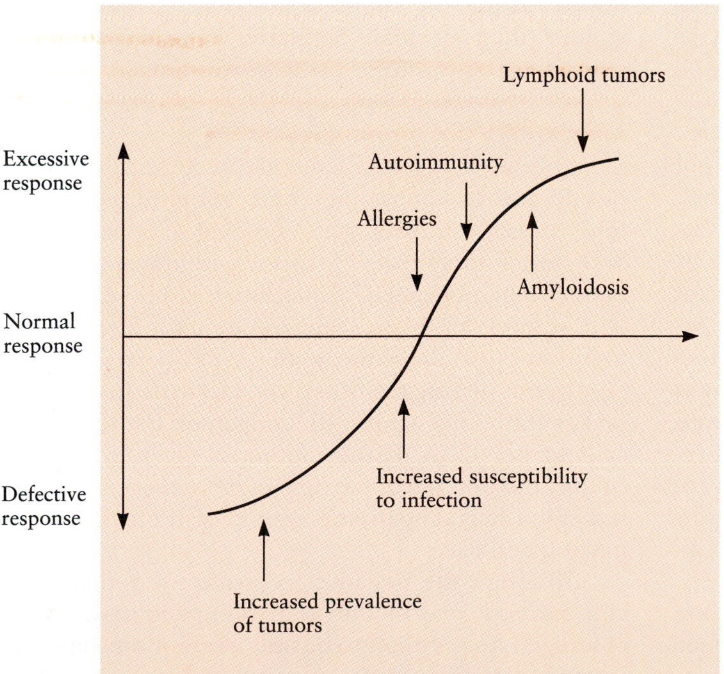

Figure 22–1 The consequences of a failure to regulate the immune system.

an appropriate time, in the correct dose, and in an appropriate manner. If the antigen is encountered at the wrong period in a cell's development, if the dose of antigen is either excessive or inadequate, or if the antigen is presented to the cells in an inappropriate fashion, then instead of responding to antigen by division and differentiation, cells may become **anergic** or may even undergo **apoptosis** and tolerance therefore results (see Fig. 21–6).

Antigen Processing and Immune Regulation

The nature of the immune response may differ in different parts of the body. Langerhans' cells seem especially suited for mediation of delayed hypersensitivity, and the dendritic cells of lymphoid follicles mediate priming of B cells. It is possible that the Langerhans' cells are optimized to present antigen to Th1 cells and the follicular dendritic cells present antigen to Th2 cells. Adjuvants also influence the type of immune response induced; they may do this through their actions on antigen-presenting cells. Lipids conjugated to protein antigens commonly induce delayed hypersensitivity rather than antibody production and localize in T-cell rather than B-cell areas of lymphoid tissues. T-independent antigens do not require T-cell help since they possess multiple epitopes and can thus strongly stimulate B cells. However, they may stimulate B cells excessively and so cause tolerance.

Regulation by Epitopes

The epitopes that trigger antibody formation are located on the surface of antigenic molecules in solution. The ability of an epitope to bind to B-cell receptors or free antibody is critically dependent on its conformation. If an epitope is treated so that its shape is altered (even though its amino acid sequence is unchanged), then it loses its ability to bind to specific antibody. Thus BCRs must bind epitopes that are not significantly modified by the activities of macrophages.

Helper T cells, in contrast, respond only to isolated antigen fragments bound to an MHC class II molecule. They arrive at this location as a result of processing and are therefore denatured. Thus, for example, antibodies formed against the enzyme ribonuclease are directed against conformational epitopes on the surface of the molecule, whereas the delayed hypersensitivity skin response, a T-cell–mediated reaction (Chapter 18), is directed against sequential epitopes (i.e., chains of amino acids in specific order rather than of specific shape). These sequential epitopes are generated by antigen processing within cells.

Some epitopes appear to selectively turn off the immune response and have been called suppressor epitopes. For example, myelin-basic protein, when injected into guinea pigs, normally provokes an autoimmune disease called experimental allergic encephalomyelitis (EAE) by stimulating the production of cytotoxic T cells against myelin. This basic protein can be cleaved into two distinct fragments. One fragment retains the ability to induce the encephalitis. Presumably, it is this fragment that stimulates the production of cytotoxic T cells. The other fragment of the molecule containing about 45 amino acids will, if given before inoculation of the basic protein, protect guinea pigs against EAE. This suppressive peptide appears to carry an epitope that provokes the development of suppressor T cells, which then act to block the immune response to myelin-basic protein.

An antigen may therefore carry up to three classes of epitope: epitopes on the surface of the molecule that stimulate B cells and thus antibody formation (e.g., most haptens), epitopes that stimulate helper cells (e.g., most carrier molecules), and epitopes that stimulate **suppressor cells.** The net response to any antigen depends on the balance between these three groups of epitopes.

ANTIBODY REGULATES IMMUNE RESPONSES

Antibodies or immune complexes generally exert a negative influence on immune responses. This influence can be shown by removing the antibody from an animal by plasmapheresis while it is mounting an immune response. (Plasmapheresis is a technique by which blood is removed from an animal, the cells and plasma are separated, and the cells returned to the donor.) As a result, the immune response proceeds indefinitely, and the total quantity of antibody produced during the immune response is many times greater than normal. Similarly, repeated adoptive transfer of responding, antibody-producing cells to syngeneic recipients permits these cells to undergo uncontrolled expansion (Fig. 22–2).

In general, IgG antibodies depress the production of IgM and IgG antibodies, and IgM antibodies tend to depress the further synthesis of IgM. Specific antibody tends to suppress a specific immune response better than nonspecific immunoglobulin. An excellent example of this is seen in the method employed to prevent hemolytic disease of the newborn (Chapter 30). In this disease a mother who lacks the Rh antigen makes antibodies against Rh antigen on the red blood cells of her fetus. If the mother is given antiserum against this antigen at the time of her exposure to fetal red blood cells at birth, she is completely blocked from making anti-Rh.

Although the negative feedback exerted by specific antibody may be due to masking and destruction of antigen, this is likely to be only a minor mechanism. Immunoglobulin molecules must have intact Fc regions to inhibit an antibody response. Both B cells

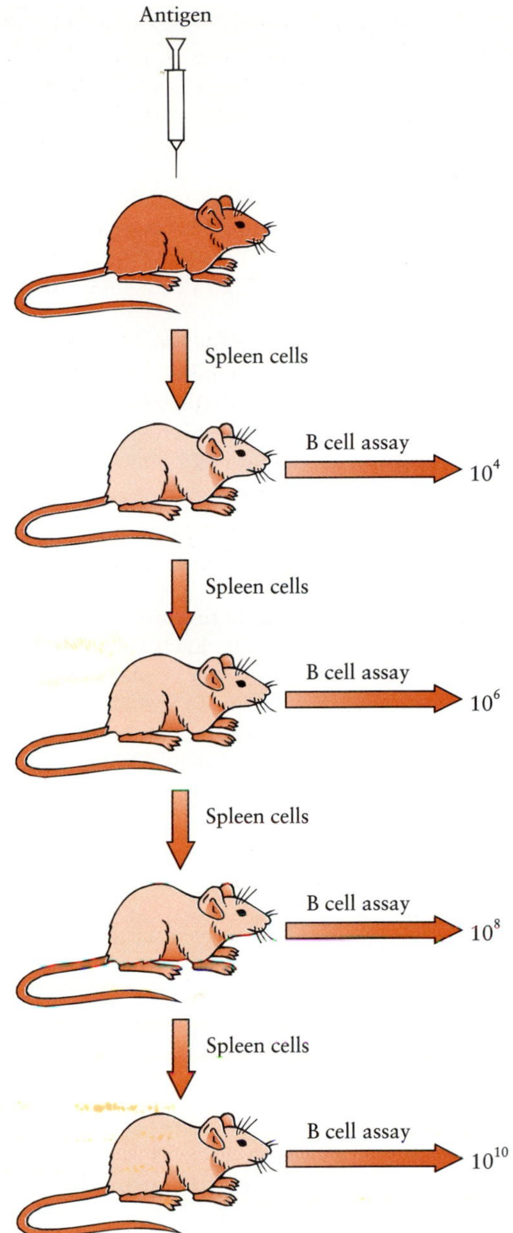

Figure 22–2 The remarkable ability of B cells to proliferate if controlling influences are removed. In this case sequential adoptive transfer of spleen cells following immunization removes the B cells from the suppressive influence of antibody. As a result, the numbers of responding B cells can reach 10^{10}.

and suppressor T cells possess Fc receptors, and the binding of antigen–antibody complexes to these receptors inhibits the antibody response (Fig. 22–3). Antigen–antibody complexes block B-cell differentiation into plasma cells but do not inhibit their proliferation. On the other hand, free Fc fragments can act as B-cell mitogens and promote polyclonal antibody formation. In addition, they bind to $CD4^+CD8^-$ cells in mice to promote the release of a nonspecific helper factor. A tetrapeptide derived from the immunoglobulin Fc region called tuftsin also promotes the macrophage-mediated presentation of antigen to T cells.

In diseases in which serum immunoglobulin levels are abnormally high, as in patients with **myelomas** (Chapter 13), these feedback mechanisms depress normal antibody synthesis. As a result, patients with myelomas are very susceptible to infections. A similar phenomenon occurs in newborn animals that acquire immunoglobulins passively from their mother. The presence of this maternal antibody, while conferring protection, effectively inhibits immunoglobulin synthesis and prevents the successful vaccination of newborn animals (Fig. 22–4).

It is also of interest that when antibody-producing plasma cells first develop, they promptly leave the lymph node cortex and the lymphoid follicles of the spleen, sites where antigen-sensitive cells normally react with antigen (Chapter 9). Presumably, if this emigration did not take place, then the production of high levels of immunoglobulin in close proximity to cells responding to antigen might turn them off and thus make an immune response terminate prematurely.

The class as well as the quantity of immunoglobulin produced during an immune response is also effectively regulated. Most unstimulated B cells have both IgM and IgD surface immunoglobulin receptors. During an immune response, these cells switch to the production of IgM, IgG, IgA, or IgE, and this switch is controlled by the activities of T cells. In animals given T-independent antigens, there is no switch from IgM to IgG production, and a persistent low-level IgM response ensues. Neonatal bursectomy may also result in a failure of the IgM-to-IgG switch, suggesting that the bursa of Fabricius is responsible for this process in birds.

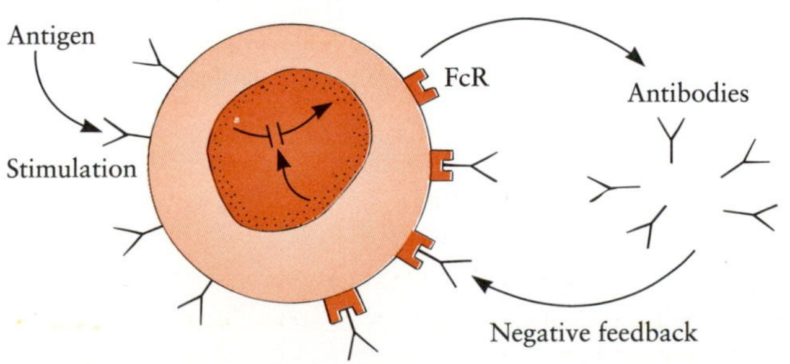

Figure 22–3 Antibody produced by responding B cells exerts a negative feedback by binding to B cells through Fc receptors such as CD32.

CLARIFICATION

How Antibody Controls the Immune Response

Henry and Jerne demonstrated the negative-feedback effect of antibody by giving antisheep red blood cell IgG to mice at various times before and after an injection of sheep red blood cells and then measuring the antisheep red blood cell response. It was clear from their results (figure) that IgG antibody given one hour before or up to four hours after injection of antigen effectively reduced the immune response. The simplest explanation for this is that the passively administered antibody prevents recognition of antigen by B cells. Gordon and Murgita conducted a similar experiment, however, in which they used not only purified immunoglobulin subclasses but also F(ab')2 fragments. They found that some subclasses, especially IgG1, were inhibitory, IgG2 enhanced the antisheep cell response, and the F(ab')2 fragments had no effect at all.

This finding implies that simple blocking of antigen binding cannot be the mechanism involved, not only because F(ab')2 was without effect, but because IgG1 and IgG2 differ only in their heavy-chain Fc region. It is this part of the immunoglobulin that must exert the controlling influence.

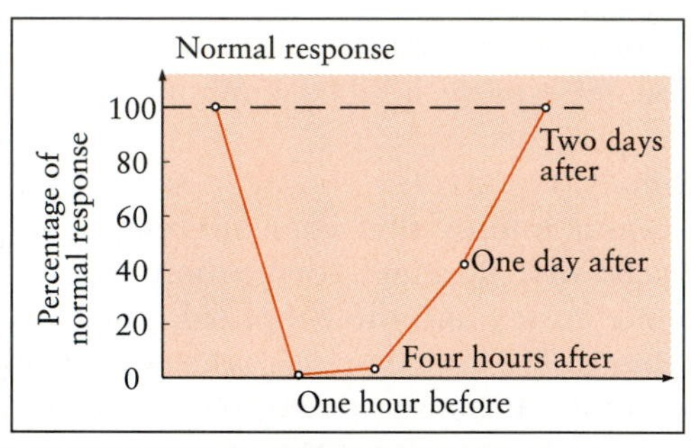

The inhibition of the antisheep erythrocyte response by administration of antisheep erythrocyte antibody. Each point represents the percentage of the maximum number of plaque-forming cells in the spleens of ten mice relative to the number in untreated mice. Another demonstration of the potent negative feedback exerted by antibody. *(From Henry, C., and Jerne, N.K. J. Exp. Med.,* 128:*133–152, 1968. With permission.)*

Idiotype Networks

In Chapter 13 it was pointed out that immunoglobulin molecules possess many variant structures. Some of these variants are located on the constant regions of these molecules and give rise to classes and **allotypes.** Major variations in structure are located on the variable regions of the light and heavy chains, where they form the antigen-binding sites. These variations, which occur between immunoglobulins of different antigenic specificity, are called **idiotopes.** The collective term for all the idiotopes on a molecule is idiotype.

Niels Jerne originally suggested that the immunoglobulins generated during an antibody response carry new idiotypes. Before immune stimulation, any

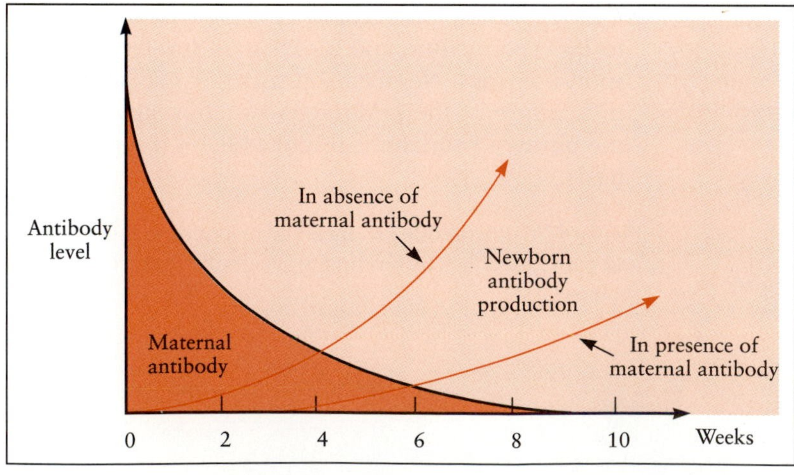

Figure 22–4 In the absence of maternally derived antibody, newborn animals produce their own antibody much earlier than they would if the maternal antibody were present. This demonstrates the inhibitory effect of immunoglobulins on the antibody response.

specific idiotype is found only on the surface immunoglobulins of a small number of B cells and will probably be present in insufficient quantity to generate tolerance. During an immune response, very large quantities of specific immunoglobulin molecules, each carrying these idiotypes, are produced. These idiotypes, not having been recognized previously, are seen by antigen-sensitive cells as foreign and stimulate the production of anti-idiotype antibodies. Anti-idiotype antibodies may be directed against idiotypes anywhere on the immunoglobulin variable regions. Some may be directed against the antigen-binding site on an immunoglobulin and hence resemble the shape of the antigen (Fig. 22–5). Other anti-idiotype antibodies may be directed against idiotopes outside the antigen-binding site and hence do not share epitopes with the antigen.

The idiotypes on the variable regions of these new anti-idiotype antibodies may also be recognized as foreign and so provoke anti-anti-idiotype antibodies. As this process continues, a complex network of interacting idiotype–anti-idiotype reactions may develop. The significance of this is that each of these reactions may enhance or suppress antibody responses. For example, anti-idiotype antibodies may enhance B-cell responses by acting like antigens. That is, they bind and so trigger the antigen receptor of B cells. Anti-idiotypes may also inhibit an immune response by triggering the production of specific suppressor cells. For example, mice challenged with the antigen phosphoryl choline (PC) produce antibodies of a single idiotype called T15. If anti-T15 antibodies are injected into these mice, they cause a transient suppression of the antibody response to PC. If mice are immunized with phosphoryl choline, then it is possible to show both an anti-PC (T15) response and an anti-T15 response developing in parallel, suggesting that some form of feedback control is occurring. The suppressive effect of anti-idiotypic antibodies probably predominates under normal conditions and thus effectively terminates an immune response.

You may also recollect that the T-cell antigen receptor contains structural variants that can act as idiotypic epitopes on its variable regions. This suggests that T cells may also be subject to regulation by anti-idiotype antibodies and cells.

IMMUNE RESPONSE GENES

The genes inherited at the class II loci of the major histocompatibility complex regulate immune responsiveness through their ability to bind specific antigen. Each individual possesses between three and six class II MHC molecules. An individual who is homozygous at all three loci would carry only three different MHC

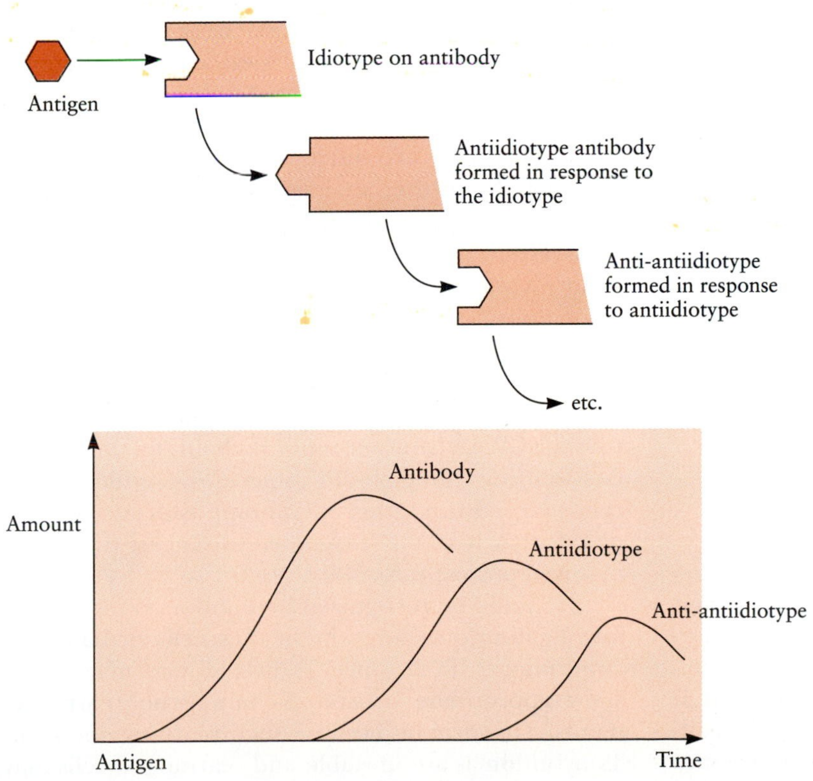

Figure 22–5 A very simple view of the idiotype network. Each antibody produced provokes the formation of anti-idiotypes against itself. The anti-idiotypic antibodies in turn provoke anti-anti-idiotypes. Anti-idiotypes exert a suppressive effect on the antibody that induces them and probably act to dampen antibody responses.

class II molecules. This would be a most unusual situation in view of the enormous polymorphism of this system. The usual situation is that each individual is heterozygous at these loci and thus has six MHC class II molecules. If a specific antigen can bind to one of these six MHC molecules, then an immune response can be mounted against that antigen. If it cannot bind, then that individual will be unresponsive to that antigen. Thus the MHC class II molecules possessed by an individual directly determine whether an immune response is mounted against a specific antigen. For this reason, the MHC genes have been called **immune response genes.** Their mode of action is discussed in Chapter 8.

Non-MHC Immune Response Genes

Biozzi noted that in outbred mice, some responded well to sheep red blood cells and others responded poorly. By selective breeding he was able to establish two strains of mice: a high-responder strain and a low-responder strain. The differences between these strains are associated with macrophage function. The macrophages from low-responder mice show enhanced lysosomal activity and hence destroy antigen effectively. The macrophages from high-responder mice have poorer lysosomal activity and, as a result, permit more antigen to persist and stimulate an immune response than the macrophages of the low-responder mice. The macrophages from low-responder mice are also better able to destroy intracellular parasites such as *Salmonella typhimurium, Mycobacterium bovis,* and *Leishmania donovani.*

REGULATORY CELL FUNCTION

The regulatory cells of the body include helper T cells, macrophages, and NK cells. (Helper cells are discussed in Chapter 11.) They may also include, and this is a matter of major controversy, a population of $CD8^+$ T cells that can suppress immune function and so are called suppressor T cells. The issue is further confused by the production of the potently suppressive cytokine, IL-10, by Th2 cells.

Suppressor T Cells

It has proved almost impossible to distinguish suppressor T (Ts) cells from cytotoxic T (Tc) cells, and as a result, many investigators have suggested that they do not exist. Nevertheless, many immunologists continue to study their properties. For example, it is claimed that it is possible to distinguish human Ts cells from Tc cells by their expression of CD11b, and clones of mouse suppressor cells without cytotoxic activity have been established. Suppressor cells are heterogeneous and include not only suppressor T cells (Ts) but also suppressor macrophages and natural suppressor (NS) cells. These cells may inhibit helper-T-cell activity, cytotoxic-T-cell activity, or even antigen-presenting-cell activity.

I-J Antigens

In the early 1980s it was suggested that mouse suppressor T cells carried a characteristic antigen on their surface that was called I-J. I-J was thought to be encoded by a gene located within the MHC class II region. It was a major shock to discover, when the class II region was mapped, that there was no I-J gene. Subsequent studies revealed that I-J is not a direct MHC gene product but a receptor-like molecule that is adoptively acquired by T cells. Its adsorption to T cells is controlled by MHC class II molecules. I-J is a protein homodimer of 84 to 90 kDa, and it is not physically associated with the TCR/CD3 complex. It is suggested that I-J is used to transmit a suppressive signal between cells. This suppressive signal is able to neutralize the positive signal induced through the TCR. Thus suppressor T cells may simply be T cells that have absorbed I-J.

Some studies have shown that mouse lymphocyte-derived, antigen-specific suppressive molecules consist of two polypeptide chains. Although the experimental evidence is inconsistent, it is likely that these suppressor molecules may be soluble forms of TCR. For example, Ts cells specific for the hapten DNP presented by a H-2K class I positive cell produce a suppressor factor that is specific for both DNP and H-2K. Thus they are functionally similar to the cell-bound TCR. Further analysis has shown that they are structurally related to TCR-α/β and that one chain bears TCR α determinants and the other bears TCR β determinants. If the two chains are separated, it can be demonstrated that the α chain may have avidity for the foreign antigen and the β chain may determine MHC restriction. Other investigators have described soluble antigen-specific suppressor factors that possess only a TCR α chain but not a β chain. In view of the fact that immunoglobulin molecules produced by B cells may exist in either membrane-associated or soluble forms, it seems reasonable to suggest that TCRs may also possess this ability.

It should be recognized here, however, that some investigators question whether Ts cells actually use conventional TCRs since Ts hybridomas may fail to show appropriate α- and β-chain gene rearrangements. This may be because such rearrangements in Ts hybridomas are unstable and rearranged cells may rapidly die. As a result, such rearrangements may fail to be detected following prolonged tissue culture. It has proved remarkably difficult to come to a firm decision on the nature of antigen-specific suppressor factors.

Macrophage-Mediated Suppression

Macrophage-derived suppressor factors include prostaglandins, nitric oxide, and reactive oxygen metabolites. The glycoproteins released by concanavalin-A treatment of T-cell cultures act on macrophages, provoking them to release prostaglandins, which cause a rise in lymphocyte cAMP and hence inhibition of blastogenesis and antibody secretion. Macrophage oxygen metabolites also act on an inactive T-cell–derived nonspecific suppressive factor (soluble immune response suppressor, SIRS) to generate its biological activity. SIRS is a 14-kDa protein that when oxidized can inhibit antibody secretion by B cells. The suppressive effects of macrophages are normally counteracted by IL-1 but enhanced by IFN-γ, which stimulates SIRS production.

Under some circumstances, T cells may act on antigen-presenting macrophages to prevent recognition of antigen. In these cases, the T cells and their targets must also be compatible at the MHC class I region. The mechanism of action of these is unclear, but it is postulated that these suppressor cells block a metabolic pathway within the antigen-presenting cells.

Natural Suppressor Cells

Natural suppressor (NS) cells are large granular lymphocytes. They secrete two factors (SFα and SFβ) that have suppressor-cell–inducing activity. They suppress B- and T-cell proliferation as well as immunoglobulin production. NS cells occur normally in the adult bone marrow and neonatal spleen. They can be found in adult mouse spleens following total lymphoid irradiation or after cyclophosphamide injection. Potent NS activity develops in mice undergoing a graft-versus-host reaction. NS cells are induced by some interferon inducers such as the polymer of inosine and cytidine (poly I:C). If these inducers are given to mice immunized two or three days previously, then the NS cells will block helper-cell activity and cause early termination of the IgM response.

When Do Suppressor Cells Work?

Suppressor-cell activities have been described as involving almost all aspects of immune reactivity. Non-antigen-specific Ts cells have been possibly implicated in antigenic competition; lack of immune responses in the newborn; immunosuppression following trauma, burns, or surgery; prevention of autoimmunity; some cases of hypogammaglobulinemia; and blocking of responses to mitogens. Antigen-specific suppressor cells are claimed to be responsible for tolerance provoked by low doses of antigen and by nonaggregated antigen and for immunological unresponsiveness mediated by some MHC class II genes. Specific suppressor cells are found in some tumor-bearing animals in which they block tumor rejection and in pregnant animals in which they block rejection of the fetus. They also occur in individuals unresponsive (anergic) to tuberculosis infection.

Cellular Regulatory Networks

Since Ts cells are themselves regulated by the interactions of other Th and Ts cells, and since T cells bear antigen receptors that could act as idiotypes, a regulatory T-cell idiotype–anti-idiotype network might exist as a functional counterpart to the antibody network. In some experimental systems a suppressor-cell cascade has been identified, in which one suppressor-cell population (Ts1) releases suppressor factors that act on a second cell population (Ts2), which in turn suppresses a third suppressor-cell population (Ts3). Each of the suppressor factors released by these subpopulations has either idiotype or anti-idiotypic specificity, and different suppressor-cell stimuli can stimulate the chain at different levels. Each suppressor-cell population can also act at different points in the immune response. The complexity of the system is further compounded by an interlocking series of positive and negative feedback loops ensuring that the system functions only at its appropriate level.

Contrasuppressor Cells

Suppressor T cells induced by oral administration of antigen (oral tolerance) may be selectively neutralized by a population of T cells found in intestinal epithelium called **contrasuppressor cells** (Tcs) (Fig. 22–6). As a result of the activities of Tcs cells, an immune response may proceed in the intestinal wall in the presence of active suppression throughout the rest of the body. Contrasuppressor T cells are not just helper cells. They possess different surface antigens. Thus in mice helper cells are I-J$^-$, and contrasuppressor cells are I-J$^+$. Helper cells are MHC class II restricted, whereas contrasuppressor cells are not. Recent evidence suggests that contrasuppressor cells possess the γ/δ form of TCR and helper cells carry TCR α/β.

When antigen is administered orally, tolerance may result and the animal may become unreactive to systemic administration of the same antigen. This tol-

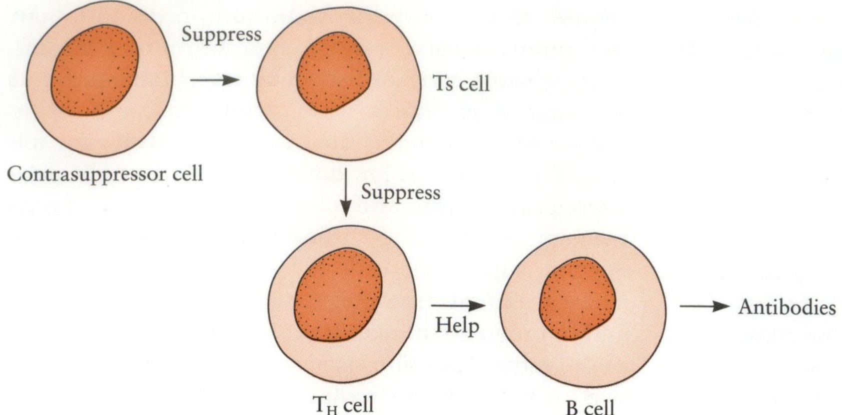

Figure 22–6 The role of contrasuppressor cells in regulating suppressor cell function. This may reflect the differing roles of Th1 and Th2 cells.

erance may be a result of the development of a suppressor-cell population. At the same time, the intestinal lymphoid tissues of this animal remain fully sensitized. The intestinal lymphoid tissues continue to be responsive to antigen since they contain a population of contrasuppressor T cells. This is probably a device to prevent a systemic response to food antigens while permitting a local protective response in the intestine. Contrasuppression may also help in recovery from the suppression associated with severe burns, and a loss of contrasuppressor cells may result in a severe immunodeficiency.

Mouse contrasuppressor T cells (I-J$^+$, CD4$^+$CD8$^-$) interact with a I-J$^+$, CD4$^-$CD8$^+$ cell to release a soluble factor (TcsF), which is an antigen-specific I-J$^+$ factor. TcsF acts on helper cells to render them insusceptible to T suppressor factors. It is suggested that the contrasuppressor factor may block suppressor factor receptors. In humans, contrasuppressor cells are CD8$^+$, MHC class II$^+$, CD3$^+$, and CD4$^-$.

Do Suppressor T Cells Exist?

Because of the difficulties encountered in characterizing suppressor-cell populations, many immunologists have suggested that a distinct suppressor-cell lineage may not exist. Certainly, some of the studies that were used to demonstrate suppressor-T-cell function were flawed. An excellent example of this is a method classically used to generate non-antigen-specific suppressor cells by stimulating them with the lectin concanavalin A. Later studies showed that the suppressive effect was mediated not by suppressor cells, but by depletion of IL-2 in the medium.

It is very likely that some of the activities attributed to suppressor cells are actually due to the antagonistic functions of the two helper-T-cell populations Th1 and Th2. For example, IFN-γ released by Th1 cells can suppress IgE production. IL-10 released by Th2 cells is very suppressive for macrophage IL-12 production and thus for the production of cytokines by Th1 cells. Likewise, IL-4 may also suppress IL-2–mediated B-cell proliferation.

Yet another possible suppressive mechanism could involve the stimulation of cytotoxic-T-cell activities by antigen presented on cells such as B cells. Studies have shown that some cytotoxic T cells are MHC class II restricted and thus destroy B cells presenting antigen in the conventional manner.

Thus for all these reasons the concept of suppressor T cells is under serious attack, although not yet dead.

NEURAL REGULATION OF IMMUNITY

It has been known for many years that mental attitudes, especially stress, significantly influence resistance to some diseases. For example, depressed individuals, students studying for examinations, and the recently bereaved show reduced lymphocyte reactivity (Fig. 22–7). Academic examinations of medical students were associated with psychological distress and depressed immune function. Thus NK activity was poorer at examination time. Students who were lonely had lower NK activity than students who were not lonely. After divorce, women who were attached to their husbands and hence were more distressed showed a depressed lymphocyte response to mitogens. A similar effect that resulted in excess mortality occurred in divorced men. Individuals caring for a patient suffering from Alzheimer's disease are also significantly immunosuppressed. NK activity is suppressed by stress as well as by sexual activity (in rats). Bereavement suppresses NK-cell activity. (Bereaved women had significantly suppressed NK activity as op-

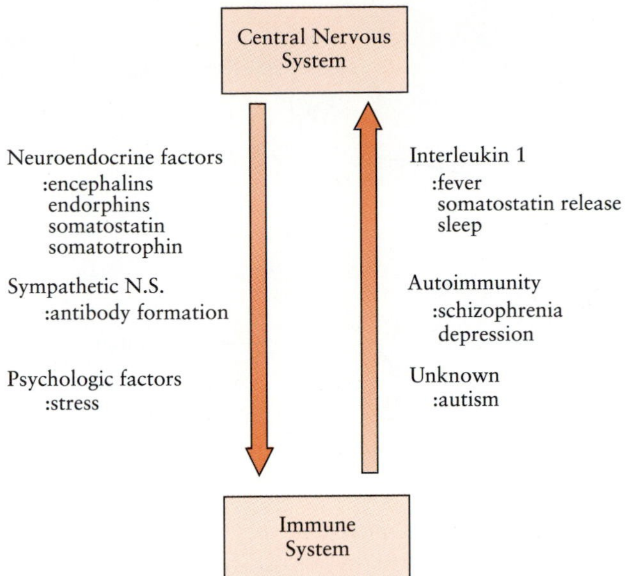

Figure 22–7 Some of the ways in which the nervous system and the immune system interact.

posed to women with healthy spouses.) Stress can depress T-cell mitogenesis, NK activity, and IL-2 production. Administration of IL-2 did not reverse this, so it is not just due to loss of IL-2. Expression of IL-2R on lymphocytes is also depressed by stress. Reduction in stress can have a reverse effect. Thus lymphocyte reactivity (as measured by lymphocyte response to phytohemagglutinin) doubled within a week after homosexual men were informed that they did not have HIV infection.

In looking for possible mechanisms of this stress effect, it is clear that neuroendocrine hormones do influence immune function. Stress-induced immunosuppression also occurs in adrenalectomized animals, so it is not due to adrenal steroids, but serum from stressed animals contains an immunosuppressive protein. Cells of the immune system possess receptors for the opioid peptides. Neuropeptides such as the enkephalins and endorphins are released during stress. These can bind to receptors on lymphocytes and influence their activity. Thus the generation of cytotoxic T cells is enhanced by [met]-enkephalin and β-endorphin. α-Endorphin suppresses antibody formation, and β-endorphin reverses this suppressive effect. Other neuropeptides that can influence the immune system include ACTH, oxytocin, vasoactive intestinal peptide (VIP), somatostatin, prolactin, substance P, and norepinephrine. VIP and somatostatin tend to inhibit T-cell function, whereas substance P is mainly stimulatory.

By controlling neurotransmitter function or the autonomic nervous system, certain sites in the brain, especially the hypothalamus, the limbic forebrain, and the brain stem, influence immune function. The pituitary influences the activity of the thymus. Thus, growth hormone injections augment the level of some thymic hormones in dogs. α-Melanocyte-stimulating hormone inhibits fever, the acute phase response, and inflammation by modulating the effects of IL-1.

Many parts of the immune system are well innervated. For example, nerves are linked to Langerhans' cells in the skin. By releasing a neuropeptide, these nerves can depress the antigen-presenting ability of these Langerhans' cells. This might explain why allergic dermatitis and psoriasis worsen with anxiety. Noradrenergic nerves innervate the thymus, the splenic white pulp, and the lymph nodes. They influence blood flow, vascular permeability, lymphocyte migration, and differentiation. Surgical or chemical destruction of sympathetic nerves to the spleen enhances antibody production and can induce changes in the distribution of lymphocyte subpopulations.

Finally, the immune system can influence nervous function. A recent study of individuals suffering from severe depression showed that some of them possessed autoantibodies directed against β-endorphin. By blocking the activities of this molecule that mediates pleasurable sensations, the autoantibodies presumably induced the depression. In addition, macrophage-derived IL-1 stimulates the synthesis of nerve growth factor in inflamed areas and so promotes innervation and nerve healing. IL-1 stimulates the synthesis of somatostatin, which in turn depresses production of growth hormone and thyroid-stimulating hormone. ACTH and endorphins are produced by cells of the immune system. Mast cells are closely associated with nerves. Nerve growth factor stimulates an increase in mast cell numbers in tissues, and mast cells influence neural function.

CONCLUSION

It should be clear to the reader who has managed to survive this far that many interacting control systems act to regulate immune reactivity (Fig. 22–8). This is an inevitable feature not only of the immune system, but of all body systems. The immune system is almost certainly not unique in this respect. It happens to be among the first of the body systems to be analyzed in detail. It may safely be predicted that other body systems that involve cell interaction will be found to be equally complex.

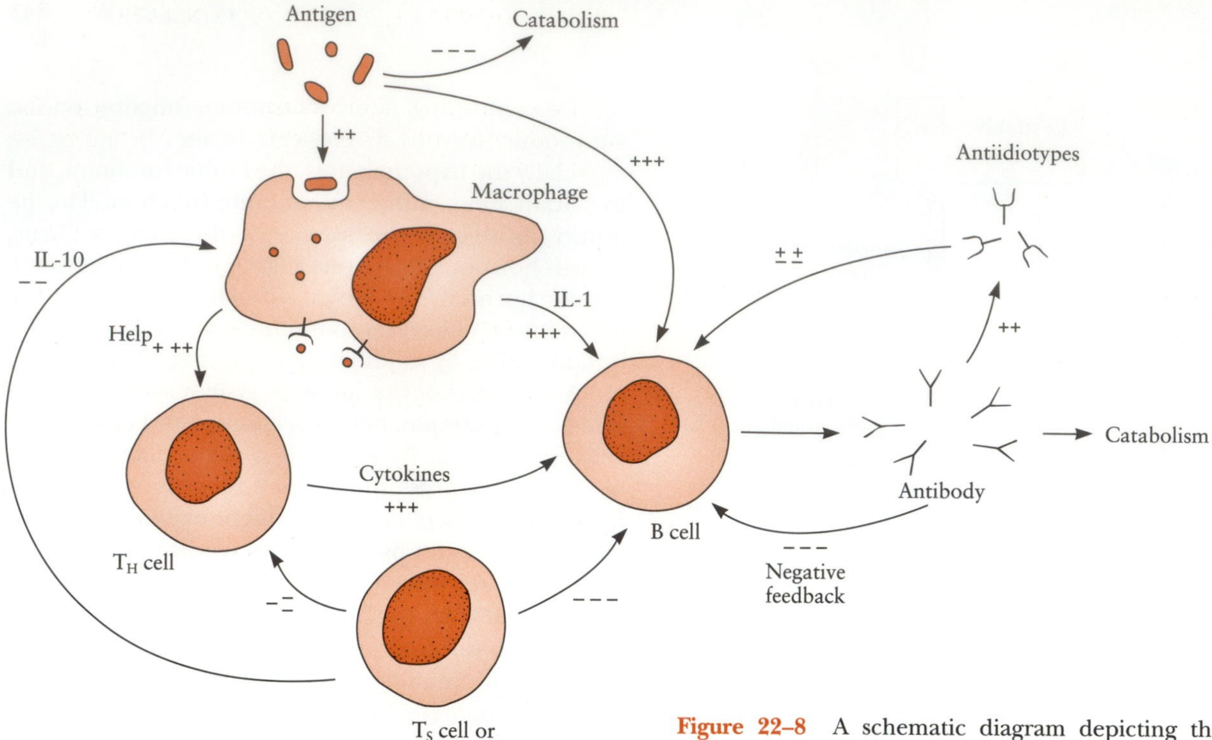

Figure 22–8 A schematic diagram depicting the major mechanisms that control immune responses.

KEY WORDS

Anti-idiotype antibodies p. 339
Contrasuppressor cell p. 341
Epitope p. 336
Experimental allergic encephalitis p. 336
Hypothalamus p. 342
Idiotype network p. 338
I-J antigen p. 340
Immune response gene p. 339
Maternal antibody p. 336
Myelin basic protein p. 336
Natural suppressor cell p. 341
Negative feedback p. 336
Regulatory network p. 341
Stress-induced immunosuppression p. 342
Suppressor cell p. 340
Suppressor epitope p. 336
Suppressor factor p. 340
Suppressor macrophage p. 341
Sympathetic nervous system p. 343

QUESTIONS

1. The conformation of an epitope is important for recognition by
   a. T cells
   b. NK cells
   c. B cells
   d. neutrophils
   e. macrophages
2. Which portion of an immunoglobulin molecule is essential for control of antibody production?
   a. Fab
   b. F (ab′)2
   c. Fc
   d. light chains
   e. V regions
3. T-independent antigens only induce antibodies of which class?
   a. IgG
   b. IgA
   c. IgM
   d. IgE
   e. IgD
4. The epitopes formed by an immunoglobulin antigen-binding site are called
   a. isotypes
   b. allotypes
   c. epitypes
   d. haplotypes
   e. idiotypes
5. I-J antigens may be characteristic features of
   a. Th1 cells
   b. Tdh cells
   c. Th2 cells
   d. Ts cells
   e. Tc cells
6. Among the suppressive factors secreted by macrophages is
   a. interleukin-1
   b. nitric oxide

c. interleukin-12
d. interleukin-6
e. interleukin-10

7. Contrasuppressor cells are mainly found in the
   a. spleen
   b. thymus
   c. tonsil
   d. gastrointestinal tract
   e. bursa
8. The effect of stress on the immune system is probably mediated by
   a. endorphins
   b. corticosteroids
   c. interleukin-1
   d. thymosin
   e. enkephalins
9. Among the immunosuppressive mechanisms induced by the fetus is/are
   a. interleukin-12
   b. immunoglobulin G
   c. interleukin-10
   d. C-reactive protein
   e. suppressor cells
10. Development of hemolytic disease in the newborn is prevented by administration to the mother of
   a. antiplacental antibody
   b. anti-Rh antibody
   c. anti-idiotype antibody
   d. anticomplement antibody
   e. antiallotype antibody
11. Construct a hypothetical antigen that might selectively stimulate suppressor-cell activity. How would you test this activity?
12. Antiserum against the Rh antigen given to a woman soon after giving birth effectively prevents her from making antibody against the Rh antigen herself. Why does this work? What mechanisms are involved? Can you think of other diseases in which such treatment might be useful?
13. The existence of suppressor cells has been questioned by many immunologists. Why is this? What do you think is the most convincing evidence for their existence?
14. What might be the function of contrasuppressor cells?
15. Given the great diversity of methods by which the body controls the immune responses, rank them in order of importance. Give your reasons.
16. Why do we not normally make antibodies against antigens in our food? Or do we?
17. Discuss the proposition that the concept of an "idiotypic network," although interesting, is of very little significance in the regulation of the immune responses.
18. Is it possible that an individual patient can improve his or her resistance to infectious disease or cancer by conscious thought?

Answers: 1c, 2c, 3c, 4e, 5d, 6b, 7d, 8a, 9e, 10b

SOURCES OF ADDITIONAL INFORMATION

Abruzzo, L.V., and Rowley, D.A. Homeostasis of the antibody response: Immunization by NK cells. Science, 222:581–584, 1983.

Berczi, I. Immunoregulation by neuroendocrine factors. Dev. Comp. Immunol., 13:329–341, 1989.

Chang, T.W. Regulation of immune response by antibodies: The importance of antibody and monocyte Fc receptor interaction in T cell activation. Immunol. Today, 6:245–249, 1985.

Cohen, J.J. Programmed cell death in the immune system. Adv. Immunol., 50:55–85, 1992.

Fairchild, R.L., Kubo, R.T., and Moorhead, J.W. DNP-specific/class I MHC-restricted suppressor molecules bear determinants of the T cell receptor α- and β-chains. The Vβ8 chain dictates restriction to either K or D. J. Immunol., 145: 2001–2009, 1990.

Gilbert, K.M., and Hoffmann, M.K. Suppressor B lymphocytes. Immunol. Today, 4:253–255, 1983.

Gordon, J., and Murgita, R.A. Suppression and augmentation of the primary *in vitro* immune response by different classes of antibodies. Cell. Immunol., 15:392–402, 1975.

Green, D.R., and Faist, E. Trauma and the immune response. Immunol. Today, 9:253–254, 1988.

Hadden, J.W., and Coffey, R.G. Cyclic nucleotides in mitogen induced lymphocyte proliferation. Immunol. Today, 3: 299–304, 1982.

Henry, C., and Jerne, N.K. Competition of 19S and 7S antigen receptors in the regulation of the primary immune response. J. Exp. Med., 128:133–152, 1968.

Jerne, N.K. Towards a network theory of the immune system. Ann. Immunol. (Paris), 125C:373–389, 1974.

Kapp, J.A., Pierce, C.W., and Sorensen, C.M. Antigen-specific suppressor T cell factors. Hosp. Pract. [off.], 11:85–98, 1984.

Koide, J., and Engleman, E.G. Differences in surface phenotype and mechanism of action between alloantigen-specific $CD8^+$ cytotoxic and suppressor T cell clones. J. Immunol., 144:32–40, 1990.

Krzych, U., et al. Induction of helper and suppressor T cells by nonoverlapping determinants on the large protein antigen, β-galactosidase. FASEB J., 2:141–145, 1988.

Moldofsky, H., et al. Effects of sleep deprivation on human immune functions. FASEB J., 3:1972–1977, 1989.

Nakayama, T., Asano, Y., and Tada, T. I-J and mechanism of immunosuppression. Immunol. Suppl., 2:16–19, 1989.

Ninnemann, J.L. Prostaglandins and immunity. Immunol. Today, 5:170–178, 1984.

Reichlin, S. Neuroendocrine-immune interactions. N. Engl. J. Med., 329:1246–1253, 1993.

Rosenthal, A.S. Regulation of the immune response—role of the macrophage. N. Engl. J. Med., 303:1153–1156, 1980.

Taniguchi, M., Takai, I., and Tada, T. Functional and molecular organization of an antigen specific suppressor factor from a T-cell hybridoma. Nature, 283:227–228, 1980.

Weigle, W.A. Immunological unresponsiveness. Adv. Immunol., 16:61–122, 1973.

CHAPTER 23

Immunity at Body Surfaces

Thomas Tomasi *In 1965, Dr. Thomas Tomasi and his colleagues were studying the structure of IgA secreted in saliva and demonstrated that it was not derived from the bloodstream by active transport but that it was locally synthesized in the salivary gland. As a result of these and other studies, Dr. Tomasi first put forward the concept of an immunological system that was characteristic of external secretions and separate from the systemic immune system. This external secretory immune system plays a very significant role in the defense of the body.*

CHAPTER OUTLINE

CHAPTER CONCEPTS

1. A wide variety of nonimmunological mechanisms is involved in ensuring that body surfaces are resistant to microbial invasion.
2. IgA is the immunoglobulin mainly responsible for protection of body surfaces against infection.
3. IgA is protected from proteolytic digestion in the intestine by secretory component. Secretory component originates as a receptor for polymeric immunoglobulins on the surface of intestinal epithelial cells.
4. IgA-producing cells emigrate from sites of antigenic stimulation such as the intestine and colonize other body surfaces.
5. IgE-producing cells are located at sites where parasites may invade the body. These include the skin, respiratory tract, and intestine.
6. IgE is produced on body surfaces when antigen gains access to tissues. This IgE can mediate an acute allergic response, which floods the surface with IgG and so prevents microbial invasion.

The body's first line of defense is on its surfaces. In this chapter you will read about the various mechanisms by which these surfaces are defended. The first part of the chapter reviews briefly the nonimmunological defenses such as the normal flora and the physical barriers to infection. The second part focuses on immunological defenses. The most important of these is IgA, which is described in considerable detail. Then the role of IgE is also described since it too is a surface immunoglobulin. The chapter concludes with a description of the surface defenses of the gastrointestinal tract, mammary gland, urogenital tract, respiratory tract, and skin.

Although mammals possess an extensive array of defense mechanisms within tissues, it is at the body surfaces that invading microorganisms are first encountered and largely repelled or destroyed (Fig. 23–1). The protective systems at body surfaces achieve this by establishing, through both physical and chemical mechanisms, local environmental conditions suitable for only the most adapted microorganisms. These surfaces are populated by an extensive microbial flora that, because it is well adapted, has low pathogenicity and effectively prevents the establishment of other, more poorly adapted and potentially pathogenic organisms. This environmental defense system is supplemented by immunological mechanisms in areas where the physical barriers to invasion are relatively weak.

NONIMMUNOLOGICAL SURFACE-PROTECTIVE MECHANISMS

One major function of the skin is to act as a barrier to invading microorganisms. The skin carries a dense and stable resident bacterial flora whose composition is regulated by factors such as continuing desquamation, desiccation, and a relatively low pH due to the presence of fatty acids in sebum. If any of these environmental factors is altered, then the composition of the skin flora is disturbed, its protective properties are reduced, and microbial invasion may result (Fig. 23–2). Thus, skin infections tend to occur in areas such as the axilla or groin, where both pH and humidity are relatively high. Similarly, individuals forced to stand in water or mud for prolonged periods show an increased frequency of foot infections as the skin becomes sodden, its structure breaks down, and its resident flora changes in response to alterations in the local environment.

The importance of the resident flora is seen to much greater effect in the digestive tract, where it is essential not only for the control of potential pathogens but also for the digestion of some foods, such as cellulose in the diet of herbivores. In addition, the natural development of the immune system depends on the continuous antigenic stimulation provided by intestinal flora. Because of the absence of a bacterial flora in the intestine, and a resulting lack of exposure to foreign antigens, germ-free animals have poorly developed **secondary lymphoid organs.**

If the **normal flora** of the intestine is eliminated or its composition drastically altered (by aggressive antibiotic treatment, for example), then dietary disturbances result and the overgrowth of potential pathogens may occur. The flora of the digestive tract normally acts competitively against potential invaders through mechanisms that supplement the other physical defenses of this system. Thus, in the mouth, the flushing activity of saliva is complemented by the generation of peroxides from streptococci. In the stomach of some animals, the gastric pH may be sufficiently low to have some bactericidal and viricidal effect, although this effect varies greatly between species and between

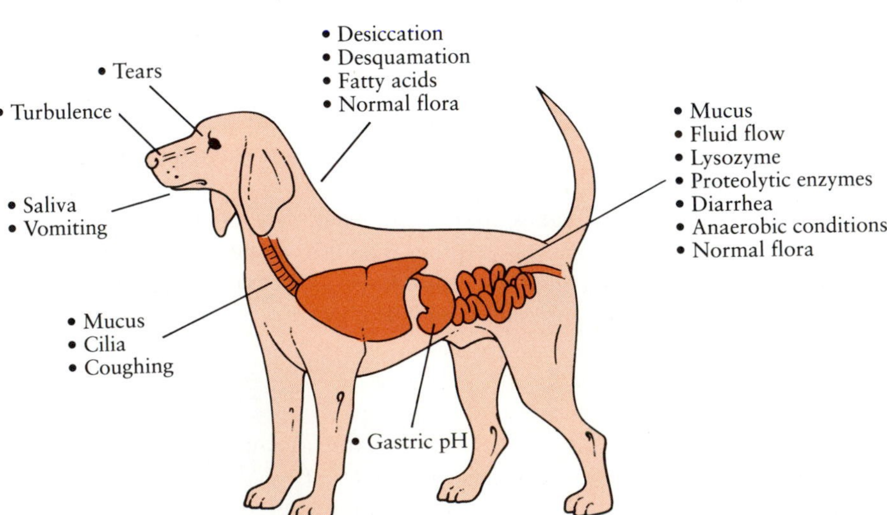

Figure 23–1 Nonimmunological surface protective mechanisms.

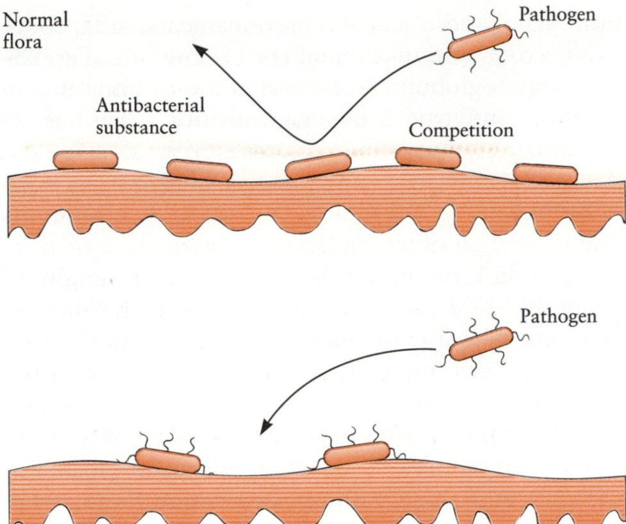

Figure 23–2 The normal bacterial flora play a major role in excluding invading microorganisms.

meals, and the pH in the center of a mass of ingested food may not necessarily drop to low levels. Some foods, such as milk, are potent buffers.

Farther down the intestine, the resident bacterial flora ensures that the pH is kept low and the contents anaerobic. The intestinal flora is also influenced by the diet; for instance, the intestine of milk-fed animals tends to be colonized largely by lactobacilli, which produce large quantities of bacteriostatic lactic and butyric acids. These acids inhibit colonization by potential pathogens, such as *Escherichia coli,* so that animals suckled naturally tend to have fewer digestive upsets than animals weaned early in life. In the large intestine, the bacterial flora is mainly composed of strict anaerobes.

Lysozyme (Chapter 5), the antibacterial and antiviral enzyme, is synthesized in the gastric mucosa and in macrophages within the intestinal mucosa. As a result, it is found in large quantities in intestinal fluid. The role of phagocytic cells in the intestine is not clear, but macrophages can migrate through the intestinal wall and may be active for a short time within the lumen.

In the urinary system, the flushing action and low pH of urine generally provide adequate protection; however, when urinary stasis occurs, urethritis resulting from the unhindered ascent of pathogenic bacteria is not uncommon. In adult women, the vagina is lined by a squamous epithelium composed of cells rich in glycogen. When these cells desquamate, they provide a substrate for lactobacilli that, in turn, generate large quantities of lactic acid, which protects the vagina against invasion. Glycogen storage in the vaginal epithelial cells is stimulated by estrogens and thus occurs only in sexually mature individuals. Because of this, vaginal infections tend to be most common before puberty and after menopause.

The mammary gland utilizes several interesting defense mechanisms. The flushing action of the milk prevents invasion by some potential pathogens, while milk itself contains bacterial inhibitors. A general term for these antibacterial substances in milk is **lactenins.** Lactenins include complement, lysozyme, the iron-binding protein lactoferrin, and the enzyme lactoperoxidase. Lactoferrin competes with bacteria for iron and therefore renders it unavailable for their growth. It also enhances the neutrophil **respiratory burst.** Milk contains high concentrations of lactoperoxidase and thiocyanate (SCN^-) ions. In the presence of exogenous hydrogen peroxide, the lactoperoxidase can oxidize the SCN^- to bacteriostatic products such as sulfur dicyanide. Hydrogen peroxide may be produced by the oxidation of ascorbic acid. Some strains of streptococci are resistant to this bacteriostatic pathway, since they possess an enzyme that reduces the SCN^-. IgA enhances the activity of lactoperoxidase. The phagocytic cells released into the mammary gland in response to irritation may also contribute to antimicrobial resistance not only through their phagocytic efforts but also by providing additional lactoferrin, hydrogen peroxide, and lysosomal peroxidases.

The respiratory tract differs from the other body surfaces in that it is in intimate connection with the interior of the body and it is required to allow unhindered access of air to the alveoli. The system obviously requires a filter. Air entering the respiratory tract is largely deprived of any suspended particles by turbu-

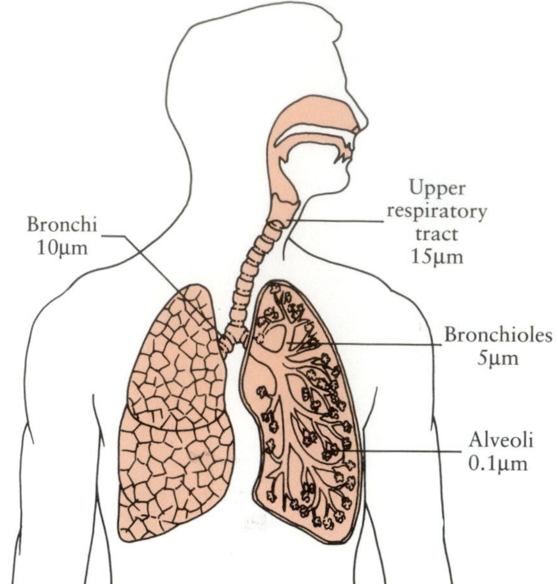

Figure 23–3 The influence of size on the site of deposition of particles within the respiratory tract: Particles greater than 10 μm in diameter will be trapped in the upper respiratory tract. Only particles of less than 5 to 6 μm diameter will reach the alveolar ducts.

lence that directs the particles onto its mucus-covered walls, where they adhere. The turbulence is brought about by the shape of the turbinate bones, the trachea, and the bronchi. This "turbulence filter" removes particles as small as 5 μm before they reach the alveoli (Fig. 23–3).

The walls of the upper respiratory tract are covered by a layer of mucus produced by goblet cells and provided with "antiseptic" properties through its content of lysozyme and IgA. This mucus layer is in continuous flow, being carried from the bronchioles up the bronchi and trachea by the cilia of the respiratory epithelial cells to the pharynx. Here the "dirty" mucus is swallowed and presumably digested in the intestinal tract. Particles smaller than 5 μm that can bypass this "mucociliary escalator" and reach the alveoli are phagocytosed by alveolar macrophages. Once these cells have successfully ingested particles, they migrate to the mucus escalator and, in this way, are also carried to the pharynx and eliminated.

IMMUNOLOGICAL SURFACE-PROTECTIVE MECHANISMS

Immunoglobulins A and E

In addition to the environmental and chemical factors that protect body surfaces, two immunoglobulin classes are found in secretions such as saliva, intestinal fluid, nasal and tracheal secretions, tears, milk, colostrum, urine, and urogenital tract secretions (Fig. 23–4). Immunoglobulin E is associated with immunity to parasites and type I hypersensitivities (Chapters 25 and 29). Immunoglobulin A, however, appears to have evolved specifically for the purpose of protecting body surfaces. In the intestine it is produced in amounts that exceed all other classes combined.

The IgA monomer has a molecular weight of about 160 kDa and is a typical four-chain, Y-shaped structure. It is usually found as a dimer with the Cα2 domain of one molecule bound by a J chain to the Cα3 domain of the other molecule. IgA has several extra cysteine residues in its heavy chains. As a result, the short interchain disulfide bonds make the chains compact and shield vulnerable peptide bonds from proteolytic enzymes. IgA dimers bind a receptor synthesized by intestinal epithelial cells and hepatocytes, termed **secretory component,** to produce secretory IgA (SIgA). The addition of secretory component renders IgA even more resistant to proteolysis by digestive enzymes.

IgA is not bactericidal, does not bind to macrophages or enhance phagocytosis, and activates complement only by the alternate pathway. It can, however, neutralize viruses, as well as some viral and bacterial enzymes, and it functions in some antibody-dependent cellular cytotoxicity (ADCC) systems. Its major function is to act as a barrier and prevent the adherence and entry of bacteria and viruses on epithelial surfaces.

IgA is synthesized by plasma cells located in the submucosa (the region immediately beneath the epithelial cells) in response to local antigenic stimulation and is thus found in high concentrations in the submucosa (Fig. 23–5). Some of this IgA binds to a receptor for polymeric immunoglobulin (pIgR) on the submucosal surface of epithelial cells. The complex of IgA and pIgR is then endocytosed and transported across the epithelial cell (Fig. 23–6). When it reaches the exterior surface, the endosome fuses with the plasma membrane and exposes the IgA to the intestinal lumen. The extracellular portion of the pIgR is then cleaved by proteolytic enzymes so that the IgA, with the receptor peptide still attached, is released. This peptide is called secretory component (SC) (Fig. 23–7). The production, transport, and secretion of secretory component occur even in the absence of IgA so that free secretory component is found in high concentrations in intestinal secretions. The complete pIgR has a molecular weight of 71 kDa. Its extracellular portion (SC) is made up of five immunoglobulin-like domains of 100 to 115 amino acids each.

Because IgA is transported through intestinal epithelial cells, it can act inside these cells (Fig. 23–8).

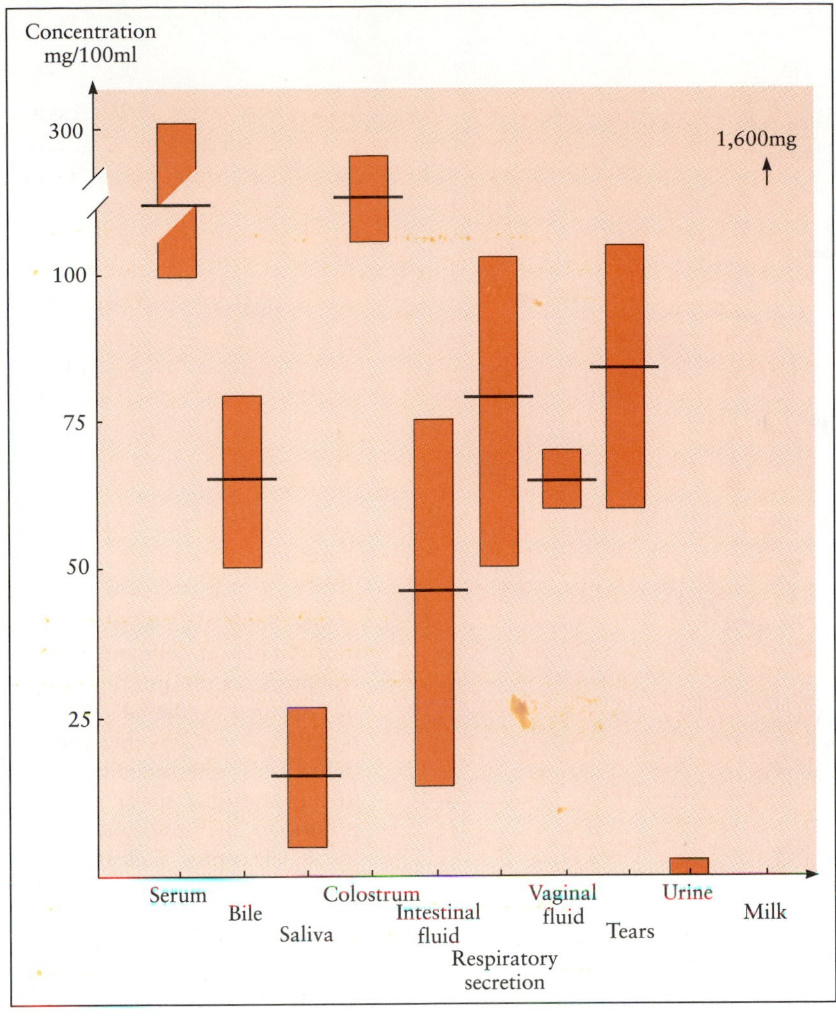

Figure 23–4 IgA levels in various body fluids in humans.

Thus IgA can bind to viral proteins inside epithelial cells and interrupt their replication. In this way, the IgA can prevent viral growth before the integrity of the epithelium is damaged. This is a unique example of an antibody acting in an intracellular location. A second function of intracellular IgA is to excrete foreign antigens. Thus IgA can bind to antigens that have penetrated to the submucosa. Once bound, the immune complexes bind to pIgR and are actively transported across the mucosal epithelial cells into the intestinal lumen. IgA can therefore act at three different levels to exclude foreign antigens: within the submu-

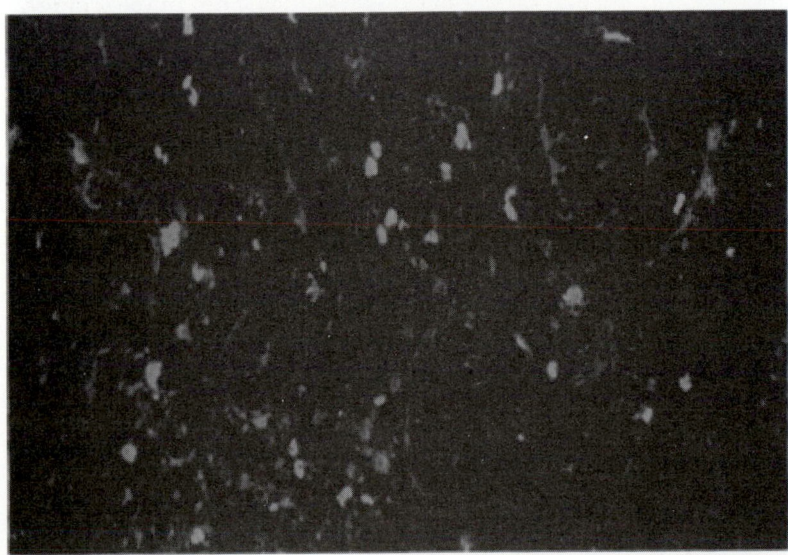

Figure 23–5 An immunofluorescence photomicrograph showing IgA-positive plasma cells in the villi and submucosa of the rabbit small intestine. *(Courtesy of Dr. D. Befus.)*

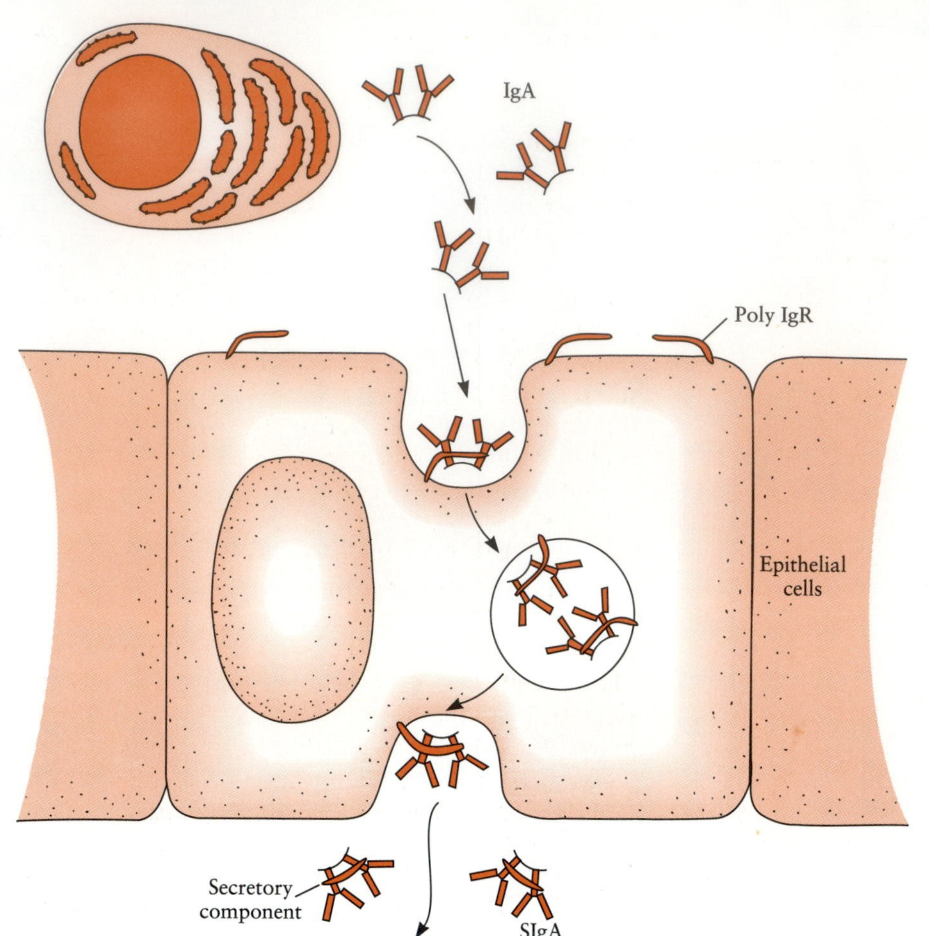

Figure 23–6 IgA is secreted by mucosal plasma cells and binds to receptors on the interior surface of intestinal epithelial cells. The bound IgA is taken into the epithelial cells, and passed in vesicles to the intestinal surface. On exposure to the intestinal lumen, the IgA receptor is cleaved from the cell and remains bound to the IgA as secretory component.

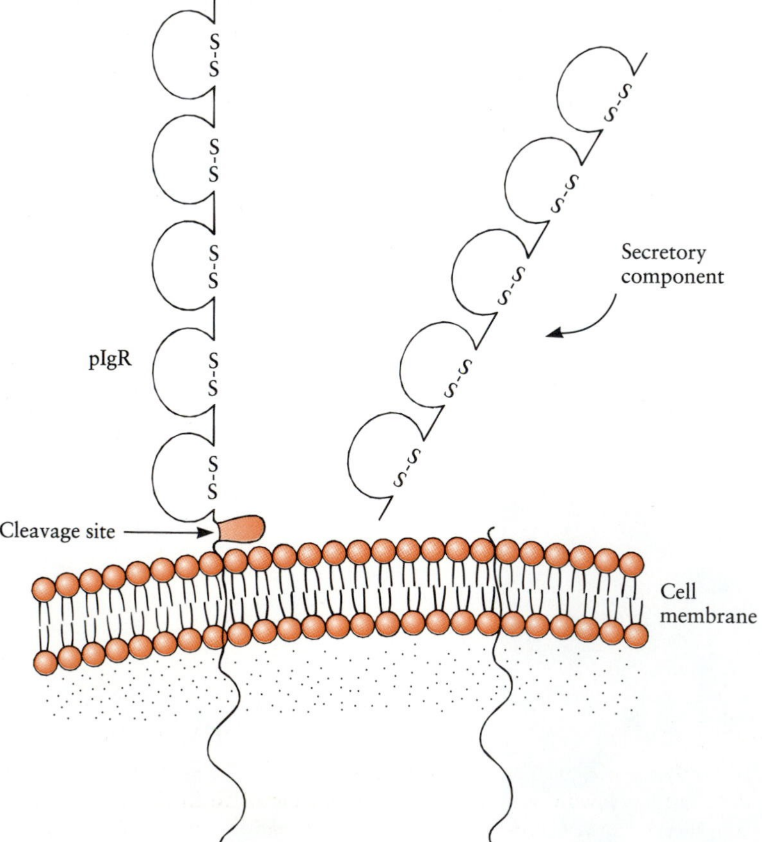

Figure 23–7 Schematic diagram showing the structure of the poly Ig receptor and its relationship to secretory component.

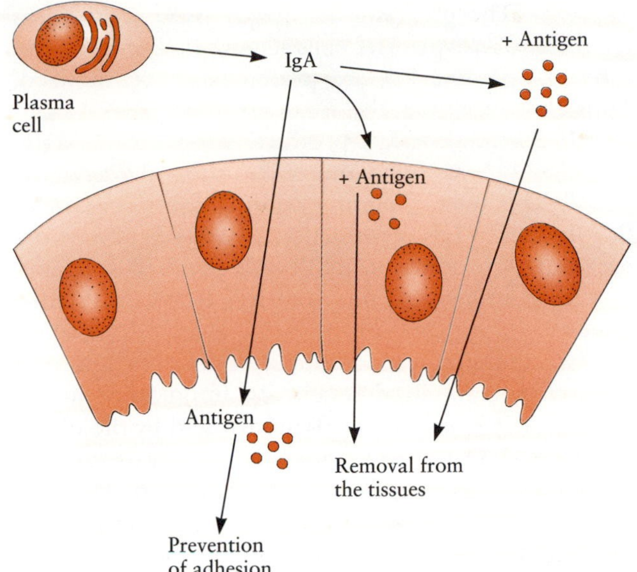

Figure 23–8 IgA acts at three levels in the intestinal wall. It can bind antigen in the submucosa and transport it to the intestinal lumen. It can bind antigen within epithelial cells and transport this to the lumen, and it can bind antigen within the lumen and prevent its adherence to the intestinal wall.

cosa, within epithelial cells, and within the mucosal secretions.

In some species, such as rats, rabbits, and chickens, between 30% and 75% of the IgA produced in the intestinal wall may diffuse into the portal blood circulation and is thus carried directly to the liver (Fig. 23–9). In these species, liver cells (hepatocytes) make pIgR and incorporate it into their membrane. The blood-borne IgA thus binds to hepatocytes and is carried across the hepatocyte cytoplasm to be released into the bile canaliculi. Bile is therefore extremely rich in IgA and is the major route by which IgA reaches the intestine in these species. It is also a route by which antigens bound to circulating IgA can be removed from the body. The reader should note, however, that the situation in ruminants, mice, dogs, and humans is different. Less than 1% of the IgA enters the bile in these species. More than 99% enters the intestinal lumen directly through epithelial cells.

The plasma cells in the gut-associated lymphoid tissue (**GALT**) arise from precursor B cells. These B cells, upon encountering antigen, respond in a manner similar to that of lymphocytes elsewhere in the body—that is, they divide and some differentiate into plasma cells. Some of these responding B cells also migrate into intestinal lymphatics, from which they reach the thoracic duct and the blood circulation. These recirculating IgA^{+}B cells have homing receptors that can bind to addressins in mucosal lymphoid tissue (Fig. 23–10). Their specific ligand is an **addressin** called MAdCAM-1. MAdCAM-1, found only on the vascular endothelial cells of mucosal lymphoid tissue, binds either L-selectin or the $\alpha4/\beta7$ integrin on IgA-producing B cells. As a result, these cells leave the bloodstream in the intestinal tract, respiratory tract, urogenital tract, and mammary gland. Thus stimulation by antigen at one surface site permits antibodies to be synthesized and secondary responses to occur at distant surfaces.

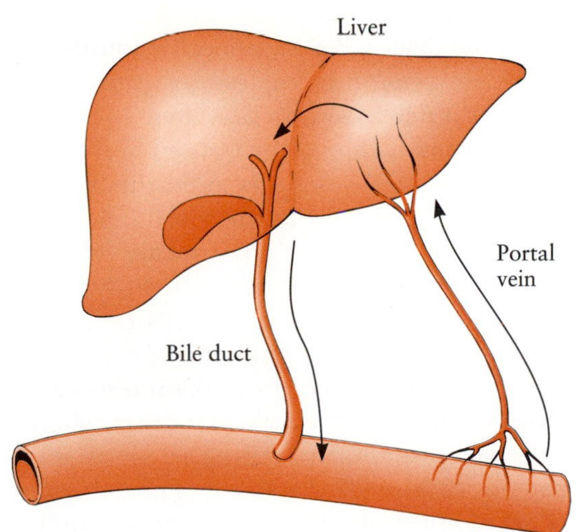

Figure 23–9 Some IgA, instead of being secreted directly into the intestine as shown in Figure 23–6, is carried to the liver where it is passed through the hepatocytes into the bile duct. Only about 1 percent of IgA reaches the intestine by this route in humans.

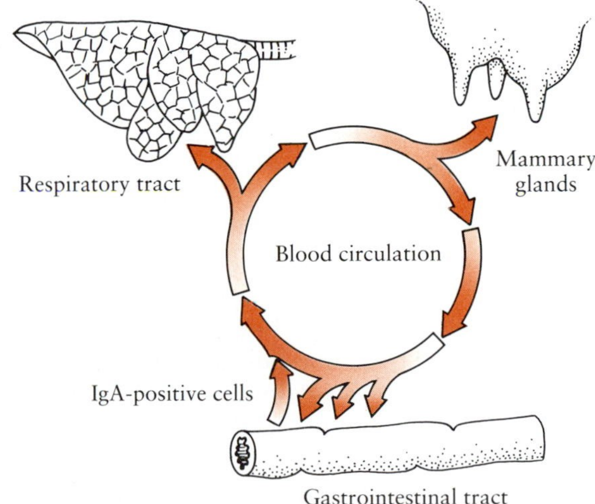

Figure 23–10 IgA producing B cells leave the intestine following antigenic stimulation, circulate in the bloodstream and eventually settle in other locations such as the lung, the mammary gland, and other regions of the gastrointestinal tract.

METHODOLOGY

How to Show That IgA Is Secreted in Rat Bile

To demonstrate the fate of serum IgA, rats were injected intravenously with purified homologous IgA labeled with ^{125}I. Blood and bile samples were taken at intervals, and the radioactivity in each of these fluids was then counted (figure).

Two hours after injection, the level of IgA in bile was up to 12 times higher than that in serum. If the bile duct was ligated, then the IgA radioactivity in serum fell very slowly, at a rate comparable to that of the other immunoglobulin classes. This presumably reflects normal protein catabolism.

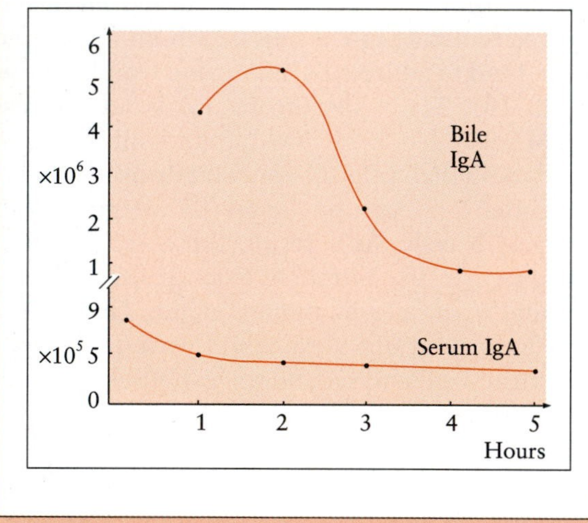

The movement of IgA-secreting B cells from the intestine to the mammary gland is of major importance, since it provides a route by which intestinal immunity can be transferred to the newborn through milk. Oral administration of antigen to a pregnant animal thus results in the appearance of IgA antibodies in its milk. In this way, the intestine of the newborn animal is flooded by antibodies directed against intestinal pathogens. T cells originating within the Peyer's patches also tend to home specifically to the intestinal mucosa, reflecting their use of specialized vascular addressins to determine lymphocyte migration patterns.

Control of IgA Production

The production of IgA in surface tissues is determined by the nature of the helper-T-cell population in these locations. The predominant helper cells on surfaces are of the Th2 type. Th2 cells secrete IL-4 and IL-5. As a result, IgA and IgE are preferentially synthesized. Thus IL-4 promotes the synthesis of IgG1, IgA, and IgE. IL-5 enhances IgA production by cells precommitted to IgA synthesis and makes them differentiate into IgA-producing plasma cells. IL-5 also enhances IL-4–induced IgE production. IL-5 selectively enhances IgA production in Peyer's patch B cells and acts as an inducer of terminal IgA B-cell differentiation.

Interleukin-6 also selectively enhances IgA production by committed B cells. Its importance in the control of IgA response is demonstrated by the observation that mice with targeted disruption of the IL-6 gene show greatly reduced IgA production. They are unable to mount IgA responses to appropriate antigenic stimulation.

Immunity in the Gastrointestinal Tract

The intestine is the most important route of entry for foreign antigens and contains a major portion of the lymphoid tissue of the body. The antigens that enter the body from the intestine include food proteins and microbial antigens from the normal flora and invading pathogens. The intestine contains, in total, more lymphoid tissue than the spleen. Antigenic particles, such as bacteria, can penetrate the intestinal mucosa relatively easily and gain access to the lacteals and portal vessels. These organisms are subsequently trapped in the mesenteric lymph nodes and liver.

Other organisms may enter the body via the surface lymphoid tissues. For instance, the tonsils are very vulnerable to invasion by microorganisms since they have a structural weakness in that their epithelial covering is particularly thin at the bottom of the tonsillar crypts (Fig. 23–11). Viruses may enter the body by this route and multiply locally in the tonsil before the development of viremia. Peyer's patches are covered with a specialized epithelium that is rich in lymphocytes and specialized antigen-sampling cells called M cells (see Fig. 9–10). These M cells can take up bacteria, viruses, or macromolecules from the intestinal lumen and transport them to the follicular areas, where they are processed by dendritic cells. The tonsils and Peyer's patches consist of relatively organized lymphoid tissues, possessing all the components required to mount an immune response—T cells, helper cells, B cells, and macrophages. Nevertheless, most IgA is formed in diffuse lymphoid nodules and in isolated plasma cells in the walls of the intestine, in salivary glands, and in the gallbladder.

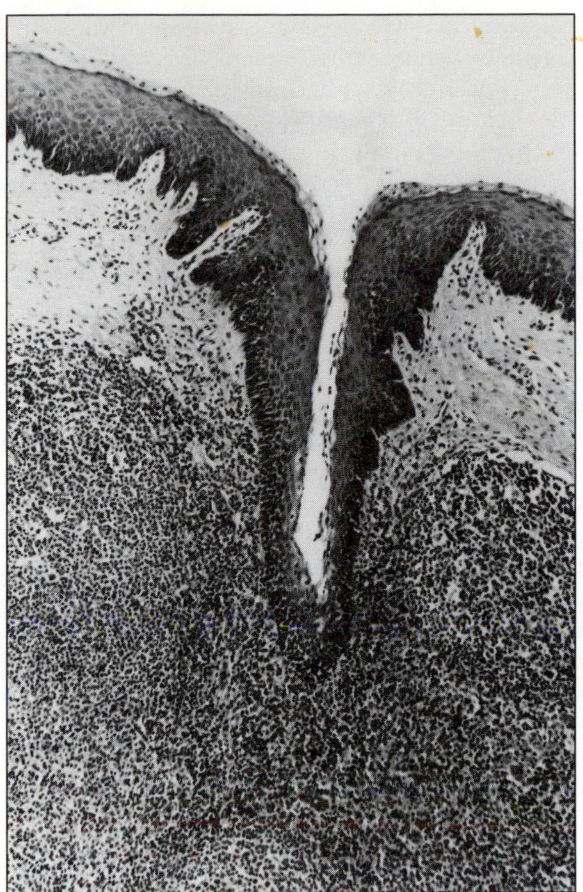

Figure 23–11 A section of tonsil showing a tonsilar crypt. Note how thin the epithelium is at the base of the crypt. An easy invasion route for many organisms. *(Original magnification × 150.)*

It is difficult to confer prolonged protection against intestinal infection by using killed organisms as a vaccine. Oral vaccination, however, may be of some use if the vaccine contains live organisms capable of colonizing the gastrointestinal tract. For example, volunteers given live cholera organisms remain resistant to cholera for at least five or six years.

Intraepithelial Lymphocytes

Cell-mediated immune responses can occur in the wall of the intestine since T cells are found beneath and between epithelial cells. They may collectively constitute up to 15% of the epithelial cell population. The phenotype of these **intraepithelial lymphocytes** (IELs) depends on their location within the intestinal tract. In the small intestine, most of these cells are $CD8^+$, implying that they may have cytotoxic functions, although $CD4^+$ cells are also present. As many as 10% of small intestinal IELs in humans and 25% in mice carry the γ/δ TCR. These cells also use an unusual $V\gamma$ gene to form the receptor (the $V\gamma7$ gene segment). This gene is not usually expressed in other lymphoid organs and suggests that these T cells are specialized for epithelial surveillance. Some of these γ/δ cells have contrasuppressor activity and can prevent the development of oral tolerance within the intestinal lymphoid tissues (Chapter 22). In the mouse large intestine, CD4 and CD8 cells are equally represented and there are very few γ/δ cells. Intraepithelial lymphocytes are uniquely positioned to encounter foreign antigens, and it is likely that these cells do have diverse functions. They may regulate B-cell IgA responses or secrete cytokines, and others are cytotoxic T cells that may attack parasites within the intestinal lumen.

Food antigens may either cause immunological tolerance or trigger formation of IgG and IgE. A measurable fraction of food proteins is not destroyed within the intestine but is absorbed into the circulation, where it may stimulate antibody formation. In contrast, bacteria that colonize the gut mucosa are much more immunogenic than food and usually stimulate a secretory antibody response consisting largely of IgA. It is not clear why food antigens stimulate a different type of immune response from that induced by microbial antigens. It is probable that the differences arise as a result of antigen processing within the gut-associated lymphoid tissues. It is clearly not due to differences in antigen uptake since these different responses can be seen when dietary proteins or microbial antigens are injected directly into Peyer's patches. The intestinal lymphoid tissue must be able to discriminate between the two types of antigen.

Immunity in the Mammary Gland

The mammary gland is protected in nonspecific fashion by the physical barrier of the teat canal, by the flushing action of the milk, and by the presence of lactenins, of which the lactoperoxidase-thiocyanate system, lactoferrin, and lysozyme are probably the most important. In addition, milk contains IgA and IgG1 in low concentrations. In simple-stomached animals, IgA predominates, whereas in ruminants, IgG1 predominates. IgG is usually synthesized locally in the mammary gland. Many of these IgA-producing cells are derived from cells originating in Peyer's patches and mesenteric lymph nodes in late pregnancy and during lactation. Once these cells colonize the mammary gland, they provide a local source of antibodies.

Human colostrum is rich in macrophages and lymphocytes. These macrophages can process antigen, and when cultured, their supernatant fluids can enhance IgA production from blood lymphocytes. The milk lymphocytes are primarily T cells. They may survive for some time in the intestine of a suckling animal and are able to penetrate the intestinal wall and reach its mesenteric lymph nodes. These cells can transfer cell-mediated immune reactions and may even provoke a transient graft-versus-host reaction in infants.

Immunity in the Urogenital Tract

Antibodies of several classes, but particularly IgA, are found in cervicovaginal mucus. Locally produced antibodies may be directed against organisms that cause infections of the cervix or vagina. IgA is present in small amounts in normal urine, produced, presumably, by lymphoid tissues in the walls of the urinary tract. If, however, a kidney infection occurs, then IgG may also be found in relatively large amounts because of the breakdown in the glomerular barrier and defects in tubular reabsorption. Urinary IgA and IgG prevent colonization of the urinary tract by preventing bacterial adherence to epithelial cells.

Immunity in the Respiratory Tract

In addition to the tonsils, the respiratory tract contains a considerable amount of lymphoid tissue in the form of nodules in the walls of the bronchi as well as lymphocytes distributed diffusely throughout the lung and the walls of the airways. The immunoglobulin synthesized in these tissues is mainly secretory IgA, especially in the upper respiratory tract. In the bronchioles and alveoli, however, the secretions contain a large proportion of IgG, the concentration of which is intermediate between the levels in the trachea and in serum. IgE is also synthesized in significant amounts in the lymphoid tissues of the upper respiratory tract. This may be of significance in the development of allergic reactions such as asthma in the respiratory tract. As on other body surfaces, IgA in the respiratory tract probably protects by preventing adherence of antigenic particles, including microorganisms, whereas IgG is probably of major importance only when acute inflammation and transudation of serum protein occur. This situation arises, for example, after an allergic reaction mediated by locally produced IgE. The combination of IgA and IgE may act in concert, so that IgA acts as a barrier that prevents antigen adherence and penetration. If, in spite of the presence of IgA, antigen gains access to the tissues, then the subsequent IgE-mediated hypersensitivity reaction will increase vascular permeability and make available large quantities of IgG in the resulting fluid exudate (Fig. 23–12). As described in Chapter 25, the IgE produced in the lung may also be of importance in controlling parasite invasion.

Many cells may be washed out of the airways of the lung by lavage with saline. These include alveolar macrophages and lymphocytes, which are mainly T cells. It is possible to demonstrate the production of cytokines by these cells and to show alveolar macrophage activation following infection with *Listeria monocytogenes*. Cell-mediated immune reactions are therefore

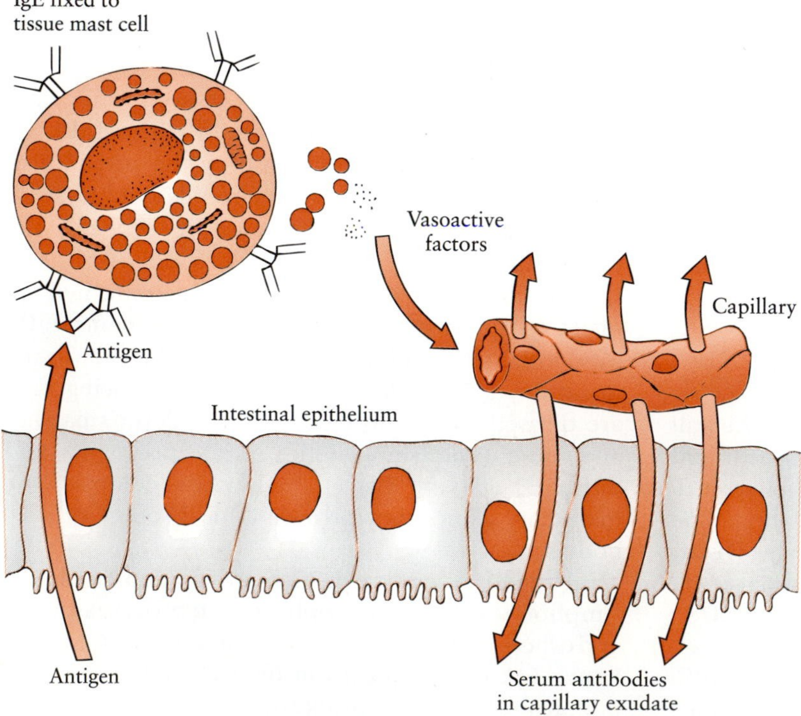

Figure 23–12 The IgE response in the intestinal wall. Antigen reaches IgE sensitized mast cells to cause their degranulation. As a result of this, vasoactive factors cause increased vascular permeability and exudation of serum IgG antibodies.

readily provoked among the cells within the lower respiratory tract.

Immunity in the Skin

The skin is the first line of defense against many microbial invaders. It carries out this function very effectively as judged by the infrequency of severe infections resulting from minor wounds. One way this is accomplished is through a local antigen-trapping system that can present antigens to lymphocytes in an efficient manner and thus provoke a rapid immune response. The antigen-trapping system consists of a network of dendritic cells in the epidermis called **Langerhans' cells.** The Langerhans' cells carry MHC class II molecules on their surface and are able to present antigen to nearby helper T cells. In mice, about 99% of the epidermal T cells carry γ/δ TCRs. The activities of Langerhans' cells are also enhanced by keratinocytes. Keratinocytes also have MHC class II molecules on their surface and are able to synthesize and secrete IL-1, so stimulating the T cells even further. There is also evidence to suggest that a T-cell subpopulation selectively homes to the skin. Thus, if an antigen is injected intradermally, such as occurs when a tick bites an animal, for example, the antigen is trapped by Langerhans' cells and presented to skin T cells, so stimulating a rapid and effective immune response. This may also involve recruitment of basophils (see Fig. 30–16). A similar reaction occurs when reactive chemicals are painted on the skin. If skin is subjected to severe ultraviolet irradiation, the Langerhans' cells are destroyed and the protective mechanisms in the skin are effectively suppressed (Chapter 22).

KEY WORDS

Bile p. 353
Colostrum p. 355
Food antigen p. 355
Gut-associated lymphoid tissue (GALT) p. 353
Immunoglobulin A p. 350
Immunoglobulin E p. 356
Intraepithelial lymphocyte p. 355
Lactenin p. 349
Lactoferrin p. 349
Lactoperoxidase p. 349
Langerhans' cell p. 357
Lysozyme p. 349
M cell p. 354
Milk p. 355
Mucus layer p. 350
Parasite p. 356
Peyer's patch p. 354
Polymeric immunoglobulin receptor p. 350
Resident flora p. 348
Secretory component p. 350
Skin p. 357
Tonsil p. 354

QUESTIONS

1. Secretory component is synthesized by
   a. neutrophils
   b. B cells
   c. plasma cells
   d. macrophages
   e. intestinal epithelial cells
2. The major disadvantage of local secretory immunization is the
   a. difficulty of administration
   b. poor cell-mediated immune response
   c. slow development of local immunity
   d. development of allergic reactions
   e. requirement for live vaccines
3. Secretory IgA contains
   a. four light chains
   b. four heavy chains
   c. one J chain
   d. secretory component
   e. all of the above
4. The two major immunoglobulin classes that protect body surfaces are
   a. IgA and IgG
   b. IgA and IgM
   c. IgA and IgE
   d. IgE and IgG
   e. IgM and IgG
5. Secretory component carries IgA across
   a. histiocytes
   b. hepatocytes
   c. macrophages
   d. neutrophils
   e. plasma cells
6. Lactoferrin found in milk inhibits bacterial growth by
   a. binding calcium
   b. binding iron
   c. acting as an opsonin
   d. stimulating IgA formation
   e. acting as a chemotactic agent
7. The predominant helper-cell activity in the intestine is
   a. Th1
   b. Th2
   c. Th0
   d. Ts
   e. Tdh
8. In which species is a high proportion of IgA secreted in bile?

a. humans
b. dogs
c. mice
d. rats
e. ruminants

9. Which of the following populations of mouse lymphocytes is rich in γ/δ TCR?
a. thymocytes
b. bone marrow
c. tonsil
d. paracortical
e. intraepithelial

10. The dendritic cell population in the skin consists of
a. Langerhans' cells
b. veiled cells
c. follicular dendritic cells
d. islet cells
e. mast cells

11. Discuss the concept that food allergies result from an IgA deficiency.

12. Outline the immunological and nonimmunological defenses in the skin. What types of treatment impair these defense mechanisms?

13. Why has it proved difficult to develop effective oral vaccines containing only inactivated organisms? Outline the properties that would be required for an effective oral vaccine.

14. How can a pathogen that invades the intestine of a female mammal provoke effective intestinal immunity in her offspring?

15. What possible benefits are conferred on an individual when antibody-producing cells migrate from the intestine to the respiratory tract?

16. Discuss the possible functions of the tonsil and the appendix, both intestinal lymphoid organs. Are they both entirely useless? Are any body organs entirely useless?

17. Why are the defense mechanisms in the urogenital tract incapable of protecting an individual from a sexually transmitted disease such as gonorrhea?

Answers: 1e, 2e, 3e, 4c, 5b, 6b, 7b, 8d, 9e, 10a

SOURCES OF ADDITIONAL INFORMATION

Bienenstock, J., and Befus, A.D. Mucosal immunology. Immunology, 41:249–270, 1980.

Brandtzaeg, P. Transport models for secretory IgA and secretory IgM. Clin. Exp. Immunol., 44:221–232, 1981.

Camerini, V., Panwala, C., and Kronenberg, M. Regional specialization of the mucosal immune system. J. Immunol., 151: 1765–1776, 1993.

Crago, S.S., et al. Secretory component on epithelial cells is a surface receptor for polymeric immunoglobulin. J. Exp. Med., 147:1832–1837, 1978.

Guy-Grand, D., Griscelli, C., and Vassalli, P. The mouse gut T lymphocyte: A novel type of cell. J. Exp. Med., 148:1661–1677, 1978.

Husband, A.J., and Gowans, J.L. The origin and antigen-dependent distribution of IgA-containing cells in the intestine. J. Exp. Med., 148:1146–1160, 1978.

Kaetzel, C.S., et al. The polymeric immunoglobulin receptor (secretory component) mediates transport of immune complexes across epithelial cells: A local defense function for IgA. Proc. Natl. Acad. Sci. U.S.A. 88:8796–8800, 1991.

Keren, D.F. Antigen processing in the mucosal immune system. Semin. Immunol., 4:217–226, 1992.

Mazanec, M.B., et al. Intracellular neutralization of virus by immunoglobulin A antibodies. Proc. Natl. Acad. Sci. U.S.A., 89:6901–6905, 1992.

McDermott, M.R., Befus, A.D., and Bienenstock, J. The structural basis for immunity in the respiratory tract. Int. Rev. Exp. Pathol., 23:47–112, 1982.

McNabb, P.C., and Tomasi, T.B. Host defense on mucosal surfaces. Annu. Rev. Microbiol., 35:477–496, 1981.

Orlans, E., et al. Rapid active transport of immunoglobulin A from blood to bile. J. Exp. Med., 147:588–592, 1978.

Ramsay, A.J., et al. The role of Interleukin-6 in mucosal IgA antibody responses in vivo. Science, 264:561–564, 1994.

Schrader, J.W., Scollary, R., and Battye, F. Intramucosal lymphocytes of the gut: Lyt-2 and Thy-1 phenotype of the granulated cells and evidence for the presence of both T cells and mast cell precursors. J. Immunol., 130:558–564, 1983.

Solari, R., and Kraehenbuhl, J.P. The biosynthesis of secretory component and its role in the transepithelial transport of IgA dimers. Immunol. Today, 6:17–20, 1985.

Tomasi, T.B., et al. Mucosal immunity: The origin and migration pattern of cells in the secretory system. J. Allergy Clin. Immunol., 65:12–19, 1980.

Underdown, B.J. Transcytosis by the receptor for polymeric immunoglobulin. *In* Metzger, H., ed. Fc Receptors and the Action of Antibodies. American Society for Microbiology, Washington D.C., 1990.

Wold, A.E., et al. Differences between bacterial and food antigens in mucosal immunogenicity. Infect. Immun., 57: 2666–2673, 1989.

Woloschak, G.E., and Tomasi, T.B. The immunology and molecular biology of the gut-associated lymphoid tissue. Crit. Rev. Immunol., 4:1–18, 1983.

CHAPTER 24

Vaccines and Vaccination

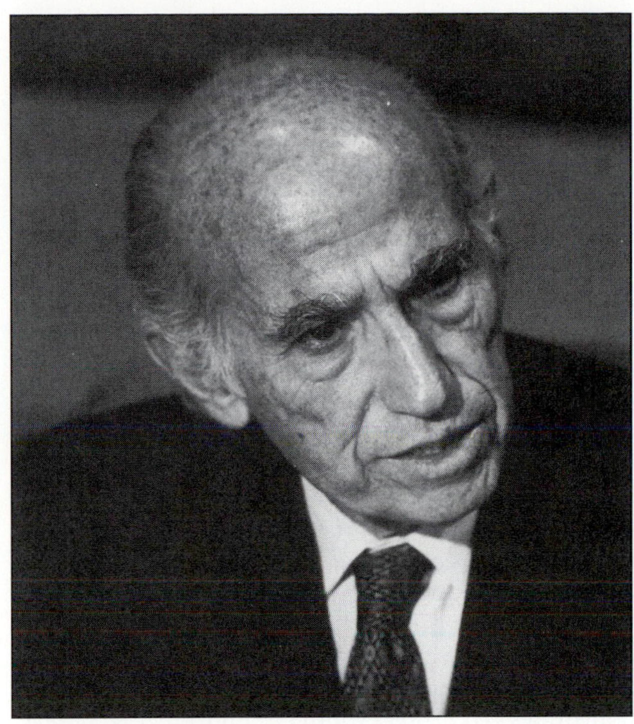

Jonas Salk *Poliomyelitis was a very devastating disease in the 1940s and '50s. The discovery that the poliovirus could be grown in tissue culture enabled Dr Jonas Salk to develop an inactivated poliovirus vaccine. Thus he adapted three strains of poliovirus to monkey cell culture. The culture fluids that contained live virus were then inactivated by treatment with formaldehyde to produce an effective vaccine. The development of this vaccine, and subsequently an attenuated live vaccine by Sabin, eventually brought poliomyelitis under control. It is estimated that this discovery alone saved many times more money than has ever been spent on medical research.*

CHAPTER OUTLINE

CHAPTER CONCEPTS

1. Protective immunity can be induced by either active or passive immunization procedures.
2. Passive immunization involves administration of preformed antibody to an individual to provide immediate protection. This passive immunity is short-lived since it wanes as the transferred antibody is metabolized.
3. Active immunization involves administration of an antigen in the form of a vaccine. Immunity provoked in this way takes several weeks to develop but may persist for a long time.
4. Vaccines containing modified live organisms generally produce better immunity than do vaccines containing inactivated organisms but are potentially more hazardous.
5. New techniques for producing antigen such as recombinant DNA techniques or the use of synthetic antigens have led to the development of several new or improved vaccines.
6. The widespread use of vaccines has had significant impact on the prevalence of many infectious diseases. Their ongoing development is hindered, however, by the issue of adverse side effects and litigation.

Vaccination is, of course, one of the major contributions of immunology to medical science. In this chapter you will first read about the two different methods of immunization, active and passive. The chapter continues with a discussion of the methods of production of vaccines and the relative merits of vaccines containing live or inactivated organisms. This is followed by a review of some newer approaches to developing more effective vaccines. The chapter concludes with the practical aspects of vaccine usage, including methods of administration, reasons why vaccines do not work all the time, some hazards of vaccination, as well as a review of the benefits of vaccination.

The observation that individuals who recovered from some infectious diseases were resistant to subsequent reinfection long preceded the development of the science of immunology and our understanding of the immune response. Indeed, the attempts made by Edward Jenner and Louis Pasteur to reproduce this phenomenon provided the impetus for the early development of immunology. Their efforts to produce immunity by artificial exposure to infectious agents were so successful that many diseases, once major scourges of mankind, were rapidly controlled. **Vaccines** were developed rapidly against smallpox, rabies, tetanus, anthrax, cholera, and diphtheria. In the United States, the administration of effective vaccines has reduced the number of reported cases of diphtheria, measles, mumps, pertussis, poliomyelitis, rubella, and tetanus by at least 97% (Table 24–1). No other form of disease control has had such an effect on the reduction of mortality. Indeed, vaccination has been responsible, in large part, for the recent phenomenal increase in the world's population.

In general, immunization procedures involve giving antigen derived from an infectious agent to an individual so that an immune response is mounted and resistance to that infectious agent is stimulated. This is known as **active immunization.** Alternatively, preformed antibodies can be administered to a susceptible recipient to confer a temporary but immediate state of immunity, a process called **passive immunization.**

Three criteria must first be satisfied in determining whether vaccination is either possible or desirable in controlling a specific disease. The first is the absolute identification of the causal organism. Although this appears to be an obvious requirement, it has not always been followed in practice, at least in some diseases of domestic animals.

Second, it must be established that an immune response can, in fact, protect against the disease in question. Thus smallpox vaccine must never be used to treat recurrent herpesvirus infections or warts. It does not work in these conditions. Other examples of nonprotective immune responses are seen in the lentivirus disease of horses, equine infectious anemia, and in the parvovirus disease of mink, Aleutian disease, in which the immune response itself is responsible for many of the disease processes and vaccination therefore increases its severity. *Haemophilus influenzae* type B capsular polysaccharide vaccine protects only children over 18 months of age, since younger children do not mount an immune response against carbohydrates. To protect younger children, a *H. influenzae* vaccine containing the polysaccharide chemically conjugated to a protein must be used.

Finally, vaccination is not without some disadvantages. Before using any vaccine, one must be absolutely certain that the risks involved do not exceed those due to the chance of contracting the disease itself. For example, smallpox has been eradicated globally since October 26, 1977; because smallpox vacci-

Table 24–1 Reported Cases of Vaccine-Preventable Childhood Diseases in the United States.*

Disease	Maximal No. of Cases (Yr)	1991	% Change
Diphtheria	206,939 (1921)	2	−99.9
Measles	894,134 (1941)	9488	−98.9
Mumps†	152,209 (1968)	4031	−97.4
Pertussis	265,269 (1934)	2575	−99.0
Poliomyelitis (paralytic)	21,269 (1952)	0‡	−100.0
Rubella§	57,686 (1969)	1372	−97.6
Congenital rubella syndrome	20,000 (1964–1965)	36	−99.8
Tetanus	1,560¶ (1923)	49	−96.9

Reprinted by permission of the New England Journal of Medicine, Vol. 327, page 1795, 1992.

*Data are from Orenstein.[1] Figures for 1991 are provisional (as of June 1992).
†Mumps first became a reportable disease in 1968.
‡Projected number of vaccine-associated cases, 5 to 10.
§Rubella first became a reportable disease in 1966.
¶Number of reported deaths.

Table 24–2 Whooping Cough Vaccine: Risks Versus Benefits

Problem	Risk of Occurrence After: Vaccination	Risk of Occurrence After: Disease
Seizures	1:1,750	1:25–1:50
Encephalitis	1:110,000	1:1,000–1:4,000
Severe brain damage	1:310,000	1:2,000–1:8,000
Death	1:1,000,000	1:200–1:1,000

nation occasionally causes encephalitis and possibly death, there is no reason whatsoever to give smallpox vaccine to normal individuals.

More recently, there has been a debate on the risks of *Bordetella pertussis* (whooping cough) vaccination relative to the hazards of whooping cough. Severe reactions to pertussis vaccine are very rare (Table 24–2), although high fever, convulsions, and collapse reactions are all well recognized. Acute encephalitis, permanent brain damage, and death are much less common. If vaccination were to be abandoned, however, it is estimated that the number of pertussis-associated deaths in the United States would climb from around 10 to between 35 to 60 annually. Clearly, the relative benefits of this and other vaccines must be carefully analyzed.

A CLOSER LOOK

Vaccine Safety

In Japan in 1974 and 1975 two children died within 24 hours of receiving pertussis vaccine. As a result, pertussis vaccine use dropped to very low levels. The number of pertussis cases in Japan climbed from less than 1000 cases in 1974 to more than 13,000 cases in 1979. In that year there were 41 deaths from pertussis.

TYPES OF IMMUNIZATION PROCEDURES

Passive immunization and active immunization are the two methods by which an individual can be made resistant to an infectious agent (Fig. 24–1). Passive im-

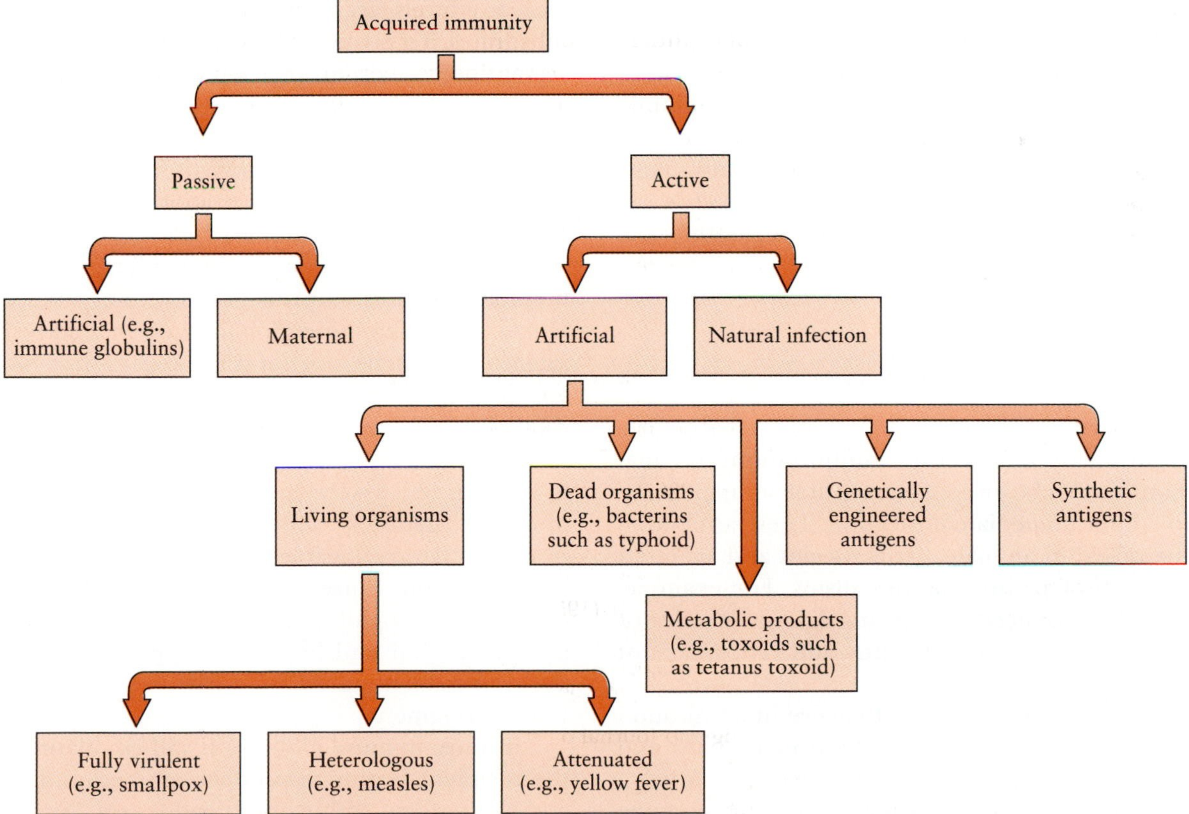

Figure 24–1 A classification of the ways in which an individual can develop immunity to an infectious agent.

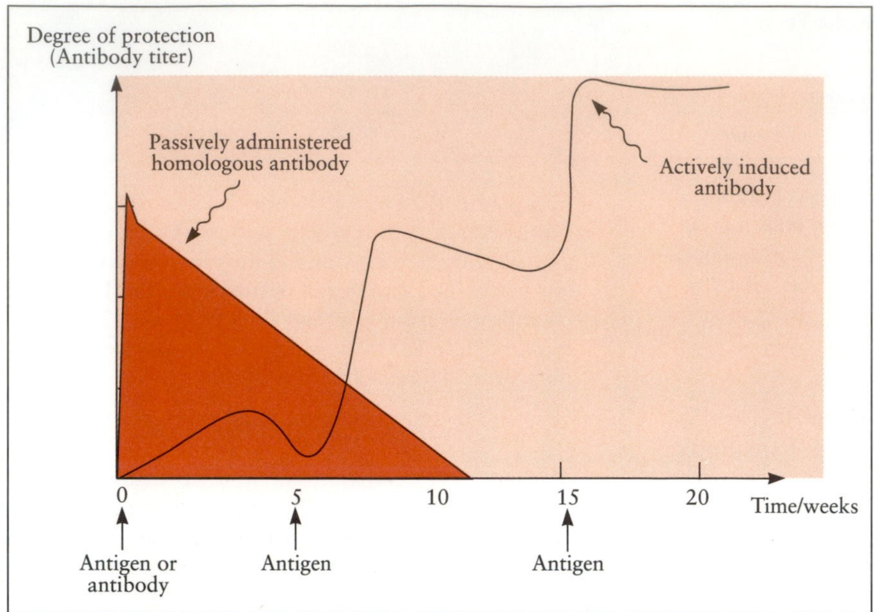

Figure 24–2 A comparison of active and passive immunization. Passive immunization gives rapid-onset but relatively short-lasting immunity. Active immunization in contrast develops slowly but gives prolonged immunity and can be readily boosted.

munization produces temporary resistance by transferring antibodies from a resistant to a susceptible individual. These transferred antibodies give immediate protection, but since they are gradually catabolized, the protection wanes and the recipient eventually becomes susceptible to infection once again (Fig. 24–2).

Active immunization has several advantages over passive immunization. This technique involves administering antigen to individuals so that they respond by mounting a protective immune response. Reimmunization or exposure to the infectious agent results in a secondary immune response. The major disadvantage of active immunization is that the protection is not immediate. Its advantage is that it is long-lasting and capable of restimulation.

PASSIVE IMMUNIZATION

Passive immunization requires that antibodies be produced in a donor by active immunization and that these antibodies be given to susceptible animals in order to confer immediate protection. These antibodies may be raised in animals of any species and against a wide variety of pathogenic organisms. For example, they can be produced in humans against measles or hepatitis, in horses against tetanus, and in dogs against canine distemper.

One of the most important passive immunization procedures has been the protection of humans and other animals against tetanus by means of antisera raised in horses. The antibodies, known as **immune globulins,** are produced in young horses by a series of immunizing injections. The toxins of the clostridia are proteins, and they can be made nontoxic by treatment with formaldehyde. Toxins treated in this way are known as **toxoids.** Initially, the horses are inoculated with toxoids, but once antibodies are produced, subsequent injections contain purified toxin. The responses of the horses are monitored, and once their antibody levels are sufficiently high, the horses are bled. Bleeding is undertaken at regular intervals until the antibody levels drop, at which time the horses are again inoculated with the antigen. Plasma is separated from the horse blood and treated with ammonium sulfate to concentrate and purify the globulin fraction. This purified antibody preparation is then dialyzed, filtered, titrated, and dispensed.

Tetanus immune globulin confers prompt immunity to tetanus, but it is not devoid of side effects. Horse tetanus immune globulin may indeed be given safely to horses, in which it is not regarded as foreign, and persists for a long time, being removed only by catabolism. If, however, this immune globulin is given to an animal of another species, such as a human, then it is regarded as foreign, an immune response is mounted against it, and it is rapidly eliminated. Although it is not possible to destroy all the immunogenicity of the horse antibodies, it is usual to treat equine immunoglobulins with pepsin so as to remove their Fc region and leave intact only the portion of the molecule required for toxin neutralization, the F(ab′)2 fragment.

If horse immunoglobulin is still circulating when the recipient begins to produce antibodies, immune complexes form within the bloodstream (Fig. 24–3). These may cause a hypersensitivity reaction known as

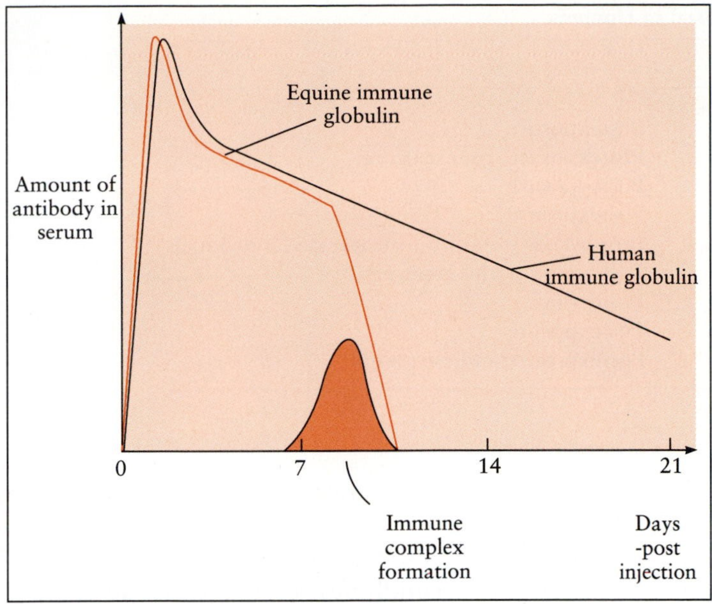

Figure 24–3 A comparison of the fate of human immune globulin and equine immune globulin when injected into a human. The equine immune globulin is antigenic and is rapidly removed once the host mounts an immune response. Immune complexes form when the foreign antigen is being removed by antibody.

serum sickness (Chapter 30) (Table 24–3). If repeated doses of horse immune globulin are given to an individual, other hypersensitivity reactions, such as **anaphylaxis,** may occur (Chapter 29).

To measure the potency of preparations of immune globulin, units of activity are calculated by comparison with an international biological standard. The international standard antiserum for tetanus toxin is a quantity held at the State Serum Institute in Copenhagen.

About 250 units of immune globulin are normally given to humans to confer immediate protection against tetanus, although the exact amount should vary with the severity of tissue damage, the degree of wound contamination, and the time elapsed since injury. About 10,000 units may be given to individuals suffering from clinical tetanus in an attempt to bind free toxin and hasten recovery, although this procedure is relatively ineffective.

Because of the problems encountered when equine antiserum is used in humans, it is preferable to use antiserum of human origin whenever possible. This human immune serum globulin (ISG) is obtained from volunteer donors who have recently been immunized against tetanus. Other common passive immunizing agents used in humans include antisera against measles, hepatitis A and B, rubella, varicella, rabies, diphtheria, group B streptococci, *Pseudomonas,* cytomegalovirus, tetanus, and botulism (Table 24–4).

Table 24–3 The Adverse Consequences of Passive Immunization

Short-lived immunity
Suppression of active immunization
Serum sickness
Anaphylaxis

Measles ISG is given to susceptible contacts of measles cases in order to prevent the disease or, alternatively, to immunodeficient individuals who may not be able to mount an active immune response. Hepatitis-A ISG can be used for preexposure prophylaxis of individuals who run the risk of exposure to this disease through either their occupation or area of residence. Hepatitis-B ISG is used for immediate protection of individuals exposed to this virus through accidental skin puncture with a contaminated needle or exposure to infected blood or body fluids. It is commonly given in conjunction with active immunization. Rubella ISG can be given to susceptible pregnant women who have been exposed to rubella (German measles). Varicella (chickenpox) immune globulin is of use in immunosuppressed children, such as leukemia patients exposed to chickenpox. Rabies immune globulin of either human or equine origin is used for postexposure prophylaxis of rabies. It is given in a divided dose, half given intramuscularly and the remainder injected around the bite wound in an effort to establish an antibody barrier to the spread of the rabies virus.

Monoclonal antibodies represent another potential source of passive protection. At present, however, these are mainly made by mouse–mouse **hybridomas** and thus consist of mouse immunoglobulins. They are therefore able to sensitize animals of other species.

Table 24–4 Passive Immunizing Agents Used in Humans

Disease	Origin	Use
Measles	Human	Postexposure
Hepatitis A	Human	Postexposure, preexposure
Hepatitis B	Human	Postexposure
Rubella	Human	Postexposure
Varicella	Human	Prophylaxis in immunosuppressed individuals
Botulism	Equine	Postexposure and treatment
Diphtheria	Equine	Postexposure
Rabies	Human or equine	Postexposure
Tetanus	Human or equine	Postexposure and treatment

ACTIVE IMMUNIZATION

The most important advantages of active immunization as compared with passive immunization are the prolonged duration of protection and the recall and boosting of this protective response by repeated injections of antigen. The ideal vaccine to be used for active immunization should be cheap, stable, and adaptable to mass vaccination. It should give prolonged strong immunity without adverse side effects and ideally should stimulate an immune response distinguishable from that due to natural infection so that vaccination does not interfere with diagnosis.

It is most important that vaccines be free of adverse side effects. This clearly precludes the use of virulent living organisms. Remember, however, that this was the method first used by the Chinese in vaccinating against smallpox (Chapter 1). The Chinese reduced the hazards of vaccination by selecting the smallpox scabs from the mildest cases available. Nevertheless, the procedure was still hazardous until Jenner introduced the use of cowpox. Unfortunately, not all infectious agents have a naturally available avirulent counterpart. For this reason, if living organisms must be used in a vaccine, they must be treated in such a way that they lose their disease-producing ability. The term used for this loss of **virulence** is **attenuation.**

The commonly used methods of attenuation involve adapting organisms to unusual environmental conditions so that they lose the ability to replicate uncontrollably in their usual host. For example, the bacillus Calmette-Guérin (BCG) strain of *Mycobacterium bovis* was rendered avirulent by growing it for 13 years on a bile-saturated medium. At the end of that time, the organisms had adapted well to growing in bile, but they had lost the ability to grow and cause disease in humans. BCG is used as a vaccine against tuberculosis. A similar situation occurred in Pasteur's original fowl-cholera experiments (Chapter 1), in which the organism was grown under conditions involving a shortage of nutrients and, as a result, lost its ability to cause disease in chickens.

Viruses can also be attenuated by growth in abnormal culture conditions. Prolonged tissue culture, especially in cells of a species that the virus does not normally infect, has the effect of reducing the virulence of viruses. Thus attenuated living poliovirus vaccine contains viruses grown in monkey-tissue culture. Rubella virus vaccine may contain virus prepared in duck-embryo cells. As an alternative to tissue culture, some viruses may be attenuated by growth in eggs. These include the vaccine strain of yellow fever and strains of influenza virus. Another interesting method of attenuation is to adapt the virus to growth at slightly lower temperatures than normal. Thus the vaccine against the herpesvirus responsible for infectious bovine rhinotracheitis uses a temperature-sensitive mutant. This vaccine is given intranasally to cattle so that the temperature-sensitive organism can replicate within the relatively cool nasal mucosa. It cannot spread within the body and so cause disease, because the body temperature is too high.

Attenuation of organisms for use in vaccines has both advantages and disadvantages. In general, vaccines containing living organisms are effective and give prolonged, strong immunity. Because, in effect, they cause transient infection, only a few inoculating doses are required and adjuvant need not be employed. As a result, there is less chance of provoking adverse hypersensitivity reactions. In addition, live virus vaccines may provoke a rapid protective response through stimulation of **interferon** production.

On the other hand, live vaccines may be difficult and expensive to produce. There exists the possibility that they may contain dangerous extraneous organisms and, more important, cause disease as a result either of residual virulence (i.e., they may not be fully attenuated) or of reversion to a more virulent form. In general, live vaccines cannot be used in patients suffering from an immunodeficiency such as AIDS.

To avoid these disadvantages, it is common to use organisms that have been killed or inactivated. In such cases, the "dead" organisms should be as antigenically similar to the living organisms as possible. Therefore, a crude method of killing microorganisms, such as heating, which causes extensive protein denaturation, tends to be unsatisfactory. If chemical inactivation is to be used, the chemicals must produce little change in the antigens that are responsible for protective immunity. Commonly used inactivating agents include formaldehyde, ethylene oxide, ethyleneimine, acetylethyleneimine, and β-propiolactone.

The relative advantages of vaccines containing **modified live** or "dead" organisms can be seen in those vaccines used against poliomyelitis (Fig. 24–4). Live poliovirus vaccines contain a mixture of three strains of attenuated poliovirus (types 1, 2, and 3) originally grown in epithelial cells from monkey kidneys. The vaccine is administered orally to children as a liquid dropped onto a sugar cube. As a result, the vaccine strains colonize the child's intestine and provoke an immune response similar to that induced by natural infection by poliovirus. This response includes the production of secretory antibodies in the intestine as well as serum antibodies. Consequently, the vaccine produces a long-lasting humoral and intestinal immunity, thus reducing the need for repeated boosters. The administration of the vaccine is simple and does not require trained personnel, and the vaccine itself is relatively cheap and can be grown on cultures of human cells, thus reducing the need to use increasingly rare and precious monkeys. On the other hand, since oral polio vaccination results in infection of recipients, the vaccine virus may spread to household contacts or nonvaccinees and cause occasional cases of paralytic polio. If the vaccinated individual is already carrying a population of other enteroviruses, a common situation in underdeveloped tropical countries, then viral interference may prevent successful colonization by the polio vaccine virus and result in failure to protect.

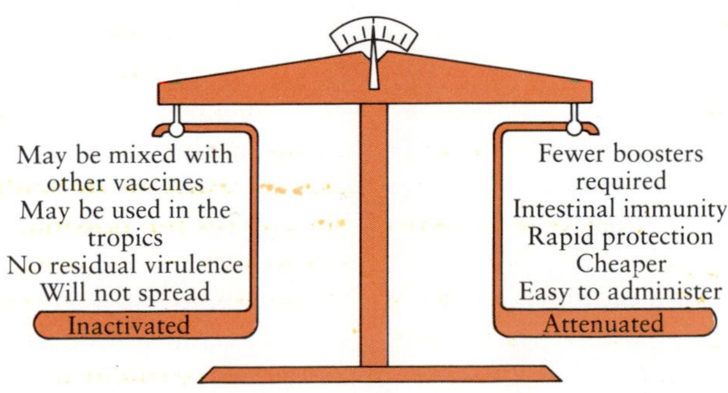

Figure 24–4 The relative merits of inactivated and attenuated poliovirus vaccine.

As with many live-virus vaccines, there is a risk that the vaccine virus may cause disease. It has been estimated that the rate of paralytic disease in polio vaccine recipients is about one case in four million doses of vaccine distributed, although there is about one case of paralytic poliomyelitis in two million doses of vaccine distributed among individuals in contact with vaccinees. Although this may be considered a very low risk, consider that paralytic polio, if not lethal, is a severely crippling disease. Thus the few individuals who develop polio as a result of exposure to vaccine may have their lives ruined. Society surely owes it to these individuals to compensate them for this; it is a small price to pay for keeping the rest of us free from this disease.

The advantages of killed poliovirus vaccines effectively reflect the disadvantages of live oral poliovirus vaccines. Thus, killed polio vaccine can be readily incorporated with other injectable vaccines as part of a single vaccination program. This method is safe in that there is no chance of reversion to a virulent form, provided it is properly inactivated. It also can be used in tropical areas where other enteroviruses interfere with the live vaccine. It is perfectly safe for immunodeficient individuals. On the other hand, to get a good protective response, multiple booster injections are required and individuals must return for boosters at regular intervals. Killed poliovirus vaccine does not prevent intestinal colonization by wild-type virus, since it does not provoke local immunity. Finally, since it must be injected, it must be pure, and because it is grown in monkey cells, it is costly.

Thus both types of vaccine have risks and benefits. The final choice as to which one to use must be made only in conjunction with an assessment of local conditions and requirements.

In the search for effective vaccines, unwanted contaminants can increase the risk of side effects without adding anything to the efficacy of the vaccine. For example, vaccines prepared from viruses or bacteria grown in culture commonly contain culture-fluid components. Likewise, many organisms carry antigens that are not protective but that may be toxic. Thus it is usual to purify vaccines as much as possible before use. The end point of this process is the isolation of pure, protective antigens (Table 24–5). The most obvious examples of such "pure" antigens are the vaccines against diphtheria and tetanus, which utilize not the whole bacterium but rather a purified preparation of exotoxin detoxified by treatment with formaldehyde. These bacterial toxoids have proved to be excellent and effective immunizing agents. Another example of this type of vaccine is that against bacterial meningitis, which contains purified bacterial cell wall polysaccharides against two types of *Neisseria meningitidis*—groups

Table 24–5 Some Subunit Vaccines Used in Humans and Other Animals

Vaccine	Subunit
Bacterial	
Tetanus	Toxin
Botulism	Toxin
Diphtheria	Toxin
S. pneumoniae	Capsular polysaccharide
Meningococci	Capsular polysaccharide
E. Coli	Adherence pili
Viruses	
Rabies	Capsid proteins
Foot-and-mouth virus	Capsid proteins
Influenza	Capsid proteins
Hepatitis B	Synthetic capsid epitopes
Foot-and-mouth virus	Synthetic capsid epitopes

A and C. Whole cell *B. pertussis* vaccines, although effective, are associated with several adverse side effects. As a result, several acellular pertussis vaccines have been produced. These contain purified bacterial subcomponents and cause significantly fewer adverse side effects. Antipneumococcal vaccine now contains a mixture of purified polysaccharides from each of 23 types of *Streptococcus pneumoniae.* These are known to cause 85% to 90% of pneumococcal pneumonia in the United States. Some strains of *Escherichia coli* that cause enteric disease of animals attach to the intestinal wall by means of specialized attachment pili. Vaccines enriched with these attachment pili reduce the severity of disease caused by pilus-bearing strains.

SOME NEW APPROACHES TO VACCINE PRODUCTION

Although conventional vaccines have been successful in controlling infectious diseases, there is always a need for improvement. Several new approaches are being studied in attempts to make vaccines more effective, cheaper, and safer (Fig. 24–5).

Genetically Modified Organisms

Attenuation may be considered a primitive form of genetic engineering. The desired result is the development of a genetically stable agent that in some way lacks the ability to cause disease. This may be difficult to achieve, and reversion to virulence is an ever-present risk. It has become possible, however, to modify the genes of organisms deliberately so that they become effectively and irreversibly attenuated. For example, a vaccine is now available against a herpesvirus that causes the disease in pigs called pseudorabies. The thymidine kinase (TK) gene has been removed from this virus. TK is necessary for herpesviruses to replicate in nondividing cells such as neurons. Viruses from which the TK gene has been removed are able to infect nerve cells but cannot replicate and cannot therefore cause disease. As a result, this vaccine not only confers effective protection, but also by blocking cell invasion by virulent pseudorabies viruses prevents the development of a persistent carrier state. Other methods of rendering viruses avirulent include gene segment reassortment in organisms with segmented genomes such as rotaviruses and influenza. Alterna-

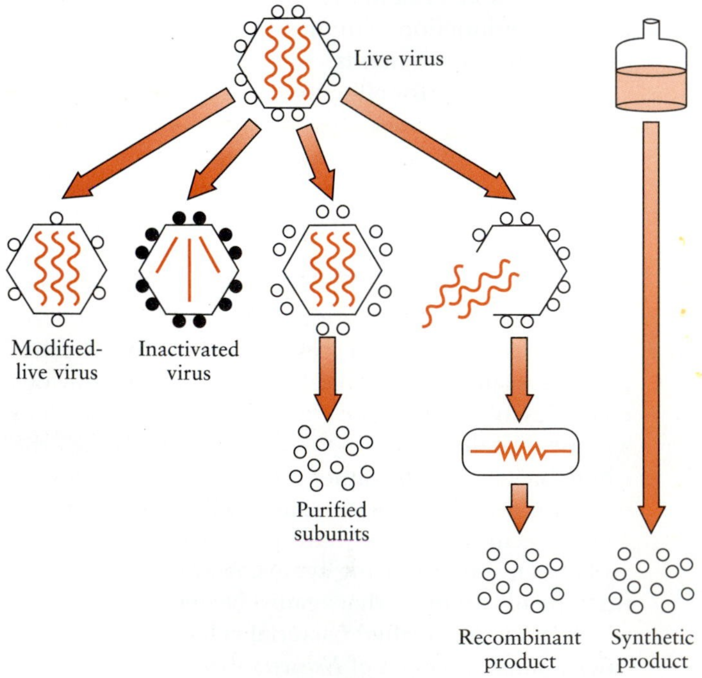

Figure 24–5 Some exciting new methods of making vaccines. Unfortunately, few of these have been approved for human use. Recombinant vaccines are used against hepatitis B, rabies, and feline leukemia. Synthetic vaccines are generally very expensive.

tively, other genes may be deleted such as in *Vibrio cholerae,* in which part of the toxin gene may be removed. A third method of causing irreversible attenuation is to insert a missense gene such as is used to generate temperature-sensitive mutants. All these approaches have worked well. It is clear that this deliberate modification of virulence is a much more effective and logical method of reducing virulence than the previous hit-or-miss methods of attenuation.

Recombinant Vaccines

Recombinant Antigens. Recombinant DNA techniques can be employed to isolate genetic material coding for a protein antigen of interest. This DNA can then be placed in a bacterium, yeast, or other cell and permitted to code for that protein. The first attempt to use gene cloning to prepare a vaccine was with foot-and-mouth disease virus (Fig. 24–6). This virus is extremely simple, the protective antigen (VP1) is well recognized, and the genes that code for this antigen have been mapped. The RNA genome of the foot-and-mouth disease virus was isolated and duplicated into DNA by means of the enzyme reverse transcriptase. The DNA was then carefully cut by restriction endonucleases so that it contained only the gene for VP1. This DNA was then inserted into an *E. coli* plasmid, the plasmid inserted into *E. coli,* and the *E. coli* grown. These recombinant bacteria synthesize large quantities of VP1 that may be harvested, purified, and incorporated into a vaccine. The process can be highly efficient since 4×10^7 doses of foot-and-mouth vaccine can be obtained from 10 liters of *E. coli* grown to 10^{12} organisms/ml.

A very successful recombinant vaccine against hepatitis B has been in use in the United States since 1984. This vaccine is produced by rDNA-transfected yeast. It is especially effective when administered together with immune globulin to newborn children of high-risk mothers.

Recombinant DNA techniques are useful in any situation in which protein antigens must be synthesized in large and pure quantities. Unfortunately, pure proteins such as these are often poor antigens because they are not effectively delivered to antigen-sensitive cells and cannot therefore provoke a sufficiently intense response. An alternative method is to clone the genes of interest into an attenuated living carrier organism.

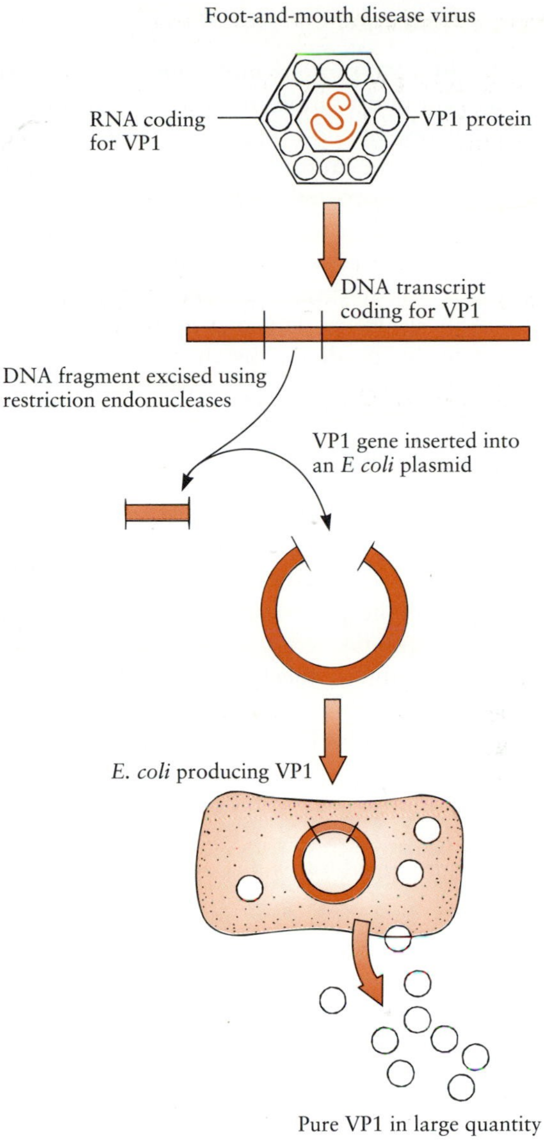

Figure 24–6 The production of large quantities of VP1 antigen from the foot-and-mouth disease virus by the use of recombinant DNA techniques. The purified VP1 derived from bulk culture of *E. coli* may be used in a foot-and-mouth disease vaccine.

Live Recombinant Organisms. Genes coding for specific protein antigens may be cloned directly into a variety of organisms. The organism that has been most widely employed for this purpose is vaccinia virus. Vaccinia virus is easy to administer by dermal scratching. It has a large genome that makes it relatively easy to insert a new gene, and it can express high levels of the new antigen. Moreover, the proteins undergo appropriate processing steps, including glycosylation and membrane transport. Individuals vaccinated with genetically modified vaccinia virus make high levels of antibodies against the introduced antigen. Because the use of vaccinia may be hazardous, alternative carriers that have been proposed include attenuated *Salmonella* strains. However, the use of attenuated carrier organisms has some intrinsic limitations and all the disadvantages of modified live vaccines.

Synthetic Peptides

Although globular protein molecules may be large, they have a limited number of epitopes on their surface. Only a few of these epitopes are important in inducing protective immunity (Chapter 22). Thus, if the structure of a protective epitope is known, it may be chemically synthesized and used in a vaccine. The procedures involved include a complete sequencing of the antigen of interest, followed by identification of the important epitopes. This may be difficult, but the epitopes may be predicted by the use of computer models of the protein or by the use of monoclonal antibodies to identify the critical protective sequences. By knowing the complete amino acid sequence of the major antigenic proteins of these organisms, it is possible to identify the sequences that are hydrophilic and thus most likely to be located on the surface of the molecule. It may be predicted that these sequences will function as epitopes. Once identified, the protective peptides can be synthesized and used in a vaccine. Experimental synthetic vaccines have been developed against several viruses, including hepatitis B and influenza A. When peptides containing these synthetic epitopes are used, they provoke some protective immunity. This technique has several significant advantages over gene splicing techniques. For example, it is much safer although considerably more expensive.

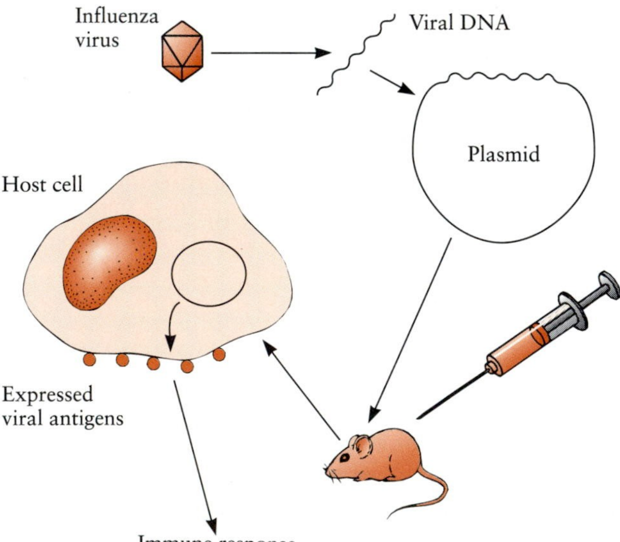

Figure 24–7 Pure nucleic acid has been successfully employed as an influenza vaccine in mice. The viral DNA, incorporated into a plasmid, can cause formation of viral antigens in host cells in the same way as the original virus. These endogenous antigens are admirably suited to provoke a protective immune response.

Naked DNA

Philip Felgner and Robert Malone have developed a unique approach to vaccination that involves injection of DNA from a virus into a host. They took the DNA coding for the nucleoprotein of an influenza virus, attached it to a plasmid, a piece of circular DNA that acts as a vector, and injected the plasmid containing the viral gene into the muscles of mice (Fig. 24–7). As a result, the mice began to make antibodies against the nucleoprotein, which indicated that the viral gene was being expressed in the recipient's cells. When they infected these mice with a lethal dose of influenza virus, 90% of the vaccinated animals survived and only 20% of the control animals did so. The gene products are treated as endogenous antigens, processed by linkage to MHC class I molecules, and displayed on the cell surface. This processed antigen then triggers a protective immune response. Even more success has been achieved in animals with DNA nose drops that stimulate mucosal immunity. The technique can be refined even further since the viral DNA can be shot directly into cells when coated onto tiny gold beads fired by a "gene gun."

Anti-Idiotype Vaccines

Antigens provoke the appearance of immunoglobulin molecules whose binding site (**idiotype**) has a complementary structure to that of the inducing epitope. Antibodies to this idiotype have the same three-dimensional structure as the inducing epitope. In other words, the binding site on an anti-idiotypic antibody has the same shape as the antigen that induced the idiotype. Anti-idiotype antibodies can be made by taking a monoclonal antibody against the epitope in question and using it to immunize an animal. The anti-idiotypes produced can, in theory, be used in turn to vaccinate an animal. The anti-anti-idiotypes formed may be protective, directed not only against the antiidiotype but also against the original antigen. This has occurred in several experimental situations.

USE OF VACCINES

Route of Administration

Most human and animal vaccines are designed to be administered by intramuscular or subcutaneous injection. Injectable vaccines should be administered in an area that affords least risk of tissue injury. Injections, however, are relatively time-consuming and painful and bear a risk of carrying unwanted organisms into an individual. Thus, when large numbers of people or animals must be vaccinated under less than ideal conditions, other methods of vaccination are employed. The use of a high-pressure jet injector can serve as a painless, sterile method of administering antigen. Some antigens may be given orally, for example, poliomyelitis or typhoid vaccines in humans. Polio vaccine is given as a solution containing live virus, and the typhoid vaccine is administered in the form of an enteric-coated capsule containing live attenuated organisms. In the poultry industry, it is common to administer vaccines to large flocks by incorporating the vaccine either in the feed or in the drinking water. An alternative method of mass vaccination used in veterinary medicine is to expose animals to a vaccine aerosol so that they will inhale the antigen. This technique is particularly useful for vaccines directed against diseases of the respiratory tract, since it provokes the production of local antibodies.

Adjuvants

Under some circumstances, as in vaccination, it may be desirable to enhance the normal immune response by administering an **adjuvant** with the antigen. Adjuvants enhance the body's immune responses to the antigen. A large variety of compounds have been employed as adjuvants. Many adjuvants act by slowing the release of antigen into the body, but in many cases their mode of action is unknown (Table 24–6). The immune system, being antigen driven, responds to the presence of antigen and terminates that response once antigen is eliminated. It is possible to slow the rate of antigen release into the tissues by first mixing antigen with an insoluble adjuvant. Injected into an animal, this forms a focus, or "depot." Examples of depot-forming adjuvants include insoluble aluminum salts, such as aluminum hydroxide, aluminum phosphate, and aluminum potassium sulfate (alum). When antigen is mixed with one of these salts and injected into an animal, a macrophage-rich **granuloma** forms in the tissues. The antigen within this granuloma slowly leaks into the body and so provides a prolonged antigenic stimulus. Antigens that normally persist for only a few days may be retained in the body for several weeks by means of this technique. These depot adjuvants influence only the primary immune response and have little effect on secondary immune responses.

An alternative method of forming a depot is to incorporate the antigen in a water-in-oil emulsion known as Freund's incomplete adjuvant. The oil stimulates a local, chronic, inflammatory response, and as a result, a thick layer of macrophages and fibroblasts, called a granuloma, forms around the site of the inoculum. The antigen is slowly leached from the aqueous phase of the emulsion. Oil-emulsion droplets may also be carried to other sites through the lymphatic system. If killed tubercle bacilli (*Mycobacterium tuberculosis*) are incorporated into the water-in-oil emulsion, the mixture, known as Freund's complete adjuvant (FCA), is extremely potent. Not only does FCA form a depot, but the tubercle bacilli contain a compound called muramyl dipeptide (MDP) (*n*-acetylmuramyl-L-alanyl-D-isoglutamine). MDP acts on macrophages to produce interleukin-1, which stimulates the cells of the immune system and so enhances immunity. Because interleukin-1 also causes side effects, including fever, muscle wastage, and depression, nonpyrogenic derivatives have now been synthesized. MDP can be covalently linked to synthetic antigens

Table 24–6 Some Common Adjuvants

Type	Adjuvant	Mode of Action
Aluminum salts	Aluminum phosphate	Slow-release antigen depot
	Aluminum hydroxide	
	Alum	
Water-in-oil emulsions	Freund's incomplete adjuvant	Slow-release antigen depot
Bacterial fractions	Anaerobic corynebacteria	Macrophage stimulator
	BCG	Macrophage stimulator
	Muramyl dipeptide	Macrophage stimulator
	Bordetella pertussis	Lymphocyte stimulator
	Lipopolysaccharide	Macrophage stimulator
Surface-active agents	Saponin	Stimulates antigen processing
	Lysolecithin	Stimulates antigen processing
Complex carbohydrates	Acemannan	Macrophage stimulator
	Glucans	Macrophage stimulator
	Dextran sulfate	Macrophage stimulator
Mixed adjuvants	Freund's complete adjuvant	Water-in-oil emulsion plus mycobacterium

to create potent, synthetic, chemically defined antigens.

Freund's complete adjuvant works best when given subcutaneously or intradermally and when the antigen dose is relatively low. It acts specifically to stimulate T-lymphocyte function and therefore enhances responses only to thymus-dependent antigens (Chapter 14). FCA promotes IgG production over IgM. It inhibits tolerance induction, favors delayed hypersensitivity reactions, accelerates graft rejection, and promotes resistance to tumors. FCA is required to induce some experimental autoimmune diseases, such as experimental allergic encephalitis and thyroiditis (Chapter 31). It also stimulates macrophages, promoting their phagocytic and cytotoxic activities.

Other bacterial products besides MDP possess adjuvant activity. For example, endotoxins enhance antibody formation if given about the same time as the antigen. They have no effect on delayed hypersensitivity, but they can break tolerance, and they have a general immunostimulatory activity, which is reflected in a nonspecific resistance to bacterial infections. Endotoxins act by stimulating macrophage production of interleukin-1. A less toxic derivative of endotoxin with excellent adjuvant properties is monophosphoryl lipid A, which, it has been suggested, acts by suppressing suppressor-T-cell activity. Similar in their effect are anaerobic corynebacteria, especially *Propionibacterium acnes* and *B. pertussis.*

Micelles can be constructed using protein antigens and a matrix of Quil A (a saponin extract derived from the bark of a South American tree, *Quillaja saponaria*). These immune-stimulating complexes (ISCOMS) appear to be highly effective adjuvants with few observed side effects.

In human vaccines, the only widely employed adjuvants are aluminum salts, especially aluminum phosphate. In domestic animals, a wider variety of adjuvants are employed. The major restrictions on adjuvants are that they should not be toxic, and in the case of food animals, they should not adversely affect the quality of the meat. Saponin, used in anthrax vaccine, is an exception to this rule, since it is required to cause local tissue damage and anaerobic conditions so that the anthrax spores in the vaccine may germinate.

Vaccination Schedules

A simple immunization schedule can be followed as children grow up (Fig. 24–8). It may be modified as appropriate if vaccination is begun at a later age than normal. In addition, certain legal requirements, such as the requirement that children be vaccinated against measles before admission to school, influence the timing of vaccination schedules.

Since newborn animals are passively protected by maternal antibodies, it is sometimes difficult to suc-

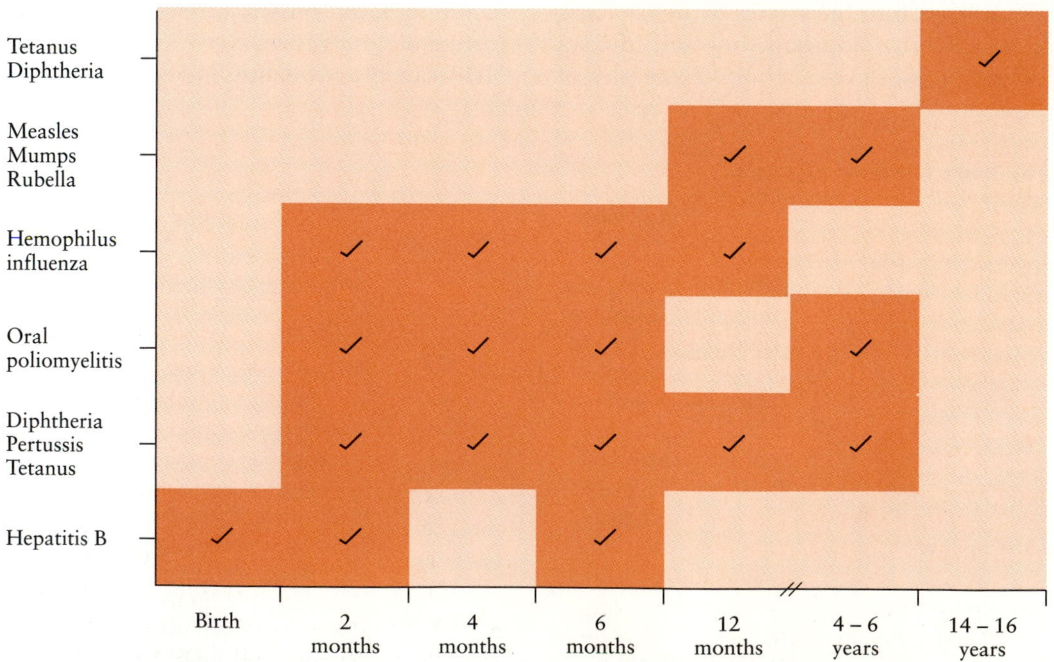

Figure 24–8 An example of a routine immunization schedule for normal infants and children.

Table 24–7 Active Immunizing Agents (Vaccines) Used Against Bacterial Disease in Humans

Disease	Antigen Used	Adjuvant
Anthrax	Purified proteins	None
Tuberculosis	Modified live (BCG)	None
Cholera	Killed organisms	None
Diphtheria	Toxoid	Aluminum phosphate
Hemophilus	Capsular polysaccharide	None
Meningococcus	Capsular polysaccharide	None
Whooping cough	Killed organisms or purified proteins	Aluminum phosphate
Plague	Killed organisms	None
Pneumococcus	Capsular polysaccharide	None
Tetanus	Toxoid	Aluminum phosphate
Typhoid	Killed or modified live organisms	None

cessfully vaccinate them in early life. Thus, if measles vaccine is given to children before 12 months of age, a second dose must be given at about 15 months to ensure adequate protection. In dogs and cats vaccines given before 10 weeks of age must be repeated at about 15 weeks; in cattle, sheep, pigs, and horses any vaccine given before six months must be repeated. In domestic animals, if protection is required early in life, it may be possible to vaccinate the pregnant mother so that peak antibody levels are achieved at the time of colostrum formation.

The interval between booster doses of vaccine varies, but **inactivated vaccines** generally produce a weak immunity that requires frequent boosters, perhaps as often as every six months (as in the case of cholera vaccine). On the other hand, live vaccines produce a much more persistent immunity. For example, BCG vaccine does not require boosting, and yellow fever vaccine requires a booster dose only every ten years (Tables 24–7 and 24–8).

Figure 24–8 Active Immunizing Agents Used Against Viral Disease in Humans

Disease	Antigen Used	Adjuvant
Measles	Modified live	None
Mumps	Modified live	None
Rubella	Modified live	None
Poliomyelitis	Inactivated or modified live	None
Influenza	Inactivated	None
Hepatitis B	Recombinant antigen	None
Rabies	Inactivated	None
Yellow fever	Modified live	None

Failures in Vaccination

The immune response, like other biological phenomena, never confers absolute protection and is never equal in all members of a vaccinated population. Since the immune response is influenced by many factors, the range of responses in a large random population tends to follow a normal distribution (Figs. 24–9 and 24–10). This means that although most individuals respond to a vaccine by mounting an average immune response, a small proportion mount a poor immune response. This group may not be protected against disease in spite of vaccination. It is therefore highly improbable that 100% of a large random population is protected by vaccination. The size of the unreactive portion of the population varies with the type of vaccine employed and the number of booster doses given. Its significance also varies. Thus, for any individual, the lack of protection is serious. However, from a public health viewpoint, less than 100% protection may be quite satisfactory if it prevents the spread of disease within the population. This phenomenon is known as **herd immunity;** its effect is due to the reduced probability of a susceptible individual encountering an infected one and so contracting disease.

A second type of apparent vaccine failure can occur when the immune response is depressed. As mentioned earlier, this effect may be due to clinical immunosuppression, or it may be due to a high-stress situation. Extremes of cold and heat or malnourishment may inhibit the immune response (Chapter 26).

Other reasons for vaccine failure include the possibility that the individual was incubating the disease prior to vaccination and the excessive use of alcohol while swabbing the skin, a procedure that may inactivate a virus vaccine.

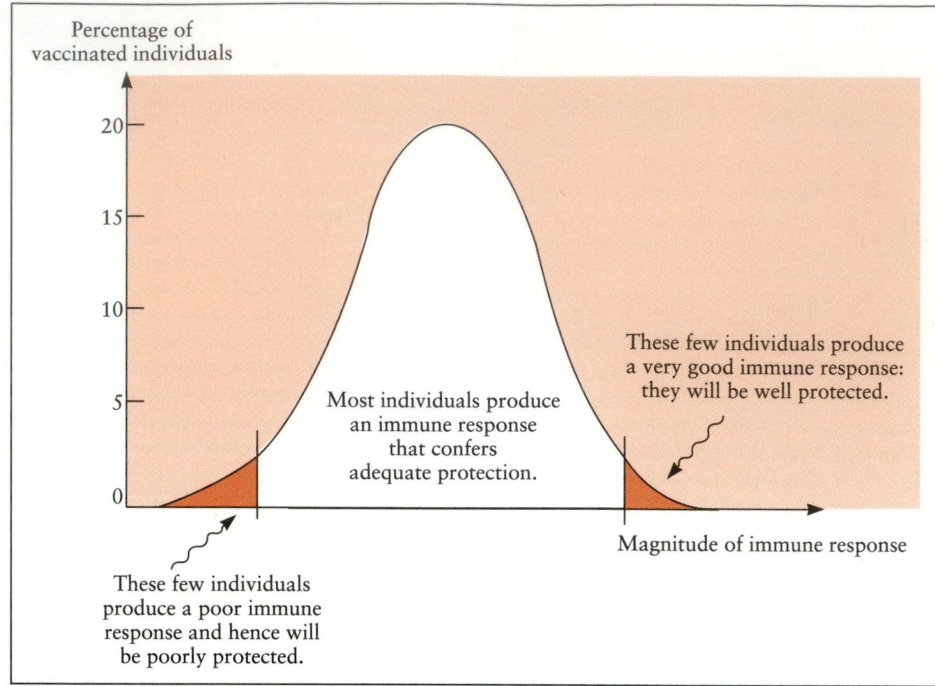

Figure 24–9 The normal distribution of immune response in a population of vaccinated individuals.

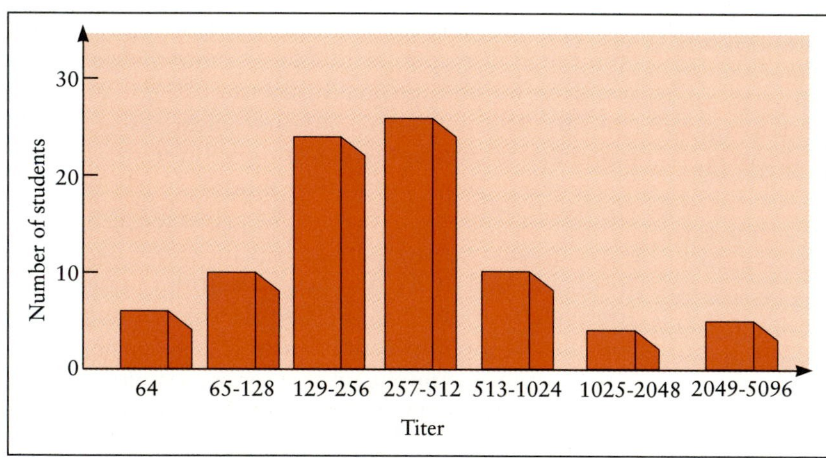

Figure 24–10 The distribution of antibody titers to rabies in 120 recently vaccinated students. Note that a few students have produced a poor antibody response and are poorly protected. This is an inevitable result of a biological response.

Adverse Consequences of Vaccination

Residual virulence, toxicity, and allergic reactions are the three most important groups of problems associated with the use of vaccines (Fig. 24–11). The most important contraindications to vaccination involve the administration of live vaccines to immunodeficient patients. Thus it is critical that no live vaccine be given to patients with immunodeficiency diseases (such as AIDS), patients with malignancies that adversely affect immunity (such as leukemias and lymphomas), and patients undergoing immunosuppressive therapy.

Although inactivated vaccines and toxoids may be given to pregnant women, live viral vaccines, especially measles and rubella, should not be given because of the potential risk to the developing fetus. On the other hand, under conditions in which the risk of disease is high, as in yellow fever or poliomyelitis, the theoretical risk may be less important and the vaccine given to pregnant women.

Vaccines containing whole killed gram-negative bacteria, such as the cholera, whooping cough, and typhoid vaccines, may be intrinsically toxic owing to their ability to stimulate cytokine release. These cytokines may cause soreness, fever, malaise, and headache. Influenza vaccines also commonly cause fever, malaise, and muscle pain. In an attempt to reduce these symptoms, the influenza virus preparation can be treated with detergents that cause its disruption and reduce the prevalence of side effects. Unfortu-

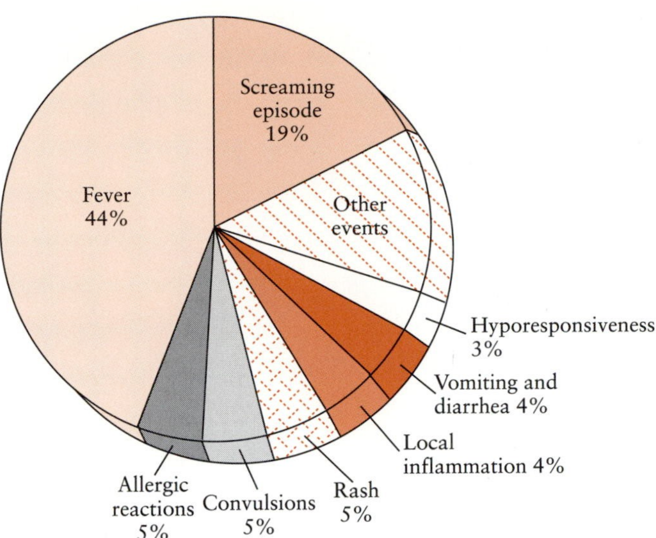

Figure 24–11 The adverse side effects of vaccination in infants and their relative prevalence. The fever is presumably due to the release of cytokines such as IL-1 and IL-6 from stimulated macrophages. Therefore, this could be considered a physiologic response.

nately, this preparation is slightly less immunogenic than the untreated vaccine. Some of the older rabies vaccines, especially those produced in duck embryos, commonly caused severe systemic toxic reactions. The newer rabies vaccines that utilize hamster kidney cells or the vaccines grown in human diploid fibroblast cultures are much less toxic.

Like any antigen, vaccine preparations have the ability to provoke hypersensitivity reactions. A type I immediate hypersensitivity can occur in response not only to the organisms but also to some of the contaminating antigens, such as egg proteins in measles, mumps, and influenza vaccines. Patients with a history of allergies to eggs should not be given these vaccines.

Some vaccines may contain trace amounts of antibiotics, for example, neomycin, and should not be given to patients with a history of allergy to this drug. As with other types of hypersensitivity, this type of reaction is more commonly associated with multiple injections of antigen and is therefore associated with the use of dead vaccines.

Under some rare circumstances, autoimmune diseases are provoked by vaccination. For example, an allergic encephalitis may be provoked by the use of vaccines such as the rabies vaccines, which contain central nervous tissue. An autoimmune polyneuritis (Guillain-Barré syndrome) has been associated with the use of certain virus vaccines (most notably swine influenza). This is a self-limiting paralytic syndrome that occurred in about ten persons per million vaccinated against swine flu in 1976. Of those affected, 5% to 10% have some residual weakness, and about 5% percent may die.

CLARIFICATION

Liability Insurance and Vaccines

Vaccines, like other biological products, are not without risk. In an era in which serious infectious diseases are of relatively minor importance in developed countries, the risks of vaccination may be perceived by many to outweigh the risks of contracting the disease. The American Academy of Pediatrics has estimated that approximately one of every 310,000 doses of diphtheria, pertussis, tetanus (DPT) vaccine results in permanent injury in sensitive individuals even when the vaccine is manufactured and administered in complete accordance with all government regulations. This vaccine must be given to all children as a condition of entering school in all 50 states. Severe reactions have also occurred as a result of polio or measles, mumps, rubella vaccination. Serious side effects caused by vaccines are no longer regarded as acceptable, and courts have found vaccine manufacturers liable for vaccine injuries even when the vaccine meets federal standards. Some of these awards have been relatively large, ranging from one million to ten million dollars. By the mid-1980s the cost of liability insurance premiums had climbed astronomically. As a result, some of the few remaining manufacturers of DPT vaccines, the required vaccine for whooping cough, had lost their liability coverage for the vaccine. Manufacturers were therefore obliged either to increase the price of the vaccine to cover the cost of health insurance or to withdraw from the market completely.

In an attempt to protect vaccine manufacturers from crippling liability while ensuring that the injured individuals were fairly compensated, The National Childhood Vaccine Injury Act was passed in 1986. It restricts claims to events directly related to vaccine usage and sets upper limits on compensation. It has not been very successful.

THE BENEFITS OF VACCINATION

Many studies have been made of the actual benefits conferred by vaccination of children. Thus, for example, in one year (1983) it has been estimated that without measles vaccination in the United States 3,325,000 cases of measles would have occurred instead of 2872 actual cases. Since measles vaccine was introduced into the United States it has been estimated that it prevented 52 million cases of measles, 5200 deaths, and 17,400 cases of mental retardation

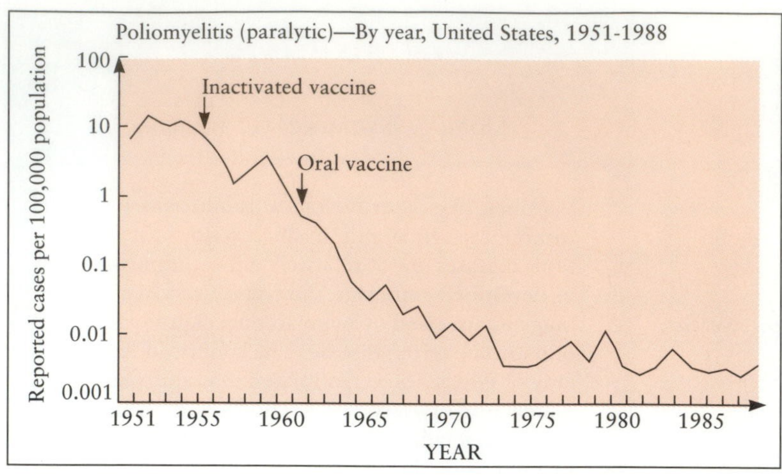

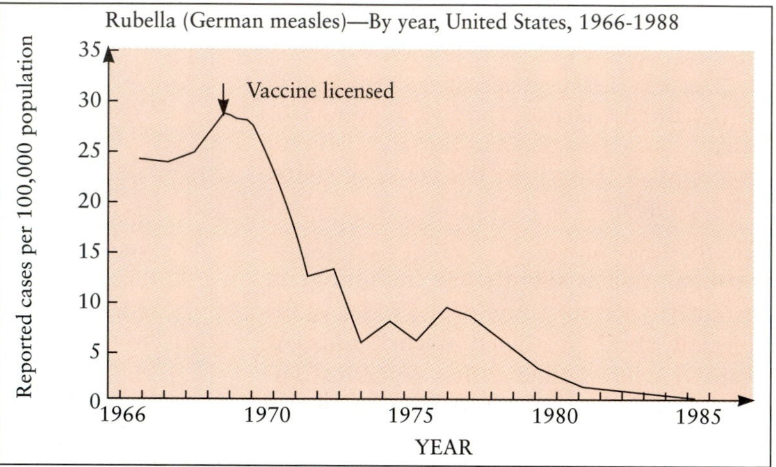

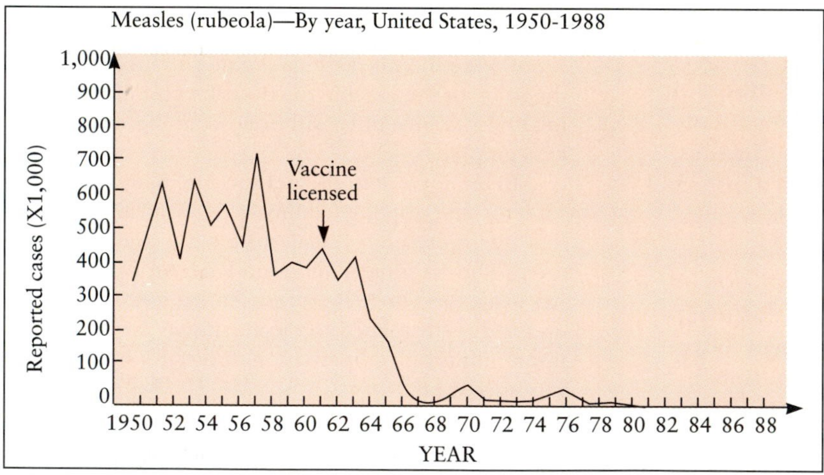

Figure 24–12 The impact of vaccination. Three graphs showing the decline in poliomyelitis, rubella, and measles in the United States following the introduction of vaccines. Note that the vertical scale on the poliomyelitis graph is logarithmic. *(From Morbidity and Mortality Weekly Reports.)*

and produced a net savings of $5.1 billion to society. The overall savings in costs to society as a result of measles, mumps, and rubella vaccination in 1983 were calculated to be $1.3 billion with a benefit-cost ratio of 14:1. There is no doubt that routine childhood vaccination confers huge benefits on society as a whole and has been largely responsible for the control of viral diseases in our society (Fig. 24–12). Few other scientific disciplines can claim to have had such an impact.

Of major significance on a global scale is the very successful Expanded Program on Immunization, organized by the World Health Organization with help from UNICEF, the World Bank, Rotary International, and developing countries themselves. This program aims to provide immunization against six major diseases to at least 80% of the children in underdeveloped countries. These diseases are tetanus, poliomyelitis, tuberculosis, diphtheria, measles, and whooping cough. Hepatitis B has recently been added to this list.

KEY WORDS

Active immunization p. 364
Adjuvant p. 369
Allergies p. 373
Anti-idiotype vaccines p. 368
Attenuation p. 364
Autoimmune disease p. 373
Booster doses p. 371
Freund's complete adjuvant p. 370
Herd immunity p. 371
Hypersensitivity p. 373
Immune globulin p. 363
Immunodeficiency p. 372
Inactivated vaccine p. 365
Modified live vaccine p. 365
Muramyl dipeptide p. 369
Naked DNA vaccines p. 368
Oral vaccines p. 369
Passive immunization p. 362
Recombinant vaccines p. 367
Serum sickness p. 363
Synthetic vaccines p. 368
Thymidine kinase p. 366
Toxoid p. 362
Vaccination p. 360
Vaccination schedule p. 370
Vaccine p. 360
Vaccine failure p. 371

QUESTIONS

1. The major disadvantage of passive immunization is that it
   a. is ineffective
   b. is expensive
   c. is liable to cause serum sickness
   d. interferes with active immunization
   e. induces short-lived immunity
2. The major disadvantage of active immunization is that it
   a. causes prolonged immunity
   b. induces rapid onset of immunity
   c. induces slow onset of immunity
   d. causes serum sickness
   e. is expensive
3. When given to humans, older forms of rabies vaccines may give rise to
   a. anaphylactoid reactions
   b. disseminated intravascular coagulation
   c. anaphylaxis
   d. serum sickness
   e. allergic encephalitis
4. Organisms suitable for use in modified live vaccines are produced by
   a. inactivation
   b. genetic recombination
   c. attenuation
   d. complement fixation
   e. hybridization
5. A suitable organism for use in recombinant vaccines is
   a. influenza virus
   b. smallpox virus
   c. *Pasteurella multocida*
   d. poliomyelitis virus
   e. vaccinia virus
6. One major vaccine component that causes allergic reactions is
   a. viral antigen
   b. bacterial antigen
   c. egg antigen
   d. tissue culture antigen
   e. endotoxin
7. Which gene is involved in one mutation that makes herpesviruses avirulent?
   a. envelope
   b. thymidine kinase
   c. nucleocapsid
   d. capsomere
   e. polymerase
8. One of the most potent adjuvants known is called
   a. alum
   b. saponin
   c. Freund's complete adjuvant
   d. liposomes
   e. endotoxin
9. One reason why vaccines fail to work in very young infants is the presence of
   a. maternal antibodies
   b. glycoproteins
   c. endotoxin
   d. adjuvant
   e. serum
10. One problem associated with the use of modified live vaccines is
    a. toxicity
    b. fevers
    c. muscle pain
    d. encephalitis
    e. residual virulence
11. What are the possible causes of a painful arm following intramuscular administration of a vaccine?
12. Why are influenza vaccines so relatively ineffective? How might you begin to improve on current influenza vaccines?
13. How would you manufacture immune globulin against rattlesnake venom?
14. Why are no vaccines available against the following: (a) the common cold; (b) gonorrhea; (c) AIDS?
15. Imagine that you discover a tropical island inhabited by a tribe that has never encountered outsiders. Outline a vaccination schedule that will protect the tribe against the diseases to which they might be exposed.
16. This society has placed severe restrictions on the development and use of vaccines containing organisms that have been modified by recombinant DNA techniques. Can these restrictions be justified?

17. If your child suffered severe side effects after vaccination, would you sue the manufacturers? On what possible grounds?
18. Why is anaphylaxis not an inevitable consequence of all vaccinations?
19. Why do we use adjuvants? What sort of adjuvants would you recommend for routine use in a vaccine to be administered to children? What side effects might you anticipate from their use?
20. Serum sickness was a significant adverse side effect of the use of tetanus antitoxin in the early years of this century. What steps were taken to reduce this problem?

Answers: 1e, 2c, 3e, 4c, 5e, 6c, 7b, 8c, 9a, 10e

SOURCES OF ADDITIONAL INFORMATION

Ada, G.L. Vaccine efficacy and the immune response. Vaccine Research, 1:17–23, 1992.

Anderson, R.M., and May, R.M. Vaccination and herd immunity to infectious diseases. Nature, 318:323–329, 1985.

Arnon, R., and Van Regenmortel, M.H.V. Structural basis of antigenic specificity and design of new vaccines. FASEB J., 6: 3265–3274, 1992.

Clancy, R.L., et al. Specific immune response in the respiratory tract after administration of an oral polyvalent bacterial vaccine. Infect. Immunol., 39:491–496, 1983.

Dick, G. Immunization. Update Books, London, N.J., 1978.

Fields, B.N., and Chanock, R.M. What biotechnology has to offer vaccine development. Rev. Infect. Dis., 11:s519–s523, 1989.

Gardner, P., and Schaffner, W. Immunization of adults. N. Engl. J. Med., 328:1252–1258, 1993.

Gershon, A.A. Immunization practices in children. Hosp. Pract. [Off.], 25:91–107, 1990.

Hennessen, V., and Huygelen, C., eds. Immunization benefits versus risk factors. Dev. Biol. Stand., 43:1–476, 1979.

Hilleman, M.R., et al. Polyvalent pneumococcal polysaccharide vaccines. Bull. World Health Organ., 56:371–375, 1978.

Horstmann, D.M. Control of poliomyelitis: A continuing paradox. J. Infect. Dis., 146:540–549, 1982.

Lerner, R.A. Synthetic vaccines. Sci. Am., 248:66–74, 1983.

Liew, F.Y. New aspects of vaccine development. Clin. Exp. Immunol., 62:225–241, 1985.

Mitchison, N.A. Rational design of vaccines. Nature, 308: 112–113, 1984.

Moss, B. Vaccinia virus expression vector: A new tool for immunologists. Immunol. Today, 6:243–245, 1985.

Peter, G. Childhood immunizations. N. Engl. J. Med., 327: 1794–1800, 1992.

Tang, D., DeVit, M., and Johnston, S.A. Genetic immunization is a simple method for eliciting an immune response. Nature, 356:152–154, 1992.

White, C.C., Koplan, J.P., and Orenstein, W.A. Benefits, risks and costs of immunization for measles, mumps and rubella. Am. J. Public Health, 75:739–744, 1985.

CHAPTER 25

Immunity to Infection

Louis Pasteur *Louis Pasteur, probably the most important figure in the history of immunology, was a chemist by training. As a result of his studies on fermentation, Pasteur firmly established that bacteria were the source of fermentative activity and it was he who finally disproved ideas of spontaneous generation of life forms. His studies on bacteria and bacterial disease led to methods of diagnosing these diseases by isolating their bacterial causes. Pasteur was a very practical man and his interest in preventing disease led to the discovery and systematic development of vaccines against fowl cholera, anthrax, and rabies. These key discoveries led to the establishment of immunology as a discipline.*

CHAPTER OUTLINE

CHAPTER CONCEPTS

1. Nonimmunological factors that influence resistance to infectious disease include the nutritional or hormonal status of an animal as well as its genetic background.
2. Immunity to bacteria may involve antibodies, complement, or activated macrophages. Bacteria may be opsonized, lysed by complement, or destroyed by activated macrophages.
3. Immunity to viruses may involve neutralization by antibodies and complement or destruction of virus-infected cells by cytotoxic T cells, by ADCC, or by antibodies and complement.
4. Immunity to protozoa involves antibodies and complement if they are circulating in the bloodstream and cytotoxic T cells and activated macrophages if they reside within cells.
5. Immunity to parasitic helminths involves the production of IgE antibodies, mast-cell degranulation, and the cytotoxic effects of eosinophils that migrate to sites of worm invasion. T-cell–mediated responses may also be involved.
6. Many organisms induce hypersensitivity reactions that can result in tissue damage and increased severity of disease.
7. Many organisms are able to evade the immune responses and so survive within infected animals. This is especially important for viruses and other obligate parasites.

The prime function of the immune system is to defend the body against invasion by infectious agents. In this chapter this enormous subject is briefly reviewed. Following an initial discussion of the nonimmunological defenses of the body, immunity to bacteria, viruses, protozoa, and helminths is described in that order. For each group of microorganisms, we review nonimmunological defenses followed by the major mechanisms by which the immunological defenses destroy (or at least hinder) invading organisms. In the case of bacterial and viral diseases, the immune responses can significantly alter the nature of disease caused and this is also discussed. Microorganisms do not, of course, allow themselves to be destroyed without fighting back. Thus each section also includes a brief review of the strategies used by organisms to evade the immunological defenses.

As pointed out at the beginning of this book, the body must defend itself from two major groups of invaders. The first group is the pathogens that invade from outside the body; these include the bacteria, viruses, fungi, protozoa, and helminths. The second group is the abnormal cells that arise within the body; these include virus-infected, chemically modified, and cancer cells. This provides a basis for the division of the immune system into its antibody-mediated and cell-mediated branches. Thus antibodies are primarily involved in immunity to extracellular infectious agents, and the cell-mediated immune responses are directed against cell-associated or intracellular agents. This is seen in immunodeficient individuals. Those who are deficient in antibodies tend to suffer from severe bacterial diseases. Those who are deficient in T cells and thus fail to mount a cell-mediated response suffer primarily from severe recurrent viral diseases.

FACTORS INFLUENCING RESISTANCE

The specific cell- and antibody-mediated immune responses are only a portion of the many defenses available to the animal. Other nonimmunological forms of protection against infection can also be identified. For example, resistance to disease can occur as a result of species insusceptibility. Thus, *Neisseria gonorrhoeae,* grows well only in humans. *Corynebacterium diphtheriae,* the cause of diphtheria, causes disease in humans and guinea pigs but not in rats; and *Brucella abortus,* an important mammalian pathogen, cannot infect chickens.

The most important of the nonimmunological factors that influence disease resistance are genetic. Under natural conditions, disease is but one of the selective pressures to act on a population. The spread of a disease through a population eliminates susceptible individuals but leaves a resistant residue to multiply and make use of the newly available resources, such as food. This was seen historically when diseases such as smallpox were introduced into susceptible native populations or when Europeans first encountered tropical diseases. The genes that influence resistance in this way are complex. Some of the most important are the MHC genes and the genes that influence macrophage activity (page 339).

A second factor that influences disease resistance is age. The very young and the very old are especially vulnerable to infection. For example, rubella infections result in severe damage to the human fetus during the first three months of pregnancy but scarcely affect the mother. Likewise, the protozoan parasite *Toxoplasma gondii* causes severe damage to the fetus, although it usually produces only a mild infection in adults. Children under six are less able than adults to localize bacterial infections. As a result, children used to develop chronic streptococcal infections, especially osteomyelitis (bone marrow infection), before the advent of antibiotics, whereas adults rarely did so. Age-related resistance is, in many cases, under hormonal influence. The resistance of the vagina to infection is controlled by hormones in such a way that vaginal infections are more common before puberty and after menopause.

Nutrition clearly influences disease resistance. Severe malnutrition, specifically protein deficiency, may impair immunoglobulin production. The stressful effect of surgery may also be magnified by the increase in protein catabolism, which occurs following surgical intervention and results in a temporary immunosuppression. It is of interest to note, however, that mild protein-calorie deficiency may enhance immunity.

IMMUNITY TO BACTERIA

Nonspecific Immunity

Convincing evidence for the ability of animal tissues to discourage bacterial invasion is readily available when we consider how infrequent bacterial infections are as a result of minor skin wounds. Some of this resistance is due to the presence of antibacterial factors in tissues (Table 25–1). The most important of these is **lysozyme,** an enzyme first recognized by Alexander Fleming, the discoverer of penicillin. Lysozyme is found in tissues and in all body fluids with the exception of cerebrospinal fluid, sweat, and urine. It is found in high concentrations in tears and egg white. Lysozyme splits the cell wall proteoglycans of some gram-positive bacteria, so killing them (see Fig. 2–5). It can also kill some gram-negative organisms in conjunction with complement (Fig. 25–1). Although

Table 25–1 Some Nonimmunological Protective Factors Found in Body Tissues and Fluids

Group	Name	Major Sources	Activity Against
Enzymes	Lysozyme	Serum; leukocytes	Gram-positive and gram-negative bacteria; some viruses
Basic peptides and proteins	β-lysin	Platelets	Gram-positive bacteria
	Phagocytin	Neutrophils	
	Leukin	Neutrophils	
	Plakin	Platelets	
Iron-binding proteins	Transferrin	Serum	Gram-positive and gram-negative bacteria
	Lactoferrin	Leukocytes; milk	
Basic amines	Spermine; spermidine	Pancreas; kidney; prostate	Gram-positive bacteria
Complement components	—	Serum	Bacteria; viruses; protozoa
Peroxide-generating mechanisms	Myeloperoxidase; xanthine oxidase	Neutrophils; milk	Bacteria; viruses; protozoa
Interferons	TFN-α-β-γ, and -ω	Most cells but not neutrophils	Viruses; some intracellular protozoa

many of the bacteria killed by lysozyme are nonpathogenic, it might reasonably be pointed out that this susceptibility could account for their lack of pathogenicity. Lysozyme is found in high concentrations in the lysosomes of neutrophils, and so it accumulates in areas of acute inflammation, including sites of bacterial invasion. The pH optimum for lysozyme activity, although somewhat low (pH3 to pH6), is easily achieved in inflamed areas and within phagosomes; as a result, it is there that it exerts its antibacterial activity. Lysozyme can also act as an opsonin, facilitating phagocytosis in the absence of specific antibodies and under conditions in which its enzymatic activity may be ineffective.

Free fatty acids also inhibit bacterial growth. In general, unsaturated fatty acids, such as oleic acid, tend to be bactericidal for gram-positive organisms, and saturated fatty acids are fungicidal. Because of this, scalp ringworm of children, which is somewhat difficult to treat, may resolve spontaneously at puberty when the amount of fatty acids in sebum increases.

Several antibacterial peptides and proteins rich in basic amino acids (lysine and arginine) have been isolated. They are derived from proteins digested by enzymes from neutrophils or platelets. β Lysin, a polypeptide active against *Bacillus anthracis* and the clostridia, is released from platelets as a result of their interaction with immune complexes.

One of the most important factors that influence the success of bacterial invasion is the level of iron in body fluids. Many bacteria such as *Staphylococcus aureus, Escherichia coli, Pasteurella multocida,* and *Mycobacterium tuberculosis* require iron for growth. Within the body, however, iron is associated with the iron-binding proteins transferrin, lactoferrin, haptoglobin, and ferritin. In addition, serum iron levels drop after bacterial invasion as a result of a reduction in intestinal iron absorption and increased production of haptoglobin

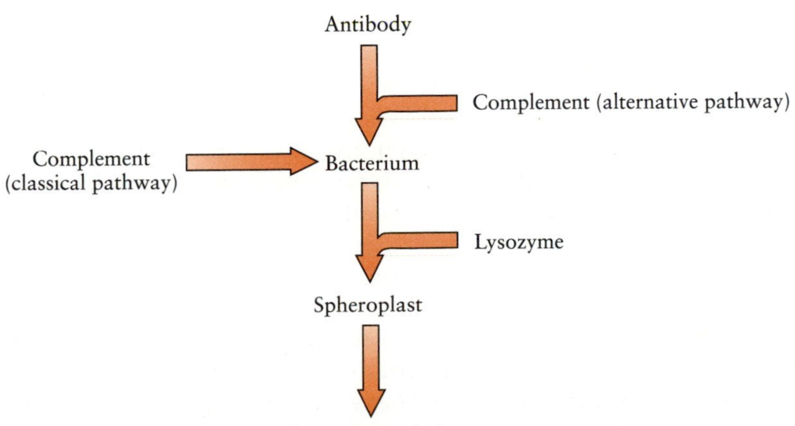

Figure 25–1 The interaction between complement and lysozyme that causes lysis of gram-negative bacteria. Although it can make an initial lesion in the bacterial cell wall, the membrane attack complex of complement is unable to penetrate the cell membrane. Serum lysozyme, however, can complete the lytic process.

and transferrin by the liver under the influence of interleukin-1. This effectively hinders bacterial invasion. A similar situation occurs in the mammary gland when, in response to bacterial invasion, neutrophils release their stores of lactoferrin and so enhance the bactericidal power of milk. In spite of the binding of iron, some bacteria, such as *M. tuberculosis* and *E. coli,* can invade the body because they produce potent iron-binding proteins called siderophores (mycobactin and enterochelin, respectively). These bacterial siderophores have such a high affinity for iron that they can withdraw it from the serum proteins and make it available to the bacteria. In diseases in which serum iron levels are elevated, such as the hemolytic anemias, individuals may be susceptible to overwhelming bacterial infection. Peroxide-generating mechanisms, which are important in neutrophils and in milk, have been discussed elsewhere (Chapters 6 and 23).

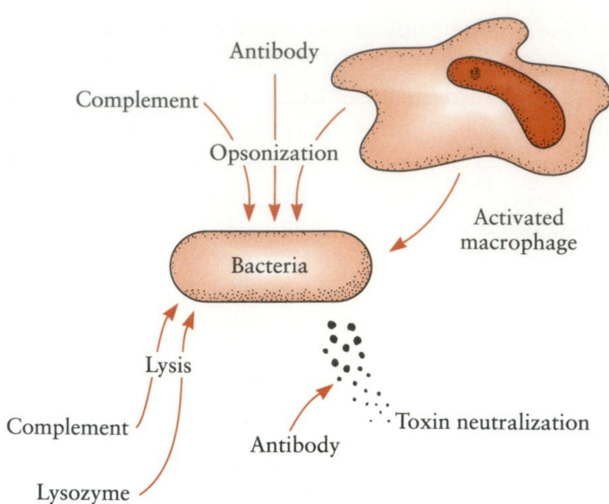

Figure 25–2 The several ways in which the immune system can protect the body against bacterial invasion. These include opsonization, lysis, toxin neutralization, and destruction within activated macrophages.

Specific Immunity

There are four basic mechanisms by which the specific immune responses combat bacterial infections (Fig. 25–2). These are the neutralization of toxins or enzymes by antibody; the killing of bacteria by antibodies, complement, and lysozyme; the opsonization and phagocytosis of bacteria by neutrophils and normal macrophages; and the phagocytosis and destruction of intracellular bacteria by activated macrophages. The relative importance of each of these processes depends on the organisms involved and the mechanisms by which they cause disease.

Immunity to Exotoxigenic Bacteria

In diseases caused by bacteria that produce **exotoxins** such as the clostridia or *B. anthracis,* the immune system must not only eliminate invading organisms, but also neutralize their toxins. Unfortunately, destruction of these bacteria may be difficult, especially if they are embedded in a mass of dead tissue. Nevertheless, antibodies can readily neutralize bacterial exotoxins by blocking the combination of the toxin with its receptor on the host cell. IgG is almost totally responsible for toxin neutralization since IgM molecules do not usually bind strongly enough. Thus, if IgM binds to the toxin, it is easily dissociated and the toxin will remain active. The neutralization process therefore involves competition between receptor and antibody for the toxin molecule. It is apparent that once the toxin has combined with the receptor, antibody will be relatively ineffective in reversing this combination. This is seen in practice when the dose of immune globulin required to produce clinical improvement in tetanus is much greater than that required to prevent the development of disease.

Immunity to Systemically Invasive Bacteria

Protection against invasive bacteria is mediated by antibodies directed against their surface antigens. Antibody directed against capsular antigens may neutralize the antiphagocytic properties of the capsule, so opsonizing the bacteria and permitting their destruction by phagocytic cells. In organisms lacking capsules, antibodies directed against cell wall antigens may serve a similar function. The fate of many bacteria depends on the immunoglobulin class produced and whether complement is activated. Thus IgG is a more effective opsonin than IgM in the absence of complement. IgM, however, is more effective at activating complement than IgG and thus is the most effective opsonin in the presence of complement. On a molar basis, IgM is about 500 to 1000 times more efficient than IgG in opsonization in the presence of complement and about 100 times more potent than IgG in sensitizing bacteria for complement-mediated lysis. Therefore, during a primary immune response, the low level of the IgM response is compensated for by its quality, so ensuring early and efficient protection.

Although some bacteria are phagocytosed by neutrophils or macrophages, others may be killed while they are free in the circulation. The bactericidal activity of serum is mediated by antibodies, complement, and lysozyme. Together, antibodies and complement can cause the development of membrane attack complexes (MACs) in the outer cell wall similar to those seen by electron microscopy on complement-lysed red blood cells. By themselves, however, the MACs may be insufficient to kill many gram-negative bacteria since they are unable to span the outer membrane and disrupt the functions of the bacterial inner membrane.

The inner cell membrane must also be disrupted. Lysozyme can act on the exposed inner cell membrane following complement activity and so kill bacteria. Gram-negative organisms may activate the alternate complement pathway by inhibiting factor-H activity; alternatively, they may directly activate the classical pathway through the interaction between lipid A and C1q or through mannose-binding protein. Complement may therefore act in the absence of antibody.

The Heat Shock Protein Response

Up to 25 new proteins are produced by stressed cells. These stresses include heat, starvation, exposure to oxygen radicals, toxins such as heavy metals, protein synthesis inhibitors, and viral infections. The heat shock proteins (HSPs) are the best understood of these new proteins. HSPs are present in all organisms at very low levels at normal temperatures. Mild heat treatment such as a low-grade fever induces HSPs in cells and increases their levels significantly. For example, they climb from 1.5% to 15% of the total protein in stressed *E. coli.* These proteins control the folding of other proteins so that the cells or organisms can function normally at a higher temperature. There are three major heat shock proteins: HSP 90, HSP 70, and HSP 60. The gene for HSP 70 is located within the MHC class III region in mammals.

When a bacterium is phagocytosed and exposed to the respiratory burst within a neutrophil, large quantities of antigenic HSP 60 are generated by the stressed organism. HSP 60 is the immunodominant antigen induced by *Mycobacterium, Coxiella burnetii, Legionella, Treponema,* and *Borrelia* infections. Antibodies against HSP 70 have been detected in patients with malaria, schistosomiasis, trypanosomiasis, filariasis, and leishmaniasis. These HSPs are highly antigenic for several reasons: (1) they are produced in abundance, (2) they are readily processed by antigen-presenting cells, and (3) there may be unusually large numbers of lymphocytes capable of responding to HSP. It is suggested that an anti-HSP response may be a major defense against pathogens.

Modification of Bacterial Disease by the Immune Response

The development of an antibacterial immune response clearly influences the course of an infection. At its best this will result in a cure. In the absence of a cure, however, the disease may be profoundly modified. This may depend on whether a cell-mediated or antibody response is generated. Thus the type of helper-cell subpopulation that develops during an immune response profoundly affects the course of disease. As described in Chapter 18, cell-mediated responses are usually required to control diseases caused by intracellular bacteria. Only activated macrophages can prevent the growth of these organisms. Macrophage activation requires that Th1 cells produce IL-2 and IFN-γ. When the macrophages are activated, they can localize or cure these infections. If, in contrast, the immune response against these organisms only stimulates Th2 cells, then cell-mediated immunity fails to develop, macrophages are not activated, and chronic progressive disease results. An example of this is seen in leprosy. Leprosy occurs in two distinct clinical forms, tuberculoid and lepromatous (Fig. 25–3). Tuberculoid leprosy, found in about 80% of cases, is associated with resistance to the disease as a result of strong cell-mediated immune responses to the leprosy bacillus. Thus patients with tuberculoid leprosy have few organisms in their tissues and therefore few lesions. The lesions are infiltrated

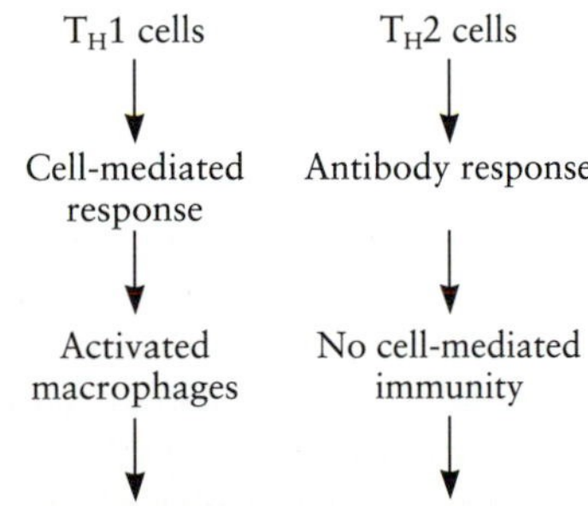

	Tuberculoid	Lepromatous
Resistance	High	Low
Presence of *M. leprae* in tissues	±	+ + + +
Lymphocyte infiltration in lesions	+ + +	–
Antibodies to *M. leprae*	≈20%	95%
Immune-complex lesions	–	+ + + +
Autoantibodies	≈10%	≈40%
Lepromin test	+ + +	–

Figure 25–3 The type of helper-cell response induced by *Mycobacterium leprae* has a very profound effect on the course of infection with this organism.

with macrophages, and the patients give a positive **delayed hypersensitivity** reaction to intradermally injected bacterial extract (called lepromin). Most of these patients do not have antibodies to *Mycobacterium leprae.*

In contrast, lepromatous leprosy, found in about 20% of cases, is associated with low resistance to the disease. Patients with lepromatous leprosy do not mount a detectable cell-mediated response but instead make high levels of antibodies against *M. leprae.* They may also develop immune complex lesions (type III hypersensitivity) and autoantibodies to normal tissues. The lesions in these patients are densely infiltrated with enormous numbers of leprosy bacilli and contain very few macrophages. The prognosis for lepromatous leprosy is much poorer than for tuberculoid leprosy.

Recent studies have clarified the mechanisms involved in the development of lepromatous leprosy. These patients mount an immune response preferentially utilizing Th2 cells. As a result, large amounts of IL-4 and IL-10 are produced. The IL-10 has a direct inhibitory effect on macrophage cytokine synthesis. It reduces the production of IL-12, TNF-α, and GM-CSF. The reduction in IL-12 production results in a decrease in Th1-derived IFN-γ. Since IFN-γ is necessary for macrophage activation, the reduction in its production directly reduces an individual's ability to control an intracellular organism such as *M. leprae.* In contrast, patients who develop tuberculoid leprosy mount a T-cell response dominated by Th1 cells. Their lesions contain IL-2 and IFN-γ as well as the macrophage-derived molecules IL-1, TNF-α, and IL-6. Presumably, these cytokines cause significant macrophage activation.

Similar examples of increased susceptibility to disease are seen in *Leishmania, Listeria, Trichinella,* and *Schistosoma* infections as a result of inappropriate Th2 responses and the immunosuppressive effects of IL-10.

METHODOLOGY

How to Detect Interferon

The most usual assays for interferon involve measurement of its antiviral effects. As can be imagined, many in vitro assay systems have been developed for this purpose. In general, for human interferon assays, a standard virus preparation, bovine vesicular stomatitis virus (VSV), Indiana strain, is grown in human fibroblast cultures. Normally, a set of fibroblast cultures is established under standard conditions. Samples containing interferon are added to these cultures at various dilutions and incubated for 18 to 24 hours. The fibroblast monolayers are then washed, and a standard quantity of VSV is added to each culture.

The VSV is allowed to adsorb to the cells for 45 minutes. Excess virus is then removed and the monolayer overlaid with melted agar. After 48 hours of incubation, the monolayers are stained and virus plaques are seen as cleared areas in the monolayer. If interferon was present in the samples, then it reduces the number of plaques formed. A greater than 50% reduction in virus plaques is considered significant. However, to confirm that the antiviral activity is indeed interferon, it is necessary to ensure that the activity belongs to a protein by showing that it is destroyed by trypsin. Interferon should not be sedimented by exposure to 100,000 g in the ultracentrifuge. (Remember, other viruses could block plaque formation by interference.) It should block the activities of other viruses, but it should be species-specific. Its sensitivity to low pH will enable it to be classified as a particular type of interferon.

Evasion of the Immune Response by Bacteria

Bacteria, like most organisms, are not well served either by the death of their host or by their own elimination. They have therefore evolved mechanisms to evade the immune responses. Some organisms such as *E. coli, S. aureus, M. tuberculosis,* and *Pseudomonas aeruginosa* secrete molecules that can depress phagocytosis by neutrophils. *Bordetella pertussis,* for example, secretes a highly active adenyl cyclase that raises intracellular cyclic AMP levels and so suppresses many cellular activities. As a result, persons infected with *Bordetella* are susceptible to other infections. *Shigella flexneri* induces apoptosis in infected macrophages. *Pasteurella haemolytica,* a major cause of pneumonia in cattle, secretes a protein that kills alveolar macrophages. Another bacterium, *P. aeruginosa,* secretes proteases that can destroy IL-2.

Many bacteria can defend themselves against the oxygen radicals produced by phagocytic cells. They can use enzymes such as superoxide dismutase and catalase, DNA repair systems, scavenging substrates, and competition with phagocytes for molecular oxygen. For example, the pigments responsible for the yellow color of *S. aureus* can quench singlet oxygen and so enable the bacteria to survive the respiratory burst. Strains of *E. coli* and salmonellae may carry plasmids that confer resistance to complement-mediated lysis. Other protective factors secreted by pathogenic bacteria include proteolytic enzymes specific for IGA

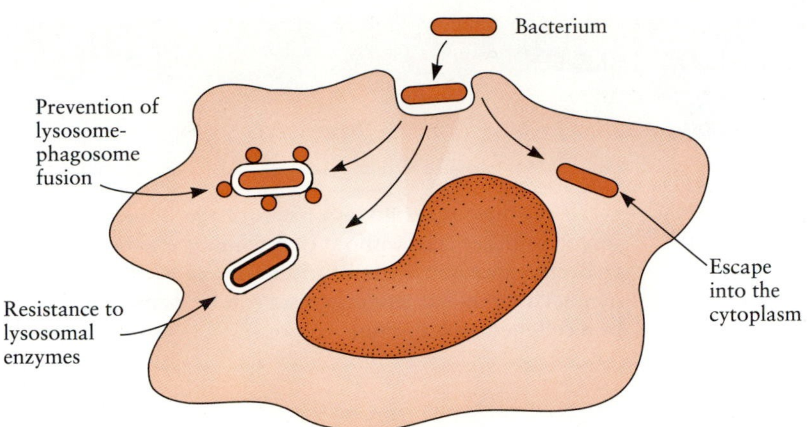

Figure 25–4 The three major ways by which bacteria and other intracellular parasites can escape destruction within macrophages.

by *N. gonorrhoeae, Streptococcus pneumoniae, Neisseria meningitidis, Haemophilus influenzae, and Streptococcus sanguis.*

Clearly, the ability to grow within cells enables an organism to avoid destruction by antibodies. In addition, intracellular bacteria such as *Listeria monocytogenes* can travel from cell to cell without exposing themselves to antibodies in the extracellular fluid. This strategy is not free of risk, however. Endocytosed organisms may be transported to acidic lysosomes and exposed to potent lysosomal enzymes. Thus these intracellular bacteria must take steps to avoid being killed by the cell. They do this in three major ways (Fig. 25–4). One way is to remain inside the phagosome but ensure that lysosomal enzymes do not enter. Thus *Chlamydia trachomatis* can inhibit both acidification and fusion between the bacteria-containing phagosome and lysosomes. This has the additional benefit that chlamydial antigens are not processed and presented on MHC class I molecules, so preventing attack by cytotoxic T cells. Other organisms that follow this intraphagosomal approach include *Legionella pneumophila* (Fig. 25–5) and *T. gondii.* Some organisms, such as *M. tuberculosis,* use a different mechanism. They can survive in the phagolysosome because they are resistant to destruction by lysosomal enzymes. The third survival technique practiced by intracellular bacteria is to escape from the phagolysosome into the host cell's cytoplasm. An organism that does this is *L. monocytogenes.*

IMMUNITY TO VIRUSES

Nonspecific Immunity

Just as nonimmunological factors influence susceptibility to bacterial disease, so they modify and control the outcome of many viral infections. Lysozyme, for example, can destroy several viruses, as can many intestinal enzymes. Bile is a powerful neutralizer of some viruses and presumably helps protect the intestine against invasion.

Interferons

Interferons are secreted by virus-infected cells within a few hours after viral invasion, and high concentrations of interferon in tissues may be achieved within a few days in vivo, when the primary immune response

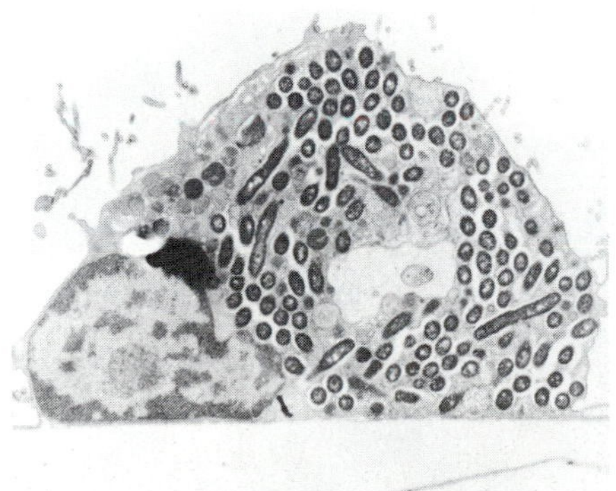

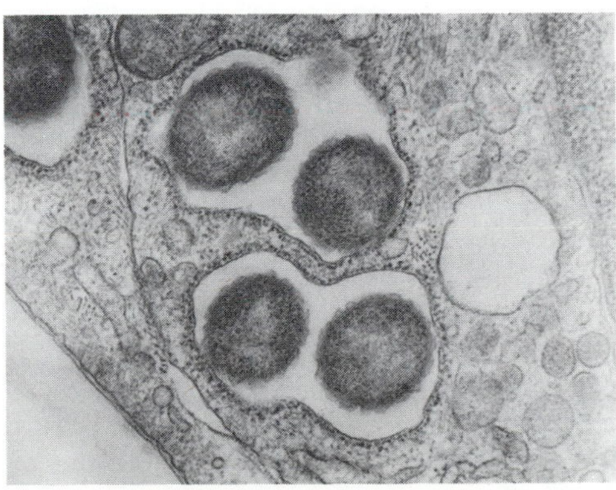

Figure 25–5 *Left,* An electron micrograph of a cultured macrophage heavily infected with *Legionella pneumophila* (original magnification × 5400). *Right, L. pneumophila* in vacuoles (phagosomes?) within the macrophage cytoplasm (original magnification ×32,400). *(Reproduced from the Journal of Clinical Investigation, 66:441, 1980. By permission of the American Society for Clinical Investigation. Courtesy of Dr. M.A. Hurwitz.)*

A CLOSER LOOK

The Curious Case of Listeria

Listeria monocytogenes, a bacterium responsible for some forms of food poisoning, exhibits some interesting mechanisms for surviving immunological attack. *Listeria* is readily phagocytosed by macrophages. Once enclosed in a phagosome, however, the *Listeria* secretes a hemolysin and a phospholipase. These enzymes break down the phagosomal membrane and enable the *Listeria* to enter the macrophage cytoplasm. Once in the cytoplasm, the *Listeria* can grow and divide. As it grows in the cytoplasm, actin filaments develop around the organism, forming a tail that extends from one pole.

When an actin-coated *Listeria* comes into contact with the plasma membrane, the organism and its actin tail form a long projection from the cell surface. When this projection comes into contact with a nearby macrophage, the second macrophage phagocytoses the projection with its contained *Listeria.* Once inside the second macrophage, the *Listeria* secretes hemolysin, and phospholipase breaks down the surrounding membranes and enters the cytoplasm.

Thus once ingested by a macrophage, *Listeria* can spread to other cells without ever coming into contact with extracellular fluid, antibodies, or complement. The only way this organism can be destroyed is by a cell-mediated immune response.

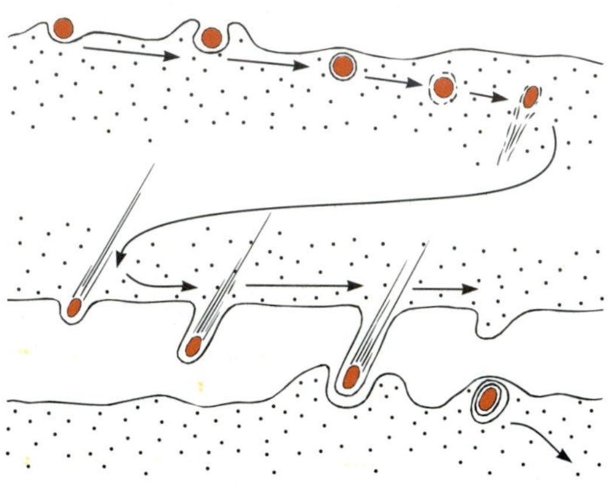

is still relatively ineffective (Fig. 25–6). For example, in individuals exposed to influenza virus, peak serum interferon levels may be reached one to two days later and then decline, although they are still detectable at seven days. In contrast, antibodies are usually not detectable in serum until five to six days after virus administration.

Interferons, as described previously, are a class of glycoproteins with molecular weights between 20 and 34 kDa. Four major classes are recognized (Table

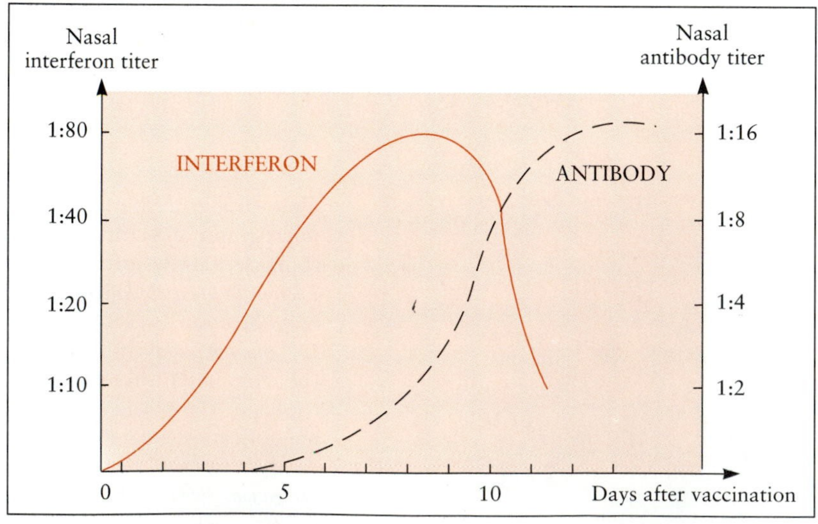

Figure 25–6 The sequential production of interferon and antibody following intranasal administration of infectious rhinotracheitis vaccine to calves. *(From data kindly provided by Dr. M. Savan.)*

Table 25–2 The Classification of Interferons

Source	Name	Target	Effect
Leukocytes	IFN-α	Virus-infected cells	Blocking of viral replication
Fibroblasts	IFN-β	Virus-infected cells	Blocking of viral replication
T, B, or NK cells	IFN-γ	Macrophages, T and NK cells	Activation and enhanced cytotoxicity
Trophoblast cells	IFN-ω	Uterine lymphocytes	Fetal survival

25–2): interferon α (IFN-α), a family of at least 20 molecules derived from virus-infected leukocytes; interferon β (IFN-β), derived from virus-infected fibroblasts; interferon γ (IFN-γ), a lymphokine derived from antigen-stimulated T cells; and interferon ω (IFN-ω), a protein secreted by the embryonic trophoblast. IFN-α, IFN-ω, and IFN-β are stable at pH 2, whereas IFN-γ is labile at a low pH.

Antiviral Activities

Interferons are produced by the association between viral nucleic acid and host-cell ribosomes, resulting in derepression of target-cell genes coding for interferon production (Fig. 25–7). Interferon is secreted by infected cells and binds to receptors on other nearby cells. IFN-α and IFN-β share the same receptor, while IFN-γ uses a different one (CDw119). Binding to the receptor results in the development of resistance to virus infection within a few minutes and peaks 5 to 8 hours later. Several mechanisms are involved in this antiviral activity.

Interferons induce tyrosine phosphorylation and activate DNA-binding proteins. These proteins activate genes and so stimulate the production of many new proteins in target cells. Some of these new proteins have antiviral activity. One of these is the enzyme 2′5′-oligoadenylate synthetase, which acts on ATP to generate 2–5A, an adenine trinucleotide; 2–5A in turn activates a latent cellular endoribonuclease (RNAase L). Once activated, RNAase L cleaves viral mRNA and so inhibits viral protein synthesis. Another protein induced by IFN-γ, especially in activated macrophages, is nitric oxide synthase. The nitric oxide produced by this enzyme has antiviral activity and so can prevent virus growth in interferon-activated macrophages. A third antiviral protein induced is a protein kinase. This protein kinase phosphorylates an initiation factor

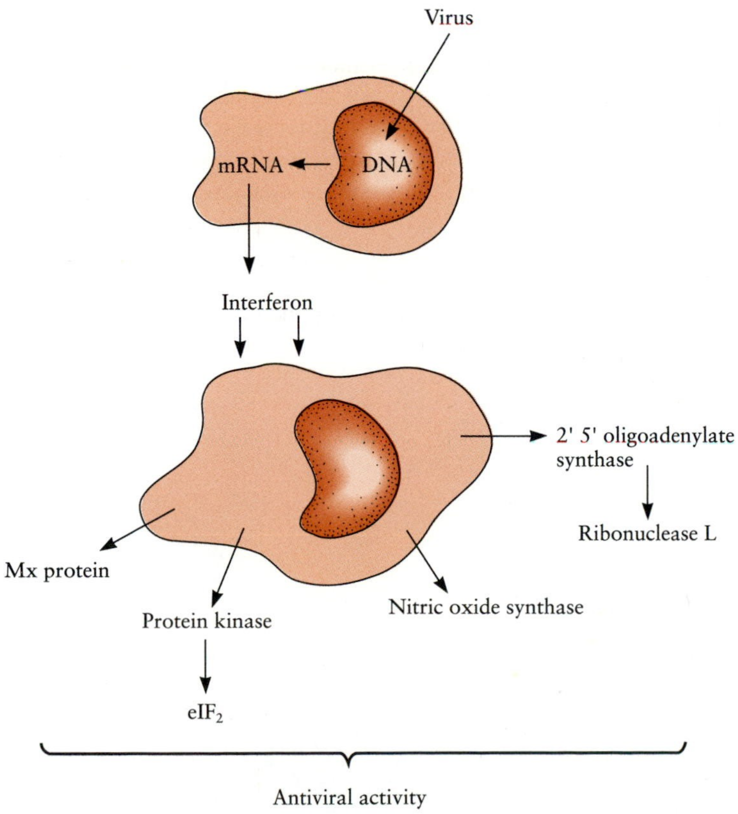

Figure 25–7 Some of the mechanisms of the antiviral effect of the interferons.

called eIF2. Phosphorylation of eIF2 causes inhibition of viral protein synthesis by preventing the elongation of viral double-stranded RNA. A fourth new protein is called Mx. Mx protein can inhibit the transcription and translation of influenza virus mRNA.

The ability of cells to produce interferon varies. Virus-infected leukocytes, especially macrophages and lymphocytes, produce IFN-α; virus-infected fibroblasts produce IFN-β; and antigen-stimulated T cells are the major source of IFN-γ (Chapter 11). Cells such as those from the kidney are relatively poor interferon producers, and neutrophils produce no interferon.

Although live or inactivated viruses are the most important stimulators of interferon production, interferons may also be produced under circumstances other than viral infection—for example, bacterial endotoxins, some plant extracts such as phytohemagglutinin, and synthetic polymers, which act by mimicking the action of viral RNA. One of the most potent of these synthetic polymers (Poly I:C) consists of inosinic and cytidilic acids.

Specific Immunity

The body employs several strategies against viruses, including both antibody- and cell-mediated immunity (Fig. 25–8). Cell-mediated immune responses, being designed to destroy intracellular parasites, are of greatest importance in antiviral immunity. Individuals who are deficient in cell-mediated immune responses commonly die from uncontrolled viral infections.

Destruction by Antibody

The capsids of viruses consist of antigenic glycoproteins, and it is against these and the viral envelope proteins that antiviral antibodies are directed. Antibodies may destroy free viruses, prevent virus infection of cells, or destroy virus-infected cells. Since viruses first must bind to receptors on cells in order to infect them, antibodies may neutralize viruses by blocking their attachment to these receptors. The efficiency of this form of virus **neutralization** varies with the virus. The T-even phages (T2, T4, and so on), which have only one critical attachment site, can be neutralized by blocking that site with a single immunoglobulin molecule. On the other hand, other viruses, such as influenza, have multiple binding sites. Neutralization of influenza therefore occurs only when all of these sites are blocked and thus requires much higher immunoglobulin levels. Small virus particles may behave as if their entire surface is critical.

The combination of antibody with virus is not lethal in itself, since splitting of virus–antibody complexes can release fully infectious virus. Nevertheless, antiviral antibodies may act as opsonins, promoting the phagocytosis of virions by macrophages; they may clump virions, so reducing the number of infectious units available for cell invasion; or they may initiate complement-mediated virus neutralization. The complement cascade may be activated by antibody bound to virus surfaces, by activation through the alternate pathway, or by direct activation of C1q by viral glycoproteins. For viruses that have a membrane-like envelope, the complement pathway may go to completion and thus cause virolysis. Other viruses, however, may be neutralized by C4a and C3b. C3b probably neutralizes the virus by blocking critical sites in the same way antibodies do.

Antibodies and complement are active, not only against free virions, but also against viral antigens expressed on the surface of infected cells. These infected cells may be destroyed by antibody-mediated opsonization or by complement-mediated lysis. Antibody

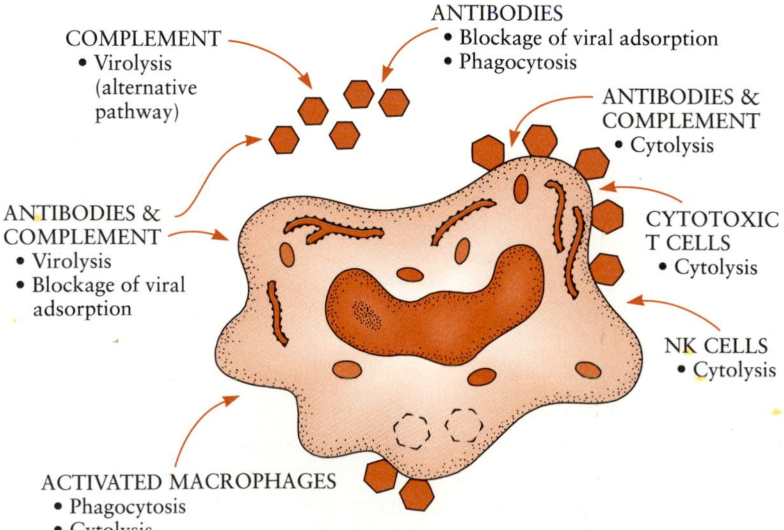

Figure 25–8 The many ways in which the immune system can defend the body against viral invasion.

binding to the surface of a virus-infected cell does not always destroy it. It may strip off viral antigens or, alternatively, suppress their production. As a result, infected cells may cease to carry viral antigen on their surfaces and become persistently infected.

Antibody-dependent cellular cytotoxicity (ADCC) is probably of major importance as a mechanism of destruction of virus-infected cells. In this process, cytotoxic cells with Fc receptors bind and lyse antibody-coated target cells. ADCC is mediated by concentrations of antibody several hundredfold less than those required to produce complement-dependent lysis of the same cells. The membrane lesions that develop on ADCC targets are similar to, but larger than, complement lesions and complement does potentiate ADCC. The cells that have been implicated in ADCC include macrophages, neutrophils, eosinophils, and NK cells.

The immunoglobulins involved in virus neutralization include IgG and IgM in serum and IgA in secretions. IgE may also play a protective role, since humans with a selective IgE deficiency show increased susceptibility to respiratory infections.

A CLOSER LOOK

Viruses Steal Genes

It is well recognized that viruses can take genes from their animal hosts and incorporate them into the viral genome. Perhaps the best recognized of these are the oncogenes (see page 82). However, some viruses, especially the poxviruses, have taken some of the genes for key surface receptors, such as cytokine receptors. Thus if a poxvirus can induce an infected cell to produce unusually large amounts of these soluble receptors, the receptors bind their cytokine ligands and effectively neutralize them. For example, a rabbit poxvirus, the myxoma virus, has a gene that codes for an IFN-γ receptor. This protein is secreted from virus-infected cells and binds IFN-γ and inhibits its activity. Another rabbit poxvirus, the Shope fibroma virus, has a gene that codes for a protein similar to one of the TNF receptors, and vaccinia and cowpox both contain a gene for a receptor for IL-1. Cowpox virus also contains a gene that codes for a protease inhibitor that prevents the proteolytic activation of IL-1 as well as a gene for a complement control protein.

Cell-Mediated Immunity

Cell-mediated immunity directed against virus-infected cells is by far the most important mechanism of resistance to viruses. The mechanisms involved in the destruction of virus-infected cells are described in Chapter 18 but can be summarized here. Virus-infected cells commonly develop new antigens on their surface. These new antigens include virus proteins exposed when viruses bud from the surface of the cell, other viral antigens expressed on the cell surface, or antigens induced by the virus on the host cells. These viral antigens are usually proteins or glycoproteins. Cell-mediated destruction of virus-infected cells is mediated through the actions of cytotoxic T cells or NK cells. The pathogenesis of some viral diseases may therefore stem from direct damage to virus-infected cells by T-cell-mediated cytotoxicity rather than from any damaging effect of the virus.

Modification of Viral Disease by the Immune Response

In some cases the immune responses directed against viruses may enhance rather than prevent the expression of disease. A good example is lymphocytic choriomeningitis (LCM). In this disease, transmission of the virus from female mice to their offspring during pregnancy causes an asymptomatic infection. The virus is not cytotoxic, and infected cells are not destroyed because of neonatal tolerance. Transmission between immunologically competent adult mice, on the other hand, leads to severe neurologic disease since cytotoxic lymphocytes attack virus-infected cells within the brain. Immunosuppressive treatment will therefore prevent the development of clinical LCM, although such treated animals will remain persistently infected.

Evasion of the Immune Response by Viruses

The relationship between host and virus must be established on the basis of mutual accommodation so that its long-term continuation is ensured. Failure to establish such a relationship results in elimination of either host or virus, an undesirable result for one or both parties. One aspect of this adaptation involves the virus's avoiding destruction by the immune system.

In contrast to the immune response against bacteria, antiviral immunity is long-lasting in many cases. This may be related to virus persistence within cells, perhaps in a slowly replicating or a nonreplicating form as typified by the herpesviruses. The persistent virus may periodically boost the immune response of the infected animal and, in this way, generate prolonged immunity. This type of infection can be eliminated only by an immune response directed against infected cells, and even this may be ineffective if the viral antigens are not presented on the cell surface. The immune responses in these cases, although not able to eliminate the virus, may prevent the development of disease and therefore be protective. This type of evasive process has been observed in measles-infected cells in humans and in LCM-infected cells in mice. Antibody against measles virus normally kills measles-infected cells. If, however, there is insufficient antibody present to be immediately lethal, then an infected cell may respond by removing the measles antigens from its surface. Once these antigens are lost, infected cells are not attacked. Removal of the measles antibody permits reexpression of measles antigen.

Immunosuppression of persistently infected animals may permit disease to occur. The association between stress and the development of some viral diseases is well recognized (for example, cold sores caused by herpes simplex virus), and it is possible that the increased ACTH production in stressful situations may be sufficiently immunosuppressive to permit activation of latent viruses or infection by exogenous ones.

Evasion of the immune response may also result from antigenic variation. This is seen among the influenza viruses. These viruses possess surface antigens of which the hemagglutinins and neuraminidases are most important. There are 13 hemagglutinins and nine neuraminidases among the type A influenza viruses, and they are identified according to a system recommended by the World Health Organization (Table 25–3). Influenza viruses in a population change their antigenic structure as mutation and selection change the amino acid sequences of their hemagglutinins and neuraminidases gradually. As a result, there is a gradual change in the antigenicity of each subtype. This **antigenic drift** permits the virus to persist in a population for many years. Influenza viruses also sporadically undergo a major change in antigenic structure (**antigenic shift**), in which a new strain develops whose hemagglutinins are very different from those of previous strains. Such a major change is due to recom-

A CLOSER LOOK

Epstein-Barr Virus and IL-10

IL-10 is normally produced by Th2 cells and has the ability to turn off macrophage production of IL-12 and thus Th1 cytokine synthesis. It is also produced by some B-cell lymphomas. IL-10 has a significant suppressive effect on cell-mediated immune responses. A segment of the Epstein-Barr virus (EBV) genome has strong sequence homology to the human IL-10 gene. As a result, EBV-infected cells can secrete a protein that has similar biological activities to human and mouse IL-10. Examination of B-cell lymphomas that produce IL-10 has shown that all are infected by EBV. If B cells that do not contain EBV are stimulated, they produce only very small amounts of IL-10, and EBV infection induces them to secrete much higher amounts. If this viral IL-10 is neutralized with a blocking anti-IL-10 antiserum, B-cell proliferation is blocked. Thus it is probable that EBV stimulates B-cell proliferation and the development of B-cell lymphomas through this IL-10 release. The benefit to the virus is a reduction in the body's cell-mediated immune defenses.

Burdin, N., et al. Epstein-Barr virus transformation induces B lymphocytes to produce human interleukin 10. J. Exp. Med., 177:295–304, 1993.

Table 25–3 Examples of Influenza A Viruses and Their Antigenic Structures

Species	Virus Strain	Antigenic Structure*
Human	A/Beijing/32/92†	H3N2
	A/Shanghai/16/89	H3N2
	A/Japan/305/57 (Asian Flu)	H2N2
	A/Bangkok/1/79	H3N2
	A/Texas/36/91	H1N1
	A/New Jersey/76 (Swine Flu)	H1N1
	A/Taiwan/1/86	H1N1
Equine	A/Equine/Prague/1/56	H7N7
	A/Equine/Miami/1/63	H3N8
Swine	A/Swine/Iowa/15/30	H1N1
Avian	A/Fowl Plague/Dutch/27	H7N7
	A/Duck/England/56	H11N7
	A/Turkey/Ontario/6118/68	H8N4

*There are many other antigens known among the human and avian influenza viruses.
†The first number is the isolate number; the second is the year of isolation.
H, hemagglutinin, N, neuraminidase

bination between two strains of virus. It is the development of these influenza viruses with a completely new antigenic structure that accounts for the periodic major epidemics of this disease.

Another form of evasion of immune responses is seen in equine infectious anemia (EIA), Aleutian disease (AD) of mink, and African swine fever. Although infected animals mount an immune response to these agents, the antibodies formed are incapable of neutralizing the virus. Virus–antibody complexes from AD-infected mink or EIA-infected horses are fully infectious. It is believed that the antibody must bind to a "noncritical" site on the virus. These viruses also show antigenic variation.

Of special interest to immunologists are the virus diseases in which profound immunosuppression occurs (Table 25–4). In many cases, the immunosuppression may be due to virus-induced destruction of lymphoid tissues. In some of these, only the primary lymphoid organs are involved. For example, mice may be infected by a herpesvirus that causes massive necrosis of the thymic cortex. This "viral thymectomy" naturally results in immunological defects. In poultry, the virus of infectious bursal disease destroys the bursa of Fabricius. The results of this infection, as might be predicted, are most evident in young birds infected immediately after hatching since they are unable to make normal levels of antibody.

Some viruses damage secondary lymphoid organs. For example, measles infection is associated with depressed delayed hypersensitivity reactions and reduced lymphocyte responses to mitogens. This immunosuppression persists for several weeks after the infected individual has recovered clinically. Its causes may include not only lymphocyte destruction and arrest of T-cell division, but also the development of suppressor cells. The most significant of these immunosuppressive viral diseases is acquired immunedeficiency syndrome (AIDS) of humans caused by human immunodeficiency virus (HIV-1) (Chapter 27).

Table 25–4 Some Viruses That Affect Lymphoid Tissues

Viruses that destroy lymphoid tissue
Herpes simplex
Dengue
Measles
Infectious bursal disease
Canine distemper
Bovine virus diarrhea
Simian immunodeficiency virus
Human immunodeficiency virus
Viruses that stimulate lymphoid tissues
Aleutian disease
Epstein-Barr virus
Visna
Viruses that cause lymphoid cancers
Marek's disease
Feline leukemia
Bovine leukemia
Murine leukemias
Human T cell leukemia virus I

A more complex interaction between lymphocytes and virus is seen in infectious mononucleosis (Fig. 25–9). The cause, Epstein-Barr virus (EBV), infects only B cells, most of which possess specific receptors for the virus (CR2, CD21). EBV activates B cells, causing them to proliferate. These abnormal, activated B cells are, however, recognized as foreign by T cells. As a result, cytotoxic T cells attack and destroy the infected B cells. It is these cytotoxic T cells that are the atypical **mononuclear cells** found in the blood of patients with infectious mononucleosis. Sometimes the T cells fail to prevent the abnormal B-cell proliferation induced by EBV. In this case, two alternative syndromes may develop: fatal infectious mononucleosis or a B cell tumor called Burkitt's lymphoma (Chapter 20). Certain genetic backgrounds predispose individuals to fatal mononucleosis, and once infection occurs the patients become profoundly lymphopenic. Burkitt's lymphoma is generally restricted to certain areas of the tropics because malaria infection predisposes to its development (Fig. 20–11). EBV may also be implicated in the development of lymphomas in patients treated with the immunosuppressive drug cyclosporine (see p. 300).

Finally, viruses commonly adopt the highly efficient strategy of suppressing the expression of MHC molecules on infected cells. They appear to be especially effective in downregulating the expression of MHC class I molecules. They suppress expression by suppressing transcription factors for MHC genes or by blocking transport of MHC molecules to the cell surface. In some cases, the viruses block the effects of interferon on MHC expression.

IMMUNITY TO PROTOZOA

Nonspecific Immunity

The nonimmunological mechanisms of resistance to protozoa are similar to those that operate in bacterial and viral diseases. Species and genetic influences are perhaps most significant; for example, *Trypanosoma lewisi* is found only in the rat and *Trypanosoma musculi* in the mouse, and neither causes disease in these animals. These species differences are a development of somewhat more subtle genetic influences. Perhaps the best analyzed case of genetically determined resistance to protozoan disease is sickle cell anemia in humans. Individuals who inherit the sickle cell trait possess he-

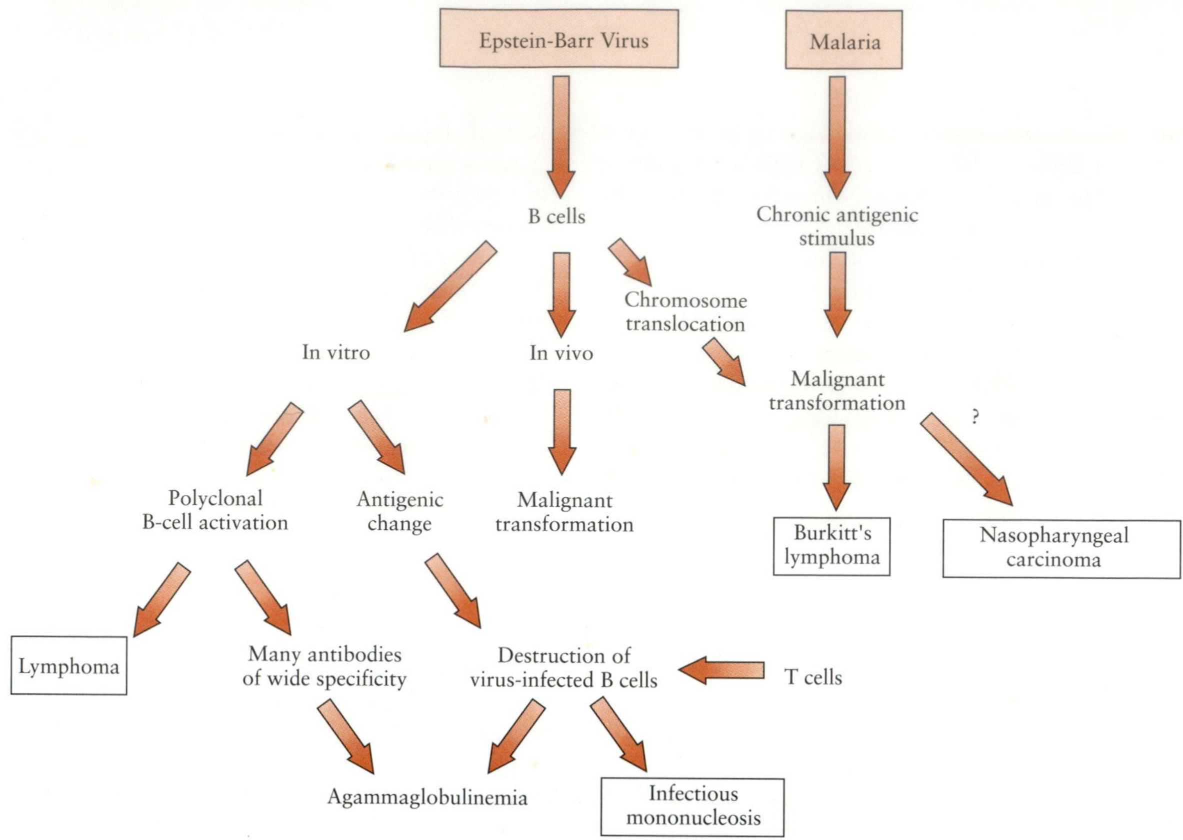

Figure 25–9 The diseases associated with infection by Epstein-Barr virus.

moglobin S (HbS), in which a residue of valine has replaced a residue of glutamic acid present in normal hemoglobin. The changes in the shape of the hemoglobin molecule induced by this substitution cause deoxygenated hemoglobin molecules to aggregate, thus distorting the shape of the red cells and resulting in increased red cell fragility and clearance. Individuals who are homozygous for the sickle cell gene may die when young from the results of severe anemia. Heterozygous individuals are also anemic, but in west central Africa, the fact that red blood cells containing hemoglobin S are not parasitized by *Plasmodium falciparum* ensures that these individuals are resistant to malaria. As a result, more of these heterozygotes tend to survive to reproductive age than normal persons. The mutation is therefore maintained in the population at a relatively high level.

Persons who lack Duffy blood group antigens (Fy [a-b-]) are also resistant to malaria. The membrane molecule that carries the Fy epitopes is also the receptor for *Plasmodium vivax*. Without it the parasite cannot invade the erythrocyte. These Fy (a-b-) individuals remain susceptible to other plasmodial infections.

Specific Immunity

The obvious inadequacies of the immune responses to many parasites led early investigators to conclude that successful parasites were, in general, poorly immunogenic. This is not the case; most parasites are fully antigenic, but in their adaptation to a parasite existence, they have developed mechanisms through which they may survive in the presence of an immune response. Therefore, like other infectious agents, protozoa can stimulate both humoral and cell-mediated immune responses. In general, antibodies control the level of parasites in the blood and tissue fluids, whereas cell-mediated immune responses are directed largely against intracellular parasites (Fig. 25–10). One mechanism of importance in immunity to intracellular protozoa such as *Leishmania* is the production of reactive nitrogen metabolites such as nitric oxide. The production of nitric oxide by these cells is promoted by TNF-α and IFN-γ. Antibodies directed against protozoan surface antigens may opsonize, agglutinate, or immobilize them. Antibodies together with complement and cytotoxic cells may lyse them, and other antibodies (called ablastins) may inhibit their division.

Evasion of the Immune Response by Protozoa

Most important protozoan parasites have evolved mechanisms to evade their host's immune responses. In general, these mechanisms resemble those evolved by other types of organisms.

Parasite-induced immunosuppression may assist parasite survival. For example, the blood parasite *Babesia bovis* is immunosuppressive for cattle. As a result

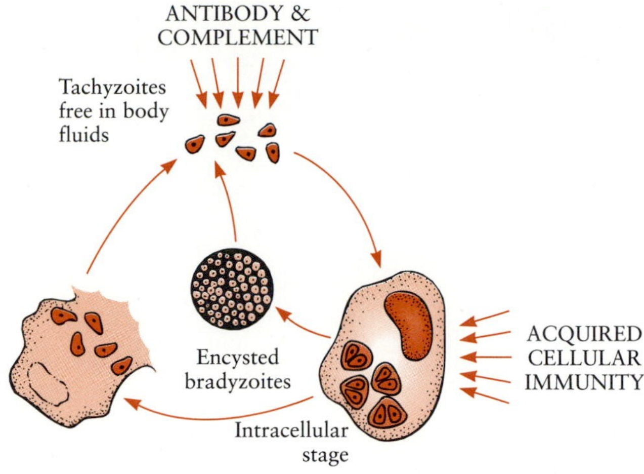

Figure 25–10 The ways in which both humoral and cell-mediated immune responses act together to destroy the protozoan parasite *Toxoplasma gondii.* (Tachyzoites are the replicating forms found in the body. Bradyzoites are the forms found within cysts.)

of this suppression, its vector, the cattle tick, is more able to survive on an infected animal. Consequently, infected cattle have more ticks than noninfected animals, and the efficiency of transmission of *B. bovis* is greatly enhanced. Recently, it has been found that *Trypanosoma cruzi,* the cause of Chagas' disease, can prevent expression of the IL-2 receptor on T cells and so induce immunosuppression. It must be pointed out, however, that parasite-induced immunosuppression can lead to the death of host animals as a result of secondary infection, so it is not always beneficial to the parasite.

In addition to immunosuppression, protozoa have evolved two other extremely effective immunoevasive techniques. One involves becoming "hypoantigenic," and the other is the ability to change surface antigens rapidly and repeatedly. An example of a hypoantigenic organism is the cyst stage of *T. gondii,* which does not stimulate a significant host response. Some protozoa can become functionally nonantigenic by masking themselves with host antigens. Examples of these are *Trypanosoma theileri* in cattle and *Trypanosoma lewisi* in rats. These are both common nonpathogenic trypanosomes that can survive in the bloodstream of infected animals because they become coated with a layer of host serum proteins and so are not regarded as foreign.

Although the absence of antigenicity may be considered the ultimate stage in the evasive process, many protozoa, especially the trypanosomes, have developed antigenic variation to a high degree of sophistication. The African trypanosomes must be able to survive in the bloodstream in high numbers since they are transmitted by biting tsetse flies. There must be lots of live organisms in the fly's blood meal if they are to be successfully transmitted. Yet the parasite does not have a thick cell wall to defend itself against immunological attack. If humans are infected with African trypanosomes, the numbers of circulating parasites in the blood fluctuate greatly. Periods of high parasitemia alternate regularly with periods of low parasitemia (Fig. 25–11). Antibodies taken from infected individuals react with trypanosomes isolated before the time of bleeding but not with those taken after. Each peak of parasitemia reflects the development of a population of trypanosomes carrying a new surface molecule called variant surface glycoprotein (VSG). The elimination of a population with one VSG leads to a fall in blood parasite levels. Some organisms, however, develop a new VSG. As a result, these survive and a fresh population arises to produce yet another period of high parasitemia. This cyclical fluctuation in parasite levels with each peak reflecting the appearance of a new VSG can continue for many months.

The sequence of VSG produced in trypanosome infections is not provoked by antibody. Trypanosomes grown in tissue culture also show spontaneous antigenic variation. By electron microscopy, it can be shown that the VSG forms a thick coat over the surface of the trypanosome. When antigenic change occurs, the glycoproteins in the old coat are shed and replaced by an antigenically different glycoprotein. Analysis of the genetics of this process indicates that the trypanosomes possess at least 1000 VSG genes and that antigenic variation occurs as a result of replacing an expressed VSG gene with another selected at random.

Since parasitic protozoa seek to evade the immune responses, it is not surprising that they also invade immunosuppressed individuals. Organisms that are normally maintained in a relatively quiescent state by the immune response, such as the cyst forms of *T. gondii,* can change to a more active form and produce severe disease in immunosuppressed animals. For this reason, acute toxoplasmosis is commonly seen in patients who are immunosuppressed for transplantation purposes or for cancer therapy (Fig. 25–12).

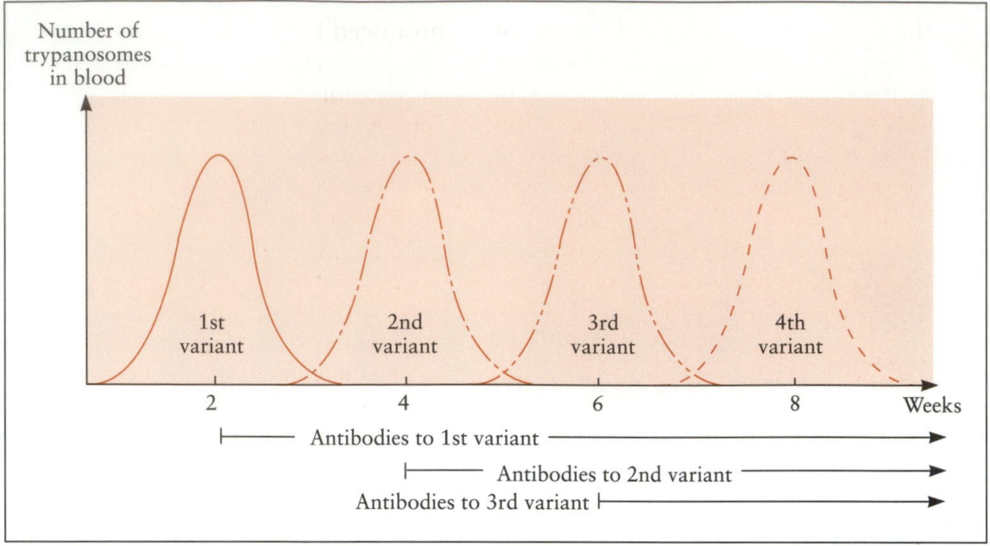

Figure 25–11 Antigenic variation in cell membrane variant-specific glycoprotein results in the cyclic parasitemia observed in trypanosomiasis. Each peak represents the growth of a population of parasites expressing new cell surface antigens.

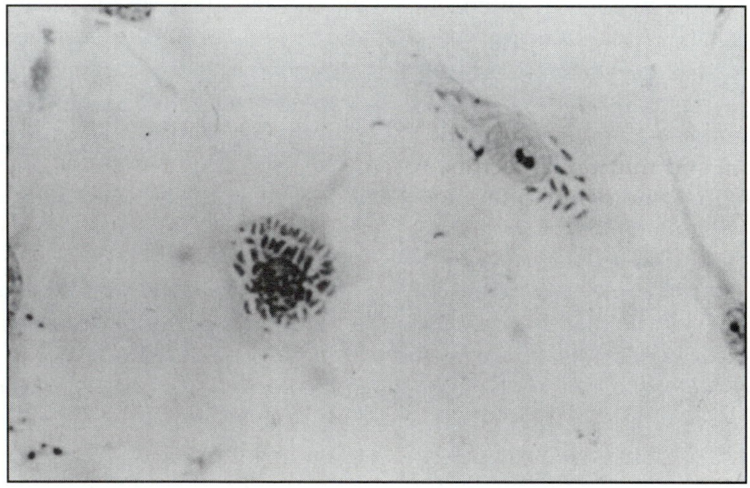

Figure 25–12 *Toxoplasma gondii* growing in a mouse macrophage culture. *T. gondii* is a facultative intracellular parasite. *(Courtesy of Dr. C.H. Lai.)*

IMMUNITY TO HELMINTHS

Parasitic worms (**helminths**) are highly prevalent in animals and in humans in developing countries. In some areas individuals may harbor worms for much of their lives. Thus these parasites can survive immunological attack. As might be anticipated, helminths use many diverse mechanisms to evade immune elimination.

Nonspecific Immunity

The factors that influence the course of parasitic worm infestations include not only the influences of host-derived factors but also of factors derived from other helminths within the same host. Thus both intraspecies and interspecies competition are known to occur. In the former case, also called concomitant immunity, it is evident in tapeworm (cestode), schistosome, and nematode infections in rodents that the presence of adult worms in the intestine delays the development of larval stages in the tissues. In interspecies competition, competition between worms for habitats and nutrients may control the numbers and composition of an animal's helminth population.

The age and sex of the host influence worm burdens through behavioral and hormonal processes. Young children tend to acquire infection much more readily than adults as a result of their behavior. In animals whose sexual cycle is seasonal, parasites tend to synchronize their reproductive cycle with that of the host. For instance, female sheep shed more worm eggs in their feces at a time that coincides with lambing. Presumably, this helps the worms spread to the newborn lambs. Similarly, the development of helminth larvae ingested by cattle in autumn is delayed until spring in a process called hypobiosis. Thus fewer eggs are shed into the environment at a time when their chances of survival are reduced. Larvae of the dog parasite *Toxocara* migrate from an infected female to the liver of the fetal puppy, resulting in a congenital infection.

One notable feature of parasitized populations is that a few individuals may be predisposed to heavy infestation and harbor the majority of worms. This is well recognized in *Ascaris, Trichuris,* and *Schistosoma* infections. This is likely caused by a combination of genetic, behavioral, environmental, and nutritional factors. It may also reflect significant immunological differences between individuals.

Specific Immunity

It is not surprising that the immune system is relatively inefficient in controlling parasitic worms. After all, these organisms have adapted to an obligatory parasitic existence, and presumably, this adaptation has involved dealing with the immune system and either overcoming or evading it. Worms are therefore not maladapted pathogenic organisms but fully adapted obligate parasites whose survival depends on reaching some form of accommodation with the host. Consequently, if an organism of this type causes disease, it is likely to be expressed either mildly or subclinically. Only when parasitic worms invade a host to which they are not fully adapted, or in unusually large numbers, does acute disease occur.

Parasitic worms can be found in the body either in tissues as larval forms or within the gastrointestinal or respiratory tract as adults. Obviously, the form of the immune response that is most effective against these stages differs considerably. Although conventional antibodies of the IgM, IgG, and IgA classes are produced in response to worm antigens, the most significant immunoglobulin class involved in resistance to parasitic worms is IgE. For example, IgE levels are usually very high in parasitized individuals; many worm infestations are associated with the characteristic signs of type I hypersensitivity, including eosinophilia, edema, asthma, and skin reactions, as well as a positive passive cutaneous anaphylaxis reaction to worm antigens (Chapter 29). Many worm antigens preferentially stimulate IgE production so that scientists who handle parasitic worms may become sensitized and suffer from asthmatic attacks or skin reactions (urticaria) on exposure to worms. Helminth antigens preferentially stimulate Th2 and possibly depress Th1 cells, so that IL-4 and IL-5 production is stimulated and IL-2 and IFN-γ production is lower than normal. IL-4 stimulates IgE synthesis, and IL-5 stimulates eosinophil differentiation and production. IL-10 production may also be enhanced in worm infestations, so reducing cell-mediated responses to these organisms. The reason for this preferential Th2 stimulation is unknown but may be due to differences in antigen processing and presentation.

Although IgE production and the allergies that result from it are considered by some to be only a nuisance, they may be of considerable benefit in controlling worm burdens. One of the best examples of this is the "self-cure" reaction in sheep infected with the gastrointestinal worm *Haemonchus contortus.* These worms, which live in the intestinal and stomach walls, secrete antigens that act as allergens. As a result, the worms provoke a local acute type I hypersensitivity reaction in the parasitized regions of the intestine (Fig. 25–13). The combination of worm antigens with mast-cell–bound IgE leads to mast-cell degranulation and the secretion of vasoactive molecules in the intestinal wall. These molecules stimulate smooth muscle contraction and increase blood vessel permeability. Thus, in the self-cure reaction, violent contractions of the intestinal muscles and an increase in the permeability of intestinal capillaries occur, allowing fluid to leak into the lumen. This combination results in dislodgement and expulsion of most of the animal's intestinal worms. In sheep that have just undergone such a self-cure, their IgE antibody levels are high (Chapter 29) and experimental administration of worm antigens will result in acute anaphylaxis, confirming the role of type I hypersensitivity in this phenomenon. IgE has other roles to play in the reduction of worm burdens in animals. For example, macrophages may bind to worm larvae such as schistosomula (immature forms of schistosomes) through an IgE-mediated pathway and cause their destruction. By causing mast-cell degranulation, IgE stimulates secretion of eosinophil

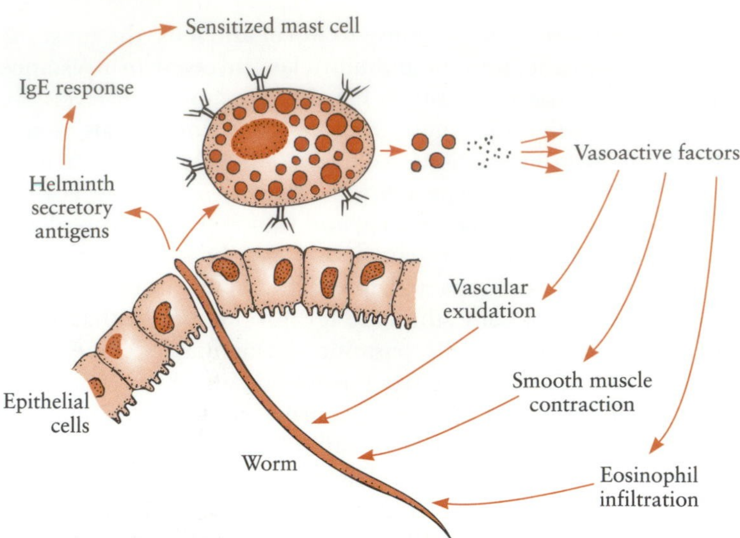

Figure 25–13 The way in which a local type I hypersensitivity reaction in the intestinal wall can discourage helminth infestation.

chemotactic factors. These peptides mobilize bone marrow eosinophils and provoke their release into the circulation. IL-5 has a similar effect. For these reasons, an eosinophilia is characteristic of worm infestations. IgE bound to eosinophils through its Fc receptor (FcεRI) can mediate the destruction of helminths such as schistosomes. Eosinophils contain many enzymes capable of causing considerable damage to helminths. (The properties of eosinophils are described in Chapter 28.)

Although the IgE-eosinophil-mediated antihelminth response is perhaps the most significant mechanism of resistance to parasitic worms, antibodies of other immunoglobulin classes also play a protective role. The mechanisms involved include complement-mediated cuticular damage (Fig. 25–14), antibody-mediated neutralization of the proteolytic enzymes used by larvae to penetrate tissues, blocking of the pores of these larvae by immune complexes formed by antibodies combined with larval excretory and secretory products (Fig. 25–15), and prevention of molting and inhibition of larval development by antibodies directed against molting enzymes.

Many parasitic worms, particularly those that undergo tissue migration, may be considered functional xenografts. It is somewhat remarkable, therefore, that they are not precipitously rejected by the cell-mediated immune system. Their survival is a reflection of

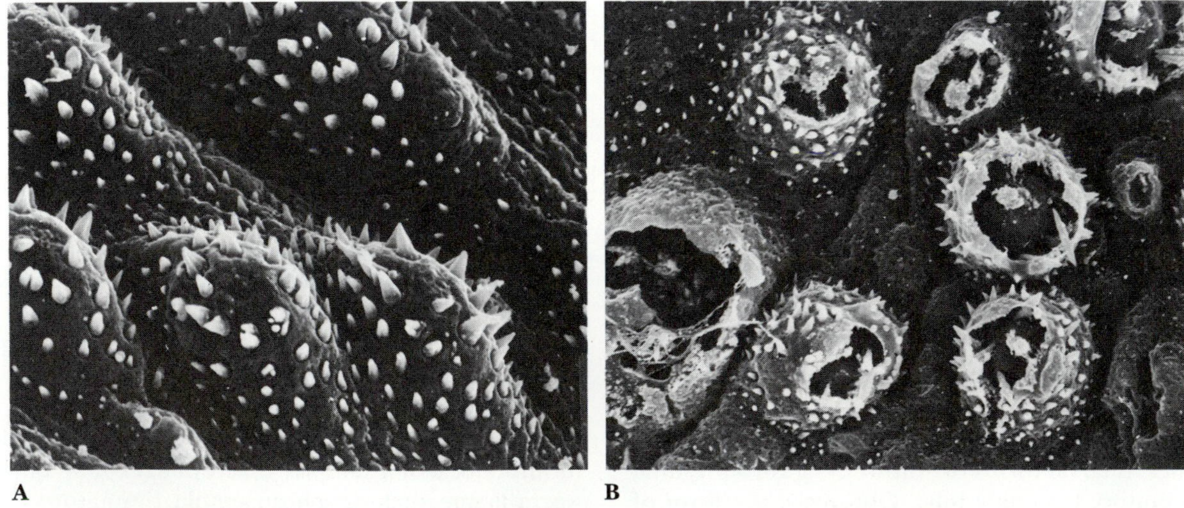

Figure 25–14 Scanning electron micrographs of the surface of an adult male *Schistosoma mansoni.* (**A**) Intact tubercles after incubation in normal mouse serum and complement. (**B**) Tubercle damage after incubation in the presence of immune mouse serum and complement (original magnification ×3000). *(Courtesy of Dr. K. Rasmussen.)*

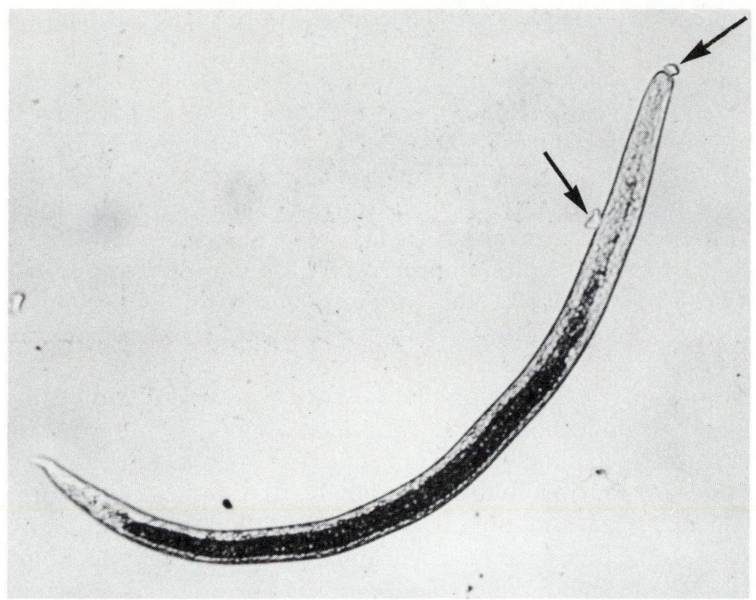

Figure 25–15 Blockage of the oral and excretory pores *(arrows)* of a *Toxocara canis* larva by immune complexes after incubation in specific antiserum.

the success of their adaptation to existence within mammalian tissues. Nevertheless, sensitized T cells and activated macrophages may successfully attack worms that are either deeply embedded in the intestinal wall or undergoing prolonged tissue migration. Sensitized T cells depress the activities of parasitic worms by three mechanisms. First, the development of an inflammatory response of the delayed hypersensitivity type tends to attract mononuclear cells to the site of larval invasion and renders the local environment unsuitable for growth or migration. Second, cytotoxic lymphocytes may be capable of causing larval destruction. It is not uncommon to observe large lymphocytes adhering firmly to migrating nematode larvae in vivo. Third, granulomas may form around parasite eggs as in schistosomiasis. These granulomas are mediated by TNF-α and do not form in mice with a deficiency in their immune system. Although they represent a defensive effort on the part of the host, the parasite has adapted to this response so well that TNF-α is required to stimulate schistosome egg production. It appears to act as an embryonic growth factor. Th1-cell–derived IFN-γ acting in association with TNF will activate macrophages. These activated macrophages kill schistosomula through the production of nitric oxide.

Evasion of the Immune Response by Helminths

Although the preceding discussion has described several mechanisms whereby animals resist parasitic worm infection, it is obvious that these responses are not very effective. Several mechanisms play a role in this adaptation. These include molecular mimicry of host antigens, adsorption of host antigens, antigenic variation, blocking of antibodies, and tolerance (Table 25–5).

The first of these mechanisms, **molecular mimicry,** is based on the concept that parasitic worms may synthesize histocompatibility or blood group antigens to match those of their host. A worm cannot synthesize all the antigens it could possibly require, since such synthesis would demand the existence of a genetic system similar in complexity to the antibody-producing system. Nevertheless, partial mimicry of host antigens is indeed possible, as shown by the fact that sheep respond to fewer antigens of *H. contortus* than do rabbits. This suggests that *H. contortus* possesses a closer antigenic similarity to sheep, its natural host, than to rab-

Table 25–5 Methods Employed by Helminths in Order to Evade the Host's Immune Response

- Hiding of helminth antigens
  - Protective coating with antibody, complement, host histocompatibility antigens, serum proteins, or blood-group antigens
- Loss of helminth antigens
  - Antigenic variation
  - Shedding of helminth surface antigens
- Release of soluble factors that:
  - Inhibit complement activity
  - Inhibit mast-cell degranulation
  - Inhibit lymphocyte proliferation
- Selective stimulation of production of:
  - Suppressor cells
  - Blocking antibodies

bits, which it does not normally infect. Also, many worms can synthesize blood group antigens, which also reduce the effective antigenicity of the worm.

Second, a migrating larval worm in tissues may be protected from the results of its host's immune response by the adsorption of host antigens onto its surface. An example of this is the adult *Schistosoma mansoni,* a worm that lives in the mesenteric blood vessels of humans and can adsorb host red blood cell antigens and MHC molecules to its surface. Tapeworm cysticerci can also adsorb MHC molecules in this way. Schistosomes can evade the alternate complement pathway by adsorbing host CD55 (DAF) onto their surfaces.

Some helminths take a more direct role in regulating the immune competence of their host by interfering with antigen presentation. Thus macrophages from schistosome-infected mice are poor antigen presenters. They may secrete molecules that block the actions of lysosomal proteases. Taeniastatin is a protease inhibitor from tapeworm larvae that inhibits neutrophil chemotaxis as well as T-cell proliferation and production of IL-2. Immunosuppression may also contribute to the survival of parasitic worms. The mechanisms involved in this are unknown. They may involve induction of suppressive cytokines such as IL-10 or selective inhibition of T-cell subsets, or alternatively, they may result from the production of blocking antibody in a manner analogous to that seen in pregnancy and in some cancers (Chapter 20). In trichinosis, for example, infected animals are nonspecifically immunosuppressed. This immunosuppression is reflected in a lowered resistance to other infections, a poor response to vaccination, and a prolongation of skin graft survival.

The techniques described previously inhibit the induction of the immune response. Other evasive techniques inhibit the effector arm of the immune system. The most important of these is antigenic variation. Although worms have not evolved a system as efficient as that seen in trypanosomes, gradual antigenic variation is recognized. For example, the cuticular antigens of *Trichinella spiralis* larvae show extensive antigenic changes following each molt. Obviously, possession of a thick, resistant cuticle that is turned over rapidly has a similar effect.

KEY WORDS

Alternative complement pathway p. 381
Antibody p. 380
Antibody-dependent cellular cytotoxicity p. 387
Antigenic drift p. 388
Antigenic shift p. 388
Antigenic variation p. 391
Bacterial immunity p. 380
Classical complement pathway p. 380
Complement p. 380
Cytotoxic T cells p. 387
Eosinophils p. 393
Epstein-Barr virus p. 389
Heat shock protein p. 381
Immunoglobulin E p. 393
Immunoglobulin G p. 380
Immunoglobulin M p. 380
Immunosuppression p. 389
Interferon-γ p. 381
Interferons p. 383
Interleukin-10 p. 382
Intracellular bacteria p. 381
Intracellular parasites p. 390
Iron levels p. 379
Leprosy p. 381
Lysozyme p. 378
Membrane attack complex p. 380
Molecular mimicry p. 395
Nitric oxide p. 390
Opsonization p. 380
Self-cure p. 393
Sickle cell anemia p. 389
Siderophores p. 380
Toxin neutralization p. 380
Virus neutralization p. 386

QUESTIONS

1. Antibodies can protect against viral diseases through which of the following?
   a. release of antibiotics from B cells
   b. blocking viral adsorption to cells
   c. T-cell–derived viricidal proteins
   d. the viricidal activity of histamine
   e. all of the above
2. "Self-cure" is a reaction that occurs when animals lose their worm burden by a
   a. hormonal change in the intestinal environment
   b. type I hypersensitivity reaction
   c. type IV hypersensitivity reaction
   d. massive eosinophil attack on the helminths
   e. type III hypersensitivity reaction
3. The major mechanism of resistance to disease caused by *Clostridium tetani* involves
   a. toxin neutralization by antibody
   b. opsonization by antibody and complement
   c. toxin neutralization by T cells
   d. toxin neutralization by activated macrophages
   e. toxin neutralization by eosinophils
4. Bacterial cell walls cause activation of which defense system?
   a. the tyrosine kinase pathway
   b. the alternative complement pathway
   c. the kinin pathway
   d. the terminal complement pathway
   e. the classical complement pathway

5. Major changes in the antigenic structure of influenza viruses are called
   a. antigenic variation
   b. signal transduction
   c. antigenic shift
   d. attenuation
   e. antigenic drift
6. The most important cells involved in the destruction of virus-infected cells are
   a. B cells
   b. macrophages
   c. NK cells
   d. helper T cells
   e. cytotoxic T cells
7. When cells, including bacteria, are stressed, they produce a major antigenic protein called
   a. endotoxin
   b. heat shock protein
   c. exotoxin
   d. hemolysin
   e. phospholipase
8. The cytokine that appears to be responsible for the diminished immune response to lepromatous leprosy is
   a. interleukin-2
   b. interleukin-4
   c. interleukin-8
   d. interleukin-10
   e. interleukin-12
9. Among the most potent bactericidal factors produced by activated macrophages is
   a. nitric oxide
   b. lysozyme
   c. transferrin
   d. complement
   e. hemolysin
10. The cytokine primarily responsible for activation of macrophages is
   a. interleukin-4
   b. interleukin-2
   c. tumor necrosis factor–α
   d. lymphotoxin
   e. interferon-γ
11. Outline how the body destroys organisms that have the ability to live within macrophages. What mechanisms do bacteria employ to ensure their survival within macrophages?
12. Bacteria, viruses, protozoa, and parasitic worms all demonstrate the phenomenon of antigenic variation. Give examples of each of these. What features do these have in common, and how do they differ?
13. How do viruses ensure their persistence for long periods of time within an animal?
14. What functions do NK cells serve in providing resistance to viral diseases?
15. How do large parasitic worms survive within the body?
16. Discuss the proposition that the immune response against viruses is more detrimental than beneficial.
17. What role does type I hypersensitivity play in protecting animals against invasive microorganisms?
18. Is it true that antibodies are primarily directed against bacteria and cell-mediated immune responses are directed against viruses? Give some examples of exceptions to the rule and comment on their significance.
19. Lysozyme is unimportant in resistance to bacterial disease because it destroys only nonpathogenic bacteria. Discuss.
20. Why have the interferons not been made widely available to the public in an attempt to prevent or treat viral diseases?

Answers: 1b, 2b, 3a, 4b, 5c, 6e, 7b, 8d, 9a, 10e

SOURCES OF ADDITIONAL INFORMATION

Acha-Orbea, H., and MacDonald, H.R. Subversion of host immune responses by viral superantigens. Trends Microbiol., 1:32–34, 1993.

Baringa, M. Viruses launch their own 'Star Wars.' Science, 258:1730–1731, 1992.

Barnes, P.F., et al. Tumor necrosis factor production in patients with leprosy. Infect. Immun. 60:1441–1446, 1992.

Burdin, N., et al. Epstein Barr virus transformation induces B lymphocytes to produce human interleukin 10. J. Exp. Med., 177:295–304, 1993.

Butterworth, A.E. Cell-mediated damage to helminths. Adv. Parasitol., 23:143–235, 1984.

Dankert, J.R. Complement-mediated inhibition of function in complement-resistant *E. coli.* J. Immunol., 142:1591–1595, 1989.

Dannenberg, A.M. Immunopathogenesis of pulmonary tuberculosis. Hosp. Pract. [Off.] 28:51–58, 1993.

Donelson, J.E., and Fulton, A.B. The pushy ways of a parasite. Nature, 342:615–616, 1989.

Elsbach, P. Degradation of microorganisms by phagocytic cells. Rev. Infect. Dis., 2:106–128, 1980.

Goodenough, U.W. Deception by pathogens. Am. Sci., 79: 344–355, 1991.

Gounni, A.S., et al. High-affinity IgE receptor on eosinophils is involved in defence against parasites. Nature, 367:183–186, 1994.

Hassett, D.J., and Cohen, M.S. Bacterial adaptation to oxidative stress: Implications for pathogenesis and interaction with phagocytic cells. FASEB J., 3:2574–2582, 1989.

Karupiah, G., et al. Inhibition of viral replication by interferon-γ-induced nitric oxide synthase. Science, 261:1445–1448, 1993.

Kay, A.B. Eosinophils as effector cells in immunity and hypersensitivity disorders. Clin. Exp. Immunol., 62:1–12, 1985.

Lafon, M., et al. Evidence for a viral superantigen in humans. Nature, 358:507–509, 1992.

Liew, F.Y., Li, Y., and Millot, S. Tumor necrosis factor-α synergizes with IFN-γ in mediating killing of *Leishmania major* through the induction of nitric oxide. J. Immunol., 145: 4306–4310, 1990.

Locksley, R.M. Interleukin 12 in host defense against microbial pathogens. Proc. Natl. Acad. Sci. U.S.A., 90:5879–5880, 1993.

Maizels, R.M., et al. Immunological modulation and evasion by helminth parasites in human populations. Nature, 365: 797–805, 1993.

Mims, C.A. The Pathogenesis of Infectious Disease, 2nd ed. Academic Press, New York, 1982.

Mims, C.A. Virus immunity and pathogenesis. Br. Med. Bull., 41:1–102, 1985.

Moore, K.W., Rousset, F., and Banchereau, J. Evolving principles in immunopathology: Interleukin 10 and its relationship to Epstein-Barr virus protein BCRF1. Springer Semin. Immunopathol., 13:157–166, 1991.

Oldstone, M.A. Viral persistence and immune dysfunction. Hosp. Pract., [Off.], 25:81–98, 1990.

Pace, J., Hayman, M.J., and Galan, J.E. Signal transduction and invasion of epithelial cells by *S. typhimurium.* Cell, 72: 505–514, 1993.

Palese, P., and Young, J.F. Variation of influenza A, B, and C viruses. Science, 215:1468–1474, 1982.

Parillo, J.E. Pathogenic mechanisms of septic shock. N. Engl. J. Med., 328:1471–1477, 1993.

Paul, W.E. Infectious diseases and the immune system. Sci. Am., 269:91–97, 1993.

Pincus, S.H., et al. Oxygen mediated killing of schistosomula of *Schistosoma mansoni:* Oxidative requirement for enhancement of eosinophil colony stimulating factor (CSF-a) and supernatants with eosinophil cytotoxicity enhancing activity (E-CEA). Cell. Immunol., 87:424–433, 1984.

Porterfield, J.S. Antibody-mediated enhancement of rabies virus. Nature, 290:542, 1981.

Rapp, F., and Cory, J.M. Mechanisms of persistence in human virus infections. Microb. Pathog., 4:85–92, 1988.

Ray, C.A., et al. Viral inhibition of inflammation: Cowpox virus encodes an inhibitor of the interleukin-1β converting enzyme. Cell, 69:597–604, 1992.

Sieling, P.A., et al. Immunosuppressive roles for IL-10 and IL-4 in human infection. J. Immunol., 150:5501–5510, 1993.

Spry, C.J.F. Synthesis and secretion of eosinophil granule substances. Immunol. Today, 6:332–335, 1985.

Tilney, L.G., and Tilney, M.S. The wily ways of a parasite: Induction of actin assembly by Listeria. Trends Microbiol., 1:25–31, 1994.

Veith, M.C., and Butterworth, A.E. Enhancement of human eosinophil-mediated killing of *Schistosoma mansoni* larvae by mononuclear cell products in vitro. J. Exp. Med., 157:1828–1843, 1983.

Young, R.A. Stress proteins and immunology. Annu. Rev. Immunol., 8:401–420, 1990.

CHAPTER 26

Defects in the Immune System

Angelo DiGeorge *Angelo DiGeorge was born in Philadelphia in 1921. An endocrinologist by training, Dr DiGeorge encountered the anomaly that bears his name when investigating newborn infants with hypoparathyroidism. Dr DiGeorge also noted that some of these children also lacked a thymus. These children suffered from what is now called the DiGeorge anomaly. Despite the absence of a thymus, Dr DiGeorge noted that these children had normal lymphocyte numbers and immunoglobulin levels despite defective cell-mediated responses. This was one of the key observations that led to the concept of a dual immune system involving both B and T cells.*

CHAPTER OUTLINE

CHAPTER CONCEPTS

1. Inherited defects in neutrophils may result either in an absence of these cells or in the development of nonfunctional neutrophils. Both conditions invariably result in a greatly increased susceptibility to bacterial infections.
2. Inherited deficiencies in the immune system vary in severity according to the precise site of the lesion. Thus individuals who fail to develop any functioning immune system whatsoever are much more severely affected than individuals who fail to produce a single immunoglobulin class.
3. As a general rule, individuals who fail to develop a functional T-cell system will die from viral infections. Individuals who fail to develop a B-cell system will die from bacterial infections.
4. Other immunodeficiencies may result from malnutrition, virus infections, or exposure to toxic chemicals or drugs.

In this chapter some of the diseases that result from defects in the immune system are described. The first group of defects (or immunodeficiencies) addressed are those that result from an inherited genetic defect and are called primary immunodeficiencies. These usually occur in children and include defects in neutrophil function as well as deficiencies in the immune system itself, including defects in stem cells, in T cells, and in B cells.

Any defect in the immune system that reduces the ability of an individual to mount either cell-mediated or antibody-mediated immune responses or interferes with the ability of phagocytic cells to ingest and destroy bacteria or process antigen results in infection. These defects are of two general types. Some defects are inherited as a result of a mutation in the parents' genes. Thus they have no obvious predisposing cause and are called **primary immunodeficiencies.** Because they are inherited, their effects are seen in infants and young children. The children suffer from recurrent, severe infections that will kill them before they get very old. The second major type of defect in the immune system is one caused by a known agent such as a drug or virus that destroys lymphocytes. This type of immunodeficiency is called a **secondary immunodeficiency** (or acquired immunodeficiency). Secondary immunodeficiencies commonly occur in adults or in the aged. Like immunodeficiencies in children, they are recognized clinically either by recurrent severe, intractable infections or by an increased susceptibility to cancer.

A special case is the DiGeorge anomaly, which is not an inherited defect but results from a failure of the thymus to develop. Because mice are a good experimental model for inherited immunodeficiencies, several of these mouse models are described next. The second group of immunodeficiencies result from secondary insult to the immune system. Thus the effects of malnutrition, exercise, trauma, and age as well as toxins on the immune system are reviewed.

DEFECTS IN PHAGOCYTOSIS AND ANTIGEN PROCESSING

Neutrophil Deficiencies

Neutrophils ingest and destroy invading foreign material, especially bacteria. If neutrophils are absent or ineffective, recurrent severe bacterial infections result. The numbers of neutrophils in the blood may be reduced by drugs, radiation, overwhelming infections, or some autoimmune diseases. Cytotoxic drugs and radiation cause a **neutropenia** because they are selectively toxic for rapidly dividing cells. Since the neutrophil precursors in the bone marrow are rapidly dividing, they are very susceptible to destruction (Table 26–1).

Severe bacterial infections commonly result from a profound neutropenia. In these cases, cause and effect are sometimes difficult to separate, since bacterial toxins released in severe infections may also depress bone marrow function. Autoantibodies to myeloid tissue also cause bone marrow depression, and the replacement of normal bone marrow by cancer cells can lead to neutropenia—a common complication of leukemia.

Some neutrophil deficiencies are inherited and fall into two distinct groups. In one group, the neutropenia occurs periodically. This "cyclic neutropenia" has been observed in children and dogs. The lesion, which presumably affects the myeloid precursor cells, has not been well defined. A second group of inherited neutropenias are progressively unremitting. As might be anticipated, individuals suffering from cyclic neutropenia suffer from repeated episodes of recurrent infection, and those suffering from noncyclic neutropenia have a more progressive disease. In all neutrophil-deficiency syndromes, affected individuals suffer from recurrent infections by bacteria such as *Streptococcus pneumoniae* and micrococci.

Neutrophil Dysfunction

If neutrophils are present but do not function correctly, the net effect is similar to that seen in a neutropenia: recurrent uncontrolled bacterial infections. Neutrophils can fail to function in many ways. Inherited defects have been described involving each of the major stages in phagocytosis: chemotaxis, opsonization, attachment, ingestion, and digestion (Table 26–2).

By far the most important group of neutrophil dysfunction syndromes is collectively known as chronic granulomatous disease. Its most common form is due to a defect in a gene located on the X chromosome; as a result, it occurs only in boys. (In girls, the presence of a second X chromosome ensures

Table 26–1 Neutrophil Deficiency Syndromes

Acquired neutropenias	Autoantibody-mediated neutropenia Drug-induced neutropenias Isoimmune neutropenia of the newborn Neutropenia in systemic lupus erythematosus
Inherited neutropenias	Cyclic and noncyclic neutropenias

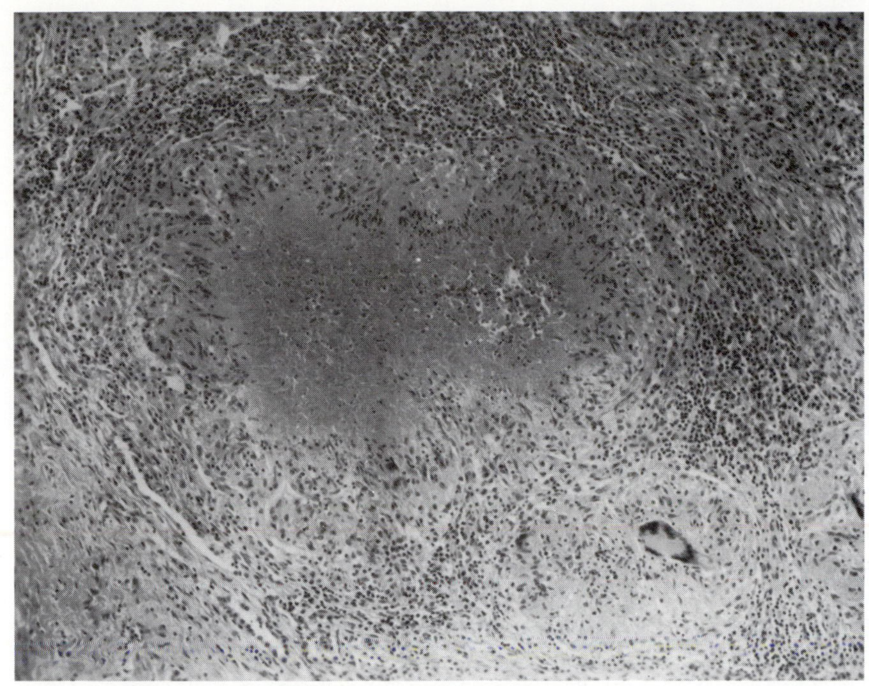

Figure 26–1 A photomicrograph of a section from a granuloma in a child with chronic granulomatous disease. There is a central area of dead cells (necrosis) surrounded by a thick layer of inflammatory cells—a granuloma (original magnification ×120). *(From Bellanti, J.A. Immunology II. W.B. Saunders, Philadelphia, 1979. With permission.)*

that a normal gene is present and the disease does not occur.) Chronic granulomatous disease usually develops in early childhood when the children begin to suffer from recurrent infections (pneumonia, dermatitis, lymphadenitis, osteomyelitis, sepsis) and multiple abscess formation caused by organisms such as *Staphylococcus aureus, Serratia marcescens,* and *Salmonella, Pseudomonas, Nocardia,* and *Aspergillus* (Fig. 26–1).

The lesions in chronic granulomatous disease are associated with an NADPH oxidase defect that results in a failure to mount a respiratory burst. NADPH oxidase is actually a complex enzyme containing four peptide chains. Mutations can occur in any of these (Fig. 26–2). As a result of any of these defects, the recycling of NADPH is prevented, which results in a deficiency of NADP and a failure to produce hydrogen peroxide during phagocytosis. NADP is involved in several other metabolic pathways in addition to the respiratory burst, including cell membrane formation. Thus this defect may also reduce the stability of lysosomal membranes and permit the release of destructive enzymes into the tissues, so causing severe damage.

The dye nitroblue tetrazolium is normally reduced by respiratory burst enzymes to a blue precipitate from a colorless precursor. Thus normal neutro-

Table 26–2 Neutrophil Dysfunction Syndromes

Depressed microbicidal activity	Chronic granulomatous disease Myeloperoxidase deficiency Chédiak-Higashi syndrome Glucose-6-phosphate dehydrogenase deficiency
Impaired opsonization	Complement deficiencies Antibody deficiencies Fibronectin deficiency
Impaired chemotaxis	Complement deficiencies Antibody deficiencies Wiskott-Aldrich syndrome Chronic mucocutaneous candidiasis Chédiak-Higashi syndrome Hyperimmunoglobulin E syndrome Chronic granulomatous disease Diabetes mellitus CD11b/CD18 deficiency

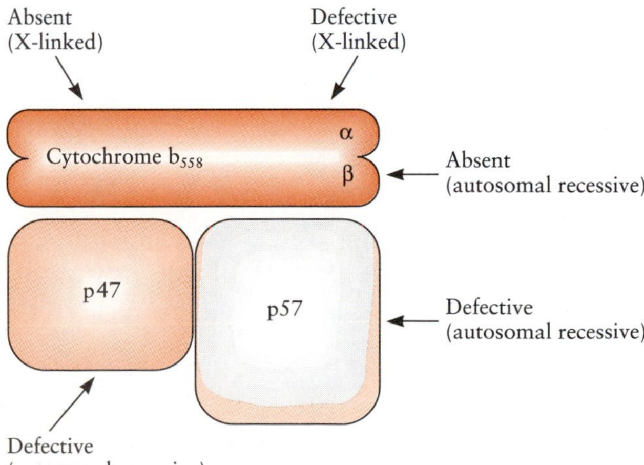

Figure 26–2 NADPH oxidase, its structure, and the recognized genetic defects that can result in chronic granulomatous disease.

Table 26–3 Methods of Evaluating Phagocytic Cell Function

Disease	Test
Chronic granulomatous disease	Nitroblue tetrazolium test Quantitative intracellular killing
"Lazy leukocyte" syndrome	Random migration
Respiratory burst abnormalities	Chemiluminescence
Various disorders	Chemotactic responsiveness

phils exposed to this dye develop deposits of an insoluble black precipitate in their cytoplasm. Neutrophils from affected children fail to do so. This procedure is a convenient aid to diagnosis (Table 26–3).

Other inherited enzyme deficiencies that lead to neutrophil dysfunction include congenital absences of glutathione peroxidase, myeloperoxidase, superoxide dismutase, and glucose-6-phosphate dehydrogenase.

A second neutrophil dysfunction syndrome is known as the Chédiak-Higashi syndrome, an inherited condition seen in cattle, beige mice, mink, Persian cats, tigers, killer whales, and humans. The melanin granules in the skin of affected animals are unusually large. As a result, coats of affected animals are an attractive pale brown color that is easily recognized. Affected animals also have a defect in their lysosomal membranes, and as a result, cellular organelles are grossly enlarged. Thus the neutrophil primary granules are large and fragile, and they rupture spontaneously, causing tissue damage (Fig. 26–3). These leukocytes are defective in chemotactic responsiveness and show reduced intracellular killing as a result of low levels of neutral proteases. The impairment seems to be related to abnormalities in cyclic nucleotide metabolism and membrane glycosphingolipids. Animals with Chédiak-Higashi syndrome also have a deficiency in NK-cell and cytotoxic–T-cell activity.

Defects in neutrophil chemotaxis may be due either to defects in the production of chemotactic factors, for example, C5a, or to an intrinsic defect in the neutrophils themselves. Thus in one condition called lazy leukocyte syndrome, neutrophils are poorly mobilized and defective in random motility.

The most severe form of leukocyte adherence deficiency (LAD-1) is associated with an inherited abnormality of CD11b/CD18. CD11b/CD18 is a member of the β_2 integrin family and a complement receptor required for chemotactic responsiveness. In CD11b/CD18-deficient individuals, neutrophils cannot respond to chemoattractants, bind to particles coated with iC3b, or bind to their ligand on endothelial cells. As a result, they cannot enter tissues to destroy invading organisms or participate in extravascular inflammation (Fig. 26–4). Affected individuals suffer from recurrent infections and fail to make pus, while, at the same time, they have large numbers of neutrophils in their bloodstream. The severity of the

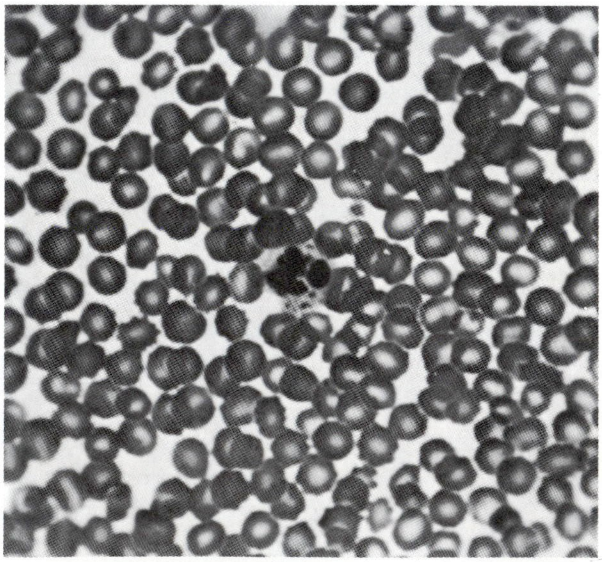

Figure 26–3 A blood smear from a mink with Chédiak-Higashi syndrome. The abnormally large dark granules within the neutrophil at the center of the photograph are characteristic of this syndrome (original magnification ×750). *(Courtesy of Dr. S.H. An.)*

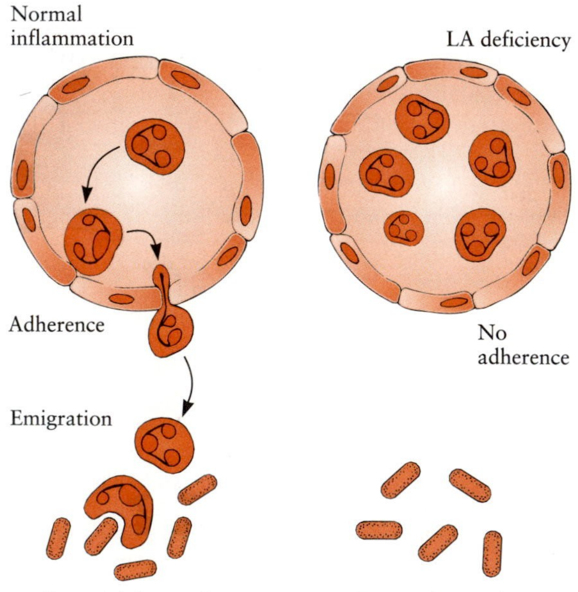

Figure 26–4 In leukocyte adherence deficiency, the inability of neutrophils to adhere to endothelial cells prevents them from entering tissues. As a result, bacteria in tissues cannot be destroyed.

syndrome depends on whether the deficiency is total or partial. If the CD11b/CD18 deficiency is only partial, it can be corrected by exposure to IFN-γ.

A milder form of LAD (LAD-II) results from a deficiency of the E-selectin ligand, sialyl Lewis^{X}. Affected children lack sialyl Lewis^{X} on their neutrophils as a result of a defect in fucose metabolism. Consequently, their neutrophils cannot decelerate by rolling on vascular endothelium (Chapter 28). If stopped, however, they can penetrate blood vessel walls and invade the tissues.

PRIMARY IMMUNODEFICIENCIES

Inherited mutations may result in defects in lymphoid cell production or function. As a rule, a defect in the production of lymphoid stem cells results in a severe deficiency of both the antibody-mediated and the cell-mediated immune responses (Fig. 26–5). A defect that occurs only in the T-cell development pathway is reflected by an inability to mount cell-mediated immune responses, although antibody production may also be affected as a result of loss of helper-cell function. Similarly, a lesion restricted to the B-cell system results in an absence of antibody-mediated immune responses.

Stem-Cell Deficiency Diseases

The most severe congenital immunodeficiency state recognized in humans is caused by a defect in the development of primordial bone marrow stem cells, as a result of which neither myeloid (neutrophils) nor lymphoid cells (T and B lymphocytes) develop. This rare condition, known as reticular dysgenesis, results in the early death of affected infants due to overwhelming infection.

Only slightly less severe than reticular dysgenesis are the combined immunodeficiencies in which lymphoid stem cells fail to develop. As a result, neither the T-cell nor B-cell system develops to maturity, and

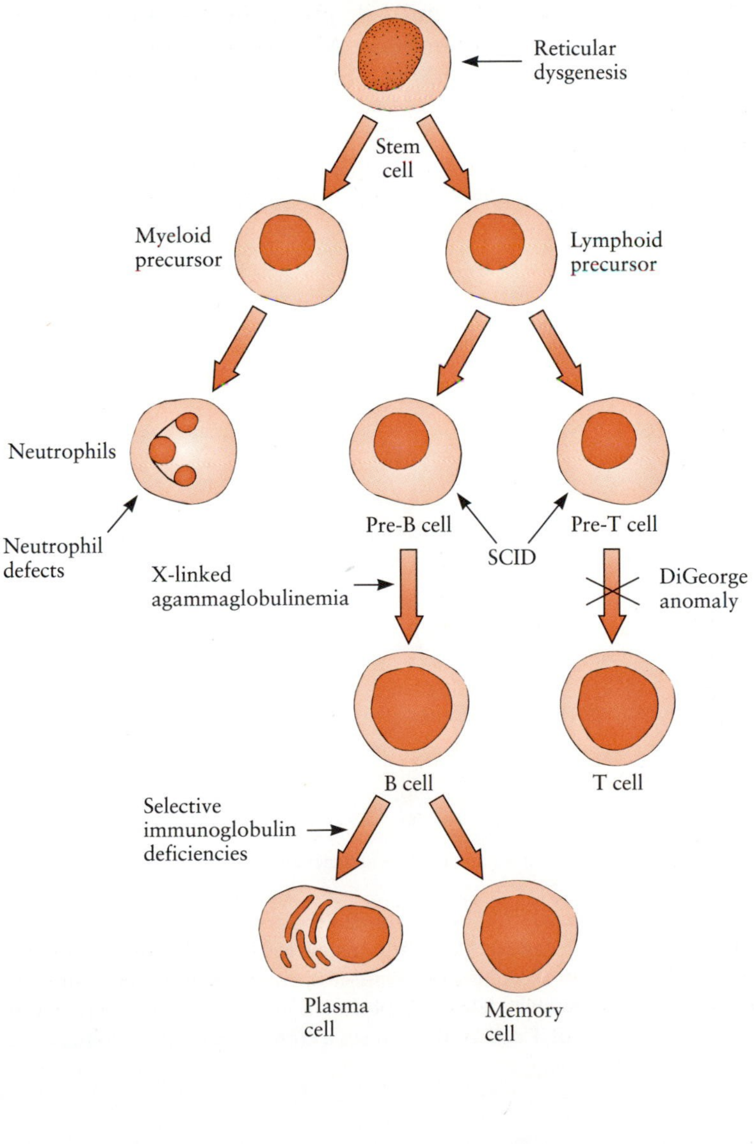

Figure 26–5 The points in the immune system at which development blocks may lead to the major immunodeficiencies.

Table 26–4 The Major Congenital Immunodeficiency Syndromes in Humans

Syndrome	Target	Mode of Inheritance
Reticular dysgenesis	Stem cells	Unknown
Combined immunodeficiency	T cells and B cells	X-linked or autosomal recessive
DiGeorge syndrome	T cells	Not inherited
PNP deficiency	T cells	Autosomal recessive
ADA deficiency	T cells and B cells	Autosomal recessive
Thymus hypoplasia (Nezelof's syndrome)	T cells (B cells)	Variable
Nucleoside phosphorylase deficiency	T cells	Autosomal recessive
Congenital hypogammaglobulinemia	B cells	X-linked
Common variable immunodeficiency	B cells	Autosomal recessive
IgA deficiency	B cells	Variable
IgM deficiency	B cells	Unknown
IgG deficiency	B cells	X-linked
Wiskott-Aldrich syndrome	T cells	X-linked
Ataxia-telangiectasia	T cells (B cells)	Autosomal recessive

no immune responses can be mounted. These congenital combined immunodeficiencies are a mixed group of disorders, with a variety of lesions, modes of inheritance, and degrees of immunologic dysfunction. They generally present as severe recurrent infections during the first weeks of life. Affected infants commonly develop oral candidiasis, pneumonias that are caused by low-grade pathogens such as *Pneumocystis carinii,* and chronic diarrhea. Unless successfully treated, these children will die by the age of two. Combined immunodeficiencies may be diagnosed by an absence of lymphocytes from peripheral blood, reduced or absent serum immunoglobulins, and an inability to mount both cell-mediated and antibody-mediated immune responses. On autopsy, both the primary and secondary lymphoid organs are devoid of lymphocytes.

Combined immunodeficiencies (CID) may be X-linked recessive or autosomal recessive diseases (Table 26–4). In some cases the nature of the lesion has been identified. These lesions include abnormal T-cell antigen receptors, abnormal cytokine receptors, failure in gene rearrangements, defective signal transduction, and defective expression of MHC class II molecules. One of the most significant autosomal recessive forms of CID is due to an inherited deficiency of the enzyme adenosine deaminase (ADA) (Fig. 26–6). ADA converts the nucleosides deoxyadenosine and adenosine to deoxyinosine and inosine, respectively. In most ADA-deficient patients, the defect in T-cell function is due to the accumulation of toxic purine metabolites. Thus deoxyadenosine phosphate accumulates within T cells, where it blocks DNA synthesis and cell division. In addition, it depletes the T cells of ATP and therefore kills resting T cells. The net effect of an ADA deficiency is the selective destruction of T cells and a loss of cell-mediated immune responses. Helper-T-cell activity is lost, so antibody production is also defective. In the T-cell membrane ADA is associated with CD26, a T-cell activation molecule, and it is possible that

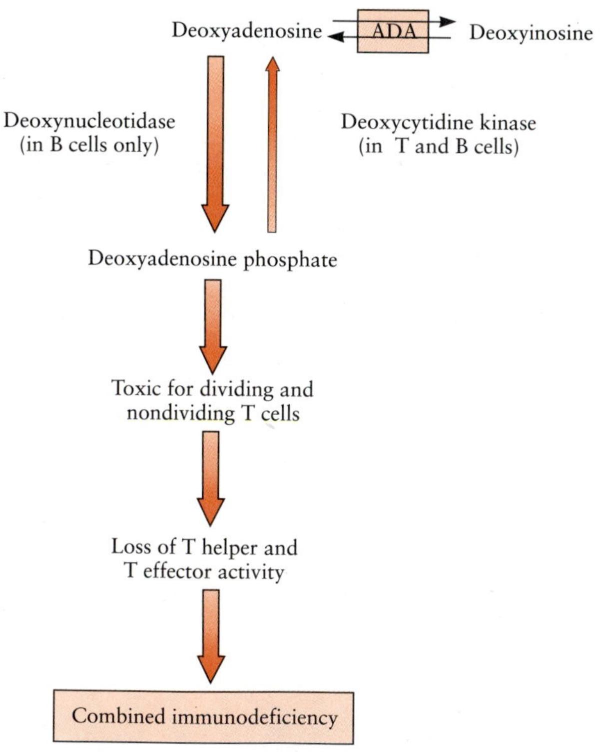

Figure 26–6 The enzyme adenosine deaminase (ADA) normally converts deoxyadenosine to deoxyinosine. In the absence of ADA, deoxyadenosine phosphate accumulates within cells. This compound is toxic for both helper and effector T cells and so causes a combined immunodeficiency.

A CLOSER LOOK

ADA Gene Therapy

Although ADA deficiency is a rare disease, it has been the prime candidate for gene therapy for several reasons. The disease itself results from the accumulation of toxic adenine metabolites. As a result, it is not necessary to replace the enzyme in every cell of the body. It is sufficient that enough enzyme be provided to ensure that the toxic metabolites are removed. ADA can be simply provided by a blood transfusion. The ADA in the transfused red cells will, for a period, remove the adenine metabolites and permit a temporary return to normal. However, the transfused red cells last for only a few weeks, and thus repeated transfusions are necessary for a significant therapeutic benefit. Purified ADA is also of significant benefit but cannot provide a cure.

To provide a lasting cure, gene therapy has been used. The gene for ADA was first inserted into a retrovirus vector (figure). Long-lived stem cells were then taken from the patient's bone marrow and cultured. Next, the genetically engineered virus was mixed with the stem cells. As a result, the new gene entered the stem cells, and they gained the ability to make the missing enzyme. Finally, the altered stem cells were transfused back into the patient in the hope that they would continue to make the enzyme and prevent the accumulation of toxic immunosuppressive metabolites.

The first recipients of this gene therapy were two young girls aged 11 and 6. They had both suffered from one severe infection after another. Respiratory infections succeeded each other with distressing regularity, and their conditions gradually deteriorated as progressive tissue damage occurred. Once their problem was identified as an ADA deficiency, they were treated with a modified bovine ADA. Within a few months their clinical condition improved as the infections stopped. The treatment saved their lives but clearly did not cure them, and it had a few adverse side effects. Gene therapy was therefore decided on. In late 1990 and early 1991, the girls received their own bone marrow cells that had been infected with a retroviral vector containing the ADA gene. The treated cells were simply injected intravenously. These initial injections, although successful, could not provide permanent cures since the transfused cells lived only for a few months and the transfusion had to be repeated. However, since there were no significant side effects, it was decided to use engineered stem cells. The stem cells could replenish themselves and thus provide a permanent cure. Following this treatment, the girls appear to be permanently cured.

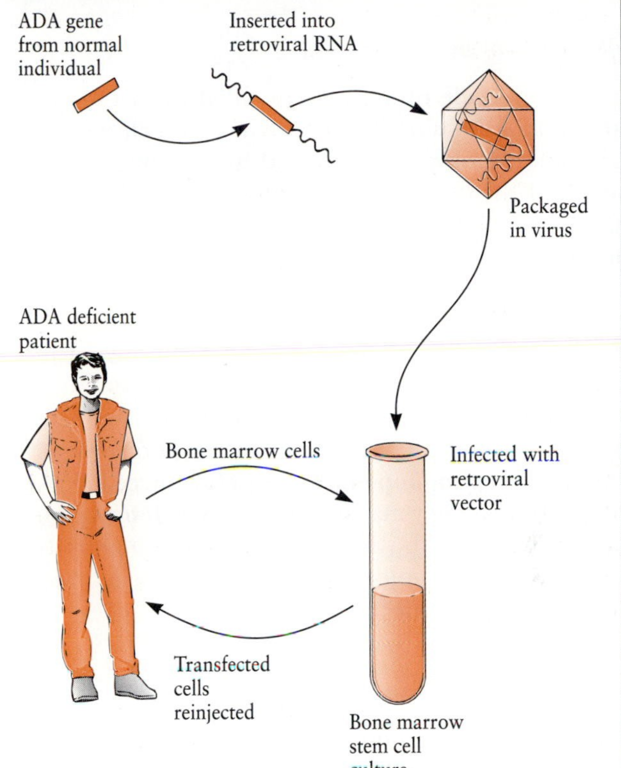

Gene transfer therapy for ADA deficiency. The gene for adenosine deaminase is inserted into the patient's bone marrow stem cells by using a retroviral vector.

some types of ADA deficiency may impair T-cell function by affecting CD26 in some way.

Other forms of CID have been described that result from a failure of development of lymphoid stem cells or from defects in thymic epithelial function. Treatment of these involves transplantation of stem cells from fetal bone marrow or liver. Graft-versus-host disease (Chapter 19) is the major complication of this therapy. A family of genetic diseases characterized by a severe combined immunodeficiency occurs as a result of defects in MHC expression. Some patients have an MHC class I deficiency, some have a class II deficiency, and others have both. All patients suffer from recurrent bacterial or viral infections and die before the age of five. The defects appear to be in the activities of regulatory proteins that control MHC expression.

Deficiencies of the T-Cell System

Purine nucleoside phosphorylase (PNP) has a similar function to adenosine deaminase in that it cleaves the nucleosides deoxyguanosine and guanosine to form guanine. PNP deficiency is a rare inherited disease that presents with immunodeficiency, an autoimmune hemolytic anemia, and neurologic lesions. As a result of a PNP deficiency, toxic deoxyguanosine accumulates in the bloodstream, where it suppresses T-cell function and cell-mediated immune responses. Antibody production is usually normal in these patients.

Many other congenital T-cell lesions have been described. Some are associated with a poorly developed thymus, **lymphopenia,** and deficient cell-mediated responses. They may be associated with **hypogammaglobulinemia** or the development of autoimmunity. Absence of IL-1 receptors on T cells is associated with severe fungal skin infections.

Deficiencies of the B-Cell System

The first recognized immunodeficiency disease was a severe hypogammaglobulinemia described by Bruton in 1952. There are several types of B-cell defect: an X-linked recessive, an autosomal recessive, and a sporadic form.

When children are born, they possess IgG acquired by transplacental passage from their mother. This passively acquired maternal IgG provides protection against infection for several months. As this maternal antibody wanes, children normally develop their own antibodies. In hypogammaglobulinemic children, however, these antibodies fail to develop. As a result, the children gradually begin to suffer from severe recurrent infections with such organisms as *S. pneumoniae, Staphylococcus pyogenes,* and *Haemophilus influenzae,* although they show normal resistance to viruses such as vaccinia, rubella, and mumps. Hypogammaglobulinemia is relatively easily diagnosed by serum electrophoresis (Chapter 13), which demonstrates reduced γ globulin levels. Hypogammaglobulinemic children usually have no B cells or plasma cells, and their lymphoid organs lack cells in the B-dependent areas. Boys with X-linked **agammaglobulinemia** do have normal numbers of pre-B cells, but these fail to develop to functional maturity. Their genetic defect lies in a tyrosine kinase gene, which is required for proper maturation of B cells into plasma cells. Treatment is through repeated administration of immune serum globulin in an attempt to passively immunize the children against infection. A secondary immunodeficiency of adults known as common variable immunodeficiency is associated with the absence of plasma cells (B-cell numbers are apparently normal) and low levels of all immunoglobulin classes. The disease appears to be due to an inability of B cells to secrete immunoglobulins and may result from a defect in T-cell–B-cell interaction. Some patients with this condition have deficiency of the enzyme 5′-nucleotidase.

There are a variety of congenital immunodeficiency syndromes in which there is a selective deficiency of some but not all of the major immunoglobulin classes. These deficiencies may be due either to the absence of a specific class or to production of nonfunctional antibodies. The most common primary immunodeficiency, occurring in about 1 in 600 Europeans, is a heterogeneous group of disorders that result from a selective IgA deficiency. This occurs as a result of a B-cell differentiation arrest. This disease may be asymptomatic, but more commonly, it leads to recurrent respiratory and digestive tract infections. IgA deficiency is also associated with severe allergies, since excessive quantities of IgE are commonly produced. In the absence of IgA, antigens may be able to penetrate mucosal surfaces and reach and stimulate B_ϵ cells. The disease is diagnosed by demonstrating an absence of IgA in serum and respiratory tract fluids. Some abnormalities of T-cell function and some IgG subclass deficiencies have also been described in these individuals.

Another interesting immunodeficiency syndrome is the X-linked hyperimmunoglobulin M syndrome. Patients with this disease have very low IgA and IgG levels associated with very high levels of IgM. The disease results from a defect in the CD40 ligand (Chapter 14). Patients suffer from severe respiratory disease, otitis, diarrhea, and autoimmunity.

A selective IgM deficiency has been described in humans, dogs, and horses. Affected humans suffer from overwhelming sepsis, and the condition is therefore treated with antibiotics. Deficiencies of each IgG subclass have also been described. These commonly occur in pairs. Thus IgG1 and IgG3 deficiency occur together, as do IgG2 and IgG4 deficiency. In many cases, affected individuals suffer from recurrent respiratory tract infections.

Complex Immunodeficiency Syndromes

The Wiskott-Aldrich syndrome is an X-linked recessive disorder characterized by diarrhea, loss of platelets (thrombocytopenia), eczema, and an immunodefi-

A CLOSER LOOK

A Failure in Cell Communication

Communication between lymphocytes is essential for the immune system to function correctly. This is well demonstrated in the rare immunodeficiency disease called X-linked hyperimmunoglobulin M syndrome (HIM). Patients with HIM produce large quantities of IgM but not of the other immunoglobulin classes. As a result, they are susceptible to recurrent infections.

Among the several key lymphocyte surface molecules that are required for helper T cells to activate B cells is the B-cell surface molecule CD40 (see Figure 14-4). B cells can be activated by exposure to monoclonal antibodies to CD40 plus IL-4. B-cell CD40 binds a ligand on the T cell. Gene mapping demonstrated that the genes for this CD40 ligand and for HIM were located at the same position on the X chromosome. It has since been shown that the CD40 ligand gene is defective in HIM patients. As a result of these defects, T cells are unable to mediate class switching and excess IgM is produced by the B cells but none of the other classes.

ciency reflected by recurrent middle ear infections. CD43 (sialophorin) is defective in these patients. Since CD43 acts as an activation receptor on T cells by binding ICAM-1(CD54), its deficiency leads to T-cell defects. It may also regulate the survival of blood cells in the circulation. Serum IgM is usually low, IgG and IgA levels are normal or slightly raised, and IgE levels may be high. This condition may gradually develop over several years as the immunologic defect gradually worsens. Treatment of this complex syndrome is difficult, but encouraging results have been achieved with a leukocyte extract.

Ataxia-telangiectasia is an autosomal recessive condition characterized by cerebellar lesions, dilation of blood vessels in the skin (telangiectases), recurrent upper respiratory tract infections, and endocrine abnormalities. The cerebellar lesions lead to ataxia. Many of these patients have no or abnormal thymus tissue. Most patients have absent or deficient IgA and IgE, and some have reduced IgG. B-cell numbers are normal but do not differentiate to IgA or IgE production. CD4 cells are depressed, so patients show delayed graft rejection or poor skin test responses. Ataxia-telangiectasia probably develops as a result of a defect in the regulation of the immunoglobulin gene locus. As a result, the expression of all proteins of this family is disordered.

Hypergammaglobulinemia E is a rare disorder associated with recurrent staphylococcal infections and chronic dermatitis. The abscesses formed in these cases may not elicit normal inflammatory responses. Affected individuals may have defective chemotaxis in their neutrophils and macrophages.

THE DIGEORGE ANOMALY

If the third and fourth pharyngeal pouches fail to develop, then affected individuals will produce neither thymic epithelium nor parathyroid glands. The resulting disease, known as the DiGeorge anomaly, is characterized by biochemical disturbances due to the absence of parathyroid glands and an absence of T-cell function. It may also be associated with cardiac and facial abnormalities.

Because of the absence of functioning T cells, individuals suffering from the DiGeorge anomaly fail to thrive and develop infections with the yeast *Candida albicans,* chronic pneumonia, and diarrhea, especially as a result of viral infection. They have few circulating lymphocytes and none with the characteristics of T cells. Immunoglobulin levels and antibody responses are relatively normal, implying that B-cell function and possibly helper-T-cell function are unimpaired. Possible treatments include the use of thymic hormones or transplantation of thymic epithelial cells, obtained either from an early fetus or from a cell culture.

IMMUNODEFICIENT MOUSE STRAINS

Our understanding of congenital human immunodeficiency diseases and the structure of the immune system has been helped considerably by analysis of spontaneous mutations leading to immunodeficiency in mice. The list of immunodeficiency mutations in mice is long. Some selected mutations are described here.

Nude Mice (nu)

The most important mouse model of immunodeficiency is the nude mouse (Fig. 26–7). Nude mice are a strain of hairless mice that fail to develop a functional thymus as a result of a mutation on chromosome 11. Because the T cells fail to develop with the thymus, these mice are deficient in mature T cells but possess a limited number of immature T cells and B cells. As a result, occasional lymphocytes are found in peripheral blood. Thymic grafts from normal mice, by restoring epithelial-cell function, permit the T cells of

Figure 26–7 A nude (*nu/nu*) mouse. In spite of an absence of T cells, these mice remain healthy if raised in clean surroundings apart from other mice.

nude mice to mature and develop immune competence. Nude mice are deficient in conventional cell-mediated immune responses, as reflected by prolonged allograft survival and lack of responses to T-cell mitogens. Their IgG and IgA levels also are depressed, presumably as a result of a loss of helper-T-cell function. *nu/nu* mice lack the ability to produce IL-3 yet have normal hematopoiesis.

Although nude mice have an enhanced susceptibility to virus-induced tumors, they fail to develop more than the normal level of spontaneous tumors. This observation was, for many years, a major objection to the **immune surveillance** theory since, if T cells destroy tumors, then T-cell–deficient animals should suffer from an increased incidence of neoplasia. However, nude mice possess normal levels of NK cells, which may protect them in the absence of T cells.

Severe Combined Immunodeficiency Mice (scid)

scid/scid mice have very low numbers of B cells and T cells. Development of B cells is halted prior to expression of cytoplasmic or cell membrane immunoglobulin. T-cell development is also arrested at a very early stage, and those lymphocytes that do reach the bloodstream are $CD4^-CD8^-$. They have no immunoglobulins and are unable to mount cell-mediated immune responses. *scid/scid* mice survive relatively well for one year in specific pathogen-free facilities but die in their second year usually as a result of *P. carinii* pneumonia. The defect in *scid* mice results from an inability to correctly rearrange their immunoglobulin or TCR genes. When T- and B-cell precursors initiate expression of their antigen receptors, a defective recombinase enzyme cuts and rejoins the receptor gene segments in an aberrant fashion. As a result, the cells cannot produce functional receptors, and no functional T or B cells are produced. The *scid* mutation also leads to an increased sensitivity to ionizing radiation since these animals cannot effectively repair DNA damage. About 15% of *scid* mice are "leaky," so that they have low levels of immunoglobulin of limited heterogeneity and can reject allografts. Antigen-presenting cells, myeloid and erythroid cells, and NK cells are normal in *scid* mice.

Motheaten mice (me)

Motheaten mice have a severely defective T-cell system but produce excessive quantities of immunoglobulin and develop autoimmune disease. Their name comes from their appearance. Within a few days of birth, neutrophils invade their hair follicles and cause patchy loss of pigment. These animals lack cytotoxic T cells and NK cells. The *me/me* mice have a very short life span and usually die as a result of an autoimmune pneumonia. The thymus of these animals shrinks unusually early, and emigration of prothymocytes into the thymus is impaired. The excessive B-cell activity may be due to overproduction of some B-cell–stimulating cytokines.

X-linked Immunodeficiency Mice (xid)

CBA/N mice have a recessive X-linked B-cell defect that makes them unable to respond to certain T-independent carbohydrate antigens. They lack some B cell subsets. It has been suggested that the lesions in CBA/N mice resemble those of the Wiskott-Aldrich syndrome in humans. The *xid* mice lack $Thy\text{-}1^+$ dendritic cells in their epidermis, although they do have $Thy\text{-}1^-$ intraepithelial lymphocytes in their intestine.

Beige Mice (bg)

Beige mice suffer from Chédiak-Higashi syndrome (page 402). As such, they lack NK cells. In addition, their cytotoxic-T-cell function is impaired and their granulocytes show reduced chemotaxis and cytotoxicity as a result of their lysosomal membrane defect.

Lipopolysaccharide Response (lps)

lps Mice do not respond to lipid A from lipopolysaccharide as a result of an almost total absence of B cells responding to lipid A. The B cells presumably lack a lipid A receptor. The macrophages of these mice also fail to develop Fc receptors.

By appropriate crossing, mice can be produced that lack any form of immune response. These *bg/nu/xid* mice are very severely immunosuppressed since they lack T, B, and NK cells. Lightly irradiated *bg/nu/*

xid mice can accept human bone marrow xenografts. These mice thus develop a human immune system and, for example, are a very convenient model for studying AIDS, a virus that does not infect normal mice.

SECONDARY IMMUNODEFICIENCIES

Malnutrition and the Immune Response

It has long been recognized that famine and disease are closely associated, and we tend to assume that malnutrition leads to increased susceptibility to infection. This is not necessarily true since the effects of malnutrition on immune functions are complex. For example, malnutrition can include not only deficiencies, but also excesses or imbalances of individual nutrients. The clinical effects of malnutrition are greatest in persons with specific nutrient requirements, for example, infants, the very old, and individuals with severe infections.

In general, severe nutritional deficiencies reduce T-cell function and therefore impair cell-mediated responses, at the same time sparing B-cell function and humoral immunity. Thus starvation rapidly induces thymic atrophy and a reduction in the level of thymic hormones. The number of circulating T cells drops, and cells are lost from the T-cell areas of secondary lymphoid tissue; delayed hypersensitivity reactions are reduced, graft rejection is delayed, and interferon production may also be impaired. In contrast, severe starvation has little effect on B-cell functions. The B-cell areas in lymphoid tissues and the number of circulating B cells remain unchanged. Serum immunoglobulin of all classes may remain normal or even rise. Secretory IgA levels commonly drop, but secretory IgE may rise, suggesting abnormal immunoregulation. Starvation will, however, result in depressed complement levels and impairment of neutrophil and macrophage chemotaxis, the respiratory burst, release of lysosomal enzymes, and microbicidal activity.

Specific nutritional deficiencies have a range of effects. Deficiencies of some B vitamins, vitamin A, and polyunsaturated fatty acids can depress immunoglobulin levels through effects on regulatory T cells. Magnesium deficiency causes a similar effect by a direct action on B cells. Deficiencies of vitamin A, vitamin B_{12}, and folic acid can depress cell-mediated immune responses. Zinc is especially critical for the proper functioning of the immune system. Zinc-deficient animals have depressed cytotoxic T-cell activity, depressed B-cell activity, and depressed NK-cell activity. If pregnant animals are deprived of zinc, their offspring have severely depressed immune function.

The effects of malnutrition may be reflected in altered resistance to infectious diseases. Because bacteria can readily survive and multiply in body tissues despite malnutrition of the host, malnutrition commonly increases the severity of bacterial diseases. Viruses, in contrast, usually require healthy host cells in which to replicate. Malnutrition, by rendering host cells unhealthy, can therefore increase resistance to viruses. Overnutrition can also influence susceptibility to viruses. For example, overfed dogs show an increased susceptibility to canine distemper virus and canine adenovirus 1. Parasites (especially intestinal helminths) usually compete with their host for nutrients. Should nutrition be reduced, the parasite may also be adversely affected; the effects of malnutrition on parasitic infections can therefore be variable.

Exercise and the Immune Response

Regular moderate exercise boosts the immune system. Thus increased antibody responses are seen in mice that take moderate exercise as compared with unexercised controls. Exercised mice also show a delay in tumor growth after administration of syngeneic tumor cells. On the other hand, strenuous exercise appears to induce enhanced susceptibility to infectious disease. For example, top runners are unusually susceptible to upper respiratory tract infection. The faster the runner, the greater the susceptibility. Lymphocyte function is significantly reduced in runners completing a marathon. This reduction correlated with increased steroid levels and may be considered the result of stress. Regular intense training in endurance athletes can also induce long-term damage to the immune system. They have lower than normal IgA and IgG levels, lymphocyte counts, and NK-cell activity. In addition, intensively exercised individuals show decreased neutrophil functions. Thus although mild exercise is good for immune function, intensive prolonged exercise may induce a functional immunodeficiency.

Trauma and the Immune Response

Severely injured individuals (wounded or burnt, for example) commonly die from sepsis as a result of an immunodeficiency. Corticosteroids, prostaglandins from damaged tissues, and a small protein called suppressive active peptide, which appears in serum following a burn, all have immunosuppressive properties. The deficiency occurs within minutes or hours and recovers as wounds heal. It affects T-cell, macrophage, and neutrophil function, but B-cell function appears to be normal. As a result, delayed hypersensitivity reactions, allograft rejection, and T-dependent antibody responses are all impaired. IL-2 and IL-2R production are reduced. $CD8^+$ cells are increased in injured individuals, suggesting that suppressor-cell function may be enhanced. Macrophages lose antigen-presenting ability as they express decreased levels of MHC class II molecules. Neutrophil and macrophage phagocytosis and respiratory burst activity are both impaired.

Age and the Immune Response

Both cell-mediated and humoral immune responses decline with advancing age.

T Cells. Aged T cells lose their ability to progress through the cell cycle. Very early events in the T-cell response to antigen such as activation of protein kinase C and the rise in intracellular calcium are impaired with age. Even after expressing IL-2 receptors and being exposed to IL-2, the aged T cell may not be able to respond effectively. Cytokine, especially IL-2 synthesis, may be impaired. Some aged T cells produce a normal amount of IL-2, but there are fewer of them. Aged T-cell populations are a mixture of fully functional and impaired cells. There is a major decline in the numbers of $CD4^+$ cells as a result of thymic involution. In mice the export of T cells from the thymus falls tenfold by six months of age. Nevertheless, the aged thymus retains the capacity to support peripheral T cells.

B Cells. The bone marrow is relatively unaffected by age, and aged bone marrow cells can reconstitute the body as well as young ones. Nevertheless, the ability of B cells to respond to signals from T cells is depressed and their affinity for antigen may be reduces. B cells tend to shift away from IgG production and produce more autoantibodies. If aged B cells are mixed with young T cells, the response is relatively normal. If the reverse is attempted (i.e., mixing young B cells with aged T cells), then the B cells respond poorly. B cells show a decline of about 20% in their mitogenic responses as they age. The comparable figure for T cells is 80% to 90%. Antigen processing and presentation are not affected by the aging process. However, aged macrophages show reduced responses to activating agents such as IFN-γ.

Toxins and the Immune Response

Many environmental toxins such as polychlorinated biphenyls, polybrominated biphenyls, iodine, lead, cadmium, methyl mercury, and DDT have a suppressive effect on the immune system. Mycotoxins may be important in situations in which animals are fed moldy grain. These include T2 toxin from *Fusarium* sp., which depresses the response of lymphocytes to mitogens and decreases the chemotactic migration of neutrophils. T2 toxin also reduces serum IgM, IgA, and C3 levels. Some of the aflatoxins may also be immunosuppressive and increase the severity of infections.

Other Secondary Immunodeficiencies

Immunodeficiencies may arise as a result of a wide variety of insults to the body. For example, immunoglobulin synthesis is generally much reduced in individuals suffering from absolute protein loss. Thus immunosuppression occurs in patients with renal failure, in heavily parasitized or tumor-bearing individuals, and in persons who have suffered severe burns or trauma. Stress may result in immunodeficiencies. For example, by chilling newborn puppies for five to ten days, it is possible to provoke an immunodeficiency syndrome similar in appearance to CID. Other stresses such as rapid weaning, prolonged transportation, and overcrowding are all recognized as effective immunosuppressants in animals. Sleep deprivation reduces the response of lymphocytes to pokeweed mitogen. Physical destruction of lymphoid tissues can result in immunodeficiencies. For example, destruction of lymphoid tissue leading to immunosuppression may occur in tumor-bearing animals, especially if the tumors themselves are lymphoid in origin.

KEY WORDS

Adenosine deaminase p. 404
Agammaglobulinemia p. 406
Ataxia-telangiectasia p. 407
Bone marrow depression p. 410
Chédiak-Higashi syndrome p. 402
Chronic granulomatous disease p. 400
Combined immunodeficiency p. 404
Cyclic neutropenia p. 400
DiGeorge anomaly p. 407
Exercise p. 409
Hypogammaglobulinemia p. 406
Immunodeficiency p. 400
Leukocyte adherence deficiency p. 402
Lymphopenia p. 406
Malnutrition p. 409
Neutropenia p. 400
Neutrophil deficiencies p. 400
Neutrophil dysfunction p. 400
Nude mice p. 407
Pneumocystis carinii p. 404
Primary immunodeficiency p. 400

QUESTIONS

1. The diseases associated with defects in the NADP oxidase system are called
   a. severe combined immunodeficiency
   b. Chédiak-Higashi syndrome
   c. Wiskott-Aldrich syndrome
   d. chronic granulomatous disease
   e. X-linked immunodeficiency
2. The main lesion that occurs in combined immunodeficiency is loss of
   a. T and NK cells
   b. B and T cells
   c. neutrophils and macrophages
   d. helper and cytotoxic T cells
   e. eosinophils and neutrophils
3. In leukocyte adherence deficiency there is a defect in which cell surface molecule?
   a. CD4
   b. CD8
   c. CD45
   d. CD11b/CD18
   e. CD23
4. The failure of the thymus and parathyroids to develop is called
   a. reticular dysgenesis
   b. combined immunodeficiency
   c. secondary immunodeficiency
   d. Wiskott-Aldrich syndrome
   e. DiGeorge anomaly
5. Which strain of mice fails to develop a thymus?
   a. New Zealand white
   b. *lpr*
   c. beige
   d. motheaten
   e. nude
6. An inherited failure to develop a functioning immune system is called
   a. primary immunodeficiency
   b. secondary immunodeficiency
   c. acquired immunodeficiency
   d. X-linked immunodeficiency
   e. genetic immunodeficiency
7. A deficiency of which of the following minerals is most likely to lead to an immunodeficiency?
   a. calcium
   b. zinc
   c. lead
   d. copper
   e. iron
8. Infection by which of the following organisms is characteristic of immunodeficiency?
   a. *Escherichia coli*
   b. Influenza virus
   c. *P. carinii*
   d. *Mycobacterium tuberculosis*
   e. *Salmonella typhimurium*
9. Successful gene therapy has been achieved for which of these immunodeficiency diseases?
   a. DiGeorge anomaly
   b. Chédiak-Higashi syndrome
   c. leukocyte adherence deficiency
   d. adenosine deaminase deficiency
   e. juvenile diabetes mellitus
10. In old age, which component of the immune system appears to be most impaired?
    a. B cells
    b. neutrophils
    c. NK cells
    d. macrophages
    e. T cells
11. How would you investigate an individual whom you believed to be suffering from an inherited deficiency in the macrophage respiratory burst? What would be the characteristics of such a deficiency?
12. LAD (leukocyte adherence deficiency) has been described in the dog. What would you predict the clinical signs to be? What sort of treatment, if any, would be appropriate for this animal?
13. Why is an increase in susceptibility to infectious disease associated with severe protein-calorie malnutrition?
14. Bone marrow transplantation is a common treatment for many immunodeficiency diseases. What are the advantages and disadvantages of this form of therapy?
15. How would you distinguish between an immunodeficiency disease caused by a defect in immunoglobulin synthesis and one caused by a defect in complement synthesis?
16. How do deficiencies in the T-cell system differ from deficiencies in the B-cell system?
17. What would be the predicted clinical signs of a deficiency in thymic hormones?
18. What histological lesions might be seen in a lymph node in an animal suffering from the following: (a) a B-cell deficiency; (b) a T-cell deficiency; and (c) a macrophage deficiency?
19. Can spending very large sums of money on treating individual children with congenital immunodeficiency disorders be justified?

Answers: 1d, 2b, 3d, 4e, 5e, 6a, 7b, 8c, 9d, 10e

SOURCES OF ADDITIONAL INFORMATION

Anderson, D.C., et al. The severe and moderate phenotypes of heritable Mac-1, LFA-1 deficiency: Their quantitative definition and relation to lymphocyte dysfunction and clinical features. J. Infect. Dis., 152:668–689, 1985.

Arnaiz-Villena, A., et al. Primary immunodeficiency caused by mutations in the gene encoding the CD3-γ subunit of the T-lymphocyte receptor. N. Engl. J. Med., 327:529–533, 1992.

Arnaiz-Villena, A., et al. Human T cell activation deficiencies. Immunol. Today, 13:259–265, 1992.

Baehner, R.L., Nathan, D.G., and Karnovsky, M.L. Correction of metabolic deficiencies in the leukocytes of patients with chronic granulomatous disease. J. Clin. Invest., 49:865–870, 1970.

Boxer, G.J., Curnette, J.T., and Boxer, L.A. Disorders of polymorphonuclear leukocyte function. Hosp. Pract. [off.], 20: 129–138, 1985.

Boxer, L.A., et al. Correction of leukocyte function in Chédiak-Higashi syndrome by ascorbate. N. Engl. J. Med., 295: 1041–1045, 1976.

Callard, R.E., et al. CD40 ligand and its role in X-linked hyper-IgM syndrome. Immunol. Today, 14:559–564, 1993.

Carson, D., Kaye, J., and Seegmiller, J.E. Differential sensitivity of human leukemic T cell lines and B cell lines to growth inhibition by deoxyadenosine. J. Immunol., 121: 1726–1731, 1978.

Clark, R.A. Genetic variation in chronic granulomatous disease. Hosp. Pract. [off.], 25:51–72, 1990.

Fitzgerald, L. Exercise and the immune system. Immunol. Today, 9:337–339, 1988.

Gupta, S. New concepts in immunodeficiency diseases. Immunol. Today, 11:344–346, 1990.

Kameoka, J., et al. Direct association of adenosine deaminase with a T cell activation antigen, CD26. Science, 261:466–469, 1993.

Koller, L.D. Effect of environmental contaminants on the immune system. Adv. Vet. Sci. Comp. Med., 23:267–295, 1979.

Marx, J. Cell communication failure leads to immune disorder. Science, 259:896–897, 1993.

Orkin, S.H. Molecular genetics of chronic granulomatous disease. Annu. Rev. Immunol., 7:277–307, 1989.

Rosen, F.S., Cooper, M.D., and Wedgewood R.J.P. The primary immunodeficiencies. N. Engl. J. Med., 311:235–242, 300–310, 1984.

Rosen, F.S., and Seligmann, M. Immunodeficiencies. Harwood Academic Publishers, Switzerland, 1992.

Snyderman, R., and Cianciolo, G.J. Immunosuppressive activity of the retroviral envelope protein p15E and its possible relationship to neoplasia. Immunol. Today, 5:240–244, 1985.

Strober, S. T and B cells in immunological diseases. Am. J. Clin. Pathol., 68(suppl.):671, 1977.

Weigel, W.O. Effects of aging on the immune system. Hosp. Pract. [off.], 24:112–119, 1989.

CHAPTER 27

Acquired Immune Deficiency Syndrome (AIDS)

Luc Montaigner *When AIDS was first recognized as a disease entity in the early 1980s strenuous efforts were made in many laboratories to identify the causative agent. This proved difficult. Nevertheless, Luc Montaigner working in Paris was able to isolate a retrovirus, not from a patient with clinical AIDS but from a non immunodeficient homosexual with lymphadenopathy. This virus was called Lymphadenopathy associated virus (LAV). Other isolates of this virus were made, for example by Robert Gallo in the United States. After a period of confusion over its name, the virus was eventually renamed Human Immunodeficiency virus (HIV).*

CHAPTER OUTLINE

The Current State of AIDS
- *United States*
- *Worldwide*

The AIDS Viruses
- *HIV-1*
- *HIV-2*
- *Genetic Variability of HIV-1*

Immunological Lesions in AIDS
- *T-Cell Abnormalities*
- *B-Cell Abnormalities*
- *NK-Cell Abnormalities*
- *Macrophage Abnormalities*
- *HIV in Other Cells*

AIDS—The Disease
- *Secondary Infections*
- *Neurologic Lesions*
- *Kaposi's Sarcoma*
- *Diagnosis of HIV Infection*
- *Treatment of HIV Infection*
- *An AIDS Vaccine?*

AIDS in Animals

Other Human Viruses Affecting the Immune System

AIDS Prevention and Its Social Implications

CHAPTER CONCEPTS

1. AIDS is a disease syndrome caused by infection with human immune deficiency virus–1, a retrovirus.
2. HIV-1 infects cells bearing CD4 on their surface. It thus infects and destroys helper T cells, although other cells of immunological importance are also affected in AIDS patients.
3. The clinical signs of AIDS are almost entirely due to secondary infections developing as a result of a loss of $CD4^+$ cells.
4. HIV-1 may attack the nervous system.
5. No vaccine is available against AIDS, and current treatment cannot ''cure'' AIDS patients.

Only one disease is described in this chapter—the virus-induced immunodeficiency syndrome called **acquired immune deficiency syndrome,** or AIDS. The chapter begins with a review of the current status of AIDS throughout the world. It is a much greater problem in other regions such as Africa than in the United States. We then go on to describe the structure of the virus that causes AIDS in some detail since this has a direct bearing on the difficulties encountered in controlling the infection. This is followed by a detailed review of the damage that this virus does to the cells of the immune system. This leads to a description of the diseases associated with AIDS. Some other viruses have similar effects on the immune system of animals, so these are briefly described. Finally, the complex social issues raised by an infection such as AIDS are mentioned to provide a relatively complete overview of this terrible disease.

The concept that viruses could destroy lymphocytes and so induce a profound **immunodeficiency** is not new. It had long been recognized that diseases such as measles in humans or distemper in dogs were accompanied by a significant **immunosuppression,** and as a result, infected individuals were more susceptible than normal to other secondary infections. Nevertheless, none of these viruses induced an immunodeficiency as severe as that induced by the human immunodeficiency virus (HIV).

Beginning in 1979, physicians began to notice a remarkable increase in cases of two unusual conditions among active male homosexuals. These conditions were a rare tumor known as Kaposi's sarcoma and a severe pneumonia caused by a parasite called *Pneumocystis carinii.* Affected individuals were found to be profoundly immunosuppressed as a result of an almost total loss of their T cells. Evidence suggested that the disease was a form of immunodeficiency transmitted by an infectious agent, so it was called the acquired immune deficiency syndrome.

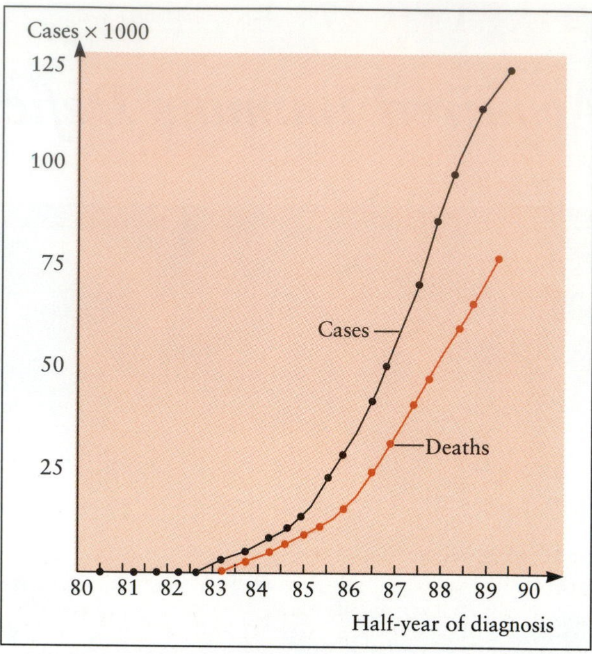

Figure 27–1 Cumulative AIDS cases and deaths in the United States by half-year since 1980.

THE CURRENT STATUS OF AIDS

United States

From June 1981 until July 1993, AIDS has been diagnosed in over 289,000 persons in the United States (Fig. 27–1). In addition, it is estimated that between one and two million individuals in the United States are infected with HIV but do not yet show symptoms. About 62% of these infected individuals are homosexual or bisexual men (Table 27–1). However, the disease also affects intravenous drug abusers (30%), heterosexual contacts of bisexual men, and infants born to women with AIDS. The disease is also transmitted by whole-blood or blood products. Before the introduction of effective screening procedures, transmission by blood was a significant problem in hemophiliacs and others who required repeated blood transfusions. This clearly points to AIDS being caused by an infectious agent.

Table 27–1 United States AIDS Statistics as of July 1, 1993*

Infected Population	Number	Percent
Adult		
Homosexual/bisexual	172,085	55
Intravenous drug abuser	73,610	24
Intravenous drug abuser (homosexual)	19,557	6
Heterosexual	21,873	7
Blood recipient	5,733	2
Hemophiliac	2,762	1
Other	15,060	5
	310,680	100
Pediatric		
From mother	3,887	87
(Intravenous drug user)	(1,768)	
(Sexual partner of intravenous drug abuser)	(761)	
Hemophiliac	194	4
Blood recipient	315	7
Undetermined	84	2
	4,480	100

*Centers for Disease Control and Prevention, July 1991–June 1993

Table 27–2 Worldwide AIDS Statistics as of July 1, 1993*

Location	Total Number of Cases	Estimated Number of Infected Individuals
United States	289,320	1 million +
Central and South America	81,766	1.5 million
Africa	246,127	9 million
Europe	92,822	500,000
Middle East	1,799	75,000
Southeast Asia	2,002	2 million
Western Pacific	5,058	25,000
Total	718,894	14 million +

**World Health Organization Global Statistics*

Worldwide

About 50% of the cases of AIDS reported worldwide have occurred in the United States (Table 27–2). However, this is very misleading since it is estimated that in the United States about 90% of AIDS cases are reported to the authorities, whereas widespread underreporting is certain in many underdeveloped countries. Taking this underreporting into account, the World Health Organization (WHO) now estimates that approximately seven million cases had occurred worldwide by the end of 1992. This implies that as many as nineteen million people may already be infected with HIV. It is also estimated that as many as three million people may already have died from AIDS. Projected estimates of the number of persons infected with HIV by the year 2000 range from 40 million to more than 100 million. Most of this increase will be in Africa and Asia (Fig. 27–2).

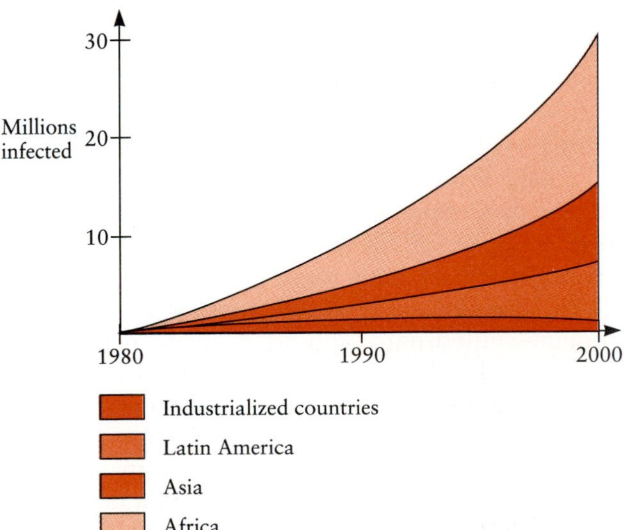

Figure 27–2 The predicted growth in the number of AIDS cases in the major regions of the world according to estimates made by the World Health Organization.

Although the first clinical cases of AIDS were reported from California in 1981, there is serological and historical evidence to suggest that it was present in central Africa at least as early as 1959. Thus antibodies to HIV have been detected in blood drawn in Zaire in 1959. In that same year a man in England died from a disease that at the time was diagnosed as leukemia. HIV has been detected in the preserved tissues of this man. Serologic evidence suggests that HIV began to spread in urban areas of Africa during the late 1970s. Isolated cases of AIDS have been diagnosed retrospectively in individuals from Central Africa in the mid-1970s. It has therefore been postulated that the virus originated in Central Africa, perhaps in monkey populations, and spread from there to Europe and North America. More recently, new AIDS-related viruses have been isolated in Africa.

The **prevalence** of AIDS has increased significantly in Africa since the mid-1970s. The epidemiology of the disease is very different from that seen in Europe and North America. For example, the disease is much more prevalent among heterosexuals in Africa than in Europe or the United States (Fig. 27–3), whereas homosexual transmission or transmission as a result of drug use remains low. The major factor contributing to transmission of the disease is heterosexual contact with many partners. Thus there is nearly an equal sex ratio among AIDS patients. As many as 88% of female prostitutes in some African countries are infected. This level reflects a significant increase from a prevalence of around 0% to 8% in 1980. The differences between the mechanism of transmission in Africa and North America or Europe might be a result of the high level of other subclinical infections in the African population. Thus genital ulcers or infection with *Chlamydia trachomatis* may promote invasion by HIV. Subclinical infection with diseases such as malaria or tuberculosis with activation of

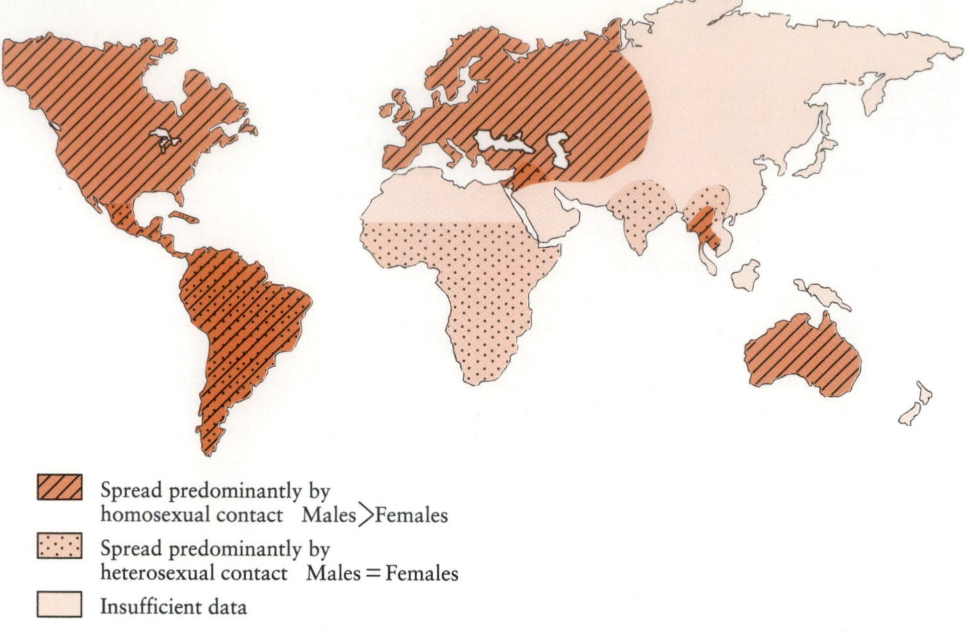

Figure 27–3 The global epidemiology of AIDS. There are two major patterns of spread. In countries such as the United States or in Europe, the disease is spread primarily by homosexual contact and drug abuse. In areas such as Africa and southeast Asia, AIDS is spread primarily by heterosexual contact. In countries such as China, there appears to be a very low incidence of AIDS at present. Such cases that do occur may be derived by homosexual conduct or by heterosexual contact during visits to southeast Asia. In South America and the Middle East, a reluctance to admit to homosexual conduct makes accurate epidemiological analysis difficult.

the immune system is more common among African adults than Americans. The progression of the disease in Africans is similar to that seen in Caucasians, with about 6% of infected individuals developing disease annually.

In populations with a high level of heterosexual transmission and a high prevalence of HIV infection in women, increased perinatal transmission is bound to occur. Thus in Africa, about 30% to 65% of infants born of HIV-infected women are also infected. Postnatal transmission through breast milk has been described in addition to congenital infection. Infant mortality in Africa will increase by 30% as a result of AIDS, and more than a million uninfected children will be orphaned because their parents will have died from AIDS by the turn of the century. It is estimated that life expectancy at birth in sub-Saharan African cities could fall by as much as six years during the next few decades as a result of the impact of AIDS.

A similar situation is now emerging in India and Thailand. In both of these countries, heterosexual prostitution appears to be the main route of transmission and the sex ratio of cases is equal.

THE AIDS VIRUSES

All responsible opinion holds that AIDS occurs as a result of infection with one of the human immunodeficiency viruses. Very rarely, an immunodeficiency syndrome may develop in humans not infected with HIV. This does not support the idea that there is another significant cause of AIDS not detected by current techniques.

HIV-1

The primary cause of AIDS is a virus called human immunodeficiency virus–1 (HIV-1) (Fig. 27–4). HIV-1 has been isolated from blood, bone marrow, lymph nodes, spleen, plasma, semen, and saliva of infected individuals, supporting the concept that AIDS is transmitted by blood, blood products, and sexual contact.

HIV-1 is an RNA virus that employs the enzyme **reverse transcriptase.** It is therefore classified as a **retrovirus.** In retroviruses, the viral RNA is first transcribed into DNA using the reverse transcriptase (Fig. 27–5). This proviral DNA can then use the cell's own protein-synthesizing ability to generate new virions. HIV-1 is related to other members of the lentivirus family of retroviruses such as visna virus of sheep, equine infectious anemia virus, and feline immunodeficiency virus. On electron microscopy, HIV-1 is seen to be a spherical virus containing a dense core surrounded by a lipoprotein envelope that is acquired as the virus buds from the surface of an infected cell (Fig. 27–6).

At the center of the **virion** are two strands of RNA, each containing about 10,000 nucleotides. Attached

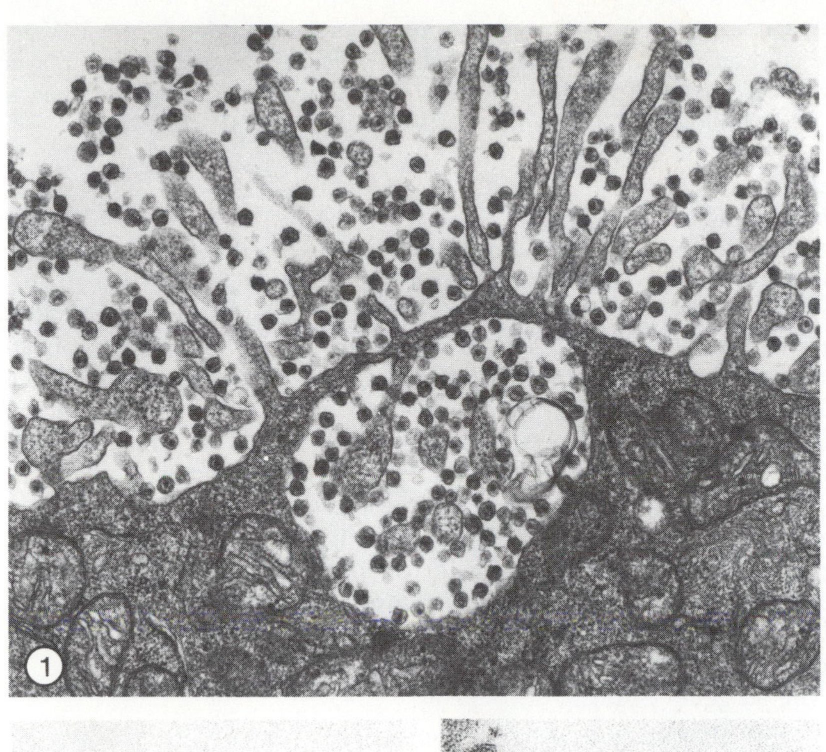

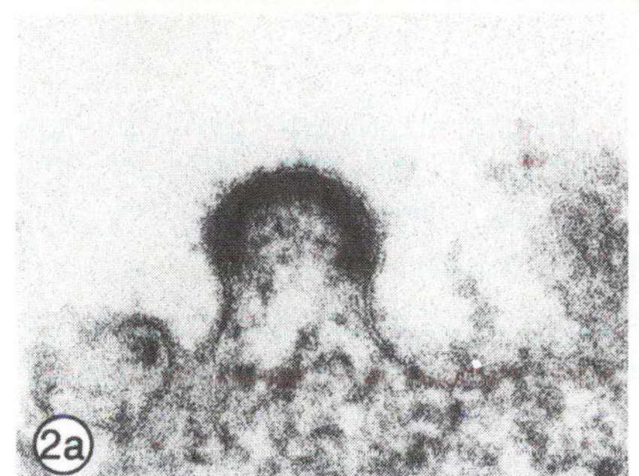

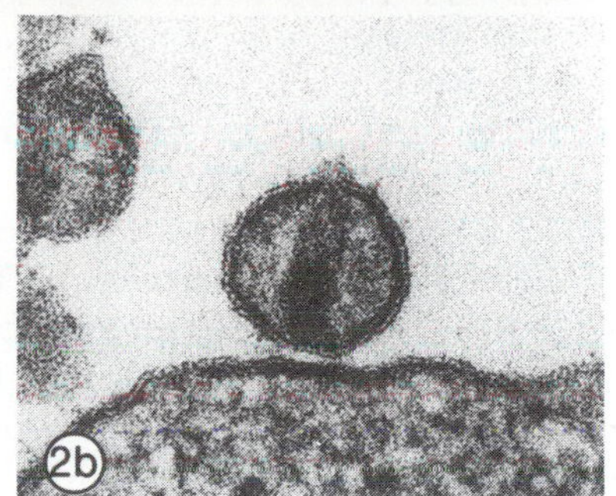

Figure 27–4 Electron micrographs of human immunodeficiency virus: (1) on a cell surface, (2a) budding from a cell, (2b) a mature virion. *(Courtesy of Dr. Robert C. Gallo.)*

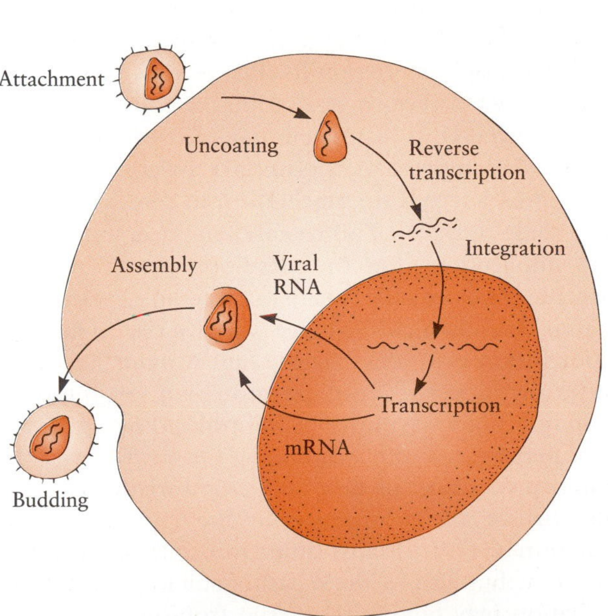

Figure 27–5 The mode of replication of a retrovirus such as HIV.

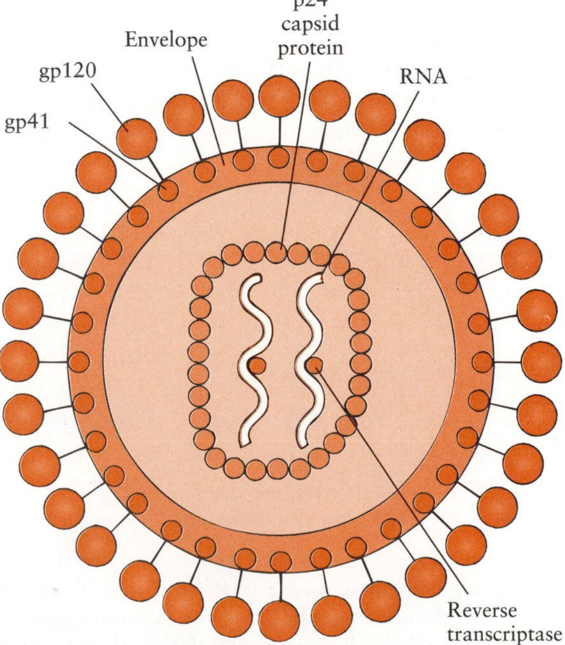

Figure 27–6 A schematic diagram showing the structure of human immunodeficiency virus (HIV).

to these strands are molecules of the reverse transcriptase as well as a protease, a ribonuclease, and an integrase. The RNA and its associated enzymes are contained within a **capsid** in the shape of a hollow truncated cone. The capsid is made of a protein called p24 (p stands for protein and the number is its mass). Surrounding the capsid is a layer of matrix protein called p17. The entire mass is covered by an envelope that consists of a lipid bilayer derived from the outer cell membrane of infected cells. Because of its origin, the envelope may also carry cell surface proteins such as MHC molecules. There are many spikes on the viral envelope formed by two glycoproteins. One glycoprotein (gp41) (gp stands for glycoprotein) is embedded in the membrane, and four molecules of another glycoprotein (gp120) form the spike.

The HIV genome is well characterized (Fig. 27–7). The viral RNA has two flanking long terminal repeat (LTR) sequences that act as binding sites for host transcription factors. There are three structural genes called *gag, pol,* and *env.* The *gag* gene codes for proteins that form the core (p24 and p17). The *pol* gene codes for a precursor protein that is cleaved to yield the four enzymes found in association with the RNA, the reverse transcriptase, the protease, the ribonuclease, and the integrase (this is endonuclease that helps integrate viral DNA into host cell DNA). The *env* gene codes for a precursor glycoprotein that is cleaved to yield the two envelope proteins gp120 and gp41.

There are three regulatory genes called *tat, nef,* and *rev.* The *tat* gene encodes a protein that attaches to the viral RNA and speeds up its transcription. The role of the *nef* gene is still unclear. Recent evidence suggests that it modifies the cell to make it more suitable for virus production. The *rev* gene also codes for a regulatory protein. This is responsible for switching the way in which viral RNA transcripts are processed once a cell has been infected for more than 24 hours.

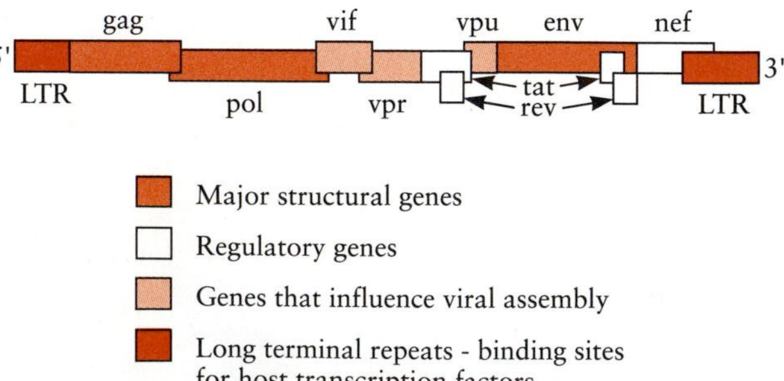

Figure 27–7 The structure of the genome of HIV-1. The genome is a single linear sequence of RNA. However, several of the genes overlap extensively. Thus, for example, *gag* and *pol* overlap. The genome is separated in the diagram to show the overlapping segments.

Three other proteins are encoded in the HIV genome. They are called *vif, vpr,* and *vpu.* They play a role in the assembly of new virions. The *vpu* gene is found only in HIV-1 and not in HIV-2.

HIV-2

HIV-2 is more closely related to the simian (monkey) immunodeficiency virus (SIV) than to HIV-1. Its genome shows about 40% homology to HIV-1 and 70% to SIV. The viral proteins of HIV-2 show variations in homology with HIV-1. Thus *gag* and *pol* show 50% **homology,** but the proteins encoded by *env, nef,* and *vif* show less than 30% homology. HIV-2 remains largely confined to West Africa. The prevalence of antibodies against the virus (seropositivity) ranges from 0.3% to 17%, but in prostitutes it may reach as high as 64%. HIV-2 spreads in a similar manner to HIV-1 in Africa, being largely heterosexually transmitted. Despite its similarity to HIV-1, it has reduced virulence as shown by a longer disease-free period.

Genetic Variability of HIV-1

HIV-1 shows extensive variability in its genetic structure. The most divergent part of the genome lies within the *env* gene, the gene for the major exterior glycoprotein on the envelope. This protein has both conserved and variable sites. The variable sites permit the structure of the envelope to vary in response to selective pressure from the immune response of an infected individual and permit the survival of viruses that can resist neutralization.

The nucleotide sequences of the *env* gene from an HIV isolate obtained from one patient at any given time vary by 2% to 3%, and new variants of HIV-1 arise continuously. The reason for this genetic variability is that the virus reverse transcriptases copy virus RNA into proviral DNA. These enzymes are very inaccurate and make on average one mistake in every 2000 nucleotides. There is no mechanism in the virus to correct these mistakes. This remarkable variation has major implications for the development of an HIV vaccine. Thus an HIV isolate put into an experimental animal such as a chimpanzee will give rise to new variants within two weeks. These new variants have antigenically different envelope proteins from the originating strain and cannot be neutralized by antibodies directed against the originating strain. Clearly, this rapid production of new antigenic variants will help the virus evade destruction by the immune system. These new HIV variants differ not only in their antigenicity, but also in their pathogenicity. The genetic variation may modify the tissue tropism of the virus and account for the great variety of clinical signs of

AIDS. The time needed for pathogenic variants to arise may contribute to the long incubation period of the disease.

Studies on the sequences of the *gag* genes of about 70 virus isolates from across the world have shown that there are seven distinct major genotypes of HIV-1 (A through G). *gag* codes for internal proteins and is thus relatively conserved. Genotype A is found in sub-Saharan Africa and Thailand. Genotype B is found in the Americas, Africa, Europe, and Asia and is thus the most widely distributed strain. Genotype C is found in Africa and India. The other genotypes are restricted to sub-Saharan Africa.

IMMUNOLOGICAL LESIONS IN AIDS

HIV infection affects the cells of the immune system, causing, over time, severe immune deficiencies. No other infectious agent is necessary, although other infections may hasten the development of an immune deficiency. The defect in the immune system is characterized by a loss of lymphocytes (**lymphopenia**) as a result of a selective deficiency of $CD4^+$ T cells (Fig. 27–8). In normal healthy individuals, the CD4:CD8 ratio is approximately 2.0. That is, there are about twice as many $CD4^+$ cells as $CD8^+$ cells. In AIDS patients the ratio drops to around 0.5; that is, there are twice as many $CD8^+$ cells as there are $CD4^+$ cells. $CD4^+$ helper T cells are responsible for successful immune responses. As a result of their $CD4^+$ T-cell deficiency, AIDS patients have major defects in their cell-mediated immune responses. Their loss of cell-mediated immunity may be demonstrated by decreased in vitro lymphocyte responses to mitogens and antigens; by decreased cytotoxic T-cell and NK-cell responses; and by a loss of **delayed hypersensitivity,** a T-cell–mediated inflammatory response to intradermally administered antigen. In addition, alterations in B-cell responses occur in AIDS patients. Thus they show elevated spontaneous immunoglobulin secretion, resulting in a hypergammaglobulinemia (Fig. 27–9).

The T-cell surface glycoprotein CD4 can serve as a high-affinity receptor for HIV-1. HIV-1 binds to the CD4 molecule through its envelope glycoprotein gp120 and probably enters the T cell by endocytosis.

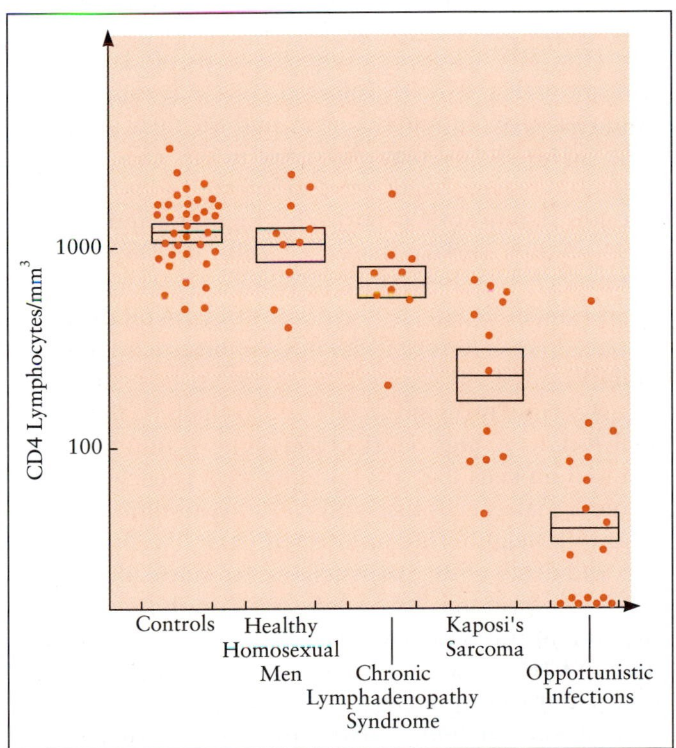

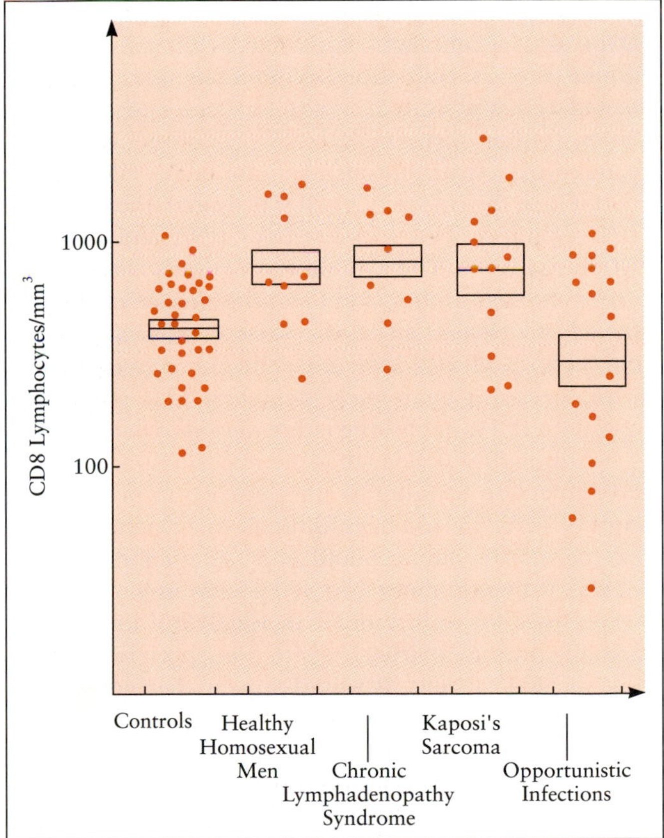

Figure 27–8 *A,* The sequential drop in the numbers of peripheral blood $CD4^+$ cells in patients with increasingly severe disease due to HIV-1 infection. Patients presenting with opportunistic infections had the lowest number of $CD4^+$ cells. *B,* The numbers of $CD8^+$ cells in patients with diseases due to HIV-1 infection. *(From Lane, H.C., et al. Am. J. Med., 78:417, 1985. With permission.)*

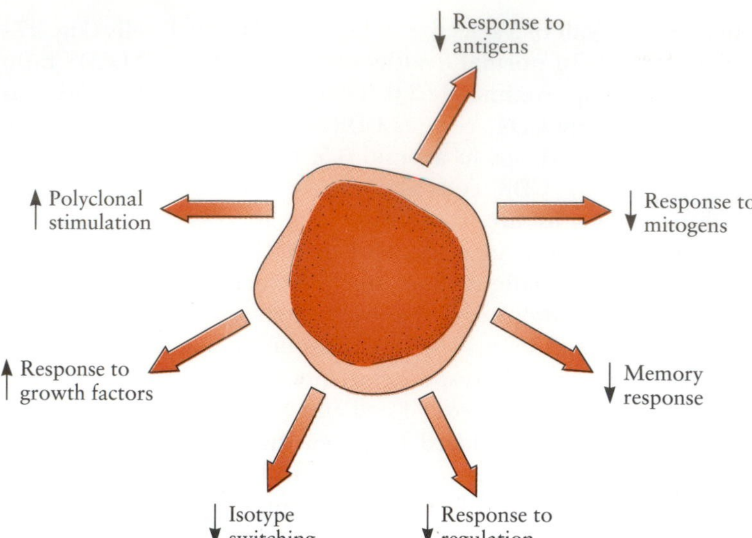

Figure 27–9 The various effects of HIV on B-cell functions.

Alternatively or additionally, the virus envelope may fuse with the T-cell membrane. Once inside cells, HIV-1 is uncoated. The viral genomic RNA is transcribed to DNA by the viral reverse transcriptase. This proviral DNA is integrated into the host cell chromosomal DNA in a process that uses the integrase coded for by the *pol* gene. Large amounts of unintegrated viral DNA also accumulate in infected cells. Once integrated, the virus infection becomes latent until the infected cell is activated by antigen, mitogen, or cytokines. Other viruses such as cytomegalovirus or Epstein-Barr virus may also activate latent HIV infection. Once the infected T cell is activated, the proviral DNA is transcribed to messenger RNA and protein synthesis occurs. The protein is processed, and virions are assembled as they bud from the surface of the infected cell. Several immune responses may have to occur before all HIV-infected cells are activated and $CD4^+$ cells are completely depleted. This could account for the variably long latent period of the disease.

Although it is clear that the major lesions seen in HIV infection are due to the destruction of $CD4^+$ T cells by HIV-1, it has been difficult to account for the extent of the immunodeficiency seen in these patients. Although about 1 in 400 T cells in blood is infected by HIV, only about 1 in 100,000 of these cells actually produces virus. Despite this, there is a major loss of $CD4^+$ T cells. It would appear that normal T-cell turnover should be able to compensate for this low rate of destruction, yet it does not do so. It is important to point out, however, that blood T cells represent only a small fraction of the total T-cell population in the body. The events in the blood are probably not typical of the events occurring elsewhere in the lymphoid system. Indeed, Giuseppe Pantaeo and Antony Fauci have shown that HIV preferentially replicates not in the blood but in the germinal centers of lymph nodes. Many of these germinal center cells contain proviral DNA but not RNA, suggesting that they are latently infected. Thus significant virus replication and T-cell destruction can occur in lymph node cells while the blood T cells appear to be relatively unaffected.

How HIV-1 kills or otherwise turns off T cells is unknown (Fig. 27–10). Infected cells may simply lyse when massive amounts of virus bud off their surface. The CD4 molecule may play a role in this since monocytes and macrophages are also infected with HIV yet are not significantly destroyed. They carry much less CD4 on their surface than do T cells. If $CD8^+$ T cells are artificially infected with HIV, they too are resistant to its cytopathic effect. The density of CD4 on the cell surface may therefore influence the destructive effect of HIV.

A second mechanism of T-cell loss may be through cell fusion. A high level of *env* gene product is expressed on cells that are budding viral particles. This protein makes cells fuse and causes **syncytium** formation with neighboring uninfected $CD4^+$ T cells. Cytolysis and death of the fused cells occur within 48 hours. The incorporation of uninfected lymphocytes into syncytia in this way may thus contribute to the depletion of $CD4^+$ cells.

A third destructive mechanism may be through immune destruction of helper T cells. $CD4^+$ T cells can capture, process, and present envelope gp120 on their surface, so making uninfected T cells a target for HIV-1–specific cytotoxic T cells. In addition, some AIDS patients produce a cytotoxic antibody reactive with activated lymphocytes.

Some investigators have suggested that antigen may trigger **apoptosis** of HIV-infected T cells. This

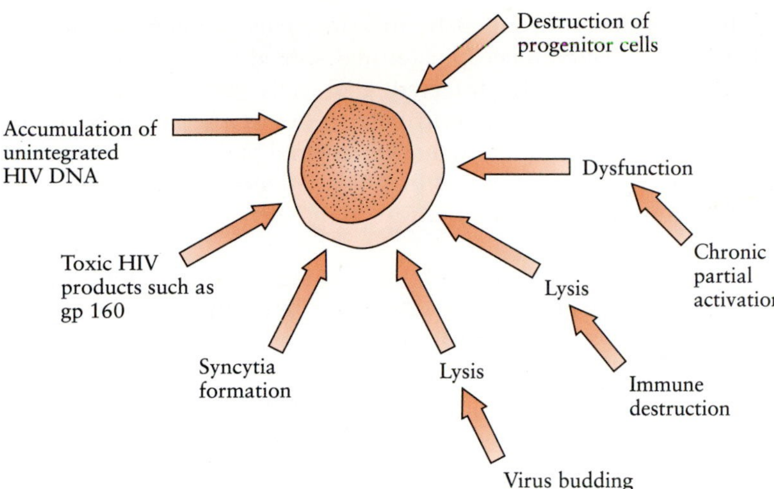

Figure 27–10 The various ways in which HIV may impair CD4 T-cell function.

could happen in two ways. Either gp120 binding to CD4 provides one signal while antigen, presented at a later time, triggers an apoptotic response. (Remember that both signals must be provided simultaneously in order to activate T cells.) Alternatively, antibody against bound gp120 may trigger apoptosis. Both mechanisms have been demonstrated to act in vitro. They imply that the T cells of HIV-infected individuals already carry one of two apoptosis-inducing signals and that almost any other pathogen may provide the second signal at a time when it would trigger apoptosis rather than an activating response.

Other postulated mechanisms that could account for major T-cell losses may involve the induction of soluble factors with toxic effects on $CD4^+$ cells; loss of a $CD4^+$ **stem cell** population; or loss of a subset of $CD4^+$ cells that are required for the production of the whole T-cell population.

T-Cell Abnormalities

In addition to destroying T cells, HIV impairs the function of surviving $CD4^+$ cells. There are several possible mechanisms for this functional defect. First, HIV-1 subunits such as the envelope are capable of directly inhibiting T-cell function. Second, HIV-1 gp120 binds avidly to CD4. In this location, it may interfere with the ability of CD4 to interact with MHC class II molecules and so blocks the ability of T cells to respond to presented antigen. Alternatively, gp120 may send a negative or desensitizing signal to the T cell, so preventing its response to antigen. T cells that are latently infected with HIV no longer express CD4 on their surface; they are defective in the synthesis and secretion of IL-2; they show decreased proliferation, decreased expression of the IL-2 receptor, and depressed production of IFN-γ. As a result of these changes, there may be a significant decline in immune function in patients whose CD4 cell numbers are normal.

HIV-1 induces a significant cytotoxic/suppressor–T-cell response in infected individuals. These $CD8^+$ cells are directed against viral epitopes and so may serve a protective role. The cytotoxic responses diminish with the progression of the disease, but they may contribute to its pathology. Some of the immunosuppression in AIDS patients may be due to stimulation of cytotoxic/suppressor activity as a result of an imbalance between $CD4^+$ and $CD8^+$.

B-Cell Abnormalities

HIV-1–infected individuals have abnormalities in B-cell function. They show polyclonal activation, hypergammaglobulinemia, immune complex deposition, and the development of autoimmunity. At the same time, their B cells show decreased proliferative responses to pokeweed mitogen and to new antigens. Although the loss of helper-T-cell function undoubtedly contributes to this defect, there is also evidence of B-cell abnormalities. For example, AIDS patients are unable to mount an adequate IgM response. Increased B-cell activity as reflected by elevated serum IgG and IgA levels is a consistent finding in AIDS. Many of the B-cell abnormalities appear to be a result of elevated IL-6 levels. B cells transformed by Epstein-Barr virus can be infected with HIV-1 since they express CD4 on their surface. The high incidence of Epstein-Barr virus and cytomegalovirus infection in AIDS patients may therefore contribute to the nonspecific stimulation of B-cell activity.

Among the many suggestions seeking to explain how HIV kills T and B cells is that the viral glycoprotein gp120 may act like a **superantigen.** Thus in HIV-infected individuals, some B-cell subpopulations, especially those bearing the immunoglobulin heavy-chain domain V_H3, are significantly reduced. It has now been shown that gp120 binds to V_H3. Indeed, gp120 can activate B cells bearing V_H3 in their BCRs. Thus it is conceivable that the binding of gp120 to V_H3 will so stimulate this subpopulation of B cells that they will eventually be depleted in the manner of the superantigens that trigger T-cell responses.

NK-Cell Abnormalities

Activity of natural killer (NK) cells is suppressed in AIDS patients, although NK-cell numbers are normal. These NK cells have reduced cytotoxic ability. This may be restored to normal by exposure of the NK cells to IL-2 or to concanavalin A. This suggests that NK cells are normally activated by T-cell–derived IL-2 and in its absence show defective function.

Macrophage Abnormalities

Monocytes and macrophages play a major role in AIDS since they are reservoirs of HIV through all stages of infection. Although CD4 is present in low concentrations on blood monocytes, the percentage of monocytes that show it detectably ranges from 5% to 90%. Monocytes expressing CD4 can be directly infected by HIV, and they may also ingest opsonized HIV. In some tissues such as brain, lymph nodes, and lung, 10% to 50% of macrophages may be infected. On the other hand, the macrophages in the liver (Kupffer cells) have not yet been shown to become infected in AIDS. HIV can replicate in macrophages, although, unlike T cells, the virions are not readily released but accumulate in large numbers in cytoplasmic vacuoles. Macrophages may constitute the major reservoir of HIV in the body since once they are infected, they can produce large quantities of virus but are not killed by the productive infection. Thus infected macrophages may spread the virus throughout the body. Infected glial cells and alveolar macrophages may contribute to the development of neurologic and lung lesions in some patients. The numbers of macrophages in AIDS patients are usually normal, even in patients who have very low or undetectable $CD4^+$ T cells.

HIV-infected macrophages show functional abnormalities (Fig. 27–11). Their adherence to surfaces is reduced, and they do not respond normally to chemoattractants. Their expression of MHC class II molecules is depressed, suggesting that their antigen-presenting ability is impaired. Their phagocytic ability is also impaired, and they show depressed clearance of particles from the blood. Persistently infected macrophages may be stimulated to release large amounts of prostaglandins, IL-1, TNF-α, and IL-6. This cytokine release may account in part for alterations in B-cell and hematopoietic-cell function. In the form of AIDS observed in Africa, patients lose so much weight that the condition is called slim disease. This may be due in part to the excessive and prolonged release of TNF-α.

It is possible that CD4-bearing dendritic cells lining mucosal surfaces are the first cells infected by sexually transmitted HIV. The follicular dendritic cells in lymph nodes are severely damaged early in HIV infection. The number of Langerhans' cells in the skin is reduced in AIDS patients, and this may contribute to the development of Kaposi's sarcoma.

HIV in Other Cells

HIV infects and affects many non–T cells. Because any $CD4^+$ cell is vulnerable to HIV infection, macrophages, monocytes, Langerhans' cells, follicular dendritic cells, microglia, and endothelial cells in the brain may

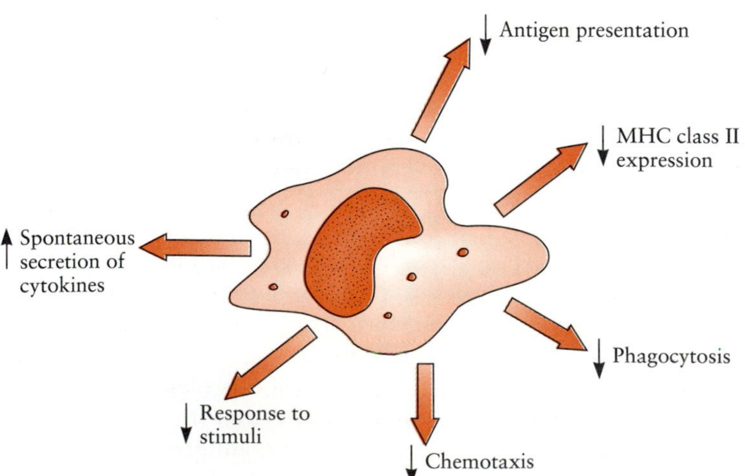

Figure 27–11 The effects of HIV-1 infection on macrophage function.

all be damaged. In addition, HIV causes a granulocytopenia, although neutrophil function is normal.

AIDS—THE DISEASE

HIV is poorly contagious and is spread either when contaminated fluids are injected or when infected lymphocytes are transmitted. The number of lymphocytes in sperm is elevated in individuals with a history of sexually transmitted diseases. The virus is known to be transmitted in blood (and in organ grafts), but blood is now screened in developed countries so that the risks of acquiring infection by this route have been significantly reduced. The fact that HIV has been recovered from saliva has given rise to fears that it might be spread by casual contact. It is highly unlikely, however, to be a natural route of virus transmission since salivary cells must first be cultured for isolation attempts to succeed. Normal household (casual) contact is also a highly unlikely mode of transmission. A small percentage of cases occur in heterosexuals with no known mode of transmission. This has given rise to fears in North America and Europe that the infection may increase in incidence in heterosexuals and spread rapidly through the population. This now appears very unlikely.

The average time from developing antibodies to HIV to the onset of disease may be as long as ten years. From the experience of those individuals who acquired the infection in the early 1980s, it is possible to estimate that each year approximately 6% of those positive for antibodies to HIV will develop clinical AIDS (Fig. 27–12). A wide range of disease syndromes is now recognized as being associated with HIV infection; they progress from mild asymptomatic infections to persistent generalized lymph node swelling (lymphadenopathy) and to AIDS itself (Fig. 27–13).

Most individuals infected with HIV develop a fever and muscle aches three to six weeks after initial infection (Fig. 27–14). This lasts for several days or weeks and is associated with **seroconversion** (antibodies to HIV begin to be produced). Large amounts of virus are in the bloodstream at this time, and the disease can be readily transmitted. At the end of this time the patient's immune response eliminates free virus and infected cells. As a result, detectable virus disappears. Unfortunately, some infected cells persist, and the virus continues to replicate in low numbers for a variably long period, commonly for several years. During this time the patient is usually fairly healthy. Eventually, however, the virus causes sufficient damage to the immune system that T cell numbers drop below the level necessary to maintain health and infections and tumors begin to develop.

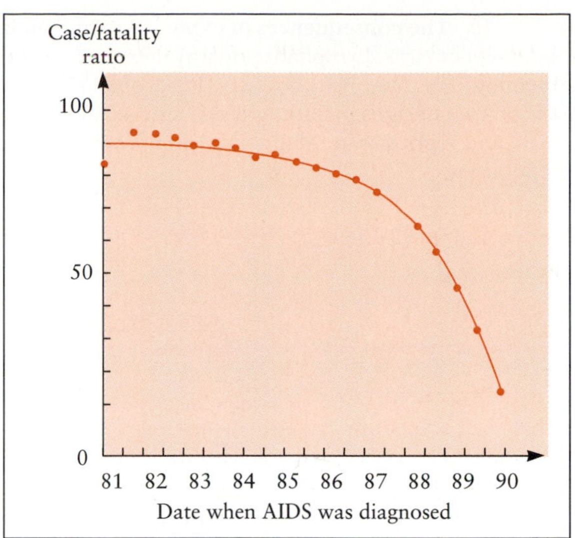

Figure 27–12 The case fatality ratio for HIV-1 infection related to the half-year when the disease was diagnosed. Thus 90% of patients diagnosed with the disease from 1980 to 1984 are now dead.

Persistent generalized **lymphadenopathy,** a more advanced disease, is characterized by the development of multiple swollen lymph nodes in the absence of other illness. The lymph nodes of these patients show hyperplasia of the follicles in the cortex and infiltration of germinal centers by $CD8^+$ cells. This eventually leads to germinal center depletion and follicular shrinkage (involution). Lymphadenopathy may persist for a long period without the patient showing clinical deterioration, but it may progress inexorably to AIDS itself.

Significant weight loss (HIV wasting syndrome) is also a common feature in AIDS patients. This may result in part from a loss of appetite, but it can also be due to increased or altered protein metabolism, intestinal malabsorption, chronic cytokine production, and endocrine abnormalities.

Secondary Infections

The full HIV-mediated disease syndrome can present in several manifestations. Some individuals simply have prolonged fever, diarrhea, or weight loss in the absence of concurrent infection. Nevertheless, numerous opportunistic infections (**secondary infections**) are usually associated with clinical AIDS (Table 27–3). These **opportunistic pathogens** include protozoa such as *P. carinii* (Fig. 27–15), *Toxoplasma gondii,* and *Cryptosporidium muris;* viruses such as cytomegalovirus and varicella zoster; bacteria such as *Mycobacterium avium;* and fungi such as *Cryptococcus, Candida,* and *Histoplasma.* These organisms, which cause few problems in normal persons, can cause severe, life-

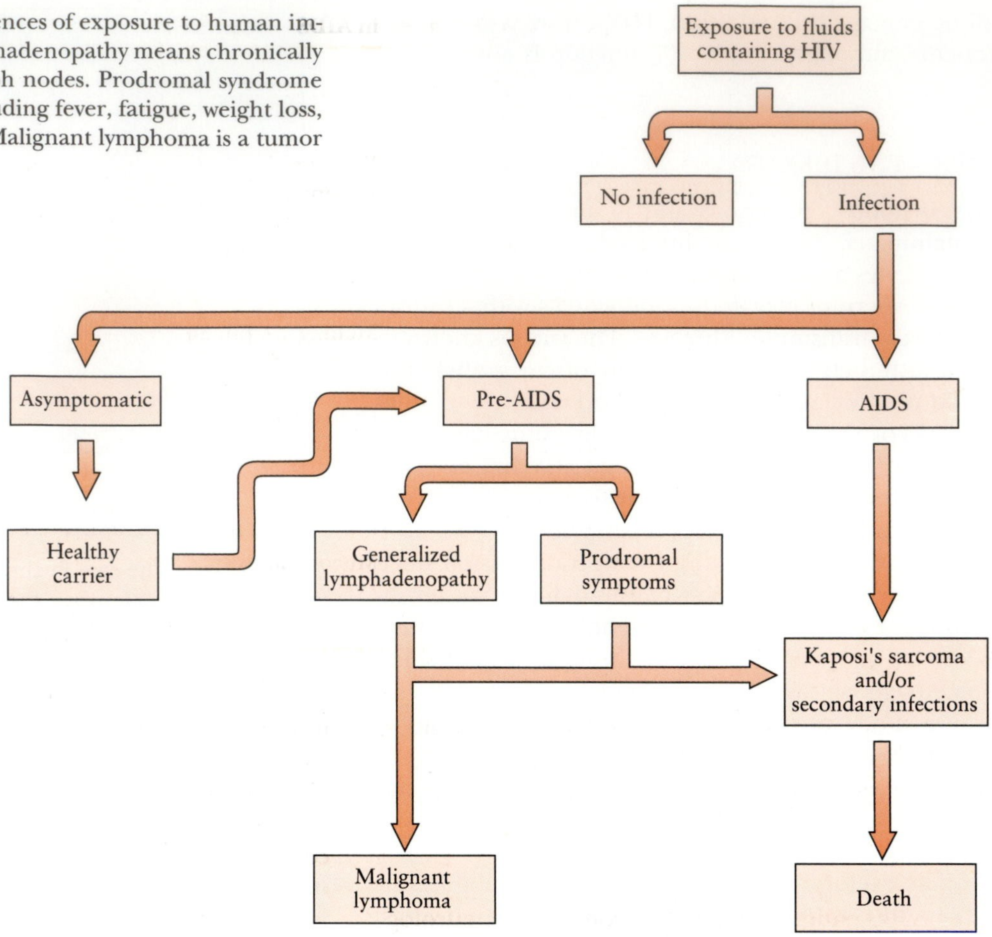

Figure 27–13 The consequences of exposure to human immunodeficiency virus. Lymphadenopathy means chronically inflamed and enlarged lymph nodes. Prodromal syndrome consists of a set of signs including fever, fatigue, weight loss, diarrhea, and night sweats. Malignant lymphoma is a tumor of lymphocytes.

Exposure to fluids containing HIV

No infection

Infection

Asymptomatic

Pre-AIDS

AIDS

Healthy carrier

Generalized lymphadenopathy

Prodromal symptoms

Kaposi's sarcoma and/or secondary infections

Malignant lymphoma

Death

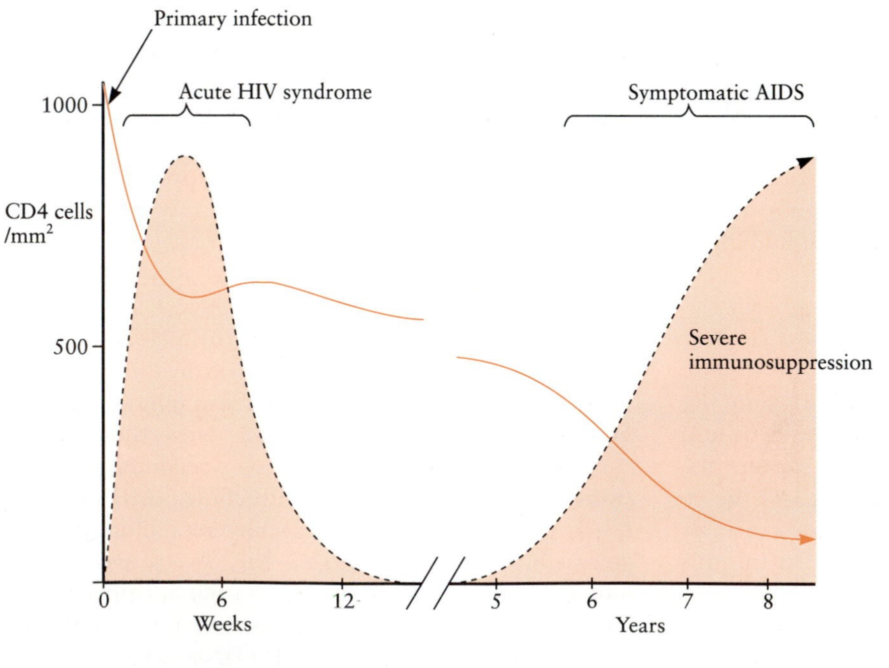

Figure 27–14 The clinical course of HIV infection. Over the period of infection there is a progressive loss of CD4 T cells that results in progressive waning of immunity. The rate of loss of these cells depends on the life-style of the patient, the patient's exposure to other pathogenic agents, the strain of HIV involved, and probably the patient's genetic background.

Table 27–3 Opportunistic Infections in AIDS

Organism	Effect
Protozoa	
Pneumocystis carinii	Pneumonia
Toxoplasma gondii	Encephalitis
Cryptosporidium muris	Gastroenteritis
Entamoeba histolytica	Gastroenteritis
Isospora belli	Gastroenteritis
Fungi	
Cryptococcus neoformans	Meningitis, pneumonia
Candida albicans	Pharyngitis, esophagitis
Bacteria	
Mycobacterium avium	Gastroenteritis
Mycobacterium intracellulare	Generalized infection
Mycobacterium tuberculosis	Generalized infection
Viruses	
Cytomegalovirus	Pneumonia, hepatitis, colitis, chorioretinitis
Herpes simplex	Mucocutaneous lesions
Varicella zoster	Disseminated herpes
Epstein-Barr virus	Pneumonitis, leukoplakia, lymphoma

threatening infections in HIV-1–infected individuals. *P. carinii* is the most common AIDS-related infection in the United States and Europe.

Neurologic Lesions

Some AIDS patients may suffer from severe neurological problems known as the AIDS dementia complex. These patients may show disturbances in cognition, movement, sensation, and behavior. Problems can develop even when there is very little brain infection with HIV, and it is not associated with secondary or opportunistic infections. CD4 is found on neurons and glial cells in several parts of the brain, so these cells may bind gp120. AIDS dementia complex results from the toxic effects of HIV-derived gp120 on neurons. The gp120 causes a rise in neuronal Ca^{2+} concentrations that eventually triggers cellular apoptosis. AIDS dementia complex has a poor prognosis, and the mean survival time after diagnosis in these cases is 11 to 15 months. About 25% of AIDS patients who show no neurological symptoms when alive have vacuoles in their myelin and macrophage infiltration in their brain on autopsy.

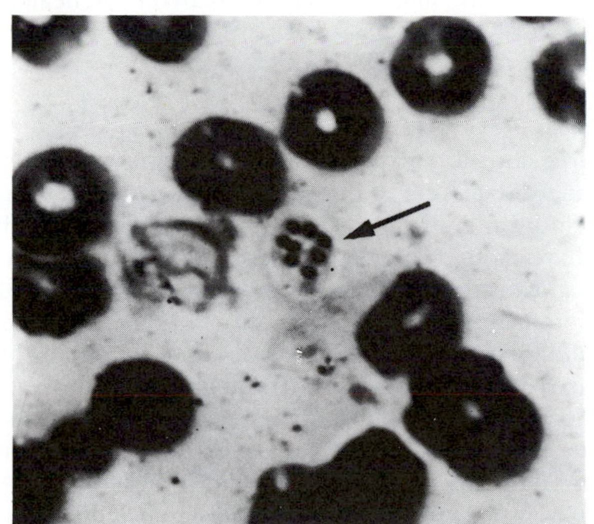

Figure 27–15 A stained preparation of a needle aspirate specimen from a case of *Pneumocystis carinii* pneumonia. This organism does not cause disease in patients with a fully functional immune system. Its presence is characteristic of a severe immunodeficiency. *(From Hughes, W.T. In Hunter's Tropical Medicine, 6th ed. W.B. Saunders Company, Philadelphia, 1984.)*

Kaposi's Sarcoma

The occurrence of a rare tumor called Kaposi's sarcoma in young homosexual men in New York and California led to the identification of the AIDS epidemic in the United States. Before that time, Kaposi's sarcoma was a rare skin cancer affecting elderly men of Mediterranean origin. It is now common in homosexuals with AIDS, occurring in about 36% of such patients (Fig. 27–16). Patients with Kaposi's sarcoma may still react to intradermally injected antigen and show less loss of $CD4^+$ cell function than AIDS patients without the tumor. Some investigators believe that Kaposi's sarcoma is a secondary disease triggered by an infectious agent different from HIV. This is because the tumor is far more prevalent among male homosexual AIDS patients than among AIDS patients from other

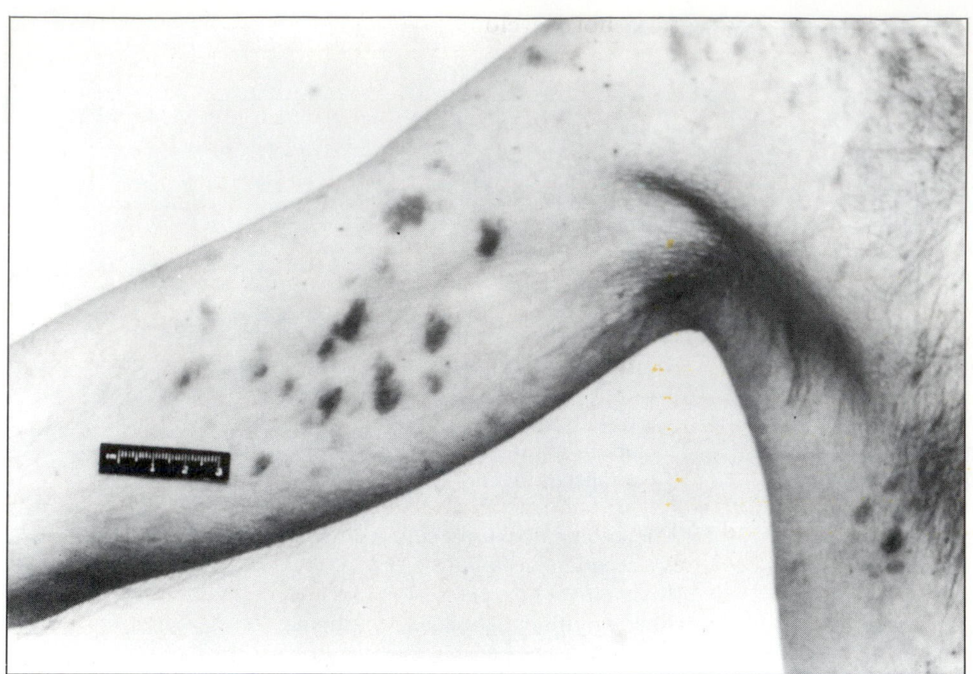

Figure 27–16 Kaposi's sarcoma on the arm of an AIDS patient. This tumor is not caused by HIV but probably results from secondary infection with another agent, possibly a form of papillomavirus. *(Courtesy of Dr. P. Mansell.)*

risk groups. One possible cause is a form of human papillomavirus. HIV promotes the development of Kaposi's sarcoma in several ways. Thus it stimulates macrophages to produce large quantities of IL-6 on which the sarcoma is dependent. Two other proteins, a T-cell–derived cytokine called oncostatin M and the *tat* gene product of HIV, are also growth factors for Kaposi's sarcoma. HIV infects Langerhans' cells in the skin, and it may therefore promote tumor growth in that region by reducing local defenses.

AIDS patients show an increased incidence of some lymphomas that may be associated with Epstein-Barr virus infection. The pathogenesis of these lymphomas is complex but may well reflect a breakdown in immune surveillance as a result of T-cell destruction. An autoimmune thrombocytopenia and bleeding disorder associated with increased platelet-bound IgG has been described in AIDS patients.

Genetic Factors. Susceptibility to HIV and the course of the disease are partially determined by genetic factors such as the MHC alleles DR5 and B35. In addition, HIV susceptibility is associated with certain alleles of a vitamin D–binding protein called Gc. HIV binds more strongly to a highly glycosylated form of Gc protein.

Diagnosis of HIV Infection

Because HIV may be transmitted by blood and blood products, it is necessary to screen blood for antibodies to this virus. The test used for this purpose is the ELISA. The sensitivity of the test is set at a very high level to ensure that no positive blood is missed. As a result, many false-positive ELISA results are obtained. These positive samples are in turn subjected either to Western blotting or to immunofluorescence assays to arrive at a definitive diagnosis. Between 1 in 6 and 1 in 20 positive ELISA samples are true positives as detected by Western blotting. Nevertheless, a carrier state can exist, and HIV may be isolated from healthy, antibody-free carriers.

Treatment of HIV Infection

Three antiviral drugs have been approved in the United States for the treatment of AIDS. These are azidothymidine (also called AZT or zidovudine), dideoxycytidine (ddC), and dideoxyinosine (ddI). All of these can block HIV replication by interfering with reverse transcription. Normally the proviral DNA chain grows by attaching each new nucleotide to a hydroxyl group located at the 3′ end of the chain to form a linkage called a phosphodiester bond. The nucleotide analogs, if present, are incorporated into the growing DNA chain. However, they lack the 3′ hydroxyl group, and as a result, no more nucleotides can be added after the chain is blocked. These compounds will therefore terminate proviral DNA chain growth and so inactivate the virus. Obviously, these compounds can also block DNA production in uninfected cells. All three drugs are therefore very toxic. A greater problem is the high mutation rate of the virus. The virus will eventually produce altered reverse transcriptases that can act even in the presence of these drugs, and their benefits are only temporary.

An AIDS Vaccine?

At present, no effective vaccine against AIDS is available. There are two main reasons why such a vaccine has not yet been developed. First, as described, HIV mutates very rapidly, producing a mixture of antigenically distinct viruses even within one infected individual. Second, HIV is transmitted via mucosal surfaces both as cell-free virions and by HIV-infected cells, and different types of response may be required to counteract this. In addition, there is a question of usage. Should the vaccine be employed to prevent infection of healthy individuals or, alternatively, to slow progression of the disease in those already infected with HIV? In experimental primate models, vaccines against HIV have been protective for only a short period and only when the vaccine strain is identical to the infecting virus. Likewise, it is clear that natural infection with HIV does not confer significant protective immunity.

Traditional approaches to vaccine development have involved immunization with either inactivated virus or recombinant HIV proteins using an adjuvant. These, it is hoped, will stimulate production of antibodies that could block HIV infection of cells. Unfortunately, the immunity induced is brief, only being effective when antibody levels are very high. An alternative approach, the use of a modified live virus in the vaccine, has a major drawback. The mutation rate of the virus is such that the virus could easily become virulent, an unacceptable result. Because of these problems, considerable attention has been paid to newer approaches to the development of an AIDS vaccine. Synthetic HIV protein epitopes have been ineffective since they generally contain too few epitopes. Live recombinant vaccines using HIV recombinants in vaccinia, mycobacteria, or enteric bacteria have not stimulated a significant protective response in animals, nor have combined approaches using live recombinant products together with recombinant vaccines been any more successful.

Some potential HIV vaccines are currently being tested for safety in humans. These include vaccines containing recombinant envelope glycoproteins, synthetic peptides, and recombinant pox viruses. Unfortunately, none has been highly protective in animal models.

AIDS IN ANIMALS

The only primates that are susceptible to HIV-1 infection in the laboratory are the chimpanzee and the gibbon. Although these animals can be infected, they do not develop signs of disease. Likewise, rhesus macaques can be persistently infected with HIV-2 but without developing disease.

There are several interesting nonprimate models of AIDS. Rabbits are susceptible to a low-level, persistent HIV infection, but like primates, they do not develop disease. *scid* mice reconstituted with human lymphocytes can harbor HIV infections, but the virus destroys their T cells very rapidly in a manner unlike the natural disease. Cats suffer severe immunosuppression as a result of infection with either the feline leukemia virus, an oncornavirus, or the feline immunodeficiency virus, a lentivirus.

Several HIV-related viruses have been isolated from nonhuman primates. These simian immunodeficiency viruses include SIV_{mac} isolated from a rhesus macaque; SIV_{agm} from an African green monkey; SIV_{sm} from a sooty mangabey; SIV_{mnd} from a mandrill; and SIV_{cpz} from a chimpanzee. All these isolates preferentially invade $CD4^+$ cells. Nucleotide homology studies suggest that HIV-2, SIV_{mac}, and SIV_{sm} are very closely related and probably arose from a common ancestor.

When SIV_{mac} infects rhesus monkeys (*Macaca mulatta*), it induces an immunodeficiency syndrome that is similar to human AIDS. It differs from the human disease in that the monkeys are immunosuppressed as a result of depletion of both B and T cells. Consequently, it is associated with a profound drop in serum IgG and IgM levels and a severe lymphopenia. Monocyte function is unimpaired, but the remaining lymphocytes do not respond to mitogens such as concanavalin A and phytohemagglutinin. The monkeys are also profoundly neutropenic. About half the infected animals survive the disease; the others die as a result of septicemia or diarrhea with wasting. In many cases, normally innocuous agents such as cytomegalovirus, *Cryptosporidium,* and *Candida albicans* cause infection. Some affected monkeys develop tumors such as fibrosarcomas. In monkey colonies, the disease is spread by the bites of carrier animals.

OTHER HUMAN VIRUSES AFFECTING THE IMMUNE SYSTEM

Six retroviruses are known to infect humans. Three of these are transforming retroviruses: HTLV-I, HTLV-II, and HTLV-V. (A transforming virus can change the growth pattern of a cell from normal to transformed. A transformed cell has the growth characteristics of a cancer cell.) These are associated with the development of T-cell neoplasms. The other three human retroviruses, HIV-1, HIV-2, and HTLV-IV, are not transforming. Both HIV-1 and HIV-2 cause AIDS. No disease has yet been associated with HTLV-IV.

HTLV-I is the cause of adult T-cell leukemia, a cancer involving $CD4^+$ lymphocytes. Only 1% of individuals exposed to this virus develop the disease, and the incubation period is as long as 20 to 30 years. HTLV-I is also associated with tropical spastic paresis, a neurologic disease characterized by difficulties in walking. HTLV-II has been isolated from several cases of hairy-cell leukemia (a B-cell tumor), and HTLV-V has been isolated from a case of mycosis fungoides (a T-cell tumor of the skin). HTLV-IV has been isolated from healthy patients in West Africa and is very similar to HIV-2.

AIDS PREVENTION AND ITS SOCIAL IMPLICATIONS

Analysis of AIDS patients based on their route of infection indicates that in the United States about 90% of cases are in men and about 90% of these men acquire their infection through certain behaviors, notably homosexual behavior and the sharing of needles when abusing drugs (Fig. 27–17). Clearly, avoidance of such behavior will go a long way toward preventing the transmission of HIV. Even in women, approximately half of the cases are associated with intravenous drug abuse, although 36% result from heterosexual contact with infected men. The risks of acquiring AIDS by either homosexual or heterosexual transmission could therefore be minimized by avoidance of sexual contact with individuals from high-risk groups of either sex. A more realistic solution is the consistent use of condoms, which prevent the transfer of HIV-infected lymphocytes in the sexual fluids to uninfected recipients. AIDS infection in intravenous drug abusers is a result of sharing of contaminated needles, resulting in the transfer of infected cells. Clearly, there is no easy solution to this problem, although some have advocated providing drug addicts with fresh needles in an attempt to prevent the reuse of contaminated ones. Screening of the blood supply and blood products has greatly reduced the risks of acquiring AIDS from transfusions. Blood for transfusion is screened by ELISA tests and Western blotting (Chapter 17).

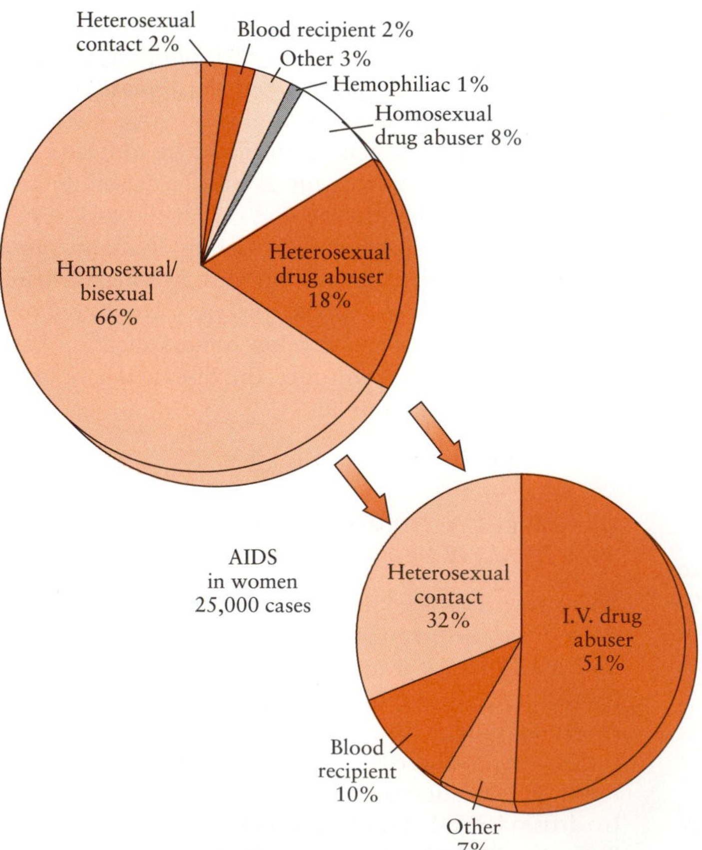

Figure 27–17 The routes of acquisition of HIV infection in the United States.

The medical and social problems associated with AIDS are unprecedented. Despite extensive and rapid research, AIDS has become a major epidemic disease of the late twentieth century. Although parasitic disease such as schistosomiasis and malaria or diarrheal diseases may be more important than AIDS as a result of their morbidity and mortality, AIDS does have a disproportionate impact on all societies where it is present. It is spreading very rapidly. There is no cure or even satisfactory treatment, and public education campaigns have had limited success. AIDS affects relatively young people during their most productive years and may lead to a significant loss of a skilled work force. The selective loss of young or middle-aged adults could lead to political destabilization in the developing world. AIDS is a disease that requires repeated hospitalization. In Africa or the inner cities of the United States, hospital services are overloaded with AIDS patients. Up to a third of the beds in some African hospitals may be occupied by AIDS patients. Treatment of these patients is usually expensive and time-consuming. AIDS also places unusually heavy demands on what are commonly the weakest areas of health care: health education, laboratory services, blood bank control, and the control of sexually transmitted diseases.

KEY WORDS

Acquired immune deficiency syndrome (AIDS) p. 414
AIDS dementia complex p. 425
AIDS vaccine p. 427
Apoptosis p. 420
Azidothymidine p. 426
B cells p. 421
CD4 p. 419
ELISA p. 426
Epstein-Barr virus p. 420
Genetic variability p. 418
gp120 p. 418
Helper T cells p. 419
HIV p. 416
Kaposi's sarcoma p. 425
Lymphadenopathy p. 423
Lymphopenia p. 419
Macrophages p. 422
NK cells p. 422
Pneumocystis carinii p. 425
Retrovirus p. 416
Reverse transcriptase p. 416
Secondary infection p. 423
SIV p. 427
Syncytium p. 420
Western blot p. 426

QUESTIONS

1. The causal agent of AIDS is
   a. EBV
   b. IDV
   c. HTLV II
   d. AIDV
   e. HIV
2. A CD4:CD8 ratio of less than 1 means that an individual
   a. is normal
   b. is immunosuppressed
   c. has activated macrophages
   d. has enhanced immunity
   e. has AIDS
3. Approximately what proportion of AIDS cases in North America are acquired by heterosexual contact?
   a. 90%
   b. 70%
   c. 50%
   d. 30%
   e. 10%
4. Which of the following species can contract a lentivirus infection that causes an immunodeficiency similar to that seen in AIDS?
   a. dog
   b. cat
   c. horse
   d. rabbit
   e. mouse
5. Which component of the HIV envelope is responsible for binding to T cells?
   a. CD4
   b. CD8
   c. gp120
   d. p24
   e. p17
6. The clinical deterioration seen in AIDS patients is due to a loss of which cell type?
   a. cytotoxic T cells
   b. macrophages
   c. nerve cells
   d. helper T cells
   e. neutrophils
7. In which continent is the AIDS epidemic most severe with the most impact on the population?
   a. South America
   b. Asia
   c. North America
   d. Europe
   e. Africa
8. Which tumor is commonly observed in AIDS patients?
   a. melanoma
   b. Kaposi's sarcoma
   c. lymphosarcoma
   d. carcinoma
   e. Burkitt's lymphoma

9. The drug that is mainly used to treat AIDS patients is called
   a. azidothymidine
   b. tetracycline
   c. cortisone
   d. cyclosporine
   e. imuran
10. The most important reason why a vaccine against AIDS will be exceedingly difficult to produce is that the virus
    a. is drug resistant
    b. has a reverse transcriptase
    c. causes immunosuppression
    d. shows antigenic variation
    e. hides within cells
11. What immunological test would you use to determine whether a patient had AIDS? Does this test detect antibody or antigen? Which test is preferable, one for antibody or one for antigen?
12. Where might HIV have originated? Can you speculate on the nature of events in Africa that led to the movement of HIV into the human population?
13. What do you think will be the eventual outcome of the present AIDS epidemic?
14. What are the major clinical features of AIDS? How can it be distinguished from other immunodeficiency disease?
15. List the reasons why an effective vaccine has not been developed against AIDS. Which are the most significant barriers to immunization?
16. Speculate on the very clear differences in the incidence and epidemiology of AIDS between the developed countries and central Africa.
17. What features of HIV-1 make AIDS an extremely difficult disease to treat?
18. What human viruses apart from HIV-1 are potently immunosuppressive? Why do these not induce an AIDS-like disease?

Answers: 1e, 2b, 3e, 4b, 5c, 6d, 7e, 8b, 9a, 10d

SOURCES OF ADDITIONAL INFORMATION

Bender, B.S., et al. Role of the mononuclear phagocyte system in the immunopathogenesis of human immunodeficiency virus infection and the acquired immunodeficiency syndrome. Rev. Infect. Dis., 10:1142–1154, 1988.

Bendinelli, M., and Ceccherini-Nelli, L. Mechanisms of retrovirus pathogenicity. Clin. Immunol. News. 9:57–60, 1988.

Bolognesi, D.P. Prospects for an HIV vaccine. Sci. Amer. Science and Medicine 1:44–53, 1994.

Chin, J. Current and future dimensions of the HIV/AIDS pandemic in women and children. Lancet, 336:221–224, 1990.

Clerici, M., et al. Restoration of HIV-specific cell-mediated immune responses by interleukin-12 in vitro. Science, 262: 1721–1724, 1993.

Desrosiers, R.C. The simian immunodeficiency viruses. Annu. Rev. Immunol., 8:557–578, 1990.

Desube, B.J., et al. Cytokine dysregulation in AIDS: In vivo overexpression of mRNA of tumor necrosis factor–α and its correlation with that of the inflammatory cytokine GRO. J. Acquir. Immune Defic. Syndr. 5:1099–1104, 1992.

Edelman, A.S., and Zolla-Pazner, S. AIDS: A syndrome of immune dysregulation, dysfunction and deficiency. FASEB J., 3:22–30, 1989.

Ensoli, B., et al. Tat protein of HIV-1 stimulates growth of cells derived from Kaposi's sarcoma lesions of AIDS patients. Nature, 345:84–86, 1990.

Fauci, A.S. CD4+ T-lymphocytopenia without HIV infection—no lights, no camera, just facts. N. Engl. J. Med., 328: 429–430, 1993.

Fauci, A.S. Multifactorial nature of human immunodeficiency virus disease: Implications for therapy. Science, 262: 1011–1018, 1993.

Gallo, R.C. The AIDS virus. Sci. Am., 256:47–56, 1987.

Gardner, M.B., and Luciw, P.A. Animal models of AIDS. FASEB J., 3:2593–2606, 1989.

Germain, R.N. Antigen processing and $CD4^+$ depletion in AIDS. Cell, 54:441–444, 1988.

Gottlieb, M.S., et al. *Pneumocystis carinii* pneumonia and mucosal candidiasis in previously healthy homosexual men. Evidence of a new acquired cellular immunodeficiency. N. Engl. J. Med., 305:1425–1431, 1981.

Greene, W.C. AIDS and the immune system. Sci. Am., 269: 99–105, 1993.

Greene, W.C. Regulation of HIV-1 gene expression. Annu. Rev. Immunol., 8:453–475, 1990.

Höllsberg, P., and Hafler, D.A. Pathogenesis of diseases induced by human lymphotropic virus type I infection. N. Engl. J. Med., 328:1173–1182, 1993.

Johnson, R.T., McArthor, J.C., and Narayan, O. The neurobiology of human immunodeficiency virus infections. FASEB J., 2:2970–2981, 1988.

Kanagawa, O., et al. Resistance of mice deficient in IL-4 to retrovirus-induced immunodeficiency syndrome (MAIDS). Science, 262:240–242, 1993.

Kornfield, H., et al. Lymphocyte activation by HIV-1 envelope glycoprotein. Nature, 335:445–447, 1990.

Letvin, N.L. Vaccines against human immunodeficiency virus—progress and prospects. N. Engl. J. Med., 329:1400–1405, 1993.

Louwagie, J., et al. Phylogenetic analysis of *gag* genes from 70 international HIV-1 isolates provides evidence for multiple genotypes. AIDS 7:769–780, 1993.

Marlink, R., et al. Reduced rate of disease development after

HIV-2 infection as compared to HIV-1. Science 265:1587–1592, 1994.

McChesney, M.B., and Oldstone, M.B.A. Virus-induced immunosuppression: Infections with measles virus and human immunodeficiency virus. Adv. Immunol., 45:335–380, 1989.

McDougal, J.S., et al. Binding of the human retrovirus HTVLIII/LAV/ARV/HIV to the CD4 (T4) molecule: Conformation, dependence, epitope mapping, antibody inhibition, and potential for idiotypic memory. J. Immunol., 137: 2937–2944, 1986.

Meltzer, M.S., et al. Role of mononuclear phagocytes in the pathogenesis of human immunodeficiency virus infection. Annu. Rev. Immunol., 8:169–194, 1990.

Meyaard, L., et al. Programmed cell death of T cells in HIV-1 infection. Science, 257:217–219, 1992.

Meyaard, L., Schuitemaker, H., and Miedema, F. T-cell dysfunction in HIV infection: Anergy due to defective antigen-presenting cell function? Immunol. Today, 14:161–164, 1993.

Morrow, W.J.W., and Levy, J.A. The viral etiology of AIDS. Clin. Immunol. News., 6:113–117, 1985.

Osborn, K.G., et al. The pathology of an epizootic of acquired immunodeficiency in Rhesus macaques. Am. J. Pathol., 114:94–103, 1984.

Pantaleo, G., Graziosi, C., and Fauci, A.S. The immunopathogenesis of human immunodeficiency virus infection. N. Engl. J. Med., 328:327–335, 1993.

Piot, P., et al. AIDS: An international perspective. Science, 239:573–579, 1988.

Price, R.W., et al. The brain in AIDS: Central nervous system HIV-1 infection and AIDS dementia complex. Science, 239: 586–592, 1988.

Rosenberg, Z.F., and Fauci, A.S. The immunopathogenesis of HIV infection. Adv. Immunol., 47:377–451, 1989.

Teminn, H.M., and Bolognesi, D.P. Where has HIV been hiding? Nature, 362:292–293, 1993.

Walker, B.D., and Plata, F. Cytotoxic T cells against HIV. AIDS, 4:177–184, 1990.

Wong-Staal, F., and Gallo, R.C. Human T-lymphotropic retroviruses. Nature, 317:395–403, 1985.

Zolla-Pazner, S., and Gorny, M.K. Passive immunization for the prevention and treatment of HIV infection. AIDS, 6: 1235–1247, 1992.

CHAPTER 28

Inflammation

Bengt Samuelsson *Dr. Bengt Samuelsson was one of the winners of the 1982 Nobel Prize in Medicine. He is a Swedish scientist who had spent his career looking at the oxidation products of the fatty acid, arachidonic acid. During these studies he discovered that some of these products, especially the thromboxanes and leukotrienes, had very significant biological activity. In fact, they together with the prostaglandins, which are also arachidonic acid metabolites, are key mediators of inflammation.*

CHAPTER OUTLINE

CHAPTER CONCEPTS

1. Acute inflammation is a rapid response to injury that results in the accumulation of immunoglobulins and cells within tissues. It is a protective response.
2. The cardinal signs of acute inflammation are heat, redness, swelling, pain, and loss of function. These may all be attributed to increased vascular permeability.
3. Hypersensitivity reactions are inflammatory responses mediated by immunological mechanisms. There are four types of hypersensitivity mechanisms.
4. Chronic inflammation is a response to prolonged injury or tissue damage. It may result in mononuclear cell infiltration of tissues and the deposition of large quantities of collagen.
5. Interleukin-1 released by macrophages is a cause of fever, production of acute-phase proteins, and many systemic responses to infection.

In this chapter we describe the response of tissues to irritation or injury, a response we call inflammation. The first part of the chapter describes the processes that occur during an acute inflammatory response. The two most important processes are an increase in blood vessel permeability and the emigration of white blood cells into the tissues. The mechanisms of both processes are described. These inflammatory changes are caused by many different vasoactive factors. These factors are described as well as the mechanisms that control inflammation.

In many cases, inflammatory responses are caused by immunological processes. There are four major types of these responses, called hypersensitivity reactions. These are described in detail since the inflammatory response is a major component of the body's defenses. A considerable section is also allocated to the role of eosinophils in the defense of the body.

A prolonged form of inflammation, called chronic inflammation, is also an important component of the defense system. Interleukin-1 plays a critical role in regulating the body's response to infectious agents, including inflammation, and its effect on body functions is described. A disease that can result from prolonged chronic inflammation, called amyloidosis, is also described here.

Inflammation is the response of tissues to irritation or injury. It is a vital protective mechanism since it provides a means by which defensive factors, such as immunoglobulins, complement, and phagocytic cells, which are normally confined to the bloodstream, can gain access to sites of microbial invasion or tissue damage. Immunoglobulins and complement components occur in normal tissue fluid at a much lower concentration than in the blood. In addition, large molecules such as IgM cannot leave normal blood vessels. Inflammation may therefore be considered a method by which the immunological protective mechanisms are focused at a localized region of tissue (Fig. 28–1).

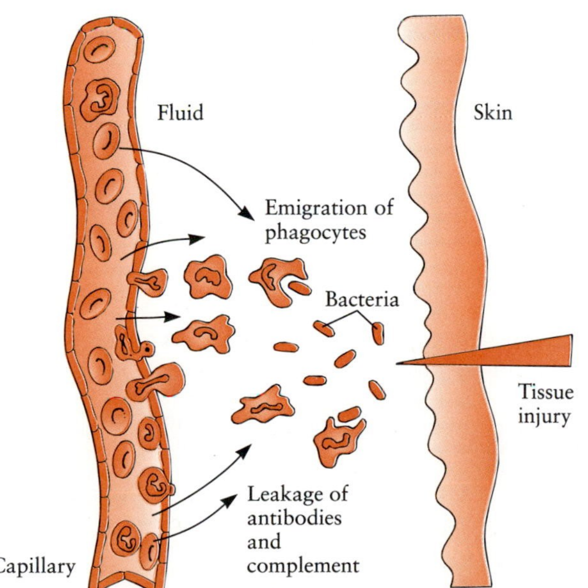

Figure 28–1 Inflammation is a process by which the effector mechanisms of the immune system, antibodies, complement, and phagocytic cells can escape into tissues and destroy invading antigen.

Inflammation is classified according to its severity and duration; recent inflammation or acute inflammation has different characteristics from the reactions seen in prolonged or chronic inflammation.

ACUTE INFLAMMATION

Acute inflammation develops within a few hours after a tissue is damaged or infected. In its classic form, acute inflammation has five cardinal signs: heat, redness, swelling, pain, and loss of function. All these signs are a direct result of changes in local blood vessels (Fig. 28–2). Immediately after injury, there is a transient constriction of local arterioles followed shortly thereafter by dilation of all the small blood vessels in the damaged area. As a result, the blood flow to the injured area increases significantly. Eventually, the flow diminishes and gradually returns to normal. While the blood vessels are dilated, their permeability also increases and protein-rich plasma exudes into the tissues, where it causes edema (swelling).

Changes in Vascular Permeability

Blood vessel permeability increases in two stages in acute inflammation. First, there is an immediate increase caused by vasoactive factors released from damaged tissues. A second phase of increased vascular permeability occurs several hours after the onset of inflammation when the leukocytes are commencing to emigrate. Permeability increases because endothelial and perivascular cells contract, so that the endothelial cells are pulled apart and fluid escapes through the intercellular spaces.

Cellular Emigration

To migrate into tissues, leukocytes must first bind to vascular endothelial cells. This attachment is caused by an increase in the adhesiveness of the endothelial cells. The adhesion process occurs in two stages. In the first stage, the neutrophils must decelerate. Blood flows through vessels at a very high speed and neutrophils must be slowed down gradually. During tissue injury or inflammation, several cytokines and pharmacological agents stimulate the endothelial cells to express P-selectin (CD62P). The P-selectin is stored

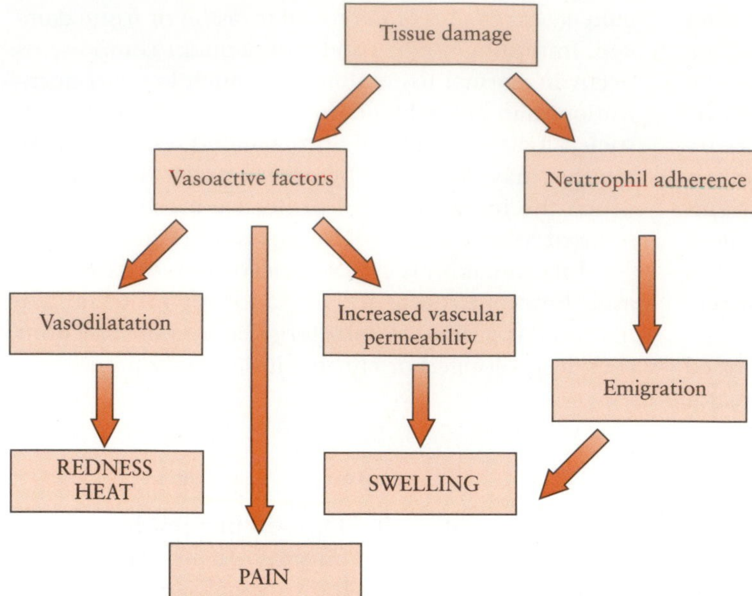

Figure 28–2 The cardinal signs of inflammation. Redness, heat, pain, and swelling all result from changes in blood vessel properties in the inflamed area.

within granules and translocated to the cell surface when the cells are stimulated by inflammatory agents such as thrombin or histamine. The P-selectin binds to an epitope called CD15s (or sialyl Lewis^x) on L-selectin located on the neutrophil surface. The adhesion is transient because the L-selectin is rapidly shed from the neutrophils. As a result of this adherence, the neutrophil rolls along the endothelial cell surface, loses speed, and eventually comes to a complete stop.

As the neutrophils are rolling along the endothelial surface, the second stage of adhesion occurs. Platelet-activating factor expressed on endothelial cells activates the neutrophils. As a result, neutrophil surface CD11a/CD18 is increased, and it undergoes a conformational change. This results in increased binding to ICAM-2 (CD102) on the endothelial cells. Similarly, α4 integrins on the neutrophil bind to VCAM-1 (CD106) on endothelial cells. This increased binding makes the neutrophil come to a complete stop despite the shearing force of blood flow (Fig. 28–3). If the endothelium is activated by IL-1 or TNF-α, it also expresses E-selectin (CD62E) and IL-8. The E-selectin enhances neutrophil adhesiveness still further, al-

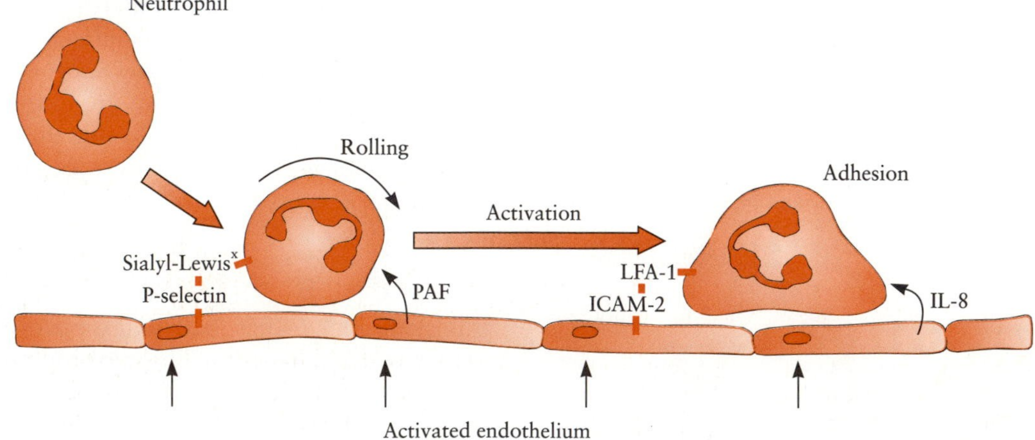

Figure 28–3 Prior to emigration from blood vessels, circulating neutrophils must be slowed down and stopped. They are slowed down by a rolling process as their sialyl-Lewis^x binds transiently to P-selectin on endothelial cells. This stimulates local release of PAF, which induces the neutrophil to express LFA-1. The LFA-1 mediates strong binding with endothelial cell ICAM-2 and brings the neutrophil to a complete stop. Local release of IL-8 then ensures that the neutrophil is firmly bound to the endothelium.

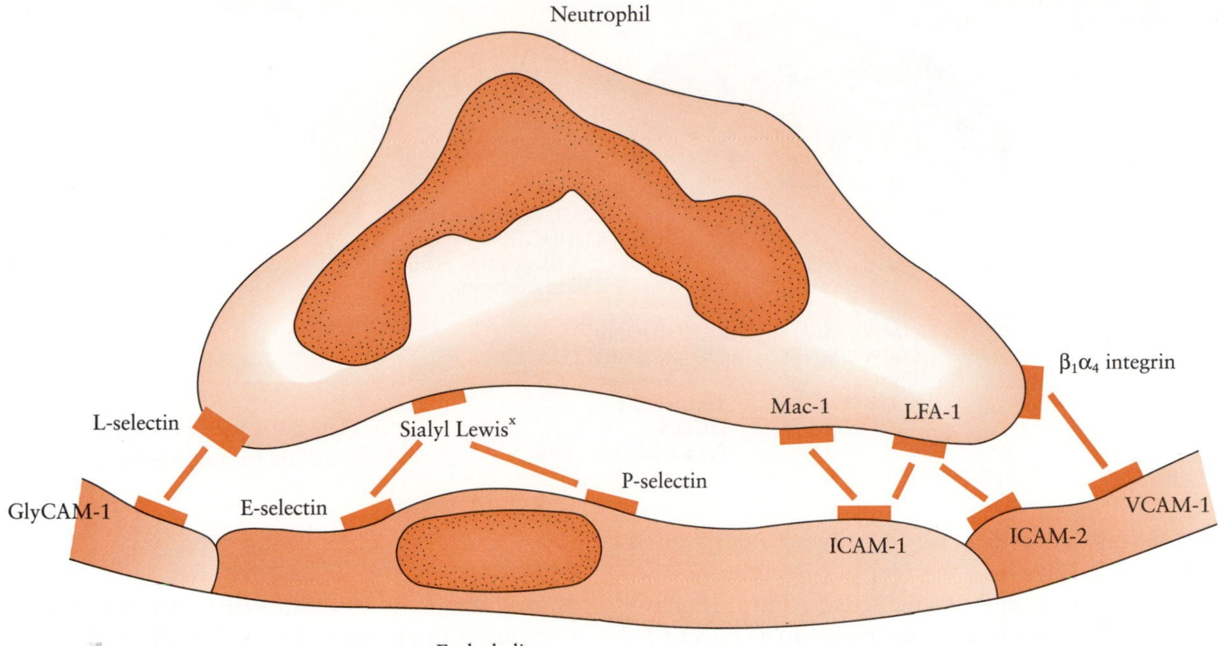

Figure 28–4 Some of the adhesion molecules that bind neutrophils to capillary endothelial cells.

though it takes up to 24 hours to appear (Fig. 28–4). The IL-8 signals to the neutrophil, causing further activation and stimulating it to migrate across the vessel wall.

Once neutrophils have bound to endothelial cells, they insert their pseudopodia between the endothelial cells and migrate between the cells and the basement membrane. Subsequently, the neutrophils move through the basement membrane into the tissue spaces under the influence of chemotactic factors (a process known as diapedesis) (Fig. 28–5). Neutrophils are the most mobile of all the blood leukocytes. They are thus the first cells to arrive in the inflamed tissues. Blood monocytes move more slowly and so arrive later. Once within tissues, the cells are attracted to sites of bacterial growth and tissue damage. There they proceed to phagocytose and destroy any foreign material and, in the case of monocytes, remove dead and dying tissue.

One of the most important chemotactic factors produced during acute inflammation is C5a. This peptide not only attracts neutrophils to sites of complement activation, but also enhances adherence to endothelial cells. C5a is unstable and readily loses its

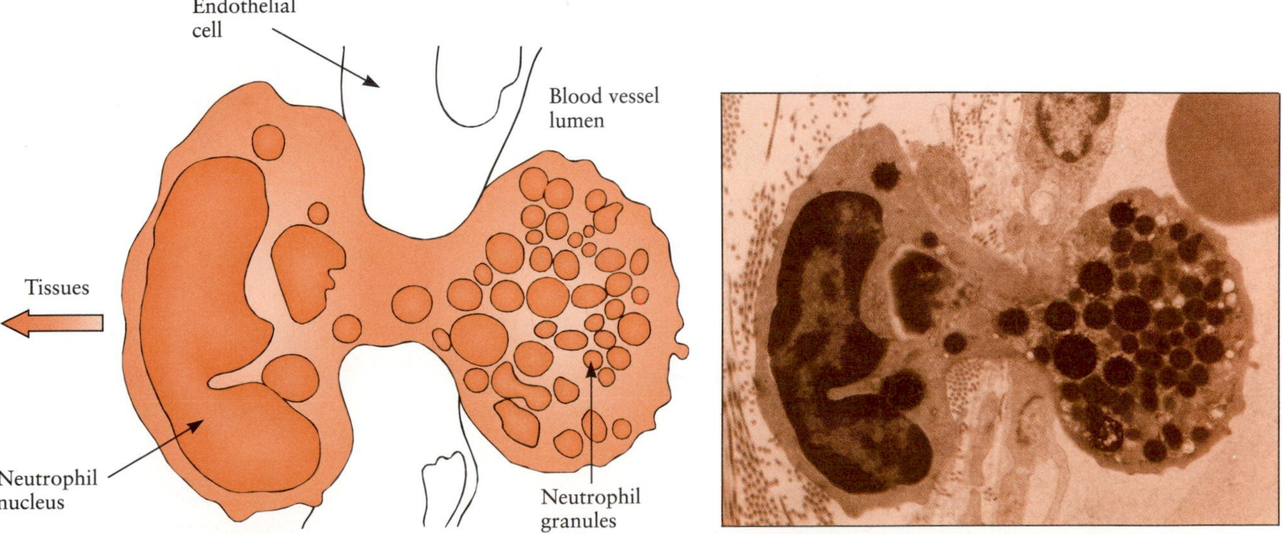

Figure 28–5 The emigration of a neutrophil through a blood vessel wall. The cell is squeezing between two endothelial cells. Its nucleus is through the wall but much of the cytoplasm containing its granules has still to get through the hole. *(Courtesy Visuals Unlimited/David M. Philips)*

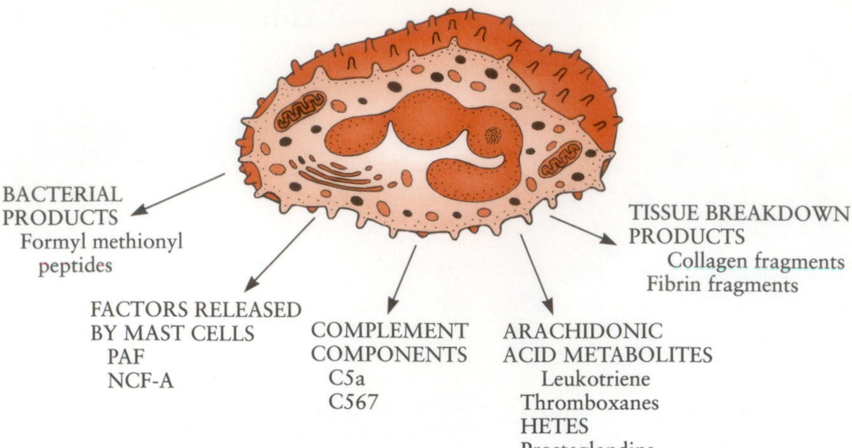

Figure 28–6 The factors that are chemotactic for neutrophils.

terminal arginine to form C5a des arg. C5a des arg has lost its ability to increase vascular permeability and reduced neutrophil chemotactic activity, but it remains as potent a macrophage chemoattractant as intact C5a (Chapter 16).

Other neutrophil chemotactic factors include collagen, chemotactic cytokines, chemotactic factors derived from macrophages, fibroblasts, and mast cells as well as factors released from neutrophils themselves (Fig. 28–6). For example, leukoegressin is a chemotactic peptide derived from cleavage of IgG by neutrophil proteolytic enzymes. Bacterial products are also chemotactic and thus attract neutrophils to sites of bacterial invasion. The most potent chemotactic factor known is leukotriene B_4, which has detectable activity at a concentration of 10^{-4}M.

Once neutrophils have reached the site of tissue damage or invasion, they ingest and destroy foreign material (Chapter 5). Most neutrophils die soon after this, but those that survive eventually become unresponsive to chemotactic agents, so enabling them to move away once their task is complete.

Chemokines

Among the many mediators that have chemotactic activity are a family of at least 14 small proteins distinguished by a characteristic sequence of four spaced cysteine residues. These mediators are called **chemokines.** They are subdivided into the C-C and C-X-C subfamilies (C-C chemokines have two linked cysteine residues while C-X-C chemokines have another amino acid sandwiched between two cysteines). Members of the C-C subfamily include macrophage inflammatory protein 1, monocyte chemoattractant protein, and RANTES protein. These tend to be chemotactic for monocytes. Interleukin-8 is a member of the C-X-C subfamily, whose members tend to be chemotactic for neutrophils.

IL-8 is a 10-kDa protein produced by fibroblasts, macrophages, endothelial cells, mast cells, hepatocytes, and keratinocytes (Fig. 28–7). Its production can be stimulated by exposure to bacterial endotoxin, viral double stranded RNA, IL-1α, or TNF–α. IL-8 is chemotactic for neutrophils, basophils, and some T

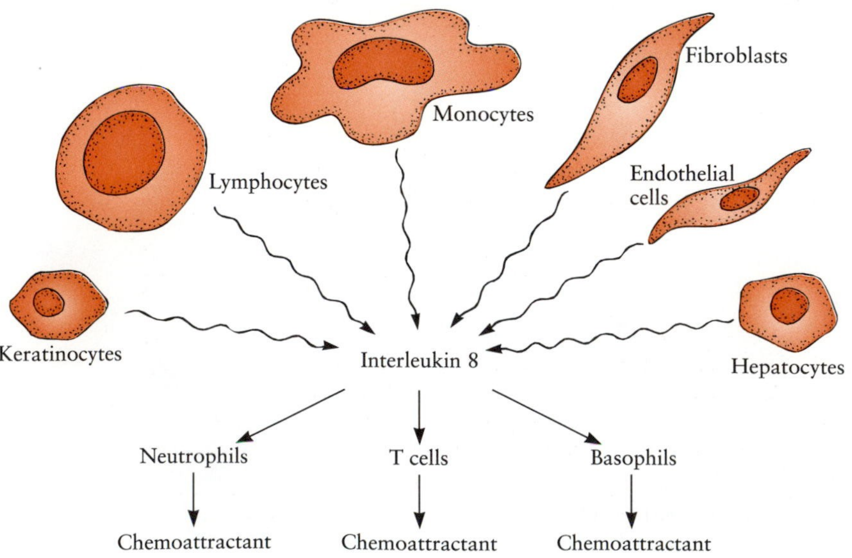

Figure 28–7 The sources and properties of the chemokine, interleukin-8.

METHODOLOGY

How to Demonstrate and Measure Chemotaxis

There are two major methods for measuring chemotaxis. In the micropore filter method, cells are placed on top of a filter and the chemotactic substance is placed on the other side. When incubated at 37° C, the cells on top migrate through the channels of the filter and, if given sufficient time, will eventually reach the bottom chamber. The technique may be quantitated either by counting the number of migrating cells or by measuring the distance they have migrated into the filter (figure).

In the agarose method, holes are punched in a sheet of agarose in a Petri dish. Cells placed in a central well are allowed to migrate under the agarose toward a well containing a chemoattractant. After incubation, the agarose can be removed and the cell pattern examined. In the absence of chemotaxis, the cells move out in a circular pattern. When stimulated by chemotactic agents, the pattern is elongated toward the well containing chemotactic factors. The specific cell types involved can also be examined easily. Although the agarose method is simple, it is about 50 to 100 times less sensitive than the micropore filter method.

Ward, P.A., and Maderazo, D.G. Micropore filter method: Leukocyte chemotaxis. *In* Manual of Clinical Immunology, 2nd ed. American Society for Microbiology, Washington, D.C., 1980.
Nelson, R.D., Quie, P.G., and Simmons, R.L. Chemotaxis under agarose: A new and simple method for measuring chemotaxis and spontaneous migration of human polymorphonuclear leukocytes and monocytes. J. Immunol., 115:1650–1655, 1975.

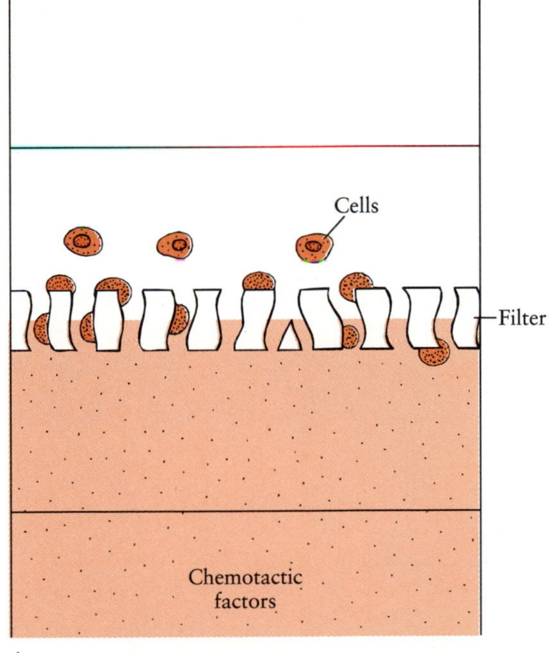

Schematic diagram showing neutrophil chemotaxis through a filter and sagittal sections of a millipore filter after placing a neutrophil suspension on the top surface. A) Using serum in the bottom chamber. B) Using chemotactic factor generated by mixing bacterial lipopolysaccharide and serum. Note how the neutrophils have migrated through the filter under the chemotactic stimulus. *(From Snyderman, R., et al. J. Exp. Med., 128:274, 1968. With permission.)*

cells. It acts on neutrophils, increasing expression of CD11a/CD18, stimulating the release of granule contents, the respiratory burst, and leukotriene release. If low doses of IL-8 are injected intradermally, they induce a T-cell infiltrate. High doses induce a neutrophil infiltrate. IL-8 receptors (CDw128) are found on neutrophils, monocytes, and T cells.

Macrophage Inflammatory Proteins (MIP–1α and 1β) are 8-kDa heterodimers produced by endotoxin-treated macrophages, activated T cells, B cells, mast cells, and fibroblasts. Their effects include activation and chemoattraction of monocytes and T cells. They produce inflammatory infiltrates when injected into the skin. MIP–1β attracts $CD4^+$ T cells, and MIP–1α tends to attract B cells and cytotoxic T cells.

Monocyte Chemoattractant Protein is an 8.4-kDa glycoprotein produced by treatment of mononuclear cells, fibroblasts, and endothelial cells with IL-1 or TNF-α. As its name indicates, it is chemotactic and activating for monocytes, stimulating the respiratory burst and lysosomal enzyme release.

RANTES Protein is an 8.4-kDa protein produced by circulating T cells. Its expression is downregulated upon T-cell activation (RANTES is an acronym for **r**egulated upon **a**ctivation, **n**ormal **T** **e**xpressed and **s**ecreted). The RANTES protein is chemotactic for blood monocytes, eosinophils, and some T-cell subsets. It also stimulates histamine release from basophils.

The Coagulation System in Inflammation

If blood vessels are damaged during microbial invasion, then platelets may bind to the vessel walls and release both vasoactive and clotting factors (Fig. 28–8). When fluid exudes from the bloodstream into the tissues, three enzyme cascades are activated: the complement system, the coagulation system, and the fibrinolytic system. The coagulation system is an enzyme cascade that is triggered following activation of the first component, Hageman factor (factor XII). Hageman factor is activated by damage to blood vessel walls. This in turn leads to a series of enzyme reactions that result in large quantities of thrombin, the main clotting enzyme, being formed. Thrombin acts on fibrinogen in tissue fluid and plasma to produce insoluble strands of fibrin, which are laid down in inflamed tissues and capillaries and form a clot, which acts as an effective barrier to the spread of infection. Activated Hageman factor, in addition to starting the coagulation cascade, initiates the fibrinolytic system. This system generates plasmin, a fibrinolytic enzyme. In de-

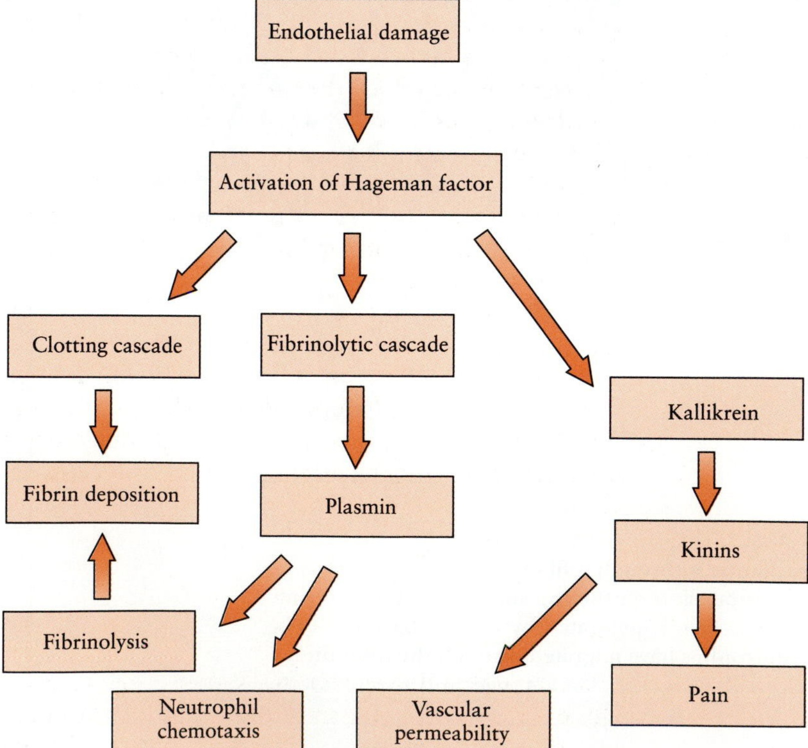

Figure 28–8 The role of the clotting system, especially Hageman factor, in mediating aspects of the inflammatory response.

stroying fibrin, plasmin breaks up blood clots and releases peptide fragments that are chemotactic for neutrophils.

VASOACTIVE FACTORS

Factors Derived from Mast Cells (Table 28–1)

Histamine. Histamine is an amine derived from the amino acid histidine and is stored within granules in mast cells. When mast-cell granules are exposed to the extracellular fluid by tissue damage, histamine is released through exchange with sodium ions. Histamine affects blood vessels, smooth muscle, and exocrine glands. It dilates most capillaries and venules but contracts certain specific vessels such as the pulmonary vessels. It causes increased permeability of small vessels so that intradermal inoculation of histamine causes swelling and redness—called a wheal and flare reaction (Chapter 29). It enhances P-selectin expression in venules and so induces neutrophil rolling. Histamine causes smooth muscle contraction, particularly in the bronchi, gastrointestinal tract, uterus, and bladder. It is a potent stimulator of secretions, stimulating bronchial mucus secretion, lacrimation, and salivation. In small quantities, histamine attracts eosinophils, which, possessing large quantities of histaminases, can readily break it down.

Serotonin (5-Hydroxytryptamine). Serotonin, a derivative of the amino acid tryptophan, is released from the mast cells of some species of rodents and the large domestic herbivores. It also exists preformed in platelets, in central nervous tissue, and in some intestinal cells. It is released from platelets by platelet-activating factor (Chapter 29). Serotonin normally causes vasoconstriction that results in a rise in blood pressure. It appears to have little effect on vascular permeability except in rats and mice.

Table 28–1 Vasoactive Factors Involved in Inflammation

Factor	Source
Histamine	Mast cell granules
Serotonin	Mast cell granules Platelets
Leukotrienes	Cell membranes
Lipoxins	Cell membranes
Prostaglandins	Cell membranes
Thromboxanes	Platelets
Kinins	Serum proteins (Kininogens)
Anaphylatoxins	Serum proteins (C3 and C5)
PF/dil	Hageman Factor
Platelet activating factor	Mast cells Platelets Leukocytes Endothelial cells

Factors Derived from Arachidonic Acid

When tissues are damaged or stimulated, cell membrane phospholipases break down the phospholipids in cell walls to release fatty acids. The most important of these fatty acids is an unsaturated 20-carbon fatty acid called arachidonic acid. Arachidonic acid is metabolized by two alternative pathways. Under the influence of an enzyme known as 5-lipoxygenase, it is converted to **leukotrienes,** A_4 B_4, C_4, D_4, and E_4. Under the influence of enzymes known as cyclooxygenases, arachidonic acid is converted to prostaglandins, prostacyclin, and thromboxanes (Fig. 28–9).

Leukotrienes. Four major leukotrienes play a central role in the inflammatory response. Leukotriene B_4 is a potent attractant for neutrophils and eosinophils. The other major leukotrienes (leukotrienes C_4, D_4, and E_4) provoke a slow contraction of smooth muscle. Leukotrienes C_4 and D_4 are up to 20,000 times more active than histamine in contracting the smooth muscle of bronchioles in certain species and are potent stimulators of increased vascular permeability.

Prostaglandins. **Prostaglandins** consist of four groups of complex lipids: the PGE series, the PGF series, the thromboxanes, and the prostacyclins. The prostacyclins are formed in vascular endothelial cells, the thromboxanes are formed in platelets, and the other prostaglandins can be generated by most nucleated cells. The biological activities of the prostaglandins vary widely, and since many different prostaglandins are released in inflamed tissues, their net effect may be complex. For example, $PGF_{2\alpha}$ and thromboxane cause smooth muscle to contract and provoke vasoconstriction. Other prostaglandins such as PGE_1, PGF_2, and prostacyclin (PGI_1) cause smooth muscle relaxation and vasodilation. PGI_2, PGE_1, and $PGF_{2\alpha}$ inhibit platelet aggregation; PGE_2 and thromboxanes promote platelet aggregation and the release of platelet mediators such as serotonin; and $PGF_{2\alpha}$ promotes mast-cell–mediator release.

Vasoactive Peptides

Several peptides play a role in inflammation. They include the **kinins** and the **anaphylatoxins.** The most important kinin is bradykinin. Kinins are derived from

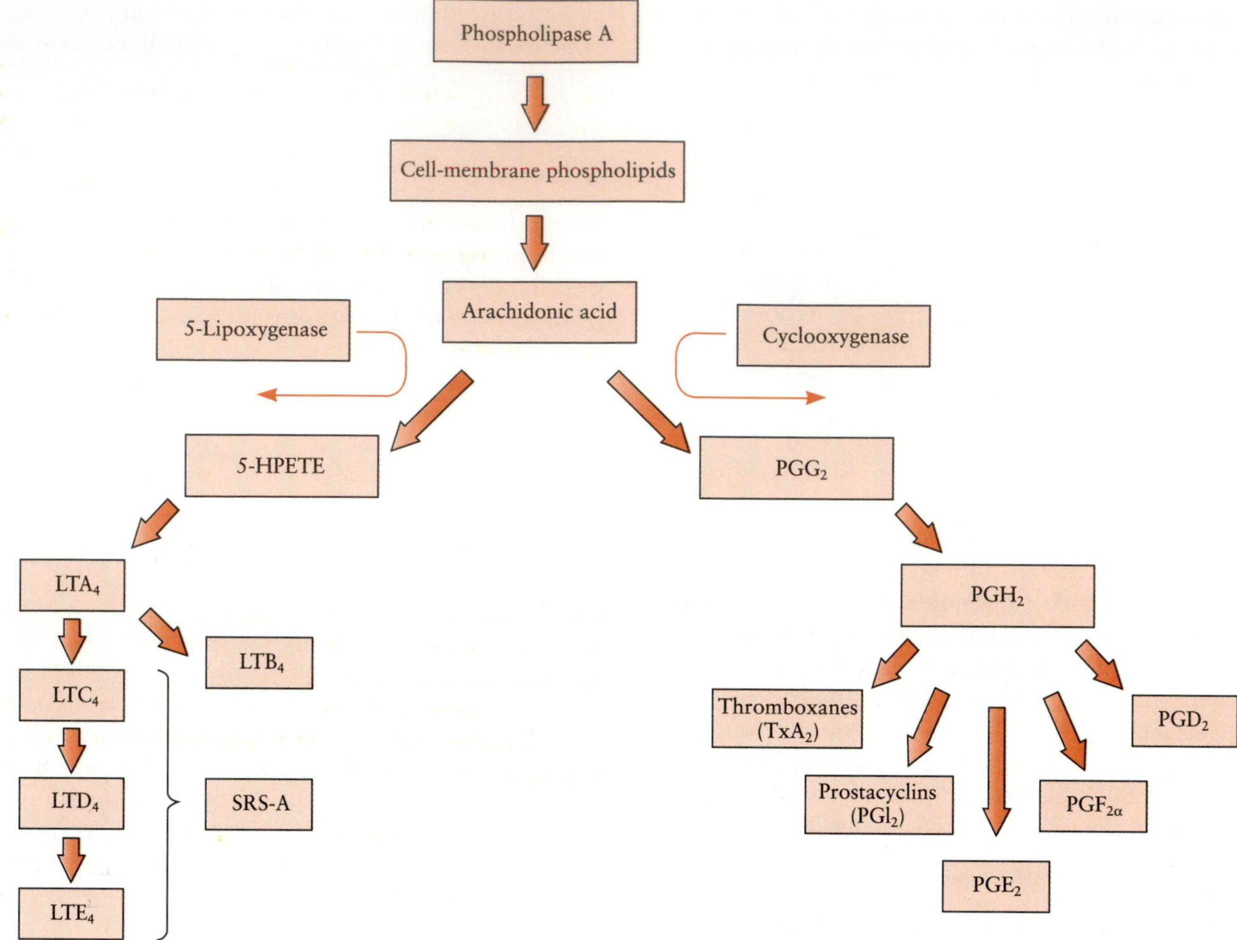

Figure 28–9 The generation of the leukotrienes, thromboxanes, prostacyclins, and prostaglandins. These compounds, collectively known as prostanoids, are very important mediators of inflammation. Slow-reacting substance of anaphylaxis (SRS-A) is a mixture of leukotrienes.

two serum proteins (kininogens) by the activity of proteolytic enzymes called kallikreins. Kallikreins may be produced directly from mast cells and basophils or indirectly from activated platelets. They may also be produced in plasma by the action of Hageman factor on an inert precursor. Kinins increase vascular permeability and stimulate smooth muscle contraction at least partly by promoting the release of leukotrienes and prostaglandins. They also stimulate pain receptors.

The anaphylatoxins are derived from the cleavage of the α chains of the complement components C3 and C5 and are known as C3a and C5a, respectively (Chapter 16). They both act indirectly on blood vessels by promoting histamine release from mast cells. In addition, C5a is a potent chemoattractant for neutrophils.

An important vasoactive peptide called PF/dil (permeability factor in diluted serum) is a fragment of the blood clotting protein, Hageman factor. PF/dil activates kallikreins, the proteolytic enzymes responsible for the generation of kinins. It is generated by damage to vascular endothelium.

Factors Derived from Neutrophils

Upon phagocytosis of particles, neutrophil primary granules fuse with the phagosomes to form phagolysosomes. Since this fusion may occur before the particle is completely ingested, lysosomal enzymes may escape into tissues. These enzymes include the kallikreins and enzymes such as superoxide dismutase, as well as free radicals from the respiratory burst.

Factors Derived from Platelets

If vascular endothelium is damaged, it provides a stimulus for local clumping of platelets. When platelets clump, they are stimulated to release many potent vasoactive agents, especially serotonin and thromboxanes.

CONTROL OF ACUTE INFLAMMATION

Since acute inflammation may cause severe tissue damage, it must be carefully controlled. Plasma contains several molecules that either inactivate inflammatory mediators or inhibit the enzymes that generate the mediators. Thus α_1-antitrypsin and α_2-macroglobulin block the proteases released from neutrophil granules as well as thrombin, plasmin, and C1 esterase.

Adenosine released from injured cells regulates the activities of neutrophils through two cell surface receptors. Adenosine A_1 receptors are stimulated by very low concentrations of adenosine. This stimulation enhances neutrophil adherence to endothelium, promotes chemotaxis, and stimulates the respiratory burst and FcγR-mediated phagocytosis. A_2 receptors are stimulated by high concentrations of adenosine and exert an opposing effect. Thus, occupation of these receptors inhibits neutrophil adherence to endothelium and their respiratory burst, although it has no effect on their chemotactic responsiveness.

Low concentrations of adenosine present in the early stages of inflammation may therefore bind A_1 receptors, stimulate neutrophil emigration and function, and promote the removal of opsonized organisms. Later, as the inflammatory reaction proceeds, neutrophil activities may be suppressed by stimulation of A_2 receptors by high concentrations of adenosine released by injured cells. This would inhibit further tissue destruction by activated neutrophils.

INFLAMMATION AS A RESULT OF IMMUNE REACTIONS

Inflammatory responses of immunological origin are known as **hypersensitivity** reactions and may be classified into four basic types, as suggested by Philip Gell and Robin Coombs (Fig. 28–10).

Type I (Immediate) Hypersensitivity

IgE and some IgG subclasses can attach to mast cells or basophils through Fc receptors. If antigen binds to this cell-fixed antibody, then the mast cell or basophil will respond by releasing the vasoactive factors, especially histamine, contained within its granules. These factors cause local acute inflammation within a few minutes (**immediate hypersensitivity**). Because mast cells also release eosinophil chemotactic factors, massive accumulations of eosinophils are characteristic of this type of hypersensitivity (Fig. 28–11).

Eosinophils

Eosinophils were first identified by Paul Ehrlich in 1879. He observed that some blood leukocytes bound acidic dyes such as eosin and so called them eosinophils. Slightly larger than neutrophils and with a bilobed nucleus, eosinophils contain two types of granule. Small granules that contain arylsulfatase and acid phosphatase and large crystalloid granules that contain four major proteins, major basic protein (MBP), eosinophil peroxidase (EPO), eosinophil cationic protein, and eosinophil-derived neurotoxin (Figs. 28–12, 28–13). Eosinophil cell membranes contain large quantities of a lysophospholipase. On their surface eosinophils have receptors for immunoglobulins (CD32, CD23, CD16, FcϵRI, and perhaps FcμR), and for complement (CD35).

Eosinophil Enzymes and Parasite Destruction

It is likely that the major function of eosinophils is the destruction of parasitic helminths (Fig. 28–14). Because they possess Fc receptors, eosinophils bind to antibody-coated parasites. Once bound, the eosinophils degranulate and release their contents onto the worm cuticle. These contents include products of the respiratory burst such as superoxide and hydrogen peroxide, potent lytic enzymes such as lysophospholipase and phospholipase D, and the eosinophil granule proteins. All of these can damage worm cuticles and may cause fatal damage to the parasite. In addition to damaging the worm cuticle at extremely low concentrations, MBP also promotes the adherence of additional eosinophils.

Inflammatory Reactions

Large numbers of eosinophils in the blood (**eosinophilia**) are characteristic of individuals with severe type I hypersensitivity. This is the result of Th2 activity. Th2 cell-derived cytokines can mobilize eosinophils

I

Mast cell with surface IgE + Antigen → Mast cell degranulation → Vasoactive factors → INFLAMMATION

II

Target cell + Antibody + Complement → Cell destruction → INFLAMMATION

III

Antigen + Antibody + Complement → Immune complexes → Neutrophil accumulation → Tissue damage → INFLAMMATION

IV

Antigen + Sensitized T cell → Lymphokines → Mononuclear cell accumulation → Tissue damage → INFLAMMATION

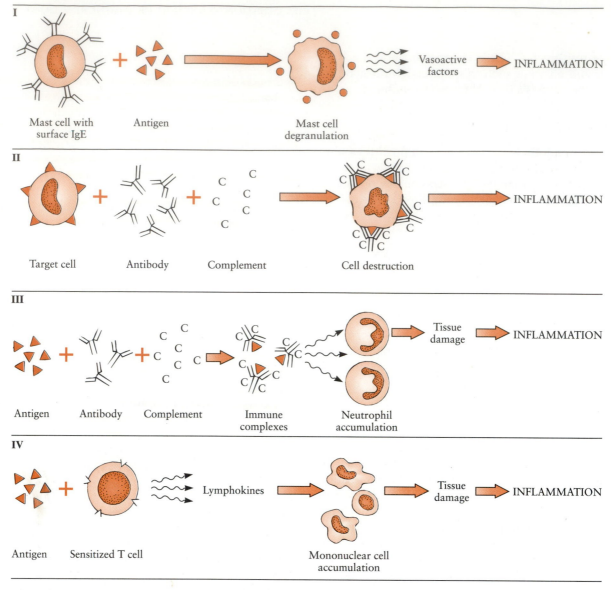

Figure 28–10 The four types of immunologically mediated inflammation or hypersensitivity as classified by Gell and Coombs.

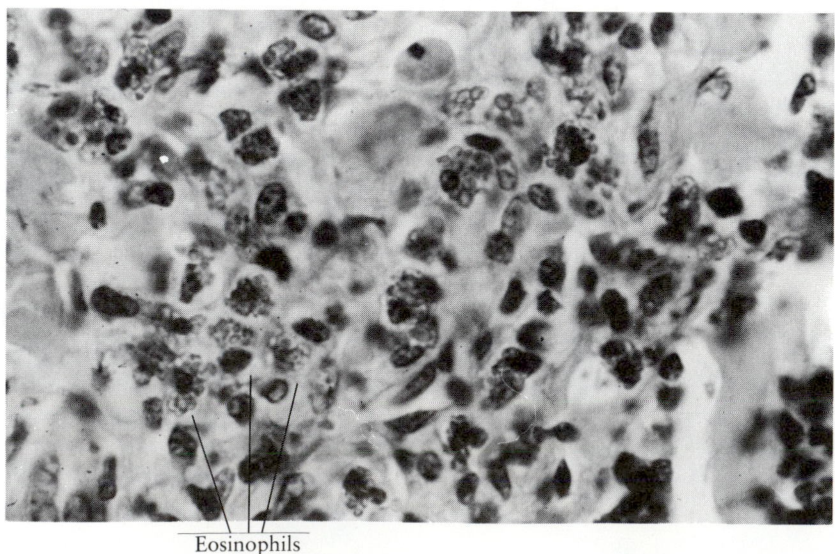

Figure 28–11 A photomicrograph of a lesion in horse skin caused by allergy to migrating parasitic helminth larvae. The extensive infiltration with eosinophils indicates the occurrence of a type I hypersensitivity reaction.

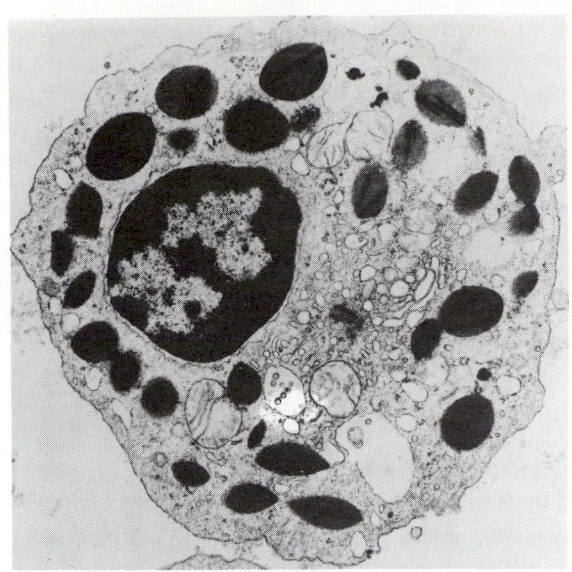

Figure 28–12 A transmission electron micrograph of a rabbit eosinophil. Note the characteristic granules containing an electron dense core. *(Courtesy of Dr. Scott Linthicum.)*

T cells
EAF
IL-5
ACTIVATION
Macrophages
IL-3
TNF
GM-CSF
IFN-β
EAF
PAF
Leukotriene C_4
Release of oxygen radicals granule proteins
Cytotoxicity
ADHERENCE
FcγRII, FcεRII
CR1, CR2
Degranulation
Cationic protein release
Cytotoxicity

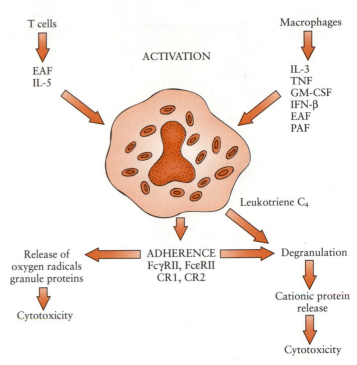

Figure 28–14 The activation of eosinophils and their functions.

from the bone marrow. Thus IL-3 stimulates the differentiation of eosinophil precursors while IL-5 causes differentiation of eosinophils in the bone marrow, causes an eosinophilia, attracts eosinophils into tissues, prolongs their life-span, and activates them. These eosinophils are then attracted to mast-cell degranulation sites by eosinophil chemotactic factors, by leukotriene B_4, by histamine, and especially by the histamine-breakdown product imidiazoleacetic acid.

When the enzymes of eosinophils were first characterized, it seemed that their prime function was to destroy the products of mast-cell degranulation. It is now clear that this is less important than their role as proinflammatory cells. For example, MBP can cause histamine release from basophils and mast cells. EPO binds strongly to mast cell granules; this EPO-granule complex has enhanced respiratory burst and parasiticidal activity, much more so than the free enzyme. Eosinophils can generate leukotrienes, particularly leukotriene C, much more effectively than neutrophils, and they can also make large amounts of platelet activating factor (Table 28–2).

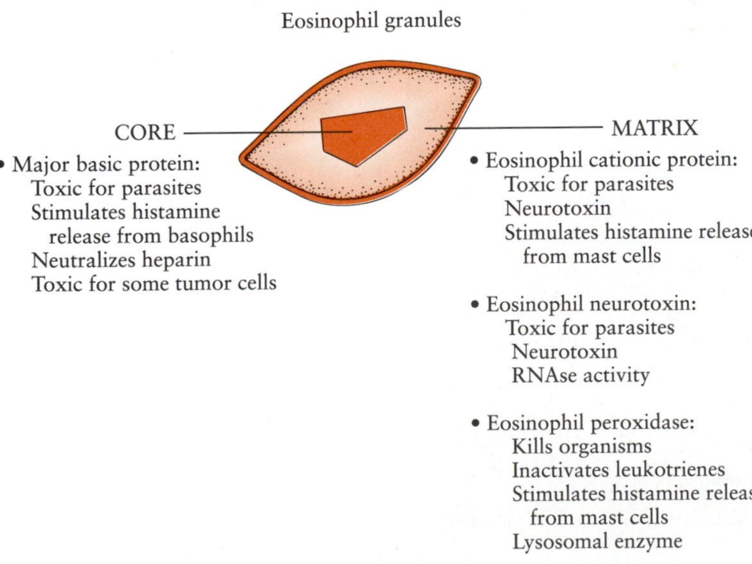

Figure 28–13 The structure and contents of an eosinophil granule.

Table 28–2 Eosinophil Enzymes

Eosinophil Enzymes
Eosinophil cationic protein (ribonuclease)
Major basic protein
Lysophospholipase
Arylsulfatase B
Phospholipid exchange protein
Phospholipase D
Peroxidase
Histaminase
Eosinophil-derived neurotoxin (ribonuclease)

Eosinophils can be activated by cytokines from macrophages or by IL-3 and IL-5 from Th2 cells. Other molecules that enhance eosinophil activity include IFN-γ, IFN-α, histamine, ECF-A, leukotriene B_4, and parasite-derived molecules.

Type II (Cytotoxic) Hypersensitivity

Antibodies may destroy antigenic cells either by activating complement or through the activities of cytotoxic cells. Cells such as neutrophils, macrophages, and some lymphocytes possess receptors for the Fc portion of immunoglobulins. These cells may kill target cells coated with immune complexes by **antibody-dependent cellular cytotoxicity** (ADCC) (Chapter 18). Cells destroyed either in this way or through complement-mediated lysis may cause acute inflammation because of the release of biologically active cell-breakdown products. This form of inflammation is observed in graft rejection (Chapter 19).

Type III (Immune Complex) Hypersensitivity

Immune complexes may activate complement when deposited in tissues. Activation of complement in this way attracts neutrophils through the production of C5a, and the neutrophils ingest the immune complexes. Neutrophils may release their free radicals and proteolytic enzymes directly into their surroundings during the phagocytic process, resulting in tissue destruction. Neutrophil-activated plasmin may in turn activate the complement system; platelet aggregation may result in the release of more vasoactive factors; and mast-cell degranulation may be mediated by anaphylatoxins. The total effect is therefore one of acute inflammation and tissue destruction (Chapter 30). This form of hypersensitivity occurs in many bacterial and viral diseases.

Type IV (Delayed) Hypersensitivity

Cell-mediated immune responses may also cause acute inflammation. If antigen is injected into an animal possessing appropriately sensitized T cells, a local inflammatory reaction may result. This reaction is known as **delayed hypersensitivity,** since it takes at least 24 hours after administration of antigen for the response to reach maximal intensity (Fig. 28–15). The inflammation results from the release of chemotactic and vasoactive lymphokines by sensitized T cells and is characterized by an accumulation of **mononuclear cells** at the site of inflammation (Fig. 28–16). Delayed hypersensitivity reactions occur in response to many bacterial antigens to virus-infected cells and in graft rejection (Chapter 18).

CHRONIC INFLAMMATION

Several hours after neutrophils have arrived at an inflammatory focus, the macrophages begin to arrive. These macrophages are attracted by many of the factors that attract neutrophils, by collagen, elastin, and fibronectin breakdown products, and by lymphokines. Once they reach the tissues, macrophages become activated and not only eliminate any invading bacteria but also phagocytose and destroy damaged cells and

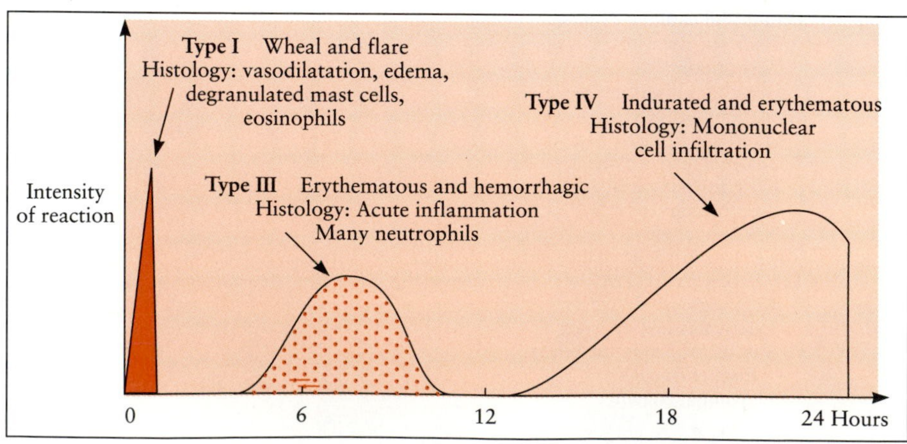

Figure 28–15 The time course, character, and histology of inflammatory skin reactions resulting from the intradermal injection of antigen and mediated by type I, III, and IV hypersensitivity reactions.

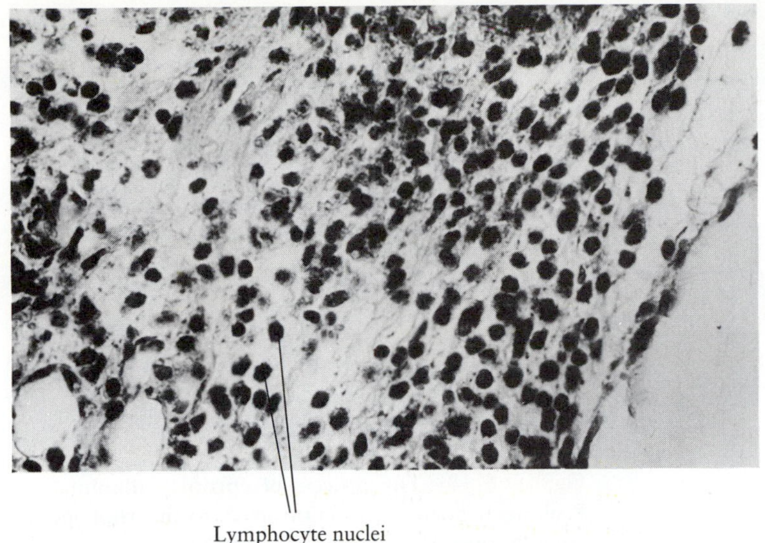

Figure 28–16 The characteristic mononuclear cell infiltration associated with a type IV hypersensitivity reaction. In this case, the response of a sensitized mouse to an extract of the protozoan parasite *Toxoplasma gondii.* *(Courtesy of Dr. C.H. Lai.)* See also Figure 25–11.

tissues, including apoptotic neutrophils. Macrophages release urokinase, collagenases, and elastases that destroy connective tissue. They also release plasminogen activator, which results in activation of plasmin. Thus macrophages destroy the local connective tissue matrix. By releasing cytokines, macrophages then attract and activate fibroblasts. Platelet-derived growth factor and tumor necrosis factor promote fibroblast proliferation and stimulate fibroblast synthesis of the collagen required to repair any tissue damage. (Despite its name, platelet-derived growth factor is produced in large amounts by macrophages.) The collagen tends to be deposited throughout the lesion and is then gradually remodeled over several weeks or months as the area returns to normal. In addition, macrophages release cytokines that promote new capillary growth. The production of these cytokines is stimulated by a lack of oxygen in the middle of a wound. Once the oxygen level is restored to normal, new blood vessel formation ceases (Fig. 28–17).

The final result of this healing process depends to a large extent on the effectiveness of the preceding acute inflammation. If the cause of the inflammation is rapidly and completely removed, then healing will follow uneventfully. If, however, the offending material or organisms cannot be destroyed because it is insusceptible to destruction by the enzymes of macrophages or neutrophils, then the inflammatory process may persist (Fig. 28–18). Thus **chronic inflammation** results from the prolonged presence of bacteria such as *Mycobacterium tuberculosis,* fungi such as *Cryptococcus,* parasite eggs such as those derived from schistosomes, or inorganic material such as asbestos crystals. In these cases, the persistent foreign material results in the continual arrival of new macrophages and fibroblasts and excessive deposition of collagen around the irritant focus. The lesion that develops in this way is known as a **granuloma** (Fig. 28–19). Granulomas consist of an accumulation of macrophages, fibroblasts, loose connective tissue, and new blood vessels called

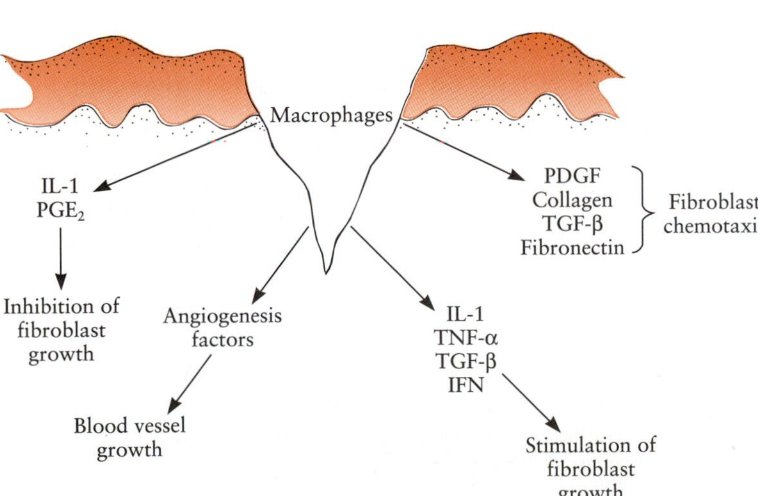

Figure 28–17 The role of macrophages in chronic inflammation and wound healing.

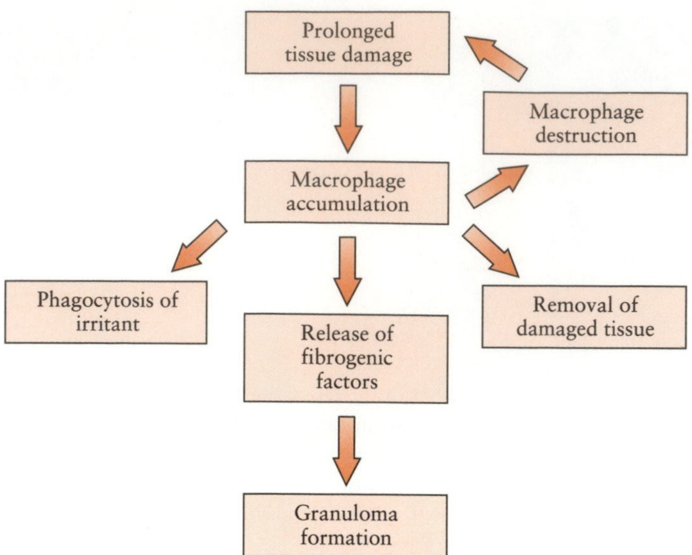

Figure 28–18 The process of chronic inflammation. Prolonged tissue destruction leads to macrophage accumulation and granuloma formation.

granulation tissue. The term "granulation tissue" is derived from the granular appearance of this tissue when cut. The "granules" are in fact new blood vessels.

If the persistent irritant is nonantigenic, for example, silica, talc, or mineral oil, then few neutrophils or lymphocytes will be attracted to the lesion. **Epithelioid** and giant cells, however, are formed in an attempt to destroy the offending material. If the material is toxic for macrophages, as is asbestos, macrophage enzymes may be released, leading to excessive tissue damage and eventually to local fibrosis and scarring.

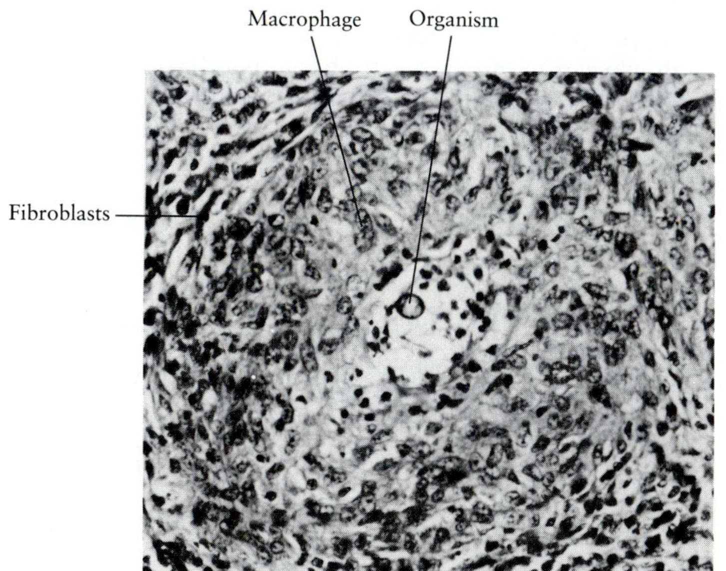

Figure 28–19 A section of liver from a patient infested by the parasitic worm, *Schistosoma mansoni.* An egg from the parasite is seen in the center of a granuloma. The cells immediately surrounding the egg are eosinophils. (Original magnification × 100.) *(From Bellanti, J.A., Immunology II. W.B. Saunders, Philadelphia, 1969. With permission.)*

If the foreign material is immunogenic, then the initial inflammatory response may be similar to that seen in a delayed hypersensitivity reaction (Chapter 18). As a result of the release of lymphokines and persistent immune stimulation, this type of granuloma will contain lymphocytes as well as macrophages, fibroblasts, and probably some neutrophils, eosinophils, and basophils. Macrophages within these granulomas may form giant cells. These chronically stimulated macrophages also release cytokines, which can stimulate collagen deposition by fibroblasts and thus eventually wall off the lesion from the rest of the body. Antigens that provoke this form of granuloma include bacteria such as *Mycobacterium tuberculosis, Brucella abortus,* and parasites such as *Microfilaria* and schistosomes.

These chronic granulomas, whether due to immunological or foreign body reactions, are important since they may enlarge and involve normal tissues. In asbestosis, for example, death eventually occurs as a result of the gradual replacement of normal lung tissues by large numbers of expanding granulomas.

CYTOKINES AND INFLAMMATION

Cytokines not only play a vital role in the local tissue response to microbial invasion and tissue damage, but also have significant systematic effects during infection, tissue injury, or inflammation. This includes the development of a fever, a neutrophilia (elevated blood neutrophils), lethargy, and eventual muscle wasting. This systemic response is mainly mediated by IL-1, IL-6, and TNF-α.

Fever. IL-1, IL-6, and TNF-α act on the thermoregulatory center in the brain to raise the body's thermostatic set-point. This causes increased heat conservation by vasoconstriction and heat production by shivering. The body temperature therefore rises until it reaches the new set-point and a fever occurs. This fever promotes T-cell proliferation in response to IL-2 and enhances the production of cytotoxic T cells and antibody synthesis. IL-1–driven secretion of IL-2 is temperature dependent, with optimal production occurring at 39° C. The alternate complement pathway also functions optimally at temperatures around 39° C (Chapter 16). These cytokines also induce sleep by promoting the release of sleep-inducing factors in the brain. Increased lethargy is commonly associated with a fever and may, by reducing the energy demands of an animal, enhance the efficiency of defense and repair mechanisms. Finally, these cytokines suppress the hunger centers of the brain and so induce the loss of appetite associated with infections (Fig. 28–20).

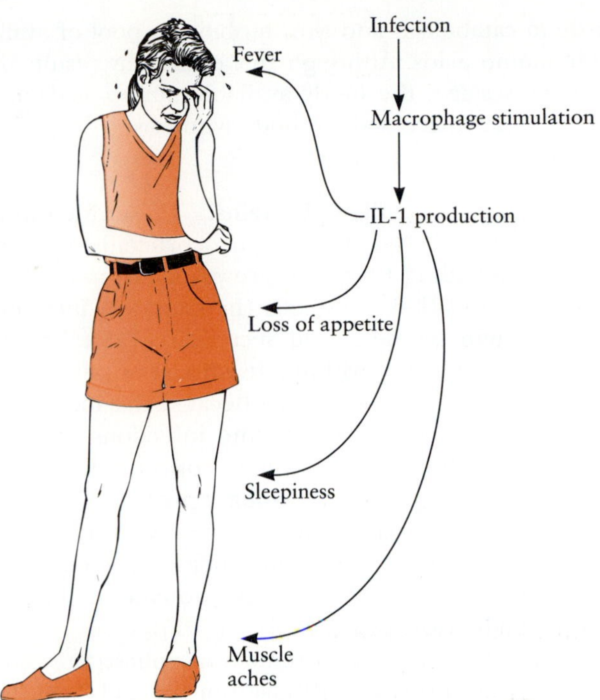

Figure 28–20 Some of the effects of interleukin-1 on the body, especially the central nervous system, that contribute to feeling "ill."

Cytokine-Induced Metabolic Changes. In addition to their effects on the nervous and immune systems, IL-1, IL-6, and TFN-α have profound metabolic effects (Fig. 28–21). They act on skeletal muscle to enhance

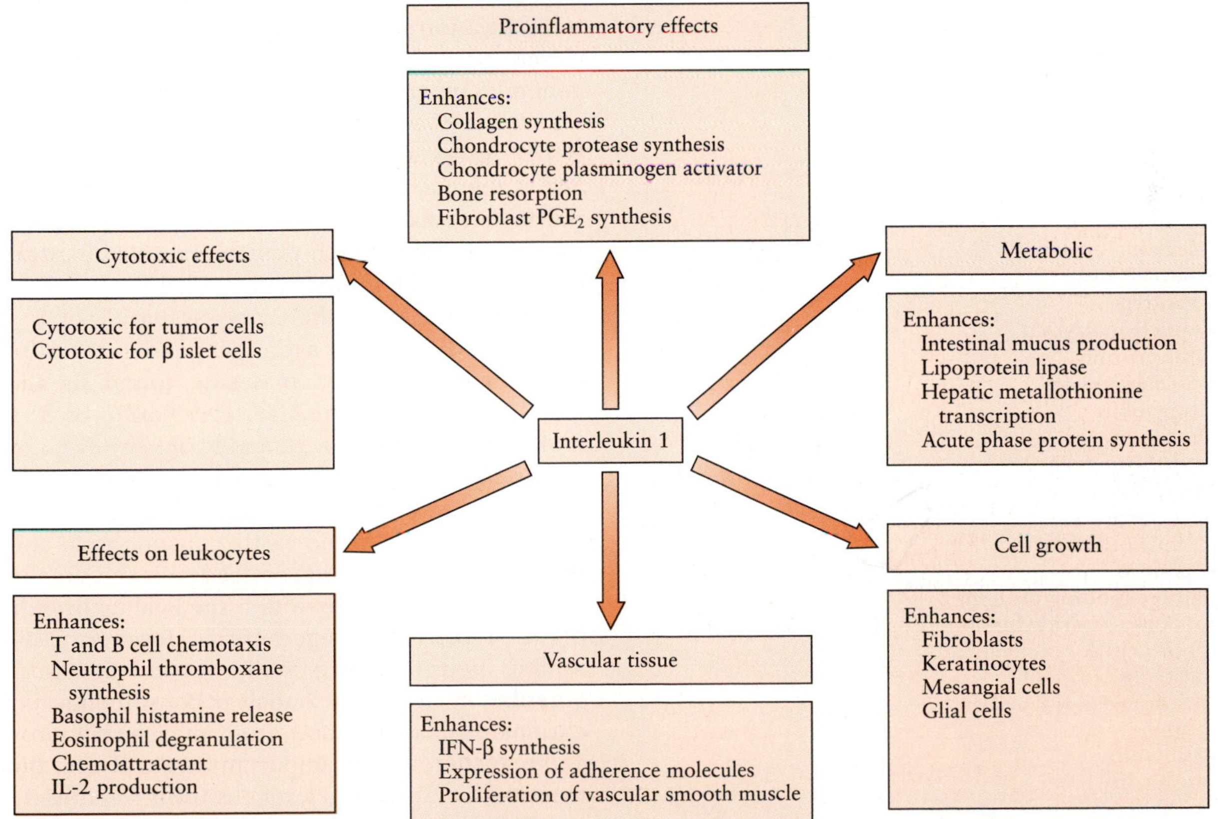

Figure 28–21 The systemic effects of interleukin-1.

protein catabolism and thus mobilize a pool of available amino acids. Although this eventually results in muscle wastage, the newly available amino acids are available for increased antibody synthesis.

Induction of Acute-Phase Proteins. Under the influence of IL-1, TFN-α, IFN-γ, glucocorticoids, growth factors (including fibroblast growth factor and TGF-β), IL-11, and IL-6, liver cells (hepatocytes) increase their protein synthesis and secretion (Table 28–3). The response begins within a few hours of injury and subsides within 24 to 48 hours. Because this increased synthesis is associated with acute infections and inflammation, those proteins whose concentrations rise so rapidly are known as **acute-phase proteins** (Fig. 28–22). The acute-phase proteins have a wide range of functions in host defense. They include complement components, coagulation factors, protease inhibitors, and metal-binding proteins.

The three most important acute-phase proteins are C-reactive protein (CRP), serum amyloid P (SAP), and serum amyloid A (SAA). The levels of these proteins may rise a thousandfold in serum following bacterial invasion and tissue damage. CRP and SAP are closely related both structurally and functionally. In general, only one of these two proteins (CRP or SAP) is an acute-phase protein in a given species. Thus CRP is the major acute-phase protein in humans while SAP is the major one in mice. Both CRP and SAP are polymers of five identical 23 kDa proteins. They belong to a family of such pentamers called pentraxins. CRP was first identified and named for its ability to bind and precipitate the C-polysaccharide of *S. pneumoniae.* CRP binds in a Ca^{++} dependent reaction to phosphatidylcholine, a molecule found in all cell membranes and to the C-polysaccharide of *S. pneumoniae.* It can bind to activated lymphocytes, to invading organisms, and to damaged tissues where it activates the classical complement pathway. CRP binds to neutrophils and promotes the phagocytosis and removal of damaged, dying, or dead cells and organisms. CRP may therefore promote tissue healing by enhancing the repair of damaged tissue. However, CRP also inhibits neutrophil superoxide production and degranulation. It binds to nuclear constituents such as chromatin, histones, and small nuclear riboproteins and may therefore be important in regulating the autoantibody response to antigens from cell nuclei.

SAP, in contrast, is a component of normal basement membranes. Like CRP it can bind nuclear con-

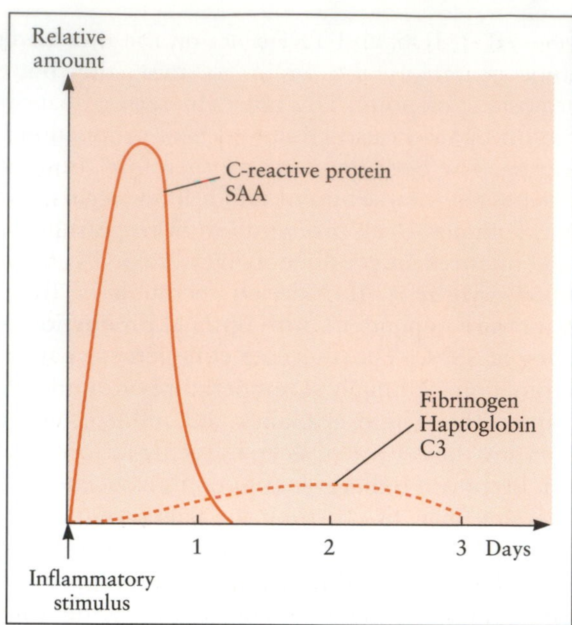

Figure 28–22 Some acute-phase responses. Proteins such as C-reactive protein and SAA rise many hundredfold. Other proteins such as fibrinogen, haptoglobin, and C3 rise only slightly.

Table 28–3 The Acute-Phase Proteins

Positive Acute-Phase Proteins
Major proteins
C-reactive protein
Serum amyloid A
Serum amyloid P
Complement components
C2
C3
C4
C5
C9
Factor B
C1-inhibitor
C4-binding protein
Protease inhibitors
α_1-antitrypsin
α_1-antichymotrypsin
β_2-antiplasmin
Metal-binding proteins
Haptoglobin
Hemopexin
Ceruloplasmin
Clotting factors
Fibrinogen
von Willebrand factor
Other proteins
Mannose-binding proteins
Lipopolysaccharide-binding proteins
Lipoprotein A
α_1-acid glycoprotein
Negative Acute-Phase Proteins
Albumin
Prealbumin
Transferrin
Apo A1
Apo A2

stituents such as DNA, chromatin, and histones. It also can bind to C1q and activate the classical complement pathway.

The other major acute-phase proteins, the SAA proteins are a family of lipoproteins whose function has not been definitely established. They bind to cell membranes and are immunosuppressive, inhibiting both B-cell and macrophage function. The SAA proteins may exert a negative feedback on IL-1 production by macrophages.

A second group of acute-phase proteins increase their concentrations twofold to threefold. These include protease inhibitors such as α_1-antitrypsin, α_1-antichymotrypsin, and α_2-macroglobulin. All of these can inhibit the tissue damage caused by the release of neutrophil proteases in sites of acute inflammation. This group also includes the iron-binding proteins, haptoglobin and hemopexin. Haptoglobin and hemopexin bind iron molecules and thus make them unavailable to invading bacteria (Chapter 25). In this way they inhibit bacterial proliferation and invasion. These proteins also reduce iron availability for red blood cell production so that anemia is commonly associated with severe or chronic infections.

Some proteins are considered negative acute-phase proteins since their levels fall during acute inflammation. Negative acute-phase proteins include albumin, prealbumin, and transferrin. Cytokines such as TNF-α acting on hepatocytes suppress transcription of their genes.

IL-1–Induced Inflammatory Changes. IL-1 stimulates the release of neutrophils from the bone marrow into the circulation and so causes a neutrophilia. IL-1 is also chemotactic for neutrophils and thus attracts them to sites of bacterial invasion. It enhances their bactericidal activity by stimulating their oxidative metabolism. IL-1 enhances inflammation by degranulating basophils and mast cells and by activating neutrophils and eosinophils so that they release their lysosomal enzymes. It acts on vascular endothelium, increasing its adhesiveness while causing vasodilation (Fig. 28–23). It may also be responsible for the endothelial proliferation seen in chronically rejected allografts (Chapter 19).

AMYLOIDOSIS

Amyloid is the name given to an amorphous, eosinophilic, waxy, extracellular substance that infiltrates organs in certain diseases (Fig. 28–24). Several forms of

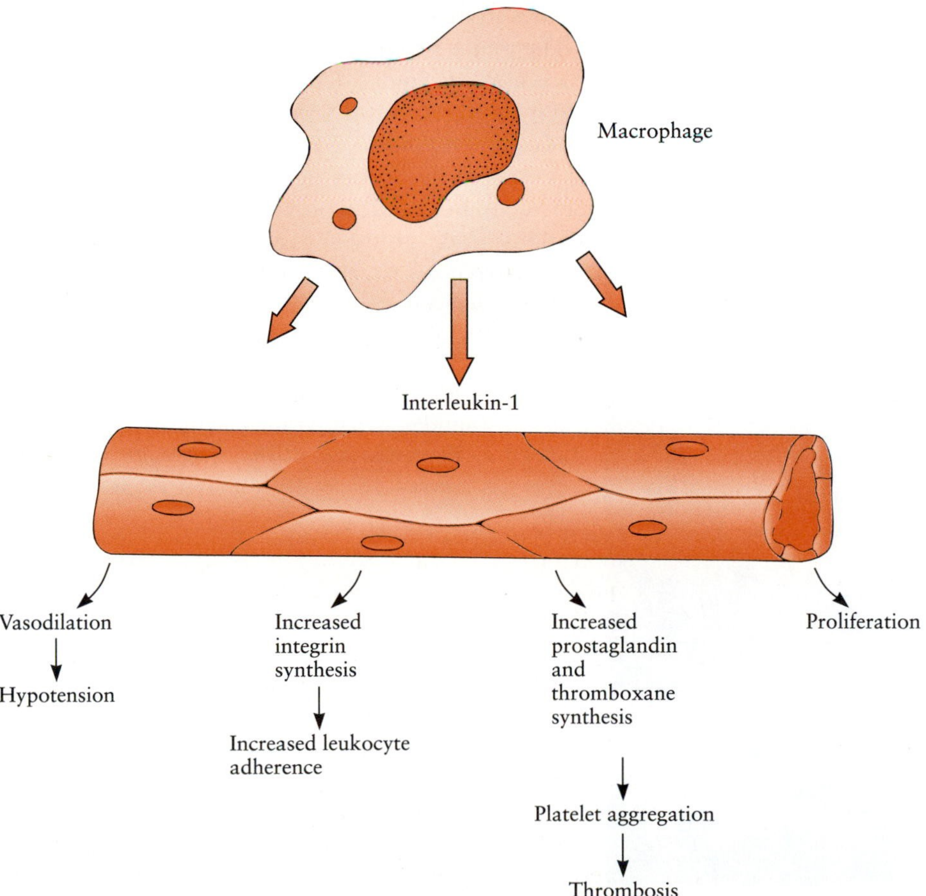

Figure 28–23 The effects of interleukin-1 on blood vessels and vascular endothelium. Thus it increases the adhesiveness of endothelial cells causing leukocytes to marginate. It may also be responsible for the proliferation of vascular endothelium seen in cardiac allografts (Fig. 19–9).

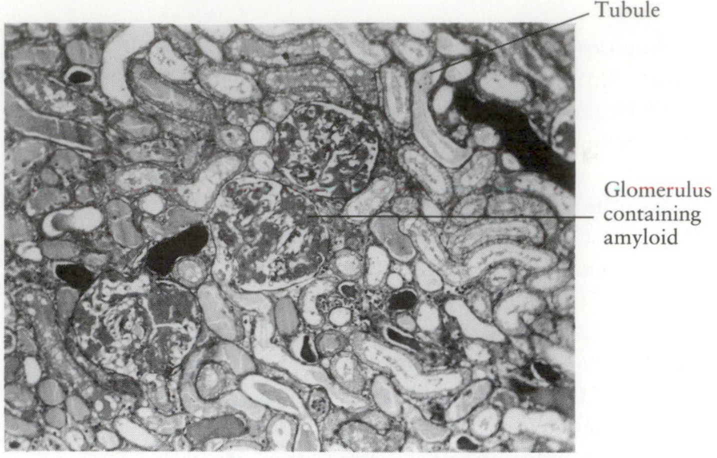

Figure 28–24 Amyloidosis of the kidney. Note the amorphous deposits within glomeruli. These will eventually result in renal failure and death.

amyloid are recognized. Thus immunocytic amyloid is associated with myelomas or other lymphoid tumors; reactive amyloid is associated with chronic inflammatory conditions such as mastitis, osteomyelitis, abscesses, and tuberculosis; and hereditary forms of amyloid are also recognized. Reactive amyloidosis is a major cause of death in animals repeatedly immunized for commercial antiserum production. It is also commonly associated with autoimmune diseases such as systemic lupus erythematosus and rheumatoid arthritis and with chronic infections such as leprosy and tuberculosis. These reactive amyloid deposits are found in the liver, spleen, and kidneys, particularly within glomeruli. Amyloidosis may be produced in laboratory mice by feeding them a diet rich in casein or by injecting large or repeated doses of bacterial endotoxin or Freund's complete adjuvant.

On electron microscopy, all forms of amyloid can be shown to consist of a feltlike mass of protein fibrils (Fig. 28–25). By x-ray crystallography, it can be further shown that all amyloid proteins have their polypeptide chains arranged in β-pleated sheets. This is a molecular conformation that renders the fibrils both extremely insoluble and almost totally resistant to normal proteolytic enzymes. Consequently, once deposited in tissues, they are almost impossible to remove. The accumulation of amyloid in tissues is therefore essentially irreversible, leading to gradual cell loss and tissue destruction. The β-pleated sheet configuration of amyloid also gives it unique staining properties; for example, it stains unusually intensely with the dye toluidine blue and binds specifically to another dye called Congo red.

Although all amyloid proteins have the β-pleated sheet configuration, biochemical analysis has shown that most are composed of one of two basic proteins (Fig. 28–26). In immunocytic amyloid and a few cases of reactive amyloid, this protein is known as amyloid L (AL). AL protein is a product of partially denatured immunoglobulin light chains. In individuals with myelomas, light chains are usually generated in excess, and limited digestion of these permits the V_L region to assume a β-pleated sheet configuration. It has been suggested that immunocytic amyloid deposits consist of light chains partially digested by proteolytic enzymes from cells such as macrophages and deposited in tissues in excessive amounts.

The major protein component of reactive amyloid is known as amyloid A (AA). AA is derived by partial proteolytic digestion of serum amyloid A (SAA) proteins (page 449). SAA proteins are found in high concentrations in the serum of humans and animals with experimental or natural amyloidosis and in the serum of patients undergoing chronic antigenic stimulation, as in tuberculosis or rheumatoid arthritis. They are found in low levels in normal serum and are produced

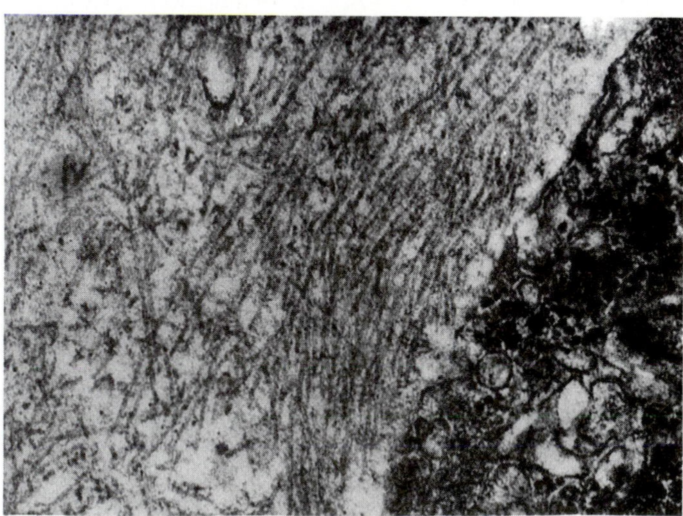

Figure 28–25 Amyloid fibrils. An electron micrograph showing bundles of paired amyloid fibrils deposited parallel to a cell membrane. *(Courtesy of the late Dr. E.C. Franklin. From Adv. Immunol., 15:258, 1972. With permission.)*

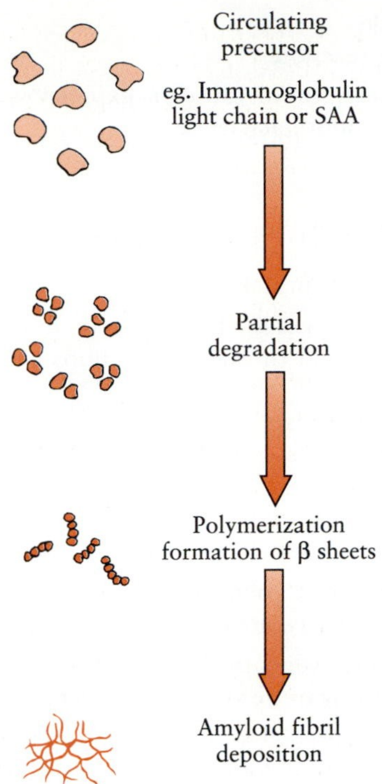

Figure 28–26 A simplified view of the probable pathogenesis of immunogenic and reactive amyloidosis.

by hepatocytes and fibroblasts in response to IL-1 and other cytokines. Since SAA proteins have been shown to be strongly immunosuppressive in mice, it has been suggested that SAA normally controls the immune response and that, under conditions of chronic antigenic stimulation, excessive production of SAA and their subsequent digestion can lead to the deposition of AA as amyloid in tissues.

SAP is found in most amyloid deposits regardless of the identity of the major fibril protein. SAP can inhibit the enzyme elastase and may therefore assist in protecting amyloid deposits against proteolytic digestion. Although AL and AA are associated with the two major forms of amyloid, it must be emphasized that any protein that can form extensive β-pleated sheets will give rise to amyloid if deposited in tissues. Many of the less common manifestations of amyloidosis may therefore be caused by proteins distinct from AL or AA. For example, chronic hemodialysis patients develop an amyloidosis due to deposition of β2-microglobulin. Alzheimer's disease is associated with deposition of an amyloid fibril in brain called A4 protein (or amyloid β protein). It has been suggested that a more appropriate collective term for the amyloidoses is β fibrilloses, since the β-pleated conformational structure is the common feature of all these conditions.

KEY WORDS

Acute inflammation p. 433
Acute-phase proteins p. 448
Adenosine p. 441
Adherence p. 434
Amyloidosis p. 449
Anaphylatoxin p. 439
Cardinal signs p. 433
C5a p. 435
Chemokine p. 436
Chemotactic factors p. 435
Chronic inflammation p. 444
C-reactive protein p. 448
Cytotoxic hypersensitivity p. 444
Delayed hypersensitivity p. 444
Diapedesis p. 435
Edema p. 433
Eosinophil p. 441
Epithelioid cells p. 446
Fever p. 447
Granuloma p. 445
Hageman factor p. 438
Histamine p. 439
Hypersensitivity reaction p. 441
Immediate hypersensitivity p. 441
Immune complex p. 444
Immunocytic amyloid p. 450
Inflammation p. 433
Integrin p. 434
Interleukin-1 p. 447
Interleukin-8 p. 436
Kallikrein p. 440
Kinin p. 439
Leukoegressin p. 436
Leukotriene p. 439
Macrophage inflammatory protein p. 438
Major basic protein p. 441
Monocyte chemoattractant protein p. 438
Monocytes p. 435
Neutrophils p. 435
Permeability p. 433
Platelet-activating factor p. 439
Prostaglandin p. 439
RANTES protein p. 438
Reactive amyloid p. 450
Serotonin p. 439
Serum amyloid A p. 448
Serum amyloid P p. 448
Vasoactive factors p. 439

QUESTIONS

1. The major protein component of reactive amyloid is known as
   a. AP protein
   b. AA protein
   c. K light chains
   d. SAA protein
   e. AM protein

2. Which of the following is an acute-phase protein?
   a. C-reactive protein

b. fibronectin
c. albumin
d. immunoglobulin
e. interleukin-1

3. Which of the following is *not* one of the cardinal signs of inflammation?
a. heat
b. redness
c. pain
d. swelling
e. pus

4. Which of the following molecules is classified as a chemokine?
a. interleukin-1
b. interleukin-2
c. interleukin-4
d. interleukin-6
e. interleukin-8

5. The key cells involved in mediating delayed hypersensitivity are
a. neutrophils
b. immune complexes
c. T cells
d. mast cells
e. complement

6. An eosinophilia (a high level of eosinophils in the blood) is suggestive of
a. parasitic disease
b. cancer
c. autoimmune disease
d. bacterial disease
e. viral disease

7. Interleukin-1 causes a fever as a result of acting on the
a. spleen
b. thymus
c. adrenals
d. liver
e. hypothalamus

8. Prolonged chronic inflammation around a foreign body such as a schistosome egg gives rise to a(n)
a. abscess
b. wheal and flare reaction
c. granuloma
d. tubercle
e. Arthus reaction

9. Inflammatory stimuli cause the expression of which molecules on endothelial cells?
a. lectins
b. cadherins
c. Fc receptors
d. selectins
e. acute-phase proteins

10. The metabolism of the long-chain unsaturated fatty acid arachidonic acid through the actions of lipoxygenases generates
a. prostaglandins
b. thromboxanes
c. prostaglandins
d. leukotrienes
e. phospholipases

11. If fever is such a good thing, why do we usually try to reduce it when we are ill?

12. What are the events that eventually give rise to asbestosis? Why do cigarette smokers not get a similar disease?

13. What do you predict would happen in an individual with a deficiency of the following: (a) mast cells; (b) neutrophils; (c) C5a; and (d) kinins?

14. Account for the cardinal signs of inflammation: heat, redness, pain, swelling, and loss of function. What are the benefits of each of these lesions?

15. List and compare the differences between acute and chronic inflammation.

16. What are the events that eventually give rise to amyloidosis? How might this condition be treated?

17. When you suffer an attack of influenza, you get ill and feel very sick. To what extent are the signs of illness attributable to interleukin-1 or other cytokines?

18. What might be the consequences of an absolute deficiency of interleukin-1?

19. What benefits may the acute-phase response confer on an individual? Are there any apparent disadvantages to this response?

Answers: 1b, 2a, 3e, 4e, 5c, 6a, 7e, 8c, 9d, 10d

SOURCES OF ADDITIONAL INFORMATION

Alam, R., et al. RANTES is a chemotactic and activating factor for human eosinophils. J. Immunol., 150:3442–3447, 1993.

Bornstein, D.L. Leukocyte pyrogen: A major mediator of the phase reaction. Ann. N.Y. Acad. Sci., 389:323–337, 1982.

Cronstein, B.N., et al. Neutrophil adherence to endothelium is enhanced via adenosine A_1 receptors and inhibited via adenosine A_2 receptors. J. Immunol., 148:2201–2206, 1992.

Dinarello, C.A. Interleukin 1 and the pathogenesis of the acute-phase response. N. Engl. J. Med., 311:1413–1418, 1984.

Dinarello, C.A. The endogenous pyrogens in host-defense interactions. Hosp. Pract. [off.], 24:111–128, 1989.

Etzioni, A. Adhesion molecules in host defense. Clin. Diagn. Lab. Immunol., 1:1–4, 1994.

Fahey, T.J., et al. Macrophage inflammatory protein 1 modulates macrophage function. J. Immunol., 148:2764–2769, 1992.

Gerritsen, M.E., and Bloor, C.M. Endothelial cell gene expression in response to injury. FASEB J., 7:523–532, 1993.

Glenner, G.G. Amyloid deposits and amyloidosis: The β fibrilloses. N. Engl. J. Med., 302:1283–1292, 1333–1343, 1980.

Jampel, H.D., et al. Fever and immunoregulation. III. Hyperthermia augments the primary in vitro humoral immune response. J. Exp. Med., 157:1229–1238, 1983.

Koch, A.E., et al. Interleukin-8 as a macrophage-derived mediator of angiogenesis. Science, 258:1798–1800, 1992.

Kushner, I. C-reactive protein and the acute-phase response. Hosp. Pract. [off.], 25:13–28, 1990.

Lawrence, M.B., and Springer, T.A. Neutrophils roll on E-selectin. J. Immunol., 151:6338–6346, 1993.

Lewis, R.A., Austen, K.F., and Soberman, R.J. Leukotrienes and other products of the 5-lipoxygenase pathway. Biochemistry and relation to pathobiology in human diseases. N. Engl. J. Med., 323:645–655, 1990.

Mantovani, A., and Dejana, E. Cytokines as communication signals between leukocytes and endothelial cells. Immunol. Today, 10:370–375, 1989.

Matsushima, K., and Oppenheim, J.J. Interleukin 8 and MCAF: Novel inflammatory cytokines inducible by IL-1 and TNF. Cytokine, 1:2–13, 1989.

Miller, M.D., and Krangel, M.S. Biology and biochemistry of the chemokines: A family of chemotactic and inflammatory cytokines. Crit. Rev. Immunol., 12:17–46, 1992.

Oppenheim, J.J., Rosenstreich, D.L., and Potter, M. Cellular Functions in Immunity and Inflammation. Elsevier-North Holland, New York, 1981.

Osborn, L. Leukocyte adhesion to endothelium in inflammation. Cell, 62:3–6, 1990.

Salmon, J.E., and Cronstein, B.N. Fcγ receptor-mediated functions in neutrophils are modulated by adenosine receptor occupancy. J. Immunol., 145:2235–2240, 1990.

Samuelsson, B. Leukotrienes: Mediators of immediate hypersensitivity reactions and inflammation. Science, 219:568–575, 1983.

Stone, M.J. Amyloidosis: A final common pathway for protein deposition in tissues. Blood, 75:531–545, 1990.

Subramanian, N., and Bray, M.A. Interleukin 1 releases histamine from human basophils and mast cells *in vitro*. J. Immunol., 138:271–275, 1987.

Weiss, S.J. Tissue destruction by neutrophils. N. Engl. J. Med., 320:365–376, 1989.

Weissmann, G., Smolen, J.E., and Korchatz, H.M. Release of inflammatory mediators from stimulated neutrophils. N. Engl. J. Med., 303:27–34, 1980.

Wolpe, S.D., and Cerami, A. Macrophage inflammatory proteins 1 and 2: Members of a novel superfamily of cytokines. FASEB J., 3:2565–2573, 1989.

Zimmerman, G.A., Prescott, S.M., and McIntyre, T.M. Endothelial cell interactions with granulocytes: Tethering and signalling molecules. Immunol. Today, 13:93–99, 1992.

CHAPTER 29

Allergies: Type I Hypersensitivity

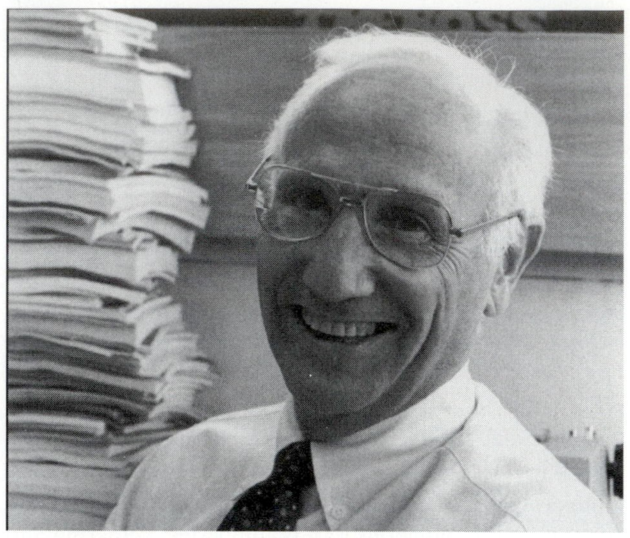

Frank Austen *As our understanding of allergies and immediate hypersensitivity increased, it became apparent that mast cell products were very complex. Dr. Frank Austen has played a prominent role in increasing our understanding of the mechanisms involved in allergy and anaphylaxis by analyzing these products. Dr. Austen's efforts have focused on the contents of mast cell granules and especially their rich content of proteases. These enzymes contribute significantly to the local inflammatory response seen in type I hypersensitivity. Their discovery has explained many puzzling features of these diseases.*

CHAPTER OUTLINE

CHAPTER CONCEPTS

1. Allergy, or type I hypersensitivity, results from the release of pharmacologically active molecules from mast cells. This release is caused by the combination of antigen with IgE bound to receptors on the mast-cell surface.
2. Type I hypersensitivity is probably a manifestation of an imbalance in helper-cell activity. Excessive Th2 activity leads to IL-4 and IL-5 release and subsequently to excessive IgE production.
3. Type I hypersensitivity is characterized by a rapid time course, by an edematous inflammatory response, and by infiltration of affected tissues by eosinophils.
4. The most important clinical manifestations of type I hypersensitivity are anaphylaxis, hay fever, and asthma. The tendency to develop this type of hypersensitivity is inherited.

In this chapter you will learn about allergies. These are a form of hypersensitivity reaction mediated by IgE and mast cells. The chapter begins with a detailed description of IgE and its properties. This is followed by a description of the structure and properties of mast cells and basophils. IgE binds to mast cells through specialized receptors and causes the release of a complex mixture of biologically active molecules. The properties of IgE receptors and the molecules released by mast cells are therefore discussed next. Because of the very serious consequences of allergic reactions, the mechanisms by which these reactions are controlled are explained. Finally, some of the most important clinical manifestations of allergic reactions in humans are addressed as well as their prevention and treatment.

Immunological responses commonly result in inflammation. If inflammation is viewed as a protective process, then these hypersensitivities can be considered to be additional protective mechanisms mediated by immune processes. In some circumstances, however, these inflammatory responses may be severe or even life-threatening.

Although hypersensitivity can be induced in many ways, the simplest way to view these reactions is through the classification scheme of Philip Gell and Robin Coombs. They noted that all these reactions fall into one of four types. (For specific details of this classification, the reader is well advised to review Chapter 28 before continuing this one.) One of these, type I hypersensitivity, popularly known as allergy, is a major clinical problem in humans. Between 10% and 20% of the population of the United States claims to suffer from some type of allergy. It has been estimated that asthma alone accounts for 1% of all health care costs in the United States.

TYPE I HYPERSENSITIVITY

Type I hypersensitivities are inflammatory reactions mediated mainly by IgE, bound to mast cells and basophils; the reactions result from the release of pharmacologically active molecules from these cells (Fig. 29–1). Because they cause so much distress and discomfort, any beneficial effects are not immediately obvious; nevertheless, certain features of type I hypersensitivity do give indications of a beneficial biological function. First, the localized acute inflammatory response that occurs in this condition probably plays a significant role in antigen elimination. Second, this type of hypersensitivity is associated with exposure to helminth antigens and plays a major role in resistance to these parasites (Chapter 25). Third, statistical evidence suggests that individuals who suffer from allergies are less likely than nonallergic people to die from cancer. The mechanism and significance of this phenomenon are unknown.

INDUCTION OF TYPE I HYPERSENSITIVITY

Type I hypersensitivity, or **allergy,** is mediated by antibodies of the IgE class. The reasons why IgE rather than IgG antibodies are produced are not clear. The antigens, or **allergens,** that induce IgE antibodies have no discernible unique biochemical features, although they must preferentially stimulate Th2 cells. Certain antigens are potent stimulators of this type of response. These include proteins of pollen grains, some helminth antigens, and some proteins in insect venoms. Freund's complete adjuvant or killed *Bordetella pertussis* organisms may preferentially stimulate IgE production in some animals.

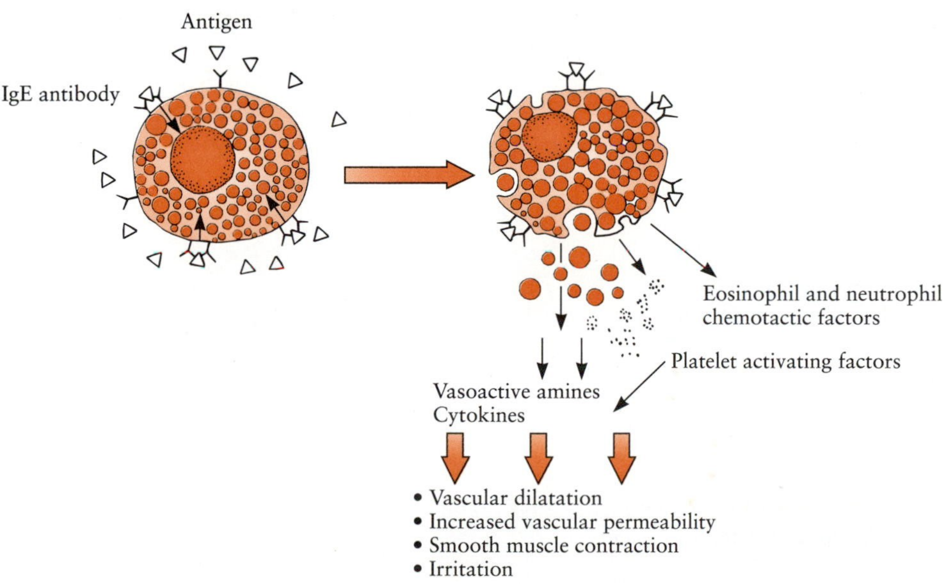

Figure 29–1 The mechanism of the type I hypersensitivity reaction.

Several general factors influence the expression of allergies. For example, there is a tendency for more males than females to be affected by allergies, and the prevalence of allergy rises to peak at about 20 to 24 years before declining gradually. Breast-feeding tends to prevent allergy by reducing neonatal infections and exposure to foreign food antigens. The most important influences on the development of allergies are genetic. Some individuals inherit a greater than normal tendency to mount an IgE response from their parents. These individuals are said to be in a state of **atopy.** In normal individuals, an IgE response can be elicited only by certain carefully designed immunization procedures. In contrast, atopic individuals make IgE constantly and have unusually high levels of this immunoglobulin. The inheritance of atopy is complex. Thus there is a major locus that influences overall IgE production on chromosome 11. It may be identical to the gene coding for FcϵRI β chain. There is also an association between certain MHC haplotypes and the occurrence of atopy. This association implies that the allergic trait is also influenced by a gene or genes within the MHC. Thus the predisposition to atopy probably depends on interaction between the gene on chromosome 11 and one or more MHC genes.

IMMUNOGLOBULIN E

IgE is a heat-labile immunoglobulin of conventional four-chain structure. As a result of the replacement of the hinge region by an extra constant domain, its heavy chain contains four C_H domains and one V_H domain. IgE thus has a molecular weight of 196 kDa, which is somewhat greater than that of IgG. IgE is produced by plasma cells, most of which are located near epithelial surfaces such as the skin, intestine, and lungs. It is found in serum in exquisitely small quantities, although much is bound through its C_H3 domain to receptors on mast cells and basophils. When attached to these cells, it has a half-life of 11 to 12 days. When free in serum, its half-life is only about two days. Some IgG subclasses may also bind to mast cells and participate in type I hypersensitivity reactions. However, the affinity of these for mast-cell receptors is considerably lower than that of IgE, and they are much less important.

The B-cell response leading to production of IgE occurs in two stages. In the first stage, a resting B cell switches from IgD/IgM production to IgE production under the influence of IL-4. Th2 cells produce IL-4 and Th1 cells produce IFN-γ, which has an opposing effect. The ratio of Th1 to Th2 cells engaged in an immune response probably determines the level of IgE produced. The second stage of the IgE response proceeds when an IgE-producing B cell develops into a plasma cell. This is controlled by cell surface protein called FcϵRII or CD23 (see later discussion). FcϵRII links the B cells to dendritic cells and so prevents their destruction by apoptosis. They are thus free to differentiate into plasma cells. FcϵRII expression is regulated by the presence of IL-4, which stimulates its production, and by IgE, which suppresses its production.

MAST CELLS AND BASOPHILS

Mast cells are large round cells (15 to 20 μm in diameter) distributed throughout the body in connective tissue (Fig. 29–2). Their most characteristic feature is a cytoplasm packed with large granules that stain intensely with basic dyes. The granules usually mask the relatively large bean-shaped nucleus (Fig. 29–3). The phenotype of mast cells is determined by their tissue environment. Thus there are two populations of mast cells in rodents and humans: connective tissue mast cells and mucosal mast cells (Table 29–1). Connective tissue mast cells arise from precursors in fetal liver and bone marrow. Their numbers increase slowly in tissues, although irritation may cause local proliferation. Connective tissue mast cells contain many uniform granules and are rich in histamine and heparin. Their life span is at least six months.

Mucosal mast cells differ from connective tissue mast cells in several respects. For example, they have few variable-sized granules. The granules of mucosal mast cells contain not heparin, but chondroitin sulfate. As a result, they have different staining properties. They contain little histamine and produce differ-

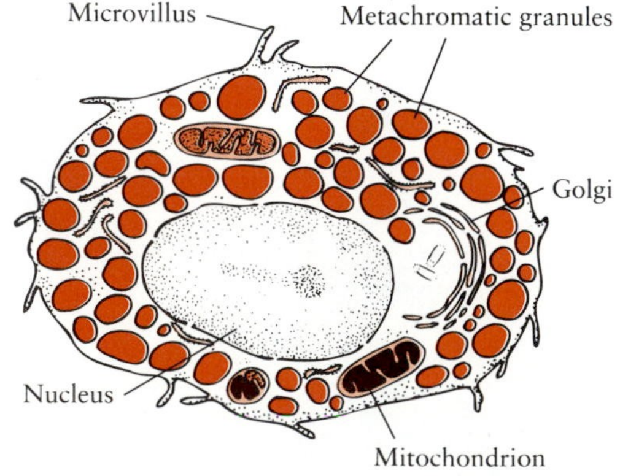

Figure 29–2 The basic features of a connective tissue mast cell.

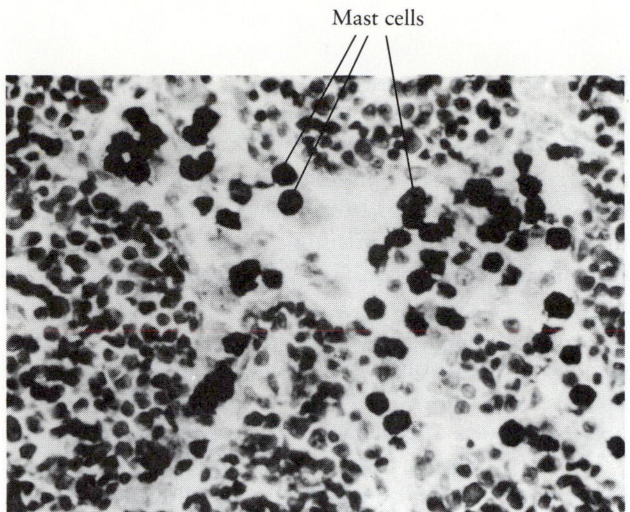

Figure 29–3 A histological section of normal human mesenteric lymph node stained to show mast cells. The intense staining of the mast cells is due to the presence of heparin within its granule.

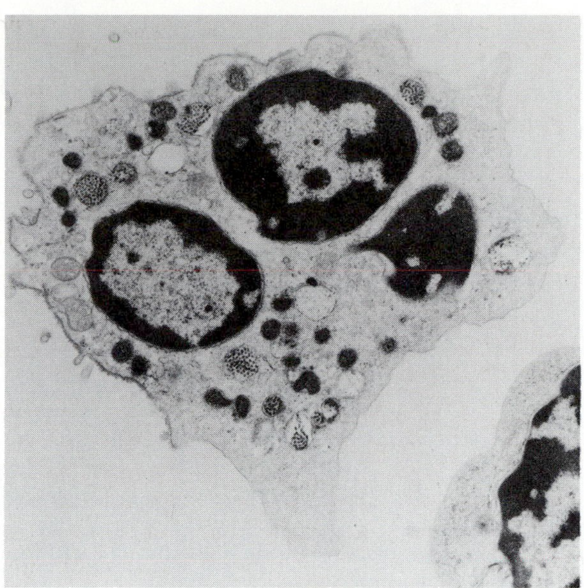

Figure 29–4 A transmission electron micrograph of a peripheral blood basophil. Note the large cytoplasmic granules. *(Courtesy of Dr. S. Linthicum.)*

ent prostaglandins and leukotrienes as well as platelet-activating factor. Whereas connective tissue mast cells remain at relatively constant levels, mucosal mast cells proliferate in response to IL-3 produced by T cells. Thus T-cell–deficient animals are also deficient in mucosal mast cells. It has been suggested that these mucosal mast cells may respond specifically to parasite antigens.

Strictly speaking, peripheral blood basophils (Fig. 29–4) should not be considered simply as circulating mast cells. They may, however, be passively sensitized with IgE, and they will respond to antigen in a manner apparently similar to that of mast cells.

IgE Receptors

There are two types of IgE receptor: high affinity (FcεRI) and low affinity (FcεRII or CD23). FcεRI is mainly found on mast cells and basophils, where there are from 10^4 to 10^5 per cell. FcεRI is a multichain receptor consisting of one α chain (45 to 65 kDa) that binds IgE, one β chain (32 kDa), and two disulfide-linked γ chains (20 kDa) that are required for signal transduction. The α and γ chains are type I integral membrane proteins. The β chain is a four-pass transmembrane protein (Fig. 29–5). The α chain belongs to the immunoglobulin superfamily and contains two immunoglobulin domains that bind to one of the CH3 domains of IgE. The γ chain (FcεRIγ) is also found associated with the other multichain receptors FcγRI, FcγRIII, and γ/δ TCR, where it also serves as a signal transducer.

The low-affinity receptor, FcεRII, is found on NK cells, macrophages, dendritic cells, eosinophils, and platelets. About 30% of B cells also possess FcεRII, and this percentage rises in atopic individuals during ragweed season. FcεRII is a selectin and is thus the only

Table 29–1 Comparison of Two Types of Mast Cell

Characteristics	Mucosal Mast Cells	Connective Tissue Mast Cells
Structure	Few, variable-sized granules	Many uniform granules
Size	9–10 μm in diameter	19–20 μm in diameter
Proteoglycan	Chondroitin sulfate E	Heparin
Histamine	1.3 pg/cell	15 pg/cell
Life span	<40 days	>6 months
Production enhanced by IL-3	+	−
Protected by disodium chromoglycate	−	+

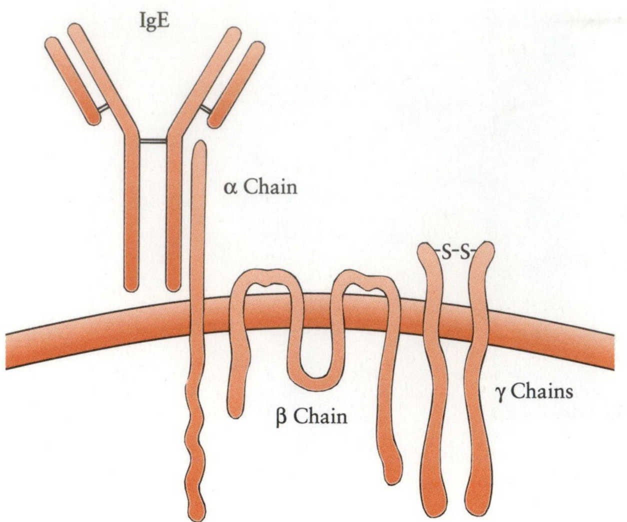

Figure 29–5 The structure of the high-affinity FcεRI found on the mast-cell surface.

known antibody receptor that does not belong to the immunoglobulin superfamily.

FcεRII is expressed only on antigen-stimulated B cells before they differentiate into antibody-secreting cells. Its expression is also inducible by IL-4. In addition to acting as a receptor for IgE, FcεRII is a ligand for the complement receptor CR2 (CD21). Thus B cells expressing FcεRII attach to CR2 on other B cells, on T cells, and, most important, on dendritic cells. By causing B cells to bind to dendritic cells, FcεRII prevents B-cell apoptosis within germinal centers. As a result, these cells are induced to become plasma cells and secrete IgE.

The membrane-bound form of FcεRII can be cleaved to form soluble IgE-binding factors that stimulate IgE synthesis after binding to B-cell CR2. Since IgE also inhibits the release of these soluble fragments, IgE thus controls its own synthesis, an interesting form of negative feedback (Fig. 29–6). B-cell FcεRII-IgE-antigen complexes are readily processed by B cells to present the antigen to T cells.

RESPONSE OF MAST CELLS TO ANTIGEN

When IgE alone binds to FcεRI on mast cells, no visible change occurs in the cell. If, however, antigen binds to mast-cell–bound IgE and cross-links two or more FcεRI structures, a series of reactions are triggered that cause the mast-cell granules to migrate to the cell

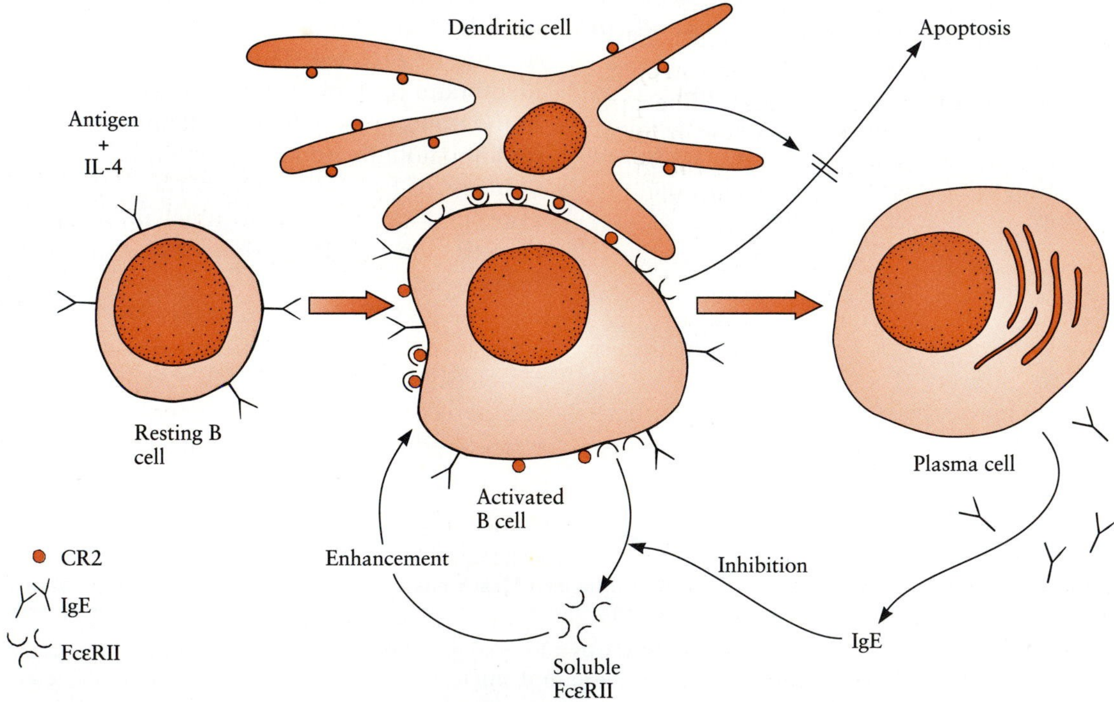

Figure 29–6 The role of FcεRII in controlling IgE production. Once B cells are activated by IL-4, they express FcεRII and membrane IgE. The FcεRII on the activated B cells binds to CR2 on dendritic cells, which then prevent the B cell from undergoing apoptosis. As a result, the B cell differentiates into a plasma cell and secretes IgE. The B cell can also release soluble FcεRII, which enhances B-cell IgE production by binding to CR2. However, the production of soluble FcεRII is inhibited by IgE. In this way, IgE controls its own production.

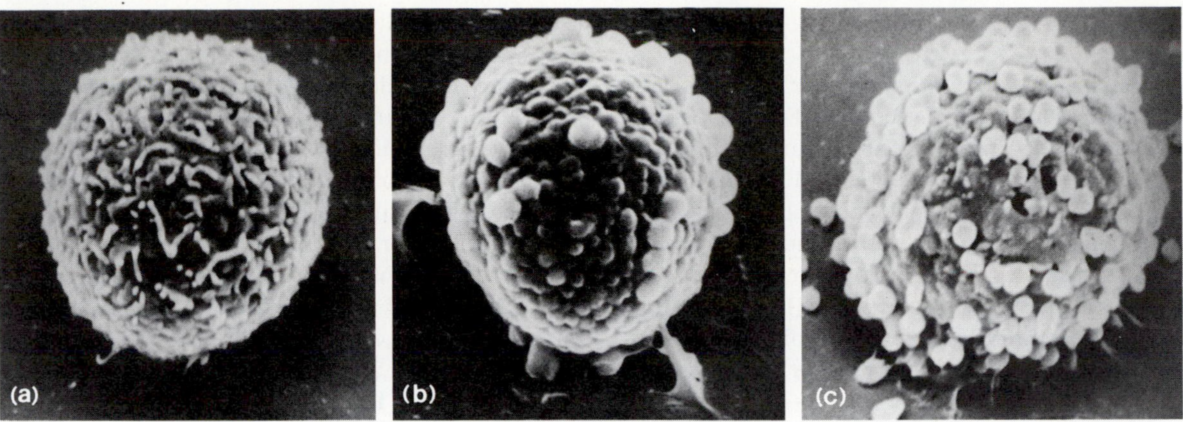

Figure 29–7 A scanning electron micrograph of (*A*) a normal rat mast cell, (*B*) a sensitized mast cell fixed 5 seconds after exposure to antigen, and (*C*) a sensitized mast cell fixed 60 seconds after exposure to antigen. (Original magnification ×3000.) *(From Tizard, I.R., and Holmes, W.L. Int. Arch. Allergy Appl. Immunol., 46:867–897, 1974. With permission of S. Karger.)*

surface, to fuse with the cell membrane, and to be extruded from the cell, releasing their contents into the surrounding tissues (Fig. 29–7). Thus, triggering of the receptor initiates several signal transduction pathways. In one pathway, a tyrosine kinase activates phospholipase C, leading to the production of diacyl glycerol and inositol trisphosphate. These in turn increase intracellular calcium levels and activate a protein kinase C. An adenylate cyclase is also activated by the receptor. This leads to a transient increase in cyclic AMP and activates another protein kinase. The protein kinases together phosphorylate myosin in intracellular filaments and on granules, causing the granules to move to the cell surface, fuse with the plasma membrane, and release their contents into the extracellular fluid. Cross-linkage of the FcεRI also activates two methyltransferases, which in turn activate a phospholipase A. This enzyme acts on membrane phospholipids to produce arachidonic acid. Other enzymes then metabolize the arachidonic acid to leukotrienes and prostaglandins. Finally, the protein kinases promote translation and expression of several cytokine genes, resulting in the synthesis and secretion of cytokines.

Some of these mast-cell responses are extremely rapid. For example, granule extrusion can occur only a few seconds after antigen binds to antibody on the cell surface (Fig. 29–8). Degranulated mast cells do not die, but they are difficult to identify because of an absence of characteristic morphological features. Mast cells may also be degranulated by drugs or chemicals. Agents that initiate mast-cell degranulation by nonimmunological mechanisms include drugs such as the antibiotic polymyxin B, morphine and the anaphylatoxins C3a and C5a (Chapter 16).

Cells that carry FcεRII such as platelets or macrophages may also be activated by antigen through IgE. Thus monocytes can produce IL-1β and TNF-α in an IgE-dependent fashion and so contribute to a type I hypersensitivity reaction.

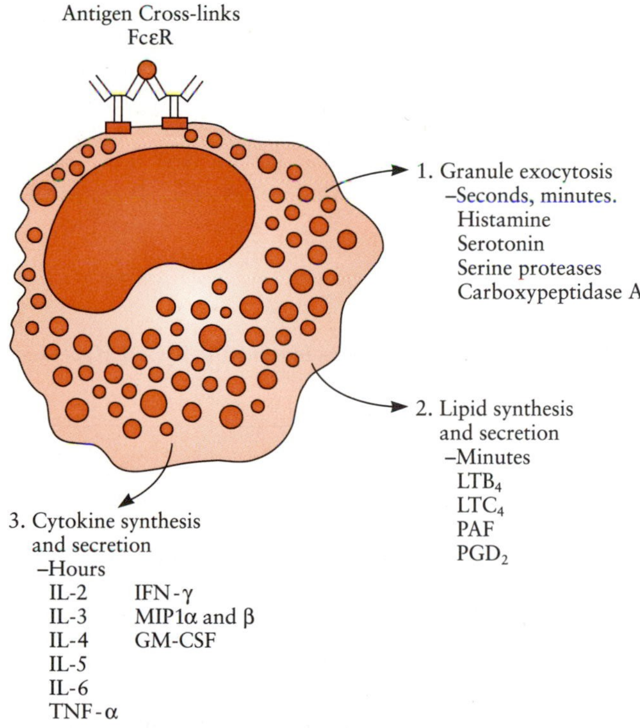

Figure 29–8 The three major groups of biologically active factors released or synthesized by mast cells. Note that the time course of this release can vary from seconds, as in the case of granule exocytosis, to hours, as in the case of cytokine synthesis.

Biologically Active Molecules Released by Mast Cells

Although mast cells can constitutively release cytokines such as IL-4, IL-6, TNF-α, and TGF-β, once stimulated by antigen they initiate the production and release of many biologically active molecules. For example, when exposed to extracellular fluid, the mast-cell granules release vasoactive amines and enzymes that cause capillary dilation and increased blood vessel permeability (see Fig. 29–8). In addition, FcεRI triggering results in the synthesis of leukotrienes and prostaglandins. Several hours after antigen triggering, mast cells transcribe and secrete a complex mixture of many cytokines. It is these molecules, both those released preformed from granules and those newly synthesized, that generate the characteristic inflammatory lesions of type I hypersensitivity (Table 29–2).

Mast-cell granules contain high concentrations of histamine and (in some species) serotonin (Fig. 29–9). Although these two compounds are important mediators of inflammation, they are only part of the total spectrum of inflammatory mediators released following mast-cell degranulation. More than half the protein in the mast-cell granules consists of trypsin or chymotrypsin-like neutral proteases. These proteases can destroy nearby cells and activate the complement components C3 and C5 to generate anaphylatoxins. Mast-cell granules also contain kallikreins (Chapter 28) and so act on kininogens to generate kinins such as bradykinin. Both the kinins and the anaphylatoxins are powerful vasoactive agents.

Platelet-activating factor (PAF) is a mixture of phospholipids closely related to lecithin. It is synthesized by mast cells following exposure to antigen or anaphylatoxins and by platelets. PAF makes platelets aggregate and release their vasoactive molecules, especially serotonin, and synthesize thromboxanes. PAFs promote neutrophil aggregation, degranulation, chemotaxis, the release of oxygen radicals, and a neutropenia, although their precise effect does depend on the molecular species of PAF involved. PAFs are major activators of eosinophils.

Activation of cell membrane phospholipases is also provoked by antigen binding to IgE on the mast-cell surface (Fig. 29–10). This leads to the release of arachidonic acid from cell membrane phospholipids. Arachidonic acid is the substrate for cyclooxygenase and for 5-lipoxygenase. The products of the cyclooxygenase pathway are the prostaglandins, prostacyclins, and thromboxanes, and the products of the 5-lipoxygenase pathway are leukotrienes (Chapter 28). All of these have major effects on blood vessel permeability. It is possible that different populations of mast cells give rise to different products from these pathways.

Mast-cell granules also release two peptides that are chemotactic for eosinophils called eosinophil chemotactic factors of anaphylaxis. Both are tetrapeptides that exist preformed within mast-cell granules. These molecules account, at least in part, for the eosinophilia so characteristic of type I hypersensitivity reactions, including helminth infestations. Mast cells secrete a mixture of cytokines, including IL-3, IL-4, IL-5, IL-6, IL-8, and GM-CSF, in response to the cross-linking of the IgE receptor. Other molecules that are found in mast-cell granules and are released on degranulation include a neutrophil chemotactic factor and a neutrophil immobilization factor. The latter

Table 29–2 Major Mediators Involved in Type I Hypersensitivity

Mediator	Major Actions
Preformed mediators	
Histamine	Smooth-muscle contraction, increased vascular permeability, itching, increased exocrine secretion
Serotonin	Vasospasm and smooth-muscle contraction
ECF-A	Eosinophil chemotaxis
Platelet activating factor	Platelet aggregation and secretion of vasoactive factors
NCF-A	Neutrophil chemotaxis
Mediators generated de novo	
Prostaglandins	Very complex; influence vascular and smooth-muscle tone, aggregation, and immune reactivity
Leukotrienes C and D	Characteristic smooth-muscle contraction and increased vascular permeability
Leukotriene B	Neutrophil and eosinophil chemotaxis
Bradykinin	Smooth-muscle contraction and increased vascular permeability
Serotonin	Smooth-muscle contraction and vasospasm

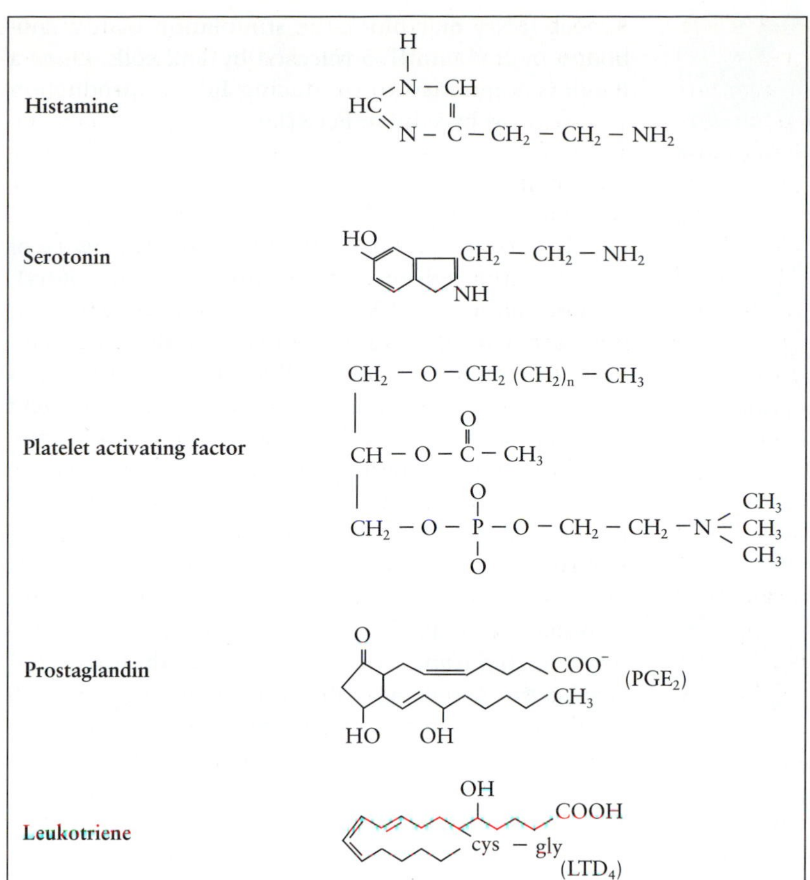

Figure 29–9 The chemical structure of some of the small biologically active factors released or synthesized by mast cells.

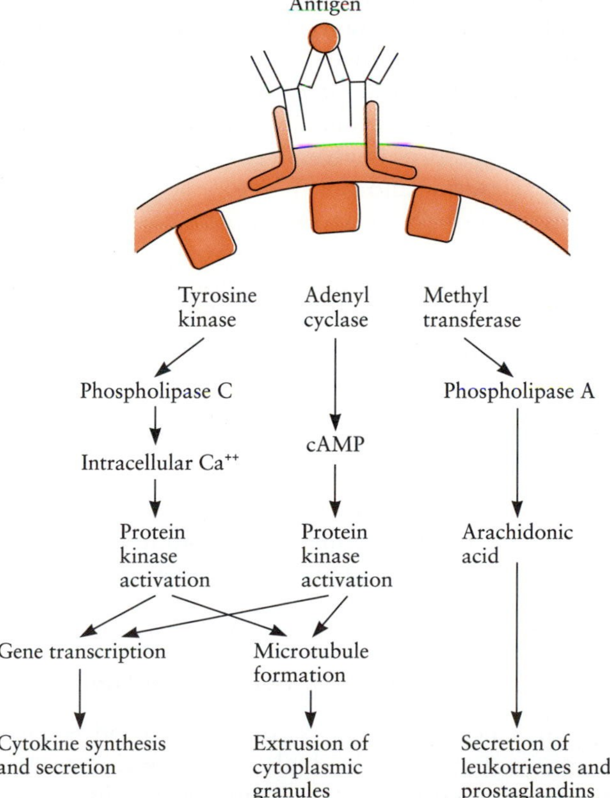

molecule, as its name suggests, stops the emigration of neutrophils once they have been attracted to an area of inflammation by the chemotactic molecule. It does not interfere with the phagocytic activities of the cells.

Proteoglycans such as heparin from connective tissue mast cells and chondroitin sulfate from mucosal mast cells are also released on mast-cell degranulation. Because of the anticoagulant properties of heparin, blood from dogs with mast-cell tumors and from animals suffering from anaphylaxis may fail to coagulate.

Mast-Cell Cytotoxicity. Mast cells are cytotoxic for certain tumor cells as well as helminth parasites. They possess multiple cytotoxic mechanisms. Thus they secrete peroxidases and proteases as well as TNF-α.

Figure 29–10 An outline of the signal transduction pathways that lead to mast-cell degranulation. The three pathways result in degranulation and the release of prostaglandins and leukotrienes. *(For details see text.)*

CLARIFICATION

Allergy "Shots"

The role of blocking antibodies was first shown in beekeepers, who, in spite of being stung frequently and having high levels of IgE in their serum, usually suffered minimal discomfort. Analysis of their serum showed that they had high levels of IgG against bee venom. If venom-allergic individuals were passively immunized with serum from these beekeepers, they could be made temporarily "immune" to allergic reactions induced by venom. In addition, this hyperimmune serum effectively prevented degranulation and histamine release from sensitized mast cells in vitro. As a result of this observation, physicians attempted to control other allergies by carefully administering low doses of antigen.

In hyposensitization therapy, antigen is administered in a form designed to stimulate the immune response while reducing, as far as possible, the risks of anaphylactic shock. Thus antigen may be administered in a suspension with alum or in a water-in-oil emulsion to form a depot. Aqueous allergens are usually rapidly absorbed but require frequent injections. The alum or oil-adjuvanted allergens are slowly absorbed, and fewer injections are required. The first injections contain only a small quantity of allergen. Over several weeks, the dose is gradually increased. If an individual's allergy is seasonal, the course of injections should be timed to reach completion just prior to the anticipated antigen exposure.

CONTROL OF TYPE I HYPERSENSITIVITY

Regulation of the IgE Response

Not all individuals are equally able to mount an IgE response. The reasons for this, as pointed out earlier, are primarily genetic. It is clear that in some individuals Th2 cells are preferentially involved in responses to antigens as a result of increased synthesis of IL-4 and IL-5. This predisposition of atopic subjects to produce IgE may be remedied by the use of desensitizing injections of antigen ("allergy shots"). The original rationale for administering antigen to atopic individuals was to stimulate the production of IgG-blocking antibodies. It was anticipated that these would compete with IgE for antigen and hence prevent the antigen from reaching the mast cells. It is now believed that these injections stimulate Th1 cells to release IFN-γ. This IFN-γ may block the stimulation of IgE antibodies by IL-4 and IL-5 released by Th2 cells. Once a B cell is committed to producing IgE, its production is controlled by soluble FcϵRII.

Regulation of Mast-Cell Degranulation

On the surface of mast cells there are two types of receptor for adrenergic agents (adrenoceptors), named α and β (Table 29–3). Compounds that stimulate the α receptor enhance mast-cell degranulation because they depress intracellular cyclic AMP. Drugs that stimulate the β receptor have the reverse effect and thus depress mast-cell degranulation. Thus β stimulants are useful in treating allergies. They include isoproterenol, epinephrine, and salbutamol. Recently, it has become clear that β-receptor impairment can also contribute to the atopic state. For example, the effects of β-receptor activation can be inhibited by certain respiratory tract pathogens, such as *B. pertussis* or *Haemophilus influenzae,* or by autoantibodies directed against the β receptor. Because of this impairment, individuals infected by these organisms are more susceptible to hypersensitivity conditions such as chronic asthma.

Regulation of the Response to Mediators

The α and β adrenoceptors are found not only on mast cells but also on secretory and smooth muscle cells throughout the body. Stimulators of α adrenoceptors mediate vasoconstriction. Consequently, α-adrenergic agents may be of use in the treatment of anaphylaxis, reducing edema and raising blood pressure. Stimulators of β adrenoceptors mediate smooth muscle relaxation and may therefore be useful in modulating the severity of smooth muscle contraction.

Pure α and β stimulants are of only limited use in the treatment of anaphylaxis because each alone is insufficient to counteract all the effects of mast-cell–derived molecules. Epinephrine, on the other hand, has both α- and β-adrenergic activity, and therefore, in addition to causing vasoconstriction in skin and viscera, its β effects cause smooth muscle to relax. This combination of effects is well suited to combat the vasodilation and smooth muscle contraction produced by histamine. Ideally, epinephrine solution should be available whenever potential allergens are administered to animals.

Function of Eosinophils in Type I Hypersensitivity

The presence of many eosinophils in blood is characteristic of atopic individuals. In severe asthma, the mucus coughed up from the lungs is rich in eosino-

Table 29–3 Effects of Stimulating α and β Adrenoceptors

System	Stimulation of α Adrenoceptor	Stimulation of β Adrenoceptor
Cyclic nucleotides	Lower cAMP Raise cGMP	Raise cAMP Lower cGMP
Mast-cell degranulation	Enhances	Depresses
Smooth muscle	Contracts	Relaxes
Blood vessels	Constrict	Dilate

phils and may even contain crystallized eosinophil phospholipase. Eosinophils are attracted to mast-cell degranulation sites by eosinophil chemotactic factors, by certain leukotrienes, especially leukotriene B_4, by histamine, and especially by the histamine-breakdown product imidiazoleacetic acid. The eosinophils exert a proinflammatory response that presumably potentiates the hypersensitivity reaction. The role of eosinophils is described in detail in Chapter 28.

CLINICAL MANIFESTATIONS OF TYPE I HYPERSENSITIVITY

All of the clinical signs of type I hypersensitivity relate to the release of vasoactive molecules from mast cells and basophils. The severity of these clinical signs depends on the number and location of the mast cells stimulated, and this in turn depends on the amount and route of antigen administration. For example, antigen injected rapidly intravenously causes generalized mast-cell degranulation. If the rate of release of vasoactive molecules is greater than the body's ability to respond to the rapid changes in its vascular system, the patient suffers from acute anaphylaxis or **anaphylactic shock** and may die. If, on the other hand, antigen is administered either locally in small quantities or slowly, then the clinical signs of hypersensitivity will be much less severe, since the individual will have time to compensate for the vascular changes provoked by the mast-cell–derived molecules.

Acute Anaphylaxis

Acute anaphylaxis, or anaphylactic shock, may be provoked in laboratory animals such as guinea pigs by first sensitizing the animal with a single injection of foreign protein. After about three weeks, IgE is produced. A large dose of antigen administered rapidly will then provoke anaphylaxis. The first signs of anaphylaxis are restlessness, ruffling of the hair on the neck, sneezing, and pawing at the nose in response to the irritant effect of released histamine. The animal will begin to cough and gasp for air. As the smooth muscle of the gastrointestinal tract and bladder contract, the animal will defecate and urinate. Finally, bronchial constriction leads to interference with respiration and asphyxiation, and death follows. On necropsy, the most marked finding in shocked guinea pigs is the overinflation of the lungs. In humans, anaphylaxis may follow the rapid administration of a drug or other allergen to which an individual is allergic (Fig. 29–11). It closely resembles the reaction seen in the guinea pig. Shortness of breath and increased heart rate precede asphyxiation as a result of bronchial constriction. Death may also result from edema of the larynx blocking the flow of air. This bronchial constriction, when less severe, is called asthma. In most mammals, anaphylaxis involves primarily the respiratory tract, either by causing bronchial constriction or by constricting the pulmonary vein and thus provoking increased pulmonary blood pressure and fluid accumulation in the lungs. In the dog, the first species in which anaphylaxis was recorded, the major target organ for anaphylaxis is not the lungs but, unusually, the hepatic blood vessels. As a result of the constriction of these blood vessels, blood remains trapped in the liver and intestine, and the animal dies from what is essentially a lack of blood. Allergens that can induce anaphylaxis include foods, drugs, insect stings, and vaccines (Table 29–4).

Table 29–4 Major Allergens in Humans

Proteins	**Foods**
Pollens	Nuts
Mold spores	Dairy products
Animal danders	Chocolate
Foreign serum	Seafood
Vaccines	Eggs
Drugs	Peas, beans
Penicillin	**Insects**
Sulfonamides	Dust mites
Local anesthetics	Bee and wasp stings
Salicylates	Ant and flea bites

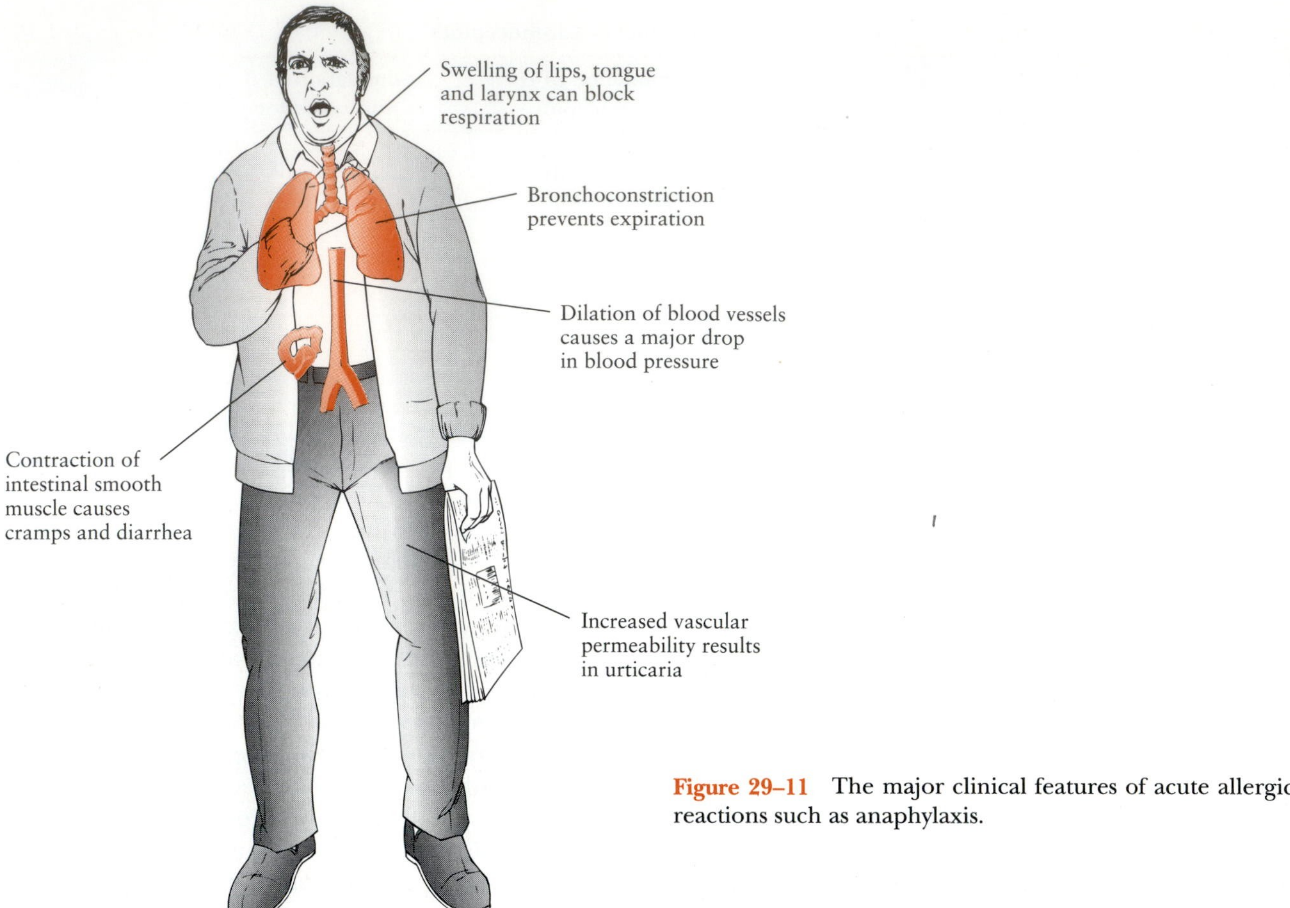
Swelling of lips, tongue
and larynx can block
respiration
Bronchoconstriction
prevents expiration
Dilation of blood vessels
causes a major drop
in blood pressure
Contraction of
intestinal smooth
muscle causes
cramps and diarrhea
Increased vascular
permeability results
in urticaria

Figure 29–11 The major clinical features of acute allergic reactions such as anaphylaxis.

Respiratory Allergy

The most important human respiratory allergens are pollens and plant spores, animal danders, and mite antigens in house dust (Table 29–4). These allergens, suspended in the air, are easily inhaled and thus come into contact with mucous membranes of the nasal and respiratory tracts as well as the eyes. In a sensitized individual, exposure to these allergens causes sneezing, wheezing, shortness of breath, and nasal discharge, conditions called hay fever and asthma. In hay fever, the local release of histamine in the nasal mucosa causes irritation, an increase of vascular permeability leading to a running nose and tears, and inflammation. In the upper respiratory tract, the local reaction leads to bronchoconstriction and subsequent difficulty in breathing—asthma. This difficulty can be very distressing for affected individuals. Because of the structure of the pulmonary airways, it is usually possible to inhale air in spite of an inability to exhale it. As a result, the lungs may become overinflated and the walls of the air sacs may break down, leading to emphysema. In addition to immunologic mechanisms, neurologic processes also contribute to the development of asthma. Thus, triggers of asthma include not only allergic reactions but also, in some asthmatics who have very sensitive airways, emotion, exercise, tobacco smoke, infection, and even cold, dry air. Asthma exacerbations are commonly associated with infections with respiratory syncytial virus, parainfluenza, or rhinoviruses.

Food Allergy

Allergies to foods are also a major class of type I hypersensitivity. In this case, the local reaction to antigens in the intestine causes smooth muscle constriction and effusion of fluid. This results in intense discomfort and diarrhea. The allergens may be absorbed and cause mast-cell degranulation elsewhere in the body. Thus local reactions in the skin can cause the development of edematous lesions commonly called hives, or urticaria. (They resemble the skin reaction caused by contact with the plant called a ''stinging'' nettle, *Urtica dioica.*) Respiratory symptoms such as asthma may also develop as a result of food allergies. Common food allergens include chocolate, cow's milk, wheat and wheat products, and fish. Avoiding these foods is generally the simplest means of controlling these problems.

A CLOSER LOOK

Lung Transplantation and Asthma

Transplantation of lungs is now a recognized form of treatment for severe lung damage. Usually healthy lungs are transplanted, but occasionally lungs from mild asthmatic donors may be used. Thus asthmatic lungs have been transplanted into nonasthmatic recipients, and nonasthmatic lungs have been transplanted into asthmatic recipients. In two reported cases, the nonasthmatic recipients of asthmatic lungs developed asthma after transplantation. In contrast, the asthmatic recipients of normal lungs, in the same study, did not develop asthma up to three years after transplantation. This clearly indicates that the asthma resulted from local conditions in the lung.

The Late-Phase Reaction

Several hours after the immediate hypersensitivity reaction is over, a second inflammatory response may develop. This late-phase reaction is due to immigration and degranulation of neutrophils and eosinophils under the influence of mast-cell–derived chemotactic molecules. Histamine, lysosomal enzymes, and reactive oxygen metabolites are released from these cells at this time. Late-phase reactions can occur in the lungs of asthmatics or in skin allergies.

DIAGNOSIS OF TYPE I HYPERSENSITIVITY

IgE antibodies bind to mast cells, including those in the skin. As a result, intradermal inoculation of antigen, even in dilute solution, provokes a local acute inflammatory reaction called a wheal and flare reaction in allergic individuals. Vasoactive molecules are released within minutes to produce local redness (erythema) as a result of capillary dilation and produce circumscribed edema (a wheal) due to increased vascular permeability. The reaction may also develop a large area of redness (an erythematous flare) due to dilation of local arterioles brought about by a nerve reflex. This type of response to antigen reaches its greatest intensity within 30 minutes and tends to fade and disappear within a few hours. A physician may inject several different allergens into the skin of a patient to determine to which antigens they may be sensitive.

PREVENTION AND TREATMENT OF TYPE I HYPERSENSITIVITY

To prevent the development of severe type I hypersensitivity reactions, it is essential that the offending allergen or allergens be identified and removed. This process may be difficult and tedious. Intradermal skin testing using dilute solutions of a variety of potential allergens may be of assistance. Alternatively (or additionally) in food allergies a change of diet and the readdition of individual components on a test basis may enable the investigator to identify a specific allergen or, more commonly, a group of allergens. When elimination of the offending allergen is not possible, as, for example, in allergies to pollens, desensitization therapy (see p. 462) may prove effective.

Asthma is usually treated using bronchodilators, although these do not affect the cause of the problem. In addition, anti-inflammatory drugs such as corticosteroids are used to reduce mucosal swelling. Inhaled steroids usually produce a satisfactory clinical response without serious side effects. Type I hypersensitivity reactions may also be treated using β-adrenoceptor stimulants such as epinephrine and isoprenaline. Salbutamol is a more selective stimulant that causes bronchodilation. α-Adrenoceptor inhibitors such as methoxamine and phenylephrine have a similar beneficial effect. Specific pharmacological inhibitors such as the antihistamines (e.g., pyrilamine, promethazine, and diphenhydramine) can effectively inhibit the activities of histamine. However, since histamine is but one of many mast-cell–derived mediators, antihistamines are of limited effectiveness in controlling hypersensitivity diseases. Another drug commonly used in the treatment of type I hypersensitivity is disodium cromoglycate, which interferes with the release of histamine and leukotrienes from connective tissue mast cells.

KEY WORDS

Chemotactic peptides p. 460
Connective tissue mast cells p. 456
Desensitizing injections p. 462
Eosinophil p. 462
Epinephrine p. 462
FcεRI p. 457
FcεRII p. 457
Food allergy p. 464
Hay fever p. 464
Heparin p. 461
Histamine p. 460
Immunoglobulin E p. 456
IgE receptors p. 457
Interleukin-4 p. 456
Kinins p. 460
Late-phase reaction p. 465
Leukotriene p. 460
Mast cells p. 456
Mucosal mast cells p. 456
Platelet-activating factor p. 460
Prostaglandin p. 460
Protease p. 460
Respiratory allergy p. 464
Th2 cells p. 456
Type I hypersensitivity p. 455
Urticaria p. 464
Wheal and flare reaction p. 465

QUESTIONS

1. Mast-cell granules contain
   a. immunoglobulin
   b. complement
   c. epinephrine
   d. acetyl choline
   e. histamine
2. The IgE-mediated degranulation of mast cells does not involve
   a. bridging two receptor-bound IgE molecules
   b. complement activation
   c. a rise in intracellular calcium
   d. release of vasoactive agents from granules
   e. synthesis of leukotrienes
3. One major function of eosinophils is to
   a. neutralize vasoactive factors from mast cells
   b. discourage biting insects
   c. inhibit mast-cell degranulation
   d. reduce prostaglandin release
   e. destroy migrating helminth larvae
4. Mast cells release one of the following on degranulation:
   a. anaphylatoxins
   b. interleukin-4
   c. interleukin-1
   d. immunoglobulin E
   e. complement
5. The overproduction of IgE in atopy is probably a result of secretion of which cytokine by Th2 cells?
   a. interleukin-2
   b. interleukin-3
   c. interleukin-4
   d. interleukin-6
   e. interleukin-12
6. FcεRII has two ligands: IgE and
   a. CR2
   b. C3
   c. IgA
   d. IL-4
   e. IFN-γ
7. During anaphylaxis, which mast-cell product stops blood clotting?
   a. histamine
   b. serotonin
   c. bradykinin
   d. platelet-activating factor
   e. heparin
8. Stimulation of β adrenoceptors on mast cells results in
   a. bronchoconstriction
   b. IgE synthesis
   c. anaphylaxis
   d. reduced mast-cell degranulation
   e. epinephrine production
9. Which of the following is the drug of choice for the emergency treatment of anaphylaxis?
   a. penicillin
   b. epinephrine
   c. cyclosporine
   d. histamine
   e. aspirin
10. Hyposensitization therapy for atopic disease is thought to work by
    a. blocking of IgE by excess antigen
    b. increased cell-mediated response to allergen
    c. induction of Th1-cell activity
    d. decrease in specific IgA antibody response
    e. all of the above
11. What possible advantages could type I hypersensitivity reactions confer on an animal?
12. List the pharmacological mediators that may be released by degranulating mast cells. How is the activity of each of these molecules controlled?
13. What is acute anaphylaxis? What are the clinical signs of this condition in humans, and how might it be treated?
14. How does the role of eosinophils in parasitic diseases relate to their role in immediate hypersensitivity?
15. What is urticaria? What is its mechanism, and how is it treated?
16. What are the differences and similarities between eosinophils and mast cells? Can they be considered to belong to a single family of cells? Why?
17. If you have a friend who suffers from allergies, find out about his or her genetically related family members and determine for yourself to what extent atopy is inherited.

Answers: 1e, 2b, 3e, 4b, 5c, 6a, 7e, 8d, 9b, 10c

SOURCES OF ADDITIONAL INFORMATION

Alber, G., Kent, U.M., and Metzger, H. Functional comparison of FcεRI, FcγRII and FcγRIII in mast cells. J. Immunol., 149:2428–2436, 1992.

Askenase, P.W. The role of basophils in health and disease. Res. Staff Phys., 32:33–41, 1986.

Austin, K.F. The heterogeneity of mast cell populations and products. Hosp. Pract. [off.], 19:135–146, 1984.

Beaven, M.A., and Metzger, H. Signal transduction by Fc receptors: The FcεRI case. Immunol. Today, 14:222–226, 1993.

Befus, A.D., Bienenstock, J., and Denburg, J.A. Mast cell differentiation and heterogeneity. Immunol. Today., 6:281–284, 1985.

Bennich, H., and Johansson, S.G.O. Structure and function of human immunoglobulin E. Adv. Immunol., 13:1–55, 1971.

Bissonnette, E.Y., and Befus, E.D. Inhibition of mast cell-mediated cytotoxicity by IFN-α/β and -γ. J. Immunol, 145:3385–3390, 1990.

Bonnefoy, J-Y., et al. A new pair of surface molecules involved in human IgE regulation. Immunol. Today, 14:1–2, 1993.

Borish, L., Mascali, J.J., and Rosenwasser, J. IgE-dependent cytokine production by human peripheral blood mononuclear phagocytes. J. Immunol, 146:63–67, 1991.

Bradding, P., et al. Interleukin 4 is localized to and released by human mast cells. J. Exp. Med., 176:1381–1386, 1992.

Conrad, D.H. FcεRII/CD23, the low affinity receptor for IgE. Annu. Rev. Immunol., 8:623–645, 1990.

Corris, P.A., and Dark, J.H. Aetiology of asthma: Lessons from lung transplantation. Lancet, 341:1369–1371, 1993.

Denburg, J.A. Basophil and mast cell lineages in vitro and in vivo. Blood, 79:846–860, 1992.

Galli, S.J. New approaches for the analysis of mast cell maturation, heterogeneity and function. Federation Proc., 46: 1906–1914, 1987.

Geha, R.S. Regulation of IgE synthesis in atopic disease. Hosp. Pract. [off.], 23:91–102, 1988.

Gleich, G.J. Current understanding of eosinophil function. Hosp. Pract. [off.], 23:137–160, 1988.

Hoffman, D.R., Wood, C.L., and Hudson, P. Demonstration of IgE and IgG antibodies against venoms in the blood of victims of fatal sting anaphylaxis. J. Allergy Clin. Immunol., 71:193–196, 1983.

Huntley, J.F. Mast cells and basophils: A review of their heterogeneity and function. J. Comp. Pathol., 107:349–372, 1992.

Ishizaka, K. Regulation of immunoglobulin E biosynthesis. Adv. Immunol., 47:1–44, 1989.

Jarrett, E.E., MacKenzie, S., and Bennich, H. Parasite-induced non-specific IgE does not protect against allergic reactions. Nature, 283:302–304, 1980.

Jarrett, E.E., and Haig, D.M. Mucosal mast cells in vivo and in vitro. Immunol. Today, 5:115–118, 1984.

Kaliner, M.A. The late phase reaction and its clinical implications. Hosp. Pract. [off.], 22:73–83, 1987.

Kay, A.B. Eosinophils as effector cells in immunity and hypersensitivity disorders. Clin. Exp. Immunol., 62:1–12, 1985.

Khalife, J., Capron, M., and Cesbron, J.Y. Role of specific IgE antibodies in peroxidase release from human eosinophils. J. Immunol., 137:1659–1664, 1986.

Kuna, P., et al. Characterization of the human basophil response to cytokines, growth factors, and histamine releasing factors of the intercrine/chemokine family. J. Immunol., 150:1932–1943, 1993.

Larrick, J.W., et al. Does hyperimmunoglobulin E protect tropical populations from allergic disease? J. Allergy Clin. Immunol., 71:184–188, 1983.

Lichtenstein, L.M. Allergy and the immune system. Sci. Am., 269:117–124, 1993.

Metzger, H., and Kinet, J-P. How antibodies work: Focus on Fc receptors. FASEB J., 2:3–11, 1988.

Peng, C., et al. A new isoform of human membrane-bound IgE. J. Immunol., 148:129–136, 1992.

Sandford, A.J., et al. Localization of atopy and β subunit of high-affinity IgE receptor (FcεRI) on chromosome 11q. Lancet, 341:332–334, 1993.

Spry, C.J.F. Synthesis and secretion of eosinophil granule substances. Immunol. Today, 6:332–335, 1985

Stevens, R.L., and Austin, K.F. Recent advances in the cellular and molecular biology of mast cells. Immunol. Today, 10:381–386, 1989.

Sutton, B.J., and Gould, H.J. The human IgE network. Nature, 366:421–428, 1993.

Ventner, J.C., Frazer, C.M., and Harrison, L.C. Autoantibodies to β_2-adrenergic receptors: A possible cause of adrenergic hyporesponsiveness in allergic rhinitis and asthma. Science, 207:1361–1363, 1980.

CHAPTER 30

Other Hypersensitivities

Robin Coombs *Dr. Robin Coombs is a veterinarian who has spent much of his career in immunology investigating and devising innovative serologic tests. One of the most significant of these is the antiglobulin (or Coombs) test. Infants with hemolytic disease of the newborn suffer from severe red cell destruction as a result of the presence of antibodies that attack their red cells. Robin Coombs recognized that these non-agglutinating antibodies could be detected by means of an antibody or antiglobulin directed against the maternal antibodies. The test that he devised is a vital aid in the diagnosis of this disease and of other immunologically mediated anemias.*

CHAPTER OUTLINE

CHAPTER CONCEPTS

1. In addition to type I hypersensitivity described in the previous chapter, there are three other major mechanisms of hypersensitivity.
2. Type II hypersensitivity results from the destruction of cells by antibodies or complement. An example of this is the destruction of foreign red blood cells as a result of an incompatible blood transfusion.
3. Blood group antigens are proteins or glycoproteins found on the surface of red blood cells. The most important human blood group antigens belong to the ABO system.
4. Type III hypersensitivity occurs as a result of the deposition of immune complexes in tissues. These immune complexes activate complement and attract neutrophils. The neutrophil enzymes cause tissue damage. Immune complexes deposited in the skin, lungs, joints, or walls of blood vessels therefore cause local inflammatory responses.
5. Immune complexes formed in the bloodstream are deposited in the glomeruli of the kidney, where they cause a glomerulonephritis and interfere with kidney function.
6. Type IV hypersensitivity reactions result from a cell-mediated response to antigen. They usually take more than 24 hours to develop and are therefore called delayed hypersensitivity reactions. They are mediated by factors released by T cells and macrophages.

In this chapter we discuss three types of hypersensitivity reaction. Type II hypersensitivity, typified by the destruction of incompatible red blood cells, is described first, followed by a brief review of blood group antigens and their structure and inheritance. An important type II hypersensitivity disease, called hemolytic disease of the newborn, is discussed next.

The second part of the chapter describes different forms of type III hypersensitivity reaction. This type of hypersensitivity is caused by the deposition of immune complexes in tissues. Its clinical manifestations therefore depend on the location of these complexes. The first set of diseases discussed are those associated with local deposition of complexes in the skin or lung. The second set of diseases results from production of complexes in the bloodstream. These tend to be deposited in sites such as the glomeruli of the kidney.

The final section of the chapter briefly describes some type IV hypersensitivity reactions. Their mechanisms have already been described in Chapter 18, but some clinical conditions are described here.

TYPE II HYPERSENSITIVITY

Type II hypersensitivity reactions are those in which tissue or cell damage is the direct result of the actions of antibody and complement. Antibodies directed against antigens on the surface of nucleated cells or red blood cells will cause their lysis in the presence of complement. If the immunoglobulin is not one that activates complement, then the target cell may still be destroyed by phagocytosis or ADCC. Thus this type of hypersensitivity is a purely destructive process. The best example of type II hypersensitivity is the destruction of foreign red blood cells following an incompatible blood transfusion.

Like nucleated cells, red blood cells possess characteristic cell surface proteins. Unlike the histocompatibility molecules of nucleated cells, however, red blood cell surface proteins, or **blood groups,** do not determine an animal's ability to mount an immune response, although they do influence graft rejection (grafts between individuals incompatible in the major blood groups are rejected rapidly). They do, however, have other important functions. For example, the glycoproteins of the ABO system are anion and glucose transporter proteins, the Kell antigen is an endopeptidase, and the Kidd antigens are urea transport proteins. Most red blood cell surface proteins are integral components of the cell membrane. However, some blood group molecules, although found on red blood cells, are synthesized at other sites within the body. These molecules are found free in serum, saliva, and other body fluids and are passively adsorbed onto red blood cell surfaces. Examples of such molecules include the Lewis antigen and the C4-related molecules, Chido and Rogers (Chapter 16).

If normal red blood cells are administered to a genetically different (allogeneic) recipient, their cell membrane proteins stimulate an immune response. The transfused red blood cells are rapidly destroyed in the bloodstream by antibody and complement and through opsonization and phagocytosis by the cells of the mononuclear-phagocytic system. This type of cell destruction is classified as a type II hypersensitivity reaction.

BLOOD GROUPS

The cell surface proteins found on red blood cells are termed blood group antigens. They vary in their antigenicity, some being more important than others. When blood is transfused into an individual, those cell surface proteins that are foreign to the recipient trigger an antibody response. By studying the pattern of these responses, it is possible to show that the expression of these blood group antigens is inherited and that many of them occur in different allelic forms. Each antigenic cell surface protein thus forms a blood group system. Within each system a variable number of alternative alleles can be recognized by specific antibodies.

In one special case, the human ABO system, in addition to possessing blood group antigens on their cells, normal animals have antibodies directed against foreign blood group antigens in their serum. Thus humans whose red cells lack the blood group antigen called A have antibodies to this molecule in their sera. These antibodies are not derived from contact with group-A red blood cells but result from exposure to cross-reacting epitopes that occur commonly in nature. The ABO epitopes are common structural components of a wide range of organisms, including plants, bacteria, protozoa, and helminths. As a result, individuals make antibodies against these epitopes from a very young age.

BLOOD TRANSFUSIONS

Red blood cells can be readily transfused from one animal to another. If the donor red blood cells carry antigens identical to those found on the recipient's red blood cells, no immune response will result. If, however, the transfused red cells are foreign to the recipient, then they will be destroyed. This can happen in two ways. If, as in the ABO system, the recipient possesses preexisting antibodies to the antigens on the foreign red blood cells, then they will be destroyed immediately. When these antibodies combine with foreign blood group antigens, they cause agglutination, hemolysis, opsonization, and phagocytosis of the transfused cells. In the absence of preexisting antibodies, the foreign red cells stimulate a primary immune response in the recipient. The transfused cells therefore circulate freely until antibody production takes place and immune elimination occurs. A second transfusion with identical foreign red cells results in their immediate destruction in the sensitized recipient.

Although the body can eliminate small numbers of aged red blood cells on a continuing basis, the rapid destruction of many foreign red blood cells from an incompatible transfusion can lead to severe tissue damage. This damage results from massive intravascular hemolysis, which triggers clotting. As a result of the blockage of capillaries by microthrombi, there are nervous system signs such as tremors, paralysis, and convulsions. In some individuals, difficulty in breathing, as well as coughing, and diarrhea may occur. Patients develop a fever, and hemoglobin is excreted in the urine. These transfusion reactions are treated by stopping the transfusion and maintaining urine flow with a diuretic, since accumulation of hemoglobin in the kidney will cause severe damage. Recovery follows elimination of all foreign red blood cells.

Blood Groups of Humans

ABO and Lewis Systems

In 1901 Karl Landsteiner first demonstrated the existence of blood group antigens on human red blood cells as well as antibodies directed against those antigens in human sera. He collected blood from members of his laboratory staff. He separated the red blood cells from the serum, and then he studied the results of mixing serum and red blood cells from different individuals. He found that some sera could agglutinate the red blood cells of some individuals but not others. In analyzing the results, he found he could group individuals. Group A individuals possessed an antigen, called A, on their red blood cells and antibodies to another antigen, called B, in their serum. Group B individuals had antigen B on their red blood cells and antibodies to antigen A in their sera. A third group, called group O, had neither A nor B on their red blood cells but had both anti-A and anti-B in their sera. Some time later, individuals were described who had both A and B antigens on their red blood cells but no antibodies to A or B in their sera. This group was called AB. Clearly, an individual could never have both a blood group antigen and its specific antibody in blood at the same time.

Although the ABO antigens were first identified as glycolipids, they are also the polysaccharide side chains of two cell surface glycoproteins. These glycoproteins are anion and glucose transporter proteins called band 3 and band 4.5, respectively. (These refer to the bands of protein seen on **electrophoresis** of red cell surface proteins.) The blood group designations A, B, AB, and O indicate the phenotype of an individual's red blood cells. These polysaccharide antigens are inherited according to a simple mendelian system involving three allelic genes called A, B, and O. A and B genes are dominant over O. Thus, an AB person is heterozygous, having inherited an A gene from one parent and a B gene from the other. An O person must be homozygous for O, and persons in groups A and B must be either homozygous (AA or BB) or heterozygous (AO or BO) (Table 30–1). In North America, approximately 45% of the population are of group O, 42% A, 10% B, and 3% AB. However, these values vary between ethnic groups, so that in blacks, for example, the distribution is 49% O, 28% A, 20% B, and 3% AB. Members of the Native American Navajo tribe are essentially 100% group O. In addition to the A and B polysaccharides, two soluble blood group antigens belong to this family of molecules. These are the Le^a (Lewis a) and Le^b (Lewis b) antigens. They are inherited at a different locus from ABO.

The genes that determine both the ABO and Lewis systems code for glycosyltransferases. These enzymes mediate the addition of carbohydrate side chains to the surface glycoproteins and glycolipids. Specifically, they attach new sugar residues to two basic precursor molecules. Both precursor molecules consist of alternating units of galactose-N-acetyl-D-glucosamine and galactose-N-acetyl-D-galactosamine. In the type 1 precursor, the terminal galactose is bound to the N-acetyl-D-glucosamine by a β-(1,3) linkage, and in the type 2 precursor it is bound by a β-(1,4) linkage. The precursor oligosaccharide is called H substance. A, B, and AB individuals express glycosyltransferase

Table 30–1 The Inheritance of the ABO Blood Group System

Genotype	Phenotype (Blood Group)	Antibodies in Serum
AA, AO	A	Anti-B
BB, BO	B	Anti-A
AB	AB	None
OO	O	Anti-A and anti-B

METHODOLOGY

How to Determine an Individual's ABO Status

To determine the ABO blood group of an individual, specific antisera are required. Anti-A is usually obtained from human volunteers of blood group B. Their natural anti-A is commonly boosted by injection of pure A substance derived from pigs. Commercial anti-A is usually dyed blue. Anti-B is similarly obtained from group A volunteers, and it is usually dyed yellow.

For rapid testing, whole blood is used as a source of cells, and two drops of blood are mixed on a slide with a drop of anti-A and anti-B. When testing blood for compatibility in transfusion, it is usual to separate it into cells and serum. The cells, after washing, are suspended in saline at 2% to 5%, mixed with the different antisera in a tube, and examined for agglutination after centrifugation. Similarly, the serum is tested against cells of known specificity.

Generally, the results of the two tests should confirm one another—that is, the results of testing the cells should be compatible with the results of testing the serum. Obviously, if discrepancies occur, the testing must be repeated and the blood not used for transfusion until the cause is known.

activities that convert the H substance into A or B antigens or both, whereas O individuals lack such activity. The A and B genes differ from each other in a few single-base substitutions. These differences result in changes in four amino acids in their products. It is these four amino acids that account for the differences in A and B glycosyltransferase specificity. In individuals of blood group O, a single-base deletion occurs in the glycosyltransferase gene, which results in a reading frame shift and the production of an entirely different protein. This new protein is incapable of modifying the H substance.

If an individual is of blood group O therefore, no further reactions occur and H substance alone is found on their red blood cells. Individuals possessing the A gene make a glycosyltransferase that links N-acetyl-D-galactosamine to the terminal galactose of H substance to create A substance. On the other hand, if the individual possesses a B gene, then the glycosyltransferase will add an additional galactose residue to the terminal galactose of H substance and so produce B substance. Possession of both A and B genes ensures that both A and B substances are produced (Fig. 30–1).

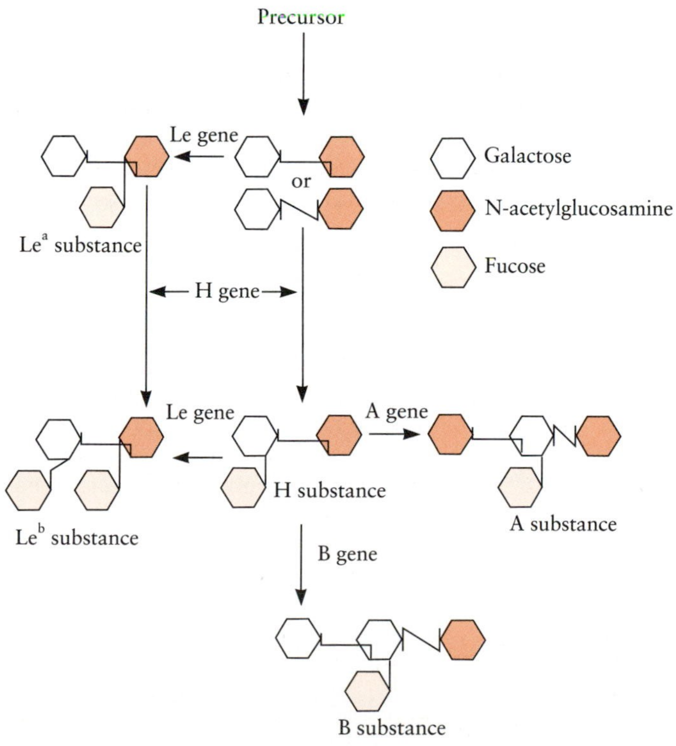

Figure 30–1 The synthesis of the ABO and Lewis blood group substances. Sugars are added in sequence to a precursor glycoprotein by glycosyltransferases controlled by the I, H, Le, A, and B blood group genes. There are two forms of precursor substances. One, with a β1,3 linkage, is the precursor of Le^{a} and Le^{b}. The other, with a β1,4 linkage, is the precursor of A and B.

Soluble Le^a substance is produced from the type 1 precursor by a glycosyltransferase coded for by the Le gene. It links a fucose molecule to the N-acetyl-D-glucosamine residue. Le^b substance is produced from Le^a through the actions of the H glycosyltransferase, which adds a second fucose molecule to the terminal galactose residue. Le^b substance may also be formed by linking fucose to an N-acetyl-D-glucosamine by the A glycosyltransferase or to a galactose by the B glycosyltransferase to produce structures with A or B epitopes.

Secretors and Nonsecretors

In about 75% of humans, A and B carbohydrates may be found not only bound to red cell surfaces but also as soluble glycolipids in body fluids such as serum, urine, and saliva. This trait is controlled by a secretor (Se) gene. The Se gene promotes the production of soluble A, B, and H glycolipids when it is present—that is, in homozygous (SeSe) or heterozygous (Sese) individuals. In the homozygous recessive state (sese), the glycolipids are not produced and, as a result, ABH glycolipids are absent from body fluids (Fig. 30–2). The glycosyltransferase of the Lewis system is functional only in Se^+ individuals. Lewis antigens are soluble glycolipids that are passively adsorbed to cells.

Rhesus System

In 1940 Landsteiner and Wiener showed that antibodies produced against rhesus monkey red blood cells agglutinated the red blood cells of 85% of a human population. The antibodies were directed against a molecule called the rhesus (Rh) antigen, and individuals possessing it were called Rh positive. The remaining 15% who did not carry it were called Rh negative. Natural antibodies against the Rh antigens do not occur. Rhesus antigens are unique nonglycosylated, very hydrophobic cell surface proteins of 32 kDa (the only nonglycosylated membrane proteins known). They are structurally related to the band 3 and band 4.5 glycoproteins, which suggests that they too may be transporter proteins (Fig. 30–3).

Immunologically, the Rh system has been shown to be immensely complex, and approximately 30 antigenic types have been identified. The system consists of five isoforms (C, D, E, c, and e), the products of three closely linked loci. The most important of these loci is called D and is responsible for the production of the D antigen, the most important of the Rh antigens. There may be an alternative allele to D called d. However, this has never been demonstrated. Thus d denotes the absence of D. A person is Rh positive who inherits D from either parent. A homozygous parent transmits D to all his or her children. A heterozygous parent transmits it to only half his or her children.

Other Human Blood Group Systems

Human red cells carry many other blood group antigens (Table 30–2). Some are relatively simple, for example, the Kidd group, which consists of two alternative alleles. Others are complex, for example, the MN system with 29 identified allelic antigens. Some antigens are found on red blood cells of individuals but are not grouped. One group, the Xg system, is sex-linked.

Blood Groups and Disease

It has already been pointed out how certain histocompatibility antigens are closely associated with disease prevalence or susceptibility. Since blood group antigens are not necessarily linked to the MHC and have no antigen-presenting ability, there is no a priori reason why they should be linked to specific disease conditions. Nevertheless, there is a clear association between stomach ulceration and nonsecretors of blood group O. (This has recently been shown to result from the fact that stomach ulcers are caused by a bacterium

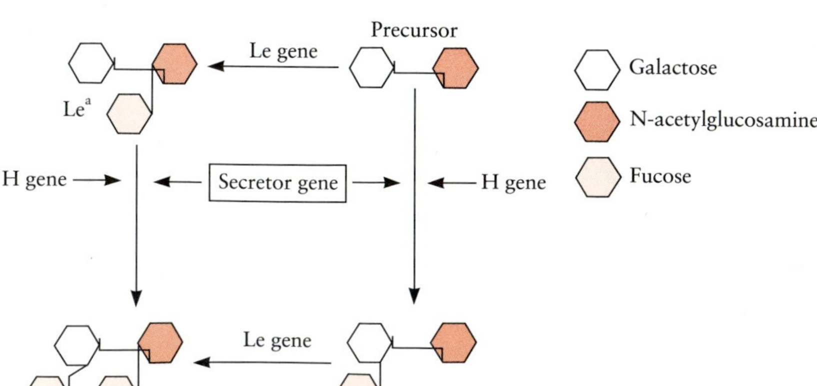

Figure 30–2 The mode of action of the secretor gene (Se): the Se gene acts on H precursor substance to produce soluble glycolipids. It is also required for the synthesis of Le^b substance. In the absence of the Se gene, H, A, and B substances bind strongly to cells and are not found free in body fluids. Le^{b+} individuals must possess the Se gene in either the homozygous or heterozygous states.

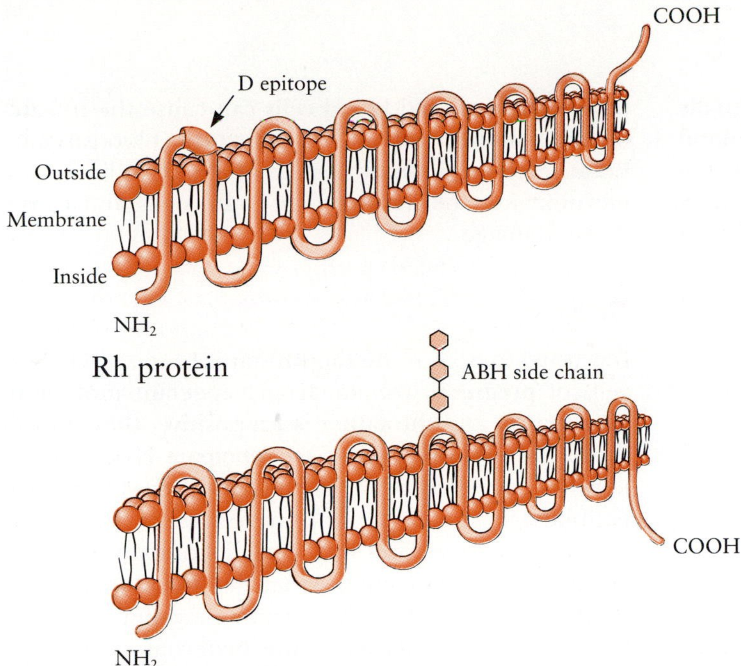

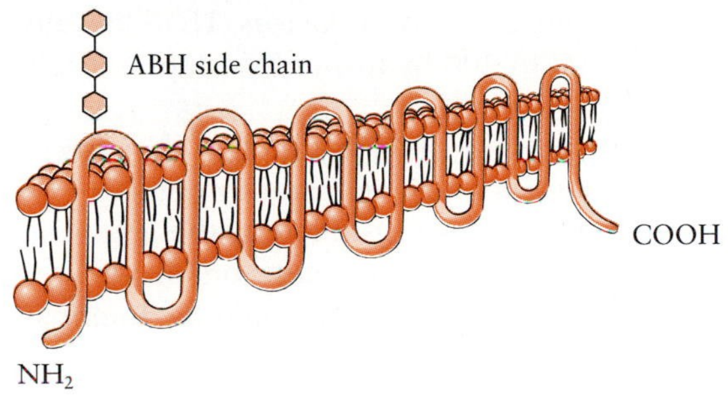

Figure 30–3 The three major proteins associated with blood group reactivity are all multipass membrane proteins that probably have a transporter function.

Table 30–2 Some of the Major Human Blood Group Antigens and Their Phenotypic Frequencies*

System	Nature	Major Antigens	Frequencies	(Percent)
ABO	Glycoprotein Glycolipid	A, B, O	A	42
			B	10
			AB	3
			O	45
Rh	Protein	C, D, E or c, d, e	D+	84
			D−	16
MN	Glycophorin A	M, N	MM	28
			NN	50
			MN	22
Kell	Endopeptidase	K, k	K	9
			k	91
Duffy	Glycoprotein	Fy^a, Fy^b, Fy	Fy^aFy^b	45
			Fy^a	20
			Fy^b	35
			Fy	<1
Lutheran	Glycoprotein	Lu^a, Lu^b	Lu^a	8
			Lu^b	99

*Other recognized human blood groups include I, Lewis, P, Kidd, Diego, Cartright, Vel, and Xg.

called *Helicobacter pylori.* This organism preferentially binds to Lewis[b] antigens that have a terminal blood group O fucose. These molecules are located on the surface of gastric epithelial cells.) In contrast, cancer of the stomach and pernicious anemia are more common in individuals of blood group A. Much more important than these associations is the direct role of blood group antigens in **hemolytic disease of the newborn** (HDN).

HEMOLYTIC DISEASE OF THE NEWBORN

HDN usually occurs as a result of rhesus incompatibility. It develops when an Rh-negative mother carries an Rh-positive fetus (Fig. 30–4). Normally, the fetal red blood cells are separated from the mother's circulation by the layer of cells in the placenta called the trophoblast. However, during late pregnancy, and especially during the process of childbirth, the fetal red blood cells may escape into the mother's circulation. (They can be detected by staining a blood smear from the mother for fetal hemoglobin [HbF], which is found only in fetal red blood cells.) Once these cells reach the mother's circulation, they are perceived as foreign and therefore provoke an antibody response.

Antibodies to fetal red blood cells are not usually made before the first childbirth. Repeated pregnancies do, however, eventually provoke high antibody levels in the mother. Maternal IgG antibodies provoked in this way can cross the placenta and reach the fetal circulation, where they react with the fetal red blood cells and cause their destruction. This red blood cell destruction may be so severe that it results in the death of the fetus. Alternatively, the fetus may survive to be born in a jaundiced or anemic state. The fetus may attempt to compensate for the loss of its red blood cells by releasing immature red blood cells (erythroblasts) from its bone marrow. HDN is therefore also called erythroblastosis fetalis. The extensive destruction of the fetal red blood cells can cause the release of toxic amounts of bilirubin into the bloodstream. Because of the immature state of the fetal liver, this bilirubin may be deposited in brain cells and cause severe damage.

Diagnosis

It is usual to determine the Rh status of the red blood cells of pregnant women. If an expectant mother is Rh-negative and the father is Rh-positive, then a physician will be in a position to anticipate HDN. HDN, however, does not occur in all cases of rhesus incompatibility. Thus not all Rh-positive fathers are homozygous, and therefore, in theory, incompatibility will occur only in 75% of pregnancies between an Rh-positive father and an Rh-negative mother. Second, fetal red blood cells reach the maternal circulation only in 20% to 70% of cases. Third, the risks of the disease occurring are very low at the time of first pregnancy but increase with each subsequent pregnancy. As a result of all these factors, HDN develops in 1 in every 200 to 400 births, unless steps are taken to prevent it.

In pregnancies with a high probability of HDN, an early diagnosis can be made by detecting fetal hemoglobin breakdown products in the amniotic fluid. After birth, diagnosis is made on the basis of clinical signs, but it must be confirmed by demonstrating antibodies directed against the fetal red blood cells. These antibodies cannot cause direct agglutination but can be detected by means of an antiglobulin or Coombs' test (Chapter 17). It is also possible to screen maternal serum during pregnancy for a rise in antibodies to antigen D.

Prevention

When the factors that influence the development of HDN were studied, it was noticed that the disease did

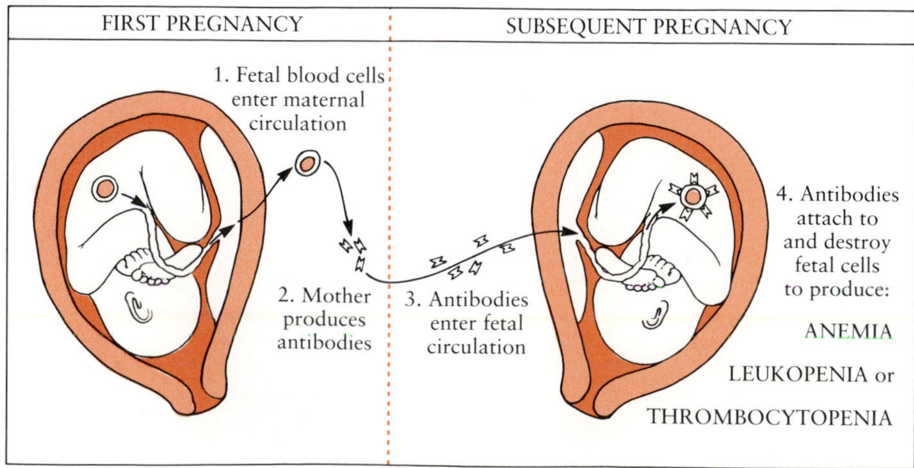
FIRST PREGNANCY
SUBSEQUENT PREGNANCY
1. Fetal blood cells enter maternal circulation
2. Mother produces antibodies
3. Antibodies enter fetal circulation
4. Antibodies attach to and destroy fetal cells to produce:
ANEMIA
LEUKOPENIA or
THROMBOCYTOPENIA

Figure 30–4 Alloimmunization due to fetomaternal blood group incompatibility. Fetal blood cells trigger an immune response in the mother. The mother's antibodies then cross the placenta and destroy the fetus's red cells. They may also destroy fetal white cells and platelets. *(From Bellanti, J.A. Immunology II. W.B. Saunders, Philadelphia, 1979. With permission.)*

not develop if the parents had differing ABO blood groups. This is because any fetal red blood cells entering the mother's circulation in these ABO-incompatible individuals are immediately destroyed by preexisting antibodies and therefore will not provoke an immune response. These natural antibodies rarely cause hemolytic disease themselves since they are of the IgM class and thus cannot cross the placenta to reach the fetus.

On the basis of this observation, it was predicted that, if the mother had preexisting anti-Rh antibodies, these would destroy any Rh-positive red blood cells entering the maternal circulation and hence prevent her sensitization. This prediction has been amply confirmed. Thus, passively administered anti-D serum is given to the mother within 72 hours of any delivery at which she might be sensitized (Fig. 30–5). (This anti-D is produced in male volunteers.) As a result of this treatment, her sensitization is effectively prevented. This treatment has reduced the prevalence of HDN to 1 in every 20,000 deliveries.

If hemolytic disease does occur in a newborn infant, the usual method of treatment is by exchange transfusion. This corrects the anemia and at the same time removes sensitized red blood cells, antibodies, and bilirubin.

BLOOD GROUPS AND HEMOLYTIC DISEASE OF ANIMALS

All mammals tested have been shown to possess blood groups similar to those seen in humans. Some are relatively simple, and some are enormously complex. The B system of cattle contains over 1000 different alleles, and it has been suggested that there are sufficient alternative combinations to provide a unique blood group for every cow in the world. HDN is also recorded in animals. In the major domestic animals, however, antibodies cannot cross the placenta. Newborn mammals therefore obtain maternal antibody

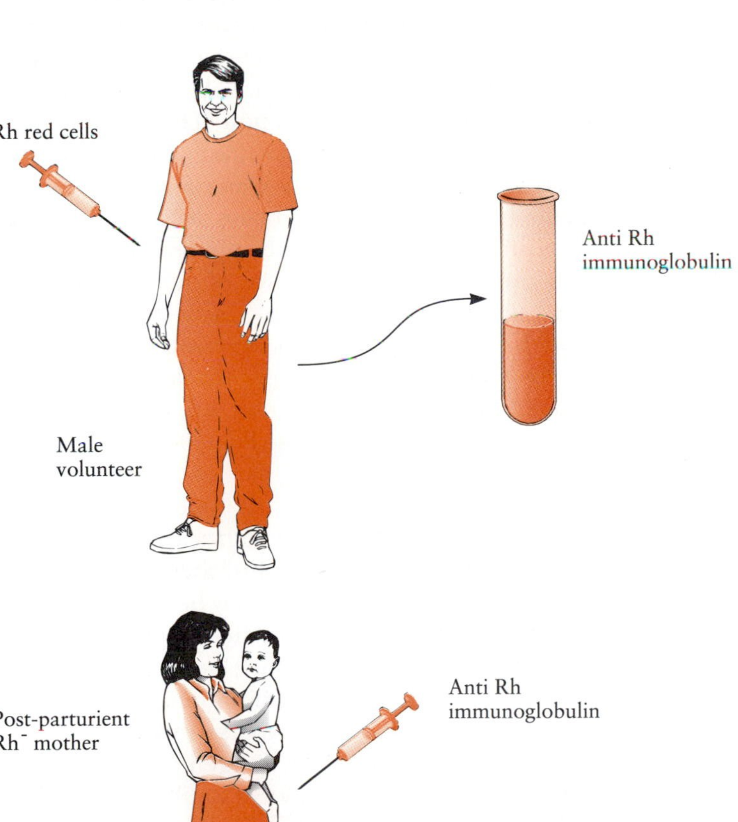

Figure 30–5 The prevention of hemolytic disease in human infants. Anti-Rh antibodies, if given to the mother within 72 hours after birth, effectively prevent sensitization at that time by inhibiting the mother's own anti-Rh response.

through the first milk, or colostrum. As a result, hemolytic disease does not develop in utero but follows immediately after a newborn animal suckles for the first time. The clinical signs of these diseases are similar to the condition in humans, but prevention is relatively simple: prevent the young animal from suckling.

One other interesting feature of hemolytic disease in cattle is that it results from human intervention. Fetal red blood cells do not cross the placenta in cattle, and as a result, cows are never naturally sensitized to fetal red blood cells. Some antiparasitic vaccines that contain bovine red blood cells are, however, given to cattle. Administration of these vaccines, therefore, may sensitize cows to other bovine blood group antigens and can lead to the development of hemolytic disease in newborn calves.

TYPE III HYPERSENSITIVITY

The formation of immune complexes through the combination of antibody with antigen is the first step in many immunological processes. One of the most significant of these processes is the complement cascade. When complement-activating immune complexes are deposited in tissues, chemotactic factors are produced and lead to a local accumulation of neutrophils. These neutrophils release their lysosomal enzymes and oxidizing radicals, and these in turn cause local tissue destruction. Lesions generated in this fashion are called type III, or immune complex–mediated, hypersensitivity reactions.

Classification of Type III Hypersensitivity

The severity and significance of type III hypersensitivity reactions depend on the amount and site of deposition of immune complexes. Two major types of reaction are recognized. One is a local reaction known as the **Arthus reaction,** named after the biologist who first described it. The Arthus reaction occurs when immune complexes are deposited in tissues. Arthus reactions may be induced in any tissue into which antigen can be deposited.

A second form of type III hypersensitivity reaction results when large quantities of immune complexes are formed in the bloodstream. This occurs when antigen is administered intravenously to a hyperimmune recipient. Complexes generated in this way are deposited in the walls of blood vessels. Local activation of complement then leads to neutrophil accumulation and the development of inflammation in blood vessel walls (**vasculitis**). Circulating immune complexes are also deposited in the glomeruli of the kidney. Therefore, the occurrence of inflammatory lesions in glomeruli (glomerulonephritis) is also characteristic of this type of hypersensitivity.

The combination of antigen with antibody always results in the generation of some immune complexes, but the occurrence of clinically significant type III hypersensitivity reactions is due to the formation of very large amounts of immune complexes. For instance, several grams of antigen may be needed to sensitize an animal, such as a rabbit, in order to produce experimental Arthus reactions or generalized immune complex disease. Minor immune complex–mediated lesions probably develop relatively frequently following the normal immune responses to many antigens, without giving rise to clinically significant disease.

LOCAL TYPE III HYPERSENSITIVITY REACTIONS

Arthus Reaction

If antigen is injected subcutaneously into an animal that has been hyperimmunized and possesses circulating antibody able to precipitate that antigen, then an acute inflammatory reaction will develop within several hours at the site of injection. The reaction starts as a reddened, edematous swelling; eventually, local hemorrhage and thrombosis occur and, if severe, the reaction culminates in local tissue destruction (necrosis). Histologically, the first change seen following antigen injection is the adherence of neutrophils to vascular endothelium, followed by their emigration through the walls of small blood vessels, especially venules. By 6 to 8 hours, when the reaction has reached maximal intensity, the injection site is densely infiltrated by many of these cells (Fig. 30–6). As the reaction progresses, destruction of blood vessel walls occurs, resulting in the escape of fluid and red blood cells into the tissues. Platelet aggregation and thrombosis are also associated with this vascular destruction. By 8 hours, **mononuclear cells** may be observed within the lesion, and by 24 hours or later, depending on the amount of antigen injected, they become the predominant cell type. Eosinophil infiltration is not a significant feature of this type of hypersensitivity.

The fate of the injected antigen may be followed by means of a direct fluorescent antibody test (Chapter 17). Antigen diffuses away from the injection site through tissue spaces. When small blood vessels are encountered, the antigen will diffuse into the vessel walls, where it comes into contact with circulating antibody. As a result, immune complexes are generated and deposited between and beneath endothelial cells. When these immune complexes activate the complement system, C5a and C567 are produced. As a result,

METHODOLOGY

How to Induce an Arthus Reaction

Monsieur Arthus produced his reaction by injecting a rabbit with horse serum daily for several weeks. It is not absolutely necessary to induce the reaction this way. All that need be done is to immunize a rabbit against an antigen so that high-titered precipitating antibodies are produced. When a small quantity (about 1 mg) of antigen is injected subcutaneously or intradermally into an animal with these antibodies, a diffuse, reddened swelling starts about 2 hours later. Small hemorrhages develop, and edema and **erythema** progressively increase. Eventually, frank hemorrhage and necrosis may develop at the injection site.

neutrophils are attracted to the site and progressively accumulate (Fig. 30–7).

Neutrophils have a C3b receptor (CR1) and so adhere to immune complexes containing this component. Neutrophils that encounter immune complexes promptly phagocytose and destroy them. During this process, however, large quantities of oxidizing radicals and enzymes are released into the tissues. These radicals and enzymes cause the tissue damage seen in the Arthus reaction. Neutrophil hydrolytic enzymes are normally stored within lysosomes. They are released into tissue through several processes, the most obvious of which is cell death; other release mechanisms, however, are probably of greater importance in the Arthus reaction. For example, when neutrophils attempt to phagocytose immune complexes attached to a nonphagocytosable structure, such as a basement membrane, they secrete their free radicals directly into the surrounding medium. Similarly, they may release lysosomal enzymes into the phagosome before immune complexes are completely enclosed so that the enzymes escape into the surrounding tissues.

The lysosomal enzymes released in this way include collagenases that disrupt collagen fibers, neutral proteases that destroy ground substances and basement membranes, and elastases that destroy elastic tissue. Normally, tissues contain antiproteinases that inhibit the neutrophil enzymes. Neutrophils can, however, subvert these antiproteinases by the release of OCl^-. The OCl^- destroys protease inhibitors and allows tissue destruction to proceed. Neutrophil proteases also act on C5 to generate C5a, which stimulates neutrophil degranulation and enzyme release and so promotes further neutrophil accumulation and degranulation. Other enzymes released by neutrophils may make mast cells degranulate or generate kinins. As a result of all this enzyme release, destruction of tissues (especially of blood vessel walls) occurs, resulting in the development of the edema, vasculitis, and hemorrhage characteristic of the Arthus reaction.

In addition to causing neutrophil accumulation, complement activated by immune complexes may make platelets clump and release clotting factors (Chapter 28). This, in conjunction with the severe vascular damage, results in extensive thrombosis. Finally, the production of **anaphylatoxins** (C3a and C5a) and of the C2 kinin, the release of kininogens from neutrophils and of vasoactive amines from neutrophils,

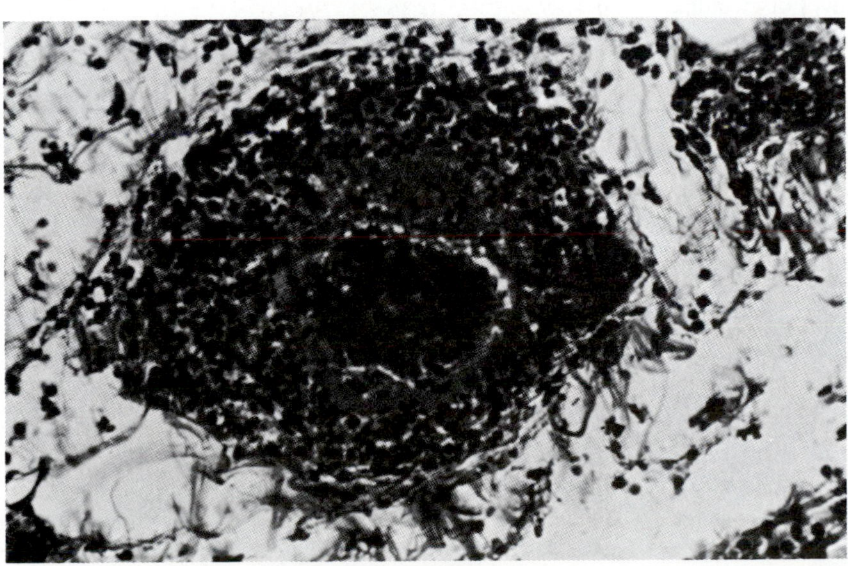

Figure 30–6 A histological section of an Arthus reaction in rabbit skin: the vessel is thrombosed and there is extensive infiltration of neutrophils around it. *(From Thomson, R.G. General Veterinary Pathology. W.B. Saunders, Philadelphia, 1978. With permission.)*

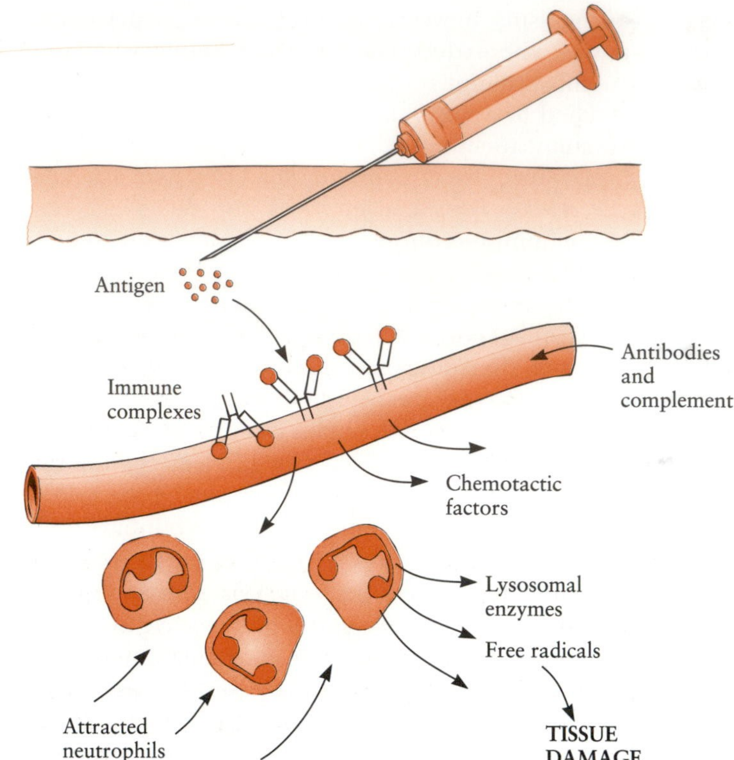

Figure 30–7 The major mechanisms involved in the Arthus reaction and other local type III hypersensitivity reactions.

platelets, and mast cells all contribute to the development of a severe local inflammatory response. The antibodies involved in the Arthus reaction are usually of the IgG class.

Although the classic direct Arthus reaction is produced by local administration of antigen to hyperimmunized animals, any technique that permits immune complexes to be deposited in tissues will stimulate a similar response. A reversed Arthus reaction can therefore be produced if antibody is administered intradermally to an animal with a high level of circulating antigen. Injected preformed immune complexes, particularly those containing a moderate excess of antigen, will provoke a similar reaction, although there is less involvement of blood vessel walls and the reaction is less severe. A passive Arthus reaction can be produced by giving antibody intravenously to a nonsensitized animal followed by an intradermal injection of antigen. Real enthusiasts can produce a reversed passive Arthus reaction by giving antibody intradermally followed by intravenous antigen.

Hypersensitivity Pneumonitis

Repeated inhalation of very small spores from fungi or actinomycetes stimulates an immune response. As a result, individuals continuously exposed to these spores develop high-titered precipitating serum antibodies to spore antigens. If exposure to these spores continues, spore antigens will encounter antibody within the lung on the surface on the alveoli (the small air sacs at the end of the airways); immune complexes will be deposited, complement will be activated, and

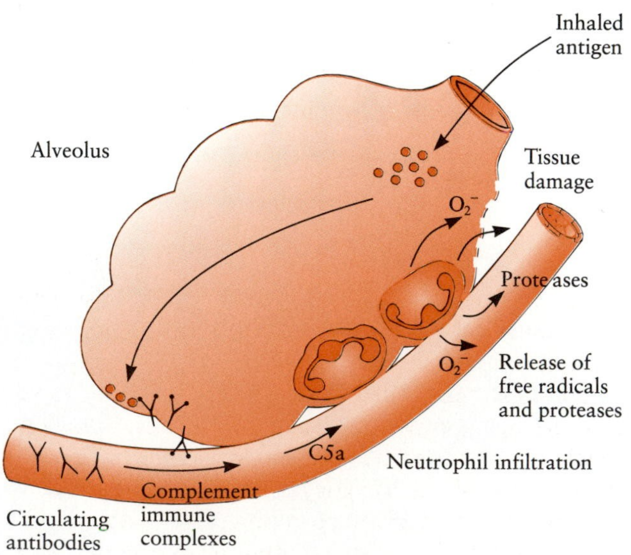

Figure 30–8 The major mechanisms involved in the pathogenesis of acute hypersensitivity pneumonitis.

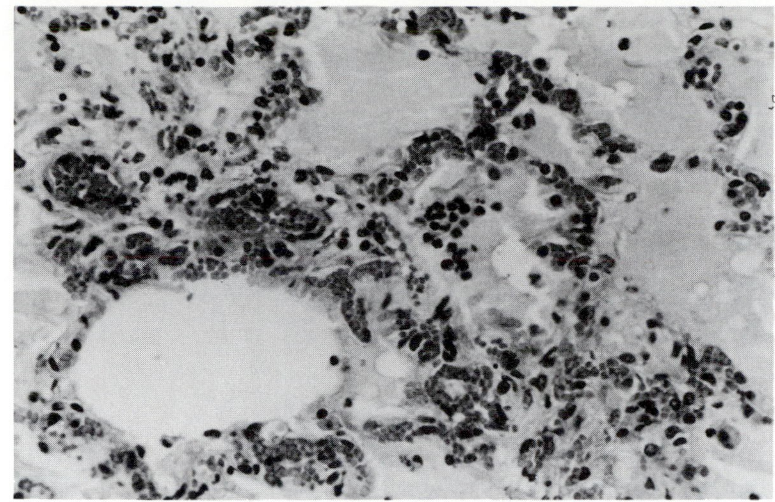

Figure 30–9 A histological section of lung from an acute case of hypersensitivity pneumonitis due to inhalation of actinomycete spores. Note that the alveoli, which should be full of air, mainly contain a pale gray substance, which is fluid. This individual clearly had difficulty breathing. *(Courtesy of Dr. B.N. Wilkie.)*

a local inflammatory response will develop (Fig. 30–8). This inflammation is characterized by a massive neutrophil accumulation and tissue damage. Fluid accumulates in the alveoli interfering with the free exchange of oxygen (Fig. 30–9). It is a form of type III hypersensitivity reaction. Clinically, this type of hypersensitivity is associated with difficulty in breathing (dyspnea) occurring 5 to 10 hours after exposure to the antigen. Once recognized, it can be prevented from recurring by removal of the source of the antigen.

Many forms of **hypersensitivity pneumonitis** in humans are mediated in this way. They are usually named after the source of the offending antigen. Thus farmer's lung develops as a result of chronic exposure to actinomycete spores released when working with moldy hay. Pigeon-breeder's lung arises following exposure to the dust from pigeon feces, mushroom-grower's disease is due to hypersensitivity to inhaled spores from actinomycetes in the soil used for growing mushrooms, and librarian's lung results from inhalation of dusts from old books and so forth.

GENERALIZED TYPE III HYPERSENSITIVITY REACTIONS

If an antigen is injected directly into the bloodstream of an animal with a high level of circulating antibodies, then immune complexes form within the circulation. Most of these complexes, especially the large ones, are removed by macrophages. Some complexes, however, especially those formed with excess antigen, are soluble and thus poorly phagocytosed. In addition, alternate-pathway complement components are able to insert themselves into large immune complexes and solubilize them. These soluble complexes may activate complement and so stimulate platelet aggregation and the release of vasoactive amines. They will therefore affect the permeability of the vascular endothelium. As a result, immune complexes may be deposited in the walls of blood vessels, particularly medium-sized arteries (Fig. 30–10). They may also be deposited in vessels where there is physiological outflow of fluid, for example, glomeruli, synovia, and the choroid plexus of the brain.

Acute Serum Sickness

When the use of antisera for passive immunization was in its infancy, it was observed that patients who had received a large single dose of foreign serum (usually horse antiserum to tetanus or diphtheria) showed a characteristic series of side effects about ten days later. The side effects consisted of a generalized vasculitis with erythema, edema, urticaria of the skin, neutropenia, lymph node enlargement, joint swelling, and proteinuria. The reaction was usually of short duration, subsiding within a few days, and was known as **serum sickness.** A similar reaction can be produced experimentally by giving rabbits a single very large intravenous dose of antigen. Its occurrence can be shown to coincide with the presence of large quantities of immune complexes in the circulation as a result of antibodies produced against circulating antigen (Fig. 30–11).

Two types of lesion occur in acute serum sickness. First, there is a transient **glomerulonephritis,** the nature of which tends to vary with the size and avidity of immune complexes involved (Fig. 30–12). Since the glomerular basement membrane is a negatively charged (anionic) barrier, it attracts immune complexes that are cationic. Some complexes penetrate the vascular endothelium but not the glomerular basement membrane, and so become deposited under the

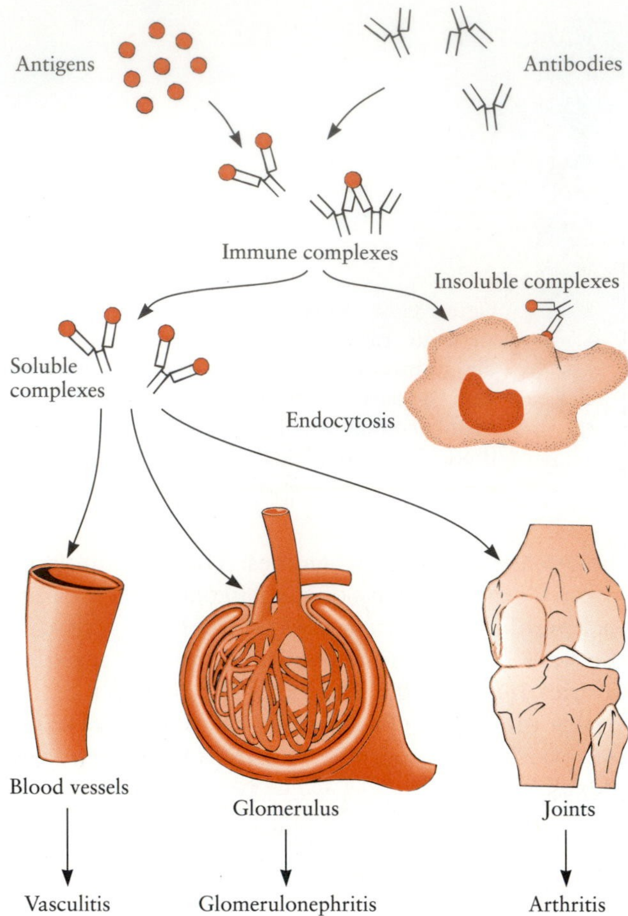

Figure 30–10 The major mechanisms involved in the pathogenesis of serum sickness and generalized immune complex–mediated disease.

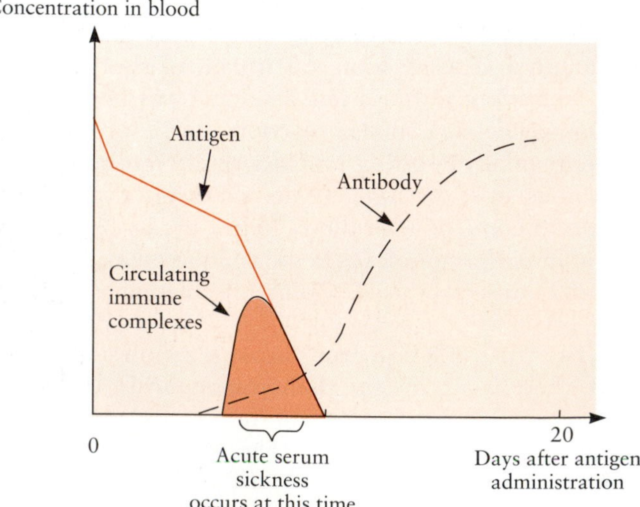

Figure 30–11 The timing of immune complex formation in acute serum sickness. It coincides with the removal of circulating antigen by newly formed antibodies.

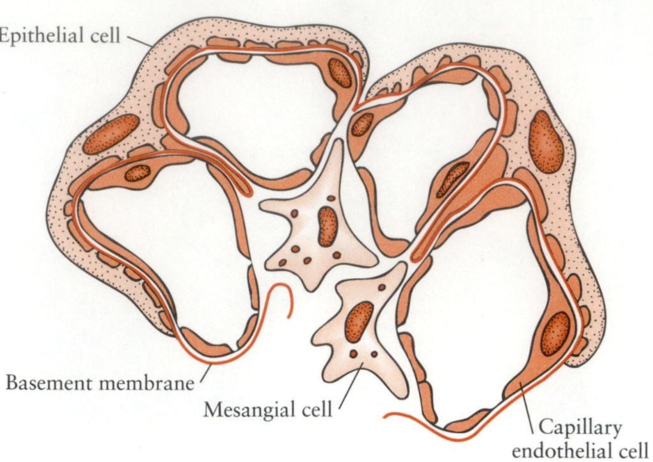

Figure 30–12 A schematic diagram showing the major features of the structure of a normal glomerulus.

endothelial cells, where they stimulate endothelial cell swelling and proliferation (Fig. 30–13). In contrast, other small complexes can penetrate both the vascular endothelium and the basement membrane and be deposited outside the blood vessels under the epithelial cells. Here they stimulate epithelial cell swelling and proliferation. Neutrophils do not normally accumulate within these glomeruli; neverth less, local damage does occur as a result of the vasoactive properties of activated complement, and this leads to the loss of protein in the urine (proteinuria). Second, lesions develop in medium-sized arteries. These include neutrophil infiltration, disruption of the internal elastic membrane, and destruction of the medial layer.

Chronic Serum Sickness

If, instead of receiving a single high dose of antigen, a rabbit is given multiple intravenous injections of small doses of antigen, then two other types of glomerular lesions may develop. Continued deposition of immune complexes under epithelial cells may lead to an increased thickness of the glomerular basement membrane, forming what is known as a wire-loop lesion or membranous glomerulonephritis. Alternatively, these immune complexes may be deposited between blood vessels within the glomeruli. In this region are found **mesangial cells.** Mesangial cells are modified muscle cells that can take up immune complexes and release prostaglandins and cytokines such as IL-1 and IL-6. They respond to the presence of immune complexes by secreting interleukin-1. The IL-1 stimulates mesangial cell proliferation (Fig. 30–14) and the secretion of proteases. Normally, this proliferation scarcely affects glomerular function unless the mesangial cells expand to completely surround the

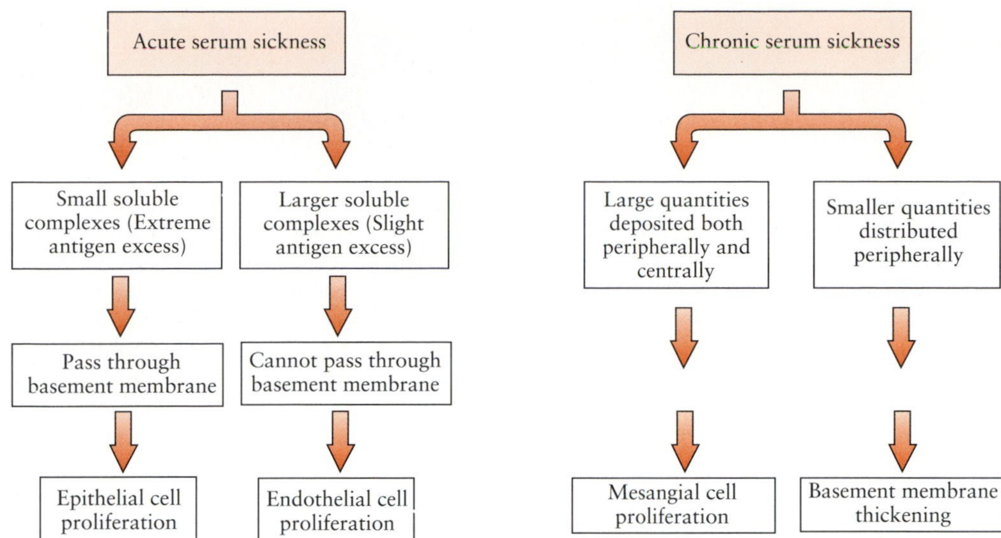

Figure 30–13 The different forms of immune complex–mediated glomerulonephritis. The lesion that develops depends the size of immune complexes formed and on whether the deposition is brief (acute serum sickness) or prolonged over many weeks or months (chronic serum sickness).

glomerular capillaries. By immunofluorescence, it can be shown that aggregates of immune complexes are deposited in capillary walls and on the epithelial side of the glomerular basement membrane (Fig. 30–15).

Chronic immune complex–mediated lesions are not restricted to the capillaries of the glomeruli. Physiological outflow of fluid is also a feature of the capillaries of the choroid plexus of the brain and of the capillaries found in the synovia of joints. Deposition of immune complexes in the walls of these vessels may also lead to the development of significant inflammation.

IMMUNE COMPLEX–MEDIATED DISEASES

Several important clinical syndromes are associated with the development of immune complex–mediated lesions in humans and animals. All are associated with

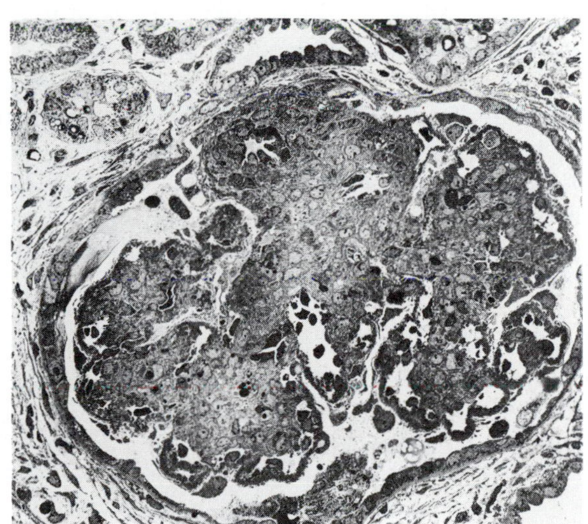

Figure 30–14 A thin section of a glomerulus with immune complex–mediated lesions. There is mesangial cell proliferation and basement membrane thickening so that the capillaries have almost disappeared. *(From Angus, K.W., et al. J. Comp. Pathol., 84:319–330, 1974. Copyright Academic Press Inc. [London] Ltd. With permission.)*

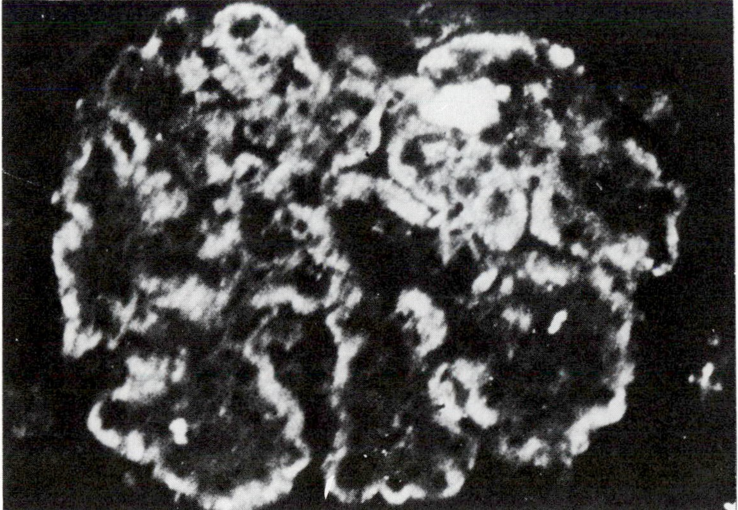

Figure 30–15 A fluorescent photomicrograph of a glomerulus with mesangiocapillary glomerulonephritis. The fluorescent antiglobulin reveals "lumpy-bumpy" deposits characteristic of immune-complex deposition. *(From Angus, K.W., et al. J. Comp. Pathol., 84:319–330, 1974. Copyright Academic Press Inc. [London] Ltd. With permission.)*

METHODOLOGY

How to Detect and Measure Immune Complexes

If a Type III hypersensitivity mechanism is suspected of being involved in a disease, it may be desirable to detect and measure the immune complexes. Several techniques are available to measure circulating immune complexes within the bloodstream. None is entirely satisfactory, and the clinical significance may not be entirely clear. Most techniques use the principle that binding of immunoglobulin to antigen causes a change in the Fc region that activates C1q and permits binding to Fc receptors on cells. Thus the immune complexes can be bound to C1q fixed to plastic tubes, and after washing, the bound immunoglobulin may be detected by means of an enzyme- or isotope-labeled antiglobulin. Alternatively, the C1q may be first radiolabeled, and after binding to immune complexes, the complexes may be precipitated and the radioactivity measured.

The Raji cell line is a cultured lymphoid cell line that possesses Fc and C3b receptors. After immune complexes are permitted to bind to these cells, they may be measured using a radiolabeled antiglobulin serum.

A third approach is to use the large immune complexes and selectively precipitate them with polyethylene glycol. The level of the immunoglobulin in the precipitate can then be measured using a radiolabeled antiglobulin serum or laser nephelometry.

the persistence of antigen in the bloodstream in the presence of antibodies. Glomerulonephritis is therefore characteristically associated with chronic viral and autoimmune diseases (Table 30–3).

Examples of chronic virus diseases that induce immune complex lesions include Aleutian disease of mink and equine infectious anemia. A glomerulonephritis has also been associated with lymphocytic choriomeningitis in mice and was, for a long time, assumed to be due to deposition of virus–antibody complexes in the glomeruli. The development of these lesions, however, is inhibited by anti-interferon serum, and pure interferon itself may also induce a severe glomerulonephritis in mice.

Table 30–3 Human Diseases Associated with Type III Hypersensitivity

Bacterial diseases
Streptococcal
Mycoplasma pneumonia
Leprosy
Syphilis
Viral diseases
Infectious mononucleosis
Hepatitis
Dengue hemorrhagic fever
Parasitic diseases
Malaria
Leishmaniasis
Trypanosomiasis
Schistosomiasis
Onchocerciasis
Tumors
Autoimmune diseases
Rheumatoid arthritis
Systemic lupus erythematosus
Autoimmune thyroiditis
Others
Serum sickness
Hypersensitivity pneumonitis
IgA nephropathy

Immune complex–mediated glomerulonephritis occurs in systemic lupus erythematosus in humans and in the autoimmune disorders of New Zealand Black/New Zealand White (NZB/NZW) hybrid mice. The immune complexes in both these cases consist largely of DNA/anti-DNA complexes. In rheumatoid arthritis, the local inflammatory joint lesion is probably also partially immune complex–mediated. The offending complexes probably consist of normal immunoglobulin complexed to an autoantiglobulin known as rheumatoid factor (Chapter 31).

One of the most common causes of renal failure in humans is a condition known as IgA nephropathy, or Berger's disease. In this condition, the origin of which is unknown, massive IgA and IgE deposits are found in the glomerular mesangia. Affected individuals have increased serum IgA and IgA^+ lymphocytes and depressed suppressor cells for IgA production.

TYPE IV HYPERSENSITIVITY

Type IV hypersensitivity reactions result from a T-cell–mediated response to antigen. Because of the need for T cells to migrate to the site of antigen deposition, the reactions usually take more than 24 hours to develop and are therefore called **delayed hypersensitivity** reactions. The best example of a type IV reaction is the tuberculin reaction. This is an inflammatory response produced in the skin in response to intradermal inoculation of an extract of the tubercle bacillus. It is used in the diagnosis of tuberculosis. The tuberculin reaction is described in detail in Chapter 18. Type IV reactions may also occur in the skin as a result of contact with reactive chemicals. In this case the reaction is directed against chemically modified skin cells. The mechanisms of type IV hypersensitivity are complex but involve T cells, macrophages, mast cells, and basophils (Fig. 30–16).

Although the intradermal tuberculin reaction is artificial in that antigen is administered by injection, a similar host response occurs if living tubercle bacilli lodge in tissues. *Mycobacterium tuberculosis* is resistant to intracellular destruction until a cell-mediated immune response has developed (Chapter 25); dead organisms are slowly removed because they contain large quantities of poorly metabolized waxes. As a result, the hypersensitivity reaction to whole organisms is prolonged, and consequently, macrophages accumulate in large numbers. Many of these macrophages attempt to ingest the bacteria and die in the process, whereas others fuse to form multinucleated giant cells. The lesion that develops around invading tubercle bacilli therefore consists of a mass of dead tissue containing both living and dead bacteria; the lesion is surrounded by a layer of macrophages, which in this location are known as **epithelioid cells.** The entire lesion is known as a tubercle. Persistent tubercles may become relatively well organized and collagen may be laid down, resulting in the formation of a **granuloma.** (Interleukin-1 stimulates collagen production by fibroblasts and hence contributes to this process.) Granuloma formation is a frequent result of local persistent inflammation. Small granulomas form if tuberculin bound to polyacrylamide beads is injected subcutaneously. This inflammation may be of immunological origin, as in tuberculosis or brucellosis in some species, but it may also occur as a result of the presence of other chronic irritants in tissues. Granulomas, for example, may arise in response to the prolonged irritation caused by talc, asbestos particles, or schistosome eggs (Chapter 28).

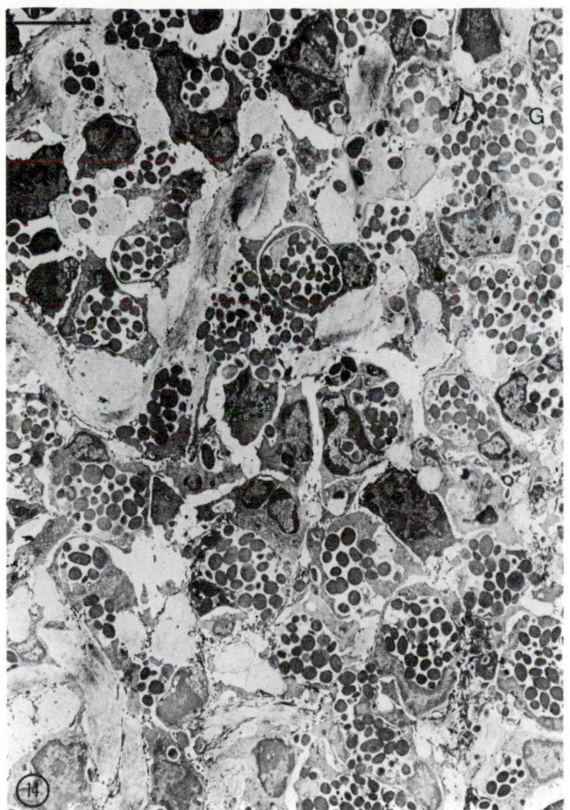

Figure 30–16 A section of guinea pig skin, 24 hours after the attachment of a tick (*Rhipicephalus appendiculatus*), showing a classic example of cutaneous basophil hypersensitivity. This guinea pig had been immunized by prior exposure to these larvae. The skin is infiltrated with many basophils. Some of these cells show degranulation sacs, indicating that they have extruded their granules. Some free granules are found in the extracellular fluid. The bar indicates 50 μm. *(With permission from D. McLaren, M.J. Worms, and P.W. Askenase, Journal of Pathology 139:299, 1983.)*

Allergic Contact Dermatitis

If reactive chemicals are painted onto the skin, they will bind to Langerhans' cells in the dermis through the formation of protein-chemical complexes. The altered Langerhans' cells form an antigenic focus for specifically sensitized lymphocytes. In addition, some Langerhans' cells emigrate from the skin through the afferent lymphatics (Chapter 8) and colonize draining lymph nodes. As a result of antigen presentation by these cells, sensitized lymphocytes migrate into the skin and attempt to destroy and remove the altered cells (Fig. 30–17). This response gives rise to a very irritating form of skin hypersensitivity known as allergic contact dermatitis.

The chemicals that induce allergic contact dermatitis are usually relatively simple; they include such compounds as formaldehyde, picric acid, aniline dyes, plant resins and oils, organophosphates, and salts of metals such as nickel and beryllium (Fig. 30–18). Allergic contact dermatitis can occur on pathologists' fingers as a result of exposure to formaldehyde, on the foot pads and ventral abdomen of dogs on exposure to some carpet dyes, on parts of the body exposed to the oils (urushiol) of the poison ivy plant (*Rhus radicans*), and around the neck of animals as a result of exposure to dichlorvos (2,2-dichlorovinyldimethyl phosphate) in flea collars. Some individuals, instead of developing the more usual type I hypersensitivity to

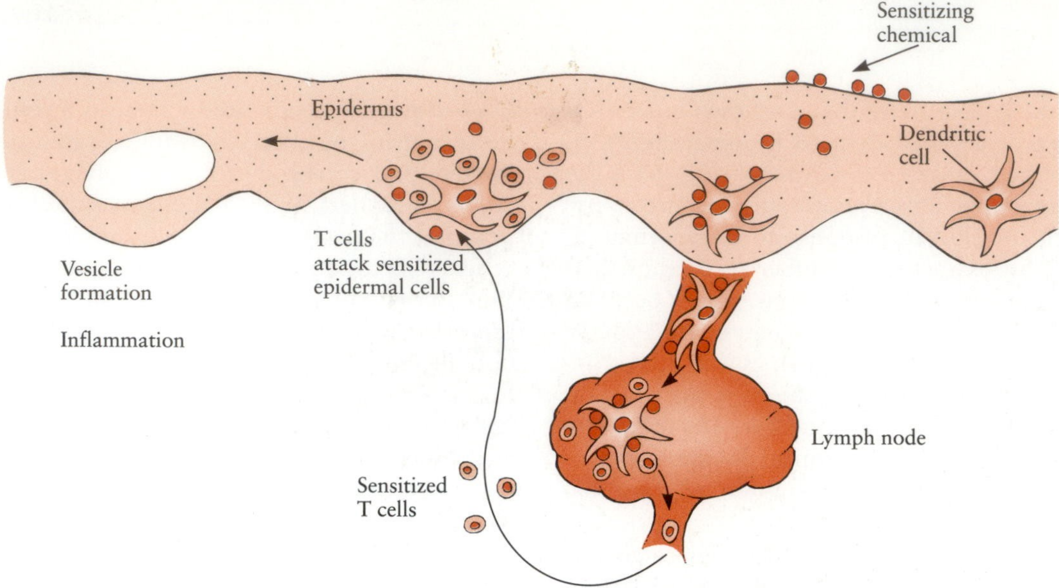

Figure 30–17 The pathogenesis of allergic contact dermatitis. When a sensitizing compound binds to epidermal dendritic cells, it is carried by them to the draining lymph node, where it initiates a T-cell response. Sensitized T cells emigrate from the node and return to the epidermis, where they destroy sensitized epidermal cells.

pollen proteins, suffer from allergic contact dermatitis as a result of a type IV hypersensitivity to pollen resins. It is unusual for allergic contact dermatitis to affect the haired areas of the skin severely unless the allergen is in a liquid form.

The lesions of allergic contact dermatitis generally vary greatly in severity, ranging from a mild erythema to severe inflammation and blister formation (Fig. 30–19). Because of the intense itching, however, self-inflicted damage and infection often mask the true nature of the lesion. In chronic lesions, severe thickening of the skin may result. Histologically, the lesion is marked by a mononuclear cell infiltration and vacuolation of skin cells under attack by cytotoxic T cells (Fig. 30–20).

Diagnosis is made by removal of the suspected antigen and by patch testing. In patch tests, a small area of skin is covered with a patch of tissue or cloth impregnated with the suspected allergen. After 24 to 48 hours, the patch can be removed. A positive reaction is indicated by local redness and blistering. Treatment consists of steroids and antibiotics to control secondary infections.

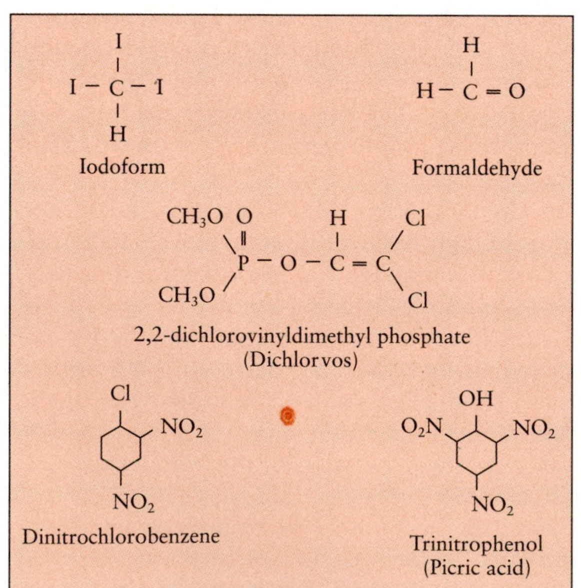

Figure 30–18 Some of the chemicals that can cause allergic contact dermatitis.

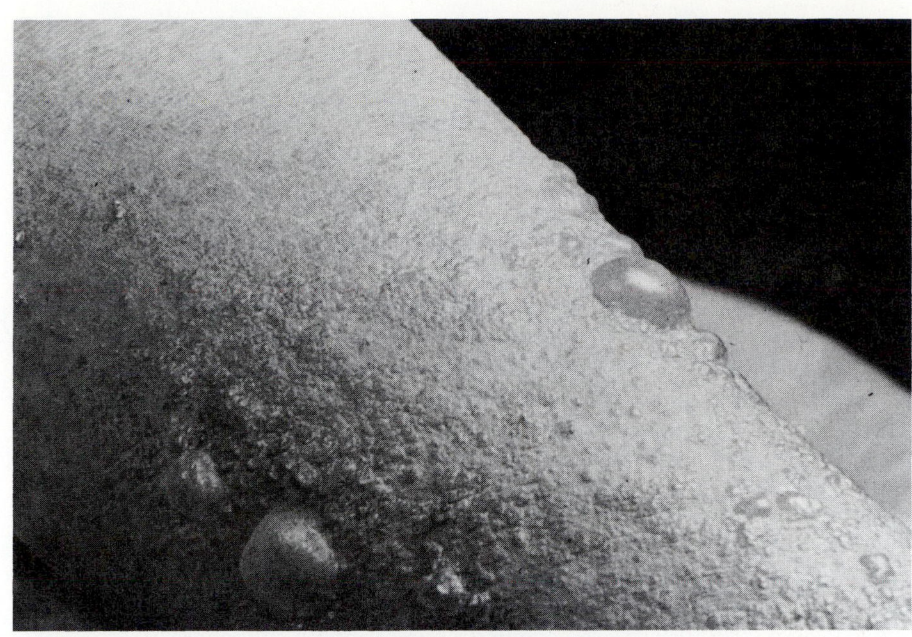

Figure 30–19 A severe case of acute allergic contact dermatitis. Note the vesicles (blisters) formed as a result of destruction of epidermal cells. *(From Parker, C.W., ed. Clinical Immunology. W.B. Saunders, Philadelphia, 1980. With permission.)*

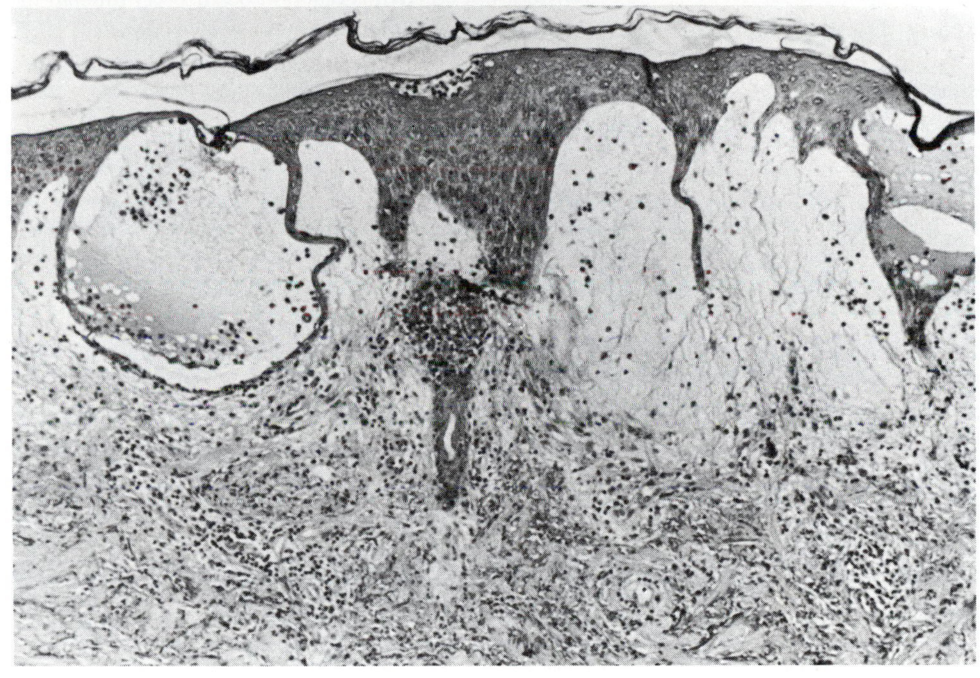

Figure 30–20 A histological section of a severe case of allergic contact dermatitis showing an extensive mononuclear cell infiltration and the development of large fluid-filled vesicles. These are in effect "holes" in the skin formed as a result of destruction of epidermal cells. *(From Parker, C.W., ed. Clinical Immunology. W.B. Saunders, Philadelphia, 1980. With permission.)*

KEY WORDS

ABO system p. 470
Acute serum sickness p. 479
Allergic contact dermatitis p. 483
Antiglobulin test p. 474
Arthus reaction p. 476
Blood groups p. 469
Blood transfusion p. 469
Chronic serum sickness p. 480
Complement p. 476
Glomerulonephritis p. 480
Hemolytic disease of the newborn p. 474
Hypersensitivity pneumonitis p. 478
IgA nephropathy p. 482
Immune complex p. 476
Incompatible transfusion p. 470
Intravascular hemolysis p. 470
Lewis system p. 470
Lysosomal enzymes p. 477
Mesangial cells p. 480
Patch test p. 484
Poison ivy p. 483
Rhesus system p. 472
Rheumatoid arthritis p. 482
Tubercle p. 483
Tuberculin test p. 483
Type II hypersensitivity p. 469
Type III hypersensitivity p. 476
Type IV hypersensitivity p. 482
Vasculitis p. 476

QUESTIONS

1. Which of the following blood group antigens is *not* determined by carbohydrate epitopes?
   a. A
   b. Lewis
   c. O
   d. rhesus
   e. AB
2. Hypersensitivity pneumonitis is a disease resulting from inhaling dust from
   a. old books
   b. pigeon droppings
   c. moldy hay
   d. sugar cane waste
   e. any of the above
3. A group O patient can safely receive
   a. type A blood
   b. type O blood
   c. type B blood
   d. type AB blood
   e. none of the above
4. Serum sickness was originally described in patients who received
   a. large doses of tetanus antitoxin
   b. large doses of penicillin
   c. human immune globulin
   d. antihistamines
   e. vaccine booster shots
5. The Arthus reaction results from local
   a. mast-cell degranulation
   b. cytokine release
   c. IgE production
   d. complement activation
   e. red cell hemolysis
6. In chronic immune complex disease, immune complexes in the glomerulus cause
   a. epithelial cell proliferation
   b. mesangial cell proliferation
   c. fibroblast proliferation
   d. T-cell proliferation
   e. macrophage proliferation
7. Immune complexes cause hypersensitivity by stimulating
   a. IgG production
   b. T cells
   c. neutrophil invasion
   d. eosinophil invasion
   e. basophil invasion
8. Allergic contact dermatitis is diagnosed by
   a. complement fixation test
   b. provocation test
   c. patch test
   d. intrapalpebral test
   e. intradermal skin test
9. Type IV hypersensitivity reactions can be differentiated from type III reactions by
   a. passive transfer with lymphocytes
   b. passive transfer with serum
   c. time course of appearance of skin reaction
   d. morphology of tissue reactions
   e. any of the above
10. Hemolytic disease of newborn infants can be prevented by
    a. allergy shots
    b. exchange transfusion
    c. administration of allergen
    d. antihistamines
    e. administration of anti-Rh antibodies
11. Why doesn't an Arthus reaction occur every time an individual receives a booster shot for vaccination?
12. Under what specific circumstances will an individual develop a hypersensitivity pneumonitis?
13. Could a type III hypersensitivity reaction occur following multiple mosquito bites? What might the signs of such a reaction be?
14. Outline the conditions that must be met for an individual to develop an immune complex–mediated glomerulonephritis.
15. How would you distinguish a skin reaction caused by type III hypersensitivity from one caused by type I hypersensitivity?
16. How might the detection and measurement of immune complexes in the bloodstream be of use to a physician?
17. What treatment would you suggest to a friend who suffered from a severe poison ivy rash?
18. Of what possible use are blood groups?
19. The tuberculin reaction is extremely useful in the diagnosis of tuberculosis. Why has it not been employed extensively for the diagnosis of other diseases?

Answers: 1d, 2e, 3b, 4a, 5d, 6b, 7c, 8c, 9e, 10e

SOURCES OF ADDITIONAL INFORMATION

Agre, P.C., and Cartron, J-P., eds. Protein Blood Group Antigens of the Human Red Cell. Structure, Function and Clinical Significance. Johns Hopkins University Press, Baltimore, 1992.

Askenase, P.W. The role of basophils in health and disease. Res. Staff Phys., 32:33–41, 1986.

Askenase, P.W., and Van Loveren, M. Delayed-type hypersensitivity: Activation of mast cells by antigen-specific T-cell factors initiates the cascade of cellular interactions. Immunol. Today, 4:259–264, 1983.

Callaghan, J.D., et al. Delayed hypersensitivity to mumps antigen in humans. Clin. Immunol. Immunopathol., 26:102–110, 1983.

Fearon, D.T. Complement, C receptors and immune complex disease. Hosp. Pract. [off.], 23:63–72, 1988.

Frigoletto, F.C., Jewett, J.F., and Konugres, A.A., eds. Rh hemolytic disease: New strategy for eradication. G.K. Hall Medical Publishers, Boston, 1982.

Godfrey, M.P., Phillips, M.E., and Askenase, P.W. Histopathology of delayed-onset hypersensitivities in contact-sensitive guinea pigs. Int. Arch. Allergy Appl. Immunol., 70:50–58, 1983.

Holdsworth, S.R. Fc dependence of macrophage accumulation and subsequent injury in experimental glomerulonephritis. J. Immunol., 130:735–739, 1983.

Lens, J.W., et al. A study of cells present in lymph draining from a contact allergic reaction in pigs sensitized to DNFB. Immunology, 49:415–422, 1983.

Mourant, A.E., Kopec, A.C., and Domaniewska-Sobczak, K. Blood Groups and Diseases. Oxford University Press, Oxford, England, 1978.

Oldstone, M.B.A., Tishon, A., and Buchmeier, M.J. Virus-induced immune-complex disease: Genetic control of C1q binding complexes in the circulation of mice persistently infected with lymphocytic choriomeningitis virus. J. Immunol., 130:912–918, 1983.

Race, R.R., and Sanger, R. Blood Groups in Man, 6th ed. Blackwell Scientific Publications, Oxford, England, 1975.

Repo, H., Kostiala, A., and Kosunen, T.V. Cellular hypersensitivity to tuberculin in BCG-revaccinated persons studied by skin reactivity, leukocyte migration inhibition and lymphocyte proliferation. Clin. Exp. Immunol., 39:442–448, 1980.

Toews, G.B., et al. Epidermal Langerhans cell density determines whether contact hypersensitivity or unresponsiveness follows skin painting with DNFB. J. Immunol., 124:445–453, 1980.

Turk, J.L. Delayed Hypersensitivity, 2nd ed. American Elsevier, New York, 1975.

Weissman, G., Smolen, J.E., and Korchak, H.M. Release of inflammatory mediators from stimulated neutrophils. N. Engl. J. Med., 303:28–34, 1980.

Yamamoto, F., Clausen, H., White, T., et al. Molecular genetic basis of the histo-blood group ABO system. Nature, 345:229–323, 1990.

CHAPTER 31

Autoimmunity: Breakdown in Self-Tolerance

Paul Ehrlich *Paul Ehrlich was born in Germany in 1854. A remarkably original and stimulating scientist, Ehrlich made major contributions to many different areas of immunology and medicine. Thus he developed methods for standardizing antigens and antitoxins, he developed the first plausible (and correct) theory to explain the immune response by suggesting that cells had receptors or "side-chains" that were shed as antibodies. He studied the hemolytic properties of blood and coined the term complement. He heads this chapter since he appears to be the first to have considered the problem of autoimmunity and self recognition. He developed the concept that the body would do everything possible to avoid an autoimmune response.*

CHAPTER OUTLINE

CHAPTER CONCEPTS

1. Not all autoimmune responses give rise to disease. For example, aged cells are naturally destroyed by autoantibodies.
2. Autoimmunity is not a rare event. All normal individuals possess lymphocytes able to react to self-antigens. These lymphocytes are normally suppressed by the control mechanisms of the immune system.
3. Autoimmune diseases may result from formation of new epitopes on normal cells, from exposure of hidden molecules, from inadvertent sharing of epitopes between infectious agents and normal molecules, from loss of control of lymphocyte responses, or from some virus infections.
4. The development and severity of autoimmune diseases are determined, in large part, by the genetic background of the affected individual.
5. Autoimmune diseases occur as a result of the tissue destruction and inflammation brought about by each of the four types of hypersensitivity mechanisms.
6. Some autoimmune diseases such as autoimmune thyroiditis, encephalitis, and diabetes mellitus are specific for a single organ.
7. Some autoimmune diseases involve many organs. An example of such a disease is systemic lupus erythematosus.

Autoimmune diseases are the subject of this chapter. The first section points out that autoimmunity is not always bad and that some physiologic functions such as the removal of aged cells may be an autoimmune process. Although the causes of autoimmune diseases are unknown, there is no shortage of ideas as to how they are induced or of the factors that control their development. These are described in some detail. This is followed by a review of how autoimmune processes actually cause disease. These generally fit the classification of the four major types of hypersensitivity. Finally, five selected autoimmune diseases, thyroiditis, multiple sclerosis, myasthenia gravis, systemic lupus erythematosus (SLE), and rheumatoid arthritis, are described in detail to give the reader an idea of the different types of disease that can result from autoimmunity.

Autoimmune diseases affect 5% to 7% of adults in Europe and North America, two thirds of whom are women. They include such important diseases as insulin-dependent diabetes mellitus, psoriasis, multiple sclerosis, and rheumatoid arthritis. More than 40 human diseases are now thought to be autoimmune in origin.

It was once believed that normal, healthy animals were unable to mount any immune response against self-antigens as a result of self-tolerance. It is now clear, however, that animals can produce autoantibodies relatively easily when their immune systems are appropriately stimulated. A small number of B cells, reactive to normal tissue antigens, are always present in lymphoid organs. Stimulation of these cells by mitogens such as bacterial lipopolysaccharide provokes the transient appearance of autoantibodies in serum. These autoantibodies are directed against common autoantigens, for example, DNA, IgG, phospholipids, red blood cells, and lymphocytes. These antibodies can react with normal tissues but usually have no adverse effects. On occasion, the regulation of these autoreactive cells may break down. When this happens, clones of lymphocytes may generate high levels of autoantibodies or autoreactive T cells.

PHYSIOLOGICAL AUTOIMMUNITY

It has been generally assumed that all autoimmune responses are bad and cause disease. This is untrue. Some autoimmune responses have important physiological functions and are therefore essential for the normal functioning of the body. These include the recognition of self-idiotypes and the recognition of senescent cells.

Recognition of Self-Idiotypes

The antigen-binding sites (idiotypes) on immunoglobulin molecules can function as epitopes and provoke the production of autoantibodies (anti-idiotypes) (Chapter 22). These newly formed anti-idiotypes also possess their own characteristic epitopes and so provoke anti-anti-idiotypes. This process can, in theory, continue indefinitely. However, since each anti-idiotype response tends to exert a negative feedback on the preceding response, the net effect is to effectively terminate antibody responses. Under some circumstances, anti-idiotype antibodies may also stimulate antibody responses. Thus the idiotype–anti-idiotype network may play a dynamic role in regulating the level of the antibody response.

Removal of Aged Cells

Red blood cells are removed from the bloodstream once they reach the end of their life span. This process is accomplished by autoantibodies. As red blood cells age, the anion transporter glycoprotein called band 3 is cleaved and a new epitope is exposed. This new epitope is recognized by a naturally occurring IgG autoantibody. Although present in low concentrations and having a low affinity for antigen, the anti–band 3 antibody has a very high affinity for C3. As a result, it is a very effective opsonin. These antibodies and complement attach to old red blood cells and promote their phagocytosis by macrophages in the spleen. Band 3 protein is also found on platelets, lymphocytes, neutrophils, hepatocytes, and kidney cells. The cleavage of band 3 in aged cells and the subsequent removal of those cells by antibodies may be a mechanism for the elimination of old cells of many different types.

INDUCTION OF AUTOIMMUNITY

The precise mechanisms of breakdown of self-tolerance are not known for any autoimmune disease. Nevertheless, several key events are likely to be involved: induction of autoimmune disease is multifactorial and polygenic (Fig. 31–1). In other words, many environmental factors and genes determine autoimmune disease susceptibility. In addition, many autoimmune diseases show spontaneous exacerbations and remissions, suggesting that there is an unstable relationship between positive and negative regulatory factors such as stimulatory and suppressive cytokines.

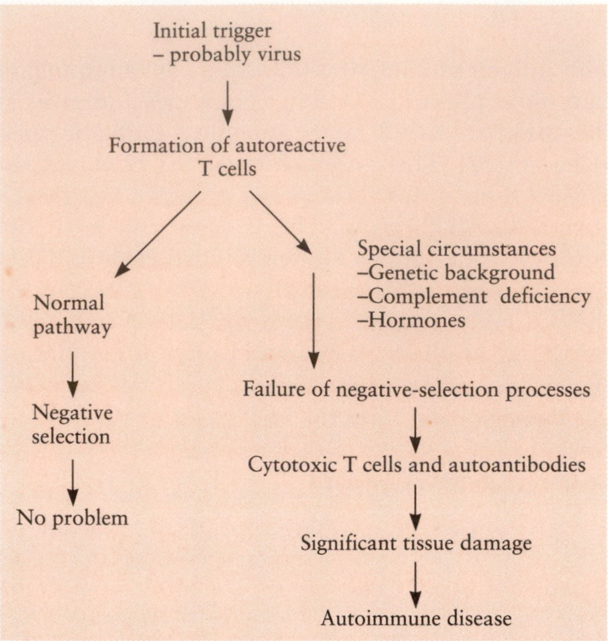

Figure 31–1 Some of the factors that influence the development of an autoimmune disease.

Exposure of Hidden Antigens

Some antigens exist within the body in locations not normally visited by circulating lymphocytes. For example, antigens may be hidden in the central nervous system and the testes, where they are not usually visited by lymphocytes. As a result, lymphocytes may not be completely tolerant to antigens in these organs. If the brain or testes are injured, then the resulting leakage of blood vessels may permit antigens released by damaged cells to reach the bloodstream, encounter antigen-sensitive cells, and stimulate an immune response. Similar considerations apply to antigens that are normally found only inside cells and to which tolerance may not be established. For example, after a heart attack (myocardial infarction), autoantibodies may be produced against the intracellular components of cardiac muscle cells such as mitochondria. In diseases such as trypanosomiasis or tuberculosis in which widespread tissue damage occurs, autoantibodies to a wide range of tissue antigens may be detected in serum at low titers.

Formation of New Epitopes

The formation of autoantibodies may be provoked by the development of new epitopes on normal proteins. An example of this type of antigen is the band 3 antigen on old red blood cells described previously. Two other examples of autoantibodies generated in this fashion are **rheumatoid factors** and **immunoconglutinins.**

Rheumatoid factors are antibodies directed against epitopes on other immunoglobulin molecules. When an immunoglobulin binds to an antigen, the Fab regions of the molecule are stabilized in such a way that new epitopes are exposed on the Fc region. These new epitopes stimulate rheumatoid factor formation. Rheumatoid factors develop in diseases in which large quantities of immune complexes are generated, such as rheumatoid arthritis and SLE.

Immunoconglutinins (abbreviated to IK after the German spelling) are antibodies directed against epitopes on the activated complement components C2, C4, and C3. The most important of these is the IK directed against C3. The epitopes that stimulate IK formation are sites on complement components that are newly revealed by complement activation. Their physiological role is unclear, but they may enhance the opsonizing function of C3.

Minor structural changes in normal body components may be generated artificially and used to induce autoantibodies. For example, chemically modified thyroglobulin can be used experimentally to stimulate the production of autoantibodies against normal thyroglobulin. It is also possible to make many normal tissues antigenic by injecting them together with Freund's complete adjuvant.

Cross-Reactivity with Microorganisms

Molecular mimicry is a term used to describe the sharing of epitopes between an infectious agent or parasite and its host. When this occurs, an immune response directed against the invader may react with normal self-antigens and cause disease (Fig. 31–2). The shared epitopes would have to be sufficiently similar for cross-reactivity to occur but sufficiently different for tolerance to be broken. Once initiated, the infectious agent could be removed while the autoimmune damage continued—a sort of "hit-and-run" process. Many examples of molecular mimicry are now recognized. One excellent example is seen in individuals infected with the protozoan parasite *Trypanosoma cruzi,* the cause of Chagas' disease. This organism possesses antigens that cross-react with mammalian neurons and cardiac muscle. In infected individuals these antigens provoke autoantibody formation, and the autoantibodies provoke nervous disorders and heart disease.

It has been suggested that the heart lesions that develop in rheumatic fever in children are a result of the production of antibodies to the cell wall M protein of group A streptococci, which cross-react with cardiac

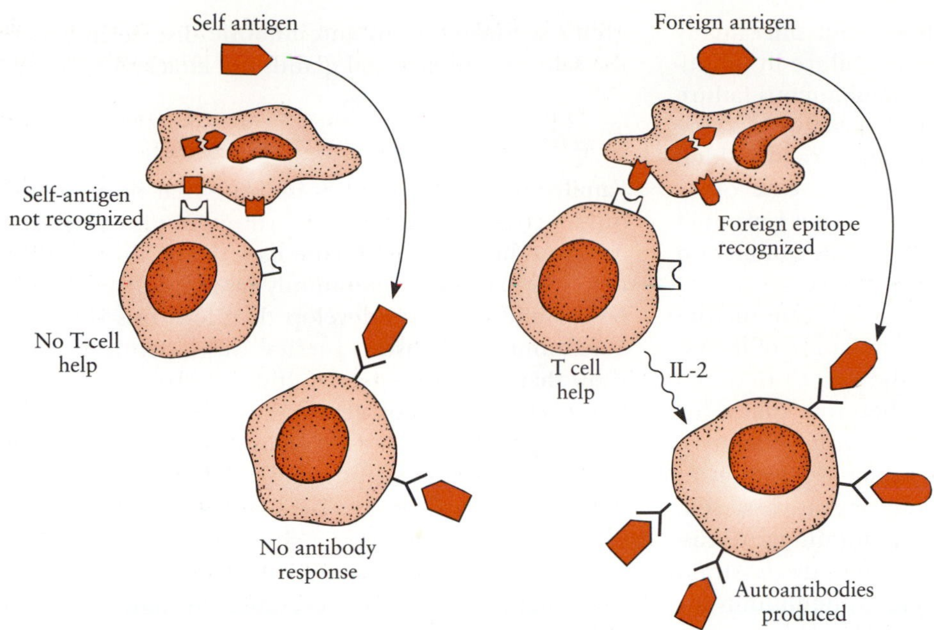

Figure 31–2 The way in which molecular mimicry antigens may initiate an autoimmune response even though the body is normally tolerant to self-antigens.

myosin. Only certain strains of group A streptococci induce antimyocardial antibodies. Other strains of these streptococci induce acute glomerulonephritis in children as a result of the production of antibodies cross-reacting with glomerular basement membranes. Epstein-Barr virus DNA polymerase cross-reacts with myelin basic protein and may be involved in the induction of multiple sclerosis. Poliovirus capsid protein VP2 cross-reacts with the acetylcholine receptor in motor end plates in muscles and may be associated with the development of myasthenia gravis.

Autoantibodies to some microbial **heat shock proteins** seem to be associated with autoimmune arthritis in humans and rats. Killed *Mycobacterium tuberculosis* in Freund's complete adjuvant can cause arthritis in rats, and T cells from these animals can adoptively transfer arthritis to normal syngeneic recipients. These T cells are specific for HSP 60, a mycobacterial heat shock protein (Chapter 25). Because heat shock proteins are highly conserved and because a clone of T cells from a human rheumatoid arthritis patient was also specific for HSP 60, it has been suggested that HSP 60 may be the ''key'' autoantigen in rheumatoid arthritis.

Ankylosing spondylitis is an autoimmune inflammatory disease that affects the sacroiliac joints, the spine, and peripheral joints. Patients also frequently develop an acute anterior uveitis (inflammation of the iris and neighboring structures in the eye). More than 90% of Caucasian patients with ankylosing spondylitis carry HLA-B27, whereas in the normal population, the frequency of this allele is less than 10%. The disease may result from molecular mimicry between HLA-B27 and bacterial antigens because antibodies to HLA-B27 cross-react with *Klebsiella pneumoniae. K. pneumoniae* is found more frequently than normal in the intestine of patients with active ankylosing spondylitis and in patients with uveitis. Patients with active disease have elevated levels of IgA antibodies to *Klebsiella* in their sera. Thus it may be that antibodies to *Klebsiella* may participate in the development of the disease.

Loss of Control of Lymphocyte Responses

Most autoimmune diseases probably occur as a result of the development of T or B cells that had previously been suppressed by the normal control mechanisms of the body. For example, when mice are injected with rat red blood cells, they not only make antibodies to the rat cells, but they also develop a self-limiting and transient autoimmune response to their own red blood cells. These autoantibodies can be detected by means of an antiglobulin test. This anti–red blood cell autoimmune response is normally rapidly controlled by suppressor cells and lasts for only a few days. If, however, suppressor cell activity in mice is depressed, as occurs in New Zealand Black (NZB) mice for example, then these autoantibodies cause red blood cell destruction and a severe anemia may result.

It is not uncommon to find autoimmune disease associated with lymphoid tumors. For example, myasthenia gravis may be associated with the presence of a thymocyte tumor. In humans, there is a fourfold increase in the incidence of rheumatoid diseases in patients with malignant lymphoid tumors, and there is evidence for a similar association in animals. The reasons for this are poorly understood, but since many lymphoid tumors may result from a failure in immunological control mechanisms, a simultaneous failure in self-tolerance may also occur. Alternatively, some tumors may represent the development of a forbidden clone of cells producing autoantibodies. One other possibility that should be considered is that lymphoid tumors may arise as a result of the chronic stimulation of the immune system by autoantigens.

The fas protein (CD95) plays a key role in lymphocyte apoptosis (Chapter 18). Stimulation of the fas protein on a cell surface causes the cell to undergo apoptosis. It has been suggested therefore that a defect in the fas protein may result in the development of autoimmunity by preventing negative selection of self-reactive T and B cells. It would permit these self-reactive cells to survive and cause autoimmune disease. This has been well demonstrated in the *lpr* strain of mice. These mice develop severe autoimmunity accompanied by lymphoproliferation. The nature of the *lpr* defect has been identified as a mutation within the *fas* gene that alters the structure of the intracellular component of the fas product. The mutation results from the insertion of a DNA sequence derived from an endogenous retrovirus into the *fas* gene. Some investigators have suggested that a mutation in *fas* may contribute to the pathogenesis of SLE.

Viruses as Inducers of Autoimmunity

A growing body of evidence has linked many autoimmune diseases to virus infections, and it has been suggested that viruses, especially those that infect lymphoid tissues, may be capable of interfering with immunological control mechanisms and so permit autoimmune disease to occur. Thus in NZB mice, persistent infection with a type C retrovirus leads to the development of autoantibodies against nucleic acids and red blood cells. SLE of dogs and humans is a similar condition in which the presence of autoantibodies to many different organs is possibly associated with either a type C retrovirus or paramyxovirus infection. Mice infected with certain reoviruses develop an autoimmune polyendocrine disease characterized by diabetes mellitus and retarded growth. These reovirus-infected mice develop antibodies against normal pituitary, pancreas, gastric mucosa, nuclei, glucagon, growth hormone, and insulin. It has been suggested that the onset of type I diabetes in children is associated with infection by a coxsackievirus. By use of the polymerase chain reaction, it has been possible to find very small numbers of the Epstein-Barr virus genome in salivary gland tissue from patients with Sjögren's syndrome. Other investigators have reported the presence of a retrovirus similar to HIV in this disease. Sjögren's syndrome is an autoimmune disease in which the salivary and lacrimal glands are attacked.

Genetic Factors and Autoimmune Disease

Family and genetic studies indicate that many genes influence the development of autoimmunity (Fig. 31–3). Thus the relatives of patients with Hashimoto's disease (see page 495) commonly have antithyroid antibodies and may also develop thyrotoxicosis or other autoimmune diseases. Studies with identical twins show that when one gets multiple sclerosis, the other has a 30% chance of developing the disease as well. Several genes may play a role in autoimmunity. They include the genes for histocompatibility molecules, for peptide transporters, for T-cell antigen receptors, for complement components, and for immunoglobulins.

MHC class II molecules regulate an individual's ability to mount immune responses through their ability to bind processed antigen peptides. As a result, they determine resistance or susceptibility to many diseases. Very few of these diseases are infectious because infectious diseases tend to affect young animals before they reach reproductive age. Therefore, there is rapid selection against genes that predispose to susceptibility to infectious agents. The MHC class II genes that animals now possess have been selected for a rapid response to most pathogens. Autoimmune diseases, in contrast, tend to affect older, post reproductive animals. As a result, they generally do not offer a significant selective disadvantage. Studies of animal populations, especially humans, have shown that almost all autoimmune diseases are linked to possession of certain MHC class II molecules (Table 31–1).

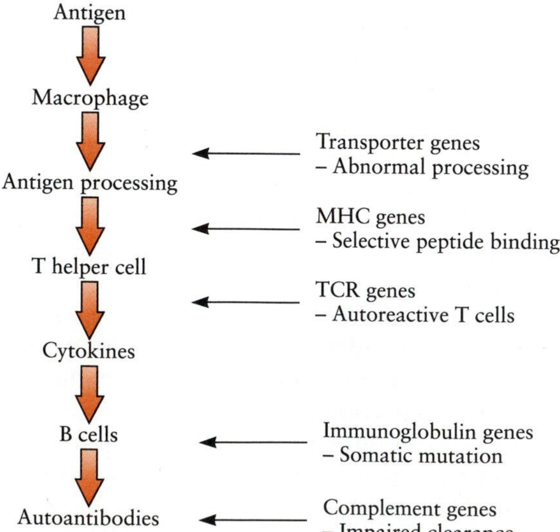

Figure 31–3 Some of the possible ways in which genes may directly influence the development of autoimmune diseases.

Table 31–1 Association Between HLA Alleles and Disease

Autoimmune Disease	Allele	Relative Risk*
Ankylosing spondylitis	B27	90
Reiter's disease	B27	36
Pemphigus vulgaris	DR4	24
Goodpasture's syndrome	DR2	15
Dermatitis herpetiformis	DR3	15
Chronic active hepatitis	DR3	14
Celiac disease	DR3	12
Membranous glomerulopathy	DR3	12
Sjögren's syndrome	DR3	10
Psoriasis	B17	6
Rheumatoid arthritis	DR4	6
Myasthenia gravis	B8	3
Multiple sclerosis	DR2	5
Pernicious anemia	DR5	5
Type I diabetes	DR3	5
	DR4	5
	DR3/DR4	20
	DR3/DQw8	100

*The number of times greater than the risk to an individual who does not possess this allele.

MHC-linked autoimmune diseases fall into two groups. One group of diseases is linked to MHC class I molecules. This includes ankylosing spondylitis linked to HLA-B27 and psoriasis, which is linked to HLA-B13, B16, and B17. The second group of associations are between MHC class II molecules and many different autoimmune diseases; for example, HLA-DR4 is associated with juvenile rheumatoid arthritis, and HLA-DR3 and HLA-DR4 are associated with type I diabetes (95% of Caucasians with type I diabetes express DR3 and/or DR4 as compared with 40% of the general Caucasian population). Individuals with HLA-DR4 are about six times more likely than the general population to develop rheumatoid arthritis, and those with HLA-DR53 have a tenfold increased susceptibility. With the notable exception of ankylosing spondylitis and HLA-B27, the possession of certain MHC haplotypes is not an absolute determinant of autoimmune disease susceptibility. Thus individuals with both HLA-DR3 and DR4 are 20 times more likely to develop juvenile diabetes mellitus than those who lack these genes. However, 5% of individuals with this disease possess neither gene.

Instead of being closely linked to a single MHC locus, some autoimmune diseases are associated with certain MHC combinations. For example, the combination of HLA-A1, B8, and DR3 is associated with increased risk of developing diabetes, myasthenia gravis, and SLE. This connection may simply mean that the gene controlling the development of these diseases is linked to the MHC. Thus the observed association of certain MHC class II molecules to SLE may be related to the inheritance of C2 or C4 null alleles in the class III region in **linkage disequilibrium** with the class I or II genes.

Inbred lines of animals have been developed that spontaneously develop autoimmune disease. Chickens of the OS strain develop an autoimmune thyroiditis. An inbred line of dogs has been used for studies on SLE. Inbred NZB mice spontaneously develop a syndrome that bears a striking resemblance to SLE (page 497). NZB mice develop an immune complex type of glomerulonephritis. They become hypergammaglobulinemic and hypocomplementemic, and they develop an autoimmune hemolytic anemia. Some mice also develop lymphoid tumors. A large variety of autoantibodies develop in these mice directed against several nucleic acid antigens, red blood cells, and T cells, while their B cells are polyclonally activated.

Genes may also confer resistance to autoimmune diseases. Thus resistance to juvenile diabetes correlates with the presence of the amino acid aspartate at position 57 of the HLA-DQβ chain. This character is common among Japanese people, who have a much lower prevalence of juvenile diabetes mellitus than white Americans. Other amino acids at this position such as serine, alanine, and valine are associated with increased susceptibility to the disease.

Hormonal Factors

Autoimmune diseases are significantly more common and more severe in women than in men. For example, 90% or more of patients with SLE are women. One possible explanation for this is that IFN-γ production is significantly enhanced by estrogens since these hormones stimulate enhancer elements for the interferon-γ genes. IFN-γ plays a major role in the pathogenesis of tissue destruction in these diseases by enhancing local MHC expression and activating macrophages.

Clinical Associations Between Autoimmune Diseases

It is not uncommon for more than one autoimmune disease to occur simultaneously in the same individual. Autoimmune thyroiditis, for example, is commonly associated with autoimmunity to gastric parietal cells (pernicious anemia) and occasionally is linked with autoimmunity to the adrenal cortex (Addison's disease). This clinical association is emphasized by a significant serological overlap between these conditions. Thus up to a third of patients with autoimmune thyroiditis may also have autoantibodies to gastric parietal cells, and half of the patients with pernicious anemia have antithyroid antibodies. The simultaneous occurrence of multiple autoimmune disorders is even more marked in SLE, in which hemolytic anemia, thrombocytopenia, and rheumatoid arthritis (see page 499) are all reported.

MECHANISMS OF TISSUE DAMAGE IN AUTOIMMUNITY

Type I Hypersensitivity

Milk allergy in cattle is an autoimmune disease in which milk α casein, normally found only in the mammary gland, gains access to the general circulation and stimulates IgE autoantibody production. This happens when the cow is not milked and intramammary pressure forces milk proteins into the circulation. Thus the milk α casein may cause acute anaphylaxis in unmilked cows (Chapter 29). Although antibodies to milk proteins are also found in human serum after rapid weaning, type I hypersensitivity is not a usual response.

Type II Hypersensitivity

Autoantibodies directed against cell surface antigens may cause lysis with the assistance of either complement or cytotoxic cells. If the autoantibodies are directed against red blood cells, then autoimmune hemolytic anemia may result; if directed against platelets, thrombocytopenia will occur; and if directed against thyroid cells, thyroiditis will result. In one form of this disease in humans, autoantibodies directed against thyroid-stimulating hormone (TSH) receptors in the thyroid stimulate thyroid activity rather than mediate thyroid destruction. Receptors are a common target of autoimmune attack. In addition to the TSH receptor, autoantibodies attack the acetylcholine receptor in myasthenia gravis and the insulin receptor in some forms of diabetes. Autoantibodies to β adrenoceptors (Chapter 29) have been detected in some patients with asthma. By blocking β receptors, these antibodies make the airways highly irritable, and affected individuals are prone to severe asthmatic attacks. It is also interesting to note that some individuals suffering from depression may have autoantibodies to β endorphin receptors in the brain.

Type III Hypersensitivity

Autoantibodies form immune complexes when bound to antigen, and these complexes may participate in type III hypersensitivity reactions. This occurs, for example, in SLE, a disease in which a wide variety of autoantibodies are produced, the most significant of which are directed against nucleic acids. DNA–antibody complexes are formed in affected individuals and are deposited in glomeruli to provoke the development of a glomerulonephritis (Chapter 30). Similarly, in rheumatoid arthritis, immune complexes formed between rheumatoid factor (the antibody) and IgG (the antigen) are deposited in joints and, by activating complement, contribute to the local inflammatory response.

Type IV Hypersensitivity

Many lesions in autoimmune diseases are heavily infiltrated with mononuclear cells, and autosensitized T lymphocytes probably contribute to the pathogenesis of disease of this type. Examples of such diseases include autoimmune thyroiditis, experimental allergic encephalitis, and multiple sclerosis, in which cytotoxic T cells can cause demyelination, and ulcerative colitis, in which cytotoxic T cells can destroy colon cells. Some cases of diabetes mellitus may also be due to a cell-mediated autoimmune response. The diseased pancreatic islets may be infiltrated by lymphocytes. In addition, lymphocytes from diabetics may be cytotoxic for pancreatic islet cells in vitro. The nonobese diabetic (NOD) mouse is a good spontaneous model of the human disease since 50% to 80% of females develop insulin-dependent diabetes mellitus by 6 months of age. The disease can be adoptively transferred to syngeneic normal mice by a mixture of $CD4^+$ and $CD8^+$ T cells from NOD mice, and depletion of T cells prevents its development. Macrophages are also involved in the development of this disease since macrophage depletion or blockade of macrophage CD11a/CD18

CLARIFICATION

Autoimmunity and Left-Handedness

In humans, left-handed individuals are more likely to develop autoimmune diseases than right-handed individuals. Thus in one study, 11% of left-handed individuals and 4% of right-handed individuals suffered from autoimmune diseases—mainly autoimmune thyroiditis, ulcerative colitis, and celiac disease.

The reasons for this are obscure. It has been suggested, however, that left-handedness and immune malfunction may both result from abnormal endocrine function in fetal life.

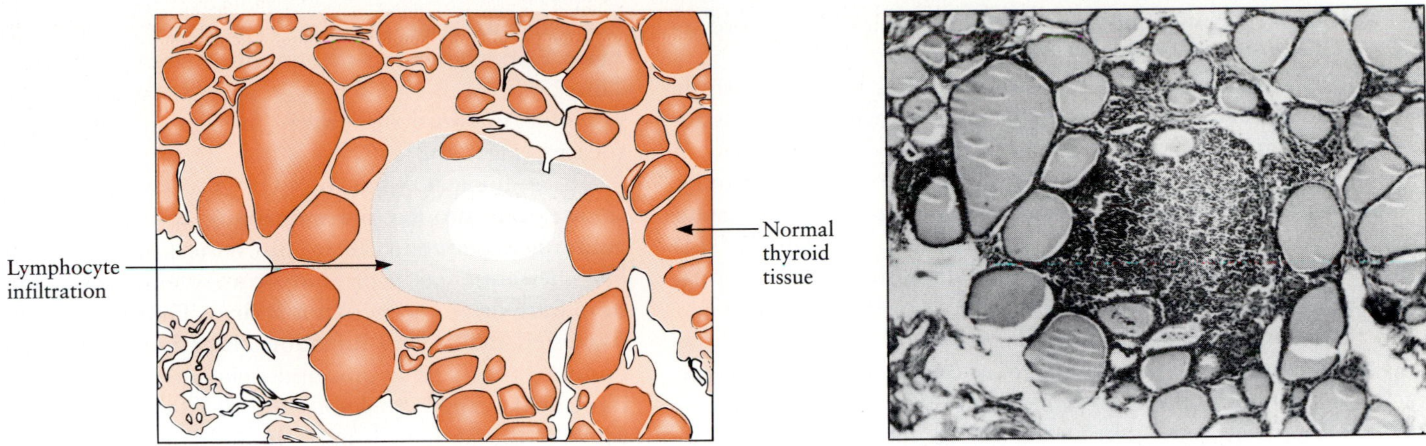

Figure 31–4 Lymphocytic infiltration in the thyroid of a dog with autoimmune thyroiditis (original magnification ×100). *(Courtesy of Dr. B.N. Wilkie.)*

(CR3) can inhibit development of diabetes in NOD mice. T-cell clones can transfer an autoimmune arthritis, and clones from mice immunized with myelin basic protein can transfer autoimmune encephalitis.

SOME SELECTED AUTOIMMUNE DISEASES

Autoimmune diseases can affect almost any organ in the body. Rather than present the reader with a catalog of these diseases, we have selected five to demonstrate the types of autoimmune diseases that occur in humans and animals.

Autoimmune Thyroiditis

Humans, dogs, and chickens can suffer from autoimmune thyroiditis (Fig. 31–4). In humans, two major forms occur (Fig. 31–5). In Hashimoto's thyroiditis, the autoantibodies formed against thyroid peroxidase can activate complement. As a result, these antibodies, together with complement, can destroy thyroid cells. The cytotoxicity probably does not account entirely for the lesions of Hashimoto's thyroiditis, since thyroid tissue fragments can grow well in the presence of cytotoxic antibodies and a T-cell–mediated process may also be involved.

Thyrotoxicosis (Graves' disease), in contrast to Hashimoto's thyroiditis, is associated with excessive thyroid activity. The autoantibodies formed in thyrotoxicosis bind to the same receptor as TSH and instead of destroying cells enhance thyroid activity. The children of women with thyrotoxicosis may be born with thyroid hyperactivity due to the passage of thyroid-stimulating antibodies across the placenta. Some patients may develop both cytotoxic antibodies and thyroid-stimulating antibodies.

Multiple Sclerosis

Multiple sclerosis is an autoimmune disease of the central nervous system. In affected individuals, the myelin sheaths that surround the nerve axons in the white matter of the brain are progressively destroyed by immunological attack. As the myelin breaks down, nerve transmission is disrupted. Patients gradually develop neurological symptoms, including changes in vision and a loss of their ability to control muscle function. The disease does not progress steadily but is characterized by unpredictable exacerbations and remissions. Thus periods of clinical disease are interrupted by variably long periods of remission. Unfortunately,

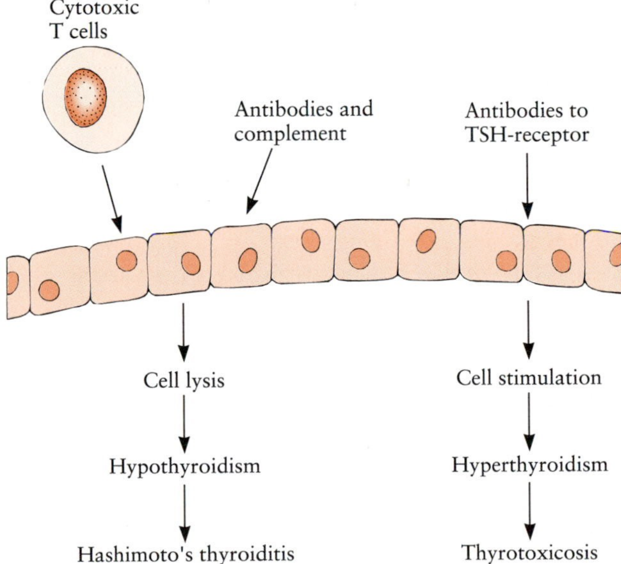

Figure 31–5 The three ways in which autoimmune reactions may attack the thyroid gland to produce very dissimilar clinical results. (For details see text.)

each clinical episode leaves the patient in a more severely affected state than previously.

Although multiple sclerosis has been extensively studied for many years, convincing evidence for its autoimmune nature has been obtained only recently. It has long been possible to induce an autoimmune brain disease in laboratory rodents by immunizing them with myelin in Freund's complete adjuvant. This disease is called experimental autoimmune encephalitis (EAE). The antigen that triggers EAE has been identified as myelin basic protein. Rodent EAE, however, is an unremitting disease, so its clinical course bears little resemblance to human multiple sclerosis.

Studies on brain tissue from multiple sclerosis patients showed, however, that some B and many T cells are present in the brain lesions. The B cells originate from a limited number of clones, and some of the antibodies they produce are directed against oligodendroglia, the cells that make myelin. Thus these antibodies together with complement can clearly contribute to the development of brain lesions. Similarly, the T cells in the lesions react against a peptide derived from myelin basic protein. This same peptide can cause EAE when used to immunize laboratory mice. Thus both the B and T cells in these lesions have the potential to cause significant damage to the myelin sheaths (Fig. 31–6).

It is unclear what triggers this antimyelin response, but two factors contribute to it. First, the amino acid sequence in myelin basic protein resembles the sequence in proteins from adenovirus 2, Epstein-Barr virus, and hepatitis B virus. Thus it has been suggested that a viral infection may trigger the autoimmune response in multiple sclerosis. Whether an individual actually develops the disease depends on whether the antigen is appropriately presented. This is controlled by the individual's MHC molecules. Thus individuals with HLA-DR2 are five times more likely to develop multiple sclerosis than those who lack this receptor. It is also clear that susceptibility to multiple sclerosis is linked to the myelin basic protein gene.

Once a specific T cell recognizes the appropriately presented antigen, it not only exerts a direct cytotoxic effect on the nerve axon but also secretes inflamma-

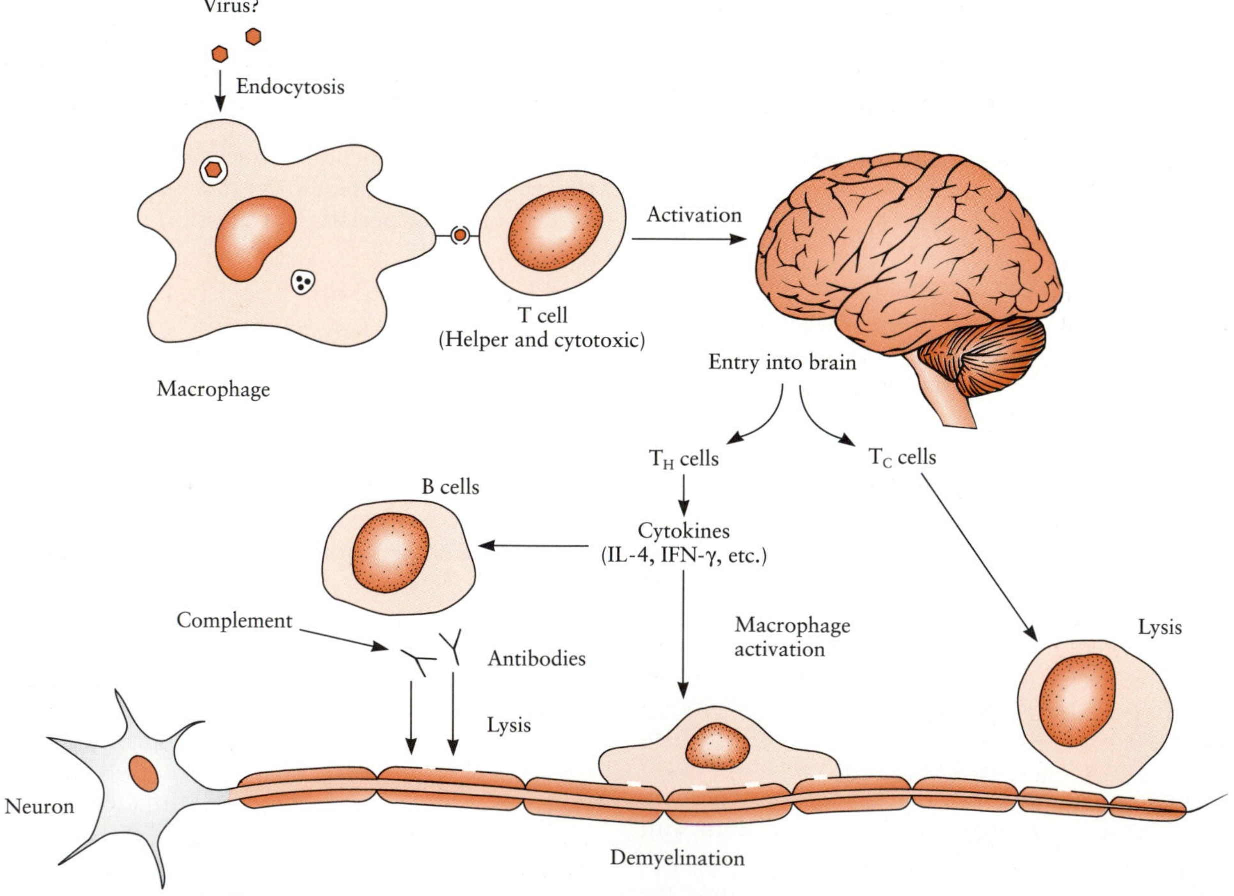

Figure 31–6 The probable pathogenesis of multiple sclerosis. Infection with a virus of unknown type triggers an immune response against the myelin in an axon sheath. As a result, the myelin is stripped from the axon.

tory cytokines, including lymphotoxin, TNF-α, and IFN-γ. These can activate macrophages, especially microglia, which then destroy the myelin on the nerve axons. In addition, IFN-γ enhances local MHC expression and so promotes local antigen presentation. For this reason, administration of IFN-γ may accelerate the course of the disease.

Interferon-β, in contrast, can inhibit MHC expression on nerve cells. As a result, administration of IFN-β reduces the number of disease episodes in multiple sclerosis patients and reduces the size of the affected area in the brain. IFN-β is now licensed for the treatment of patients with this disease. More conventional immunosuppressive agents have not been very useful.

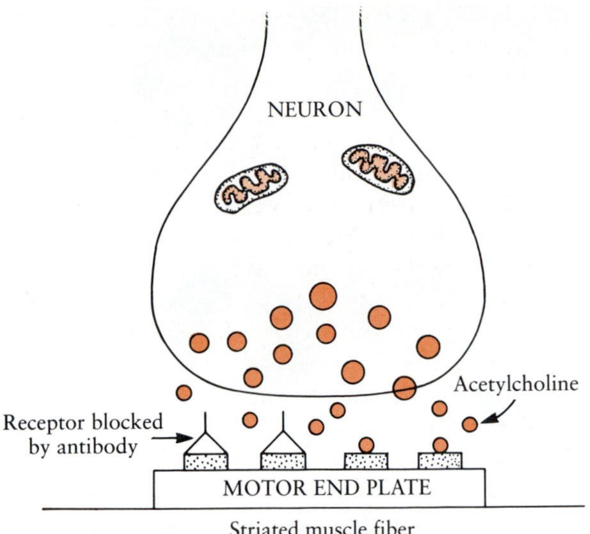

Figure 31–7 The pathogenesis of myasthenia gravis. Blockage of receptors on the motor end plate by autoantibodies reduces the number of receptors that are available to be stimulated by acetylcholine. Anticholinesterase drugs prevent acetylcholine destruction and so ensure that the remaining receptors are stimulated more effectively.

Myasthenia Gravis

Myasthenia gravis, which afflicts humans, dogs, and cats, is a disease of skeletal muscle characterized by abnormal fatigue and extreme weakness after relatively mild exercise. An individual with myasthenia gravis, for example, will collapse exhausted after walking for only a few yards. The muscles holding up the eyelids tire, making it difficult for affected individuals to keep their eyes open. The muscles of the esophagus relax, and swallowing is difficult. Respiratory difficulty may also occur. Myasthenia gravis results from destruction of acetylcholine receptors on the motor end plates of striated muscle. This destruction is caused by IgG autoantibodies against the acetylcholine receptor protein (Fig. 31–7). These autoantibodies may block binding sites. They can activate complement, and they can accelerate endocytosis and degradation of receptors. As a result, the number of acetylcholine receptors is reduced, and the end plate potentials induced at the neuromuscular junctions fall below threshold levels and therefore fail to trigger muscle contraction. Repeating the stimulus is ineffective, since all available receptors are saturated with acetylcholine.

In nearly 80% of patients with myasthenia gravis, the thymus is abnormal. These abnormalities include hyperplasia of the medulla, the presence of activated B cells, germinal center formation, and the presence of thymic tumors (thymomas) in 15% of patients. Since normal thymus tissue contains myoid cells (striated muscle cells that possess acetylcholine receptors of their own), the thymic changes may result from an immunological attack on myoid cells. It has been suggested that T cells within the thymus may respond to myoid cell antigens on thymic antigen-presenting cells. These T cells may then leave the thymus and interact with B cells specific for the acetylcholine receptor, leading to autoantibody formation. This hypothesis is supported by the observation that total thymectomy usually, but not always, leads to significant remission of myasthenia gravis. More than 90% of patients with autoimmune myasthenia gravis respond to treatment with inhibitors of acetylcholinesterase.

Systemic Lupus Erythematosus

SLE is a generalized immunological disorder that has been described in humans, mice, dogs, and cats. About 90% of cases occur in women. It results from a loss of control of the B-cell system. B cells also express increased amounts of IL-6R (CD126) while at the same time secrete IL-6. They are therefore activated by their own IL-6 in an autocrine process. As a result of this loss of control, affected individuals make autoantibodies against a range of antigens found in normal organs and tissues. These multiple autoantibodies in turn give rise to a wide spectrum of pathological lesions and clinical manifestations.

One consistent feature of SLE is the development of autoantibodies against nucleic acids. Several antinuclear antibody systems have been described in the disease, the most important of which are antibodies to DNA. These autoantibodies cause tissue damage by several mechanisms. They can combine with free DNA to form DNA–anti-DNA immune complexes, which may then be deposited in glomeruli (Chapter 30), causing a membranous glomerulonephritis. The complexes may also be deposited in the walls of arterioles, where they result in tissue destruction and fibrosis, or in joint synovia, where they provoke arthritis. Antinuclear antibodies also bind to the nuclei of degen-

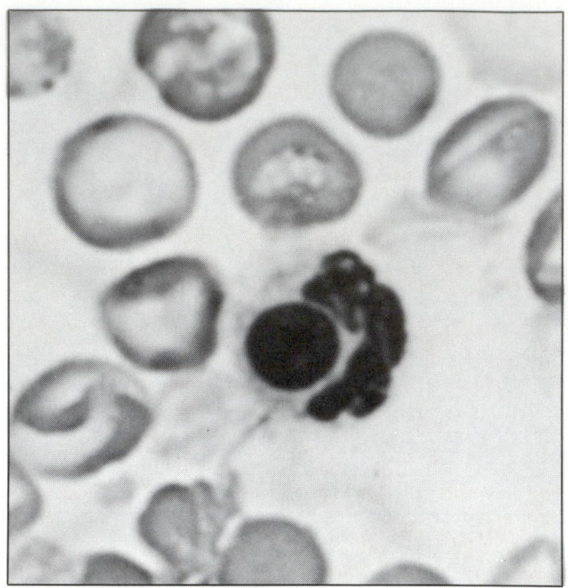

Figure 31–8 The cell in the center of this photograph is an LE cell from a case of systemic lupus erythematosus. The round basophilic body within the LE cell is probably a phagocytosed nucleus. *(Courtesy of Dr. B.N. Wilkie.)*

erating cells. These opsonized nuclei may be phagocytosed, forming structures known as lupus erythematosus (LE) cells (Fig. 31–8). LE cells are found mainly in bone marrow and less commonly in blood.

Although antibodies to nucleic acids are characteristic of SLE, many other autoantibodies are also produced. Autoantibodies to red blood cells, for example, cause an antiglobulin-positive hemolytic anemia. Antibodies to platelets give rise to an immunologically mediated thrombocytopenia. Antilymphocyte antibodies may be present, and it is suggested that they may selectively destroy suppressor cells, thus enhancing immune reactivity. About one third of SLE patients have antibodies to CD45, and these too may alter immune function. Antimuscle antibodies may provoke muscle inflammation (myositis), and antibodies to heart muscle may provoke myocarditis or endocarditis. Antibodies to skin components give rise to a characteristic symmetrical rash on the bridge of the nose and the area around the eyes (Fig. 31–9). SLE patients also make antibodies to heat shock proteins. Their grossly excessive immune reactivity is also reflected in a polyclonal hypergammaglobulinemia, enlargement of lymph nodes, and thymic enlargement.

Although it is clear that SLE involves a loss of control of the specificity of the B-cell response with resulting multiple autoimmune disorders, the initiating causes remain obscure. There is good evidence for a genetic predisposition in humans and mice (Fig. 31–10). SLE is associated with an increased incidence in blood relatives and the presence of certain MHC molecules in humans. Much evidence suggests that viral or microbial infections may be responsible for initiation of the condition. Various viruses have been implicated. Thus individuals with SLE commonly have high levels of antibodies to parainfluenza 1 and measles. Myxovirus-like structures have been observed within kidney endothelial cells from SLE patients. Similarly, certain type C retroviruses have been isolated from SLE patients and associated with the disease. Humans with SLE may have antibodies to micrococcal and staphylococcal DNA, and these appear to be very specific. Perhaps bacterial DNA induces cross-reactive antibodies that trigger SLE.

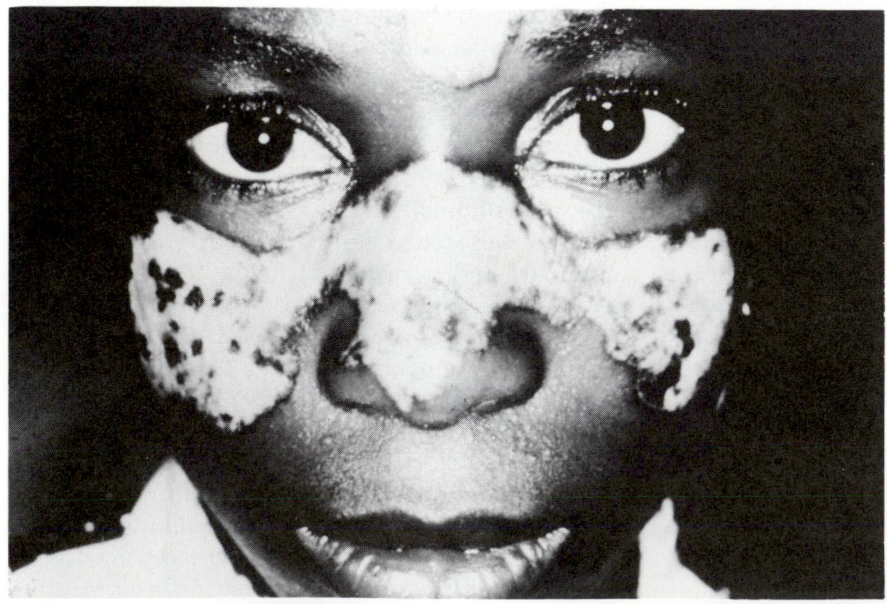

Figure 31–9 The characteristic facial lesions of SLE. This rash extending over the bridge of the nose and across the cheeks is exacerbated by sunlight, which probably accounts for its location. *(From Moschella, S.L., and Hurley, H.J. Dermatology. W.B. Saunders, Philadelphia, 1985. With permission.)*

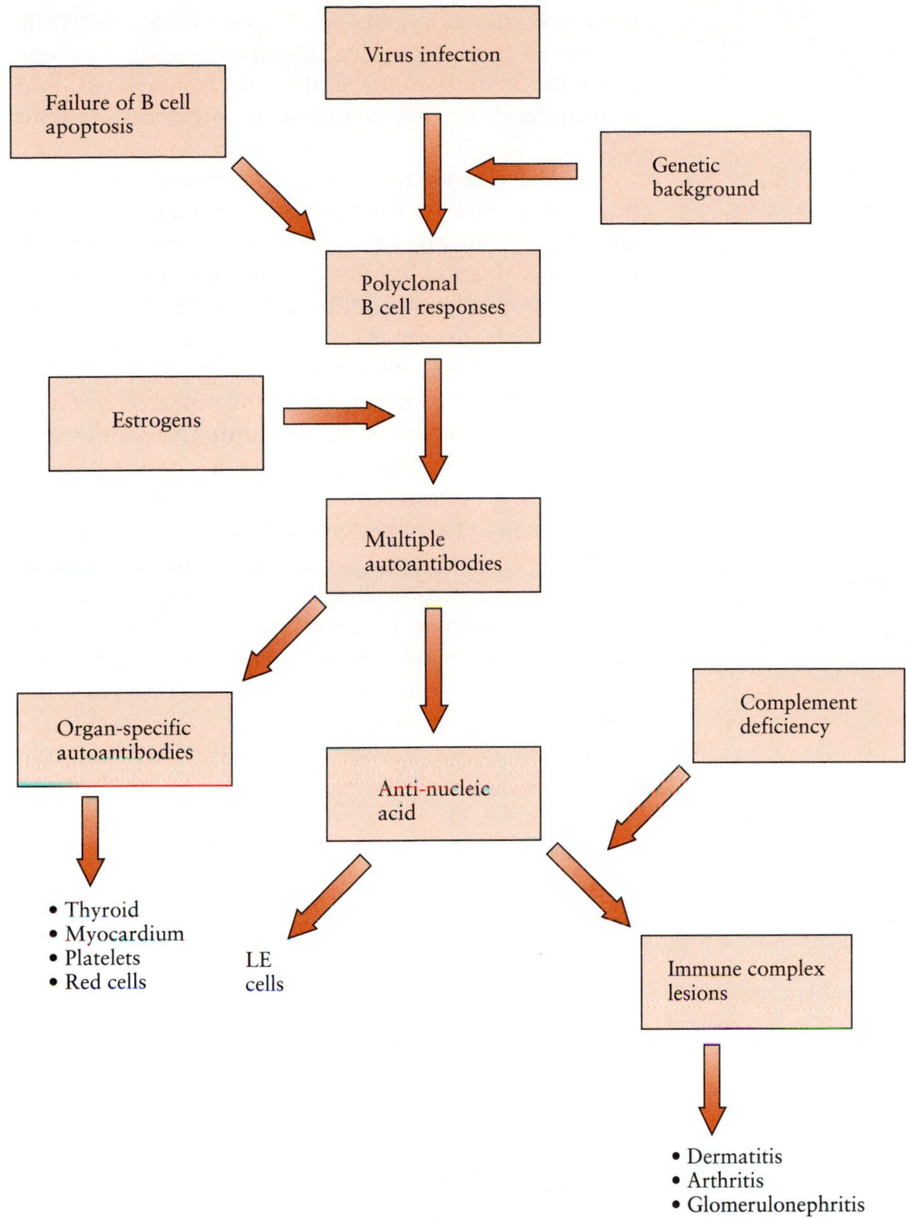

Figure 31–10 Some of the factors that influence the pathogenesis of SLE.

It is important to point out that many SLE patients are defective in their ability to remove immune complexes from their circulation. Thus clearance of antibody-coated particles (Chapter 6) is defective, and the association of SLE with certain complement component deficiencies also reflects the nonremoval or failure of solubilization of immune complexes (Fig. 31–11).

Rheumatoid Arthritis

Rheumatoid arthritis is a common, crippling disease in humans (Fig. 31–12); it is also seen in domestic animals, especially dogs. Although, as its name suggests, it typically involves the joints, other body systems are commonly affected. The disease commences when B lymphocytes in the synovium are activated and produce autoantibodies, including rheumatoid factor (see page 490). As a result, immune complexes are deposited in joint cartilage. Some of these antibodies may be directed against collagen. They may also be bound by the glycosaminoglycans in the cartilage. The immune complexes activate complement and release chemoattractants. As a result, neutrophils are attracted to the sites of immune complex deposition, and hydrolytic enzymes are released. Factors derived from platelets induce endocytosis, enhancement of cell migration, and an increase in membrane synthesis. Capillaries and venules proliferate as a result of the release of cytokines by activated macrophages. Free

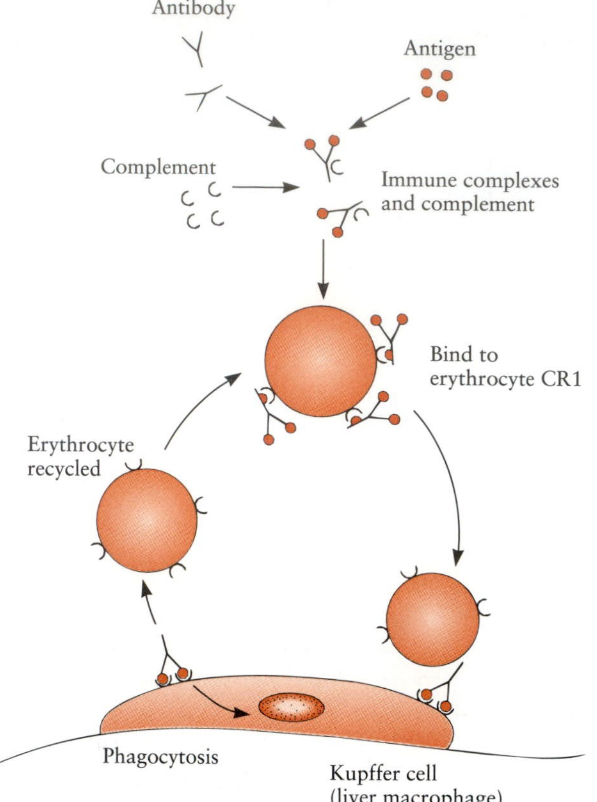

Figure 31–11 In primates, immune complexes containing C3b bind to erythrocyte CR1. The coated erythrocytes bind to Kupffer cells in the liver, where they are phagocytosed. The immune complexes may also dissociate from the red cells, permitting the cells to reenter the circulation while the complexes are endocytosed. In nonprimates, platelets bind immune complexes and may thus remove them from the circulation.

soluble IL-2R, IL-6, GM-CSF, IFN-γ, TGF-β, and TNF-α have all been found in the synovial fluid of rheumatoid arthritis patients, and T cells from rheumatoid arthritis patients are defective in their response to IL-2.

As the disease progresses, the synovia swell and begin to proliferate. Outgrowths of these proliferating synovia extend into the joint, where they are known as pannus. Pannus consists of fibrous vascular tissue that, as it invades the joint cavity, releases proteases that erode the articular cartilage and the neighboring bony structures. As the arthritis progresses, the infiltrating neutrophils may be partially replaced by $CD4^+$ lymphocytes, which can form lymphoid nodules and germinal centers. Another feature of rheumatoid arthritis that is commonly encountered is the development of subcutaneous nodules, foci of dead tissue surrounded by fibrous connective tissue containing lymphocytes and plasma cells.

The immediate cause of rheumatoid arthritis is unknown. Several investigators have proposed that the initiating cause is a microorganism. Although none of these theories has been confirmed, recent evidence suggests that Epstein-Barr virus (EBV) may be implicated in some way. EBV can promote polyclonal B-cell expansion, which, if not fully controlled, may give rise to multiple autoantibody production. Rheumatoid arthritis is also very similar to the arthritis caused by *Borrelia burgdorferi,* the cause of Lyme disease.

Development of autoantibodies to IgG is characteristic of rheumatoid arthritis. These autoantibodies, called rheumatoid factors, are of the IgM class and are directed against epitopes in the C_H2 domains of anti-

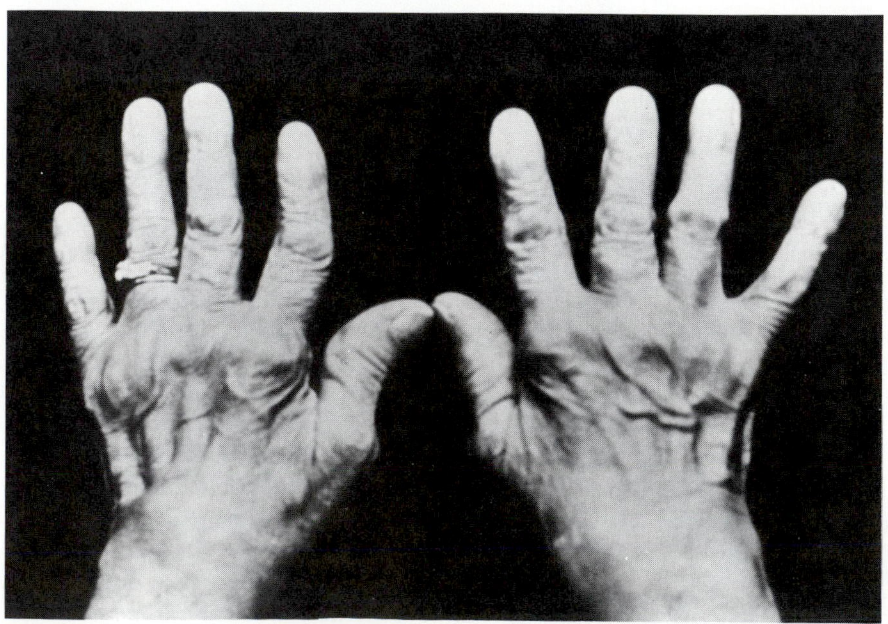

Figure 31–12 Severe rheumatoid arthritis destroys bone and articular cartilage. It produces characteristic deformities in the hands of affected individuals. *(From Parker, C.W. Clinical Immunology, vol 1. W.B. Saunders, Philadelphia, 1980. With permission.)*

gen-bound IgG. The presence of rheumatoid factor is associated with more destructive disease and with complications outside joints. Rheumatoid factors are found not only in rheumatoid arthritis but also in SLE and other conditions in which extensive immune complex formation occurs. It is possible to induce an arthritis in experimental animals by immunizing the animals against their own type II collagen. This experimental arthritis has many of the features associated with human rheumatoid arthritis. Rheumatoid arthritis and SLE patients may also have autoantibodies to IgD. It is conceivable that this may have an effect on the regulation of the immune system in these individuals.

TREATMENTS FOR AUTOIMMUNE DISEASES

Conventional treatment of autoimmune diseases has been by administration of immunosuppressive drugs. The same drugs used to prevent graft rejection have been given to patients in order to suppress their immune responses. Thus methotrexate is now probably the best available treatment for rheumatoid arthritis. Generally, the doses required are considerably lower than are needed to suppress graft rejection or prevent the growth of cancer. Nevertheless, these drugs may have some adverse side effects. For this reason, considerable effort is being put into the development of much more specific treatments, designed, as far as possible, to selectively deplete only the self-reactive cells.

One way in which autoimmune diseases may be treated is to suppress T cell responses to autoantigens. This might be done by immunizing patients against their own T cells in order to induce an immune response against the specific T cell antigen receptors. This anti-idiotype response could, in theory, suppress the autoimmune T cells and so reduce the severity of the disease.

Experimentally, animals have been injected with either killed autoimmune T cell clones or low doses of living autoimmune T cells. The T cells were killed by irradiation or chemicals after first being activated with mitogens. This vaccination effectively prevented the development of experimental autoimmune diseases such as encephalitis, adjuvant arthritis, or thyroiditis. The mechanism of this response is somewhat more complex than described here, however, since investigators have also obtained positive responses when activated, nonspecific T cells were used for immunization. Thus the underlying mechanism of T cell vaccination remains unclear.

One novel approach is to feed large quantities of the appropriate autoantigen to patients. Thus feeding myelin basic protein has induced tolerance to myelin proteins in animals and reversed the paralysis seen in experimental autoimmune encephalitis. This oral tolerance therapy makes use of the normal processes that prevent us from mounting immune responses to food proteins (Chapter 23). Attempts to treat multiple sclerosis by feeding myelin to patients have not yet yielded significant results. Other techniques being studied are to destroy helper T cells with monoclonal antibodies to CD4 or to prevent them from entering tissues with an autoantibody to VLA-4. These have been reported to be of benefit in rheumatoid arthritis patients.

KEY WORDS

Ankylosing spondylitis p. 491
Antinuclear antibodies p. 497
Autoantibodies p. 489
Autoimmune disease p. 489
Autoimmune thyroiditis p. 495
Band 3 protein p. 489
Cross-reactivity p. 490
Diabetes mellitus p. 494
Experimental autoimmune encephalitis p. 496
fas/Apo-1 protein p. 492
Hashimoto's thyroiditis p. 495
Heat shock proteins p. 491
Hidden antigens p. 490
Idiotype p. 489
Immunoconglutinins p. 490
Interferon-γ p. 497
LE cells p. 498
MHC molecules p. 492
Molecular mimicry p. 490
Multiple sclerosis p. 495
Myasthenia gravis p. 497
Myelin basic protein p. 496
New Zealand Black mice p. 491
Rheumatoid arthritis p. 499
Rheumatoid factors p. 490
Systemic lupus erythematosus p. 497
Thyrotoxicosis p. 495
Type I hypersensitivity p. 494
Type II hypersensitivity p. 494
Type III hypersensitivity p. 494
Type IV hypersensitivity p. 494

QUESTIONS

1. Autoimmune diseases are caused by
   a. a defect in the cell-mediated immune system
   b. a defect in thymus development
   c. an immune response against self-antigens
   d. a defect in the antibody-mediated immune system
   e. T cell deficiency
2. Systemic lupus erythematosus is most characteristically associated with

a. rheumatoid arthritis
b. amyloidosis
c. severe combined immunodeficiency
d. lymphoid cell tumors
e. antinuclear antibodies

3. The experimental animal model of the human disease multiple sclerosis is
a. the Arthus reaction
b. experimental autoimmune encephalitis
c. the nude mouse
d. passive cutaneous anaphylaxis
e. the complement fixation test

4. Aged red cells are removed by autoantibodies against
a. complement
b. type O blood group
c. idiotypes
d. band 3 protein
e. immunoconglutinin

5. Rheumatoid factor is an
a. antinuclear antibody
b. antibody to synovial membranes
c. antibody to glomerular basement membrane
d. antibody to complement
e. antibody to immunoglobulin

6. The best example of an autoimmune disease mediated by type I hypersensitivity is
a. systemic lupus
b. milk allergy in cattle
c. food allergy in dogs
d. myasthenia gravis
e. diabetes mellitus

7. Some autoimmune diseases may result from cross-reactions between organisms and tissues, a phenomenon called
a. molecular mimicry
b. antigenic variation
c. tissue sensitization
d. autosensitization
e. clonal anergy

8. Thyrotoxicosis results from autoimmune attack against the
a. acetylcholine receptor
b. insulin receptor
c. thyroid-stimulating hormone receptor
d. thyroglobulin receptor
e. thymic hormone receptor

9. Ankylosing spondylitis may result from cross-reactions between normal tissues and which organism?
a. *Escherichia coli*
b. *Mycobacterium tuberculosis*
c. Epstein-Barr virus
d. *Borrelia burgdorferi*
e. *Klebsiella pneumoniae*

10. Which of the oncogenes appears to regulate apoptosis, and whose defect leads to autoimmunity?
a. *src*
b. *jun*
c. *fos*
d. *fas*
e. *myc*

11. Does autoimmune disease automatically result from the formation of autoantibodies? If not, why not?

12. Speculate on the "cause" or "causes" of rheumatoid arthritis, considering that about 2% of the U.S. population suffers from it.

13. Systemic lupus erythematosus now appears to be a much less serious disease than when it was first described in the 1920s. Why might this be?

14. Discuss the "physiological" autoimmune responses. How do you think they might have evolved?

15. Compare milk allergy in cows with milk allergy in humans. What might account for these differences between species?

16. Suggest appropriate forms of treatment for autoimmune thyroiditis or myasthenia gravis.

17. Many autoimmune diseases involve the development of immune responses against endocrine tissues. Why might this be?

Answers: 1c, 2e, 3b, 4d, 5e, 6b, 7a, 8c, 9e, 10d

SOURCES OF ADDITIONAL INFORMATION

Barnett, L.A., and Fujinami, R.S. Molecular mimicry: A mechanism for autoimmune injury. FASEB J., 6:840–844, 1992.

Carson, D.A. Genetic factors in the etiology and pathogenesis of autoimmunity. FASEB J., 6:2800–2805, 1992.

Castaño, L., and Eisenbarth, G.S. Type I diabetes: A chronic autoimmune disease of humans, mouse and rat. Annu. Rev. Immunol., 8:647–679, 1990.

Chantry, D., and Feldmann, M. The role of cytokines in autoimmunity. Biotech. Ther., 1:361–409, 1990.

Chàrriere, J. Immune mechanisms in autoimmune thyroiditis. Adv. Immunol., 46:263–334, 1989.

Charron, D. Molecular basis of human leukocyte antigen class II disease associations. Adv. Immunol., 48:107–159, 1990.

Doherty, P.C., et al. Heat shock proteins and the $\gamma\delta$ T cell response in virus infections: Implications for autoimmunity. Semin. Immunopathol., 13:11–24, 1991.

Eisenbarth, G.S. Type I diabetes: Clinical implications of autoimmunity. Hosp. Pract. [off.], 22:167–184, 1987.

Hang, L., et al. Induction of severe autoimmune disease in normal mice by simultaneous action of multiple immunostimulators. J. Exp. Med., 161:423–428, 1985.

Harris, E.D. Rheumatoid arthritis: Pathophysiology and

implications for therapy. N. Engl. J. Med. 322:1277–1289, 1990.

Janeway, C. Beneficial autoimmunity. Nature, 299:396–397, 1982.

Kahn, R. Autoimmunity and the aetiology of insulin-dependent diabetes mellitus. Nature, 299:15–16, 1982.

Kantor, F.S. Autoimmunities: Diseases of dysregulation. Hosp. Pract. [off.], 23:75–84, 1988.

Khansari, N., and Fudenberg, M.H. Immune elimination of autologous senescent red cells by Kupffer cells in vivo. Cell. Immunol., 80:426–430, 1983.

Kirkland, H.H., Mohler, D.N., and Horwitz, D.A. Methyldopa inhibition of suppressor-lymphocyte function: A proposed cause for autoimmune hemolytic anemia. N. Engl. J. Med., 302:825–832, 1980.

Kroemer, G., and Wick, G. The role of interleukin 2 in autoimmunity. Immunol. Today, 10:246–251, 1989.

Lahita, R.G. Sex steroids and autoimmunity. Adv. Inflamm. Res., 8:143–164, 1984.

Lopate, G., and Pestrone, A. Autoimmune myasthenia gravis. Hosp. Pract. [off.], 28:109–131, 1993.

Lutz, H.U., Nater, M., and Stammler, P. Naturally occurring anti–band 3 antibodies have a unique affinity for C3. Immunology, 80:191–196, 1993.

Marx, J.L. Autoimmunity in left-handers. Science, 217:141–144, 1982.

Schwartz, R.S. Viruses and systemic lupus erythematosus. N. Engl. J. Med., 307:1499–1507, 1982.

Shoenfeld, Y., and Schwartz, R.S. Immunologic and genetic factors in autoimmune diseases. N. Engl. J. Med., 311:1019–1029, 1984.

Sinha, A.A., Lopez, M.T., and McDevitt, H.O. Autoimmune diseases: The failure of self tolerance. Science, 248:1380–1388, 1990.

Steinman, L. Autoimmune disease. Sci. Am. 269:107–114, 1993.

Steinman, L., and Mantegazza, R. Prospects for immunotherapy in myasthenia gravis. FASEB J., 4:2726–2731, 1990.

Stewart, T.A., Hultgren, B., Huang, X., et al. Induction of type I diabetes by interferon-α in transgenic mice. Science, 260:1942–1944, 1993.

Stuart, J.M., et al. Type II collagen-induced arthritis in rats: Passive transfer with serum and evidence that IgG anticollagen antibodies can cause arthritis. J. Exp. Med., 155:1–16, 1982.

Tienari, P.J., et al. Genetic susceptibility to multiple sclerosis linked to myelin basic protein gene. Lancet, 340:987–991, 1992.

Watanabe-Fukunaga, R., et al. Lymphoproliferation disorder in mice explained by defects in fas antigen that mediates apoptosis. Nature, 356:314–317, 1992.

Wu, J., et al. Autoimmune disease in mice due to integration of an endogenous retrovirus in an apoptosis gene. J. Exp. Med., 178:461–468, 1993.

CHAPTER 32

The Phylogeny of the Immune System

Bruce Glick *Dr. Glick is a poultry scientist who made a key contribution to immunology in 1955 when he identified the function of the avian bursa of Fabricius. Dr. Glick was raising antisera to* Salmonella typhimurium *in poultry and, to save money, injected some birds that had previously been bursectomized. These birds either died or survived and produced no antibodies. Nonbursectomized birds survived and produced a normal level of antibodies. In a subsequent study he confirmed that birds bursectomized when young failed to make antibodies. Thus through his insight, the second key arm of the immune system was identified.*

CHAPTER OUTLINE

CHAPTER CONCEPTS

1. The ability to differentiate between self and not-self and to respond to microbial invaders is found in both invertebrates and vertebrates.
2. Invertebrates possess protective mechanisms that are analogous to vertebrate phagocytosis, humoral immunity, and cell-mediated immunity.
3. Although all vertebrates possess lymphocytes, the thymus and spleen are found only in vertebrates more highly evolved than cyclostomes. Bone marrow appears only in reptiles and more evolved classes, and true lymph nodes first emerge in marsupial and eutherian mammals.
4. IgM antibodies are produced by all vertebrates except for the cyclostomes. Antibodies similar to IgG first develop in anuran amphibians, and IgA and possibly IgE and IgD appear in birds.

This chapter describes the evolutionary development of the immune system. The first section describes some of the defense mechanisms employed by invertebrates. These have many similarities to the mechanisms employed by vertebrates. This is followed by a review of the ways in which vertebrates of different types have solved the problem of defending themselves against invasion. In general, the reader will find that there has been a progressive increase in the complexity of the immune system as vertebrates have evolved. As a result, many of the features described for the mammalian system are absent or very different in other vertebrates.

All animals, regardless of their complexity or evolutionary origin, must be able to exclude invading organisms that might cause disease or death. Likewise, should abnormal cells arise, there must be mechanisms for identifying and eliminating them. The reader should be aware, however, that although the mechanisms employed by invertebrates and vertebrates to protect themselves against invaders have a superficial similarity, they may have a different evolutionary origin. Thus, the protective mechanisms of invertebrates are functional **analogs** of vertebrate immune mechanisms insofar as they have similar effects but are of totally different origin or chemical composition. In contrast, the various vertebrate immune systems are usually **homologs** of each other, having not only similar functions, but also the same origins.

IMMUNITY IN INVERTEBRATES

For simplicity, invertebrates can be divided into a "simple" group and a "complex" group based on possession of a body cavity, or coelom. The simple group consists of the acoelomates, which include the sponges and coelenterates. The complex group consists of the coelomates. The coelomates themselves are divided into two evolutionary lines. One line includes the annelids, mollusks, and arthropods and is called the Protostomia. The other group, which includes the echinoderms, protochordates, and chordates, is called the Deuterostomia (Fig. 32–1). It is from the Deuterostomia that the vertebrates eventually evolved.

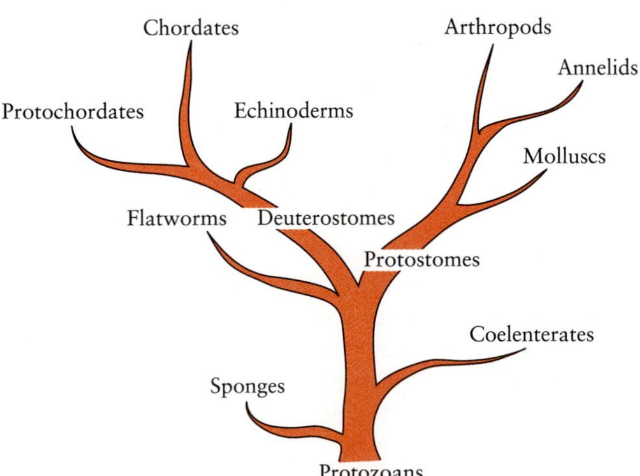

Figure 32–1 A simplified phylogenetic tree showing the relationships of the invertebrates.

Invertebrates protect themselves against invasion by processes analogous to phagocytosis, humoral immunity, and cell-mediated immunity. Unfortunately, the diversity of the invertebrates is so enormous that we possess only a set of disconnected observations on immunity in many different species. As a result, descriptions of invertebrate immune capabilities as yet lack an underlying pattern.

Phagocytosis

In 1884 Elie Mechnikoff discovered the process of phagocytosis when examining starfish larvae. He showed that rose thorns introduced into the body cavity of these larvae were attacked by mobile cells. Two types of phagocytic cell are now recognized in coelomate invertebrates: the hemocytes (blood leukocytes) and the coelomocytes (body cavity cells). Hemocytes have been described in protostomes (mollusks, annelids, and arthropods); coelomocytes are found in both protostomes and in deuterostomes (chordates and echinoderms). These phagocytic cells behave in a manner similar to that described for mammalian phagocytes and demonstrate chemotaxis, adherence, ingestion, and digestion.

Humoral Immunity

Soluble protective factors have been detected in the body fluids of both acoelomates and coelomates. When sea anemones (coelenterates) are injected with bovine serum albumin (BSA), they produce a protein in their tissue fluid that can bind to the BSA. This protein binds more strongly to BSA than it does to human serum albumin, suggesting some specificity of the reaction. Certain snails (mollusks) can produce proteins that bind and immobilize invading schistosome miracidia (larval forms of schistosome parasites). These proteins appear to be specific for the schistosome. Although proteins may be important in protecting the snail, it should be pointed out that the invading miracidia also trigger a great increase in the number of circulating hemocytes in these snails.)

If insects are injected with heat-killed bacteria, proteins appear in their **hemolymph** that can either prevent bacterial growth or cause bacterial cell lysis.

For example, following infection of tomato hornworms, their hemocytes and hemolymph plasma enzymes degrade the cell wall of the bacteria and release peptidoglycan fragments. These peptidoglycans in turn elicit increased synthesis of several antibacterial proteins that accumulate in the hemolymph. In some insects large proteins are induced by toxins and mimic the activity of antibody molecules. Among the antibacterial proteins that are induced in this way is the enzyme **lysozyme.** The other proteins that are induced are named after the insect of origin: for example, cecropins are found in moths and diptericins in flies. These antibacterial peptides appear about two hours after exposure to the bacteria and reach peak levels at 24 hours. In some insects, the activity is short-lived and disappears by four days; in others, it may last for several months. The activity is non-antigen-specific since it may also be induced by physical damage to the insect cuticle. Passive immunity can be conferred on recipient insects by transfer of hemolymph from immune insects. Little is known about the mechanisms involved in the production of these antibacterial proteins. It is suggested, however, that when hemocytes phagocytose bacteria, they release a cell wall component that stimulates other tissues to secrete the antibacterial peptides.

Natural agglutinins are found in lobsters, scorpions, crayfish, horseshoe crabs, and mollusks. These agglutinins are lectins with carbohydrate-binding specificity. Horseshoe crab agglutinin binds N-acetyl-D-glucosamine, and scorpion and lobster agglutinins bind sialic acid and its derivatives. When these molecules bind to carbohydrate residues on the surface of invading microorganisms, they may act as opsonins.

Echinoderms such as the sea star, *Asterias rubens,* possess an axial organ. This organ contains cells that can bind antigens, respond to mitogens, and secrete lymphokine-like substances. Antibody-like proteins are also released from the axial organ. The production of these proteins requires cooperation between two different cell populations. If sea stars are injected with the hapten trinitrophenol (TNP) attached to a carrier protein, the proteins produced by the axial organ will bind and lyse TNP-coated red blood cells in the presence of mammalian complement.

Proteins that resemble the terminal components of the complement pathway have been found in hemolymph of some protostomia. For example, the horseshoe crab and sipunculid worms have a lytic activity in their hemolymph that is inducible after treatment with cobra venom factor (an enzyme that activates mammalian C3). The coelomic fluid of earthworms can lyse vertebrate erythrocytes and digest some vertebrate serum proteins.

The Prophenoloxidase System

The prophenoloxidase system is a hemolymph-based complex of enzymes that, when activated, generate peptides and adhesive proteins, which then mediate many of the defense functions in arthropods. Activation of the system occurs by cuticular and hemolymph proteases as well as by microorganisms and their cell wall extracts. The system generates phenoloxidase, an adhesive enzyme that binds to foreign surfaces. The system has been likened to a primitive complement system. The end-point of the system is generation of a black pigment that is assumed to be melanin around immune defense reactions. In addition, the system results in enhanced phagocytosis of particles, plasma coagulation, and killing of bacteria and fungi.

Cell-Mediated Immunity

Analogs of cell-mediated immunity are identifiable in simple organisms such as the sponges. The primitive defense cells of acoelomates (sponges and coelenterates) are ameboid mesenchymal cells. These have evolved into primitive lymphoid cells found in protochordates and echinoderms. True lymphocytes are seen only in vertebrates.

A cell-mediated graft rejection process is seen in sponges, coelenterates, annelids, and echinoderms. For example, sponges show allogeneic incompatibility. When two identical sponge colonies are placed side by side and cultured in contact with each other, no reaction occurs. If, however, sponges from two different colonies are made to grow in contact, local tissue destruction occurs along the area of contact as each sponge attempts to destroy the other. This reaction is controlled by histocompatibility proteins.

Allograft reactions have not been demonstrated among mollusks or arthropods, although these Protostomia can reject xenografts. Their failure to mount an allograft response may be because the experimental populations employed were very closely related. Annelids such as earthworms can reject both allografts and xenografts. The rejection of xenografts (from other species of earthworms) takes about 20 days. The graft is invaded by lymphocyte-like and phagocytic cells. The grafted tissue turns white, swells, becomes edematous, and eventually dies. If the recipient worms are grafted with a second piece of skin from the same donor, the second graft is rejected faster than the first. This ability to reject second grafts rapidly may be adoptively transferred by coelomocytes from sensitized animals.

Echinoderms and tunicates are also able to reject allografts. Certain colonial tunicates have an analog of a histocompatibility locus that determines whether a new individual can be accepted into a parabiotic colony. (A parabiotic colony is one in which all the individuals are linked and so share vascular and nutrient functions.) Urochordates reject allografts using lymphocyte-like cells. First-set allografts were rejected in about 40 days. Second-set allografts were rejected in 20 days. Specific memory lasted for at least 50 days after the first graft was rejected.

Many invertebrates have been shown to possess either genes or antigens that are related to those found in vertebrates. Thus tunicates possess an antigen that is a homolog of the lymphocyte surface antigen CD5 as well as antigens coded by genes related to the J_H segments of mammalian immunoglobulin heavy chains. Thy-1 antigen (CDw90), a highly conserved member of the immunoglobulin superfamily associated with vertebrate T cells and nervous tissue, has been identified in the squid, in oysters, and in locusts, as well as in tunicates. Tunicate Thy-1 shows close homology to the μ chains of fish immunoglobulin. β_2-microglobulin, a component of MHC class I molecules, has been identified in the leukocytes of earthworms and in cultured *Drosophila* cells.

THE EVOLUTION OF VERTEBRATES

There are only seven classes of living vertebrates: the Agnatha, the jawless fishes; the Chondrichthyes, the fish with cartilaginous skeletons; the Osteichthyes, the bony fishes; the Amphibia, the frogs and salamanders; the Reptilia, the snakes, alligators, turtles, and lizards; the Aves, the birds; and the Mammalia, the mammals. Because of the great range of complexity within each of these classes (Fig. 32–2), it is not always possible to generalize about the immune system of each.

The most primitive living vertebrates belong to the Agnatha. These are cyclostome fishes such as the lamprey and hagfish. Somewhat more advanced than the cyclostomes are the Chondrichthyes. These fish have cartilaginous skeletons and include the skates, rays, and sharks. The most complex fish are the bony fish of the class Osteichthyes, which include the overwhelming majority of modern fish, the teleosts. Major differences exist between the immune systems of each of these three groups.

There are two major orders of amphibians: the relatively less complex urodeles, which include the

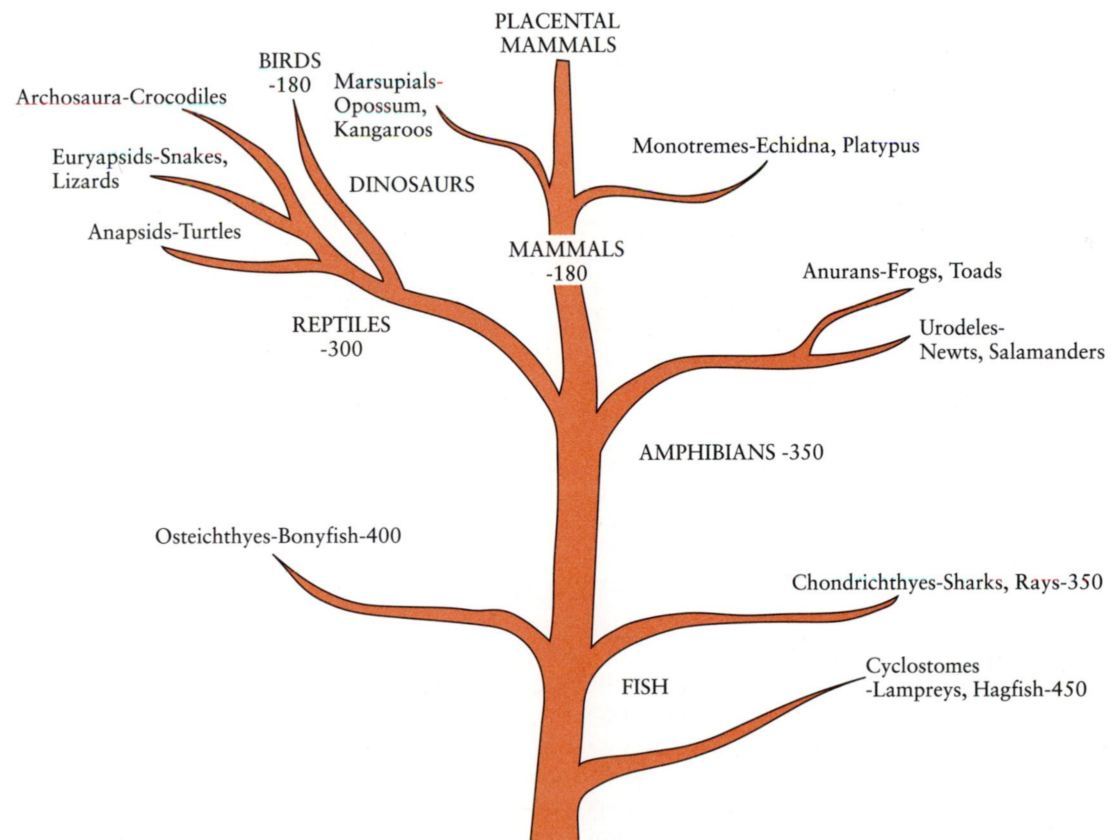

Figure 32–2 A simplified phylogenetic tree for the vertebrates. The figures are an estimate of the number of million years since these groups emerged.

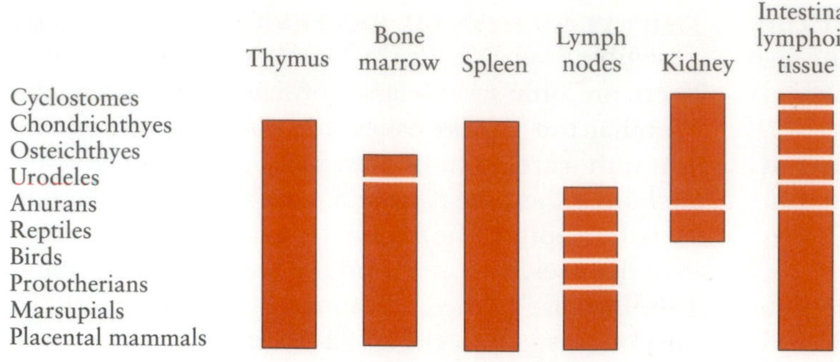

Figure 32–3 The lymphoid tissues of vertebrates. Discontinuous shading indicated that the organ has not assumed its fully evolved form in these groups.

tailed amphibians such as the salamanders and the newts, and the anurans, an advanced order that includes the frogs and toads.

Three subclasses of reptiles currently exist: the Anapsida, which contains the turtles; the Euryapsida, which includes the lizards and snakes; and the Archosaura, which includes the crocodiles and alligators.

The dinosaurs, although related to the reptile Archosaura, were sufficiently different from true reptiles to be classified in a class of their own, the Dinosaura. Although the great majority of dinosaurs disappeared 65 million years ago at the end of the cretaceous period, their modern descendants are probably the birds, now classified as members of the class Aves. Unlike the reptiles, birds are, and dinosaurs probably were, endothermic (warm-blooded). As a result of their endothermia, birds share with mammals all the benefits that come from greatly increased physiological and biochemical efficiency.

The mammals consist of two subclasses: the Prototheria, or egg-laying mammals, and the Theria, or mammals bearing live young. The Prototheria are represented by two Australian mammals, the platypus and the echidna. The Theria are divided into the marsupials, the pouched mammals such as the opossum and kangaroos, and the eutherians, or placental mammals. All of this book (except this chapter) has been devoted to the immunology of eutherian mammals.

THE EVOLUTION OF LYMPHOID ORGANS

As vertebrates have evolved, they show a progressive increase in the number and complexity of their lymphoid organs (Fig. 32–3).

Bone Marrow. The development of bone marrow depends on the development of a suitable bone structure. As a result, marrow is found only in vertebrates adapted to a terrestrial existence. Thus fish are totally lacking in bone marrow. Some lungless salamanders may have a small amount of lymphoid tissue within their long bones. A fully functional bone marrow is present in anuran amphibians, reptiles, birds, and all mammals.

Thymus. There is no detectable thymus in cyclostomes, although hagfish can destroy allografts. It does, however, appear in the cartilaginous and bony fish. The thymus is located underneath the epithelium in the bronchial cavity and arises from the first gill arches (Fig. 32–4). Thymectomy in fish can lead to prolongation of allograft survival. Antibodies or antigen-binding cells may be detected in the fish thymus during an immune response, suggesting that it contains both T-like and B-like cells.

The tailed amphibians such as the newts and salamanders also possess a thymus, which consists of four pairs of nodules arising from the second, third, fourth, and fifth pharyngeal pouches. It develops slowly, only appearing at the seventh week of life. Thymectomy delays or blocks rejection of skin allografts. The thymus of fish and urodele amphibians is not divided into cortex and medulla.

The thymus of anuran amphibians arises as a bud from the second pharyngeal pouches and involutes by about one year of age. It lies just below the skin posterior to the middle ear. In contrast to the more primitive vertebrates, it has an outer cortex and central medulla. The thymic cortex is full of proliferating lymphocytes. The medulla contains fewer lymphocytes, but thymic corpuscles and myoid cells are present. Surface immunoglobulins can be found on about 80% of these thymocytes. Larval thymectomy in the toad reduces their response to foreign red blood cells,

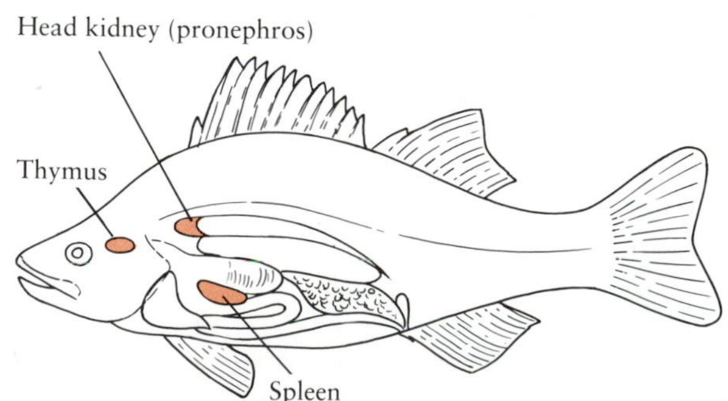

Figure 32–4 A schematic diagram showing the major lymphoid organs of a bony fish.

A CLOSER LOOK

An Antimicrobial Steroid from Sharks

Fish, like other animals, possess nonimmunological defenses to prevent local infections. Thus in the dogfish shark stomach, there is a very potent antifungal agent called squalamine. Squalamine is a steroid linked to an amide called spermidine. It is suggested that squalamine may be a component of an initial defense mechanism that prevents microbial invasion. It may be especially important in cold water, where the conventional immune response is very slow to respond to antigens. Because of its potent antifungal activity, squalamine may have potential uses in the treatment of human fungal infections.

but the response to bacterial lipopolysaccharide is unaffected, suggesting this is a T-independent response. Thymectomy slows but does not completely prevent graft rejection in toads.

The reptilian thymus develops from the pharyngeal pouches and is histologically similar to that seen in other classes of vertebrates. Age involution of the reptile thymus has been reported as well as a seasonal involution; the thymus shrinks in winter and enlarges in summer.

The thymus in birds, monotremes, and marsupials is essentially similar to that seen in eutherian mammals.

Spleen. The spleen appears first in cartilaginous and bony fish. It is absent in cyclostomes. Its overall structure and location in fish are similar to that in mammals, although the red and white pulps are not separate. The spleen of urodele amphibians is also not clearly divided into red and white pulp, but the spleen of anuran amphibians is. Thus in frogs and toads, for the first time, the red pulp and the periarteriolar white pulp are separated by boundary layer cells. The reptilian spleen also shows a clear separation between red and white pulps. Germinal centers are absent from fish, amphibian, and reptile spleens, which raises interesting questions about their B-cell responses. In contrast, the germinal centers of birds are large and well defined.

Lymph Nodes. Lymph nodes are absent from fish and urodele amphibians, but structures that resemble lymph nodes are seen in some anurans. These proto–lymph nodes consist of a mass of lymphocytes surrounding blood sinusoids. As a result, they filter blood rather than lymph. Lymphomyeloid nodes that closely resemble lymph nodes are seen for the first time in reptiles. They have a relatively simple structure consisting of a lymphoid parenchyma with phagocytes and intervening sinusoids. Primitive lymph nodules surrounding the aorta, vena cava, and jugular veins are seen in some reptiles.

Although birds are commonly considered not to possess lymph nodes, they do possess structures that can be considered their functional equivalent. These avian lymph nodes consist of a central sinus that is the main lumen of a lymphatic vessel. It is surrounded by a sheath of lymphoid tissue that contains germinal centers. Avian lymph nodes have no external capsule.

The prototheran mammals have lymph nodes with a distinctive structure (Fig. 32–5). They consist of several lymphoid nodules, each containing a germinal center, suspended by its blood vessels within the lumen of a lymphatic plexus. Thus, each nodule is bathed in lymph. There is usually just one germinal center per nodule.

Intestinal Lymphoid Tissue. Lymphoid aggregates are found in the intestinal wall of some cyclostomes.

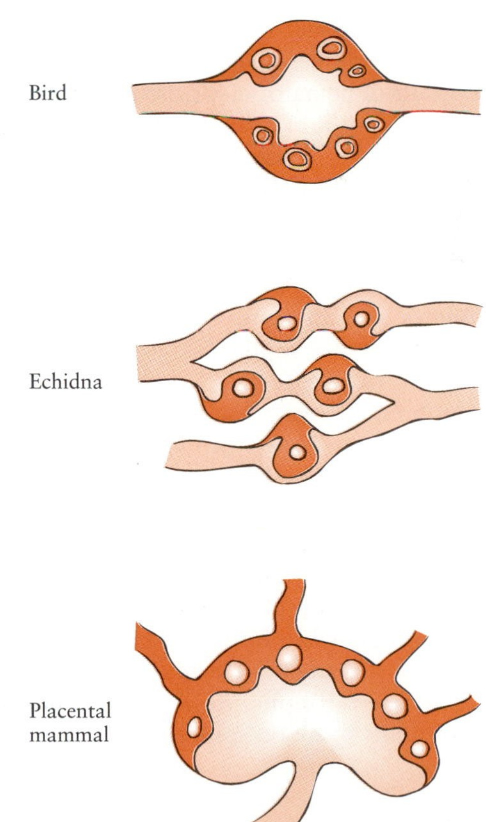

Figure 32–5 The structure of the "lymph nodes" from three groups of vertebrates. All serve to "filter" tissue fluid and permit antigens in that fluid to encounter antigen-trapping and presenting cells.

These aggregates may become quite prominent in the more complex fishes. Lymphocytes and plasma cells are found in nodules in the intestinal wall of all the more complex vertebrates. Some turtles and snakes (but not alligators) have lymphoid aggregations that project into the cloacal lumen. These aggregates are larger in adults than in young turtles and are therefore not primary lymphoid organs and cannot be regarded as a primitive bursa.

The bursa of Fabricius, found only in birds, is increasingly regarded as little more than a specialized area of intestinal lymphoid tissue. Its structure has been described in Chapter 9. Bursectomy results in a loss of antibody production, although bursectomized birds can still reject foreign skin grafts. These results have been interpreted to suggest that the bursa is a primary lymphoid organ whose function is to serve as a maturation and differentiation site for the cells of the antibody-forming system. However, recent studies have suggested that bursectomy permits excessive suppressor-cell activity. The bursa also functions as a secondary lymphoid organ since it can trap antigen and undertake some antibody synthesis. Birds also have extensive lymphoid accumulations in the cecal tonsils and in the skin.

Renal Lymphoid Tissue. The kidney of a fish differentiates in two sections. The pronephros, or anterior kidney, consists of hematopoietic tissue containing antibody-forming cells and phagocytes. It thus performs a function analogous to mammalian bone marrow and lymph nodes. The opisthnephros, or posterior kidney, retains its excretory function. The kidney retains this lymphopoietic function in amphibians. Thus, stem cells arise from the intertubular areas of the kidney in both urodeles and anurans. A few lymphocytes are found in the kidney of reptiles.

Other Lymphoid Organs. Larval amphibians such as the bullfrog tadpole have lymphomyeloid organs in their branchial region called ventral cavity bodies. Sinusoids in the organ are lined with macrophages that effectively remove injected particles. Removal of these organs renders tadpoles incapable of making antibodies to soluble antigen. They disappear at metamorphosis. Lymphocytes are found in large numbers in the subcapsular region of the liver in fish, amphibians, and reptiles. These lymphocyte accumulations occur in close proximity to blood sinuses.

Lymphocytes

Fish. Hagfish have two types of leukocyte in their blood. One population is monocyte-like; the other population is lymphocyte-like. These lymphocytes originate from the anterior kidney. About 70% of them possess surface immunoglobulins. Plasma cells have not been observed in cyclostomes.

Cartilaginous and bony fish possess lymphocytes and plasma cells. These cells have properties similar to those described in mammals. Thus lymphocytes carrying immunoglobulin on their surface can be found in the thymus, kidney, spleen, and blood, and they are assumed to be homologous to B cells. This lymphocyte surface membrane immunoglobulin acts as an antigen receptor since anti-IgM antisera inhibit antigen binding by fish lymphocytes.

Fish lymphocytes can respond to bacterial lipopolysaccharide (LPS) and PPD tuberculin by dividing. Lymphocytes from the anterior kidney respond only to LPS, whereas thymocytes will not respond to either LPS or PPD. T cells can be detected in fish by using antiserum directed against the fish homolog of Thy-1. These fish T lymphocytes respond to phytohemagglutinin and concanavalin A, although they do not form spontaneous rosettes with foreign erythrocytes (a reaction mediated by CD2). Fish T cells can perform a helper function and stimulate B cells in order to provoke antibody formation. For example, a carrier effect has been demonstrated in bony fish such as the flounder and goldfish. (That is, a secondary antibody response to a hapten will be elicited only in fish that have been primed with the same carrier.) NK cells have been described in bony fish.

Amphibians. Both adult and larval amphibians have circulating lymphocytes. Although some investigators have suggested that these lymphocytes arise in the thymus, it is more likely that they originate in the ventral cavity bodies or the liver. The thymic lymphocytes and about 80% of circulating lymphocytes carry surface IgM. Frogs possess NK-like and T cytotoxic-like killer cells.

Reptiles and Birds. Lymphocytes are well recognized in reptiles. Birds possess lymphocytes that originate in the yoke sac and migrate either to the bursa or to the thymus. Immature lymphocytes that enter the thymus mature under the influence of factors derived from thymic epithelial cells, and cells with recognizable T-cell markers emigrate from the thymus. In avian blood, T cells constitute between 60% and 70% of the total lymphocyte population. Immature lymphocytes that enter the bursa emigrate as immunoglobulin-positive B cells.

THE EVOLUTION OF HUMORAL IMMUNITY

Immunoglobulins

Fish. Cyclostomes can make antibodies. In the hagfish, these antibodies are large (23.8S) macroglobu-

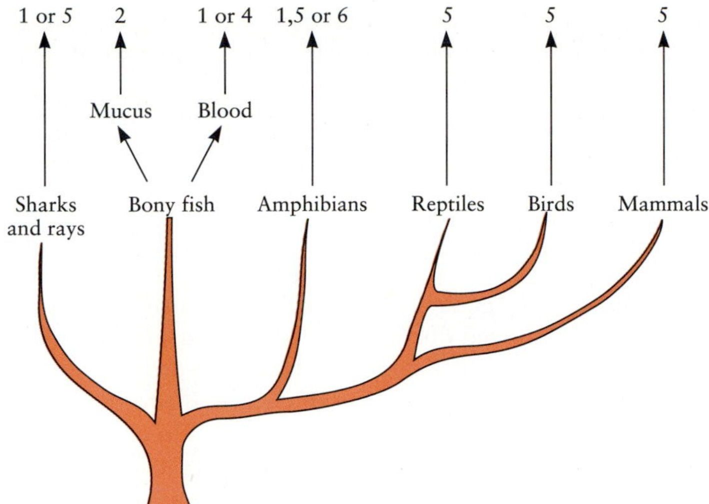

Figure 32–6 The relative sizes of IgM polymers found in different groups of vertebrates.

lins. In the lamprey, three antibody molecules of 7, 9, and 14S are formed. The 7S molecule has both light and heavy peptide chains. These heavy chains resemble the μ chains of higher vertebrates, but they have no interchain disulfide bonds. Bony and cartilaginous fish produce only IgM. Cartilaginous fish usually have serum IgM consisting of pentameric (19S) and monomeric (7S) forms. Bony fish have tetrameric (17S) and monomeric forms of IgM in their serum (Fig. 32–6). Peptides homologous to the J chain have also been described in fish.

Blood vessels of fish are permeable to serum immunoglobulins. As a result, immunoglobulins are found in most fish tissue fluids and in plasma, lymph, and skin mucus. In blood they constitute about 40% to 50% of the total serum proteins. Dimeric IgM is found in the mucus of bony fish; it is not derived by exudation from the serum immunoglobulin pool but must be locally synthesized (Fig. 32–7). Fish antibody in the presence of normal serum as a source of complement is capable of lysing target cells. Likewise, fish antibodies are effective at agglutination and are not difficult to detect.

It is possible that the evolution of multiple immunoglobulin classes arose to meet the demands of a more efficient vascular system. In animals with a higher blood pressure, there would be a need for tight junctions between vascular endothelial cells in order to reduce the exudation of plasma fluids and proteins. This would, of course, result in the inability of immunoglobulins to escape into tissue fluids and would favor the production of low-molecular-weight immunoglobulins. Fish do not have to deal with these problems, since their metabolic needs can be met by a vascular system of low hydrostatic pressure.

The way in which immunoglobulin molecules are coded for by light- and heavy-chain genes and the structures of the V, J, and C regions is conserved throughout the vertebrates. C-region domains are also well conserved as opposed to whole C regions. The V and J segments of TCRs are also conserved and related to the equivalent segments in immunoglobulins, although the C regions of immunoglobulins and TCR are distinct. Nevertheless, there are differences in the precise arrangement of the immunoglobulin genes. For example, dogfish have clustered immunoglobulin

CLARIFICATION

Sedimentation Rates

It is possible to determine the approximate molecular weight of a protein by measuring its sedimentation rate in the ultracentrifuge. This sedimentation rate is measured in **Svedberg units (S).** A 19S molecule has a molecular weight of approximately 900 kDa, and a 7S molecule has a molecular weight of approximately 160 kDa.

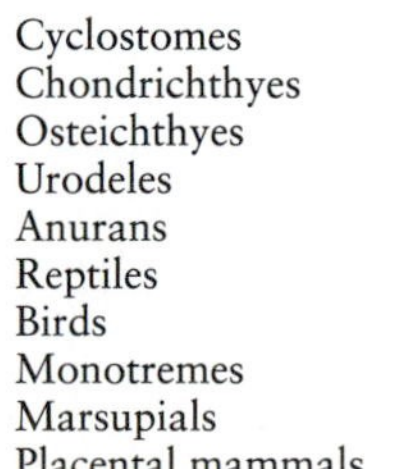

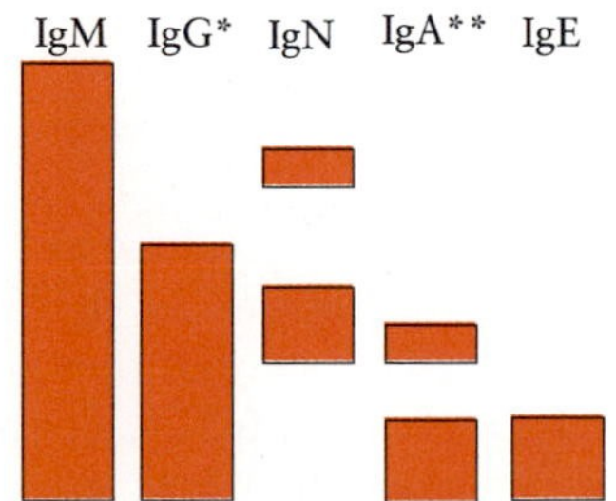

Figure 32–7 The types of immunoglobulins found in different vertebrate groups. *This includes other 7S immunoglobulin molecules irrespective of their name. **This includes all related secretory immunoglobulins.

genes where one V, D, J segment is linked to its own constant segment, thus

-VDJ-C—VDJ-C—VDJ-C—VDJ-C—VDJ-C—

There are more than 500 of these clusters, and each cluster is about 16 kilobases in size. (This arrangement is somewhat similar to that seen in the TCR-β genes in mammals.) This is different from teleost fish, which have a mammalian immunoglobulin gene arrangement with multiple Vh genes. Coelocanths also have an immunoglobulin gene arrangement similar to that seen in mammals.

Urodele Amphibians. Only IgM is produced by urodeles. Monomeric antibody is produced, but as in fish, it has μ-like heavy chains. Urodeles give a good but slow antibody response against bacterial antigens, but they do not respond to soluble antigens such as serum albumin or ferritin.

Anuran Amphibians. Anuran amphibians have two serum immunoglobulin classes: a high-molecular-weight form, classified as IgM and consisting of either pentamers or hexamers (in *Xenopus*), and a low-molecular-weight (170-kDa) molecule. The heavy chain of this small molecule is larger than that of mammalian IgG. Frogs given bacteria or foreign erythrocytes will produce only IgM. Bacteriophages or soluble foreign proteins induce both IgM and IgG formation. Anuran amphibians possess secretory immunoglobulins in bile and the intestine (but not in skin mucus). They consist of IgM and IgG but not IgA.

Reptiles. The reptiles that have been studied possess IgM and at least one other distinct class homologous to IgG. The IgM of turtles is comparable to mammalian IgM in size, chain structure, and carbohydrate content. The major low-molecular-weight antibody is of 120 kDa with two light and two heavy chains. Turtles may also have an immunoglobulin of 180 kDa. Alligators have been shown to possess two forms of immunoglobulin light chain, perhaps homologs of kappa and lambda in the mammal. The very primitive reptile the tuatara (*Sphenodon punctatum*) has both 19S and 7S immunoglobulins. These have identical light chains but different heavy chains.

Birds. Chickens can produce at least three antibody classes, immunoglobulins G (IgY), M, and A (IgB). There is some evidence that they possess an avian homolog of IgD. Chicken IgM is a pentamer with similar structure to mammalian IgM. Chicken IgG is similar to mammalian IgG but is somewhat larger; because of this and other differences, it has been termed IgY by some workers. Likewise, chicken IgA has been deemed so different from mammalian IgA that it has been called IgB by some investigators. The biliary immunoglobulins of birds appear to be distinctly different from their mammalian counterparts.

Birds generate antibody diversity in a manner that is quite distinct from that seen in mammals. Thus the germline contributes little to diversity since there is only one V1 subgroup, one Vh subgroup, one J gene, and one functional Vλ gene. There are 25 Vλ pseudogenes that serve as sequence donors to diversify the Vλ gene by gene conversion. This diversification process is induced during the clonal expansion of B cells in the bursa of Fabricius. The B-cell progenitors rearrange their light chain gene between 10 and 15 days of embryogenesis. During recombination of the V and J segments single nonrandom bases are added to each segment (N-region addition) and joining occurs at random. Chicken light chains are further diversified by somatic point mutations and imprecise V-J joining. Usually only one allele is rearranged. The frequency of somatic mutations is the same as in the mouse. As a result of all this, the chicken can generate about 10^6 different antibody molecules. This is one order of magnitude less than the mouse can generate.

Primitive Mammals. The monotremes synthesize only IgM and IgG. Marsupials can synthesize IgA and IgE as well. Since IgD has been described in birds, it seems reasonable to assume that this immunoglobulin class will eventually be found in all mammalian groups.

Immunoglobulin N. The Australian lungfish possesses a second immunoglobulin class in addition to IgM. This second immunoglobulin is a 5.9S molecule with a molecular weight of 120 kDa. Its small size is due to an antigenically distinct heavy chain of 38 kDa. This chain is called the ν (nu) chain, and the entire molecule is called IgN. IgN has been described in the lungfish, the giant grouper, some turtles, and geese and ducks. It has also been recorded in the quokka, a marsupial (*Setonix brachyurus*). The major serum immunoglobulin class in ducks is IgN. IgN activates complement.

The Evolution of Complement

Although there are many mammalian complement components, they fall into a limited number of protein families (for example, C3, C4, and C5 or C1, C2, and B). Thus is is reasonable to suggest that the present mammalian system has evolved from a simpler system with fewer components. Cyclostomes such as the hagfish have a serum protein that is structurally similar to mammalian C4 but functionally similar to C3. This protein has no hemolytic activity but is a potent opsonin. It may be activated by an alternative-like pathway. A complement system that has both classical and alternative components probably evolved with the jawed fishes. For example, nurse shark serum has been shown to contain six functionally distinct complement components. The relationship of these components to their mammalian homologs is unclear. Nurse shark complement generates a membrane attack complex similar to the complement system of mammals. However, it does work at a lower optimum temperature, generally around 25° C.

Amphibians, both urodeles and anurans, possess complement similar to that of mammals, but it is more effective at 16° C. *Xenopus* has a classical and alternate complement pathway.

The Antibody Response

As a broad generalization, vertebrates respond better to particulate antigens than to soluble ones, since IgM antibodies are produced in response to particulate antigens such as bacteria and IgG antibodies are associated with the response to soluble antigens.

Fish. Hagfish can make antibodies to mammalian erythrocytes and to keyhole limpet hemocyanin. (Keyhole limpet hemocyanin is a common experimental antigen. It is fairly readily available and potently antigenic.) Lampreys have been shown to make antibodies to *Brucella,* mammalian red blood cells, and bacteriophage. Soluble protein antigens are poorly immunogenic in fish, whereas particulate antigens such as bacteria or foreign erythrocytes are highly immunogenic. There is evidence to suggest that many cartilaginous fish show seasonal effects on antibody production, in that under constant conditions of light and temperature immune responses are poorer in winter than in summer. Social interactions can also influence their immune response since fish kept at high population density are immunosuppressed.

Amphibians. Soluble antigens and bacteriophage can induce the production of both IgG and IgM in toads. The IgG takes up to a month to appear, and its level is very low. Amphibians do not mount a secondary immune response to erythrocytes and bacteria, but memory does occur in response to the antigens that stimulate an IgG response. Studies on immunologic memory are complicated by the fact that antigen may persist in the circulation for several months following injection.

Reptiles. Turtles and lizards immunized with BSA, pig serum, or red blood cells can mount both primary and secondary antibody responses. The antibody produced in the primary response is 19S; the antibody produced in the secondary response is a 7S molecule. Secondary responses and 7S antibody production do not occur in response to certain bacterial antigens such as *Salmonella adelaide, Brucella abortus,* and *S. typhimurium.*

Birds. Birds produce primary and secondary responses in a manner similar to mammals. The predominance of IgM production in the primary immune response and of IgG in the secondary response is less marked than in mammals.

Primitive Mammals. Monotremes, like other mammals, produce predominantly IgM in the primary immune response and IgG in secondary immune responses. The opossum, a marsupial, resembles more primitive vertebrates in that it responds well to particulate antigens such as bacteria, while responding poorly to soluble antigen. When opossums were injected with sheep red blood cells, the primary immune response was long-lived and reasonably strong. The secondary response was weaker than the first and lasted for a much shorter period.

THE EVOLUTION OF CELL-MEDIATED IMMUNITY

Fish. Hagfish kept under good conditions in a warm environment can reject skin allografts. First-set grafts take about 72 days at 18° C to be rejected; second-set grafts are rejected in about 28 days. This suggests that immunologic memory has developed. Lampreys can also reject skin grafts slowly, and their lymphocytes divide in the presence of cells from a second lamprey and phytohemagglutinin.

Cartilaginous fish reject allografts slowly as in lampreys, but bony fish reject them rapidly. Repeated grafting leads to accelerated rejection. The rejected grafts are infiltrated by lymphocytes and show destruction of blood vessels and pigment cells. As in all ectotherms, graft rejection is slower at lower temperatures. MHC molecules have been identified in both cartilaginous and bony fish. They possess cell surface molecules homologous to MHC class I α chains, β_2-microglobulin, and MHC class II β chains. Monocytes of teleosts can release a molecule very similar to interleukin-1.

Amphibians. It takes about 28 to 42 days for a skin graft to be rejected in urodele amphibians. The graft looks healthy for about three weeks, and it is then slowly rejected. Destruction of the pigment cells makes destruction of the graft readily visible since it turns white. Second-set rejection takes about 8 to 20 days in the newt, and the alloimmune memory lasts at least 90 days.

Anurans such as bullfrogs and toads show a fairly rapid graft rejection. A first-set reaction takes about 14 days at 25° C. The graft shows capillary dilation, lymphocyte infiltration, and disintegration of pigment cells. Second-set grafts do not even become vascularized and are destroyed within a few days. If these amphibians are kept in the cold, a skin graft may take 200 days to be rejected.

By using specific antisera, it is possible to show that toad tadpoles (*Xenopus*) possess MHC class II molecules but not class I molecules. Adult toads have both class I and class II molecules. It is probable that larval amphibians lack MHC class I molecules.

During amphibian metamorphosis from larval stage to adult, a temporary immunosuppression occurs as shown by a slowing in graft rejection. Some grafts may even be tolerated at this time. As tadpoles change into frogs or toads, the thymus shrinks and there is a drop in the number of immunoglobulin-positive lymphocytes, a drop in serum globulins, and a drop in antibody levels.

Toad lymphocytes respond to phytohemagglutinin and in the mixed lymphocyte culture. The intensity of this reaction depends on the degree of incompatibility between the cell populations. *Xenopus* lymphocytes possess a receptor resembling IL-2R and can be stimulated by human IL-2 in vivo. IL-1–like activity is generated by *Xenopus* peritoneal cells.

Reptiles. As in other ectotherms, the rate of graft rejection is temperature dependent. Allogeneic skin grafts are rejected by turtles, snakes, and lizards at about 40 days at 25° C. Graft-versus-host disease can be induced by injection of parental cells into newborn turtles and can lead to death. The severity of the disease depends on the genetic disparity between the turtles. Mortality, however, is greater at 30° C than at 20° C.

Birds. Chickens possess T cells that can respond to phytohemagglutinin and concanavalin A. Avian T cells can induce delayed hypersensitivity, graft-versus-host disease, and allograft rejection. Avian homologs of mammalian γ/δ and α/β TCRs have been identified (TCR-1 and TCR-2) as well as a third population (TCR-3). TCR-3 is a heterodimer of 48 and 40 kDa and may be a subfamily of TCR α/β. Birds can reject skin allografts in about 7 to 14 days. Histologic examination shows infiltration of the grafted tissue with lymphocytes. These cells are believed to be T cells, since neonatal thymectomy results in a loss of the ability to reject grafts. If T cells are placed on the chorioallantoic membrane of 13- to 14-day-old chick embryos, they attack the chick tissues, resulting in pock formation on the membrane and a greatly enlarged spleen. The grafted cells attack primarily the hematopoietic cells of the recipient. A few days after hatching, chicks become resistant to this form of graft-versus-host attack.

TEMPERATURE AND THE IMMUNE RESPONSE

Fish, being ectothermic, respond faster and with a greater immune response at higher temperatures. At low temperatures, the lag period may be long or there may be a complete absence of an immune response. Only certain phases of the antibody response are temperature dependent. Thus secondary immune responses may be elicited at low temperature, provided primary immunization is carried out at a high temperature. There is some evidence that the cells that are most sensitive to cold temperature in fish are helper T cells. The poor response of chilled fish to antigens may be due to a loss of T-cell membrane fluidity and reactiveness to interleukins. Acclimatization to low temperatures can also occur; goldfish that are acclimatized at a low temperature may be able to produce a similar number of antibody-forming cells to those that have remained at a warmer temperature. The nature of the antigen is also critical in that certain T-cell–dependent mitogens are ineffective at low temperatures, again implying that the target cell is a helper T cell. The rejection of allografts in all ectotherms is influenced by the temperature at which the animal is kept.

Some mammals, most notably bears, some rodents, and bats, are able to hibernate. At this time, their body temperature may fall significantly. If bats are cooled to around 8° C, they cease antibody production, but rewarming permits rapid resumption of antibody synthesis. This cessation of the antibody response in hibernating bats may be of significance in allowing them to act as persistent carriers of viruses such as rabies. It is also of interest to note that mammalian immune responses function optimally at a slightly higher temperature than normal so that infections induce a fever as a result of the release of interleukin-1.

INFLAMMATION IN FISH

Fish possess macrophages in many tissues, especially in the mesentery, spleen, kidney, and atrium of the heart. These cells are highly phagocytic, and in histochemical staining and ultrastructure they closely resemble mammalian macrophages.

Fish neutrophils are similar in morphology to mammalian neutrophils and are frequently seen in inflammatory lesions. There is little evidence that they are actively phagocytic, and there is no evidence that

> **CLARIFICATION**
>
> ***Fever in Ectothermal Animals?***
>
> Reptiles, being ectotherms, are unable to change their body temperature by physiologic mechanisms. As a result, they will not develop a fever if maintained in a constant-temperature environment. If maintained in an environment with cool and warm areas, they will cycle between these areas and maintain their body temperature within well-defined limits. Vaugn and her colleagues infected desert iguanas (*Dipsosaurus dorsalis*) with the bacterium *Aeromonas hydrophila*. They observed that uninfected iguanas maintained their temperature between 37.4° and 41.1° C. Infected iguanas, however, modified their behavior so that they spent more time in the warm environment. Their temperatures cycled between 39.7° and 42.7° C. Once the bacterial infection was cured, the iguanas resumed their normal behavior. Thus the iguanas effectively induced a fever by their behavior.
>
> Vaugn, L.K., Bernheim, H.A., and Kluger, M.J. Nature, 252: 473–474, 1974.

they can ingest foreign material. They do, however, possess most of the enzymes of mammalian neutrophils. It has been suggested that the fish neutrophil may carry out its bactericidal function extracellularly rather than intracellularly. This may cause severe tissue damage if free oxygen radicals are released. The fat of fish is highly polyunsaturated as an adaptation to low temperatures. Polyunsaturated fats are prone to oxidation, and free radicals have the potential to cause severe host tissue damage and oxidation of tissue lipids. Fish therefore require a powerful means of modulating a response like this. One suggested mechanism is through the use of melanin and related pigments. In the lymphoid tissues of most bony fish as well as in inflammatory lesions are found many cells containing dark-colored pigment, which is mainly melanin. Because melanin has the ability to quench free oxidizing radicals, it has been suggested that it protects tissues against free radicals produced by phagocytic cells such as neutrophils.

Hypersensitivity

Anaphylaxis in fish has been well described, although histamine has not been found and the response of fish to histamine can be variable. Intradermal injection of some fungal extracts produces immediate hypersensitivity reactions in marine teleosts such as flatfish. Fish do not contain typical mammalian mast cells, although a granular leukocyte may be the homolog of the mammalian mast cell. In flatfish and salmon, an eosinophilic cell has been suggested as the responsive cell. Localized degranulation of this eosinophilic cell has been described in fish undergoing hypersensitivity responses. Anaphylactic (or anaphylactoid) reactions have been described in amphibians, reptiles, and birds.

KEY WORDS

Allograft rejection p. 506
Analog p. 505
Axial organ p. 506
Bone marrow p. 508
Bursa of Fabricius p. 510
Cell-mediated immunity p. 513
Coelomocyte p. 505
Cyclostome p. 507
Ectotherm p. 514
Hemocyte p. 505
Hemolymph p. 505
Hibernation p. 514
Homolog p. 505
Immunoglobulin p. 511
Immunoglobulin N p. 512
Inflammation p. 514
Invertebrate p. 505
Lymph node p. 509
Lymphocyte p. 510
Natural agglutinin p. 506
Phagocytosis p. 505
Renal lymphoid tissue p. 510
Spleen p. 509
Temperature effects p. 514
Thymus p. 508
Ventral cavity body p. 510
Vertebrate p. 507

QUESTIONS

1. In which invertebrate does allograft rejection first appear?
   a. sponges
   b. annelids
   c. insects
   d. tunicates
   e. echinoderms

2. The axial organ, a primitive lymphoid organ, is found in
   a. crustaceans
   b. insects
   c. starfish
   d. tunicates
   e. cyclostomes

3. Organs having a similar origin and function are considered
   a. analogs
   b. isotypes
   c. allotypes
   d. symbiotic
   e. homologs
4. In which of these species is IgN found?
   a. chicken
   b. platypus
   c. duck
   d. *Xenopus*
   e. echnida
5. In which vertebrate group does IgA first appear?
   a. birds
   b. reptiles
   c. amphibians
   d. bony fish
   e. cartilaginous fish
6. Ventral cavity bodies are immunological organs found in
   a. chickens
   b. starfish
   c. agnathans
   d. marsupials
   e. laryal amphibians
7. In which vertebrate group does bone marrow first appear?
   a. bony fish
   b. reptiles
   c. amphibians
   d. cartilaginous fish
   e. monotremes
8. Cecropins are protective proteins found in
   a. starfish
   b. tunicates
   c. annelids
   d. moths
   e. sponges
9. What is the pronephros?
   a. a primitive spleen
   b. the anterior kidney
   c. a thymus
   d. a ventral cavity body
   e. an axial organ
10. Which T-cell surface antigen has been identified in many invertebrates?
   a. TCRα
   b. MHC class II
   c. Qa1
   d. CD8
   e. Thy 1
11. If most vertebrates can function well without possessing either IgA or IgE, are these immunoglobulins of any use?
12. Why do birds and mammals apparently not require kidney-associated lymphoid tissue? Might this have something to do with an adaptation to a terrestrial environment?
13. On what basis are the phagocytic cells of invertebrates considered to be analogs rather than homologs of vertebrate phagocytes?
14. What properties of IgM suggest that it evolved before IgG?
15. Justify the existence of the bursa of Fabricius. Why might it be found only in birds?
16. What is the biological significance of the finding that the genes for proteins such as Thy-1, which are well recognized in mammals, have also been found in some invertebrates?
17. The development of the bone marrow is associated with adaptation to a terrestrial existence. Comment on possible reasons for or advantages of this.
18. It has been suggested that insects do not need an immune system since they are short-lived and can adapt very rapidly to pathogenic microorganisms. Is this a valid argument? If not, why not?

Answers: 1a, 2c, 3e, 4c, 5a, 6e, 7c, 8d, 9b, 10e

SOURCES OF ADDITIONAL INFORMATION

Char, D., et al. A third sublineage of avian T cells can be identified with a T cell receptor-specific antibody. J. Immunol., 145:3547–3555, 1990.

Delmotte, F., et al. Purification of an antibody-like protein from the sea star, *Asterias rubens* L. Eur. J. Immunol., 16: 1325–1330, 1986.

Du Pasquier, L., Schwager, J., and Flajnik, M.F. The immune system of Xenopus. Annu. Rev. Immunol., 7:251–275, 1989.

Edwards, B.F., and Ruben, L.N. Aspects of amphibian immunity. *In* Hay, J.B., ed. Animal Models of Immunological Processes. Academic Press, London, 1982.

Grossberger, D., and Parham, P. Reptilian class I major histocompatibility complex genes reveal conserved elements in class I structure. Immunogenetics, 36:166–174, 1992.

Kluger, M.J. Fever, its biology, evolution and function. Princeton University Press, Princeton, N.J., 1979.

Manning, M.J., and Tatner, M.F. Fish Immunology. Academic Press, London, 1985.

Manning, M.J., and Turner, R.J. Comparative Immunobiology. John Wiley & Sons, New York, 1976.

Marchalonis, J.J. Immunity in Evolution. Harvard University Press, Cambridge, Mass., 1977.

Marchalonis, J.J., and Schluter, S.F. Evolution of variable and constant domains and joining segments of rearranging immunoglobulins. FASEB J., 3:2469–2479, 1989.

Marchalonis, J.J., and Schluter, S.F. Immunoproteins in evolution. Dev. Comp. Immunol., 13:285–301, 1989.

McCormack, W.T., and Thompson, C.B. Somatic diversification of the chicken immunoglobulin light chain gene. Adv. Immunol., 48:41–67, 1990.

Nonaka, M., and Takahashi, M. Complete complementary DNA sequence of the third component of complement of lamprey. J. Immunol., 148:3290–3295, 1992.

Reinisch, C.L., and Litman, G.W. Evolutionary immunobiology. Immunol. Today, 10:278–281, 1989.

Robertson, M., and Postlethwait, J.H. The humoral antibacterial response of *Drosophila* adults. Dev. Comp. Immunol., 10:167–179, 1986.

Sato, K., et al. Evolution of the MHC: Isolation of class II β chain cDNA clones from the amphibian *Xenopus laevis*. J. Immunol., 150:2831–2843, 1993.

Solomon, J.B., and Horton, J.D. Developmental Immunobiology. Elsevier/North Holland Biomedical Press, Amsterdam, 1977.

Stone, R. Déjà vu guides the way to new antimicrobial steroid. Science, 259:1125, 1993.

Glossary

acquired cell-mediated immunity An immune state mediated by T cells and characterized by the development of activated macrophages.

acquired immune deficiency syndrome (AIDS) A disease caused by the human immunodeficiency virus, which destroys key components of the immune system. As a result, infected individuals become very susceptible to infections and cancers.

activated macrophage A macrophage in a state of enhanced metabolic and functional activity.

active immunity Immunity produced as a result of administration of an antigen.

acute inflammation Rapidly developing inflammation of recent onset. It is characterized by tissue infiltration by neutrophils.

acute-phase protein A protein secreted by hepatocytes whose levels in the blood rise rapidly in response to acute inflammation or infection.

addressin A protein on lymphocyte surfaces that binds to homing receptors on blood vessel walls and so regulates lymphocyte emigration from blood.

adjuvant A substance that, when given with an antigen, enhances the immune response to that antigen.

adoptive immunity The development of immunity as a result of the transfer of cells from an immunized animal to an unimmunized recipient.

affinity The strength of binding between two molecules such as an antigen and antibody. Usually expressed as an association constant (Ka).

affinity maturation The progressive increase in antibody affinity for antigen that occurs during an immune response.

agammaglobulinemia The absence of gamma globulins in blood.

agglutination The clumping of particulate antigens by antibody.

AIDS The acronym for acquired immune deficiency syndrome. A disease caused by human immunodeficiency virus.

albumin Strictly speaking, those serum proteins that remain in solution in the presence of half-saturated ammonium sulfate. In practice it is the name given to the major serum protein of 60 kDa.

alleles An alternative form of a gene from a single locus.

allelic exclusion The ability of a cell from a heterozygous individual to synthesize only one of two possible alleles.

allergen An antigen that provokes type I hypersensitivity (allergy).

allergic contact dermatitis An inflammatory skin condition occurring as a result of a type IV hypersensitivity reaction to skin cells modified by exposure to foreign chemicals.

allergy Immediate (type I) hypersensitivity. An immune response characterized by the release of pharmacological agents as a result of mast-cell and basophil degranulation. This degranulation is usually immunologically mediated. However, many clinicians and laypersons use the term "allergy" to describe any unpleasant reaction of immunological or quasi-immunological origin.

allogeneic Genetically dissimilar individuals of the same species.

allograft A tissue or organ graft between two genetically dissimilar animals of the same species.

allotype Genetically coded differences between the proteins of different individuals of a species.

alternative complement pathway A series of enzyme reactions triggered by interactions on activating surfaces leading to activation of the complement system.

amyloid An extracellular amorphous, waxy protein deposited in the tissues of individuals suffering from chronic inflammation or a myeloma.

analog An organ or tissue that has the same function as another but is of different evolutionary origin.

anamnestic response A secondary immune response.

anaphylatoxin A complement fragment with the ability to stimulate mast-cell degranulation and smooth muscle contraction.

anaphylaxis A sudden, systemic, severe immediate hypersensitivity reaction occurring as a result of rapid generalized mast-cell degranulation.

anergy The failure to respond to an antigen in a sensitized animal.

antibiotic A chemical compound, usually obtained from microorganisms, that can prevent growth or kill bacteria. Do not confuse this with an antibody.

antibody An immunoglobulin protein molecule synthesized on exposure to antigen that can combine specifically with that antigen.

antibody-dependent cellular cytotoxicity (ADCC) A cytotoxic process mediated by effector cells linked to target cells by cell-membrane-bound antibody.

antigen A foreign substance that can induce an immune response.

antigenic determinant *See* epitope.

antigenicity The ability of a molecule to be recognized by an antibody or lymphocyte.

antigenic variation Progressive change in surface antigens exhibited by viruses, parasites, and some bacteria. This is an effective device used to evade destruction by the immune system.

antigen-presenting cell A cell that can ingest antigen, process it, and present this processed antigen to antigen-sensitive cells in association with histocompatibility molecules.

antigen processing The series of events that modify antigens so that they can be recognized by antigen-sensitive cells.

antigen-sensitive cell A cell that can bind and respond to specific antigen.

antiglobulin Antibody made against an immunoglobulin, usually by injecting immunoglobulin into an animal of another species.

antiserum Serum that contains specific antibodies. Synonymous with immune globulin.

antitoxin Antiserum directed against a toxin and used for passive immunization.

apoptosis A form of cell death exhibiting a distinctive morphology and mediated by physiologic pathways.

Arthus reaction A local inflammatory reaction due to a type III hypersensitivity reaction; it is induced by the repeated injection of antigen into the skin of an immunized animal.

asthma A type I hypersensitivity disease characterized by a reduction in airway diameter, leading to difficulty in breathing (dyspnea).

atopy A genetically determined predisposition to develop clinical allergies.

attenuation The reduction of virulence of an infectious agent.

autoantibody An antibody directed against an epitope on normal body tissues.

autoantigen A normal body component that acts as an antigen.

autograft A tissue or organ graft made between two sites within the same animal.

autoimmune disease Disease caused by an immune attack against an individual's own tissues.

autoimmunity The process of mounting an immune response against a normal body component.

benign tumor A tumor that does not spread from its site of origin.

blast cell A cell that has large amounts of cytoplasm immediately prior to division.

blocking antibody An antibody that, by binding to a target cell, protects it from immune destruction.

blood group An antigen found on the surface of red blood cells. Blood group expression is inherited.

B lymphocyte, or B cell A lymphocyte that has undergone a period of processing in the bursa or bursa equivalent. B cells are responsible for antibody production.

bursectomy Surgical removal of the bursa of Fabricius.

capping The clumping of surface structures such as antigens or receptors in a small area on the surface of a cell.

capsid The protein coat around a virus.

carcinoma A tumor originating from cells of epithelial origin.

carrier An immunogenic macromolecule to which a hapten may be bound, so making the hapten immunogenic. This part of the molecule is recognized by T cells.

cascade reactions An interlinked series of enzyme reactions in which the products of one reaction catalyze a second reaction, and so forth.

CD molecule A cell surface molecule classified according to the internationally accepted CD (Cluster of Differentiation) system. They are identified by monoclonal antibodies.

cell-mediated cytotoxicity Destruction of target cells induced by contact with lymphocytes or macrophages.

cell-mediated immunity A form of immune response mediated by T lymphocytes and macrophages; it can be conferred on an animal by adoptive transfer.

cestode A parasitic tapeworm.

chemokine A family of proinflammatory and chemotactic cytokines with a characteristic sequence of four cysteine residues.

chemotaxis The directed movement of cells under the influence of an external chemical gradient.

chimera An animal that contains cells from two or more genetically different individuals.

chromosome translocation A form of mutation in which portions of two chromosomes switch position.

chronic inflammation Slowly developing or persistent inflammation characterized by tissue infiltration with macrophages and fibroblasts.

class The five major forms of immunoglobulin molecules common to all members of a species.

class switch The change in immunoglobulin class that occurs during an immune response as a result of heavy-chain gene rearrangement.

classical complement pathway A series of enzyme reactions triggered by antigen–antibody complexes, leading to activation of the complement system.

clonal deletion The elimination of self-reactive T cells in the thymus.

clonal selection A key concept in immunology suggesting that an individual possesses many lymphocyte clones that can each react with only one epitope. These clones are generated at random, but cells responsive to self-epitopes are destroyed during fetal life.

clone The progeny of a single cell.

clonotype A clone of B cells with the ability to bind a single epitope.

cluster of differentiation (CD) *See* CD molecule.

coated pit A structure employed to endocytose ligand attached to cell surface receptors while effectively recycling the receptors.

coelomocyte A phagocytic cell found in the coelomic cavity of invertebrates.

colostrum The secretion that accumulates in the mammary gland in the last weeks of pregnancy. It is very rich in immunoglobulins.

combined immunodeficiency A deficiency in both the T-cell- and B-cell-mediated components of the immune system.

complement A group of serum proteins that is activated by factors such as the combination of antigen and antibody and results in a variety of biological consequences, including cell lysis and opsonization.

complementarity-determining region An area within the variable regions of antibodies and T-cell antigen receptors that binds to antigen and determines the molecule's antigen-binding specificity. Synonymous with hypervariable regions.

concanavalin A (ConA) A lectin extracted from the jack bean, which makes T cells divide.

conglutinin A bovine mannose-binding protein that combines with the activated third component of complement.

constant domain A structural domain with little sequence variability found in antibodies and TCR.

constant region The portion of immunoglobulin and TCR peptide chains that consists of a relatively constant sequence of amino acids.

contrasuppression The suppression of suppressor cells by a population of contrasuppressor T cells. Contrasuppressor cells are distinct from helper cells.

convertase A protease that acts on a complement component to cause its activation.

cortex The outer region of an organ such as the thymus or lymph node.

corticosteroid Steroid hormone released from the adrenal cortex that has profound effects on the immune system. Some corticosteroids may be synthetic in origin.

costimulator A molecule required to stimulate an antigen-sensitive cell simultaneously with antigen in order to initiate an effective immune response.

cross-reaction The reaction of an antibody directed against one antigen with a second antigen. This occurs because the two antigens possess epitopes in common.

cutaneous basophil hypersensitivity A form of delayed hypersensitivity associated with an extensive basophil infiltration.

cytokine A protein that mediates cellular interactions and regulates cell growth and secretion. As a result, cytokines regulate the immune response.

cytolysis Destruction of cells by immune processes.

cytophilic antibody Immunoglobulin that has the ability to bind spontaneously to cellular Fc receptors.

cytotoxic cell A cell that can injure or kill other cells.

delayed hypersensitivity A cell-mediated inflammatory reaction in the skin, so called because it takes 24 to 48 hours to reach maximum intensity.

dendritic cell A macrophage-like cell that possesses long cytoplasmic processes and functions as an antigen-trapping and antigen-presenting cell.

desensitization The prevention of allergic reactions through the use of multiple injections of allergen.

diapedesis The emigration of cells from intact blood vessels during inflammation.

disseminated intravascular coagulation Activation of the clotting cascade within the circulation.

disulfide bonds Bonds that form between two cysteine residues in a protein. They may be either interchain (between two peptide chains) or intrachain (joining two parts of one chain).

domain Discrete structural units by which protein molecules are organized.

dysgammaglobulinemia The abnormal production of gamma globulins in blood.

effector cell A cell that is able to "effect" an immune response. These include cytotoxic T cells and plasma cells.

electrophoresis The separation of the proteins in a complex mixture by subjecting them to an electrical potential.

ELISA Enzyme-linked immunosorbent assay. An immunological test that uses enzyme-linked antiglobulins and substrate bound to an inert surface.

endocytosis The uptake of extracellular substances by cells.

endogenous antigen Foreign antigen synthesized within body cells. An example of this is a virus protein.

endosome A cytoplasmic vesicle formed by invagination of the outer cell membrane.

endothelium The cells that line blood vessels and lymphatics.

endotoxin The lipopolysaccharide component of gram-negative bacterial cell walls.

enhancement Improved survival of grafts or tumor cells induced by some antibodies.

eosinophilia Increased numbers of eosinophils in the blood.

epithelioid cell A macrophage that accumulates within a tubercle and resembles an epithelial cell in histological section.

epitope An area on the surface of an antigen that stimulates a specific immune response and against which that response is directed. Synonymous with antigenic determinant.

Epstein-Barr virus A herpes virus that is the causal agent of Burkitt's lymphoma and infectious mononucleosis.

erythema Redness due to inflammation.

eukaryotic organism An organism characterized by cells possessing a distinct nucleus and containing both DNA and RNA.

exocytosis The export of material from a cell by the fusion of cytoplasmic vesicles with the outer cell membrane.

exogenous antigen A foreign antigen that originates at a source outside the body, for example, bacterial antigens.

exon A segment of DNA that contains expressed genes.

exotoxin A soluble protein toxin, usually produced by gram-positive bacteria, that has a specific toxic effect.

Fab fragment The antigen-binding fragment of a partially digested antibody. It consists of light chains and the N-terminal halves of heavy chains.

facultative intracellular organism An organism that can, if necessary, grow within cells.

Fc receptor A cell surface receptor that specifically binds antibody molecules through their Fc region.

Fc region That part of an immunoglobulin molecule consisting of the C-terminal halves of heavy chains; it is responsible for the biological activities of the molecule.

fibrinolytic Breakdown of fibrin.

first-set reaction The initial rejection of a foreign tissue graft.

fluorescent antibody An antibody chemically attached to a fluorescent dye.

framework region The part of a variable region of immunoglobulins and TCRs that has a relatively constant amino acid sequence and so forms a structure on which the hypervariable, complementarity-determining regions may be constructed.

gamma (γ) globulin A serum protein that migrates toward the cathode on electrophoresis. Gamma globulins contain most of the immunoglobulins.

gammopathy An abnormal increase in gamma globulin levels.

gel diffusion An immunoprecipitation technique that involves letting antigen and antibody meet and precipitate in a clear gel such as agar.

gene complex A cluster of related genes occupying a restricted area of a chromosome.

gene conversion The exchange of blocks of DNA between different genes.

genes A unit of DNA that codes for the amino acid sequence of a polypeptide chain.

gene segment An exon that codes for a portion of a peptide chain.

germinal center A structure characteristic of many lymphoid organs in which rapidly dividing B cells form a pale-staining spherical mass surrounded by a zone of dark-staining cells. The location of somatic mutation and memory cell formation.

globulin A serum protein precipitated by the presence of a half-saturated solution of ammonium sulfate.

glomerulonephritis Pathological lesions in the glomeruli of the kidney.

glycoform Differing molecular forms of a protein resulting from differences in glycosylation.

glycoprotein A protein that contains carbohydrate.

G-protein A GTP-binding protein that acts as a signal transducer for a cell surface receptor.

graft-versus-host reaction (or disease) Disease caused by an attack of grafted T cells on the cells of a histoincompatible and immunodeficient recipient.

granulocyte A myeloid cell containing prominent cytoplasmic granules. Granulocytes include neutrophils, eosinophils, and basophils.

granuloma A localized nodular inflammatory lesion characterized by chronic inflammation with mononuclear cell infiltration and extensive fibrosis.

granzyme A family of proteolytic enzymes found in the granules of cytotoxic T cells.

growth factor A biologically active molecule that promotes cell growth.

haplotype The alleles at all loci within a gene complex.

hapten A small molecule that cannot initiate an immune response unless first bound to an immunogenic carrier molecule.

heat shock protein A protein synthesized by cells in response to physiologic stress.

helminth A worm, many of which are parasites and so stimulate immune responses.

helper T cell A class of T cells that promotes immune responses by releasing cytokines such as interleukin-2 or interleukin-4.

hemagglutination The agglutination of red blood cells.

hematopoietic organ An organ in which blood cells are produced.

hemocyte Phagocytic cells found in invertebrate hemolymph.

hemolymph The fluid that fills the body cavities of invertebrates. It has functions analogous to those of blood.

hemolysin An antibody that can lyse red blood cells in the presence of complement.

hemolytic disease Disease occurring as a result of the destruction of red blood cells by antibodies transferred to the young animal from its mother.

herd immunity Immunity conferred on a population as a whole as a result of the presence of some immune individuals within that population.

heterodimer A molecule consisting of two different subunits.

heterophile antibody An antibody that reacts with epitopes found on a wide variety of unrelated molecules.

high endothelial venule A specialized blood vessel lined with high epithelium, found in the paracortex of lymph nodes.

hinge region The region between the first and second constant domains in some immunoglobulin molecules that permits them to bend freely.

histiocyte A connective tissue macrophage.

histocompatibility molecule A cell membrane protein that is required to present antigen to antigen-sensitive cells.

HIV Human immunodeficiency virus, the cause of AIDS.

homing receptor A lymphocyte cell surface receptor that enables lymphocytes to bind to ligands on endothelial cells as a preliminary step to leaving the bloodstream.

homodimer A molecule consisting of two identical subunits.

homolog A part similar in structure, position, and origin to another organ.

homology The degree of sequence similarity between two genes (nucleotide sequences) or two proteins (amino acid sequences).

humoral immunity Immunity that can be transferred in body fluids—antibody-mediated immunity.

hybridoma A cell line formed by the fusion of a myeloma cell with a normal antibody-producing cell.

hypersensitivity An immunologically mediated damaging inflammatory response to a normally innocuous antigen.

hypersensitivity pneumonitis An inflammatory reaction in the lung caused by a type III reaction to inhaled antigen.

hypervariable region A small area within an immunoglobulin or TCR variable region where the greatest variations in amino acid sequence occur.

hypogammaglobulinemia Low levels of gamma globulins in blood.

Ia antigen A mouse MHC class II molecule.

iccosome An immune complex body. A fragment of dendritic cell that carries antigen to antigen-sensitive cells.

idiotope An epitope formed by the variable amino acid sequences located in or close to the antigen-binding site of an immunoglobulin.

idiotype The collection of idiotopes on an immunoglobulin molecule.

idiotype network The series of reactions between idiotypes, anti-idiotypes, and anti-anti-idiotypes that play a role in controlling immune responses.

immediate hypersensitivity The hypersensitivity reaction mediated by IgE and mast cells, otherwise known as type I hypersensitivity.

immune complex Another term for antigen–antibody complexes.

immune elimination The removal of an antigen from the bloodstream by circulating antibodies and phagocytic cells.

immune globulin An antibody preparation containing specific antibodies against a pathogen and used for passive immunization.

immune response gene An MHC class II gene, so called because it regulates the ability of an animal to respond to specific antigens.

immune stimulant A compound, commonly bacterial in origin, that stimulates the immune system by promoting cytokine release from macrophages.

immune surveillance The concept that lymphocytes survey the body for cancerous or abnormal cells and then eliminate them.

immunity The state of resistance to an infection.

immunization The administration of an antigen to an individual to confer immunity.

immunoconglutinin Autoantibodies directed against activated complement components.

immunodeficiency Disease conditions in which immune function is defective.

immunodiffusion Another name for the gel-diffusion technique.

immunodominant Describing the epitope on a molecule that provokes the most intense immune response.

immunoelectrophoresis A procedure involving sequential electrophoresis and immunoprecipitation; it is used to identify the proteins in a complex solution such as serum.

immunofluorescence Immunological tests that use antibodies conjugated to a fluorescent dye.

immunogenetics That portion of immunology that deals with the direct effects of genes on the immune system.

immunogenicity The ability of a molecule to elicit an immune response.

immunoglobulin A glycoprotein with antibody activity.

immunoglobulin superfamily A family of proteins that contain characteristic immunoglobulin domains.

immunological paralysis A form of immunological tolerance in which an ongoing immune response is inhibited by the presence of large amounts of antigen.

immunoperoxidase An immunological test that uses antibodies chemically conjugated to the enzyme peroxidase.

immunosuppression Inhibition of the immune system by drugs or other processes.

inactivated vaccine A vaccine containing an agent that has been treated in such a way that it is no longer capable of replication in the host.

incomplete antibody An antibody that can bind to a particulate antigen but is incapable of causing its agglutination.

indurated Hardened.

inflammation The complex series of responses of tissues to injury. These responses generally enhance tissue defenses and initiate repair processes.

integral membrane protein Cell surface proteins that are integral components of the cell membrane as opposed to proteins that are passively adsorbed to cell surfaces.

integrins A family of adhesion proteins found on cell membranes that bind to connective tissue proteins such as collagen or fibronectin.

interchain bond A bond between two peptide chains, usually formed by a disulfide linkage between two cysteine residues.

interdigitating cells A form of dendritic cell found within lymphoid organs.

interferons A protein that can interfere with viral replication. Some interferons play an important role in the development of cellular immunity.

interleukin A protein that acts as a growth and differentiation factor for the cells of the immune system.

intrachain bond A bond between two cysteine residues on a single peptide chain. Because disulfide bonds are short, its effect is to produce a loop in the peptide chain.

intraepithelial lymphocyte A small lymphocyte, probably a T cell, located among the epithelial cells in the intestinal wall.

intron A segment of DNA inserted between exons that does not contain expressed genes. Introns are transcribed into RNA but are not translated into protein.

isoform Different molecular forms of a protein that are generated by differential processing of RNA transcripts obtained from a single gene.

isogeneic (syngeneic) Genetically identical.

isograft A graft between two genetically identical animals.

isotype The structural variants of a protein that are present in all members of a species. Immunoglobulin classes and subclasses are examples of isotypes.

isotype switching The change in immunoglobulin isotype that occurs during the immune response as a result of heavy-chain gene switching.

J chain A short peptide that joins two monomers in the polymeric immunoglobulins IgM and IgA.

joining (J) gene segment A short gene segment that is located 3′ to the V gene segments in immunoglobulin and TCR V genes.

kinin A vasoactive peptide produced in injured or inflamed tissue.

Kupffer cell Macrophages lining the sinusoids of the liver.

lactenin A bactericidal factor in milk.

lag period The interval between administration of antigen and the first detection of antibody.

Langerhans' cell A dendritic cell found in the skin. These cells are very effective antigen-presenting cells.

lectin A protein, usually of plant origin, that can bind specifically to a carbohydrate. Many lectins can induce lymphocytes to divide.

leukemia A cancer consisting of white cells that proliferate within the blood.

leukocyte A white blood cell. This general term covers all the nucleated cells of blood.

leukotriene A vasoactive metabolite of arachidonic acid produced by the actions of lipoxygenase.

ligand A generic term for the molecules that bind specifically to a receptor.

linkage disequilibrium A situation in which a pair of genes is found in a population at an unexpectedly high frequency when compared with the frequency of the individual genes.

linked recognition The necessity for lymphocytes to receive two simultaneous signals in order to be activated.

locus The location of a gene on a chromosome.

looping out A method of excising a segment of intervening DNA (intron) in order to join two gene segments (exons).

lymph The clear tissue fluid that flows through lymphatic vessels.

lymphadenopathy Literally, disease of the lymph nodes. In practice it is used to describe enlarged lymph nodes.

lymphoblast A dividing lymphocyte.

lymphocyte A small mononuclear cell with a round nucleus containing densely packed chromatin. Most have only a thin rim of cytoplasm.

lymphocyte trapping The trapping of lymphocytes within a lymph node during the node's response to antigen.

lymphokine A cytokine released by lymphocytes.

lymphokine-activated killer (LAK) cell A lymphocyte activated by exposure to cytokines such as IL-2 in vitro.

lymphopenia Abnormally low numbers of lymphocytes in blood.

lymphotoxin A cytotoxic cytokine secreted by lymphocytes.

lysosomal enzyme One of the complex mixture of enzymes, many of which are proteases found within lysosomes.

lysosome A cytoplasmic organelle found within phagocytic cells that contains a complex mixture of potent proteases.

lysozyme An enzyme present in tears, saliva, and neutrophils. It attacks the mucopeptides in the cell wall of gram-positive bacteria.

macrophage A large phagocytic cell containing a single rounded nucleus.

major histocompatibility complex The gene region that contains the genes for the major histocompatibility molecules as well as for some complement components and related proteins.

malignant tumor A tumor whose cells have a tendency to invade normal tissues and spread by lymphatics or blood to distant tissue sites.

maternal antibodies Antibodies that originate in the mother but enter the bloodstream of her offspring either by transport across the placenta as in primates or by adsorption of ingested colostrum in other mammals.

medulla The region in the center of lymphoid organs such as the thymus or lymph nodes.

membrane attack complex The complement protein structure that is embedded in target-cell membranes, resulting in their lysis.

memory cell A lymphocyte formed as a result of exposure to antigen. These cells have the ability to mount an enhanced response to antigen as compared with lymphocytes that had not previously encountered antigen.

memory response The enhanced immune response that is triggered as a result of exposing a primed animal to antigen.

mesangial cell A modified muscle cell found within a glomerulus.

MHC molecules A protein coded for by genes located in the major histocompatibility complex.

MHC restriction The need for a T cell to recognize an antigen in association with a MHC molecule. It is required for helper and cytotoxic T cells to recognize antigen and for helper T cells to cooperate with B cells.

microglia Macrophages resident within the brain.

mitogen Any substance that makes cells divide.

mixed lymphocyte reaction Lymphocyte division induced by the presence of foreign cells.

modified live virus A virus whose virulence has been reduced so that it can replicate in the host but cannot cause disease in normal animals.

molecular mimicry The development by parasites of molecules whose structure closely resembles molecules found in their host. In this way the parasites may be able to evade destruction by the immune system.

monoclonal Originating from a single cell.

monoclonal antibody Antibody derived from a clone of cells and hence chemically homogeneous.

monoclonal gammopathy The appearance in serum of a high level of a monoclonal immunoglobulin. This is commonly, but not always, associated with the presence of a myeloma.

monocyte An immature macrophage found in the blood.

monokine A cytokine secreted by macrophages and monocytes.

monomer The basic unit of a molecule that can be assembled using repeating subunits.

mononuclear cell A leukocyte with a single round nucleus, for example, a lymphocyte or a macrophage.

mononuclear-phagocytic system The cells that belong to the macrophage family and their precursors.

myeloid system All the granulocytes and their precursors. These precursor cells are found in the bone marrow.

myeloma A tumor of plasma cells.

myeloma protein The immunoglobulin product secreted by a myeloma cell.

natural antibody An antibody against foreign antigens found in the serum of normal, unimmunized individuals. Most probably arise as a result of exposure to cross-reacting bacterial antigens.

natural killer (NK) cell A non-T, non-B large granular lymphocyte found in normal individuals and capable of killing some tumor cells and some virus-infected cells.

natural suppressor cell A cell found in unimmunized individuals that has the ability to suppress some immune responses.

necrosis Cell death due to pathological causes.

negative feedback A control mechanism whereby the products of a reaction suppress their own production.

negative selection A euphemism for the killing of cells that have the potential to react to self-antigens. It is a key mechanism in the prevention of autoimmunity.

nematode A roundworm.

neoplasm A synonym for a tumor.

neutralization Blockage of the activity of an organism or a toxin by antibody.

neutropenia Low numbers of neutrophils in blood.

neutrophil A polymorphonuclear neutrophil granulocyte.

neutrophilia High numbers of neutrophils in blood.

NK cell A natural killer cell.

noncovalent bond A chemical bond such as a hydrogen or hydrophobic bond that can reversibly link peptide chains. These play a key role in the binding of antigen with antibodies or with T-cell antigen receptors.

normal flora The microbial population consisting mainly of bacteria that colonize normal body surfaces. They play a key role in preventing invasion by pathogenic organisms.

nucleocapsid The key structural component of a virus, consisting of the viral nucleic acid and its protective capsid coat.

nude mice A mutant strain of mice that have no thymus and are hairless. This strain of mouse has proved very useful in immunological research.

oligosaccharide Chains of at least two covalently linked sugars.

oncogene A gene whose protein product plays a key role in cell division. As a result, its uncontrolled production leads to excessive cell growth and tumor formation. Oncogenes may be found in normal cells as well as in cancer-causing viruses.

oncogenic virus A virus that causes cancer.

ontogeny The embryonic development of an organ or animal.

opportunistic pathogen An organism that, although unable to cause disease in a healthy individual, may invade and cause disease in an individual whose immunological defenses are impaired.

opsonin A substance that facilitates phagocytosis by coating foreign particles.

optimal proportions When antigen and antibody combine, this is the ratio of reactants that generates the largest immune complexes.

paracortex The region located between the cortex and medulla of lymph nodes in which T cells predominate.

paratope The antigen-combining site on an immunoglobulin.

passive agglutination The agglutination of inert particles by antibody directed against antigen bound to their surface.

passive immunization Protection of one individual conferred by administration of antibody produced in another individual.

pathogenesis The mechanism of a disease.

pathogenic organism An organism that causes disease.

perforin A family of proteins made by T cells and NK cells that when polymerized can insert themselves into target-cell membranes and provoke cell lysis.

phagocyte A cell whose prime function is to eat bacteria (phagocytosis). These include macrophages and related cells, neutrophils, and eosinophils.

phagocytosis The ability of some cells to ingest foreign materials. Literally, eating by cells.

phagolysosome A structure produced by the fusion of a phagosome and a lysosome following phagocytosis.

phagosome The cytoplasmic vesicle that encloses an ingested organism.

phenogroup A set of blood group alleles that are consistently inherited as a group.

phylogeny The evolutionary history of a plant or animal species.

phytohemagglutinin (PHA) A lectin derived from the red kidney bean. It acts as a T cell mitogen.

pinocytosis The endocytosis of small fluid droplets.

plaque-forming cell An antibody-secreting cell capable of secreting anti-red cell antibodies and so forming a plaque in a layer of red blood cells in the presence of complement.

plasma The clear fluid that forms the liquid phase of blood.

plasma cell A fully differentiated B cell capable of synthesizing and secreting large amounts of antibody.

point mutation A mutation resulting from an alteration in a single base in a gene.

pokeweed mitogen (PWM) A lectin derived from the pokeweed plant that stimulates both T and B cells to divide.

polyclonal gammopathy The appearance in serum of a high level of immunoglobulins of many specificities originating from many clones.

polymorphism Inherited structural differences between proteins from allogeneic individuals as a result of multiple alternative alleles at a single locus.

polymorphonuclear neutrophil granulocyte A blood leukocyte possessing neutrophilic cytoplasmic granules and an irregular lobed nucleus.

positive selection The enhanced proliferation of cells within the thymus that can respond optimally to foreign antigen.

prevalence The number of cases of a disease.

primary binding test A serologic assay that directly detects the binding of antigen and antibody.

primary immune response The immune response resulting from an individual's first encounter with an antigen.

primary immunodeficiency An inherited immunodeficiency disease.

primary lymphoid organ An organ that serves either as a source of lymphocytes or in which lymphocytes mature.

primary pathogen An organism that can cause disease without first suppressing an individual's immune defenses.

primary structure The amino acid sequence of a protein.

privileged site A location within the body where foreign grafts are not rejected. A good example is the cornea of the eye.

prokaryotic organism An organism composed of cells whose genetic material is free in the cytoplasm. As a result, these cells do not contain a recognizable nucleus.

prostaglandin A biologically active lipid metabolite of arachidonic acid produced by the actions of the enzyme cyclooxygenase.

protein kinase An enzyme that phosphorylates proteins.

proteasome A complex cytoplasmic structure consisting of several proteases. It can act on proteins to cleave them into multiple fragments.

proto-oncogene A normal cellular gene that, when mutated, can result in a cell's becoming malignant.

prozone The inhibition of agglutination by the presence of high concentrations of antibody.

pseudogene A gene that cannot be expressed.

pyrogen A fever-causing substance.

pyroninophilic Stained by the dye pyronin. This stain preferentially binds to RNA, and thus a cell whose cytoplasm stains intensely with pyronin is rich in ribosomes and is probably therefore a protein-synthesizing cell.

quaternary structure The three-dimensional structure of a polymeric protein.

radioimmunoassay An immunological test that requires the use of an isotope-labeled reagent.

receptor A structure on cell membranes that binds specifically to ligands in the surrounding fluid.

recombinant vaccine. A vaccine containing antigen prepared by recombinant DNA techniques.

respiratory burst The rapid increase in metabolic activity that occurs in phagocytic cells while particles are being ingested.

reticuloendothelial system All the cells in the body that take up circulating colloidal dyes. Many are macrophages. This term is best avoided.

retrovirus An RNA virus that employs the enzyme reverse transcriptase to convert its RNA into DNA.

reverse transcriptase An enzyme that reversely transcribes RNA to DNA. It is found in retroviruses such as HIV.

rheumatoid factor Autoantibodies directed against epitopes on the immunoglobulin Fc region.

rosette The structure formed when several red blood cells bind to the surface of another cell in culture.

SALT, GALT, BALT, and MALT Abbreviations for skin-, gut-, bronchus-, and mucosa-associated lymphoid tissue, respectively.

sarcoma A tumor arising from cells of mesodermal origin.

secondary binding tests Serologic tests that detect the consequences of antigen–antibody binding such as agglutination and precipitation.

secondary immune response An enhanced immune response that results from second or subsequent exposure to an antigen.

secondary immunodeficiency An immunodeficiency disease resulting from a known nongenetic cause.

secondary infection Infections by organisms that can only invade a host whose defenses are first weakened or destroyed by other organisms.

secondary lymphoid organ An organ in which effector lymphocytes are located.

secondary structure The way in which a peptide chain is made up of structures such as α-helices and β-pleated sheets.

second-set reaction The rapid rejection of an organ or tissue graft by a previously sensitized host.

secretory component A protein found on intestinal epithelial cells that functions as an IgA receptor and, on binding to IgA, protects IgA against proteolytic digestion in the intestine.

selectin A family of cell surface adhesion proteins that bind cells to glycoproteins on vascular endothelium.

self-cure The elimination of helminth parasites by a localized hypersensitivity reaction in the intestinal tract.

sensitization The triggering of an immune response by exposure to an antigen.

septic shock A disease that results from the massive release of cytokines such as TNF as a result of infection with large numbers of gram-negative bacteria.

seroconversion The appearance of antibodies in blood indicating onset of infection.

serology The science of antibody detection.

serum The clear yellow fluid that is expressed when blood has clotted and the clot contracts.

serum sickness A type III hypersensitivity response to the administration of foreign serum as a result of the development of immune complexes in the bloodstream.

signal transduction The transmission of a signal through a receptor to a cell by means of a series of linked reactions.

skin test A diagnostic procedure that induces a local inflammatory response following intradermal inoculation of an antigen or allergen.

somatic antigen An antigen associated with bacterial bodies.

somatic mutation Gene mutation occurring in non germ-like cells.

specificity A term that describes the ability of a test to give true-positive reactions.

splice The joining of two DNA or RNA segments.

stem cell A cell that can give rise to many differentiated cell lines.

substrate modulation A method of controlling enzyme activity seen in the complement system in which a protein cannot be cleaved by a protease until it first binds to another protein.

superantigen A molecule that, as a result of its ability to bind to certain TCR variable regions, can cause T cells to divide. It is distinguished from a mitogen by its specificity for T cells bearing these variable regions. Mitogens make all T cells divide.

superfamily A grouping of molecules that share certain distinguishing common structures. Thus the members of the immunoglobulin superfamily all contain characteristic immunoglobulin domains.

suppressor cells A lymphocyte (usually a T cell) that is claimed to suppress the response of other cells to antigen. Their existence is disputed.

Svedberg unit (S) A measure of the sedimentation velocity of a molecule. It is determined by several factors, including molecular weight.

syncytium (*plural:* syncytia) The fusion of many cells into one large cytoplasmic mass containing multiple nuclei. This is usually the result of viral action.

syndrome A group of symptoms that together are characteristic of a specific disease.

syngeneic (isogeneic) Genetically identical.

tertiary binding test A serological test that measures the protective ability of an antibody in living animals.

tertiary structure The way in which the peptide chains of a protein are folded together.

thoracic duct The major lymphatic vessel that collects the lymph draining the lower portion of the body.

thymectomy Surgical removal of the thymus.

thymocyte A developing lymphocyte in the thymus.

thymus-dependent antigen An antigen that requires the assistance of helper T cells to provoke an immune response.

thymus-independent antigen An antigen that can provoke an antibody response without help from T cells.

titer The reciprocal of the highest dilution of a serum that gives a reaction in an immunological test.

titration The measurement of the level of specific antibodies in a serum by testing increasing dilutions of the serum for antibody activity.

T lymphocyte A lymphocyte that has undergone a period of processing in the thymus and is responsible for mediating cell-mediated immune responses.

tolerance The specific absence of an immune response to an antigen.

tolerogen A substance that induces tolerance.

toxic shock A disease resulting from the uptake of large amounts of staphylococcal superantigen.

toxoid Nontoxic derivatives of toxins used as antigens.

transcription The conversion of a DNA nucleotide sequence into an RNA nucleotide sequence by complementary base pairing.

transduction The conversion of a signal from one form to another.

translation The conversion of the RNA nucleotide sequence into an amino acid sequence in a ribosome.

transporter protein Proteins that bind fragments of endogenous antigen and carry them to newly assembled MHC class I molecules in the endoplasmic reticulum.

trematode A helminth known as a fluke. These include important human parasites such as schistosomes.

tubercle A persistent inflammatory response to the presence of mycobacteria in the tissues.

tuberculin An extract of tubercle bacilli that is used in a diagnostic skin test for tuberculosis.

tumor necrosis factor A macrophage or lymphocyte-derived cytokine that can exert a direct toxic effect on neoplastic cells.

tyrosine kinase An enzyme that phosphorylates tyrosine residues in proteins. They play a key role in signal transduction.

vaccination The administration of an antigen (vaccine) to stimulate a protective immune response against an infectious agent. The term originally referred to protection against smallpox.

vaccine A suspension of living or inactivated organisms used as an antigen to confer immunity.

variable region That part of the immunoglobulin or TCR peptide chains where the amino acid sequence shows significant variation between molecules.

variolation An early method of protecting an individual against smallpox by inoculation with live smallpox virus.

vasculitis Inflammation of blood vessel walls.

virgin lymphocyte A lymphocyte that has not previously encountered antigen.

virion A virus particle.

virulence The ability of an organism to cause disease.

xenograft A graft between two animals of different species.

Index

Italic page numbers indicate figures; t indicates table.